AF864006

Springer Proceedings in Business and Economics

Springer Proceedings in Business and Economics brings the most current research presented at conferences and workshops to a global readership. The series features volumes (in electronic and print formats) of selected contributions from conferences in all areas of economics, business, management, and finance. In addition to an overall evaluation by the publisher of the topical interest, scientific quality, and timeliness of each volume, each contribution is refereed to standards comparable to those of leading journals, resulting in authoritative contributions to the respective fields. Springer's production and distribution infrastructure ensures rapid publication and wide circulation of the latest developments in the most compelling and promising areas of research today.

The editorial development of volumes may be managed using Springer Nature's innovative EquinOCS, a proven online conference proceedings submission, management and review system. This system is designed to ensure an efficient timeline for your publication, making Springer Proceedings in Business and Economics the premier series to publish your workshop or conference volume.

This book series is indexed in SCOPUS.

Vicky Katsoni
Editor

Tourism, Travel, and Hospitality in a Smart and Sustainable World

9th International Conference, IACuDiT, Syros, Greece, 2022—Vol. 1

Springer

Editor
Vicky Katsoni
University of West Attica
Athens, Greece

ISSN 2198-7246 ISSN 2198-7254 (electronic)
Springer Proceedings in Business and Economics
ISBN 978-3-031-26828-1 ISBN 978-3-031-26829-8 (eBook)
https://doi.org/10.1007/978-3-031-26829-8

© The Editor(s) (if applicable) and The Author(s), under exclusive license to Springer Nature Switzerland AG 2023
This work is subject to copyright. All rights are solely and exclusively licensed by the Publisher, whether the whole or part of the material is concerned, specifically the rights of translation, reprinting, reuse of illustrations, recitation, broadcasting, reproduction on microfilms or in any other physical way, and transmission or information storage and retrieval, electronic adaptation, computer software, or by similar or dissimilar methodology now known or hereafter developed.
The use of general descriptive names, registered names, trademarks, service marks, etc. in this publication does not imply, even in the absence of a specific statement, that such names are exempt from the relevant protective laws and regulations and therefore free for general use.
The publisher, the authors, and the editors are safe to assume that the advice and information in this book are believed to be true and accurate at the date of publication. Neither the publisher nor the authors or the editors give a warranty, expressed or implied, with respect to the material contained herein or for any errors or omissions that may have been made. The publisher remains neutral with regard to jurisdictional claims in published maps and institutional affiliations.

This Springer imprint is published by the registered company Springer Nature Switzerland AG
The registered company address is: Gewerbestrasse 11, 6330 Cham, Switzerland

Editorial

"Tourism, Travel and Hospitality in a Smart and Sustainable World"

Tourism continues to be one of the sectors that most acutely and pervasively was affected by the COVID-19 pandemic, curtailing the growth and development of the tourism ecosystem, with serious spill-over effects on the economic outlook worldwide. Before the pandemic, the travel and tourism sector accounted for 10.6% of all jobs (334 million) and 10.4% of global GDP (USD 9.2 trillion); in 2019, the global travel and tourism industry directly contributed 2.9 trillion USD to the global GDP. Employment in the hospitality industry has never been as high as in the last few years. However, as a result of the COVID-19 crisis and the significant reduction in the mobility of tourists, in 2020, this sector generated USD 4.7 trillion, thus reducing its share in GDP to 5.5% and the number of jobs in tourism decreased by 18.5% or 62 million (WTTC, 2021).

The adverse impacts of the COVID-19 pandemic have dramatically changed the hospitality industry, shaking political systems and responses, damaging or breaking supply chains. At the same time, awareness and understanding of sustainability have been raised by the crisis, creating momentum to rebuild a more resilient, sustainable and smart tourism ecosystem. Crisis-related events have always created new opportunities and encouraged people to look for new development paths, to undertake entrepreneurial activities and to introduce innovation (Kraus et al, 2020). Thus, the crisis needs to be used as an opportunity for the tourism industry to follow green and digital transition pathways, in order to adapt to a post-pandemic "new normal". Solutions will differ from country to country, and the pace and scope of recovery will of course depend on global developments.

Promoting tourism all year round, shifting to niche markets thereby fighting seasonality and over-tourism, would have a major impact on economic development and job growth (Kaushal & Strivastava, 2020). It would make tourism a more economically sustainable business and investment, with clear repercussions on its resilience.

In parallel, as Micro and SMEs make up for more than 90% of tourism enterprises globally, an updated support strategy targeting their specific challenges can contribute to unleash extended benefits both for the tourism ecosystem by focusing to resiliency, sustainability, technology and interconnectedness among diverse stakeholders in the sector.

Development trends and the adoption of smart destinations and tourist services, along with artificial intelligence, robotics and other digital achievements and innovative solutions, (such as mobility monitoring applications, non-contact provision of services, digital health passports and identity checks, social distance technologies, crowd control), make it likely that the tourism industry will experience long-term structural changes (Shin and Kang, 2020; Sigala, 2020).

IACuDIT's mission is to contribute to the paradigm changes needed, which closely links to this year's conference's theme "Tourism, Travel and Hospitality in a Smart and Sustainable World". The aim of the conference to promote constructive, critical and interdisciplinary conversations on the above-mentioned challenges emerging in tourism by bringing together researchers, communities, industry and government stakeholders, is hereby fully achieved.

An array of 34 interdisciplinary conference topics was included in the call for papers. Papers submitted through virtual and poster presentations were considered for publication. All papers for the conference were double-blind peer reviewed in two phases. All the abstracts submitted for the conference were firstly peer reviewed by two experts in the field for relevance to the conference theme and contribution to the academic debate. A total of 171 abstracts were reviewed, and 124 were found acceptable. All accepted papers were then given the opportunity to submit to full conference proceeding papers utilizing the comments from the first phase of double-blind peer review for the improvement and submission of a full conference paper. 88 full conference proceedings were submitted for inclusion in the Springer proceedings in Business and Economics. These full paper submissions were again double-blind peer reviewed by two experts in the field as well as the editor. During the second review process, each submission was reviewed for: (a) relevance to the conference theme, (b) quality of the paper in terms of theoretical relevance and significance of the topic, and (c) contribution to the academic debate. The contributions were thematically selected for each group and are arranged in order of presentation in the proceedings. The subthemes to be covered by the conference were further classified into six distinctive parts, which are:

Part I Tourism Sustainability
Part II Tourism Technology and Innovation
Part III Tourism Destinations, Hospitality and Diversification
Part IV Tourism Economics, Management and Education
Part V Cultural Heritage Tourism
Part VI Tourism Consumer Behavior

Part I—Tourism Sustainability

The *first part* of the book consists of *sixteen chapters* and aims at presenting the sustainable practices and guidelines applicable in tourism. More specifically:

Chapter "Sustainability Assessment of Master Commercial European Ports Through Environmental Prismas", elaborated by Vassiliki Balla and Koskeridi A., compares Master ports of Europe based on their environmental responsibility and performance. Healthy environment, safety issues, wind and solar energy, waste management, are some of the many axes that current assessment is based on.

Chapter "Perspectives of Sustainable Tourism Development in the Region—Case Study", by Lubica Sebova, Klara Chovanova, Izabela Lazurova and Adrian Svec, focuses on tourism development in the High Tatras—Foothills, one of the most visited regions in Slovakia. The aim of the paper is to propose further options for the sustainable development of tourism in the region on the basis of the analysis and evaluation of the current state of tourism in the region.

Chapter "Environmental Sustainability Application in Athenian Hotels", by Ioanna Samartzi, tries to identify the environmental sustainability practices of 50 urban hotels in Athens, capital of Greece. After the data collection, the results estimated a lack of environmental practices in the majority of the hotels, such as insufficient waste management, recycling, water conservation and lack of environmental training among the employees.

Chapter "Sustainability Tools for Beach Management: Awareness of Integrated Coastal Zone Management and Current Compliance with Blue Flag Criteria at Eastern Beach in East London, South Africa", by Ntuthuzelo Headman Sayedwa and Dorothy Ruth Queiros, aims to investigate sustainability tools for beach management. Through proper management tools, the beach could be uplifted as a coastal tourist attraction, be accredited with Blue Flag status and managed sustainably for the enjoyment of present and future generations.

Chapter "The Prospects of Renewables for Electricity Production in Greece: Diversification Between Island and Mainland Greece", by Grigorios Kyriakopoulos, Garyfallos Arabatzis and Miltiadis Chalikias, focuses on the utilization of renewable energy sources (RES) in terms of large-scale technical infrastructure and their emerging technologies. Their research unveiled constructive outcomes and multifaceted challenges that emerge in energy utilization in the tourism sector, while considering the ongoing diffusion of RES in the energy mix of Greece.

Chapter "Fostering Sustainability in Tourism Destinations Through Residents' Assessment", by Pedro Liberato, Dália Liberato and Ricardo Cerqueira, investigates the characteristics of residents in two cities in the North of Portugal, as well as their perception about the tourism impacts at economic, sociocultural and environmental level. By analyzing residents' perceptions about the tourism impacts in the region, their results identify which factors need to be improved to influence residents' opinions over-tourism and leverage their intention to support the sector's development, in advance of strategies addressed to DMOs.

Chapter "Digital Innovation and Sustainability Practices in Tourism: An Overview", by Artemis Giourgali, seeks to develop a theoretically oriented reflection on sustainability digital innovations within the context of tourism and the potentials of this relationship. To this end, a literature review analysis was chosen for the collection, selection and critical evaluation of the literature on this topic. The author asserts that there is an ongoing process of evolutionary sustainability practices in tourism, indicating a theoretical approach rather than specific actions. The significant progress on the acceptance of the basic principles of sustainability from businesses and individuals still lacks the application of the appropriate instruments in the tourism industry so as to maximize the efficiency and successful development of sustainable digital innovations.

Chapter "Transformative Tourism as a Mean of Region's Sustainable Development", by Ekaterina Buzulukova, shows the possibilities of transformative tourism as a tool to overcome the crises in the tourism industry for sustainable regions' development. Based on transformative tourists experience in Russia, South Africa and Turkey, content analysis of feedbacks, qualitative and quantitative research, the authors discovered great potential for the travel industry recovery by means of the transformative tourism.

Chapter "Sustainability Practices in Events' Organization in Lisbon. Empirical Study of the Rock in Rio Music Festival", by Dália Liberato, Elga Costa, Cláudia Moreira, Pedro Liberato and Joaquim Ribeiro, aims to study music events and to identify main sustainable practices. The main results suggest that the Rock in Rio event impacts Lisbon, positively and negatively, as a destination, with the most evident positive impacts related to the city destination's promotion, supply development and the image and reputation creation. Even though the sustainability approach is already in everyday life, sustainable practices are not as evident in events as in destinations, particularly in the hotel sector. However, they are already defined as a valued element for choosing destinations and events.

Chapter "Sustainable Tourism and Degrowth: Searching for a Path to Societal Well-Being", by Zuzana Gajdošíková., examines the truly essence of sustainable tourism, its real application into praxis and its ability to ensure a more sustainable tourism sector. According to the literature review focused on sustainable development, tourism degrowth and local inhabitants in tourism destinations, the article analyzes the crossing needs and opinions of local inhabitants and visitors in a mature destination—the High Tatras in Slovakia with the intention to find out how tourism development should change in order to create a better living space for local inhabitants and be aware of the principles of sustainable development.

Chapter "The Sustainable Strategic Innovation Adopted by the Archaeological Museum of Taranto "MArTA" Which Drives the Repositioning of Taranto as a Cultural Tourist Destination", by Nicolaia Iaffaldano, aims to shed light on the path of sustainable strategic innovation undertaken by museum organizations. The document provides policy indications to public decision-makers operating in the cultural sector and in the local administration in order to design and develop a management model that responds to a systemic perspective.

Chapter "Experience Tourism as a Smart and Sustainable Form of Tourism in the Twenty-First Century" by Lubica Sebova, Radka Marcekova, Ivana Simockova and Kristina Pompurova, examines the supply and demand of experiential tourism products in a selected region with respect to smart and sustainable tourism development. Promoted visitor experience can be an important tool for destination marketing communication in the twenty-first century. If the destinations want to be competitive in a dynamically developing tourism market, the must adapt to this situation and offer the visitors an unforgettable authentic experience in smart and sustainable way. The goal is to create an offer that allows the visitors to actively engage in tourism activities and thus to intensify their emotions.

Chapter "The Needs of the Hospitality Industry in Its Transition to the Circular Economy", by Marival Segarra-Oña, Angel Peiró-Signes, Blanca De-Miguel- Molina and María De-Miguel-Molina, analyzes the needs of companies in the tourism industry regarding its transition to a circular economy. Three main conclusions are drawn from the authors' analysis. First, size, management and financial resources are the main barriers to the circular economy. Second, companies do not receive real support from the public administration, and they feel penalized in relation to those that do not apply these measures or to suppliers in the value chain that transfer their own waste. Third, there is a lack of a circular economy culture, which indicates the need for training at every level, so that the transition can be shared by all.

Chapter "Internal Sustainability Reporting in the Hotel Industry", by Lahorka Halmi and Katarina Poldrugovac, tests the de facto level of internal sustainability reporting harmonization of the hotel companies on the subject of (1) measuring and disclosing environmental and social key performance indicators (KPIs), and (2) the assessment of the importance of environmental and social KPIs. Data were collected via online questionnaire and explored by qualitative methods, where measurement and disclosure of KPIs was measured with C index and KPIs importance assessment was measured with non-parametric Kendall W test of concordance. Research results indicate quite low level of internal sustainability reporting harmonization on the two chosen topics. The novelty of the research is in applying qualitative methods in accounting research and in discovering the path to connect internal sustainability reporting processes to the external sustainability disclosures.

Chapter "Food Tourism Sustainability in Portugal: A Systematic Literature Review", by Teresa Mendes, Pedro Liberato, Dália Liberato, Teresa Dieguez and Hugo Barreira, delivers a systematic literature review on food tourism sustainability concerning tourism destination Portugal, based on a specific articles' selection protocol. The selected articles, analyzed through a visual bibliometric networks' software, produced three main clusters, which presented a research trend, stressing critical aspects concerning the role of regional governance in food tourism and the still scarce academic production to this matter.

Chapter "Scientific Narratives on Creative Tourism: A Theoretical Framework", Cristina Rodrigues, Dália Liberato, Elsa Esteves and Pedro Liberato provides a current scientific framework on creative tourism, from a conceptual point of view, presenting the most relevant issues, with focus on sustainability. To this end, a large body of scientific literature was analyzed through a combination of two techniques, namely: content analysis and systematic literature review. These techniques were applied as a research methodology to identify keyword trends understanding its meaning and analyze the main links between creative tourism and tourism sustainability, respectively. Main results provide a critical analysis of its characteristics as well as the correlations between creative tourism and the sustainable development of tourism activity, highlighting the sustainable matrix of this new model of tourism.

Part II—Tourism Technology and Innovation

Part II of the book consists of *eleven chapters* and aims at exploring the role of technology and innovative practices in the tourism industry. More specifically,

Chapter "Exploring the Role of Technology in Adventure Tourism", by Tatiana Chalkidou and George Skourtis, discusses how the adventure tourism industry is implementing and managing technology, addressing the changes that technology has brought and identifying the opportunities that can be exploited. In order to better understand the role of technology in adventure tourism management, we suggest the adoption of service-dominant logic and its ecosystem perspective (Vargo and Lusch, 2008). More specifically, we propose a transition of adventure tourism management from the perspective of S-D logic, reconsidering the conclusions that have been drawn from traditional perspectives (G-D logic). Importantly, a service-ecosystems view broadens the scope of adventure tourism management and enables researchers and practitioners to zoom out beyond dyadic exchange encounters and to view technology as a necessary resource for value co-creation.

Chapter "Dynamic Areas of Interest Inside an Urban Destination Using Visitors' Geolocation", by Ioannis A. Nikas, Athanasios Koutras and Alkiviadis Panagopoulos, assumes that travelers' choices have an impact on their footprint and propose an approximate spatio-temporal representation model to describe their decisions. This model is based on the implicit trace left by them, as it is captured in photo-sharing platforms and social networks. The study of these mappings can be used to reveal to policy and decision-makers a dynamic picture of the areas of interest inside destinations, as well as any other areas that can be further developed in the immediate future. Their proposed model is applied and tested in the city of Athens, an urban destination, using data from the photo-sharing platform Flickr.

Chapter "Towards a Framework for Participative Innovation in Tourism", by Susanne Marx and Lorena Arens, develops a framework for describing participation in innovation activities in tourism to help address this challenge. The research process follows the method framework for Design Science Research with a design- and development-centered entry point. The research advances the understanding of design opportunities for participative innovation activities in tourism from the DMO perspective.

Chapter "A Conceptual Framework for Applying Social Signal Processing to Neuro-Tourism", by Constantinos Halkiopoulos, Evgenia Gkintoni, Garefalea Kakoleres, Giorgos Telonis and Basilis Boutsinas, claims that neuro-tourism in combination with social signal processing techniques more efficiently explores the tourists' behaviors/preferences when they are browsing webpages or even when they are traveling. The authors propose a conceptual framework for applying social signal processing to neuro-tourism. The proposed conceptual framework takes into consideration various application areas within tourism field and constitutes a roadmap of analyzing/exploiting social signals produced during social interactions or human computer interactions.

In Chapter "Applying Big Data Technologies in Tourism Industry: A Conceptual Analysis" by Constantinos Halkiopoulos, Dimitris Papadopoulos and Leonidas Theodorakopoulos, the adoption of big data technologies in the field of tourism is examined through a review of the existing literature, with the main goal of gaining a deeper knowledge and understanding of the requirements of tourists in our country and improving the way of decision making. The paper also mentions the advantages and benefits of using big data technologies and popular methods used for sorting and contracting Big Data

Chapter "Integration of Blockchain Technology in Tourism Industry: Opportunities and Challenges", by Constantinos Halkiopoulos, Hera Antonopoulou and Nikos Kostopoulos, supports the view that as a nascent technology, blockchain has the potential to revolutionize the tourism sector since it provides a secure platform connecting the tourism business and its customers (employees and tourists). Through this endeavor, a comprehensive literature assessment on blockchain technology in smart tourism was conducted. In addition, the Internet of things (IoT) and large-scale analytics can be utilized to enhance the tourist experience. From the perspective of smart tourism, blockchain can serve as a platform that reliably and efficiently connects travelers with tourism products

Chapter "Application of Importance–Performance Analysis in Determining Critical Smart Hotel Technology Amenities: An Examination of Customer Intentions", Maria Tsourela, Dimitris Paschaloudis and Georgiadis Mauroudis, explores the smart hotel technology amenities identified in the literature, in an international tourist hotel. The study collected customers responses and applied an importance-performance analysis (IPA), using the self-explained matrix, to indicated which the order they should be addressed. Based on the research results, the authors discuss some useful implications and contributes to raising awareness of the importance of technological amenities and services for the future visiting intentions of hotel consumers.

Chapter "Implementation of Digital Marketing Techniques in Smart Tourism", Constantinos Halkiopoulos, Hera Antonopoulou and Konstantinos Giotopoulos, highlights the importance of digital marketing techniques in the field of tourism. Through their research, specific digital marketing techniques such as websites, promotional videos and informative blogs appear to be reliable tools for the promotion of the tourist product and related services as well as for the preference of new tourist applications/platforms in terms of savings of time. Therefore, the results, among others, highlight the importance of modern digital marketing techniques for their systematic application in the tourism industry

Chapter "The Trade Fair Industry in Transition: Digital, Physical and Hybrid Trade Fairs. The Case of Thessaloniki" by Dimitris Kourkouridis, Eleana Kostopoulou and

Spyros Avdimiotis, investigates the transition of exhibition activity from physical to digital and hybrid exhibitions, with particular emphasis on how the COVID-19 pandemic will affect the future of trade fairs. The results show that exhibitors recognize the benefits of digital trade fairs and digital tools that can be used but do not want to replace physical trade fairs.

Chapter "Role of Instagram Influencers on Destination Image: A Netnographic Study", by Hulisi Binbasioglu, Yasemin Keskin Yilmaz and Mehmet Baris Yilmaz, reveals how the posts of popular travel influencers who produce special content for Cappadocia determine the image and perception of Instagram users about the region. For this purpose, user comments (emotions, thoughts, expectations, etc.) on the posts of influencers about Cappadocia on Instagram were examined by the netnography method. Considering the findings obtained from the social media user' comments, the posts of the travel influencers about Cappadocia activated the emotions of the users and contributed positively to the perception of the region

Chapter "Gamification and Storytelling Enhancing Successful Wine and Food Tourism Products", by Dália Liberato, Pedro Liberato and Marta Nunes, focuses on the proposed tourism offer gamification and storytelling in the Douro region. The main results highlight the development of the region's tourism resources, which are crucial in a process of international promotion, increasing tourism demand by proposing a game as a driver of tourism and gambling associated with storytelling, and a key element in involving the player, telling the story and spreading the word about the destination. Studies and projects dealing with gamification and storytelling in tourism are unknown, particularly in low-density inland destinations, being crucial to propose projects on this topic, suitable to similar destination typologies.

Part III—Tourism Destinations, Hospitality and Diversification

Part III of the book consists of 19 chapters and aims at presenting issues regarding tourism destination policies, accommodation and various forms of special kinds of tourism, such as medical and wellness tourism, film tourism, agritourism, accessible tourism, sports tourism, coastal and maritime tourism.

Chapter "A Review of Destination Personality Literature: Focus on Articles Published in the Last Decade (2013–2022)", by Athina Nella, provides a review of the most recent literature in the field of destination personality (DP) or destination brand personality (DBP). DP has been linked with destination image, self-congruence and behavioral intentions, with a positive relationship being supported in most cases. Due to increased research interest in the field, an updated literature review of destination personality literature was considered as valuable.

Chapter "Agritourism: A Lever for the Development of the Greek Countryside. The Case of Nymphaeum of Florina/Greece", by Konstantina Boulouta and Georgios Karagiannis, attempts to map agritourism and outline its process that has been taken up today in Greece. Their research sets questions regarding agritourism development issues and their relationship with the area under investigation, which is Nymfaio in Florina, Greece.

Chapter "Consumers' Perceptions Regarding Film Tourism at the Level of the Tourist Destination Transylvania", by Diana Foris, Florina-Nicoleta Dragomir and Tiberiu Foris, attempts to identify the opinions of young consumers regarding the travel and the tourism generated by the film at the level of the tourist destination Transylvania. In this respect, a quantitative marketing research was conducted. The results of the study highlighted the positive attitude of young consumers toward film tourism and the potential benefits of this form of tourism regarding the development of tourist destinations. The results of the study are released for the managers of tourist destinations in order to develop attractive and successful destinations.

Chapter "Projecting an Image: TV Series Amor Amor and Lua de Mel and the Impacts of Film-Induced Tourism in Penafiel", by Ana Ferreira, Maria Carlos Lopes and Elisa Alen, highlights the importance of the films and television series, as they are an excellent tool for promoting tourism for the territories where they are filmed. The appearance of a particular area in a film or television series can have a huge effect on the number of visitors of an already existing place and create a new kind of tourism to the area and generate a boost for the local economy.

In Chapter "Tourism Destinations *"Where You Can Engage in Battles with the Romans"*: The Role of Reenactment in Shaping a Destination Image", Simona Malaescu examines both the direct and indirect role of reenactors in developing a heritage site as a tourism destination. Her study also revealed the reenactors' perspective of the importance of the context in which they act. The practical implications of the study reside in helping the local authorities to design new levers of developing a heritage tourism destination and to understand the long-term implications of making from reenactment performances funding a priority

Chapter "Accessible Tourism Businesses in the Cross-Border Region of Greece—Republic of Northern Macedonia: The Case of Greece", by Dimitris Kourkouridis, Vicky Dalkrani and Asimina Salepaki, evaluates the performance of the existing accessible tourist services and facilities of the tourism businesses in five regional units of Northern Greece. The most encouraging finding of the survey is the fact that those involved in tourism in the study area are willing to be educated and trained about accessibility, and also, they are willing to invest money in their businesses' accessibility. The article concludes with some suggestions for improving the accessibility of tourism businesses in the study area, based on the barriers that emerged from the research.

Chapter "Predicting the Impact of Cruise Ships on Island Destinations", Ángel Peiró-Signes, Oscar Trull Domínguez and César Gómez-Palacios use the human pressure indicator (HPI), that makes it possible to know the instantaneous situation of the population in highly touristic places, and especially on islands, including both tourists and residents. In their article, they implement a new HPI model that includes the influence of cruise ships arriving at the Hawaiian Islands, where they evaluate the real pressure exerted on them, so that they can appreciate the influence of cruise ships on them.

In Chapter "What Motivates Qualified Scuba Divers to Visit the Maltese Islands?", Simon Caruana and Francesca Zammit examine the factors that motivate divers to come and dive in the Maltese waters and connect their research with sustainability issues. Their research suggests that water quality (in terms of pollution and consequently fish life) is a determining factor. Interestingly, divers place much more importance toward sustainability today hence becoming a significant motivating factor in selecting the region to visit.

Chapter "The Authentic Marathon Swim: Understanding Swimmers' Experiences and Behavior in a Small-Scale Sport Tourism Event", by Stella Leivadi and Eugenia Tzoumaka, reveals that a successful event contributes not only to the image of the location but it also has a significant economic impact. In addition, their results also confirm previous research, which claims that positive word of mouth and the behavioral intentions of active tourists are affected positively by the destination image.

Chapter "Systematic Revision of the Literature on Satisfaction in Marinas and Nautical Ports", by Fernando Toro Sánchez, Pedro Palos Sánchez and Nélson Manuel da Silva de Matos, presents a systematic literature review (SLR) with a selection of 60 studies and references from 2005 to August 2022 using the bases of World of Science and Scopus. The authors aim at finding homogeneous directions of study that facilitates analysis based on the use of advanced statistical methodologies, in order to understand the nautical tourism sector.

In Chapter "Implication of Ukraine War on EU Migration Flows: Perspectives and Challenges", Mirela Ionela Aceleanu and Ioana Manuela Mîndrican investigate the outbreak of war in Ukraine and its effect on various areas of the economic, social and political life of European states, including tourism. The main objective of this paper is to highlight the situation of the migrant crisis and the implications that this phenomenon has on the destination states.

Chapter "Perception of Security by Tourism Service Providers: The Case of Acapulco, Mexico", by Karla Rosalba Anzaldúa Soulé, Mirella Saldaña Almazán, Arely Adriana Almazán Adame and Carmen Exchel Cliche Nájera, tries to determine the perception of tourist security in Acapulco based on the opinion of tourist service providers and to implement strategies that allow them to recover their competitiveness.

Chapter "Sports Tourism an Opportunity for Urban Areas: Classifying Sports Tourists According to Their Level of Sports Engagement and Consumer Behavior", by Maria Vrasida, Tatiana Chalkidou and Despina Gavrili, try to identify and classify sports tourists according to their different level of engagement in the sporting activity. Furthermore, a hypothesis is made and tested that consumer behavior will present differences both in the motivational (deciding on a destination) but also in the experiential phase (visiting and experiencing a destination). A multi-input matrix will allow to determine proximity to the culturally oriented or health-oriented part of the scale and define and identify consumer behavioral patterns accordingly.

Chapter "Quality of Life and Health Tourism: A Conceptual Roadmap of Enhancing Cognition and Well-Being", by Evgenia Gkintoni, Giorgos Telonis,

Constantinos Halkiopoulos and Basilis Boutsinas, aims to investigate the contribution of wellness tourism to enhancing the quality of life. In addition, cognitive and emotional characteristics of the tourist traveler's personality that can be enhanced through wellness tourism and contribute to the general improvement of life quality are investigated. The article is an overview review of research conducted over the past ten years. The results underlined, among other things, the role of wellness tourism in enhancing the quality of life in its many facets, including physical and mental health, well-being and social environment.

Chapter "Pilot Survey to Investigate Participation in Exercise for All Programs and Sport Events and Their Impact on the Evaluation of the Attica Region", by Christina appl, Georgia Yfantidou and Charilaos Kouthour, is attempt to estimate and understand the relationship between the image of a small-scale sport tourism event, its impact on destination image and the profile of participants. The results of this research could help to provide destination marketers and sport event marketers with insight to how various entities should be best combined according always to the participants' profile.

Chapter "Wellness Tourism in the Post COVID-19 Era in Blue Zones and Insular Areas. The Case of the Island of Ikaria", by Vakoula Eleonora and Constantoglou Mary, aims to examine how wellness tourism evolved during the COVID-19 pandemic, how it is located on the island of Ikaria and also what is the perspective of potential and current consumers of tourist services on this island. The research indicated that although the pandemic damaged severely the tourism sector, tourists are willing to travel, showing also a preference in the country.

Chapter "Brand and Competitiveness in Health and Wellness Tourism", by Eleonora Santos, Cátia Crespo, Jacinta Moreira and Rui Alexandre Castanho, intends to draw inferences about the effect of the brand on the competitiveness of the wellness tourism supply, across NUTS II regions of the continent, during the period of ten years, with a special focus on the recent period of the COVID-19 pandemic. The correlation analysis between brand and competitiveness indicators suggests a strong positive and significant relationship between brand and market share, in the Centro and Lisbon regions, starting in 2017 and increasing over time. Finally, some recommendations on tourism management are made to improve the competitiveness of the supply.

Chapter "Investigation into the Contribution of Social Tourism Programmes to the Satisfaction of Entrepreneurs in the Hospitality Sector", by Maria Doumi and Georgia Despotaki, investigates the value of the contribution of Greek Social Tourism programs given to the hospitality sector by evaluating the level of satisfaction of entrepreneurs according to the effectiveness of their implementation, and with respect to both their sociodemographic and business' profile. Through this research, the significance of the implementation of social tourism programs in hospitality businesses was highlighted. This research is the springboard for future planning of social tourism programs not only in Greece but also worldwide.

Chapter "Hybrid Events and Congresses and Their Impact on the Greek Meetings Industry", by Athina Papageorgiou and Alexandra Chalkia, investigates how hybrid conferences affected the Greek Meetings Industry during the COVID-19 pandemic.

For this, we conducted both a primary and secondary research; primary data were collected by interviewing the administrative personnel of Hotels, PCO's, DMC's and Audiovisual companies and secondary data were obtained by reviewing journals, relevant studies conducted by tourism associations and also UNWTO statistics. Their results confirmed that hybrid events are currently the preferred type of the Hellenic Conferences industry and will play a major role in the years to come.

Part IV—Tourism Economics, Management and Education

Part IV of the book consists of twenty-one chapters and aims at presenting economic and management practices in tourism, as well as educational considerations in the tourism digital era. Gender equality issues were all discussed.

Chapter "Economic Indicators and Cost Analysis in the Hotel Industry of Crete", by Efstathios Velissariou and Georgedaki Ioanna, presents the results of a survey in 160 hotels in the region of Crete in Greece, aiming to investigate the use of indicators as a management tool in hotels and the ability to calculate the cost, but also the break-even point of hotel businesses. The survey was conducted in the spring of 2021, and the results were disappointing, since only the half of the hotels in Crete use indicators to measure their performance. Also, one in four hotels in Crete and one in three hotels that operate all year round cannot exactly calculate their costs. Hotels use financial indicators but not entirely effective as they are mainly used for annual comparisons, while the most popular indicators used are those associated with the daily performance of the business.

In Chapter "How Can We Measure the "Success of Change Management? An Exploratory Factor Analysis in a Sample of Employees in the Greek Hotel Industry", Dimitrios Belias and Nikolaos Trihas have assessed an instrument of research which measures the change success as perceived from hotel employees. The instrument of research derives from the work of Zand and Sorensen (1975). The pilot test took place in a sample of 372 hotel employees from 4- and 5-star hotels in Greece. The result indicated that the employees' perception is the following: the hotel's management is not ready to implement with success a change program.

Chapter "The Efficiency of the Spa Enterprises in Slovakia", by Radka Marčeková, examines the efficiency of selected spa enterprises in Slovakia, by data envelopment analysis, and determines the most efficient and the least efficient spa enterprises. The paper uses the radial output output-oriented efficiency DEA model under variable returns to scale assumptions and the Malmquist productivity index. This paper refers to the application of DEA analysis to tourism business praxis.

Chapter "How Can We Measure the "Resistance to Change"? An Exploratory Factor Analysis in a Sample of Employees in the Greek Hotel Industry", by Dimitrios Belias and Nikolaos Trihas, elaborates on an instrument of research on resistance to change among hotel employees; more precisely on four-star and five-star hotels in Greece. The research instrument was developed by Oreg (2003); it is made by four variables (Routine seeking, emotional reaction to change, short-term focus and cognitive rigidity). The findings indicate that the questionnaire was reliable, though the cognitive rigidity dimension as marginally accepted by the Cronbach Alpha test. The outcome of the pilot study indicated that the resistance to change can be avoided if the hotel management tries to implement a well-programmed change where the employees will participate on the decisions.

Chapter "Pricing Strategy for Green Hotel Industry", by Flávio Ferreira, Fernanda Ferreira and Cristina Rodrigues, studies the effects of corporate social responsibility policies in the hotel industry. To do that, they model a non-cooperative hotel competition using game theory concepts. The model consists in a competition between a CSR hotel and a state-owned (SO) hotel that set room rates. The choice of room rates can be made either simultaneously or sequentially. Their main result is that the CSR hotel profits more than the SO hotel, regardless of the order of movements.

Chapter "Residents' Perception of Yachting Tourism amid COVID-19: A PLS-SEM Approach framed by SET, Revised SET and PMT", by Stelios Ioannidis and Maria Doumi, investigates the local community's perception of yachting tourism amid the epidemic threat of COVID-19. The theoretical base of social exchange theory (SET), revised SET and protection motivation theory (PMT) were utilized for the purpose of this study. The conceptual model built within this study includes four exogenous variables (place community attachment, economic gain, environmental impacts and COVID-19 Threat) and their impact on Support Yachting Tourism, with the usage of two moderators (positive and negative impacts). The survey took place at a well-established yachting area in Greece, consisting of Volos, Skiathos, Skopelos and Alonissos. The study confirmed the residents' support to yachting tourism amid COVID-19 and clarified the reasons for this support.

Chapter "Economic and Social Security", by Andreea Claudia Serban, Ionuț Jianu and Vicky Katsoni, elaborates on the complexity of the concept of economic security and concludes that it requires separate approaches for its different dimensions, being closely related to social aspects. The authors have identified five relevant dimensions/pillars interacting with economic and social aspects: demographic, globalization and technology, respectively environment and, based on a set of relevant indicators for each of them, they propose a composite index of economic and social security for countries in Euro-Atlantic area.

Chapter "Leadership Styles and Group Dynamics in the Tourism Industry", by Angelos Ntalakos, Dimitrios Belias, Nikolaos Tsigilis, provides the current literature review concerning the relationship between organizational leadership styles and group dynamics in general, as well as in tourism industry, providing a useful methodology through selection of papers and articles retrieved from online bibliographic databases.

Chapter "Modeling the Elements of Organizational Culture that Formulate the Design of the Internal Environment of Tourist Enterprises: A Literature Review", by Christos Kakarougkas, attempts to record, model and synthesize the most important elements of organizational culture that determine the design of the internal environment of tourist enterprises. His findings are characterized by originality and scientific and practical contribution, since, at an international level, a corresponding attempt to record, model and synthesize the elements of organizational culture that determine the design of the internal environment of tourist enterprises has not been identified yet.

Chapter "The Relationship Between Leadership Styles & Communication—Effect on Team Efficiency on the Tourism Industry", by Angelos Ntalakos, Dimitrios Belias and Athanasios Koustelios, examines the connection between leadership styles & communication on the tourism industry providing the theoretical and the empirical models that can shed some light on the above relationship. The methodology used is a literature review of papers, studies and articles retrieved from online bibliographic databases.

Chapter "Total Quality Management in Hotel Businesses Depending on Their Category and Size", by Aristides Katsaitis, Fragiskos Bersimis and Paris Tsartas, investigates the degree of implementation of the total quality management of the hotels in terms of the quality category of the services provided (5*–4*) and the size (small–large), in a sample of hotel businesses in the Prefecture of Attica. A survey was carried out using an electronic questionnaire that included appropriate questions addressed to the HR managers of the hotel units. The results show that 5-star and larger hotels show a higher degree of agreement in the attitudes that express the application of the principles of total quality management, compared to 4-star and smaller hotels.

Chapter "Gravity Models in Tourism: The "Crises" Factor" by Ioulia Poulaki, proposes a gravity model production for predicting spatial interactions and travel flows in times of crises that, in addition to its contribution to the literature of the tourism academia with a new (sophisticated) model of Newtonian gravity applied to tourism, it may significantly contribute to tourism professionals (organizations and businesses) as it is expected to predict changes in travel flows, thus supporting decision-making process regarding crisis management and tourism recovery, and producing multiple economic, social and environmental benefits for international popular Mediterranean destinations

Chapter "Silver Tourism in Greece. Basic Elements of Marketing and Branding", by Rentzi Konstantina and Constantoglou Mary, records the opinions and preferences of senior tourists regarding the choice of a tourist destination in general, and especially in the current COVID-19 period, which factors influence their decision, which media they use when searching for information and then the goal is to investigate their opinion on the recognition of Greece as a silver tourism destination, through the existing marketing and branding. The goal of the work is also to study the ability of Greece to emerge as a health tourism destination, specifically medical tourism for international elderly patients. Research has been contacted through online questionnaires on a random sample of elderly people living abroad during spring 2021

Chapter "Features of Foreign Working Force for Cypriot Hotel Industry", by Volha Yakavenka, Dimitrios Vlachos, Agis M. Papadopoulos, Tomáš Kliestik, Yianna Orphanidou, Leonidas Efthymiou and Danae Psilla, presents the results of the industry- based research focusing on the Greek and Slovakian hotel labor market, including but not limited to: (i) elements that impede staff mobility to Cyprus; and (ii) employer and job market characteristics that the ongoing human capital seeks to find in another country in order to engage in professional mobility. The aim of the research is to understand the features (skills and needs) of the potential employees from those pools which affect the willingness and easy to work in Cyprus. The results have been achieved by receiving information from the supply side, that is from the potential candidates who are interested in working in Cyprus hotels, coming from Greece and Slovakia (the second as a typical Eastern European country).

Chapter "Developing and Enhancing Emotional Intelligence in the Tourism Sector: The Role of Transformative Learning", Sofia Kallou, Aikaterini Kikilia, Michail Kalogiannakis and Nick Z. Zacharis, investigates the development and enhancement of emotional intelligence and the role of transformative learning theory on learners' outcomes focusing to the tourism sector. According to the findings, the authors conclude that transformative learning is an effective and valuable teaching and learning adult theory for the tourism education and training, since it has a positive impact on emotional intelligence improvement and on the personal development of tourism employees and executives.

Chapter "Online Collaborative Platforms: Analysis of Challenges to the Hospitality and Tourism Industry and Governments", by Mary Mutisya, tries to determine the impact of online collaborative economy platforms to hotels; secondly, the study examines the challenges to governments with regard to the collection of levies. The study adopted the cross-sectional survey design. The target population comprised of government agencies concerned with tourism levy, homes listed as Airbnb accommodation listings and hotel manager. The unit of observation were customers and employees from the management level. The study used a stratified random technique to select the respondents. Questionnaires, interview and Focus Group Discussion guides were used to collect the data. The analysis of the data was done descriptively and by content analysis. The study identified factors behind the success of online collaborative platforms and the strategies that can be adopted by hotels to compete with these platforms effectively. The study further identifies levy collection avenues

Chapter "Olympic Games and Mega Events Legacy Planning as a Tourism Initiation Strategy: Developments and Implications", by Ourania Vrondou, examines host city experiences and sport bodies' legacy policies to advice on applicable sport tourism destination enhancement techniques embedded in the basic city's event selection and hosting process. The analysis of available material produced positive indications of legacy programs encompassing tourism initiation tools as part of a holistic regenerating strategy for local communities, evidently impatient to experience the promised positive returns of a mega event that didn't fail to create a sustainable positive image for the locality.

Chapter "Does it Matter? The Influence of Leadership, Motivation and Empowerment on the Career Progression of Women in Tourism—A South Africa Case", by Nasreen Tisaker and Magdalena Petronella Swart, investigates how leadership, motivation and empowerment affect women's career progression in the South African tourism sector. The study was quantitative and descriptive, allowing researchers to evaluate the impact of leadership, motivation and empowerment on women's career progression in the tourism sector in South Africa. This study is expected to address a gap in the body of knowledge on women in tourism in the South African tourism sector, as well as contribute to the rising global discourse on women's motivation and empowerment in tourism.

Chapter "Exploring Gender Differences in Hotel Choice Safety Factors in the Early Post-COVID-19 Era", by Stavros Mamalis, Irene Kamenidou and Aikaterini Stavrianea, aims to explore Greek tourists' gender differences in the required safety features of tourist accommodation in the early post-pandemic era. Seven statements were used for choosing a tourist accommodation referring to safety measures required by them. Factor analysis reduced the seven statements to three dimensions-variables tested for gender differences. Independent sample t-tests revealed gender differences for two out of three dimensions, based on which discussion, suggestions and marketing communication techniques for accommodation providers are offered.

Chapter "Women in Tourism Industry: From Glass Ceiling Effect to #metoo Movement", by Poulaki Ioulia, Vassiliki Karioti and Charalampia Tsoulogianni, highlights two specific phenomena that afflict women during their professional careers: glass ceiling effect and sexual harassment. These phenomena are perpetuated within societies since stereotypes, prejudices and taboos lead to the so-called gender discrimination, especially in the work environment. The purpose of their paper is to develop the theoretical and research background on glass ceiling effect and sexual harassment that involve the women of the tourism industry, while it discusses sociopsychological solutions that may contribute to the efforts of HRM to eliminate these two phenomena".

Chapter "Factors Influencing Urban Tourism Demand:—The City of Tshwane Case", by Medalto Gotore, Magdalena Petronella Swart and Katlego Oliphant, illustrates the correlation between cognitive tourist demand, affective tourist demand and conative tourism demand in the City of Tshwane. The study was quantitative and descriptive in nature. This study is expected to address a gap in the body of knowledge on how cognitive, affective and conative tourism demand are related in the South African context and how they can influence urban tourism demand.

Part V—Cultural Heritage Tourism

Part V of the book consists of eleven chapters and aims at presenting issues regarding cultural routes, heritage and authenticity.

Chapter "Prospects for Heritage Tourism Branding in Eastern Arabia", by Adriaan De Man, provides an analysis on place branding, in the specific context of the United Arab Emirates and Oman. It does so through a qualitative insight into the heuristics and into the effective potential of cultural landscapes, relating to both policy and practice. It also challenges the notion that tangible heritage functions as a predictable accelerator for economic growth and favors more moderated takes on the effectiveness of geographical uniqueness as an instrument. The paper concludes by arguing that branding a regional product in its real diversity cannot insist excessively on history and heritage, as it needs to forge a cohesive message that embraces technology, hospitality, travel and the many other elements that appeal to the senses of the consumer.

Chapter "Meet West in East: New Routing of Old Footprints in the Heritage Tourism in Macao", Yuanyuan Cao and Eusebio Leou, explores the possible undiscovered inbound visitors' source to Macao, by introducing the theme of "Portuguese Origins (Origem Portuguesa)". Through the field measurement and consultative interview methods, this study confirmed that the concept of "Portuguese Origins" can demonstrate the cultural uniqueness in Macao, which would be a positive impact on the attractiveness of non-Chinese-speaker inbound tourists. To the finding, four theme routes of "Portuguese origins" are designed by linking the relevant locations in Macao, in order to serve as the tourist attractions with these optimal paths.

Chapter "Digital Activities as Online Getaways to Cultural Heritage: The Paradigm of Greek Museums During the COVID-19 Era", Sofia Boutsiouki and Maria Damou, examines the museum management in Greece in the midst of the COVID-19 pandemic and present museum initiatives that enabled the digital consumption of heritage and the increased virtual attendance of wider audiences via extrovert, online, interactive activities and participatory experiences. The authors also outline the existing gaps and the arising challenges and opportunities for the museums' digital strategy, resilience and sustainability in the post-coronavirus world.

Chapter "A Legal Approach of Diving Tourism and the Protection of Marine Cultural Heritage. The Case of Greece", by Dimitrios Mylonopoulos, Polyxeni Moira and Aikaterini Kondoukaki, based on the new institutional framework for the development of diving tourism in Greece, examines whether the new provisions for the development of diving tourism meet the needs of the protection of marine archeological resources but also the need for sustainable tourism development.

Chapter "Authenticity as an Antecedent of the Tourist Experience in Measuring the Gap Between Tourist Experiences at an Accommodation Establishment and A Tourist Attraction in Lesotho", by Shale Johannes Shale, Sello Samuel Nthebe and Magdalena Petronella Swart, explores whether authenticity can be applied as an antecedent of the tourist experience in measuring the gap between tourist experiences at an accommodation establishment and a tourist attraction in Lesotho. The researchers adopted qualitative content analysis, following content-logical guidelines and step-by-step paradigms before providing qualification in the form of tables, to show that the dimensions local culture, involvement, meaningfulness and social interaction to be the most appropriate for measuring authenticity.

Chapter "Turning Cultural and Creative Challenges into Local Development Opportunities", by Andreea Claudia Şerban, Huseyin Uzunboylu, Gheorghe-Alexandru Stativă, Ana-Maria Bocăneală, focuses on the cultural and creative index as a useful tool to compare cities' performance and to identify recommendations for increasing the capability of cultural and creative favorable context to support local development. The authors identified three dimensions that need to be considered and explored to make possible the objectives of the EU: building cohesive societies through culture and education and creating a more inclusive union, supporting innovation, creativity and sustainable jobs and growth, in order to provide a scheme for turning cultural creative challenges into opportunities for local development.

Chapter "Much More than a Museum: Motivations and Experiences of Young Visitors at Urban Cultural Clusters", by Daria Kondrateva, Alexander Pakhalov and Natalia Rozhkova, identifies factors that influence the motivations and the experiences of young visitors at urban cultural clusters and centers. The authors' results show that for Gen Zers information from social media is a key factor in shaping the initial intention to visit a cultural cluster or center. In turn, satisfaction and revisit intention increase if an attraction provides young visitors with more interactive and immersive experiences.

Chapter "Exploring the Role of Toponymy in Wine Naming Strategy. A Greek Case Study", by Theodosios Tsiakis, Eleni Anagnostou and Giuseppe Granata, aims at deciphering the toponymy of Greek wine labels, providing valuable insights about description of the name, the cause of selection and the language. The choice of wine label raises questions whether it intends to communicate historical concepts, identity or quality, since there are elements that can influence the buying operation, either in terms of purchase decision or based on the perception of quality–price.

Chapter "Cultural Tourists and Their Intention to Visit Greece During the Pandemic Era", Vasiliki Papaioannou, Pavlogergatos Gerasimos and Mary Constantoglou, studies the intention of people to visit Greece for cultural reasons. Questionnaires were given to answer to people traveling on board to flights to Greece, UK, Belgium and Italy. On board passengers had the time and comfort to answer a fourteen questions long questionnaire aiming at examining the intention to undertake cultural trips. The research took place during the first lockdown at the beginning of the COVID-19 pandemic, and it contributes to the assessment of cultural tourism dynamics.

Chapter "Urban Mega-Projects, Cultural Flagships and Their Effect on the Destination Image and Visitor's Economy: The Case of the Waterfront of the Metropolitan Area of Athens", Nicholas Karachalis and Efthymia Sarantakou, elaborates on how cultural infrastructure and emblematic buildings are strongly associated with efforts to change the image and fame of a city with the use of city branding techniques, often leading to dilemmas regarding their effectiveness and sustainability. The aim of the paper is to critically discuss the contradictions within this discussion. While Athens has an opportunity to develop a new identity through the developments at its waterfront, a coherent strategy mission and goals toward this direction are still missing.

Part VI—Tourism Consumer Behavior

The last part of the book consists of eleven chapters and aims at presenting different approaches of tourism behavior.

Chapter "Gastronomy Tourism in Athens, Motivations and Resources. A Push and Pull Approach", by Charilaos Kalpidis, Fragiskos Bersimis and Paris Tsartas, aims to utilize the push-pull factors approach to address the internal motives of tourists and their preferred gastronomy-related resources in the capital city of Greece, Athens. The research results add to the comprehension of food travel motivations and attractions from a tourist's and the destination's viewpoint, with relevant implications for restaurant entrepreneurs, destination management organizations (DMOs) and tourism policy planners

Chapter "Gastronomy Tourist Behavior in Macao as the "UNESCO Creative City of Gastronomy", Valeria Wenlan Jiang and Eusebio Leou, investigated tourists' perception and its effect on loyalty through questionnaires and interviews, aiming for a deeper mining to visitors' real experience during their stay in Macao. Moreover, based on Hofstede's cultural dimension theory, the authors also explored the connection among tourists' behaviors, culture impressions and the holistic impressions of Macao. The results have revealed that the perception of gastronomy tourism is multi- dimensional, which could create cultural value and work as guidance for further promotion.

In Chapter "Behavior Analysis of Glamping as a Novel Tourism Marketing Trend" by Ioanna Giannoukou, Eirini Fafouti and Constantinos Halkiopoulos, the concept of alternative tourism is analyzed, along with its characteristics and philosophy, with an emphasis on camping as well as a profile analysis of campers. In addition, the concept of glamping, a trend in alternative tourism that blends living in nature with the conveniences of a luxury hotel, is examined.

Chapter "Behavioural Intention of the South African Youth Towards Birds, Bird Habitat and Avitourism", by Nicolene Conradie, investigates behavioral intention of the South African youth toward birds, bird habitat and avitourism. The research poses clear challenges to professionals and educators within the tourism industry of South Africa to increase learners' willingness and motivation to act pro-environmentally through dedicated education. This supports the need to introduce an intervention program to promote behavioral intention and ultimately pro-avi and environmental behavior among learners.

Chapter "Exploring the Relationship Between Tourist Safety and Tourist Experience: Theories from Accommodation Establishments and Attractions", by Shale Johannes Shale, Sello Samuel Nthebe and Magdalena Petronella Swart, attempts to determine if there is a relationship between tourist safety and tourist experiences supported by the literature, specifically to investigate the gap between these concepts at accommodation establishments and attractions in Lesotho. This method adhered to the interpretive and applied content analysis as an analytical approach to produce results and put them into context. The literature shows that the relationship between tourist safety and tourist experience can be supported specially to investigate the gap between these concepts at accommodation establishments and attractions in Lesotho.

Chapter "Satisfaction of Visitors With Folklore Festivals in Slovakia", by Andrej Malachovský, examines consumer behavior and product satisfaction with four selected folklore festivals in Slovakia. A structured questionnaire was used for data collection, which was placed on social networks. The obtained data were processed by selected methods of descriptive statistics and graphically interpreted. The results of the research indicate reserves in terms of comprehensive provision of the festival product, the aim of which is to achieve the satisfaction of visitors.

Chapter "The Benefits of Tourism and Travel to the Tourist's Well-Being: A Conceptual Discussion" by John Ebejer, explores how the engagement with different elements of the destination enables tourists to enhance their physiological and psychological well-being. The tourism activities considered in this paper are (i) engagement with nature (ii) engagement with the physical environment (iii) engagement with culture (iv) engagement with the spiritual (v) engagement with other people. For each, the implications for the tourist's well-being are discussed.

In Chapter "Understanding Behavioural Intentions in a Spa Experience: Insights from Contemporary Literature", Cristina Rodrigues, Fernanda Amelia Ferreira, Vânia Costa, Maria José Alves, Márcia Vaz, Paula Odete Fernandes, Alcina Nunes try to understand, from the existing literature, the consumption behavior toward a spa experience, from the consumers' point of view. This paper was developed based on content analysis and literature review, a systematic and narrative approach. The main results highlight the most critical aspects affecting satisfaction, service quality and perceived consumer value, and identify three key moments that can systematize the spa experience: expectations, consumption and consequences.

Chapter "To Read or Not to Read? Dilemmas of Potential Tourism Visitors: Quantitative Study of Online Review Management from Slovakia", Kristína Pompurová, Ivana Šimočková, Ľubica Šebová, examines the extent to which the inhabitants of Slovakia as potential tourism visitors read online reviews and to analyze the reasons for reading reviews. The results of their survey confirmed the importance of reviews in the consumer decision-making process and underline the fact that tourism businesses should pay greater attention to electronic word of mouth if they want to be successful.

Chapter "Profiling Coastal Tourists to the Democratic Republic of Congo", by Nadine Kisema, Elricke Botha and Nicolene Conradie, creates a profile using descriptive statistics and identified demographic and behavioral characteristics. In addition, travel motivations were identified using exploratory factor analysis. Based on the results, recommendations for the public and private tourism sectors were made. This is the first coastal tourism study based on the DRC. The profile of coastal tourists can assist the DRC in effectively communicating and improving the priorities in the development of tourism in the country.

Chapter "Profiling Visitors of Four Selected Urban Green Spaces in the City of Tshwane", by Nicolene Conradie, Anna Maria Magdalena Erasmus and Ciné van Zyl, intends to provide a profile of visitors to four selected urban green spaces (i.e., botanical garden, zoo, nature reserve and bird park) in the City of Tshwane (South Africa). Visitor motivation, environmental awareness, subjective well-being and revisit intention are investigated. Tourism managers may use the information to promote the benefits experienced by visiting urban green spaces.

The editor and IACUDiT anticipate that readers of this volume will find the papers informative, thought provoking and of value to their niche research areas.

References

COM (2021) 350 final. https://ec.europa.eu/info/sites/default/files/communication-sme-strategy-march-2020_en.pdf

Kaushal, V., & Srivastava, S. (2020). Hospitality and Tourism Industry amid COVID-19 Pandemic: Perspectives on Challenges and Learnings from India. International Journal of Hospitality Management, 102707-102707.

Kraus, S., Clauss, T., Breier, M., Gast, J., Zardini, A., & Tiberius, V. The economics of COVID-19: initial empirical evidence on how family firms in five European countries cope with the corona crisis. IJEBR, 26, 5.

Shin, H.; Kang, J. Reducing perceived health risk to attract hotel customers in the COVID-19 pandemic era: Focused on technology innovation for social distancing and cleanliness. Int. J. Hosp. Manag. 2020, 91, 102664.

Sigala, M. (2020). Tourism and COVID-19: Impacts and implications for advancing and resetting industry and research. Journal of Business Research, 117, 312–321. https://doi.org/10.1016/j.jbusres.2020.06.015

Stankov, U., Filimonau, V., & Vujičić, M. D. (2020). A mindful shift: an opportunity for mindfulness-driven tourism in a post-pandemic world. Tourism Geographies, 22(3), 703-712.

UNWTO. (2020). UNWTO World tourism barometer, May 2020: Special focus on the impact of COVID-19. https://webunwto.s3.eu-west-1.amazonaws.com/s3fs-public/2020-05/Barometer%20-%20May%202020%20-%20Short.pdf

Wachyuni, S.S.; Kusumaningrum, D.A. The Effect of COVID-19 Pandemic: How are the Future Tourist Behavior? J. Educ. Soc. Behav. Sci. 2020, 33, 67–76.

WTTC. Economic Impact Reports. Available online: https://wttc.org/Research/Economic-Impact (accessed on 11 November 2022).

Contents

Tourism Sustainability

Sustainability Assessment of Master Commercial European Ports Through Environmental Prismas 3
Archontoula Koskeridi and Vassiliki Balla

Perspectives of Sustainable Tourism Development in the Region – Case Study 23
Lubica Sebova, Klara Chovanova, Izabela Lazurova, and Adrian Svec

Environmental Sustainability Application in Athenian Hotels 39
Ioanna S. Samartzi

Sustainability Tools for Beach Management: Awareness of Integrated Coastal Zone Management and Current Compliance with Blue Flag Criteria at Eastern Beach in East London, South Africa 55
Ntuthuzelo Headman Sayedwa and Dorothy Ruth Queiros

The Prospects of Renewables for Electricity Production in Greece: Diversification Between Island and Mainland Greece 73
Grigorios L. Kyriakopoulos, Garyfallos Arabatzis, and Miltiadis Chalikias

Fostering Sustainability in Tourism Destinations Through Residents' Assessment 93
Pedro Liberato, Dália Liberato, and Ricardo Cerqueira

Digital Innovation and Sustainability Practices in Tourism: An Overview 115
Artemis Giourgali

Transformative Tourism as a Mean of Region's Sustainable Development 127
Ekaterina Buzulukova

Sustainability Practices in Events' Organization in Lisbon. Empirical Study of the Rock in Rio Music Festival 139
Dália Liberato, Elga Costa, Cláudia Moreira, Pedro Liberato, and Joaquim Ribeiro

Sustainable Tourism and Degrowth: Searching for a Path to Societal Well-Being 159
Zuzana Gajdošíková

The Sustainable Strategic Innovation Adopted by the Archaeological Museum of Taranto "MArTA" Which Drives the Repositioning of Taranto as a Cultural Tourist Destination 175
Nicolaia Iaffaldano

Experience Tourism as a Smart and Sustainable Form of Tourism in the Twenty-First Century 189
Lubica Sebova, Radka Marcekova, Ivana Simockova, and Kristina Pompurova

The Needs of the Hospitality Industry in Its Transition to the Circular Economy 201
Marival Segarra-Oña, Blanca De-Miguel-Molina, Angel Peiró-Signes, and María De-Miguel-Molina

Internal Sustainability Reporting in the Hotel Industry 221
Lahorka Halmi and Katarina Poldrugovac

Food Tourism Sustainability in Portugal: A Systematic Literature Review 237
Teresa Mendes, Pedro Liberato, Dália Liberato, Teresa Dieguez, and Hugo Barreira

Scientific Narratives on Creative Tourism: A Theoretical Framework 253
Cristina Rodrigues, Dália Liberato, Elsa Esteves, and Pedro Liberato

Tourism Technology and Innovation

Exploring the Role of Technology in Adventure Tourism 281
Tatiana Chalkidou and George Skourtis

Dynamic Areas of Interest Inside an Urban Destination Using Visitors' Geolocation 291
Ioannis A. Nikas, Athanasios Koutras, and Alkiviadis Panagopoulos

Towards a Framework for Participative Innovation in Tourism 307
Susanne Marx and Lorena Arens

A Conceptual Framework for Applying Social Signal Processing to Neuro-Tourism 323
Evgenia Gkintoni, Garefalea Kakoleres, Giorgos Telonis, Constantinos Halkiopoulos, and Basilis Boutsinas

Applying Big Data Technologies in Tourism Industry: A Conceptual Analysis 337
Leonidas Theodorakopoulos, Constantinos Halkiopoulos, and Dimitris Papadopoulos

Integration of Blockchain Technology in Tourism Industry: Opportunities and Challenges 353
Constantinos Halkiopoulos, Hera Antonopoulou, and Nikos Kostopoulos

Application of Importance–Performance Analysis in Determining Critical Smart Hotel Technology Amenities: An Examination of Customer Intentions 369
Georgiadis Mauroudis, Maria Tsourela, and Dimitris Paschaloudis

Implementation of Digital Marketing Techniques in Smart Tourism 381
Constantinos Halkiopoulos, Hera Antonopoulou, and Konstantinos Giotopoulos

The Trade Fair Industry in Transition: Digital, Physical and Hybrid Trade Fairs. The Case of Thessaloniki 399
Eleana Kostopoulou, Spyros Avdimiotis, and Dimitris Kourkouridis

Role of Instagram Influencers on Destination Image: A Netnographic Study 417
Hulisi Binbasioglu, Yasemin Keskin Yilmaz, and Mehmet Baris Yilmaz

Gamification and Storytelling Enhancing Successful Wine and Food Tourism Products 433
Dália Liberato, Pedro Liberato, and Marta Nunes

Tourism Destinations, Hospitality and Diversification

A Review of Destination Personality Literature: Focus on Articles Published in the Last Decade (2013–2022) 455
Athina Nella

Agritourism: A Lever for the Development of the Greek Countryside. The Case of Nymphaeum of Florina/Greece 469
Konstantina Boulouta and Georgios Karagiannis

Consumers' Perceptions Regarding Film Tourism at the Level of the Tourist Destination Transylvania 483
Diana Foris, Florina-Nicoleta Dragomir, and Tiberiu Foris

Projecting an Image: TV Series Amor Amor and Lua de Mel and the Impacts of Film-Induced Tourism in Penafiel 493
Ana Ferreira, Maria Carlos Lopes, and Elisa Alen

Tourism Destinations *"Where You Can Engage in Battles with the Romans"*: The Role of Reenactment in Shaping a Destination Image 505
Simona Mălăescu

Accessible Tourism Businesses in the Cross-Border Region of Greece—Republic of Northern Macedonia: The Case of Greece 519
Dimitris Kourkouridis, Vicky Dalkrani, and Asimenia Salepaki

Predicting the Impact of Cruise Ships on Island Destinations 537
Oscar Trull Domínguez, Cesar Gómez-Palacios, and Angel Peiró-Signes

What Motivates Qualified Scuba Divers to Visit the Maltese Islands? 551
Simon Caruana and Francesca Zammit

The Authentic Marathon Swim: Understanding Swimmers' Experiences and Behavior in a Small-Scale Sport Tourism Event 571
Stella Leivadi and Eugenia Tzoumaka

Systematic Revision of the Literature on Satisfaction in Marinas and Nautical Ports 587
Fernando Toro Sánchez, Pedro Palos Sánchez, and Nélson Manuel da Silva de Matos

Implication of Ukraine War on EU Migration Flows: Perspectives and Challenges 613
Mirela Ionela Aceleanu, Ioana Manuela Mîndrican, and Javier Ordóñez Monfort

Perception of Security by Tourism Service Providers: The Case of Acapulco, Mexico 629
Carmen Exchel Cliche Nájera, Karla Rosalba Anzaldúa Soulé, Mirella Saldaña Almazán, and Arely Adriana Almazán Adame

Sports Tourism an Opportunity for Urban Areas: Classifying Sports Tourists According to Their Level of Sports Engagement and Consumer Behavior 641
Maria Vrasida, Tatiana Chalkidou, and Despina Gavrili

Quality of Life and Health Tourism: A Conceptual Roadmap of Enhancing Cognition and Well-Being 651
Evgenia Gkinton, Giorgos Telonis, Constantinos Halkiopoulos, and Basilis Boutsinas

Pilot Survey to Investigate Participation in Exercise for All Programs and Sport Events and Their Impact on the Evaluation of the Attica Region 667
Christina Avloniti, Georgia Yfantidou, and Charilaos Kouthouris

Wellness Tourism in the Post COVID-19 Era in Blue Zones and Insular Areas. The Case of the Island of Ikaria 685
Vakoula Eleonora and Constantoglou Mary

Brand and Competitiveness in Health and Wellness Tourism 707
Eleonora Santos, Cátia Crespo, Jacinta Moreira, and Rui Alexandre Castanho

Investigation into the Contribution of Social Tourism Programmes to the Satisfaction of Entrepreneurs in the Hospitality Sector 723
Despotaki N. C. Georgia and Doumi P. Maria

Hybrid Events and Congresses and Their Impact on the Greek Meetings Industry 737
Athina Papageorgiou and Alexandra Chalkia

Tourism Sustainability

Sustainability Assessment of Master Commercial European Ports Through Environmental Prismas

Archontoula Koskeridi and Vassiliki Balla

Abstract The functionality of European ports is the central interest of worldwide industry in current times. After the pandemic, ports must reassure that they are ready again to create economic and social value to the global economy as they manage to sustain their functions and continue their operations in an environmentally friendly manner. Healthy environment, safety issues, wind and solar energy and waste management are some of the many axes that current assessment will be based on. Master ports of Europe will be compared based on their environmental responsibility and performance. At the same time, climate change demands modifications to port operations and functionality. What is the role of robots in nowadays transportation? How automated processes affect in overall sustainability? Digitization of processes and innovation in operations are key elements of assessment in transport industry. Still each port follows specific rules, and native management style let alone must be aligned to current government policies. Does this in line mean that power countries play crucial role to the maintenance and further development of environmental management system that master ports follow?

Keywords Sustainability management · Environmental economics · Transport · Tourism

JEL Classifications Q01 · Q56 · Q5 · Z30

1 Introduction

Ports' operations offer to humanity by contributing to the direct and indirect employment of people, and as such, they play an essential role in the economic development

A. Koskeridi
University of the Aegean, Mytilene, Greece

V. Balla (✉)
University of Patras, Patras, Greece
e-mail: vballa@upatras.gr

© The Author(s), under exclusive license to Springer Nature Switzerland AG 2023

V. Katsoni (ed.), *Tourism, Travel, and Hospitality in a Smart and Sustainable World*, Springer Proceedings in Business and Economics,
https://doi.org/10.1007/978-3-031-26829-8_1

of each country (Roh et al., 2016). One could say that on the one hand, they act as social caretakers for workers and communities, reinforcing and supporting socio-economic priorities. In Europe, 2200 port operators employ more than 110.000 workers involved in loading and unloading ships and port services such as warehousing and logistics (Van Hooydonk, 2014). Moreover, ports are inevitable stations in maritime supply chains (Notteboom et al., 2020). They are characterized as numerous gateways for international trade. Eleven billion tons of maritime trade were carried by almost 98.140 ships as of 2018, accounting for about 60–70% of the value of global trade and 80% of trade volume (United Nations Conference on Trade and Development, 2020a). Shipping and ports have been the spearhead of global transport even in the worst crises, especially in pandemic COVID-19, transporting necessities such as food, medical supplies, primary materials, and manufactured products all over the world (UNCTAD, 2019a, 2020a). Since ports are hubs in the supply chain and, in combination with the scale of their port activities, they cause environmental and social externalities (Dinwoodie et al., 2012).

On the other hand, they bring negative impact to the environment even though they contribute significantly to the economy (Walker et al., 2019). Their different functions cause environmental impacts, that is, connectivity with maritime and land transport networks, cargo handling, logistics and distribution activities, industrial and semi-industrial activities, and energy production and distribution (Notteboom et al., 2020). Serious impacts on the environment are also created by the expansion and operation of ports and by shipping activities (Acciaro et al., 2014), for example, by emissions from combined transport serving inland ports and from ship berthing (Lam & Notteboom, 2014). Seas and oceans are affected by the negative impacts caused by ports and exacerbate the already existing problem (Darbra et al., 2009). They also cause local air pollution, noise and light pollution, traffic congestion, import of invasive species, and impacts from marine accidents and spills (Walker et al., 2019).

This research aims at building to contribute to the existing literature by providing a comparative overview of the largest commercial European ports and then to analyze the environmental sustainability of each port and the whole sample, using specific criteria. The rest of the paper is organized as follows: In the second section, the importance of sustainability for ports is analyzed; then in the third section, the data used for the research and given useful information about them is analyzed; and in the next section, the criteria and their subcategories are discussed. The paper concludes with the fifth section in which presents the results. The research is based around the following questions: Q1—Are all ports at the same level of sustainability and if not is there a difference between them in this direction? Q2—Is there a difference according to their location? Q3—Are measures for the sustainability of the ports in the sample being implemented?

2 Sustainability

In World's Association for Waterborne Transport Infrastructure (PIANC, 2022) definition of sustainability, a sustainable port is one in which the port authority, alongside with port users, proactively and responsibly adopts a green growth strategy. It ensures that through this strategy the long-term vision of the port is achieved promoting simultaneously the stakeholder's participation to meet the contemporary needs of the region it serves. Port sustainability includes addressing social needs, the economic development management system, and the concepts of minimizing port environmental externalities (Cheon, 2017; Cheon et al., 2017; Laxe et al., 2017).

In addition to the social and economic dimensions, one of the three dimensions of sustainability that have been specified is environmental sustainability (Giddings et al., 2002; Souza & Alves, 2018). Also, the public authorities and the wider community put strong pressure on ports to fulfill their social responsibility (De Grosbois, 2016). Consequently, it is becoming more and more important for ports to improve their "corporate responsibility (CR) profile" in order to appear environmentally conscious and sustainable and to respond to local community pressures and regulatory requirements (Acciaro, 2015).

The issue of an environmentally sensitive and sustainable attitude presents several challenges because it involves various issues, such as limiting emissions caused by current and upcoming port activities. Therefore, ports, in the spirit of sustainability, seek to increase their efficiency and performance and, consequently, their ability to compete as well as their impact to the wider economy. Meanwhile, the ports are taking a more socially active policy toward their local societies and a stronger environmental management, respecting the current rules (Sislian et al., 2016; Wang et al., 2021). Several ports are trying to become more "green" by incorporating and taking proactive practices and measures to reduce their impact on their environmental footprint (Davarzani et al., 2016), with users' and stakeholders' demands more and more concentrated on a more complete and holistic perspective so as achieve a balance among environmental, social, and economic interests.

The ports enclose various enterprises involved in a range of activities and provide a large number of services (Hakam, 2015). These enterprises are environmentally hazardous because of their characteristics, that is they produce a large amount of waste, and they release large scale of emissions and the noise pollution (Darbra et al., 2005). To minimize this damage, the sustainability comes first as the most significant aspect (Broesterhuizen et al., 2012).

3 Data

The data of this research have been derived from Sustainability Reports of the ports, as well as from recent Port Environmental Reports and recent Port Environmental Reviews. The data dates range from 2019 to 2021. Additional data were also retrieved from port's official websites. However, some ports created Master Plans—in which their environmental issues are mentioned—or other small issues for some additional information in an environmental category. The ports selection was based on Eurostat table from the archive image of the 20 largest container handling ports, 2008–2018.[1] This list was chosen as a guide because container handling ports provide high energy consumption and as such this is associated with the greatest need for energy consumption. Even though, in the corresponding list, the investigation ports amount to 20 only 17 were included to the research sample, as ports of Gioia Tauro, Izmit, and Ambarli—did not provide sufficient information. Therefore, given that 85% of the Eurostat list is gathered, the sample is considered adequate to provide secure results about the sustainability of European ports.

4 Criteria

According to the Word Port Sustainability Program (WPSP), their seventeen Sustainable Development Goals (SDGs) are grouped into five categories:

- Climate and Energy.
- Resilient Infrastructure.
- Safety and Security.
- Community Outreach and Port-City Dialogue.
- Governance and Ethics.

However, a report by European Sea Ports Organisation (ESPO) gives another comparison perspective such as:

- The instruments for achieving environmental ambitions (e.g., Port Vision, Spatial planning and infrastructure management, Port charges, etc.).
- Environmental management framework/environmental management standards (e.g., EcoPorts, ISO 14001, etc.).
- The development of environmental and sustainability management in ports.
- Elements of environmental management progress.
- Assistance tools and methods (e.g., Port Environmental Review System (PERS)).
- Established systems.
- Approaches to addressing environmental priorities (Air quality management, Energy conservation and climate change, Noise management, Waste management, Water management (consumption and quality, etc.).

Previous research dealing with similar issues has reported water and air quality as the main environmental problems together with waste disposal, noise of the areas, and conservation of their habitats (Comtois & Slack, 2007). Some other research included soil emissions, sediment and water discharges, and resource consumption (Puig et al., 2015). Many studies have been conducted on air quality (Winnes et al.,

[1] See Appendix, Fig. 10 "Top 20 ports handling containers, 2008–2018."

2015; EcoPorts, 2016; Lam & Notteboom, 2014), noise (Khoo & Nguyen, 2011; Mustonen, 2013; Schenone et al., 2014; Witte, 2016), and water pollution (Grifoll et al., 2011; Kröger et al., 2006). Water pollution is caused by waste disposal, ballast water, fuel oil, cargo, and oil spills (Ng & Song, 2010; OECD, 2011; Lam and Notteboom, 2014).

Although in literature review the ports' reports have a different pattern, which means that they were not organized in the same order of topics and with the same themes, in current research the criteria were organized into subcategories, in such a way so as to highlight the ports' similarities and differences.

1. *Safe and Healthy Environment*

All port data were compared among each on a basis of "Safe and Healthy Environment," which exposes the quality of flood risk management, as it is being affected directly by climate change. In addition to that, this criterion includes organized truck stops, which provide the rest necessary facilities such as toilets, showers, and security surveillance. Moreover, there is a discount for clean shipping that some ports may offer to ships that maintain all the compliance protocols for safe and healthy environment. In this criterion, the surrounding area of each port weighs significantly. Also, the measures that a port can take have a special weight, as with specific actions and measures it could attract more species or it could assist in the appropriate reproduction and co-existence of existing ones. Some examples are the Bird Valley in Rotterdam's port, that many different species of coastal birds and singing birds can be found in the valley. At the port of Antwerp, one of the measures taken is created of a "spawning zone" for the reproduction of fish and makes smooth quay walls rougher to create a refuge where small aquatic animals could live. Also, in the port of Bremen-Bremerhaven there is the Luneplate (an official EU bird habitat), and the near river Billerbeck is suitable as a habitat for plants and animals. Figure 1 shows the percentages of all ports and their contribution to this criterion. It can be observed that the ports of Rotterdam, Antwerp, and Bremen-Bremerhaven have a high contribution of 14%, while there are many ports with zero contribution, such as Gdańsk, Mersin, and La Spezia, or a minimal contribution of 3%, such as Le Havre and Algeciras. This result is considered rather logical given that these ports are bigger and as such are expected to demonstrate better safety regulations (Fig. 2).

2. *Climate and Energy*

The next criterion under assessment is "Climate and Energy." This represents the energy that each port uses, whether it is solar, wind, or electric. The outcome may be twofold. The port may have carried out an agreement to provide a different form of heat source such as steam some ports have carried out this agreement such as the ports of Rotterdam and Antwerp, or shore power, which allows ships ashore to shut down their engines or generator and connect to the grid. The latter reduces emissions (NOx, Sox, and particulate matter in ports' air), improves air quality, and reduces noise pollution. LED lights and upgraded terminals are included in this category alongside with the carbon capture and storage. Additionally, port's carbon footprint

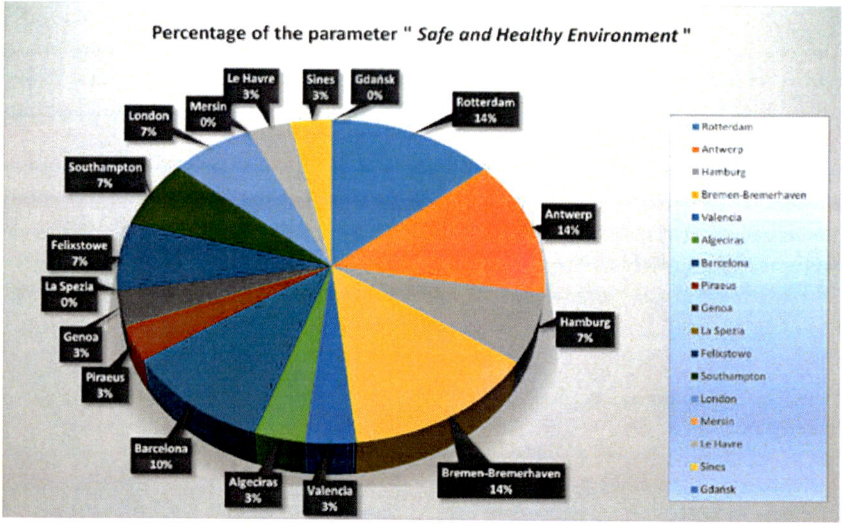

Fig. 1 Percentage of the parameter "Safe and Healthy Environment"

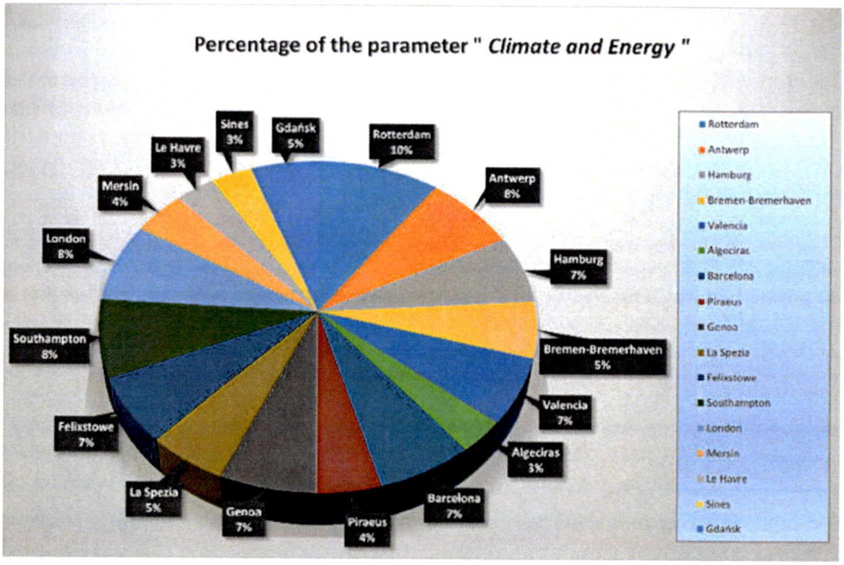

Fig. 2 Percentage of the parameter "Climate and Energy"

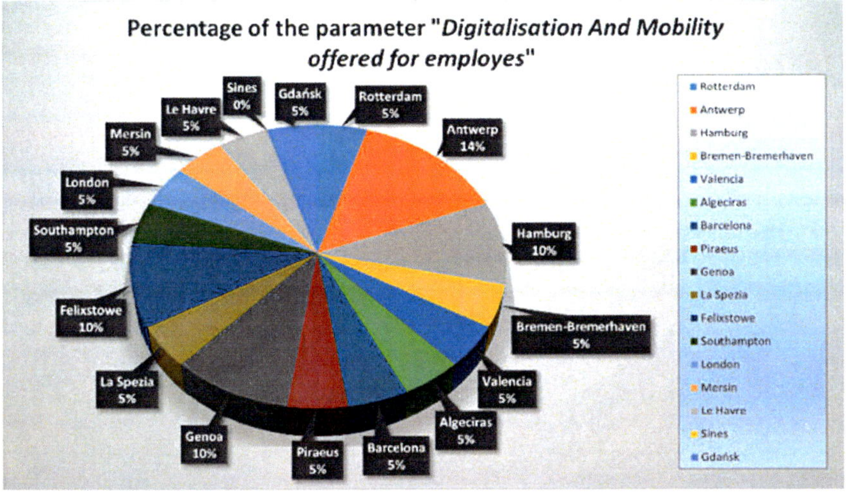

Fig. 3 Percentage of the parameter "Digitalization and Mobility offered for employees"

is also included in this criterion. Most importantly, this criterion assesses each port's energy management which is measured by environmental review tools such as ISO[2] and PERS[3] certifications.

Figure 3 provides percentage levels of all ports. The port with the highest percentage for this criterion is the port of Rotterdam with 10%, followed by the port of Southampton, London, and Antwerp with 8%. However, the ports with the lowest percentage are Algeciras, Le Havre, and Sines with 3%.

3. *Digitalization and Mobility Offered for Employees*

Digitalization is a rather powerful tool for ports as it provides timely mistake and omission prevention and information analysis. For example, it can be used for the estimation of the project's duration as well as the assessment of its infrastructure damage. Furthermore, digitization can provide an organization in the port area, either in terms of organizing traffic in the larger area of the port or in terms of sharing (live) data and information with several people at the same time. This can be achieved

[2] ISO certification is a seal of approval from a third-party body that a company runs to one of the international standards developed and published by the International Organization for Standardization (ISO). The ISO are an independent, non-governmental international organization who brings together experts to share knowledge and develop international standards that support innovation and provide solutions to global challenges. ISO Quality Services LTD, Source: https://www.isoqsltd.com/faq/.

[3] The Port Environmental Review System (PERS) does not only incorporate the main general requirements of recognized environmental management standards (e.g., ISO 14001), but also takes into account the specificities of ports. PERS builds upon the policy recommendations of ESPO and gives ports clear objectives to aim for. A PERS certification is valid for a period of 2 years. EcoPorts Tools, Source: https://www.ecoports.com/pers.

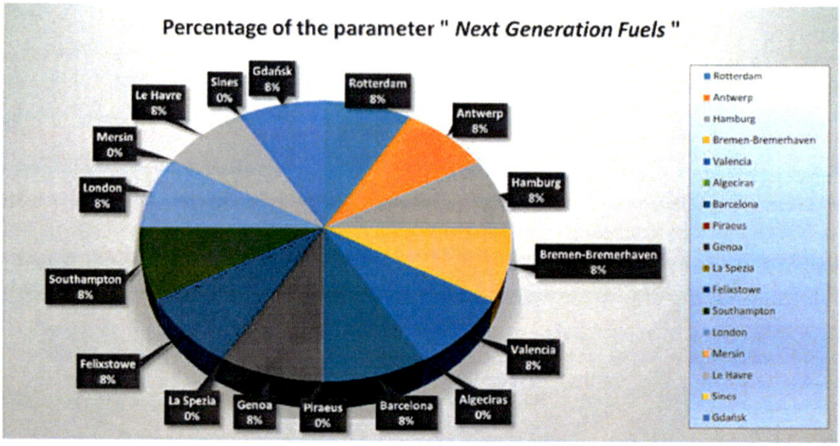

Fig. 4 Percentage of the parameter "Next Generation Fuels"

through technological programs and simulations of the whole port (such as buildings simulations, ones of bridges, and seabed) and with the assistance of the autonomous vehicles, the autonomous inland waterway vessels, and the drones. In combination with digitization, this criterion also reveals the projects and actions of some ports that have implemented eco-transport for their employees. For example, the Port of Antwerp has created the Bike Bus, which is a bus that takes its employees from one side of the port to the other, together with their bicycles through the tunnels and the Waterbus that takes them from the city to the port and vice versa together with their bicycles. Similar practice the Port of Bremen-Bremerhaven follows providing its employees with a subsidized "job ticket." Examining Fig. 3, Antwerp accounts for 14% of the total, with the ports of Hamburg, Genoa, and Felixstowe coming in second place. However, the smallest percentage is the 0% belonging to Sines, concluding that for this criterion no condition is satisfied.

4. *Measures*

Through this criterion, extra measures are assessed related to green policies. In more detail, such measures may include:

1. Energy-neutral buildings (triple glazing, heat and cold storage, and underfloor heating).
2. New types of asphalt.
3. Reducing energy consumption by running the fleet at lower speeds.
4. 24-h air quality monitoring stations,
5. SWOT analyses.
6. Additional bicycle parking and a charger for e-bikes.

Among all seventeen ports, no port was found that has not created additional measures to protect the environment, either more complex or simpler, so in this case there was no differentiation in percentages.

5. *Next Generation Fuels*

Fuel is one of the first sources of CO_2 release, so throughout the years many ports have made it their goal for their fuel source to release zero emissions. The main new generation of fuel found in ports is Liquefied Natural Gas (LNG). However, some ports also provide other types of fuel such as green hydrogen, blue hydrogen, gas to liquids (GTL), and biokerosene. Hydrogen is a fuel that has the advantage of being "burnt" without emitting carbon dioxide. On the other hand, blue hydrogen is produced by steam reforming of methane, whereas the carbon dioxide emissions are captured and stored with appropriate technology. Green hydrogen is produced via electrolysis of water and energy from renewable sources, such as the sun and wind. GTL is an alternative diesel fuel derived from natural gas, which burns cleanlier than conventional crude oil-based diesel.[4] Nonetheless, several ports are using other fuel sources, such as advanced diesel. In Fig. 4, these ports are listed with a percentage of 0%, while ports that have moved to the next generation of fuel are listed with a percentage of 8%.

6. *BIO*

The criterion "BIO" includes four parameters. The first one is the '**Bio-based cluster**', which reuses products (e.g., reuse and exchange of raw materials, such as a processing plastic waste, reuse of water) or supplies bio-based chemicals to factories. The next one is the '**Recycling Hub**', which includes actions and equipment, such as plastic and waste cleaning action organizations, waste and plastic collection vessels, and land-based waste management systems. Also, it includes waste separation and recycling of waste. Moving on to the '**Clean Port**' parameter, this refers to allocate everything a port does to be cleaner and greener. For example, Rotterdam's port has created 250-ha depot for contaminated dredged material, while the port of Antwerp researches and remediates soil, processes the contaminated sludge, makes regular checks in the waterbed quality, and makes durable cleaning of hulls and propellers. One more example is that of Bremen-Bremerhaven's port which introduced regulations for dealing with waste water from ships, such as the ballast water, the waste water from the scrubbers, and the domestic waste water. Finally, the fourth parameter '**Economic**' indicates ports that have a circular economy or circular chemistry. Figure 5 shows that most ports have a percentage in the 3% criterion, but the port of Antwerp has the highest percentage with 13% and the port of Barcelona has the lowest percentage with 0%. But these results were expected. Usually, the ports with the highest percentages for this criterion (13 and 10%) are also the largest ports which are able to ensure spatial and economic and other environmental protection measures.

[4] Shell GTL Fuel, Source: https://www.shell.com/business-customers/commercial-fuels/shell-gtl-fuel.html.

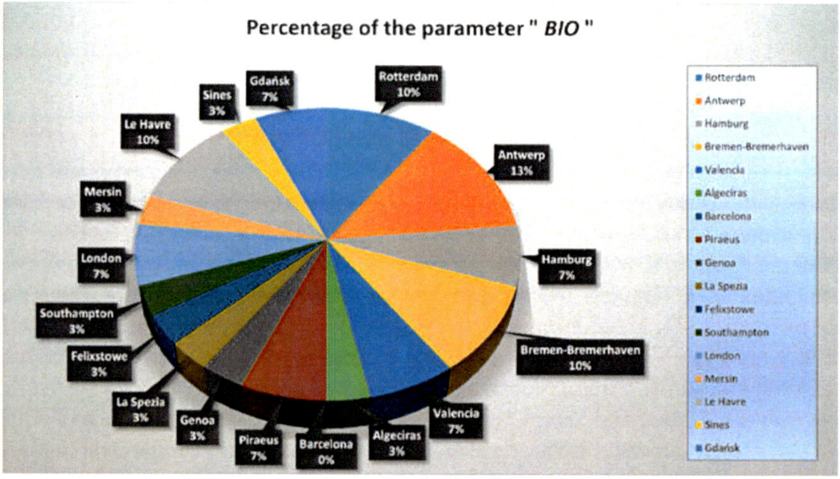

Fig. 5 Percentage of the parameter "BIO."

7. *Environmental Performance Indicators*

This criterion promotes the necessity to address the main environmental problems created by ports, such as air emissions, noise, and odors. In more detail, this criterion includes ports that have indexes for accident prevention, port cleanliness, and indicators related to climate change and energy transition. However, only the port of Piraeus and port of Bremen-Bremerhaven uses the QM Port indicator (Quality Management System indicators) and PERS. None of the rest ports have developed the exact environmental performance indicators. Furthermore, the ports evaluate their water quality, but the indicators used are not clearly mentioned and therefore removed from the generation of the figures. For example, the port of Piraeus for its seawater quality specifies that it measures pH, Salinity, BOD, COD, Enterococci, E-Coli, Total Coliforms, TDS, and Heavy Metals. Under these conditions, the percentages generated in the respective ports are presented in Fig. 6. With the highest percentages being in the ports of Rotterdam and Antwerp at 24%, the next highest percentage is 10% in the ports of Sines, Gdansk, Barcelona, and London, while the remaining 11 are at 5 and 0%.

8. *Database-Programs*

In the criterion "Database-Programs," all the programs that assist the ports to cope with their daily operations have been added together with the databases that each port handles. For example, Digital twin is a program that visualizes the port's facilities and can assist ports in reducing emissions, improving security procedures, and sharing data efficiently. However, all ports have their respective criterion resulting in a 6%

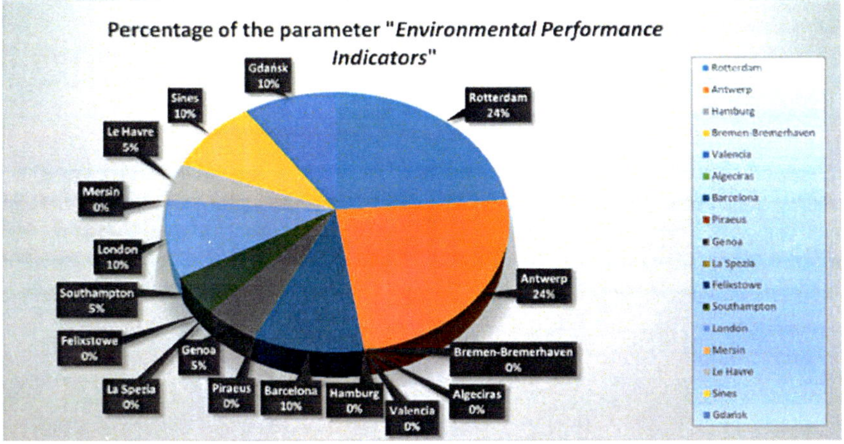

Fig. 6 Percentage of the parameter "Environmental Performance Indicators"

share in each apart from La Specia. However, each may have different programs and bases.

9. *Climatology Programs*

Each port has projects and programs to improve their ecological footprint, so that each one can contribute in its own way to a cleaner environment. For example, the port of Antwerp participates in the project "Operation Clean Sweep," where the main objective is to avoid the discharge of pellets into the environment, and the port of Bremen-Bremerhaven runs the research project called "Port Klima," which develops education and training modules for integrating adaptation to climate change. All ports, apart from La Specia, take significant actions through climatological programs, ensuring the community of their green concerns.

10. *Sustainable Development Goals (SDG'S)*

"The Sustainable Development Goals (SDGs), also known as the Global Goals, were adopted by the United Nations in 2015 as a universal call to action to end poverty, protect the planet, and ensure that by 2030 all people enjoy peace and prosperity.[5]" These 17 goals are integrated. They support the idea that the ports must balance between environmental sustainability and economic prosperity. The importance of this criterion is therefore quite significant for the progress and optimization of both the environment and the society. However, some ports did not provide information or when this was provided it was minimal and therefore the percentage is zero or very low. Figure 7 presents the highest percentage of the criterion is 12% belonging to the port of Antwerp and next with 11% is the port of Valencia.

[5] United Nations Development Programme, Source: https://www.undp.org/sustainable-development-goals.

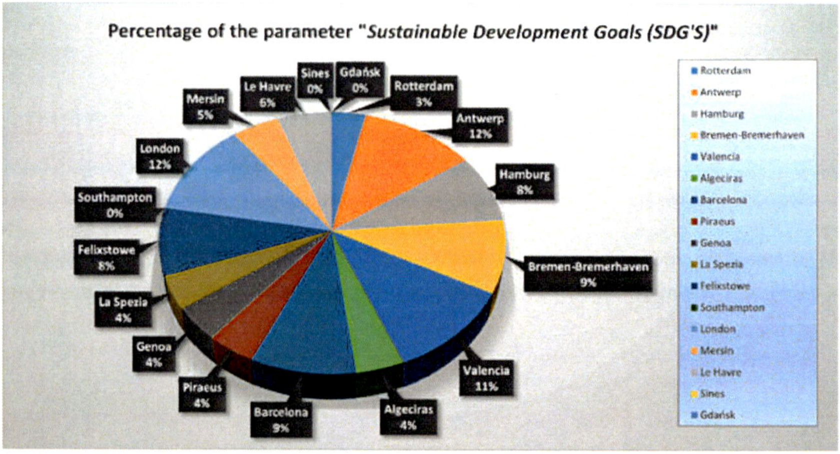

Fig. 7 Percentage of the parameter "Sustainable Development Goals (SDG'S)."

11. *GRI (Global Reporting Initiative)*

The GRI Standards are a modular system of interconnected standards. They allow organizations to publicly report the impacts of their activities in a structured way that is transparent to stakeholders and other interested parties.[6] Table 1 presents the categories of the GRI parameter for which ports.

In Fig. 8, it can be identified that several ports do not record the GRI standards in their reports; therefore, they have been assigned the 0% rate. However, the port of Bremen-Bremerhaven has the highest percentage with 18%, followed by the port of Piraeus with 17%.

12. *MARPOL*

The International Convention for the Prevention of Pollution from Ships (MARPOL) is the main international convention covering prevention of pollution of the marine environment by ships from operational or accidental causes.[7] It is therefore an important criterion as it links pollution to the marine environment and ships. Table 2 shows the criteria for the full protection of the marine environment under the MARPOL Convention, which was also another criterion for comparing port reports.

[6] https://www.globalreporting.org/media/wtaf14tw/a-short-introduction-to-the-gri-standards.pdf.

[7] https://www.imo.org/en/KnowledgeCentre/ConferencesMeetings/Pages/Marpol.aspx.

Table 1 GRI categories

GRI (Global Reporting Initiative)	
102—General disclosures	402—Labor/management relations 2016
103—Management approach	403—Occupational health and safety
201—Economic performance	404—Training and education
202—Market presence 2016	405—Diversity and equal opportunity
205—Anti-corruption	406—Non-discrimination
203—Indirect economic impacts	407—Freedom of association and collective bargaining
204—Procurement practices	408—Child labor
206—Anti-competitive behavior 2016	409—Forced or compulsory labor 2016
207—Taxes 2019	410—Security practices 2016
301—Materials	411—Rights of indigenous peoples 2016
302—Energy	412—Human rights assessment
303—Water and effluents	413—Local communities 2016
304—Biodiversity	414—Supplier social assessment 2016
305—Emissions	415—Public policy 2016
306—Waste	416—Customer health and safety 2016
307—Environmental compliance	417—Marketing and labeling 2016
308—Supplier environmental assessment	418—Customer privacy 2016
401—Employment	419—Socioeconomic compliance

Figure 9 shows the ports that mentioned the MARPOL criteria in their latest reports, with the highest percentage being the port of Bremen-Bremerhaven with 16%, followed by the port of Piraeus with 14%. However, 35% of ports show a rate of 0% for this criterion.

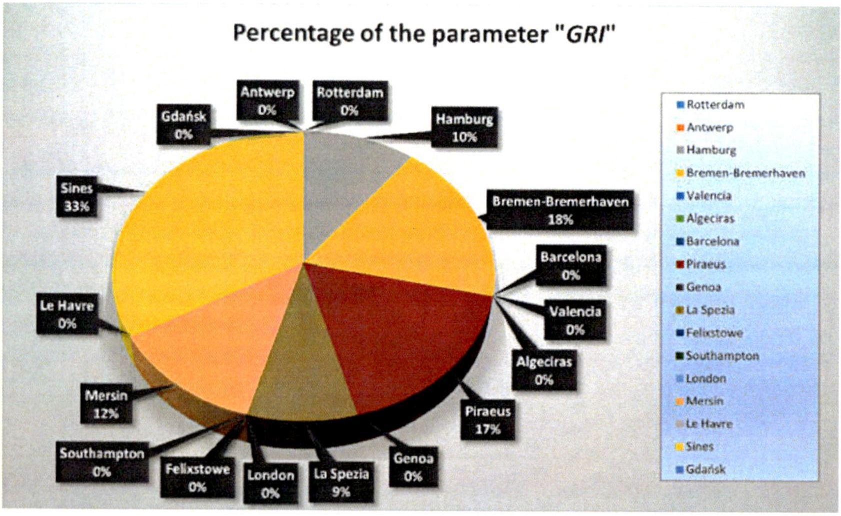

Fig. 8 Percentage of the parameter "GRI"

Table 2 MARPOL categories

MARPOL (International Convention for the Prevention of Pollution from Ships)
Annex I—Prevention of pollution by oil
Annex II—Carriage of chemicals by ship
Annex III—Prevention of pollution by harmful substances carried by sea in packaged form
Annex IV—Regulations for the prevention of pollution by sewage from ships
Annex V—Prevention of pollution by garbage from ships
Annex VI—Prevention of air pollution from ships

Results

This paper concludes that Europe's largest container ports differ from each other in terms of their environmental character, as presented in Table 3. This table breaks down the overall percentages by adding all the criteria and their parameters analyzed in the current research in an ascending order. All the criteria are 12 so the highest percentage for the ports that could be evaluated in this table is with the percentage 12. In other words, in the northwest of Europe there is an upward trend, ports to be more technologically, ecologically, and environmentally advanced than others. This in line enhances the idea that these ports present themselves as being more environmentally conscious. However, this cannot be considered entirely attainable as in the development of a business, the economy of the respective country it belongs play a major role to its sustainability goals. That is, a better economy also corresponds

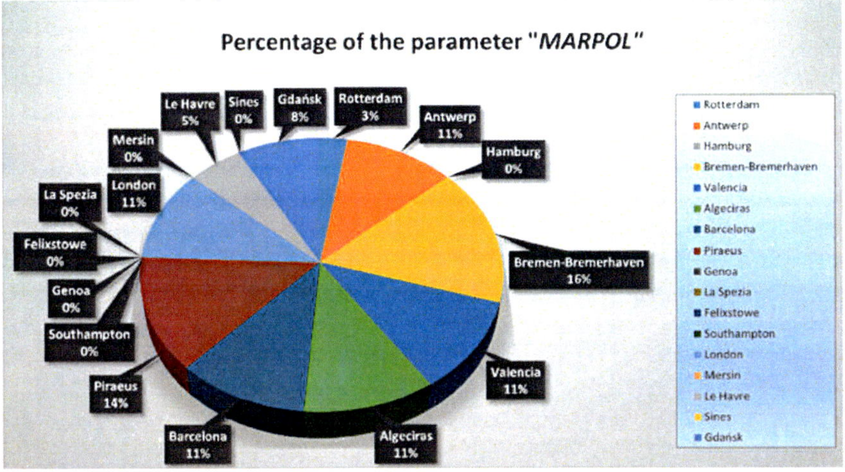

Fig. 9 Percentage of the parameter "MARPOL"

to the greater provision of money for appropriate technological development, such as the creation of smartports, and in combination with the environmental awareness of all those involved gives the result of creating a greener port. However, this does not leave the other ports behind as they too enter the game of competition and cooperation to sustainable world. Although they do not fulfill all the facilities of larger ports, still they are following significantly at their own pace. Having all the necessary measures and means for the environmental safety and protection of their environment and waters, they are struggling to manage in a greener manner.

Although various intersections of daily life were affected during the years of the pandemic, this did not significantly affect the environmental duties of several ports, as their upward trend from the rift and reefs of previous years is evident. Still, their adaptability to the new circumstances throughout these years is worthy of mention, in addition to some ports made sure to add the necessary equipment to deal with a possible new pandemic.

Table 3 Final percentage by adding all the criteria of the parameters in the research

No	Ports	Final percentage
1	Antwerp	10.2
2	Bremen-Bremerhaven	8.9
3	Rotterdam	8.3
4	London	8.0
5	Barcelona	7.4
6	Valencia	7.3
7	Hamburg	7.3
8	Felixstowe	6.7
9	Le Havre	6.6
10	Genoa	6.3
11	Piraeus	6.2
12	Gdańsk	6.2
13	Southampton	6.0
14	Algeciras	5.2
15	Sines	5.1
16	Mersin	4.8
17	La Spezia	2.7
18	Gioia Tauro	0.0
19	Istanbul	0.0
20	Izmit	0.0

Appendix

See Fig. 10.

Top 20 ports handling containers, 2008-2018
(thousand TEUs)

Rank 2018	Port	*	2008	2009	2010	2011	2012	2013	2014	2015	2016	2017		2018		Change 2018/2017 (%)	
			Total	Total	Total	Total	Total	Total	Total	Total	Total	Total	of which empty	Total	of which empty	Total	of which empty
1	Rotterdam (NL)	=	10 631	9 579	11 017	11 340	11 418	11 621	11 634	11 577	11 675	12 892	2 485	**13 598**	**2 635**	**5.5**	**6.0**
2	Antwerpen (BE)	=	8 379	7 014	8 144	8 317	8 174	8 256	8 812	9 370	9 891	10 032	1 495	**10 830**	**1 560**	**7.9**	**4.4**
3	Hamburg (DE)	=	9 767	7 031	7 906	9 035	8 891	9 302	9 775	8 848	8 929	8 860	1 191	**8 741**	**1 104**	**-1.3**	**-7.3**
4	Bremerhaven (DE)	=	5 451	4 552	4 858	5 911	6 111	5 822	5 731	5 467	5 510	5 458	858	**5 442**	**729**	**-0.3**	**-15.0**
5	Valencia (ES)	=	3 606	3 654	4 211	4 332	4 471	4 328	4 407	4 609	4 693	4 814	1 050	**5 169**	**1 211**	**7.4**	**15.3**
6	Piraeus (EL)	+2	437	667	850	1 681	2 815	3 199	3 493	3 360	3 736	4 120	805	**4 886**	**1 070**	**18.6**	**32.9**
7	Algeciras (ES)	-1	3 291	2 947	2 773	3 583	4 113	3 988	4 556	4 516	4 762	4 381	646	**4 773**	**770**	**9.0**	**19.2**
8	Gioia Tauro (IT)	+1	3 165	2 725	3 897	3 307	3 725	3 652	3 708	3 030	3 796	3 391	311	**4 005**	**347**	**18.1**	**11.5**
9	Felixstowe (UK)	-2	3 131	3 021	3 415	3 249	3 368	3 434	4 072	4 043	4 016	4 160	1 172	**3 781**	**985**	**-9.1**	**-16.0**
10	Barcelona (ES)	+1	2 567	1 846	1 928	2 006	1 745	1 717	2 056	1 950	2 225	2 998	756	**3 422**	**767**	**14.2**	**1.4**
11	Ambarli (TR)	-1			2 464	2 625	3 024	3 318	3 445	3 062	2 781	3 123	718	**3 170**	**685**	**1.5**	**-4.6**
12	Le Havre (FR) (*)	=	2 512	2 257	2 369	2 222	1 997	2 186	2 433	2 560	2 480	2 799	460	**2 866**	**467**	**2.4**	**1.6**
13	Genova (IT)	=	1 462	1 311	1 020	1 277	1 578	1 546	2 014	2 079	2 356	2 332	13	**2 554**	**29**	**9.5**	**115.5**
14	Southampton (UK)	=	1 617	1 385	1 567	1 591	1 489	1 489	1 894	1 956	2 040	2 008	541	**1 970**	**562**	**-1.9**	**3.9**
15	Sines (PT)	=	220	253	382	447	553	931	1 228	1 332	1 513	1 669	131	**1 750**	**151**	**4.9**	**15.2**
16	Gdansk (PL)	+2	183	233	510	685	933	1 189	1 232	1 041	1 559	1 473	255	**1 736**	**315**	**17.8**	**23.4**
17	London (UK)	+2	983	646	733	737	687	944	1 059	1 185	1 492	1 375	459	**1 680**	**417**	**22.2**	**-9.3**
18	Mersin (TR)	-1			1 016	1 127	1 251	1 367	1 484	1 428	1 406	1 554	338	**1 662**	**379**	**7.0**	**12.3**
19	La Spezia (IT)	-3	1 186	840	1 181	1 205	1 181	1 207	1 262	1 579	1 605	1 612	93	**1 653**	**191**	**2.5**	**105.6**
20	Izmit (TR)	+1			416	508	630	808	899	989	1 143	1 316	236	**1 598**	**290**	**21.4**	**22.6**
Total top 20 ports (*)			**64 363**	**54 386**	**63 093**	**67 267**	**69 569**	**70 181**	**75 483**	**74 247**	**77 710**	**80 413**	**13 988**	**85 285**	**14 665**	**6.1**	**4.8**

( ) not available
Note: TEU: Twenty-foot Equivalent Unit (unit of volume equivalent to a 20 foot ISO container)
(*) column indicates number of positions lost or gained compared to 2017
(*) 2012-2013: partially estimated by Eurostat
(*) Total figure for the ports being part of the top 20 ports of the countries reporting data during the reference year concerned. Turkish ports are not included in 2008 and 2009
Source: Eurostat (online data code: mar_mg_am_pvh)

eurostat

Fig. 10 Top 20 ports handling containers, 2008–2018 (thousand TEUs). *Source* Eurostat

References

Acciaro, M., Vanelslander, T., Sys, C., Ferrari, C., Roumboutsos, A., Giuliano, G., Lam, J. S. L., & Kapros, S. (2014). Environmental sustainability in seaports: A framework for successful innovation. *Maritime Policy and Management, 41*(5), 480–500.

Acciaro, M. (2015). Corporate responsibility and value creation in the port sector. *International Journal of Logistics Research and Applications, 18*, 291–311.

Broesterhuizen, E. F. M., Vellinga, T., Docters van Leeuwen, Linda, Zwakhals, J. W., Taneja, P., & Nijdam, M. (2012). Sustainability as a procurement criterion for port investments. CESUN2012, Delft.

Cheon, S. (2017). The economic-social performance relationships of ports: Roles of stakeholders and organizational tension. *Sustainable Development, 25*(1), 50–62.

Cheon, S., Maltz, A., & Dooley, K. (2017). The link between economic and environmental performance of the top 10 US ports. *Maritime Policy and Management, 44*(2), 227–247.

Comtois, C., Slack, B. (2007). Restructuring the maritime transportation industry: global overview of sustainable development practices, Québec. Retrieved from March 26 2017, from http://www.bv.transports.gouv.qc.ca/mono/0938424.pdf 26.03.17.

Darbra, R. M., Pittam, N., Royston, K. A., Dabra, J. P., & Journee, H. (2009). Survey on environmental monitoring requirements of European ports. *The Journal of Environmental Management, 90*(3), 1396–1403.

Darbra, R. M., Ronza, A., Stojanovic, T. A., Wooldridge, C., & Casal, J. (2005). A procedure for identifying significant environmental aspects in sea ports. *Marine Pollution Bulletin, 50*(8), 866–874.

Davarzani, H., Fahimnia, B., Bell, M., & Sarkis, J. (2016). Greening ports and maritime logistics: A review. *Transportation Research Part D: Transport and Environment, 48*, 473–487.

De Grosbois, D. (2016). Corporate social responsibility reporting in the cruise tourism industry: A performance evaluation using a new institutional theory based model. *Journal of Sustainable Tourism, 24*, 245–269.

Dinwoodie, J., Truck, S., Knowles, H., Benhin, J., & Sansom, M. (2012). Sustainable development of maritime operations in ports. *Business Strategy and the Environment, 21*, 111–126.

EcoPorts. (2016). ESPO/EcoPorts Port Environmental Review 2016, Brussels. Retrieved from August 06, 2017, from https://www.ecoports.com/laravel-filemanager/files/common/publications/ESPO_EcoPorts_Port_Environmnetal_Review_2016_v1.pdf.

Giddings, B., Hopwood, B., O'brien, G. (2002). Environment, economy and society: Fitting them together into sustainable development. *Sustainable Development, 10*, 187–196.

Grifoll, M., Jordà, G., Espino, M., Romo, J., & García-Sotillo, M. (2011). A management system for accidental water pollution risk in a harbour: The Barcelona case study. *Journal of Marine Systems, 88*, 60–73.

Hakam, M. H. (2015). Nordic container port sustainability performance: A conceptual intelligent framework. *Journal of Service Science and Management, 8*(01), 14–23.

Khoo, I. H., Nguyen, T. -H. (2011). Study of the noise pollution at container terminals and the surroundings, long beach, California, Retrieved August 18, 2017, from https://www.metrans.org/sites/default/files/research-project/0909_Khoo_METRANS_final_report_0_0.pdf.

Kröger, K., Gardner, J. P. A., Rowden, A. A., & Wear, R. G. (2006). Long-term effects of a toxic algal bloom on subtidal soft-sediment macroinvertebrate communities in Wellington Harbour, New Zealand. *Estuarine, Coastal and Shelf Science, 67*, 589–604.

Lam, J. S. L., & Notteboom, T. (2014). The greening of ports: A comparison of port management tools used by leading ports in Asia and Europe. *Transport Reviews, 34*(2), 169–189.

Laxe, F. G., Bermúdez, F. M., Palmero, F. M., & Novo-Corti, I. (2017). Assessment of port sustainability through synthetic indexes: Application to the Spanish case. *Marine Pollution Bulletin, 119*(1), 220–225.

Mustonen, M. (2013). *Noise as an environmental challenge for ports, Stockholm*. Retrieved September 17, 2017, from http://projects.centralbaltic.eu/images/files/result_pdf/PENTA_result4_noise.pdf.

Ng, A. K. Y., & Song, S. (2010). The environmental impacts of pollutants generated by routine shipping operations on ports. *Ocean and Coastal Management, 53*, 301–311.

Notteboom, T., Van Der Lugt, L., Van Saase, N., Sel, S., Neyens, K. (2020). The role of seaports in green supply chain management: initiatives, attitudes, and perspectives in Rotterdam, Antwerp, North Sea Port, and Zeebrugge. Sustainability.

OECD. (2011). *Environmental impacts of international shipping: the role of ports*. OECD Publishing.

PIANC. (2014). Sustainable Ports—A Guide for Port Authorities. Report N∘ 150; Environmental Navigation Commission: Brusells, Belgium. Retrieved March 22, 2022, from http://www.pianc.org.

Puig, M., Wooldridge, C., Casal, J., & Darbra, R. M. (2015). Tool for the identification and assessment of environmental aspects in ports (TEAP). *Ocean and Coastal Management, 113*, 8–17.

Roh, S., Thai, V. V., & Wong, Y. D. (2016). Towards sustainable ASEAN port development: Challenges and opportunities for Vietnamese Ports. *The Asian Journal of Shipping and Logistics, 32*(2), 107–118.

Schenone, C., Pittaluga, I., Repetto, S., Borelli, D. (2014). Noise pollution management in ports: a brief review and the EU MESP project experience. 21st International Congress on Sound and Vibration. Beijing, China. Retrieved August 18, 2017, from https://www.researchgate.net/profile/Corrado_Schenone/publication/265216678_NOISE_POLLUTION_MANAGEMENT_IN_PORTS_A_BRIEF_REVIEW_AND_THE_EU_MESP_PROJECT_EXPERIENCE/links/5405b9fb0cf2c48563b185f9?origin=publication_detail.

Sislian, L., Jaegler, A., & Cariou, P. (2016). A literature review on port sustainability and ocean's carrier network problem. *Research in Transportation Business and Management, 19*, 19–26.

Souza, J. P. E., & Alves, J. M. (2018). Lean-integrated management system: A model for sustainability improvement. *Journal of Cleaner Production, 172*, 2667–2682.

Sustainable Development Practices, Québec (2007). Retrieved March 26, 2017, from http://www.bv.transports.gouv.qc.ca/mono/0938424.pdf.

UNCTAD. (2019a). Review of maritime transport 2019a. In *United nation conference on trade and development*. Geneva, Switzerland.

UNCTAD. (2020a). Ports in the fight against COVID-19. https://tft.unctad.org/ports-covid-19/.

Van Hooydonk, E. (2014). *Port labour in the EU: Labour market, qualification and training, health and safety*. European Commission.

Walker, T. R., Adebambo, O., Del Aguila Feijoo, M. C., Elhaimer, E., Hossain, T., Edwards, S. J., Morrison, C. E., Romo, J., Sharma, N., Taylor, S., & Zomorodi, S. (2019). *Environmental effects of marine transportation* (pp. 505–530). World Seas: An Environmental Evaluation, Academic Press.

Wang, C., Haralambides, H., & Zhang, L. (2021). Sustainable port development: The role of Chinese seaports in the 21st century Maritime Silk Road. *International Journal of Shipping and Transport Logistics, 13*, 205–232.

Winnes, H., Styhre, L., & Fridell, E. (2015). Reducing GHG emissions from ships in port areas. *Research in Transportation Business and Management, 17*, 73–82.

Witte, J. R. (2016). Noise in ports. GREEN4SEA Conf. Award, 2016 Retrieved from August 8, 2017, from https://www.green4sea.com/noise-in-ports-2/.

Perspectives of Sustainable Tourism Development in the Region – Case Study

Lubica Sebova, Klara Chovanova, Izabela Lazurova, and Adrian Svec

Abstract Tourism development can be defined as the process of creating strategies and plans to increase, support, and promote tourism in a destination. A tourism region is a destination with primary and secondary offer in which the visitor has the opportunity to experience tourism activities. In our paper, we focus on tourism development the selected region, which is the High Tatras—Foothills, as one of the most visited regions in Slovakia. The aim of the paper is to propose further options for the sustainable development of tourism in the region on the basis of the analysis and evaluation of the current state of tourism in the region. The intention is to examine and analyze the tourism offer in the region, to find out what is the demand for tourism in the region and to identify the perspectives for the development of sustainable tourism in the region. The economic principle and securing income from tourism are one of the decisive factors, but at the same time, the main emphasis in the development should be placed on preserving the natural environment, i.e., respecting the environmental principle, and at the same time, also the local population should benefit from the development of tourism, i.e., the social principle. The result of this strategy is to satisfy all stakeholders and increase the region's competitive position on the international tourism market, in line with respect for the protection of natural and cultural resources.

Keywords Destination · Region · Sustainability · Tourism

JEL Classification L83 · Z32 · M31

1 Introduction

Tourism development is an impulse for destination development. At a basic level, tourism development can be defined as the process of developing strategies and plans to support and promote tourism in a particular destination. In tourism development,

L. Sebova (✉) · K. Chovanova · I. Lazurova · A. Svec
Matej Bel University, Banska Bystrica, Slovakia
e-mail: lubica.sebova@umb.sk

© The Author(s), under exclusive license to Springer Nature Switzerland AG 2023

V. Katsoni (ed.), *Tourism, Travel, and Hospitality in a Smart and Sustainable World*, Springer Proceedings in Business and Economics,
https://doi.org/10.1007/978-3-031-26829-8_2

it is necessary to examine the demand and supply of tourism and the perspectives for its development. Tourism development is a subject of interest of various tourism organizations and subjects.

In this paper, we discuss the perspectives of tourism development using the example of the Slovak region of High Tatras—Foothills. Visitors to the Tatra region mainly perceive resorts directly in the High Tatras and explore nowadays the foothills of the High Tatras. The High Tatras are overcrowded in some resorts, and the foothills of the High Tatras offer the beauty of the Tatra region without the crowds of visitors and are a space for the development of sustainable tourism. The region is not industrially oriented, and tourism is a great opportunity for its overall economic development. In this paper, we focus on the region perspectives, how it could exploit its potential and increase visitor numbers and competitiveness under the principle of sustainability. The destination management organization plays an important role in the development of sustainable tourism in the region.

2 Literature Review

A number of authors have addressed the theoretical knowledge of examining tourism development in a destination. They theoretically define tourism development, analyze tourism supply and demand, and address what are the perspectives for tourism development. They define the tourism destination and the organizations that coordinate tourism development in the territory. The focus is on its sustainability. Peace and a favorable international climate are essential conditions for tourism development. Its development is also influenced by natural, ecological, technical, demographic, economic, professional, administrative, and other factors.

Tourism development is the process of creating and sustaining a tourism industry in a locality. At a basic level, tourism development can be defined as the process of developing strategies and plans to increase, further develop, and promote tourism in a particular destination (Abir & Khan, 2022).

Oliveira (2022) refers to tourism development as a strategy of cultural, economic, social, and environmental development to meet the needs of visitors and residents. Saner-Yiu et al. (2015) describe tourism development as the activity of planning and implementing strategies to develop the industry.

In the global economy, tourism is one of the most prominent and growing industries (Buhalis, 2000). This sector plays an important role in supporting the national economy. Increasing the flow of tourism can bring positive economic outcomes to nations, especially in gross domestic product (GDP) and employment opportunities (Manzoor et al., 2019).

The focus is on the destination. The authors define destination as follows. Oliveira (2022) as a place that has a set of cultural and natural heritage resources, infrastructure, and services that attracts tourism visitors. According to Demir et al. (2021), a destination place is a place that is visited when traveling or on a trip. Ramos et al. (2019) refer to a destination as an area that relies heavily on the primary tourism supply to generate income. Gómez-Suárez and Martínez-Ruiz (2016) refer to a destination as a geographic area consisting of a set of resources and attractions that are visited by tourism visitors.

Destinations are strongly influenced by globalization and high competition in the tourism market, which implies a competitive struggle for the destination (Valeri, 2021). As a consequence of the high competition, small operators have a harder chance to succeed, and mergers are occurring. Pooling of operators is characteristic mainly for small destinations that cannot compete with fully developed destinations. The creation of a destination, with a view to increasing competitiveness at the same time, should be in line with the principles of sustainable tourism development. This method also follows the intentions of the state tourism policy, which emphasizes increasing competitiveness, sustainable development, employment development, regional development, unified presentation, and promotion of Slovakia (Strategy of sustainable tourism development to the year 2030, 2020).

Issues of tourism sustainability and urban development have become top priorities for public policy makers around the world (Keller, 2015). Today, perhaps more than ever, there is a need to manage sustainable tourism development, which cannot be achieved without taking into account environmental issues and their global dimension. The various problems and demands of society and urban development can be addressed by transforming cities into attractive tourism destinations (Zamfir & Corbos, 2015).

In defining sustainable development, the UNWTO defined three basic prerequisites for sustainable tourism. The first is the optimal use of environmental resources, which forms a key element in tourism development, maintaining essential ecological processes and helping to protect natural heritage and biodiversity. The second is to respect the sociocultural authenticity of host communities, to protect their cultural heritage and traditional values, and to contribute to intercultural understanding and tolerance. The third prerequisite for sustainable tourism is ensuring viable, long-term economic operations, providing socio-economic benefits to all stakeholders that are equitably distributed, including stable employment and income-generating opportunities and the use of social services for host communities, and contributing to poverty alleviation (www.unwto.org, 2022a, 2022b).

Janusz and Bajdor (2013) revealed the essence of sustainable tourism and described the main framework of sustainable tourism which consists of twelve elements—economic capacity, local prosperity, employment quality, social balance, visitor satisfaction, local control, local community satisfaction, cultural richness, physical integration, biodiversity, resource efficiency, and clean environment.

Another model of sustainable tourism was also limited to present key elements—carrying capacity, local community, experience, destination management organization, seasonal management, demand management, and smart destinations (Adillon, 2019).

Ryglová et al. (2011) divide the preconditions for tourism development into three main categories, namely selective (incentive) factors, locational, and implementation conditions. The first category, which includes selective factors, is intended to stimulate the emergence and development of tourism, which is linked to consumer behavior. Selective factors are further divided into objective and subjective factors. Objective factors include political and economic factors, demographic realities, and administrative conditions. Subjective factors mainly include psychological factors that influence consumer behavior. This includes, for example, advertising, promotions, and trends. The second category is locational conditions, which are determined by the territory. We divide locational conditions into natural conditions, which include, for example, climate and terrain and social attractions, which include customs, traditions, monuments, etc. The third category is implementation factors, which enable visitors to access and use services in the destination, and these include transport, accommodation, and food services.

Sustainability is the perspective of long-term tourism development. Given the need for sustainable development of tourism destinations, it has become critical to design destinations in a way that mutually facilitates experiences to enhance the perceived value of the visitor experience and the sustainability of the destination. However, it is only recently that tourism research has turned its attention to sustainable experiences as a potential means of gaining competitive advantage while enhancing destination sustainability and perceived experience value (Poudel & Nyaupane, 2013).

Sustainable development represents a meeting point for environmentalists and developers. In other words, sustainable development can be seen as a combination of two processes, development, and sustainability (Dresner, 2008).

Actions toward the development of a region's tourism should be implemented taking into account the global trends that shape the future of tourism and the local conditions and specificities that determine the current and potential state of development. Today, new trends in tourism are emerging in the aftermath of the blockage caused by the COVID-19 pandemic (Ianioglo & Risannen, 2020).

Tourism destination managers need to understand what visitors perceive as engaging, intense, and memorable. However, they also need to be mindful of how residents perceive local tourism development and how it affects their community. This is one of the fundamental aspects of tourism development, as host communities that support tourism development tend to be more hospitable to visitors, which influences visitor satisfaction and loyalty. In addition, interaction with host community residents is a key component of the quality of the experience that contributes to the long-term success and sustainability of destinations (Pinto & Guerreiro, 2020). Slovakia has rich natural and cultural potential, which are prerequisites for tourism development. This wealth of the country combined with the activities of tourism entities creates real preconditions for the prospective development of this sector.

Therefore, the creation of a tourism product with a view to its sustainability is an important task for the entities involved in the development of tourism in a destination. According to the UNWTO definition, a tourism product is a combination of tangible and intangible elements such as natural, cultural, and man-made resources, attractions, facilities, services, and activities around a specific center of interest that form the core of the destination's marketing mix and creates the overall visitor experience, including the emotional aspects for potential visitors. The tourism product is valued and sold through distribution channels and has a life cycle (www.unwto.org, 2022a, b).

For a destination to be successful and attract visitors, it needs to create a full-fledged tourism organization. The success of a destination requires balanced tourism development and combines understanding with research and concerted efforts to sustain desired tourism initiatives and maintain competitiveness. Destination management ensures the optimal allocation and utilization of resources for the best use and in the most sustainable manner (Varghese, 2021).

Nowadays, destination management organization should consider social media as one of the main tools to improve their competitiveness, for example, by engaging with visitors and thus understanding visitor behavior and needs (Carvalho et al., 2021).

Magas (2010) discusses tourism trends and new challenges such as sociocultural aspects, consumption changes, and technological aspects, along with integral tourism product planning, toward a tourism experience. Destination management organization can thus be the strategic leader and developer of the destination, facilitating partnership and collaboration in line with the shared vision of the destination. To ensure the quality of services and facilities at the destination, it is critical to develop destination management organization that can monitor the quality aspects of the resources at the destination.

Destination marketing requires strategic planning, decision-making, and organization. Effective positioning will result in a strong brand that develops an emotional and productive two-way relationship. Regardless, destination managers should have relevant knowledge and understanding of traditional and current marketing channels to better reach potential visitors (Camilleri, 2018).

Tourism consumers are becoming more discerning and are seeking to base their travel decision-making process on relevant and credible tourism information. In recent years, user-generated content on social media, travel bloggers' opinions, and entertainment programs in the media have influenced the public's travel buying behavior and act as a driving force for the development of tourism products. On the other hand, tourism organizations and destination management organizations face major challenges in communicating tourism product attributes as they cannot be experienced before consumption. Therefore, they need to know their visitors and know-how and through what means or platforms of communication they can inform potential visitors (Dinis et al., 2021).

3 Methodology

The subject of the paper is sustainable development of tourism, and the object of study is a selected tourism region, namely the High Tatras—Foothills region in Slovakia. The aim of the paper is to define further perspectives of sustainable tourism development in the High Tatras—Foothills region on the basis of the analysis and evaluation of the current state of tourism in the region. The main objective has been divided into the following sub-objectives:

O1: Examine the tourism offer in the region.
O2: Find out what is the demand for tourism in the region.
O3: Indicate the prospects for sustainable tourism development in the region.

In the article, we use a combination of theoretical and empirical research methods, analysis and synthesis, and induction and deduction. As a tool of information gathering, we use the method of questioning. The questionnaire was distributed electronically to visitors to the region. The survey was conducted in the months of March–April 2022. 287 respondents participated in the survey. The collected data were evaluated. The questionnaire consisted of 15 questions, the structure of which was designed to identify the visitor and their preferences in the region, while giving room for their own opinions and suggestions. 11 questions were focused on the High Tatras—Foothills region, and 4 questions were focused on the personal profile of the respondent. To evaluate the questionnaires, we used mathematical-statistical methods, namely the method of arithmetic mean. We have also used graphical representation of the results of the questionnaire survey by means of tables and graphs, which we will use for the concrete presentation of the data obtained from the processing of the questionnaires. In the text, we present numerical data in relative or absolute terms.

For the proposal of further development of sustainable tourism in the region, we assess the current situation with SWOT analysis. The SWOT analysis assesses the strengths, weaknesses, opportunities, and threats that exist in the market. The importance of tourism in the region can be determined through various indicators, we will focus on the tourism function in the High Tatras—Foothills region. The tourist function, the so-called Baretjes and Defert index, captures the number of inhabitants in the region and the number of beds in accommodation facilities. We calculate the tourist function and then the density of beds.

We use primary and secondary sources to process the work. Secondary sources include domestic and foreign literature sources, internal documents of the regional tourism organization High Tatras—Foothills, and data drawn from the Internet. Primary data were obtained through the evaluation of questionnaires.

4 Results

High Tatras—Foothills are a tourism destination located in the north of Slovakia near the highest mountain of the country, the High Tatras. Slovakia has a total of 21 tourism regions, the territory of the High Tatras—Foothills falls under the Tatra Tourism Region. The Tatra Region is the most touristically interesting region in Slovakia. It consists of the sub-regions of the Tatra Mountains and Northern Spiš. The region includes the districts of Poprad, Kežmarok, and Stará L'ubovňa.

The High Tatras—Foothills region is characterized by its rich nature, its territory extends in the Poprad and Hornád basins, and it extends into the Tatra National Park, the Low Tatras National Park, and the Slovak Paradise National Park. The Čierny Váh flows through the area, and the Hornád River rises near the village of

Liptovská Teplička. The tourism offer in the High Tatras—Foothills region consists of 9 villages and their natural and cultural-historical potential. In a small area, visitors to the region are offered an aquapark and a thermal swimming pool, a golf course, campsites, wellness centers, monumental reserves of folk architecture, relax hotels, congress and conference hotels and facilities. In winter, it offers excellent conditions for cross-country skiing, ski mountaineering, winter hiking, winter adrenaline activities, and events in ski resorts, of which there are several in the region. In summer, it is an ideal place for hiking, mountain hiking, cycling, stays in rural homesteads, farms and guesthouses, combined with economic activities, folk traditions, regional gastronomy, and folklore events. In terms of social events, modern and folklore festivals, as well as gastronomic events, are organized in the region. The wide natural wealth, extensive meadows and pastures, also create suitable conditions for farmers who produce and sell their products in organic quality.

The importance of tourism in the region can be determined through various indicators, and we focus on the tourist function in the High Tatras—Foothills region. The tourist function captures the number of inhabitants in the region and the number of beds in accommodation establishments in the region, which gives us a picture of the region's accommodation capacity, which is essential for residential tourism (Table 1).

The calculation of the tourism function provided us with data on how many beds are per 1,000 inhabitants of a region or state. The tourist function of the High Tatras—Foothills region results in 46 beds per 1,000 inhabitants. To better interpret the result of the region's tourism function, we calculated the tourism function of Slovakia in the same way, which is approximately 34 beds per 1,000 inhabitants in 2020. Based on these calculations, we can compare the High Tatras—Foothills region with Slovakia and conclude that the tourist function in the region is 37.19% higher than the average tourist function in Slovakia.

The index of the tourist function can also be supplemented by the employment indicator, which assesses employment in a particular industry or sector in relation to the population of the region or the state. Due to the lack of information on the exact number of people employed in tourism in the High Tatras—Foothills region, we have omitted this indicator because the interpreted data would not have the correct predictive value.

Table 1 Tourist function of the region in 2020

Destination	Number of the inhabitants	Capacity of accommodation facilities (beds)
High Tatras—Foothills	14,319	6,714
Slovak Republic	5,459,000	181,588

Source Own elaboration to datacube.statistics.sk, 2022

Calculation of tourism function (Tf):

High Tatras—Foothills = (6, 714)/(14, 319)*1 000 = 46,89

Slovak Republic = (186, 588)/(5, 459 000)*1 000 = 34,18

Table 2 Bed density in 2020

Destination	Area (100 km^2)	Capacity of accommodation (beds)	Bed density (beds/100 km^2)
High Tatras—Foothills	2,23	6,714	3,010
Slovak Republic	490,35	181,588	370

Source Own elaboration based on data from datacube.statistics.sk, 2022
Calculation of bed density:
High Tatras—Foothills = 6 714/2,23 = 3 010
Slovak Republic = 181 588/490,35 = 370

Another indicator that we looked at is bed density. This indicator provides data on the number of beds in accommodation per 100 km^2; in layman's terms, it shows how many beds are per 100 km^2 (Table 2).

The average number of beds per 100 km^2 in Slovakia in 2020 was approximately 370 beds. In the High Tatras—Foothills region, it was 3 010 beds per 100 km^2, which is more than 8 times higher than in Slovakia. On this basis, we can conclude that there is a really high concentration of beds per 100 km^2 in the region, which is well above the Slovak average.

From the point of view of visitors, the region is mainly aimed at the segment of families with children. Accommodation facilities are adapted to this type of visitor, and families with children are the predominant clientele in both the winter and summer seasons. There are a sufficient number of facilities for tourism visitors in the region.

Visitors to the region are mainly families with children and people traveling for sightseeing and tourism. Visitors are mostly from Slovakia, followed by visitors from Poland, the Czech Republic, Hungary, and Germany. The largest number of foreign visitors is from Poland, which is only 35 km away from Vel'ká Lomnica, and from the Czech Republic, where the High Tatras are considered to be one of the most popular holiday destinations for Czech inhabitants. According to the Statistical Office of the Slovak Republic, the region was visited by more than 355,000 visitors between 2017 and 2021, of which 25.61% were foreign visitors, representing more than 90,000 visitors. The average figures speak of about 27% of foreign visitors in the current year. Between 2017 and 2019, the number of visitors to the region showed an upward trend. The highest number of visitors was recorded in 2019 at 101,505 visitors. The number of visitors in 2020, due to the impact of the COVID-19 pandemic, decreased significantly. Accommodation establishments, but also other tourism service providers, had to significantly limit their activities for more than 6 months. Due to the very short winter season and the inter-season, the number of visitors in 2020 was 65 057. The pandemic affected tourism in 2021 to an even greater extent than in 2020. This year, accommodation establishments and other tourism service providers, had to limit their activities for more than 5 months, even during the top seasons, which include the Easter and Christmas holidays and the New Year's Eve to Three Kings Day. In 2021, only 34,109 visitors to the region were recorded as a result of these developments. The number of domestic and international and also the total number of visitors to the region in each year, from 2017 to 2021, are shown in Table 3.

The number of domestic and international overnight stays as well as the total number of overnight visitors in the region in each year, from 2017 to 2021, are shown in Table 4.

The demand and attitude of visitors to tourism products and services in the High Tatras—Foothills region were surveyed by means of a structured questionnaire. The structure of respondents in terms of gender consisted of 181 women (63.07%) and 106 men (36.93%). The largest group of respondents, in terms of age, was the category of respondents aged between 41 and 50 years (36.59%). The age categories 21–30 years (18.47%), 31–40 years (17.77%), and 51–60 years (16.38%) were approximately equally represented. For the employment status, most respondents gave the answer employee (67.25%).

At the beginning of the questionnaire survey, we asked the respondents whether they had ever visited the High Tatras—Foothills region. Out of a total of 287 respondents, 280 (97.56%) answered yes to this question, which means that they have already visited the region and 7 respondents answered no, that they have not visited the region yet. The aim of this question was to divide the respondents into existing and potential visitors.

We also asked respondents how often they visit the High Tatras—Foothills region. The aim of the question was to find out whether regular or occasional and infrequent visitors predominate in the region. The higher the frequency of visits, the more objectively the respondents can evaluate how the level of tourism development in the region is increasing over time and recommend further opportunities for improvement. The

Table 3 Number of visitors in the region High Tatras—Foothills

Year	Domestic visitors	International visitors	Total
2017	52,366	19,986	72,352
2018	59,672	22,319	81,991
2019	73,939	27,569	101,508
2020	49,375	15,682	65,057
2021	28,731	5,378	34,109

Source Statistical Office of the Slovak Republic (2022a, b)

Table 4 Number of overnight stays in the region High Tatras—Foothills

Year	Domestic visitors	International visitors	Total
2017	129,266	64,839	194,105
2018	138,235	71,180	209,415
2019	181,130	85,646	266,776
2020	135,616	48,009	183,625
2021	81,257	15,748	97,005

Source Statistical Office of the Slovak Republic (2022a, b)

largest group is made up of regular visitors, with exactly 50.36% of the respondents. 35.71% of the respondents visit the region occasionally and only 12.86% of them rarely. When it comes to visiting the region, only 1.07% of the respondents said that they do not visit the region anymore. They justified their choice of this option by dissatisfaction with services and lack of free time to visit. Considering the data obtained, we can conclude that the region is dominated by regular visitors.

In the next question, existing visitors we were interested in the motive for visiting the High Tatras—Foothills region. Respondents were given a choice of several options, and they also had the opportunity to indicate more than one answer. The main purpose of visiting the region was recreation, which was indicated by 47.14% of the respondents. Right after recreation, the second most frequently indicated answer was tourism with 37.5%, and the third most frequently indicated answer was visiting relatives with 32.14%. A significant part of the respondents indicated that they visit the region for the purpose of attending various events (18.21%), in the wintertime for the purpose of skiing (29.64%) and in the summertime for the purpose of cycling (16.07%). 9.64% of the respondents indicated that they visit the region for the purpose of health and use of wellness services. Only 5.36% of the respondents travel to the region for the work purpose. 5.71% of the respondents indicated the other option. On the basis of the obtained data, we can evaluate that the region needs to develop mainly recreational and sports tourism.

Regarding the length of stay, most of the respondents, up to 34.64%, indicated that they will stay in the region for 2–3 days during their visit, mainly for weekend stays. This means that with this length of stay, visitors are likely to make use of accommodation, catering, and also other tourism services during their stay. In second place were day trippers with a percentage of 22.86%. In third place, by not much difference to day trippers (22.14%), were visitors who spend 3 days or more in the region. Similar to visitors who stay for 2–3 days, visitors who come to the region for 3 or more days represent the largest tourism revenue for the region, and the ambition in developing tourism in the region should be to attract as many visitors as possible who spend 3 or more days in the region. The remaining 20.36% of respondents indicated the option of being locals.

Tourism demand is influenced by various factors, and respondents were given the opportunity to rate the different criteria and express their satisfaction with the services in the High Tatras—Foothills region. Visitors rated the criteria on a scale from 1 (best) to 5 (worst), if they did not know how to rate a criterion, they could mark option N (do not know how to rate). Table 5 captures the results in absolute value.

Respondents rated a total of 9 criteria. The best rated criterion was location and environment, with an average score of 1.56. We consider this criterion to be one of the strongest points of the High Tatras—Foothills region, its territory is located in several national parks, and we see the greatest potential for tourism development in the region. In second place, with an average mark of 1.85, it was the respondents' satisfaction with the quality of accommodation services. Quality accommodation services determine the length of a visitor's stay in the region, so we take this rating as positive. In third place with an average score of 1.87, visitors rated satisfaction with

Table 5 Visitors' evaluation of the region High Tatras—Foothills

Criteria	Evaluation						
	1	2	3	4	5	N	Average
Location and environment	201	34	21	7	15	2	1,56
Quality of accommodation services	135	65	38	13	11	18	1,85
Marking of hiking trails	121	68	43	12	7	29	1,87
Sports facilities	132	76	36	19	11	6	1,91
Cycle routes	73	88	47	13	5	54	2,07
Quality of catering services	116	68	51	20	15	10	2,07
Cultural events	83	71	56	31	5	34	2,20
Transport accessibility	70	104	72	18	6	10	2,21
Museums and expositions	45	73	77	27	11	47	2,51

Source Own elaboration, 2022

the marking of hiking trails, and close behind with a score of 1.90 was the criterion of the possibility of sports activities. The criterion of cycle paths received a rating of 2.07. Only 80.71% of the surveyed respondents commented on the evaluation of the cycle routes, the remaining respondents apparently do not use the cycle routes and thus cannot assess their level. The quality of catering services was rated 2.07, cultural events 2.20, and transport accessibility 2.21. The worst rated criterion was museum and exhibition, which was rated with an average mark of 2.51. In terms of the number and quality of museums and expositions, the High Tatras—Foothills region really lags behind, with most museums in the region being small and with irregular opening hours. Also the promotion of museums and expositions is very poor, 16.79% of the respondents said that they could not assess this criterion, which may be due to the fact that they do not register museums and expositions in the region.

We further investigated whether the visit to the region met visitors' expectations, with the majority of visitors being very satisfied (62.86%) and satisfied (29.64%). Visitors also perceive the region competitive to other larger tourism regions. 60.36% perceived the activities of the Destination Management Organization High Tatras—Foothills mainly in the area of marketing and promotion and rated them as beneficial for the development of the region.

As already mentioned, the development of tourism in the area is the responsibility of the Destination Management Organization High Tatras—Foothills. The main objective of the organization was to link the services and attractions in the High Tatras—Foothills area into a more coherent whole and thus create a more interesting tourism offer. They saw the greatest potential in the development of local and regional cooperation, which can better promote local attractions, exploit the potential of agrotourism, and offer space for even the smallest service providers to promote their services and products in the region.

In order to propose the further development and perspectives of sustainable tourism in the region, it is necessary to assess the current situation. A SWOT analysis assesses the strengths, weaknesses, opportunities, and threats that exist in the market. The SWOT analysis method was also used in the study of the High Tatras—Foothills region in terms of sustainable tourism development. The Table 6 shows the findings.

We found that the region has more strengths than weaknesses. Among its strengths, we include its large area, which is 222,609 km^2. Another strength associated with the size is that the region is spread over the territory of up to three national parks—NAPANT, TANAP, and the Slovak Paradise National Park. The combination of the vast location and the national parks makes the region a great place for hiking. National parks are protected areas where certain rules apply to protect nature, which may restrict visitors in some ways—for example, a ban on fishing—but the role of these restrictions is to preserve a clean environment, which is another strength of the region. The High Tatras—Foothills region is located within the Tatra region, which is characterized by overcrowding in some resorts, especially directly in the town of High Tatras and its surroundings, despite the overcrowding one of the most sought-after regions of Slovakia. The Foothills of the High Tatras offer the beauty of the Tatra region without the crowds of visitors. Another advantage that could make the High Tatras—Foothills region compete with larger regions is the ski resorts, which offer groomed ski trails of all difficulty levels, from easy to moderate to difficult, and cross-country ski trails. There are even two ski resorts in the village of Liptovská Teplička, and another ski resort in the region is located in the village

Table 6 SWOT analysis of the tourism development in the region

Strengths	Opportunities
Large area of the region	Suitable conditions for the development of agrotourism
Protected area of NAPANT, TANAP,	Potential of the area for recreation
and Slovak Paradise National Park	Extension of cycling routes and winter cross-country skiing trails
Excellent location within the Tatra	Year-round usable offer
Region	Development of a tourist center
Numerous accommodation facilities	Improvement of marketing and promotion
Clean environment	
Ski resorts	
Weaknesses	**Threats**
Irregular opening hours of monuments	COVID-19 pandemic
Transport accessibility of the region	Shortage of skilled labor
Quality of roads	Competition from larger regions
Lack of contact between the local population and visitors	Environmental pollution
Absence of tourism services in the evening	Disappearance of crafts and traditions after the death of elders

Source Own elaboration, 2022

of Vernár. There are several cross-country skiing trails in the region, for example, in the winter months, the Black Stork Golf Resort turns into a 10 km long cross-country skiing trail. Traditional events are held in the region all year round in an effort to preserve customs and traditions. Some of the villages also have their own folklore groups and organize folklore festivals. The Websites of the municipalities and destination management organization regularly inform about the events and all those who want to experience the atmosphere of the former village life are invited. Traditional agriculture has also been preserved in the villages, with agricultural cooperatives still farming the surrounding pastures and selling their products. By offering accommodation facilities, the region can appeal to a wide range of visitors, whether families with children seeking the services of hotels and guesthouses, or adventurers camping in a campsite.

Despite the numerous strengths, we should focus more on the weaknesses and eliminate them. Cultural opportunities in the form of visits to museums, galleries, and open-air museums are unfortunately not widely represented in the region. The offer of these facilities is limited, and another disadvantage associated with this is that they do not have regular opening hours. The region is accessible by bus and train, but unfortunately, the train service does not cover all the villages in the region, and the frequency of bus services is insufficient. Weak transport accessibility makes the region less attractive, especially for environmentally oriented visitors who exclusively use public transport when traveling. The rules of the protected area and the difficult natural conditions result in an underdeveloped technical infrastructure. The strict measures do not affect the whole region, but in some parts, for example, there is a ban on winter salting of roads, and the poor condition of the roads is also due to the difficulty of road maintenance, as the rules do not allow frequent intervention by machinery. Another obstacle to tourism development is the reluctance of the local population to engage with visitors and thus add value to their experience and impression of the region. In the absence of tourism service provision in the evening, it is mostly catering services, with restaurants closing too early, leaving the visitor severely constrained.

Opportunities for tourism development derive from the region's strengths. The municipalities of the High Tatras—Foothills region have excellent conditions for the establishment and development of agrotourism. Mutual cooperation of accommodation facilities with agricultural cooperatives, and the creation of agrotourism products is one of the greatest opportunities for the region. The offer of activities in the region is not only seasonal, but active leisure is also offered in every season, and this is followed by the opportunity to extend cycling routes and winter cross-country skiing trails. The more activities a region offers, the more interesting it is and the more visitors it can attract, and possibly contribute to visitors spending more time in the region. For example, a visitor can combine a skiing holiday with a visit to an organized event and spend more days in the region. On the other hand, the lack of marketing and promotion, which could be seen as a negative, gives the region room for development. Proper marketing and promotion can reach more visitors than the product of the region itself. The High Tatras—Foothills region lacks an information center, and we see its completion as another opportunity to improve the overall visitor impression of the region.

The biggest threat to tourism, as shown in 2020, is the COVID-19 pandemic. Tourism revenues were at their highest level in decades when travel was almost completely halted by the pandemic. Tourism is an industry that will not disappear, but the impact of the pandemic on its operations has been fatal, and further, similar events could be destructive for the industry. There was already a shortage of skilled labor in the tourism industry before the pandemic, and this situation has only exacerbated it. Unskilled staff can mean a reduced quality of service and therefore a reduced overall visitor impression of the region. Another threat to the region is competition from larger regions. State funding of regions also depends on their size; the High Tatras—Foothills region is small and poorly funded, so it has difficulty competing with large regions such as the High Tatras or Liptov, which are heavily funded by the state and also generate large tourism revenues. The development of tourism is also linked to the construction of new facilities such as hotels and restaurants, the construction of which has a negative impact on the environment. The key to sustainable tourism is to strike a balance between construction and maintaining a clean environment. The threat that most directly affects the High Tatras—Foothills region is the disappearance of crafts and traditions after the death of older inhabitants. The region presents itself and creates the image of a traditional region, focused on traditions, their disappearance would mean that the region would lose its authenticity. The solution is to pass on traditions to the next generation.

5 Conclusion

The development of tourism in the region does not only have a positive impact on this sector, thanks to the multiplier effect, it has an impact on the overall economic development of the region. Through tourism development, the region can also meet objectives such as economic growth and sustainable development.

The paper dealt with the perspectives of tourism development in the High Tatras—Foothills region. In the paper, we fulfilled the main objective, which was to present the perspectives of sustainable tourism development in the region based on the analysis and evaluation of the current state of tourism in the High Tatras—Foothills region. We fulfilled the main objective as well as the sub-objectives, which were to examine the tourism offer in the region, to find out what is the demand for tourism in the region, and to indicate the perspectives of sustainable development of tourism in the region. We calculated the tourism function of the region. We conducted an analysis of strengths, weaknesses, opportunities, and threats, which yielded interesting facts. We have emphasized the important role of the regional tourism organization in the development of sustainable tourism in the region. Priority is given to the cooperation of stakeholders in sustainable tourism development and the creation of a comprehensive product that will increase visitation and competitiveness of the region.

Acknowledgements This research was funded by the Scientific Grant Agency of the Ministry of Education, Science, Research and Sport of Slovak Republic VEGA, grant number 1/0368/20 "Sharing economy as an opportunity for sustainable and competitive development of tourist destinations in Slovakia".

References

Abir, T., & Khan, M. Y. H. (2022). Importance of ICT advancement and culture of adaptation in the tourism and hospitality industry for developing countries. In C.M.Q. Ramos et al. (Eds.), *ICT as innovator between tourism and culture* (pp. 30–41). IGI Global.

Adillon, R. (2019). Diamond model: A theoretical framework for the sustainable development of tourism. *Arts and Humanities Open Access Journal, 3*(1), 11–23. https://doi.org/10.15406/ahoaj.2019.03.00099

Buhalis, D. (2000). Marketing the competitive destination of the future. *Tourism Management, 21*(1), 97–116. https://doi.org/10.1016/S0261-5177(99)00095-3

Camilleri, M. A. (2018). *Strategic perspectives in destination marketing.* IGI Global.

Carvalho, R. C. (2021). *Handbook of research on multidisciplinary approaches to entrepreneurship, innovation, and ICTs.* IGI Global.

Demir, M., Dalgic, A., & Ergen, F. D. (2021). *Handbook of research on the impacts and implications of COVID-19 on the tourism industry.* IGI Global.

Dinis, M. G. et al. (2021). *Impact of new media in tourism.* IGI Global.

Dresner, S. (2008). *The principles of sustainability.* Earthscan.

Gómez-Suárez, M., & Martínez-Ruiz, M. P. (2016). *Handbook of research on strategic retailing of private label products in a recovering economy.* IGI Global.

Ianioglo, A., & Risannen, M. (2020). Global trends and tourism development in peripheral areas. *Scandinavian Journal of Hospitality and Tourism, 20*(5), 520–539. https://doi.org/10.1080/15022250.2020.1848620

Janusz, G. K., & Bajdor, P. (2013). Towards to sustainable tourism—Framework, activities and dimensions. *Procedia Economics and Finance, 6*, 523–529. https://doi.org/10.1016/S2212-5671(13)00170-6

Keller, P. (2015). A pioneer of a sustainable tourism policy. In H. Pechlaner & E. Smeral (Eds.), *Tourism and leisure: Current issues and perspectives of development* (pp. 41–50). Springer Gabler.

Magas, D. (2010). *Why the destination management organization?* Retrieved from https://www.proquest.com/docview/763422402

Manzoor, F. et al. (2019). The contribution of sustainable tourism to economic growth and employment in Pakistan. *International Journal of Environmental Research and Public Health,* 16(19). https://doi.org/10.3390/ijerph16193785

Ministry of Transportation and Construction of Slovak Republic. (2020). *Strategy of sustainable tourism development to the year 2030.* Ministry of Transportation and Construction of Slovak Republic.

Oliveira, L. (2022). *Handbook of research on digital communications, internet of things, and the future of cultural tourism.* IGI Global.

Pinto, P., & Guerreiro, M. (2020). *Handbook of research on resident and tourist perspectives on travel destinations.* IGI Global.

Poudel, S., & Nyaupane, G. P. (2013). The role of interpretative tour guiding in sustainable destination management a comparison between guided and nonguided tourists. *Journal of Travel Research, 52*(5), 659–672. https://doi.org/10.1177/0047287513478496

Ramos, C. M. Q., Almeida, C. R., & Fernandes, P. O. (2019). *Handbook of research on social media applications for the tourism and hospitality sector.* IGI Global.

Ryglová, K., Burian, M., & Vajčnerová, I. (2011). Cestovní ruch - podnikatelské principy a příležitosti v praxi. Praha : Grada.
Saner-Yiu, L., Saner, R., & Lee, M. R. (2015). Lesotho, a tourism destination: An analysis of Lesotho's current tourism products and potential for growth. In A.A. Camillo (Ed.), *Handbook of research on global hospitality and tourism management* (319–339). IGI Global.
Statistical Office of Slovak Republic. (2022a). *Capacities and outcomes of the accommodation facilities*. Retrieved April 16, 2022a, from http://datacube.statistics.sk
Statistical Office of Slovak Republic. (2022b). *Number of inhabitants*. Retrieved April 16, 2022b, from http://datacube.statistics.sk
UNWTO. (2022a). *Product development*. Retrieved January 8, 2022a, from https://www.unwto.org/tourism-development-products
UNWTO. (2022b). *Sustainable development*. Retrieved January 8, 2022b, from https://www.unwto.org/sustainable-development
Valeri, M. (2021). *New governance and management in touristic destinations*. IGI Global.
Varghese, B. (2021). *Evolving paradigms in tourism and hospitality in developing countries*. Apple Academic Press.
Zamfir, A., & Corbos, R. A. (2015). Towards sustainable tourism development in urban areas: Case study on Bucharest as tourist destination. *Sustainability, 7*, 12709–12722. https://doi.org/10.3390/su70912709

Environmental Sustainability Application in Athenian Hotels

Ioanna S. Samartzi

Abstract Environmental sustainability is indeed a buzzing subject in a variety of fields. At its' core, it is an essential tool for the protection of future generations against environmental challenges, such as climate change and natural disasters that put human well-being in danger. Likewise, the tourism field is a highly dynamic sector, while for many countries, it is their main capital source but is also a sensitive one. The connection between the tourism sector and environmental sustainability occurs in the necessity for a harmonious coexistence between tourism activities and environmental protection. Urban destinations, such as hotel facilities, are especially important for environmental sustainability, because of two factors. The first one is the increased number of tourists in urban areas, and the second is the high amount of waste due to human activities in hotels. Tourism activities pressure the environment, making sustainability now crucial more than ever. The aim and purpose of this research are to identify the environmental sustainability practices of 50 urban hotels in the capital of Greece. The questionnaires were delivered on 3 February and were completed until 21 February 2020, and they enclosed 15 questions in total. After the data collection, the results estimated a lack of environmental practices in the majority of the hotels, such as insufficient waste management, recycling, water conservation, and lack of environmental training among the employees. Nonetheless, most of the hotel managers stated a willingness to adopt environmental practices in future.

Keywords Environmental sustainability · Tourism · Hotel management · Environmental practices

JEL Classification Z32 · Q56

I. S. Samartzi (✉)
Department of Sports Management, University of Peloponnese, Sparta, Greece
e-mail: ioannasamartzh@gmail.com

© The Author(s), under exclusive license to Springer Nature Switzerland AG 2023

V. Katsoni (ed.), *Tourism, Travel, and Hospitality in a Smart and Sustainable World*, Springer Proceedings in Business and Economics,
https://doi.org/10.1007/978-3-031-26829-8_3

1 Introduction and Background (Theoretical Considerations)

Sustainable tourism is a complex concept to define (Ruhanen, 2008), because of the different aspects and orientations given by many sectors involved in its examination. For Holden (2016), sustainability "is an integral component of tourism policy and strategy", while for Moyle et al. (2021), it "is a vital strategic objective for tourism destinations worldwide to protect the environment". In the past years, as Almeida et al. (2020) stated, "sustainability has been actively considered in social and scientific fields". In the tourism sector, the research regarding sustainability issues is expanding rapidly, but the content remains almost the same (Moyle et al., 2020). However, Abdou et al. (2020) noted that "sustainability in the hospitality industry has been addressed from different points of view". From a different perspective, tourism is vital for the economy of many countries and essential for a big part of the worldwide population. The report of the World Travel and Tourism Council (2020) claimed that "in 2019, the hospitality industry contributed US$8.9 trillion to the world's GDP, 10.3% of the global GDP, and 1 in 10 jobs worldwide". In addition, sustainable tourism, in the past few years, has been a necessity for businesses, especially tourism companies (Boboc et al., 2019), which are using sustainable development goals as a guideline for adopting a sustainable point of view (Torelli, 2020), but many scholars pinpoint the need to acquire a better knowledge of sustainable business decisions in the hotels, due to the lack of theoretical frameworks in the literature (Calisto et al., 2021), but they estimate that focusing on sustainability is given an advantage in the market share of hotels (Oxenswärdh, 2020). Therefore, tourists have expressed a preference to stay in green hotels (Ogbeide, 2012). In order to travel, tourists need hotels and other services, creating in this way a crucial environmental impact, which needs to be reduced (Khatter et al., 2019).

Concerning the environmental impacts of tourism, the outcome is negative, and as Azam et al. (2018) pinpointed, "tourism is the most environmentally harmful sector, alongside the industrial one", while the hotels specifically are consumers of a variety of resources daily (Verma & Chandra, 2018). For some scholars, sustainable tourism, despite the criticism, is the main tool for eliminating the negative aspects of tourism in the environment (Sharpley, 2020); thus, it is crucial to urgently achieve environmentally sustainable practices in the tourism sector. On the other hand, for some others, sustainable tourism is often ineffective, and not well managed (Hens, 2015). In parallel hotels are responsible for a high amount of waste (Sloan et al., 2013). The negative effects of tourism have led to environmental degradation of its components (Tsekouropoulos et al., 2022), not only in rural areas but also in mass tourism and niche tourism (World Tourism Organization, 2021). Exemplifying this, urban destinations have been established in central cities as a state of growth and innovation, making them a very important factor to ensure environmental practices nowadays (Krassmira, 2007), but in fact, mass urbanization and mass tourism have led to unsustainable environmental actions and outcomes (Riffat et al., 2016). The urban destination tourists could also differ according to their motivations, from business trips to school holidays and leisure travelers (Buhalis, 2000). Expectedly, most urban tourism research focuses on the social and economic aspects of sustainability, and less on the environment, thus hospitality sustainable practices research is restricted in research contexts (Aall & Koens, 2019).

In fact, the hotel industry is one of the main sectors of tourism (Pirani & Arafat, 2015), besides that, hotel service is also considered a contributor to environmental pollution, because of its human-oriented characteristics (Dogru et al., 2020), while daily operations of hotels require a high number of resources (Verma & Chandra, 2018). Regarding the outcome of environmental sustainability in hotels, in general, as UNTWTO (2017) states, minimizing negative environmental impact is gaining ground in the tourism field. Specifically in European hotels, 80% of them are involved in environmental protection activities and procedures, with the main efforts in energy, water-saving, and waste decrement.

2 Environmentally Sustainable Practices in Hotels

Over the last few years, there has been a significant interest in investigating the topic of sustainable tourism, and how it could be achieved (Niñerola et al., 2019), while many consider it a key tool for the competitiveness between hospitality businesses (Goffiet et al., 2019). This growing interest from media, the public sector, and consumers, is driving many hotels to adopt environmentally friendly practices (Kapera, 2018), but there are a number of main issues in achieving sustainability in hotels, such as the building design, and energy consumption rates, but they are not enough. Due to its anthropocentric nature, human resource management is also an essential key tool for green hotels to achieve (Jauhari, 2014). Environmental sustainability in the hospitality industry refers to the healthy balance between tourism activity, hotel operations, and the environment (Lim, 2016).

Hotels with an environmental focus, otherwise named "green hotels", highlight environmental protection through environmental operations and reasonable consumption (Wu, 2021). Manaktola and Jauhari (2007) define a green hotel as "a lodging facility committed to ecological practices" such as saving water and energy conservation, waste management, and use of recycled materials. Green hotels, in order to reduce their environmental impacts, implement into their strategy environmentally responsible practices, to protect the environment, and to consider "environmental criteria" (Tzschentke et al., 2008). In a quite different perspective, many scholars declared that the adoption of environmental practices for hotels, results in cost reduction, improvement of hotel image, and customer satisfaction thus is a strong asset for competitiveness (Pereira et al., 2021). This is the reason why consumer behavior is a "key driver" in hotel strategies and makes the transition of hotels into green hotels, more smoothly (United Nations Environment Programme, 2018). Consumers are now, more than ever, aware of the environmental problems that pollute natural ecosystems, and humans indeed. Therefore, they are willing to pay more for "green tourism", instead of choosing hotels with standard services

(Cingoski & Petrevska, 2018), while "information technology practices" by the hospitality industry could play a vital role in the implementation of sustainability (Apostolakis et al., 2020).

In general, the tourism industry, especially mass tourism destinations, and hotels have been connected negatively to damaging the environment through their activities, requiring them to concentrate on environmental issues' solutions (Chan et al., 2018), but they are always solutions. A way to ensure the development of green practices in hotels is the installation of environmentally friendly equipment (Luo et al., 2021), and the promotion of green human resources management, thus, is essential for them to motivate the hotel staff to involve and be trained, in sustainable activities, such as recycling (Irani et al., 2022). The hotel industry has also introduced eco-labels to "reveal their willingness to support sustainability" and to adopt environmental strategies (Abdou, 2021).

However, there are also many barriers to overcome to ensure green hotel practices. A barrier for hotels in adopting environmental practices is employee involvement (Pham et al., 2019), while from the perspective of the hotel managers of small and medium-sized hotels, as Satchapappichit et al. (2015) claimed barriers as the limited "resource availability", and the lack of motivations, and knowledge of the positive effects of environmental management, while in the same wavelength, according to Wang et al. (2018): "The higher is the star-level of a hotel, the higher enthusiasm on energy-saving a hotel has". Nowadays, the hospitality industry seems to understand the importance of eco-friendly activities and environmental sustainability more than in the past (Trang et al., 2019), with regards to the latter, "81% of travelers" were willing to stay "at least once in an eco-friendly facility", in 2021 (Lock in Statista, 2021). All the above observations lead to three major key issues for sustainable tourism: solid waste management, water waste management, and energy efficiency, which are analyzed below.

2.1 Solid Waste Management

A significant environmental issue concerning the tourism industry is the waste management of solid materials (Ioannidis et al., 2021), and it is the main contributor to greenhouse gas emissions (Singh et al., 2014). According to Sarkis and Dijkshoorn (2005), "waste is the unused materials of inadequate practices of production and consumption", which leads to economic and environmental harm. In the tourism sector, most researchers have examined the volume of waste which is generated by hotels and their services (Bhat et al., 2014). Quaintly, smaller hotels also produce a notable amount of solid waste, despite their size (Radwan et al., 2012), while larger hotels found it easier to implement environmental management practices (Sánchez-Medina et al., 2016). The largest percentage of waste from tourism activities is the enormous amount of plastic waste (Maione, 2019), especially during the peak seasons of tourism, depending on the destination. Therefore, the higher number of tourists, the higher the number of waste hotels generate, making the procedures more complicated than already (Chaabane et al., 2018). Hotels have a tendency to make use of a high number of single-use plastics, and according to United Nations Environment Programme and World Travel and Tourism Council (2021): "The production, consumption, and disposal of additional single-use plastic products may increase air pollution, greenhouse gas emissions, waste, and littering". In the same wavelength, a World Wildlife Fund (2019a, 2019b) report highlighted "that plastic waste produced in the Mediterranean increased by up to 30% in the summer months", when most of the tourists travel to these locations. Thus, plastic usage is one of the main contributing factors to environmental pollution, and it must be reduced urgently (Ramirez & George, 2019. The solution to this problem is the fact that almost 30% of solid waste in a hotel can be sorted, reused, recycled, or recovered, according to Mohan et al. (2017). Nowadays, hotel operators use environmental practices such as hotel waste separations, soil composting (Han et al., 2018), reduction of plastics, and recycling systems adoption (United Nations Environment Programme and World Travel and Tourism Council, 2021).

2.2 *Water Waste Management*

Concerning the importance of water in the tourism industry, according to DeSouza (2016), tourism accounted for "less than 1% of worldwide water use", thus water services in hotels required double, or triple amounts of water than normal (Styles et al., 2015), while the report by the World Bank suggested that "by 2050, 3.40 billion tons of waste will be generated around the world every year" (Kaza et al., 2018); therefore, water conservation is a pivotal issue for humans and the environment. In more depth, most of the hotels have adopted some water conservation strategies, but for cost-saving reasons (Becken & Dolnicar, 2016), while most tourism activities are taking place, according to Gössling (2005): "in water-poor countries", such as Mediterranean destinations. The amount of water consumption and wastewater in hotels, such as "wastewater from bathing, laundry, and toilets, are generated in hotels through guest use, cleaning, swimming pool, and other operating services" according to Sadi & Adebitan, 2014), in an amount that is impossible to be ignored, and may increase in future (Yu et al., 2021), causing environmental problems, such as GHS, pollution, and contribution to climate change (Leverenz et al., 2021). Some scholars have proposed solutions such as the separation of the waste and its recycling for further use (Atanasova et al., 2017), the installation of water-efficient devices, and reuse programs to reduce water consumption (Han et al., 2018).

2.3 *Energy Efficiency*

Regarding energy resources, hotels use energy for all their services, and all year round in an intensive way (Parpairi, 2017), in fact, hotels are "one of the most energy-intensive facilities" according to Cingoski and Petrevska (2018). Heating and cooling account for almost half of the total energy consumption in hotels (Parpairi, 2017), thus electricity consumption is the main source of GHCs in hotels (Lai, 2015). Noteworthy, according to Santiago (2021) "depending on the tourism destination, carbon emissions per occupied room have been reported from 7.2 $kgCO_2$-e up to 199.1 $kgCO_2$-e". The solution here, given by Bohdanowicz (2005), is the implementation of "renewable energy programs", eco-friendly devices, LED lights, and more technological solutions. Renewable energy sources in hotels provide many possibilities, such as increment in energy efficiency and cost reduction, with the form of wind power systems, solar panels (Cerovic et al., 2014), and even solar water heaters (Chan et al., 2013). From a different perspective, sustainable tourism has introduced certain indicators, which are implemented environmental protection, among its operations (Guizzardi et al., 2017), many of them are voluntary environmental management tools, such as "third party certified eco-labels", which have an informative role for tourists, and they verify the green management development for hotels (Geerts, 2014). At the same time, they are used as environmental certifications (Esparon et al., 2014).

3 Methodology and Sample

This study sought to examine the environmental practices of the hotel sector of Athens, Greece. In more detail, the sample of the research enclosed the hotel managers of 50 Athenian hotels. Notably, 130 questionnaires were distributed at first, but 62% of hotel managers refused to answer. The questionnaires were delivered to the hotel managers of these 50 hotels, who in front of the researcher, filled up the questions, and then the researcher received them back. The questionnaire included in total of 15 questions, 4 introductory, 8 about the environmental practices in the hotel, such as waste management strategies, water conservation, and energy efficiency, and 3 questions that examined the perceptions of hoteliers about the importance of environmental strategy now and in future. In addition, the questionnaires encompassed 1 open question, 8 close questions, and 3 linear scale questions, and for the purpose of the research, primary and secondary data, through quantitative research, were collected. Therefore, using quantitative research was considered important to the examination of the connection between variables, (Tharenou et al., 2007). While regarding primary and secondary data, the primary data were collected via questionnaires, and the responses were obtained from the hotel managers of Athenian hotels, as the study is based on the city hotel sector regarding environmental practices and perspectives. In parallel, the secondary data were collected from the examination of the literature review of sustainable tourism, from the latest and historical references available from journals, research publications, conferences, and other relevant academic sources.

In addition, the main objective of the research is to investigate the environmental sustainability practices, and perceptions in urban hotels in the center of Athens,

Table 1 Hypothesis

Hypothesis 1	50% of hotels are using recycling systems
Hypothesis 2	10% of hotels are using renewable energy systems
Hypothesis 3	30% of hotels adopt water conservation strategies
Hypothesis 4	Hoteliers claim, on average of 4, willingness to adopt an environmental strategy in future

Greece. The barriers to the study were identified as the high percentage of hotel managers that refused to fill up the questionnaire, and the lack of knowledge of the hotel managers, regarding environmental sustainability strategies. Exemplifying this, the majority of the participants had questions about the environmental part of the questionnaires, asking for clarifications. In the bargain, the high percentage of refusal has related to the worldwide outbreak of coronavirus (COVID-19), which affected all humanity deeply. The outbreak began in December of 2019 in Wuhan, a city in China (Yang et al., 2020), and became an international problem in January 2020, and finally a pandemic 2 months later, on March 11 of 2020 (Farzanegan et al., 2020), thus, the tourism was deeply affected also in Greece.

Regarding the structure of the study, and based on the purpose of investigating environmental sustainability practices in hotels, the research design focused on the following research hypotheses, which are also illustrated in the following research table (see Table 1).

4 Presentation of Research Findings

As has been mentioned above, the findings were retrieved by the method of questionnaires, which were distributed to hotels in the center of the Attica region, in Athens, Greece, and specifically in the locations of Syntagma, Monastiraki, Omonia, and Metaxourgio. The collection of the data began on 3 February and was completed on 21 February 2020. Not many days after, on 28 February 2020, the first case of COVID-19 was detected in Greece. This led to the first measures for restricting COVID-19 in early March 2020 (Vatavali et al., 2020), which had a huge impact on Greek hospitality and on the Greek tourism industry in general and in the tourism industry worldwide (Chang et al., 2020).

However, after the collection, the data were analyzed and are presented below: First of all, ten questionnaires were filled up by hotel managers in Syntagma, eight in Monastiraki, twenty-five in Omonia, and seven in Metaxourgio, which counts for fifty answered questionnaires in total. Regarding the hotel star rating systems, which according to Vagena and Manoussakis (2021): "provide accurate and objective accommodation assessment data according to defined criteria and guidelines", in Syntagma, 70% of the hotels that took part in the survey have three stars, and 30% four stars, in Monastiraki 37.5%, are ranked with three stars and 62.5% with four. In the region of Omonia, 4% ranked with one star, 48% with two stars, 32% with three stars, and 16% with four stars, while in Metaxourgio, 28.5% of the hotels ranked with two stars, 28.5%with three stars, 28.5% with four stars, and 14.5% with five stars. Remarkably only in Metaxourgio, a hotel was ranked with five stars. Concerning the capacity, of hotels' guest's rooms, the results showed that 20% of the hotels have less than 20 guest rooms, 42% have 20–40 guest rooms, 14% have 40–70 guest rooms, 8% have 70–100 guest rooms, 6% 100 to 120 guest rooms, 8% have 120–150 guest rooms, and 2% more than 150 guest rooms. The last question of the introductory ones was related to the number of employees, in which the average number was calculated as 15.

The next questions were designed to identify the application of hotels' environmental practices. According to the responses, only 5 hotels from the 50 had hired an employee responsible for environmental management, but their tasks were not named; thus, 90% of the hotels did not have an environmental manager, or at least one employee responsible for environmental practices. Moreover, only 33% of the hotels have developed waste management in their hotels, half of the hotels in Syntagma, 43% in Monastiraki, 12% in Omonia, and 28.50% in Metaxourgio, while 33.5% of the hotels are recycling. Overall, 50% of Syntagma hotels are recycled, 35% in Monastiraki, 20% in Omonia, and 29% in Metaxourgio. Regarding the purchase of environmentally friendly products for hotels' daily operations, the average percentage was 12.25%, in Syntagma 10%, in Monastiraki 12.50% same as Omonia, and 14% in Metaxourgio. Analyzing the results about water conservation practices, the average percentage was 5.5%, and for energy efficiency, it was 5%, while surprisingly none of the hotels used renewable energy programs. Another important question examined the number of hotels with an environmental certification, the results showed that only one in the 50 hotels is certificated in environmental sustainability standards. In the following table (Table 2), the most important results regarding the environment are summarized.

The last part of the questionnaire examined the intention of hoteliers and tourists regarding hospitality environmental practices. The first question was whether the hotels consider it useful to develop environmental strategies, with one as the least, and five as the most. The average number was calculated as 3.1. For the hotels of Syntagma, the average was 3.3, for Monastiraki 3.6, for Omonia 2.7, and for Metaxourgio 3.4. The second question examined whether tourists are asking for

Table 2 Environmental hotel practices

	Environmentally friendly products (%)	Waste management (%)	Water conservation (%)	Energy efficiency (%)
Syntagma	10	50	8	6
Monastiraki	12.50	43	2	4
Omonia	12.50	12	6	4
Metaksourgio	14	28.50	6	6

Table 3 Environmental perspective

Description	Average number (1 the least, 5 the most)
Importance of environmental practices	3.1
Tourists demanding environmental practices	2.7
Development of environmental strategies in future	3.4

Table 4 Hypothesis confirmed or failed

	Confirmed	Failed
Hypothesis 1	No	Yes
Hypothesis 2	No	Yes
Hypothesis 3	No	Yes
Hypothesis 4	No	Yes

more environmental practices, with one as the least and five as the most. The average of all respondents reached 2.7. More specifically, in Syntagma, it reached 3.5, in Monastiraki 2.3, in Omonia 2.4, and in Metaxourgio 3.1. The last question was whether hotels are interested in developing more sustainable strategies in future, and the average of all hotels reached 3.4. More specifically, in Syntagma, the average was 3.5, in Monastiraki 3.9, in Omonia 3, and in Metaxourgio 3.9 (see Table 3).

In line with all of the above results, the hypotheses failed. In more depth, hypothesis 1, regarding recycling systems in the hotels, was 50%, but it was calculated as 33.5%, while hypothesis 2 about the existence of renewable energy systems was 10% but calculated as 0. Moreover, hypothesis 3 estimated that 30% of hotels had adopted water conservation management techniques, but the results showed that only 6% did, while hypothesis 4 also failed. Hypothesis 4 stated that the willingness of hotels to adopt an environmental strategy in future, the average number was 4, but the numbers showed the number 3.4 (Table 4).

5 Conclusion

The collection and analysis of the data showed a negative outcome regarding environmentally sustainable practices in Athenian hotels, and none of the hypotheses were confirmed. Regarding the issue of the hoteliers regarding environmental practices, only 1 in 10 hotels had an employee in this sector, while only 33% of the hotels had developed waste management techniques, with 33.5% choosing recycling. Meanwhile, regarding water conservation, the average percentage of hotels that had implemented them was 5.5%, and the energy efficiency was 6%. Remarkable, none of the hotels had developed renewable energy solutions in their operations.

Table 5 Potential environmental management mechanisms and their actions

Potential mechanisms	Actions and characteristics
Solid waste management	Reduction of the amount and toxicity of waste sent to the landfill
Recycling	The hotels should have recyclable products with the least possible environmental impact and to implement recycling programs for customers and employees
Eco-friendly products	Replacing toxic products with non-toxic ones by choosing environmentally safe cleaning products
Renewable energy programs	Solar power
Water management	Adoption of water-saving strategies
Energy efficiency management	LED lights and other

It is important to pinpoint that the hotel sector in Greece "is responsible for 10% of the country's total energy consumption", with heating and cooling as the main contributors (HES, 2021), however, Greece in the past was ranked, among other countries, as the top spots for the total installation of solar water heating technologies in the world, according to Sawin (2011).

Moreover, this study brought to light the potential mechanisms responsible for solid waste management, recycling, purchasing of eco-friendly products, renewable energy programs, water management, and energy efficiency, as illustrated in Table 5.

Continuously, the most important barrier for this research was the "fear" of COVID-19 in the hotel industry. International travel became more complicated and challenging, especially with travel restrictions (Gössling & Schweiggart, 2022). In the meantime, COVID-19 resulted in economic and health crises all over the world (Anderson et al., 2020), and according to Asquith (2020), "more than 90% of people worldwide were living under travel restrictions" for months.

Nowadays, sustainability has been a priority among younger generations. They are more worried about environmental protection, and they are also more well-informed about them (Palazzo et al., 2022). This perspective has been transferred also to the tourism industry. Particularly, according to Karantzavelou in Sustainable Travel Survey (2022): "78.3% of the travelers are more willing to pay more for their vacations in order to lower their carbon footprint", and 35% claimed that sustainable travel is a highly important factor when choosing their travel destinations, facilities, and activities. The willingness of travelers is critical because the tourism system occurs at multiple embedded scales affecting many (Hall, 2022).

For further research, it is essential to analyze the same hotels regarding environmental sustainability now, two years later, in order to examine how COVID-19 affected their willingness to adopt more green actions, and if they have adopted environmental strategies. Moreover, an examination regarding food waste management could be also useful. Environmental sustainability in the hotel industry is an urgent issue to be solved, due to its human-centered approach, and its effect on the environment. It is crucial to examine further, all the daily operations of a hotel, in order to design and manage the most eco-friendly systems in this environment.

References

Aall, C., & Koens, K. (2019). The discourse on sustainable urban tourism: The need for discussing more than overtourism. *Sustainability, 11*(15), 4228. https://doi.org/10.3390/su11154228

Abdou, A. H., Hassan, T. H., & El Dief, M. M. (2020). A description of green hotel practices and their role in achieving sustainable development. *Sustainability, 12*(22), 9624. https://doi.org/10.3390/su12229624

Almeida, M., Barbosa, R., & Malheiro, R. (2020). Effect of embodied energy on cost-effectiveness of a prefabricated modular solution on renovation scenarios in social housing in Porto, Portugal. *Sustainability, 12*(4), 1631. MDPI AG. Retrieved from https://doi.org/10.3390/su12041631

Anderson, R. M., Heesterbeek, H., Klinkenberg, D., & Hollingsworth, T. D. (2020). How will country-based mitigation measures influence the course of the covid-19 epidemic? *The Lancet, 395*(10228), 931–934. https://doi.org/10.1016/s0140-6736(20)30567-5

Apostolakis, A., Jaffry, S., & Kourgiantakis, M. (2020). Examination of individual preferences for Green Hotels in Crete. *Sustainability, 12*(20), 8294. https://doi.org/10.3390/su12208294

Asquith, J. (2020). Coronavirus travel update-over 90% of people live under travel restrictions. Forbes. Retrieved June 12, 2022, from https://www.forbes.com/sites/jamesasquith/2020/04/02/april-update-coronavirus-travel-update-90-of-countries-have-travel-restrictions

Atanasova, N., Dalmau, M., Comas, J., Poch, M., Rodriguez-Roda, I., & Buttiglieri, G. (2017). Optimized MBR for greywater reuse systems in hotel facilities. *Journal of Environmental Management, 193*:503–511.https://doi.org/10.1016/j.jenvman.2017.02.041.

Azam, M., Ramanathan, A., Wagnon, P., Vincent, C., Linda, A., Berthier, E., Pottakkal, J. (2016). Meteorological conditions, seasonal and annual mass balances of Chhota Shigri Glacier, western Himalaya, India. *Annals of Glaciology, 57*(71), 328–338. https://doi.org/10.3189/2016AoG71A570

Becken, S., & Dolnicar, S. (2016). Uptake of resource efficiency measures among European small- and medium sized accommodation and food service providers. *Journal of Hospitality and Tourism Management, 26*, 45–49. https://doi.org/10.1016/j.jhtm.2015.11.001

Bhat, R. A., Nazir, R., Ashraf, S., Ali, M., Bandh, S. A., & Kamili, A. N. (2014). Municipal solid waste generation rates and its management at Yusmarg forest ecosystem, a tourist resort in Kashmir. *Waste Management and Research, 32*(2), 165–169. https://doi.org/10.1177/0734242X13518089

Boboc, C., Ghita, S., & Vasile, V. (2019). Patterns in romanian tourism activity—A factorial analysis. In *Caring and sharing: The cultural heritage environment as an agent for change* (pp. 91–101). Cham, Switzerland. ISBN: 978-3-319-89468-3.

Bohdanowicz, P. (2005). European hoteliers' environmental attitudes. *The Cornell Hotel and Restaurant Administration Quarterly, 2005*(46), 188–204. https://doi.org/10.1177/0010880404273891

Buhalis, D. (2000). Marketing the competitive destination of the future. *Tourism Management, 21*(1), 97–116. https://doi.org/10.1016/S0261-5177(99)00095-3

Calisto, M. D. L., Umbelino, J., Gonçalves, A., & Viegas, C. (2021). Environmental sustainability strategies for smaller companies in the hotel industry: Doing the right thing or doing things right? *Sustainability, 13*(18), 10380. https://doi.org/10.3390/su131810380

Cerovic, L., Drpic, D., & Milojica, V. (2014). Renewable energy sources in the function of sustainable business in tourism and Hospitality Industry. *Turizam, 18*(3), 130–139. https://doi.org/10.5937/turizam1403130c

Chaabane, W., Nassour, A., & Nelles, M. (2018). Solid waste management key indicator development for hotels: A Tunisian case study analysis. *Recycling, 3*(4), 56. https://doi.org/10.3390/recycling3040056
Chan, E., Okumus, F., & Chan, W. (2018). Barriers to environmental technology adoption in hotels. *Journal of Hospitality and Tourism Research, 42*, 829–852. https://doi.org/10.1177/1096348015614959
Chan, W. W., Li, D., Mak, B., & Liu, L. (2013). Evaluating the application of solar energy for Hot water provision: An action research of Independent Hotel. *International Journal of Hospitality Management, 33*, 76–84. https://doi.org/10.1016/j.ijhm.2013.01.008
Chang, C. L., McAleer, M., & Ramos, V. (2020). A charter for sustainable tourism after COVID-19. *Sustainability, 12*, 3671. https://doi.org/10.3390/su12093671
Cingoski, V., & Petrevska, B. (2018). Making hotels more energy efficient: The managerial perception. *Economic Research-Ekonomska Istraživanja, 31*(1), 87–101. https://doi.org/10.1080/1331677x.2017.1421994
Cucculelli, M., & Goffi, G. (2016). Does sustainability enhance tourism destination competitiveness? Evidence from Italian destinations of excellence. *Journal of Cleaner Production, 111*, 370–382. https://doi.org/10.1016/j.jclepro.2014.12.069
DeSouza, M. (2016). Coping with water scarcity in agriculture: A global framework for action in a changing climate (Job No. 15604E). Retrieved from http://www.fao.org/documents/card/en/c/8dd680fd-70d3-4725–8d9f-30f9a02455a0
Dogru, T., Bulut, U., Kocak, E., Isik, C., Suess, C., & Sirakaya-Turk, E. (2020). The nexus between tourism, economic growth, renewable energy consumption, and carbon dioxide emissions: Contemporary evidence from OECD countries. *Environmental Science and Pollution Research, 27*(32), 40930–40948. https://doi.org/10.1007/s11356-020-10110-w. PMID: 32681326.
Esparon, M., Gyuris, E., & Stoeckl, N. (2014). Does ECO certification deliver benefits? An empirical investigation of visitors' perceptions of the importance of ECO certification's attributes and of operators' performance. *Journal of Sustainable Tourism, 22*, 148–169. https://doi.org/10.1080/09669582.2013.802325
Farzanegan, M. R., Gholipour, H. F., Feizi, M., Nunkoo, R., & Andargoli, A. E. (2020). International tourism and outbreak of Coronavirus (COVID-19): A cross-country analysis. *Journal of Travel Research, 60*(3), 687–692. https://doi.org/10.1177/0047287520931593
Geerts, W. (2014). Environmental certification schemes: Hotel managers' views and perceptions. *International Journal of Hospitality Management, 39*, 87–96. https://doi.org/10.1016/j.ijhm.2014.02.007
Goffi, G., Cucculelli, M., & Masiero, L. (2019). Fostering tourism destination competitiveness in developing countries: The role of sustainability. *Journal of Cleaner Production, 2019*(209), 101–115. https://doi.org/10.1016/j.jclepro.2018.10.208
Gössling, S. (2005). Chapter 20. Tourism's contribution to global environmental change: Space, energy, disease, water. *Tourism, Recreation and Climate Change*, 286–300. https://doi.org/10.21832/9781845410056-022
Gössling, S., & Schweiggart, N. (2022). Two years of covid-19 and tourism: What we learned, and what we should have learned. *Journal of Sustainable Tourism, 30*(4), 915–931. https://doi.org/10.1080/09669582.2022.2029872
Guizzardi, A., Mariani, M., & Prayag, G. (2017). Environmental impacts and certification: Evidence from the milan world expo 2015. *International Journal of Contemporary Hospitality Management, 29*(3), 1052–1071. https://doi.org/10.1108/IJCHM-09-2015-0491
Hall, C. M. (2022). Sustainable tourism beyond bau (Brundtlund as usual)?: Shifting from paradoxical to relational thinking? Frontiers. Retrieved June 12, 2022, from https://www.frontiersin.org/articles/https://doi.org/10.3389/frsut.2022.927946/full
Han, H., Lee, J., Trang, H. L. T., & Kim, W. (2018). Water conservation and waste reduction management for increasing guest loyalty and green hotel practices. *International Journal of Hospitality Management, 2018*(75), 58–66. https://doi.org/10.1016/j.ijhm.2018.03.012

Hens, L. (2015). Urban sustainability in theory and practice: Circles of sustainability. *Environment, Development and Sustainability, 17*(3), 681–682. https://doi.org/10.1007/s10668-015-9643-0

Holden, A. (2016). *Environment and tourism.* Routledge. ISBN 9781138785762.

Hotel Energy Solutions. (2021). *Analysis on energy use by European hotels: Online survey and desk research.* HES publications. Retrieved June 14, 2022, from http://www.nezeh.eu/assets/media/fckuploads/file/Reports/10.HESreserch.pdf

Ioannidis, A., Chalvatzis, K. J., Leonidou, L. C., & Feng, Z. (2021). Applying the reduce, reuse, and recycle principle in the hospitality sector: Its antecedents and performance implications. *Business Strategy and the Environment, 30*(7), 3394–3410. https://doi.org/10.1002/bse.2809

Iwanowski, K., & Rushmore, C. (1994). Introducing the Eco-Friendly Hotel: There are lots of reasons to pay attention to eco-tourism and let's face it, the main ones have to do with money. *Cornell Hotel and Restaurant Administration Quarterly, 35*(1), 34–38. https://doi.org/10.1177/001088049403500104

Jauhari, V. (2014). Managing sustainability in the hospitality and tourism industry: Paradigm and directions for the future. *Canadan Journal of Cleaner production,* 1–17. (Apple Academic Press: Toronto, ON). https://doi.org/10.1201/b16789.

Kapera, I. (2018). *Sustainable development in the hotel industry: Between theory and practice in Poland., 28,* 23–30. https://doi.org/10.2478/tour-2018-0011

Karantzavelou, V. (2022). Sustainable Travel Survey 2022—87% say it is important; 78% would pay more to lower carbon footprint. TravelDailyNews International. Retrieved June 15, 2022, from https://www.traveldailynews.com/post/sustainable-travel-survey-2022-87-say-it-is-important-78-would-pay-more-to-lower-carbon-footprint.

Kaza, S., Yao, L., Bhada-Tata, P., & Van Woerden, F. (2018). *What a waste 2.0: A global snapshot of solid waste management to 2050.* World Bank Publications.

Khatter, A., McGrath, M., Pyke, J., White, L., & Lockstone-Binney, L. (2019). Analysis of hotels' environmentally sustainable policies and practices. *International Journal of Contemporary Hospitality Management, 31*(6), 2394–2410. https://doi.org/10.1108/ijchm-08-2018-0670

Krassmira, A. P. S. (2007). New paradigms in city tourism management: Redefining destination promotion. *Journal of Travel Research, 46*(108), 109. https://doi.org/10.1177/0047287507302394

Lai, J. H. K. (2015). Carbon footprints of hotels: Analysis of three archetypes in Hong Kong. *Sustainable Cities and Society, 14,* 334–341. https://doi.org/10.1016/j.scs.2013.09.005

Legrand, W., Sloan, P., & Chen, J. S. (2017). *Sustainability in the hospitality industry.* Routledge.

Leroux, E., & Pupion, P.-C. (2018). Factors of adoption of eco-labelling in hotel industry. *Technological Forecasting and Social Change, 129,* 194–209. https://doi.org/10.1016/j.techfore.2017.09.018

Leverenz, D., Hafner, G., Moussawel, S., Kranert, M., Goossens, Y., & Schmidt, T. (2021). Reducing food waste in hotel kitchens based on self-reported data. *Industrial Marketing Management, 93,* 617–627. https://doi.org/10.1016/j.indmarman.2020.08.008

Lim, W. M. (2016). Creativity and sustainability in hospitality and tourism. *Tourism Management Perspectives, 2016*(18), 16. https://doi.org/10.1016/j.tmp.2016.02.0011-167

Lock, S. (2021). *Global travelers that want to use green lodging in the next year 2021.* Statista. Retrieved June 18, 2022, from https://www.statista.com/statistics/1055777/sustainable-travel-travelers-staying-in-green-lodging-in-the-next-year/

Luo, J. M., Chau, K. Y., Fan, Y., & Chen, H. (2021). Barriers to the implementation of green practices in the integrated resort sector. *SAGE Open, 11*(3), 215824402110302. https://doi.org/10.1177/21582440211030277

Maione, C. (2019). *Emergence of plastic pollution on tourism beaches in Zanzibar, Tanzania.* Master's thesis, School for Environment and Sustainability, University of Michigan. Retrieved June 11, 2022, from https://deepblue.lib.umich.edu/bitstream/handle/2027.42/148801/Maione_Carol_Thesis.pdf?sequence=1&isAllwed=y

Malheiro, A., Sousa, B. B., Liberato, D., & Liberato, P. (2020). Sustainability in tourism and hospitality: Trends and challenges. In the 35th IBIMA conference education excellence and innovation management: A 2025 vision to sustain economic development during global challenges (pp. 15545–15555). IBIMA.

Manaktola, K., & Jauhari, V. (2007). Exploring consumer attitudes and behavior towards green practices in the lodging industry in India. *International Journal of Contemporary Hospitality, 19*(5), 364–377. https://doi.org/10.1108/09596110710757534

Mohan, V., Deepak, B., & Mona, S. (2017). Reduction and management of waste in hotel industries. *International Journal of Engineering Research and Applications, 07*(07), 34–37. https://doi.org/10.9790/9622-0707103437

Moyle, B. D., Weaver, D. B., Gössling, S., McLennan, C. L., & Hadinejad, A. (2021). Are water-centric themes in sustainable tourism research congruent with the UN sustainable development goals? *Journal of Sustainable Tourism, 1–16*, 1–16. https://doi.org/10.1080/09669582.2021.1993233

Moyle, B., Moyle, C.-l., Ruhanen, L., Weaver, D. & y Hadinejad, A. (2020). Are we really progressing sustainable tourism research? A bibliometric analysis. *Journal of Sustainable Tourism, 29*(1), 106–122. https://doi.org/10.1080/09669582.2020.1817048.

Niñerola, A., Sánchez-Rebull, M. V., & Hernández-Lara, A. B. (2019). Tourism research on sustainability: A bibliometric analysis. *Sustainability, 11*, 1377. https://doi.org/10.3390/su11051377

Obradovi´c, S., Stojanovi´c, V., Kova˘ci´c, S. Jovanovic, T., Panteli´c, M. & Vuji˘ci´c, M. (2021). Assessment of residents' attitudes toward sustainable tourism development-A case study of Ba˘cko Podunavlje Biosphere Reserve, Serbia. *Journal of Outdoor Recreation and Tourism, 35*, 100384 https://doi.org/10.1016/j.jort.2021.100384.

Ogbeide, G. C. (2012). Perception of green hotels in the 21st century. *Journal of Tourism Insights, 3*(1), 1. https://doi.org/10.9707/2328-0824.1032

Oxenswärdh, A. (2020). Sustainability practice at hotels on the Island of Gotland in Sweden–an exploratory study. *European Journal of Tourism, Hospitality and Recreation, 10*(3), 203–212. https://doi.org/10.2478/ejthr-2020-0018

Palazzo, M., Gigauri, I., Panait, M. C., Apostu, S. A., & Siano, A. (2022). Sustainable tourism issues in European countries during the Global Pandemic Crisis. *Sustainability, 14*(7), 3844. https://doi.org/10.3390/su14073844

Parpairi, K. (2017). Sustainability and energy use in small scale Greek hotels: Energy saving strategies and environmental policies. *Procedia Environmental Sciences, 38*, 169–177. https://doi.org/10.1016/j.proenv.2017.03.099

Pereira, V., Silva, G. M., & Dias, A. (2021). Sustainability practices in hospitality: Case study of a luxury hotel in Arrábida Natural Park. *Sustainability, 13*(6), 3164. https://doi.org/10.3390/su13063164

Pham, N. T., Thanh, T. V., Tuckova, Z., & Thuy, V. T. N. (2019). The role of green human resource management in driving hotel's environmental performance: Interaction and mediation analysis. *International Journal of Hospitality Management, 10*, 1–10. https://doi.org/10.1016/j.ijhm.2019.102392

Pirani, S., & Arafat, H. (2015). Reduction of food waste generation in the hospitality industry. *Journal of Cleaner Production.* https://doi.org/10.1016/j.jclepro.2015.07.146

Radwan, H. R., Jones, E., & Minoli, D. (2012). Solid waste management in small hotels: A comparison of green and non-green small hotels in Wales. *Journal of Sustainable Tourism, 20*(4), 533–550. https://doi.org/10.1080/09669582.2011.621539

Ramirez, A., & George, B. (2019). Plastic recycling and waste reduction in the hospitality industry: Current challenges and some potential solutions. *Economics, Management and Sustainability, 4*(1), 6–20. https://doi.org/10.14254/jems.2019.4-1.1.

Riffat, S., Powell, R., & Aydin, D. (2016). Future cities and environmental sustainability. *Future Cities and Environment, 2*, 1. https://doi.org/10.1186/s40984-016-0014-2

Ruhanen, L. (2008). Progressing the sustainability debate: A knowledge management approach to sustainable tourism planning. *Current Issues in Tourism, 11*(5), 429–455. https://doi.org/10.1080/13683500802316030

Sadi, I., & Adebitan, E. (2014). Waste water recycling in the hospitality industry. *Academic Journal of Interdisciplinary Studies*. https://doi.org/10.5901/ajis.2014.v3n7p87

Sánchez-Medina, P. S., Díaz-Pichardo, R., & Cruz-Bautista, M. (2016). Stakeholder influence on the implementation of environmental management practices in the hotel industry. *International Journal of Tourism Research, 18*(4), 387–398. https://doi.org/10.1002/jtr.2056

Santiago, D. E. (2021). Energy use in hotels: A case study in Gran canaria. *International Journal of Low-Carbon Technologies, 16*(4), 1264–1276. https://doi.org/10.1093/ijlct/ctab048

Sarkis, J., & Dijkshoorn, J. (2005). An exploratory evaluation of economic and environmental efficiency of solid waste management in Welsh small and medium-sized enterprises. Working paper series no. 31. Cardiff: BRASS. https://doi.org/10.1117/12.631778

Satchapappichit, S., Hashim, N. A., & Hussin, Z. (2015). *Factors influencing adoption of green practices by small and medium sized hotels in Thailand*. http://repo.uum.edu.my/20687/1/JBMA%203%202013%201%2015.pdf

Sawin, J. (2011). Renewable 2011 Global Status Report. REN 21 Secretariat, Paris.

Sharpley, R. (2020). Tourism, sustainable development and the theoretical divide: 20 years on. *Journal of Sustainable Tourism, 28*(11), 1932–1946. https://doi.org/10.1080/09669582.2020.1779732

Singh, N., Cranage, D., & Lee, S. (2014). Green strategies for hotels: Estimation of recycling benefits. *International Journal of Hospitality Management, 43*, 13–22. https://doi.org/10.1016/J.IJHM.2014.07.006

Sloan, P., Legrand, W., & Chen, J. (2013). *Sustainability in the hospitality industry: Principles of sustainable operations*. Routledge.

Styles, D., Schoenberger, H., & Galvez-Martos, J. L. (2015). Water management in the European hospitality sector: Best practice, performance benchmarks, and improvement potential. *Tourism Management, 46*, 187–202. https://doi.org/10.1016/j.tourman.2014.07.005

Taherdoost, H. (2017b). Measurement and scaling techniques in research methodology; Survey/questionnaire development. *International Journal of Academic Research in Management, 6*(1), 1–5. ISSN: 2296–1747

Tharenou, P., Donohue, R., & Cooper, B. (2007). *Management research methods*. Cambridge University Press.

Torelli, R. (2020). Sustainability, responsibility and ethics: Different concepts for a single path. *Social Responsibility Journal*. ahead-of-print. https://doi.org/10.1108/srj-03-2020-0081.

Trang, H. L. T., Lee, J. S., & Han, H. (2019). How do green attributes elicit pro-environmental behaviors in guests? The case of green hotels in Vietnam. *Journal of Travel and Tourism Marketing, 2019*(36), 14–28. https://doi.org/10.1080/10548408.2018.1486782

Tsekouropoulos, G., Gkouna, O., Theocharis, D., & Gounas, A. (2022). Innovative sustainable tourism development and entrepreneurship through sports events. *Sustainability, 2022*(14), 4379. https://doi.org/10.3390/su14084379

Tzschentke, N., Kirk, D., & Lynch, P. (2008). Going green: Decisional factors in small hospitality operations. *International Journal of Hospitality Management, 27*(1), 126–133. https://doi.org/10.1016/j.ijhm.2007.07.010

United Nations Environment Program. (2018). Mapping of global plastics value chain and plastics losses to the environment: With a particular focus on marine environment. https://wedocs.unep.org/handle/20.500.11822/26745.

United Nations Environment Programme and World Travel & Tourism Council. (2021). Rethinking single-use plastic products in travel & tourism—impacts, management practices and recommendations. Nairobi.

United Nations World Tourism Organization. (2017). The 3th UNWTO International congress on ethics presents the recommendations on the responsible use of ratings and reviews. http://media.unwto.org/press-release/2017-05-04/3th-unwto-internationalcongress-ethics-presents-recommendations-responsibl

Vagena, A., & Manoussakis, G. (2021). Group analysis of official hotel classification systems: A recent study. *Current Approaches in Science and Technology Research, 4*, 1–17. https://doi.org/10.9734/bpi/castr/v4/2173f

Vatavali F, Gareiou Z, Kehagia F, Zervas E. (2020). Impact of COVID-19 on Urban everyday life in greece. perceptions, experiences, and practices of the active population. *Sustainability, 12*(22), 9410. https://doi.org/10.3390/su12229410.

Verma, V. K., & Chandra, B. (2018). An application of theory of planned behavior to predict young Indian consumers' green hotel visit intention. *Journal of Cleaner Production, 172*, 1152–1162. https://doi.org/10.1016/j.jclepro.2017.10.047

Wang, X., Wu, N. Q., Qiao, Y., & Song, Q. B. (2018). Assessment of energy-saving practices of the hospitality industry in Macau. *Sustainability, 10*(1), 255. https://doi.org/10.3390/su10010255

World Tourism Organizations: A United Nations Specialized Agency. (2021). Sustainable development. https://www.unwto.org/sustainable-development.

World Wildlife Fund. (2019a). *Plastic waste and climate change—what's the connection?*. www.wwf.org.au/news/blogs/plastic-waste-and-climate-change-whats-the-connection#gs.hei9mm

World Wildlife Fund. (2019b). *Stop the flood of plastic: Effective measures to avoid single-use plastics and packaging in hotels.* https://www.wwf.de/fileadmin/fm-wwf/Publikationen-PDF/WWF_Plastikstudie_Hotelma%c3%9fnahmen_eng.pdf

World Travel and Tourism Council. (2020). World Travel and Tourism Council (WTTC). In Economic Impact Reports.

Wu, C. (2021). Exploring green hotel competitive strategies by using the hybrid method for complex data analysis. *Mathematical Problems in Engineering, 2021*, 1–10. https://doi.org/10.1155/2021/9982387

Yang, Y., Zhang, H., & Chen, X. (2020). Coronavirus pandemic and tourism: Dynamic stochastic general equilibrium modeling of infectious disease outbreak. *Annals of Tourism Research.* https://doi.org/10.1016/j.annals.2020.102913

Yu, J., Park, J., Lee, K., & Han, H. (2021). Can environmentally sustainable development and green innovation of hotels trigger the formation of a positive brand and price premium? *International Journal of Environmental Research and Public Health, 18*, 3275. https://doi.org/10.3390/ijerph18063275

Sustainability Tools for Beach Management: Awareness of Integrated Coastal Zone Management and Current Compliance with Blue Flag Criteria at Eastern Beach in East London, South Africa

Ntuthuzelo Headman Sayedwa and Dorothy Ruth Queiros

Abstract Marine and coastal tourism continues to be documented globally as a vital niche of the tourism industry. In South Africa, various statutory beach management tools such as Integrated Coastal Zone Management (ICZM) and the Blue Flag ecolabel exist, to facilitate the management of beaches for environmental sustainability. South Africa hosts a few well-managed beaches that have Blue Flag accreditation. Yet, several beaches lack resources, inclusive development and effective implementation of sustainable beach management tools. This is a widespread problem, which calls for urgent attention to sustainable marine and coastal tourism management. The focus of this research is on Eastern Beach in East London, which suffers from the abovementioned problems. This paper aimed to investigate sustainability tools for beach management in the context of Eastern Beach, particularly the awareness of ICZM and current compliance with Blue Flag criteria (with a view to attaining this status in future). Via virtual semi-structured interviews with twenty participants, who were tourism stakeholders, various closed and open-ended questions were posed. A few participants voiced that ICZM is useful for environmental conservation, planning and local beach upgrade programmes. Moreover, most participants voiced challenges regarding insufficient implementation of ICZM at Eastern beach. In terms of Blue Flag compliance, findings illustrate noncompliance concerning environmental education and information criteria, with no information to guide and educate users on appropriate use of the beach. As a result of challenges with sewage discharge to the sea, non-compliance with water quality criteria was reported, along with noncompliance regarding environmental management criteria, due to limited rest rooms and inadequate ablution facilities. Participants reported compliance with safety and services criteria, despite lacking emergency plans to cope with pollution risks. Through these management tools, this beach could be uplifted as a coastal

N. H. Sayedwa (✉) · D. R. Queiros (✉)
University of South Africa, Pretoria, South Africa
e-mail: headmansayedwa@outlook.com

D. R. Queiros
e-mail: queirdr@unisa.ac.za

© The Author(s), under exclusive license to Springer Nature Switzerland AG 2023
V. Katsoni (ed.), *Tourism, Travel, and Hospitality in a Smart and Sustainable World*, Springer Proceedings in Business and Economics,
https://doi.org/10.1007/978-3-031-26829-8_4

tourist attraction, be accredited with Blue Flag status and managed sustainably for the enjoyment of present and future generations.

Keywords Sustainability · Marine and coastal tourism · Beach management · Integrated Coastal Zone Management · Blue Flag criteria · Blue Flag status

JEL Classification Fishing tourism · Maritime policy · Cultural ports

1 Introduction

Globally and locally, the tourism industry is continuously growing and has become an economically important sector for both developing and developed nations (León-Gómez et al., 2021). Tourism has economic, social and cultural significances and offers real prospects for sustainable and inclusive coastal and marine development (United Nations World Tourism Organisation, 2021). Marine and coastal tourism therefore continues to be documented as a vital niche (León-Gómez et al., 2021). Administrations are globally encouraging it due to its high multiplier effects with respect to the generation of employment, increase in foreign exchange earnings and stimulation of supply sectors such as transport, hospitality and gaming. These spin-offs can alleviate poverty and improve the living standards for local communities (Khan et al., 2021). Coastal and marine regions are, however, fragile and complex ecosystems, with a variety of land uses (Sakellariadou, 2007). The Ecosystems Approach (EA) has been globally presented and emerged as a leading paradigm for managing coastal and marine ecosystems (Sardá et al., 2015). This approach presents new prospects for sustainable use of the sea and improves understanding of how marine socio-ecological systems operate, how they generate goods and services, as well as how these benefits are captured and can be sustained (Farmer et al., 2012). The intense use of beaches has forced beach administrators to introduce sustainable beach management tools, such as the use of Performance Awards (for example, Blue Flag Ecolabel) and Integrated Coastal Zone Management (ICZM).

The Blue Flag ecolabel remains a sought-after ecolabel (Foundation for Environmental Education, 2006), dating back to the 1980s (World Commission on Environment and Development, 1987). It applies environmental law (Coetzee, 2016) and focuses on the sustainable ecological management of beaches, marinas and boating tourism operators (Klein & Dodds, 2017). As a beach management tool, it plays two important roles—to guide the tourism industry to improve its performance by supporting operational guidelines; and to communicate to beach users the importance of and compliance with environmental, educational, safety and accessibility standards (Wildlife and Environment Society of South Africa, 2021).

ICZM, on the other hand, is a tool to develop and use coastal natural resources to protect the best interests of the public and economy, while ensuring ecological sustainability (McLean & Glazewski, 2009; South Africa, 2016). Over the years,

ICZM has become a key model for sustainable development in coastal destinations. It has emerged in different contexts and been adopted for various purposes (Ameersingh, 2016; Phillips & Jones, 2005).

In South Africa, various statutory beach management tools, including ICZM and Blue Flag ecolabel exist, to facilitate the management of beaches for environmental sustainability (WESSA, 2021). The country currently hosts a few well-managed beaches that have Blue Flag status. As of 2021, 51 beaches across the provinces of the Western Cape, Eastern Cape and KwaZulu Natal have been accredited with Blue Flag status. The Eastern Cape, however, contains only seven of these (Business Tech, 2021; WESSA, 2021), in spite of beach tourism being a strong component of the province's economic growth, drawing domestic and international tourists (Klein & Dodds, 2017). The province's coastline is rich with natural biodiversity and beautiful beaches such as Eastern Beach in East London, which currently does not have Blue Flag status. Moreover, several beaches lack resources, inclusive development and effective implementation of sustainable beach management tools. This is a widespread problem, which calls for urgent attention to sustainable marine and coastal tourism management. Therefore, the focus of this research is on Eastern Beach in East London, which suffers from the abovementioned problems. This paper aimed to investigate sustainability tools for beach management in the context of Eastern Beach, particularly the awareness of ICZM and current compliance with Blue Flag criteria (with a view to attaining this status in future).

2 Literature Review

2.1 Statutory Tools for Sustainable Tourism Development in South Africa

South Africa implemented a compulsory Tourism White Paper of 1996 to administer sustainable tourism development as a key driver to strengthen the country's tourism sector (Department of Environmental Affairs and Tourism, 1996). Post the inception of this policy, the country continued to implement coastal and marine tourism policies, which aimed at improving coastal sustainable tourism (Coetzee, 2016; National Department of Tourism, 2016). Amongst these, the National Environmental Management Act (NEMA) 107 of 1998, the National Tourism Sector Strategy of 2011, as well as the Integrated Coastal Management Act (ICMA) 24 of 2008 are noteworthy in the context of this paper and supplement the White Paper agenda (South Africa, 2016). Furthermore, the integrated environmental management (IEM) agenda in the country encourages coastal tourist destinations to emphasise the impacts of human activities on coastal or marine resources (Department of Environmental Affairs and Tourism, 2017). The National Development Plan 2030 was also implemented and currently presents a similar notion of coastal or marine sustainability, to ensure that coastal ecosystems are conserved for current and future generations (South Africa,

2016). An Integrated Coastal Management programme has been developed over the past few decades to address multiple-use conflicts, plan for new uses and protect vulnerable ecosystems and marine biodiversity. The ICMA is classified as an adaptable multi-sectoral and governmental method that has proved to strike a balance between the development, use and protection of coastal surroundings (South African Institute of International Affairs, 2015).

2.2 *Integrated Coastal Zone Management as a Sustainability Tool*

Post the adoption of the ICMA, the country's approach towards coastal management transformed from being a conservation focused and bureaucratic approach, into a more human inclusive development based on sustainable development goals (SAIIA, 2015). In addition, the ICMA was later amended to ICM Act 36 of 2014, to clarify responsibilities for the organs of state, with reference to coastal areas (DEA, 2012). It has since been a baseline for ICZM and sustainable coastal development for coastal areas (DEA, 2012; SAIIA, 2015). ICZM relates to the effective conservation and utilisation of the coast, as a dynamic ecosystem at the interface between land and sea (Ameersingh, 2016). It is also considered as a beach management tool, aiming to ensure that development in the coastal zone is regulated to protect intrinsic environmental features and ecological processes (Glavovic, 2006). In addition, the Blue Flag programme is another valued instrument within the Integrated Environmental Management toolbox for sustainable tourism development and management. Due to its importance, South Africa adopted it and becomes the first country outside the European continent to be accorded Blue Flag accreditation rights for its beaches in 2001 (Blue Flag South Africa, 2020).

2.3 *Blue Flag Ecolabel as a Sustainability Tool*

Awareness of the Blue Flag ecolabel remains limited in most developing countries, like South Africa. Yet, the concept of beach and marine tourism in the country continues to be directly framed around sustainable standards of seashores, with the intention to promote and enhance coastal tourism (DEA, 2012; Lucrezi et al., 2016). This is possible with working beach management tools such as Blue Flag ecolabel. The literature positions the Blue Flag ecolabel as a sensitive beach management tool, which reinforces environmental sustainability (BFSA, 2020). The country adopted this ecolabel to utilise it as a sustainable approach, which ensures that local beaches can be promoted for their sustainable management, international standard of cleanliness, safety and environmental responsiveness (BFSA, 2020; NDT, 2016). The Blue

Flag criteria provide guidelines regarding the processes to obtain certification, as well as the related administrative processes to maintain the status (WESSA, 2021).

The Blue Flag programme requires beaches to meet 33 criteria, grouped under environmental education and information; water quality; environmental management; and safety and service (WESSA, 2021). It is vital that beach management or municipal representatives ensure that water quality aligns with internationally recognised standards (McKenna et al., 2011) and with the Blue Flag necessities regarding the microbiological parameter Escherichia coli (faecal coli bacteria) and intestinal enterococci (streptococci) (FEE, 2006). The Blue Flag award is given to all beaches that pass the criteria for a specific season (which officially opens on 1 December annually). Post the award, beaches must continue to comply with the standards set (Silwana, 2015) or risk losing this status. While South Africa's Blue Flag beaches do increase the attractiveness and standing of South Africa as an international destination (BFSA, 2020; Slatter & Mearns, 2018), since the award is moderately new in the country, a consideration of sustainability tools for better development and management of these local beaches is vital. If not, it may take more time for international tourists to visit the country due to its accredited beaches and for locals to notice the value of the award and its associated improvement in beach quality standards (Nahman & Rigby, 2008).

3 Methodology

3.1 Sampling Technique, Sample Size, Inclusion Criteria, Participants and Ethical Considerations

The primary researcher approached public and private stakeholders who are managers within their organisations. These participants are involved in tourism planning, coastal development and environmental management of East London's coastal attractions. Out of 52 active tourism stakeholders (Buffalo City Metropolitan Municipality, 2018), purposive sampling was used to create a final sample of 20 (ten public and ten private stakeholders). Purposive sampling is used to select individuals or groups of individuals that are knowledgeable about the phenomenon of interest (Creswell & Plano Clark, 2011). This approach assists in ensuring the validity and reliability of the data gathered (Palys, 2008). The inclusion criteria were that participants needed to be between the ages of 18 and 65 and be employed as a public or private tourism stakeholder in East London. Participants provided informed consent and anonymity was assured. Prior to the fieldwork, ethical clearance was obtained from the Ethics Review Committee of the Department of Applied Management at the University of South Africa.

3.2 Research Instrument

Primary data collection was done via virtual semi-structured interviews, based on an interview guide, whereby various closed and open-ended questions were posed. The questions were informed by literature and finalised by the primary researcher. The virtual nature of interviews was triggered by the COVID-19 pandemic. The restrictions imposed compelled the use of virtual communication tools and provided an opportunity to grow knowledge and understanding regarding using information technology effectively in fieldwork (Sah et al., 2020). Microsoft Teams (MS) was utilised to this end, with each interview being recorded with the consent of the participant.

3.3 Study Area

Eastern Beach in East London, which falls with the Eastern Cape province of South Africa, was the study site for this research (Fig. 1). The city of East London is found on the southeast coast of South Africa, within the Buffalo City Metropolitan Municipality (BCMM). East London's population is over 267 000 and covers 168,86 km^2 (Buffalo City Metropolitan Development Agency, 2016). BCMM is the local tourism district of East London (Ebomah et al., 2019).

3.4 Data Analysis

To assist the analysis, ATLAS.ti, version 9.0.18 was used. The use of ATLAS.ti allowed transcribed text to be linked back to original recordings and facilitated the organisation of codes, sub-codes and comment writing (Queiros & Mearns, 2019) for the open-ended questions. The recorded interviews were transcribed into word documents, the data was cleaned and then coded with codes and sub-codes developed by the primary researcher based on the themes emerging from the data (Maguire & Delahunt, 2017). Quotations are used to substantiate codes and are cited according to the participant number and the line within the transcript where the quote comes from. For example, "P9:54" refers to Participant 9, line 54. Quantitative questions are reported as per the number of responses, followed by a discussion of these findings.

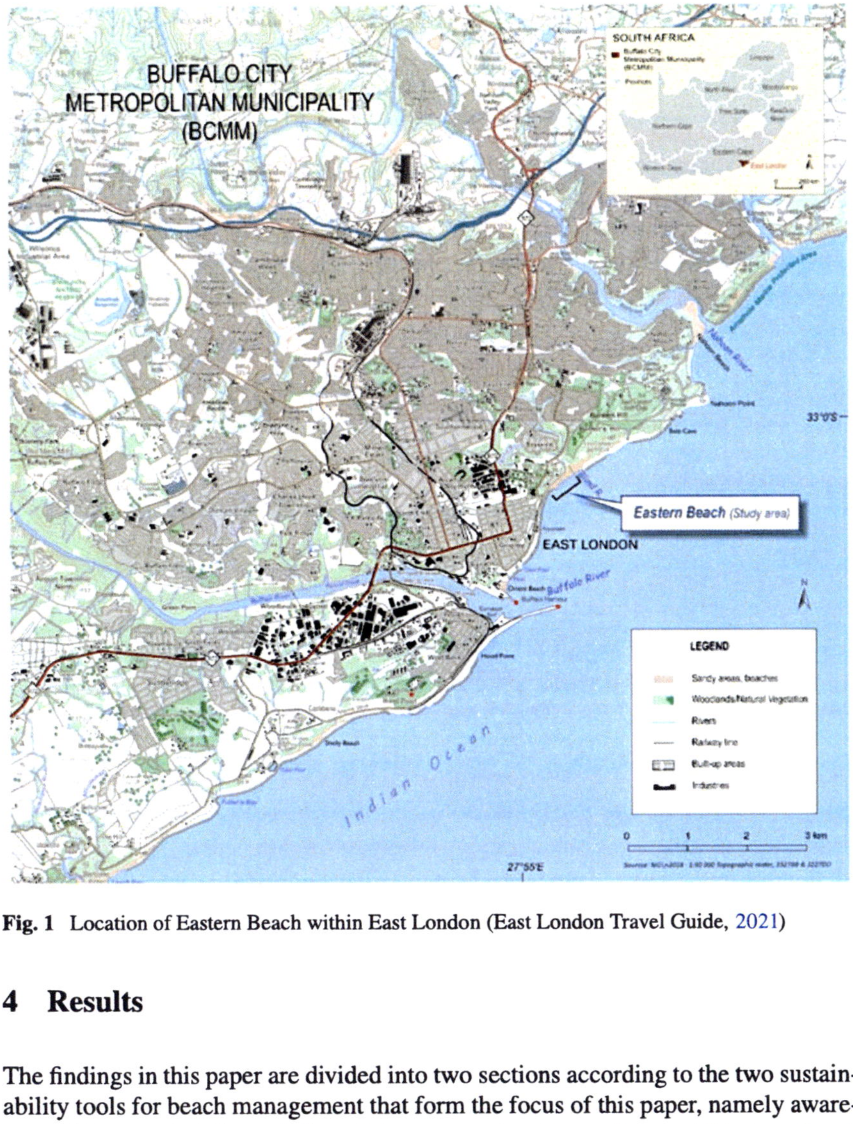

Fig. 1 Location of Eastern Beach within East London (East London Travel Guide, 2021)

4 Results

The findings in this paper are divided into two sections according to the two sustainability tools for beach management that form the focus of this paper, namely awareness of **Integrated Coastal Zone Management** and compliance of Eastern Beach with **Blue Flag** criteria. These findings are interpreted below, where relevant, findings are also compared with those of previous studies.

Awareness of Integrated Coastal Zone Management (ICZM)

Regarding awareness of ICZM, participants were asked a closed question on whether they had come across the concept of ICZM before. Their responses are depicted in Fig. 2. The majority (eleven) of these participants were unfamiliar with ICZM. Nine

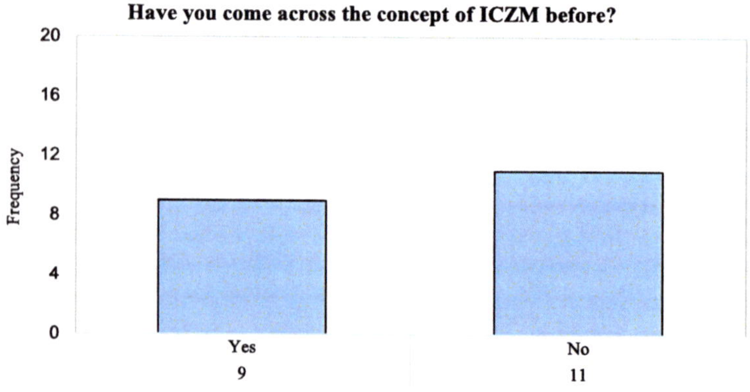

Fig. 2 Stakeholder awareness of ICZM

Table 1 Code frequencies for Integrated Coastal Zone Management

Code: Integrated Coastal Zone Management (ICZM)	
Sub-codes	Public and private stakeholders
ICZM: Framework use/application	14
ICZM: No ICZM implementation	12
Total	26

indicated that they had come across the concept before and were familiar with it from the organisations they work for.

In addition to the quantitative question above, participants were also asked a qualitative question relating to ICZM, namely "If yes, how is it used in your work?" Findings from this question are demonstrated in Table 1. Under the main code of 'Integrated coastal zone management (ICZM)', two sub-codes developed: 'Framework use/application' with 14 quotes and 'No ICZM implementation' with 12 quotes. These sub-codes capture how public and private stakeholders use ICZM in their organisations as well as responses regarding a lack of ICZM implementation.

4.1 Framework Use/Application

The sub-code of 'Framework use/application' contained 14 quotations and responses were obtained from participants who had come across the concept of ICZM before. One confirmed that ICZM was adopted for a **beach development programme**: "*... so, from my office, what we've also done has to do with what is called an* ***Integrated Beach Programme*** *(IBP) for all the beaches from East to West of East London*"

(P3:123). Results revealed that the use of ICZM relates to environmental conservation and planning, as evidenced in the following quotations:

> When I am developing management plans, then we have to look at … what is going on in that protection zone, to see what activities are appropriate or not appropriate in that protected zone, so it's almost proactive planning (P4:77–79).
>
> … I have used it in the planning side of things (P4:76).
>
> I think it is a good tool for planning … (P14:71),
>
> The municipality is updated with that, and it is being used (P21:33).
>
> Our organisation has a development unit, and we have a biodiversity and conservation unit, so you know it might have affected people within those units or they would have been part of that, but not from the marketing perspective (P18:40).
>
> … it has actually been on a superficial level, I think in previous years, work was done to protect the shoreline in Nahoon Beach … (P20:28).
>
> I have heard about it in the areas where we manage the national reserves and the identified one was East London Coast Nature Reserve (P19:47).

The ICZM framework remains a recognised science-driven management approach for capacity development and sustainability, for all coastal tourist destinations like East London (Glavovic, 2006; Quesada et al., 2018; Republic of South Africa, 2003; United Nations Environmental Programme, 2010).

4.2 No ICZM Implementation

The sub-code of 'No ICZM implementation' contained 12 quotations. Implementation challenges emerged strongly in the data, as cited in the following quotations:

> … I think implementation has been a challenge and I don't think its high on the priority list for Buffalo City (P2:97).
>
> I have heard of it, but the plans in the metro, in my opinion, possibly have failed to implement that appropriately (P3:128).
>
> … I think it's very limited in terms of actual implementation, so there are challenges with implementation (P14:71).
>
> I've seen such plans and I believe BCMM should have that plan, so it is applicable, but it is not implemented … (P10:31).

Results revealed that a gap exists concerning ICZM [*"I think there is a gap there"* (P4:81)] and things are fragmented ["*Perhaps things are fragmented and incoherent, I cannot exactly pinpoint as to what is the problem*" (P6:38)]. Even after 40 years of ICZM introduction around the world, there are still numerous challenges to be overcome with regard to its implementation (SAIIA, 2015). If a beach is missing ICZM, it is most likely that public awareness regarding complexities of the coastal area and active participation in the management of the coast becomes disjointed too (DEA, 2012; Republic of South Africa, 2003, 2008).

Compliance of Eastern Beach with Blue Flag Criteria

The current section provides results regarding compliance of Eastern Beach with Blue Flag criteria. The questions posed to participants in this section were closed questions in the form of statements derived from the mandatory Blue Flag criteria (Tables 2, 3, 4 and 5). The results are followed by discussions. Participants were asked to answer 'Yes', 'No' or 'Not sure'.

Table 2 Environmental education and information criteria

Environmental education and information criteria		Response frequency		
		Yes	No	Not sure
1	There is information presently displayed regarding the Blue Flag ecolabel at the Eastern Beach	0	17.0	3.0
2	Eastern Beach is promoting environmental education activities for the beach users	4.0	15.0	1.0
3	A beach map indicating the different facilities available is currently displayed at the Eastern Beach	10.0	10.0	0.0
4	Eastern Beach is currently offering a Code of Conduct, which is displayed and guiding beach users about the appropriate use of the beach and surrounding areas	11.0	8.0	1.0

Table 3 Water quality criteria

Water quality criteria		Response frequency		
		Yes	No	Not sure
1	Eastern Beach is currently complying with water quality sampling and frequency requirements for the Blue Flag beaches	4.0	11.0	5.0
2	There is perceived industrial, wastewater or sewage-related discharges, currently affecting Eastern Beach	17.0	2.0	1.0
3	Eastern Beach is fully complying with the standards and requirements for water quality analysis	0.0	11.0	9.0

Table 4 Environmental management criteria

Environmental management		Response frequency		
		Yes	No	Not Sure
1	A beach management committee appointed by the local authority/beach operator for the Eastern Beach already exists	3.0	13.0	4.0
2	Eastern Beach has clean facilities, which are meeting required standards for Blue Flag beaches	1.0	17.0	2.0
3	Facilities for the separation of recyclable waste materials (glass, tins, dry and wet waste), exist at Eastern Beach	5.0	15.0	0.0
4	At Eastern Beach, there is an adequate number of restrooms and ablution facilities for beach users	2.0	17.0	1.0
5	Eastern Beach has strict and controlled measures that are guiding the access of dogs and other domestic animals	6.0	14.0	0.0

Table 5 Safety and services criteria

Safety and services		Response frequency		
		Yes	No	Not sure
1	Appropriate public safety control measures currently exist for beach users at Eastern Beach	13.0	5.0	2.0
2	Essential facilities and access for physically disabled visitors or residents already exist at the Eastern Beach	8.0	11.0	1.0
3	Emergency plans to cope with pollution risks currently exist at Eastern Beach	6.0	13.0	1.0

4.3 Environmental Education and Information Criteria

Table 2 presents the results obtained from participants regarding the compliance of Eastern Beach with the environmental education and information criteria.

As indicated in Table 2, the majority of stakeholders (17), indicated that information regarding the Blue Flag label is not on display, while the other three participants were not sure. Eastern Beach is also not promoting environmental education activities, based on the majority of participants' (15) replying 'no'. These findings might mean that there is a lack of awareness of any environmental public initiatives at Eastern Beach.

Regarding the display of a beach map showing facilities, ten participants felt Eastern Beach complied with this, while ten felt it did not comply. Eleven participants agreed that a code of conduct was on display at the beach, while eight said 'no'. This could suggest a link between prevailing negative ecological impacts of beach users and limited sustainable information, which beach users should be cognisant of at Eastern Beach. Therefore, much effort is required in terms of environmental education and information criteria, to improve the current status of this beach. Such ecological efforts should be aimed at meeting South Africa's international obligations related to coastal and educational matters for beach users at Eastern Beach (Klein & Dodds, 2017; Republic of South Africa, 2001).

4.4 Water Quality Criteria

Table 3 presents the results relating to compliance with water quality criteria.

With respect to compliance with water quality sampling and frequency, the majority (11) of interviewed participants selected 'no', indicating that this beach does not currently comply. It is concerning that the majority of participants (17) indicated that there are problems regarding industrial or sewage water waste at Eastern Beach. Such findings are concerning, especially when environmental legislation in South Africa puts emphasis on the proper execution of regulations, to avoid impaired coastal planning, land-use and unwanted sewage-related spills on its beaches (BFSA, 2020; WESSA, 2021).

As expected, most participants (11) subsequently disagreed that Eastern Beach is complying with the standards and requirements for water quality analysis. Other participants (9) were unsure about this criterion, indicating a concern or a gap which perhaps requires further research. Currently, and due to lack of Blue Flag status at this beach, there is no official encouragement from the city to abide by the Bathing Water Directive (BWD) policies. These policies are concerned with beach certification, water quality, beach safety, cleanliness and reliability (BFSA, 2020; WESSA, 2021). These results indicate that Eastern Beach is currently not complying with the standards for swimming water, which should be in line with the Blue Flag requirements for the microbiological parameters regarding faecal coli bacteria and intestinal enterococci (FEE, 2006).

4.5 Environmental Management Criteria

Table 4 displays the results regarding Eastern Beach's compliance with environmental management criteria.

Interviewed participants disclosed a noteworthy negative response in relation to the existence of a beach management committee, which should be appointed by the beach operator or by the BCMM, to look after this beach. This finding reveals a lack of a joint coastal management effort, which should exist to ensure that there is a shared responsibility for planning, development, management and monitoring of this natural attraction, towards Blue Flag status. A high frequency of responses was received (17, 0) (the majority of stakeholders) for criterion 2, where participants indicated that Eastern Beach does not have clean facilities which is a required standard for a Blue Flag ecolabel. The results also indicate that this beach is not complying with the criterion related to the availability of facilities for the separation of dry and wet waste, with most (15, 0) participants responding with a 'no' answer.

Restrooms and ablution facilities for beach users are also not adequate; 17 participants responded with a 'no' answer to the criterion. This reveals a need for facilities at this beach to be sufficiently developed and then maintained. These results suggest that a degree of noncompliance exists at Eastern Beach, which oppose the required national standards and regulations on acceptable beach facilities. These results also signify that beach authorities, in conjunction with the BCMM, must ensure that facilities and activities under their responsibility comply with environmental criteria. While facilities are concerning at this beach, the management of the beach, its facilities and immediate surrounding areas must also comply with appropriate coastal development plans and monitoring, as per the planning regulations of the Republic of South Africa (2014).

Finally, Table 4 indicates that the majority of participants (14, 0) felt that Eastern Beach lacks strict and controlled measures to guide access of dogs and other domestic animals. Non-adherence to Blue Flag beach requirements on pets, as well as inadequate beach facilities, reveals a lack of a broad coastal public consciousness of policies and regulations, which should be positively influencing sustainable tourism practices at Eastern Beach.

4.6 *Safety and Services Criteria*

Table 5 presents three criteria relating to safety and services, where participants were asked if they feel Eastern Beach is complying or not complying.

In spite of incidents related to people drowning, robbery and other types of crime at this beach, appropriate public safety measures do exist. A high frequency of responses (13, 0) supported this. However, further research is needed to determine the extent of public safety control measures towards the prevention of previously mentioned safety issues at Eastern Beach.

In terms of facilities and access for physically disabled visitors, responses are mixed, with eight participants saying 'yes', 11 saying 'no' and one being unsure. The public and private stakeholders gave negative responses when it comes to the existence of emergency plans to cope with pollution risks at this beach (13 said 'no', six said 'yes'). This is therefore an area for future action, as beach management should have contingency plans to manage oil, hazardous or toxic waste spills. As per the criteria, this beach should have defined plans for events that could lead to large-scale pollution impacts on the beach or bathing water (BFSA, 2020; WESSA, 2021). With respect to safety and services criteria, these results signify that there is a need for Eastern Beach to be managed based on the implemented safety plans of the Republic of South Africa (1998).

5 Conclusion

This paper aimed to investigate sustainability tools for beach management in the context of Eastern Beach, particularly awareness of ICZM and current compliance with Blue Flag criteria (with a view to attaining this status in future). Through virtual semi-structured interviews, public and private stakeholders were questioned to contribute to the research gaps identified. These gaps indicated that despite the country's legislative beach management tools for sustainability, it still hosts few accredited beaches. Only seven are found in the Eastern Cape province, which hosts rich biodiversity and beautiful coastlines. Most beaches lack resources, inclusive development and effective implementation of sustainable beach management tools. Furthermore, there is limited research on sustainable marine and coastal tourism management, particularly at Eastern Beach in East London.

The paper first considered ICZM as a sustainability tool for beach management. It emerged from the data that the use of ICZM is linked to beach development programmes, environmental conservation and planning. Moreover, challenges regarding insufficient implementation of ICZM at Eastern beach were revealed. Considering that interviews conducted were with public and private stakeholders, it was surprising that the majority could not relate to the concept of ICZM. The results portray that the ICZM framework application is limited to a few organisations. Several references were made regarding the lack of ICZM implementation at Eastern Beach, and BCMM was professed as the main holder of this management tool.

This paper also provided findings concerning the Blue Flag ecolabel as a sustainability tool for beach management at the study area. Findings focused on the criteria required to achieve this status. This research firstly stressed noncompliance concerning *environmental education and information criteria*, with no information to guide and educate users on appropriate beach usage. In terms of the second criteria (*water quality*), most participants indicated noncompliance pertaining to industrial or sewage water waste entering the sea. The majority disagreed that Eastern Beach is complying with water quality criteria. Regarding *environmental management criteria*, participants were negative about the current lack of a beach management committee, which should be appointed by the beach operator (BCMM) to look after Eastern Beach. Furthermore, challenges were identified in terms of inadequate restrooms and ablution facilities. In terms of the *safety and services criteria*, the majority agreed that this beach has appropriate safety and security and hence complies with the criterion. Finally, most participants disagree that Eastern Beach has emergency plans to cope with pollution risks.

This research highlighted findings regarding limited awareness of ICZM and limited compliance with Blue Flag criteria, as sustainable beach management tools. Through this research, guidelines can be developed to improve beaches lacking resources and sustainable and inclusive tourism development. Findings could be used by involved beach management stakeholders, providing a holistic picture of mandatory sustainability tools to be considered, to take Eastern Beach towards its desired future as a Blue Flag accredited beach. If this beach could be improved and sustained using these tools, it would be a victory for the disadvantaged East London and Eastern Cape province.

Acknowledgements The authors thank the public and private tourism stakeholders in East London, who participated in this research. Thank you to the University of South Africa for funding provided through its Masters and Doctoral Research Bursary Programme.

References

Ameersingh, S. (2016). *The South African coastal zone: a critical assessment of whether the manner in which the coastal zone is defined in the National Environmental Management: Integrated Coastal Management Act 24 of 2008 facilitates an integrated, and especially ecosystem based, approach to managing the South African coast*. Master's Dissertation, Durban, South Africa, University of KwaZulu-Natal.

Blue Flag South Africa. (2020). *About blue flag in South Africa*. Retrieved June 13, 2022, from https://blueflag.org.za/index

Buffalo City Metropolitan Municipality. (2018). *Integrated development Plan—2018–2011*. East London, Buffalo City Metropolitan Municipality.

Business Tech. (2021). *The safest, cleanest beaches in South Africa*. Retrieved July 24, 2022 from, https://businesstech.co.za/news/lifestyle/546766/51-of-the-safest-cleanest-beaches-in-south-africa/.

Coetzee, J. H. (2016). *Sustainable development in South African environmental law and its relationship with the national development plan*. Master's dissertation, Environmental Law and Governance, North-West University.

Creswell, J. W., & Plano-Clark, V. L. (2011). *Designing and conducting mixed methods research* (2nd ed.). SAGE Publications.

Department of Environmental Affairs and Tourism. (1996). *White paper on development and promotion of tourism*. Government Printers.

Department of Environmental Affairs & Tourism. (2017). *Annual report 2017/18*. Retrieved July 09, 2021, from https://www.deat.gov.zal/Documents/PreviousAmmualReports/2017Jan/annual review

Department of Environmental Affairs. (2012). *Overview of national strategy for sustainable development*. Department of Environmental Affairs.

East London Travel Guide. (2021). *East London maps*. Retrieved December 09, 2021, from http://soileiragusgonta.com/map-of-east-london/map-of-east-london-4-travelguide-accommodation-tourist-information/

Ebomah, K., Sibanda, T., Adefisoye, M., Nontongana, N., Nwodo, U., & Okoh, A. (2019). Evaluating Nahoon Beach and canal waters in Eastern Cape, South Africa: A public health concern. *Journal of Environmental Studies, 28*(3), 1115–1125.

Farmer, A., Mee, L., Langmead, O., Cooper, P., Kannen, A., Kershaw, P. & Cherrier, V. (2012). *The ecosystem approach in marine management*. ISBN 0-9529089-5-6. European Union KNOWSEAS Project.

Foundation for Environmental Education. (2006). *Awards for improving the coastal environment: The example of the Blue Flag*. Foundation for Environmental Education

Glavovic, B. (2006). The evolution of coastal management in South Africa: Why blood is thicker than water. *Ocean Coast Management, 49*(2), 889–904.

Khan, A., Sughra Bibi, S., Ardito Lorenzo, A., Lyu, J. & Babar, Z. U. (2021). Tourism and Development in Developing Economies: A Policy Implication Perspective. *Sustainability, 12*(16), 2–21.

Klein, L., & Dodds, R. (2017). Blue Flag beach certification: An environmental management tool or tourism promotional tool? *Tourism Recreation Research, 10*(3), 1–14.

León-Gómez, A., Ruiz-Palomo, D., Fernández-Gámez, M. A., García-Revilla, M. R. (2021). Sustainable tourism development and economic growth: Bibliometric review and analysis. *Sustainability, 1*(13), 2270.

Lucrezi, S., Saayman, M., & Van der Merwe, P. (2016). An assessment tool for sandy beaches: A case study for integrating beach description, human dimension, and economic factors to identify priority management issues. *Ocean and Coastal Management, 9*(4), 21–22.

Maguire, M., & Delahunt, B. (2017). Doing a thematic analysis: A practical, step-by-step guide for learning and teaching scholars. *All Ireland Journal of Teaching and Learning in Higher Education Journal, 1*(3), 3352–3359.

McKenna, J., Williams, A. T., & Cooper, J. A. G. (2011). Blue Flag or Red Herring: Do beach awards encourage the public to visit beaches? *Tourism Management, 32*(3), 576–588.

McLean, B. & Glazewski, J. I. (2009). Marine systems. Environmental management: An introduction. In H.A. Strydom & N.D. King (Eds.), *Environmental management in South Africa* (2nd ed.). Juta Law.

Nahman, A., & Rigby, D. (2008). Valuing Blue Flag status and estuarine water quality in Margate, South Africa. *South African Journal of Economics, 76*(4), 721–737.

National Department of Tourism. (2016). *National Tourism Sector Strategy (NTSS): Eastern Cape State of the environmental report*. Government Printers.

Palys, T. (2008). Purposive sampling. In L.M. Given (Ed.), *The SAGE encyclopaedia of qualitative research methods*. SAGE Publications.

Phillips, M. R., & Jones, A. L. (2005). Erosion and tourism infrastructure in the coastal zone: Problems, consequences and management. *Tourism Management, 27*(3), 517–524.

Quesada, G. C., Klenke, T., & Mejía-Ortíz, L. M. (2018). *Regulatory challenges in realizing integrated coastal management—lessons from Germany, Costa Rica, Mexico and South Africa*. University of Oldenburg.

Queiros, D., & Mearns, K. (2019). Khanyayo village and Mkhambathi Nature Reserve, South Africa: A pragmatic qualitative investigation into attitudes towards a protected area. *Journal of Sustainable Tourism, 27*(6), 750–772.

Republic of South Africa. (1998). *National environmental management act, (Act No.107 of 1998)*. Cape Town: Government Printer.

Republic of South Africa. (2001). *First edition environmental implementation plan for defence, in government gazette, 401–22022*. Pretoria: Government Press.

Republic of South Africa. (2003). *National environmental management: Protected areas act (Act No. 57 of 2003)*. Cape Town: Government Printers.

Republic of South Africa. (2008). *National environmental management: integrated coastal management amendment act (Act 24 of 2008)*. Pretoria: Government Printer.

Republic of South Africa. (2014). *National environmental management: integrated coastal management amendment act (Act 36 of 2014)*. Pretoria: Government Printer.

Sah, L. K., Singh, D. R. & Sah., R. J. (2020). Conducting Qualitative Interviews using Virtual Communication Tools amid COVID-19. *Journal of Nepal Medical Association, 58*(232),1103–1106.

Sakellariadou, F. (2007). Marine tourism development and sustainability. Case study of the Saronicos Gulf Marinas. Greece. In *CIESM Congress Proceedings (No.38)*. CIESM, Monaco.

Sardá, R., Pintó, J., Lozoya, J. P., & Sansbellò, R. M. M. F. (2015). Towards a new integrated beach management system: The ecostem-based management system for beaches. *Ocean and Coastal Management, 118*(1), 167–177.

Silwana, H. L. S. (2015). *Blue flag beaches in the eastern cape: Implications for tourism, the environment and socio-economy*. Masters dissertation, East London, South Africa: University of Fort Hare.

Slatter, R., & Mearns, K. (2018). Perceptions and activity profiles of Blue Flag beach users in South Africa. *African Journal of Hospitality, Tourism and Leisure, 7*(4), 1–15.

South Africa. (2016). *The National Development Plan 2030: Our future—make it work*. Retrieved July 27, 2022, from https://www.gov.za/issues/national-development-plan-2030

South African Institute of International Affairs. (2015). Promoting the integrated governance of South Africa's coastal zone. Cape Town: Governance of Africa's resources programme. *Occasional Paper, 218*, 1–5.

United Nations Environmental Programme. (2010). *Marine and coastal ecosystems and human well-being: A synthesis report based on the findings of the millennium ecosystem assessment*. Retrieved October 14, 2019, from http://www.lme.noaa.gov/

United Nations World Tourism Organisation. (2021). *Tourism and Covid-19—Unprecedented economic impacts*. Retrieved March 25, 2022, from https://www.unwto.org/tourism-and-covid-19-unprecedented-economic-impacts

WESSA. (2021). Wildlife and Environment Society of South Africa. *Blue Flag programme—What we do*. Retrieved August 09, 2022, from https://www.wessa.org

World Commission on Environment and Development. (1987). *Our common future*. Oxford University Press.

The Prospects of Renewables for Electricity Production in Greece: Diversification Between Island and Mainland Greece

Grigorios L. Kyriakopoulos, Garyfallos Arabatzis, and Miltiadis Chalikias

Abstract Global research interest has recently focused on the utilization of renewable energy sources (RESs) in terms of large-scale technical infrastructure and their emerging technologies. In parallel RES-based electricity production at existing large-scale power stations, it is determined by the current availability of exploitable/remote energy sources. In a technological point of view, the discrete utilization of biomass, hydroelectric, wind and solar energy is prevailing globally. In Greece, meeting national energy needs, it has been remarkably bounded on high level of conventional fuels' exploitation, including abundant stocks of lignite, as well as imports of hydrocarbons and natural gas. In systematically approaching RES in Greece, an hierarchical cluster analysis was applied on data of the number and the rated power of RES plants installed at mainland, insular, and peripheries. Then, the K-means method was used as the distance criterion. The results of the descriptive statistical analysis of the variables supported grouping of peripheral units into clusters. At the same time, the typology results of all grouped areas were visually mapped. Subsequently, the cluster analysis was performed on different pairs of variables in order to find the differences among similar groups produced. The statistical analysis was performed using data for the years 2015 and 2019. The number of formed clusters groups was from 3 to 6, while—according to the ANOVA analysis—the optimal possible and good data separation in the aforementioned groups was achieved on the RES: wind (in all 5 cases hierarchical groups), being followed by the small hydroelectric in 2 cases: (one for 2015 and one for 2019) and the photovoltaic in 2 cases: (one for 2015 and one for 2019). The research unveiled constructive outcomes and multifaceted

G. L. Kyriakopoulos
School of Electrical and Computer Engineering, National Technical University of Athens, 15780 Athens, Greece

G. Arabatzis
Department of Forestry and Management of the Environment and Natural Resources, Democritus University of Thrace, Pantazidou 193, Orestiada, Greece

M. Chalikias (✉)
Department of Accounting and Finance, University of West Attica, 12244 Egaleo, Greece
e-mail: mchalik@uniwa.gr

© The Author(s), under exclusive license to Springer Nature Switzerland AG 2023

V. Katsoni (ed.), *Tourism, Travel, and Hospitality in a Smart and Sustainable World*, Springer Proceedings in Business and Economics,
https://doi.org/10.1007/978-3-031-26829-8_5

challenges that emerge in energy utilization, while considering the ongoing diffusion of RES in the energy mix of Greece.

Keywords Renewable energy sources · Hierarchical cluster analysis · K-means analysis · Electricity consumption · Greece · Energy consumption

JEL Classification Q42 · Q43 · Q57

1 Introduction

Recently, there is abundant international literature focused on the utilization of RES for electricity generation, either on an experimental or a pilot scale of analysis (Zhou et al., 2015; Kyriakopoulos & Arabatzis, 2016; Castagneto Gissey et al., 2018). Besides, there is a widespread geographical coverage of the respective studies: Europe (Moser et al., 2020), New Zealand (Crossland et al., 2020), the island of Mauritius (Timmons et al., 2020), Kenya (Lai et al., 2017). Off-grid RES technological systems comprise a critical infrastructure for providing electricity to small communities, especially in remote and rural areas where there are no connection points to the mainland grid (Arabatzis et al., 2010; Koutroumanidis et al., 2009; Lai et al., 2017). These studies highlighted the high sensitivity and the economic efficiency of energy storage facilities: on the one hand to investment costs and on the other hand to the individual systems of installation, operation and maintenance (Moser et al., 2020). In such a study, it was also been denoted that existing hydropower storage potential could be considered to achieve almost overly energy autonomy from the use of coal for electricity generation in New Zealand (Crossland et al., 2020).

Another research aspect of many studies regards the cost of electricity storage that confuses the cost of charge/discharge power in kW with the cost of potential energy storage in kWh. This multi-parametric approach to renewable energy sources (RES) technologies results in fundamentally different cost scenarios for different storage technologies. In such studies, it was stated that although large battery price reductions make these RES technologies competitive, they cannot significantly reduce overall costs (Timmons et al., 2020). On the contrary, the reservoir-type storage pumped hydroelectric (PHES) system is considered a fully mature technology having the potential to wider energy availability than the existing one. To this end, energy policy-makers argue that commitments to RES development are not functioning unambiguously: either on the development of new storage technology or on the reduction of battery prices (Timmons et al., 2020).

Furthermore, energy network companies are able to determine the value of the storage system, but their ownership can only be acceptable if the operation of the system becomes independent of the operation of the energy network. Indicative initiatives to address these issues include flexible connection agreements and the development of improved responsiveness of relevant rapid energy storage services (Castagneto Gissey et al., 2018). However, in order to remove the identified obstacles, a market structure is needed that can value the energy flexibility offered by storage, considering energy storage systems as a complementary (not competing) capital asset with respect to other energy grid and generation assets (Castagneto Gissey et al., 2018). The current assessment of storage is considered a capital component of electricity generation, despite the fact that energy storage does not provide a positive net flow of electricity, but it is used to justify a doubling of grid usage charges (Castagneto Gissey et al., 2018).

In Greece, but also internationally, electricity generation is commonly linked to industrial-scale chemical processes as well as to the inducing techno-economic and environmental problems of utilizing raw materials with a high energy content, such as coal. In particular, it is mentioned the post-combustion process where porous adsorbents proved to be very effective for carbon sequestration (Vatalis et al., 2012). To this end, the respective research focuses mainly on carbon dioxide (CO_2) treatment after combustion. From environmental and economic perspectives, potential applications of new and low-cost adsorbents can be considered to achieve an energy-efficient separation and capture of flue gases. To this end, the optimal adsorbent materials are characterized by the capacity of increased pollutant adsorption, the increased separation capacity, as well as the stable thermo-mechanical properties which achieve fast kinetics of the adsorbent. Also, an important feature of the process is the affordable economic cost, making positive the utilization prospects in large-scale technology applications. One such case is the exhaust gases produced in the Greek regions of Western Macedonia and in particular the prefectures of Florina-Ptolemaida-Kozani, where electricity is produced at thermal stations using the underground lignite deposits. In these deposits, almost 58% of the total electricity is consumed by Greek energy end users of all energy production of the Greek territory (Vatalis et al., 2012).

Among the available RES for electricity generation is nuclear energy, while biomass has the least significant impact on climate change policy costs (Romano & Scandurra, 2011). Moreover, the levels of nuclear and bioenergy development are comparably high in all baseline model scenarios, while the level of RES development increases significantly under all climate change policy alternatives (Tavoni et al., 2012). Important implications of these policies relate to the "best technology push," such as regulatory and normative standards. Therefore, it is important to evaluate and rank each RES technology with energy utilization in terms of power generation and regarding the mitigation of carbon emissions and climate changes in the sectors of transport, electricity generation, and industry (Tavoni et al., 2012). The assumption of full availability of technologies is the first step toward pursuing climate change policies in a "second" (in order of feasibility) optimal environment. This research framework of RES for electricity generation is important tool of assessing the economic value of low-carbon technologies and investments in research and development (R&D) and, subsequently, contributing to the design of policies that ensure efficient technological development, and, prioritizing investments in key fields of technology (Tavoni et al., 2012; Ntanos et al., 2018a, 2018b, 2019).

2 Literature Review

2.1 Socio-economic and Legislative Framework for Electricity Production

The socio-economic context of running technologies for electric energy storage (EES) for power generation is determined by the fact that the reduction of electricity costs is related to energy storage in off-peak electricity consumption hours. During the peak period, energy production can be achieved with the complementary contribution of cost-effective forms of energy supply based on coal and nuclear technology. During peak periods, cost-effective forms of supply must compete with less efficient but more flexible forms of generation such as oil and natural gas. Contrarily, during the off-peak period, less electricity is consumed; therefore, those of the expensive types of power generation that operate during peak hours can now close their operation and, consequently, the pricing of the generated electricity it is reduced (Roodenrijs et al., 2020). At the same time, certain consideration of the large-scale research of EES technologies regards the efficiency, the lifetime, and the cost of environmentally friendly RES technologies in an energy clean future of power generation. To this end, the factors of environment and climate bring greater social awareness compared to the recent past, regarding the future prospects of RES. Tracking the research progress focuses on the interaction between the decision-making process and social factors (Roodenrijs et al., 2020).

At an international level of analysis among RES technologies for electricity generation, it is noteworthy that the installed capacity of hydroelectric projects worldwide reaches 127 GW, a percentage which corresponds to 99% of the total installed capacity of energy storage systems used in electricity networks and which (regarding the potential of hydroelectric projects) corresponds to about 3% of the global capacity. The second largest energy storage technology is the installed capacity of compressed air electric generating stations (CAES), but in this case, only few systems are operating. Besides, the two energy storage technologies of PHS and CAES are economically feasible but with little application in practical operation of electricity generation, although they are characterized by long discharge times (up to months) and huge energy storage potential (in the order of TWh). At the same time, further development among long-term energy storage technologies for power generation, such as hydrogen and SNG, it is realistic with a full-cycle electricity conversion efficiency (of these technologies) ranging at 50–75% (Kyriakopoulos & Arabatzis, 2016).

At the Greek level of analysis, Directive 2009/28/RE is evaluated as very important legislative tool for the RES promotion in energy uses. This directive aims to create a common framework since it was for the first time that RES utilization was foreseen for all energy uses: electricity generation, cooling/heating, transport/biofuels. Subsequently, the Greek RES legislation for power generation includes "Υ.Α.Π.Ε./Φ1/οικ.24839" (Government Gazette B 1901/03.12.2010) regarding "Guaranteeing for signing of Connection Agreements in the distribution networks of power plants using RES which are exempt from the obligation to obtain a production license." Furthermore, the pricing of electricity produced by different RES-typed power plants is important. Overall, it is noteworthy that the particular characteristic that constitutes a technology as "good" refers mainly on the perspective of the decision maker, while the criteria of EES technologies efficiency and lifetime are not always considered as important factors in the functionality of EES technologies (Sundararagavan & Baker, 2012).

2.2 *RES Technologies for Electricity Production*

2.2.1 Biomass

The technological and the socio-economic effects of biomass sources are particularly popular among developing countries, such as in West African countries (West African Monetary and Economic Union, in French: UEMOA, 2019), but also in countries with extensive geographical coverage of forest biomass (Arabatzis et al., 2015; Kyriakopoulos et al., 2010, 2015; Chalikias et al., 2010; Kolovos et al. 2011; Chalikias & Kolovos, 2013). The available and commercially viable technologies for bioenergy conversion are either in the pilot phase, in research and development (R&D), or in large-scale applications. Bioenergy consists of solid, liquid, or gaseous fuels. Liquid fuels can be used directly in existing road, rail, and air transport networks, as well as in engine and turbine power generators. Solid and gaseous fuels can be used to produce electricity from on-site power plants (UEMOA, 2019).

During the pre-treatment of biomass, the process of pelletization leads to a homogenized and uniform feed, improving the controllability of the process (it is a significant advantage for waste and waste-derived raw materials, maintaining high levels of raw material variability). Pelletization can further minimize feedstock losses during storage, as biomass pellets can sometimes be easily integrated into existing fossil fuel infrastructure (such as co-installation in mills and large coal-fired power plants). However, pelletization requires drying, grinding, and compressing the primary biomass source, applying sufficient pressure to melt the lignin content in the wood. In some cases, a binder may be added to improve the physical properties of the particle-pellet. After pelletization, the raw material becomes more resistant to mechanical handling, thus ensuring less damage throughout the supply chain process (Adams et al., 2017).

Biomass energy technologies can convert renewable biomass fuels into heat and electricity using processes similar to those used for fossil fuels. Currently, the primary choice for producing electricity from biomass is direct combustion. Combustion systems to produce electricity and heat are similar to most fossil fuel power plants. Biomass fuel is burned in a boiler to produce high pressure steam. The steam is then fed into a steam turbine and flows over an arrangement of turbine blades, causing the turbine to rotate. The turbine is connected to an electric generator while the air flows and comes back to the turbine. Therefore, the electric generator produces electricity making the above energy technology as one of the most widely available commercial technologies (UEMOA, 2019).

The efficiency of the combustion boiler process is determined by the amount of fuel energy that is transferred to the water having the final uses of heating or electricity generation. Common side-effects of the process are heat losses through flue gases, conduction, and radiation to the environment, as well as losses due to incomplete fuel combustion. It is also noteworthy that the lowest efficiency is reported in the biomass pellet boiler, which holds a correspondingly high level of PM10 particle emissions. Therefore, very low burn levels could be improved by better air/fuel mixing, higher combustion temperatures, or longer fuel residence time. The overall efficiency of an already operating commercial power plant is lower than the combustion efficiency, as the commercial plant incurs additional significant losses from both electrical conversion and fuel processing. The non-burnt quota of carbon that remains in the ash represents the efficiency reduction that adversely affects the balance of greenhouse gases (GHGs) originating from the combustion system (Adams et al., 2017).

In the case of satisfying the energy needs from biomass through cogeneration of electricity and heat (bio-CHP) (including cooling and air conditioning), then, electricity pricing is an important factor in the implementation of such a large-scale energy projects, as the chosen electricity generation technology can be co-evaluated through sensitivity analysis between alternative scenarios and with a price/costing (cost–benefit) level determined by follow-up studies (Holmgren et al., 2018). The second-generation outline of liquid biofuels is presented in Table 1.

Table 1 Production and uses of second-generation liquid biofuels (edited by authors, modified by UEMOA, 2019)

Second generation of biofuels				
Type of biofuel	Nomination	Plant origin of biomass	Process	Applications
Biodiesel	Hydrotreated biodiesel	Vegetable fuels and animal fats	Hydrotreatment	Internal combustion engines for motor transport
Bioethanol	Cellulosic bioethanol	Lignocellulosic matter	Hydrolysis and fermentation	
Synthetic biofuels	Biomass-to-liquids, BTL Fischer–Tropsch (FT) Bio-methanol Bio-dimethylether (Bio-DME)	Lignocellulosic matter	Gasification and synthesis	
Bio-hydrogen		Lignocellulosic matter	Gasification and synthesis	

2.2.2 Hydroelectricity

Hydroelectric power can contribute substantially to a cleaner environment. Indicatively, it is reported that approximately 2800 TWh of hydropower worldwide can replace either 1200 Mt of coal or 4.5 billion barrels of crude oil. In fact, more than 1300 Mt CO_2 emissions are avoided due to hydroelectric power (Kaldellis, 2008). In economic terms, hydroelectric plants into operation are more efficient compared to other RES, although they entail higher construction costs, low operating costs, and relatively low maintenance costs, provided that there is favorable annual rainfall (Manzano-Agugliaro et al., 2017). Indeed, the energy conversion efficiency of the waterfall is more than 90% (when the best/optimum efficiency of fossil fuel power plants is only 50%). At a socio-environmental level of analysis, hydropower offers multi-use advantages as a RES technology of large initial capital costs but also with a long lifetime, balancing out over time the high initial investment costs (Tortajada & Saklani, 2018).

Small power plants have attracted the attention of energy policy-makers because of their high potential to electrify rural areas and contributing to the development of them. These areas are characterized by low costs and per capita income. To this end, small hydropower plants provide affordable electricity in remote areas where there is no connection to the mainland grid (Bilgili et al., 2018), with the advantage of rapid responding to short-term changes in electricity demand (Manzano-Agugliaro et al., 2017).

Western and Northern Greece sustain significant hydroelectric potential which is still partially exploitable. The configuration (topography) of the ground combined with the relatively high rainfall facilitates the applications of similar power plants (Kaldellis, 2008). In Greece, there are fifteen large hydroelectric power plants (LHP) with a total capacity of 2950 MW and almost fifty small hydroelectric power plants (SHP) with a total nominal capacity of 70 MW. The first two power plants are in operation since 1954, named that of SHP hydroelectric plants of Agra (I, II) and the SHP factory of Louros. Since then, several large hydroelectric plants have been built, such as in Kremasta (440 MW) and Kastraki (320 MW) (West Central Greece), as well as in Polyphyto (375 MW) (Central Macedonia) (Kaldellis, 2008). Greece includes the operation of 32 units with a nominal power of 60 MW. Eleven SHP plants sustain a rated power of 1–10 MW which represent 82% of the total installed capacity. In addition, there are very small hydroelectric plants with a nominal power of less than 1 MW and a total capacity of 10.7 MW. Six of the eleven SHP units (42.2 MW) belong to public electric grid (PPC), while the remaining five were created by private investors. All the private SHP units–together with the Glafkos power station–are located in the Peloponnese (Kaldellis, 2008).

Although LHP plants represent the large majority of installed hydroelectric plants, SHP plants are considered more important infrastructure of electricity generation maintaining a higher degree of utilization of its hydroelectric potential with more installed units (having interest in further development) comparing to the number of installed LHP hydroelectric plants. Hydroelectric power generation is not the major national water management policy, so most hydroelectric plants are mainly used to meet peak demand (Kaldellis, 2008). It is important to emphasize that the gross theoretical hydroelectric potential is a necessary but not a sufficient precondition for understanding the historical evolution of hydropower in different countries. Besides, being naturally endowed with vast water or mineral resources does not necessarily mean that a country will use that resources (del Mar Rubio & Tafunell, 2014).

2.3 *Wind Energy*

Over the past two decades, alongside China's rapid industrialization and urbanization, prospects for reducing pollution and carbon emissions have been increased. At the same time, one of the most important approaches to achieve economic development and environmental protection is the shift of coal-based electricity generation by RES. In the past fifteen years, the promotion and the development of wind power in China were implemented supporting policies regarding either the tariff policy or the non-tariff policy. China's options for achieving an environmentally friendly economic development in the country are consistent with the strengthening of energy production from RES. The power generation industry in China is covered by thermal energy for 70% of the total rated power and 80% of the total electricity generation. At the same time, fossil coal consumption for electricity generation in China accounts for 60% while carbon emissions from the national electricity generation industry account for 40% of total carbon emissions. Therefore, it becomes imperative need to adjust China's energy mix by increasing the RES applicability. Over the past decade—based on existing technology—China's onshore wind resources supported the theoretical production capacity of more than 1100 GW of wind power. Therefore, China's corresponding statistical data collection showed that if the wind energy power of 1100 GW is achievable, then wind energy is feasible to account for more than 90% of the total power. Furthermore, since the power factor of wind energy is about 30%, China used (a decade ago) only 1.98% of its total available wind resources/capacity. From a technological point of view, it can be said that the overall available/energy-absorbable resource includes wind potential at a height of 70 m with a density of at least 300 W/m^2, as well as potential offshore energy potential in the 5–50 m depth areas. Subsequently, China's rapid growth of wind-based electricity generation accounts for only 2.78% of the total, although the share of wind power remained at a low percentage of installed capacity, namely 7% of the total power generation. However, both the increasing research production and the technological construction of wind projects in China indicate a prospect of great dynamic development of wind energy in the near future (Voutsinas, 2007). This dynamic growth in China is based on the industrialization of the country as well as the developmental opportunities of both wind power and other RES technologies. This energy "pluralism," in alignment with the population growth, supports China to improve its national power generation mix by preventing long-term and irreversible negative impacts on environmental protection and sustainable economic development (Zhao et al., 2016).

2.4 Technological Background—Design Guides for RES Projects

2.4.1 Wind Energy-Grid Compatibility

Regarding the evolution of wind turbine design standards, it is clear that wind energy technology has been developed very rapidly, and due to the accumulated knowledge, many problems have been successfully solved. However, the rapid development of the wind energy market confronts new challenges (Willis et al., 2018). In particular, the energy supply of a wind turbine (AG, in Greek) is not constant due to the wind speed/direction fluctuations. Therefore, in case the installed wind power is small compared to the size of the electricity grid, then fluctuations due to wind-related variability can be easily harnessed by the grid. Otherwise, instabilities may occur in the electrical network. Therefore, in order to achieve high RES penetration, it is important to develop procedures that will minimize disruptions to the power grid. To this end, an important process is the adaptation of variable speed wind turbines, as this type of AG allows the operator to define a specific operating point, making it possible to operate modern wind turbines as if they were power plants. Furthermore, this practice may lead to significant exploitation of the available wind energy, and therefore, additional actions must be taken by the electricity grid operator in order to minimize wind energy losses (Willis et al., 2018).

2.4.2 Solar Energy Materials of Photovoltaics (PVs) Construction

CdTe material contains the highly toxic metal cadmium. Although the current cost to install CdTe PV solar panels is very low (Hermerschmidt et al., 2015), the economic criterion should not be the only one at selecting and constructing of a PV. Cadmium must be recycled after use, and this must be taken into account when calculating the techno-economic feasibility of the whole PV farm. The last decade has shown an exponential growth in the energy content of different kinds of PV systems, with a combined annual growth rate of about 50% (Hermerschmidt et al., 2015). In 2010, the production of PV cells was estimated at a level of the order of 27 GigaWatt-peak (GWp). The majority of manufacturing PV cells is constructed by multicrystalline silicon (mc-Si). The expected PV-cell electricity generation in 2021 is more than 50 GWp. The vast majority of PV cells are based on silicon cost (such as the CdTe thin films), where it is expected that this cost will decrease in the near future making PV technologies competitive and cost-effective (Hermerschmidt et al., 2015). At this point, it is noteworthy that although small PV electricity producers may own grid-connected systems, the aggregation of several such small energy suppliers into an expanded grid can result in significant energy production, which can be comparable to similar electricity generation from central power plants, reducing energy losses during electricity transmission and distribution, as well as black-outs case (Pacca, 2003).

2.4.3 Biomass

It is also noteworthy that coal emissions from biomass burning for domestic use have decreased, as a result of the widespread demographic trend toward urbanized populations, with urbanization and urbanization beginning in the 1990s. Further reductions were observed in industrial coal emissions and the use of coke ovens, encouraging the transition to more efficient power plants, compared to using coal for power generation (Sweerts et al., 2019). A notable characteristic of biomass utilization is the wide geographical dispersion of energy crops and, subsequently, the wide variation of biofuels production in quantities, costs, and applications of energy interest.

2.4.4 Hydroelectric Energy

Regarding the technological background of hydroelectric energy, the prevailing technology involved is that of turbines, being valued as strong and credible. Despite this technological maturity of hydropower among other RES technologies, the research production focuses on the adaptation of different types of turbines and installations to the type of waterfall and flow in order to exploit the huge untapped potential and optimal efficiency of power generation with low maintenance costs (Manzano-Agugliaro et al., 2017). Other research approaches focus on efforts to improve efficiencies through redesigning the water impingement fins, improving energy conversion efficiency, either through intermittent generation due to rainfall, or through varying generator voltage or frequency (Manzano-Agugliaro et al., 2017), or through the utilization of low-altitude waterfalls to utilize the embodied energy in small-scale systems for domestic energy production in a way that is economically feasible and technologically exploitable, especially with regard to the hydroelectric resources of low-flow waterfalls (Bilgili et al., 2018).

Overall the development and implementation of alternative scenarios of RES systems can support energy policy-makers to choose appropriate RES technologies for electricity generation. When appointing weighted criteria for alternative RES technologies, negative values indicate that the corresponding parameters that have an inhibiting contribution, while positive values indicate that the corresponding parameters that have a positive contribution. The remaining parameters of such a multi-criteria scenario affect either positively or negatively the intertwined technologies and implementation processes of each a large-scale RES (Papageorgiou et al., 2020).

3 Methods and Results

For the statistical analysis, SPSS Statistical Package was used (Chalikias, 2015). Primary data on the number of appointed units and the corresponding capacity (production) of electricity generation per RES type were utilized for 13 geographical regions in Greece (For the years 2015 and 2019, power values are placed in MW). An investigation of the variables was carried out with descriptive statistics, in order to examine whether the normal distribution was followed. These results were presented in Table 2.

However, both normality tests clearly rejected the null hypothesis that the data follow a normal distribution ($p = 0$), Table 3.

Indeed, both the Kolmogorov–Smirnov and Shapiro–Wilk normality tests, as well as the set of generated figures, they concluded that the normality condition does not hold. Therefore, the parametric test was not applied, but the Mann–Whitney U Test was applied, which coincides with the Kruskal Wallis test in the two samples.

In particular, the Mann–Whitney U test showed a statistically significant difference between island regions and mainland regions in the two RES: biomass, small hydropower ($p = 0.000$), while the same test did not show differences between island and mainland regions for the other two RES: wind, biogas, with significance rates of $p = 0.650$ and $p = 0.205$, respectively.

At this point, it is noteworthy to emphasize that the percentages of significance, p-values, for the 4 RES in 2015 and 2019, were exactly the same. This finding was not surprising because the Mann–Whitney U test does not take into account the values of the variable (like the t-test) but their order (rank), which (order) coincides in both cases (years 2015 and 2019). Therefore, the entire set of results (in the forms of Tables, Figures) for the year 2015 is presented in this study, while the entire set of results for the year 2019 was omitted for narrative simplicity reasons. Then, for the year 2015, the following Table 4 refers to the wind RES in alignment with its sub-figures Normal Q-Q plot (NQQ), Detrended Normal Q-Q plot, and histogram (Fig. 1), while similar set of results was examined for the other three RES: biomass, small hydroelectric-SHP, as well as photovoltaic-PV. Table 4 presented the statistics and the corresponding set of results for the non-parametric tests with independent samples with the Mann–Whitney U test for separation in mainland and island areas for wind. In addition, for the other three RES biogas, biomass, small hydro, respectively (at the reference year 2015), the statistical controls were listed.

Nonparametric tests with independent samples (Nonparametric Tests: Independent samples) were performed for the 4 RES: wind, biogas, biomass, small hydro, since the normality hypothesis was rejected for the comparison of island and mainland areas, so a Mann–Whitney U test was performed. Based on Table 4 and Fig. 1, it was shown, on the one hand, a statistically significant difference between island regions and mainland regions in biomass and small hydropower ($p = 0.000$) and, on the other hand, a non-statistically significant difference between wind and biogas ($p = 0.650$ and $p = 0.205$, respectively).

Table 2 Descriptive statistics of normal distribution

Descriptive				
			Statistic	Std. error
Wind	Mean		502,1050	85,76,195
	95% Confidence interval for mean	Lower bound	329,8471	
		Upper bound	674,3630	
	5% Trimmed mean		423,4123	
	Median		374,0000	
	Variance		375,110,743	
	Std. deviation		612,46,285	
	Minimum		0,00	
	Maximum		3768,24	
	Range		3768,24	
	Interquartile range		501,55	
	Skewness		3,326	0,333
	Kurtosis		15,664	0,656
Biomass	Mean		7,8286	1,42,375
	95% Confidence interval for mean	Lower bound	4,9689	
		Upper bound	10,6883	
	5% Trimmed mean		6,6916	
	Median		4,9900	
	Variance		103,381	
	Std. deviation		10,16,764	
	Minimum		0,00	
	Maximum		37,72	
	Range		37,72	
	Interquartile range		10,00	
	Skewness		1,593	0,333
	Kurtosis		1,806	0,656
Small hydropower (SHP)	Mean		17,6514	4,04,536
	95% confidence interval for mean	Lower bound	9,5260	
		Upper bound	25,7767	
	5% trimmed mean		13,5393	
	Median		5,2600	
	Variance		834,610	
	Std. deviation		28,88,963	
	Minimum		0,00	
	Maximum		162,90	

(continued)

Table 2 (continued)

Descriptive

			Statistic	Std. error
	Range		162,90	
	Interquartile range		27,60	
	Skewness		3,034	0,333
	Kurtosis		12,202	0,656
Photovoltaics (PVs)	Mean		58,2250	16,17,192
	95% confidence interval for mean	Lower bound	25,7427	
		Upper bound	90,7072	
	5% trimmed mean		37,7907	
	Median		19,5737	
	Variance		13,338,088	
	Std. deviation		115,49,064	
	Minimum		0,00	
	Maximum		612,48	
	Range		612,48	
	Interquartile range		55,93	
	Skewness		3,539	0,333
	Kurtosis		13,722	0,656

Table 3 Normality test

Tests of normality

	Kolmogorov–Smirnov[a]			Shapiro–Wilk		
	Statistic	df	Sig	Statistic	df	Sig
Wind	0,206	51	0,000	0,690	51	0,000
Biomass	0,221	51	0,000	0,767	51	0,000
SHP	0,271	51	0,000	0,645	51	0,000
PVs	0,316	51	0,000	0,527	51	0,000

[a] Lilliefors Significance Correction

- The null hypothesis is that the distribution for wind is the same in all categories: Separation into mainland and island regions with the Mann–Whitney U Test, and it was concluded to maintain the null hypothesis ($p = 0.650$).
- The null hypothesis is that the distribution for biogas is the same in all categories: Separation in mainland and island areas with the Mann–Whitney U test, and it was concluded to maintain the null hypothesis ($p = 0.205$).

Table 4 Separation profile into mainland and island regions by Mann–Whitney U test for wind RES

Total N	51
Mann–Whitney U	237,500
Wilcoxon W	940,500
Test statistic	237,500
Standard error	47,377
Standardized test statistic	−0,454
Asymptotic Sig. (2-sided test)	0,650
Sig	0,650

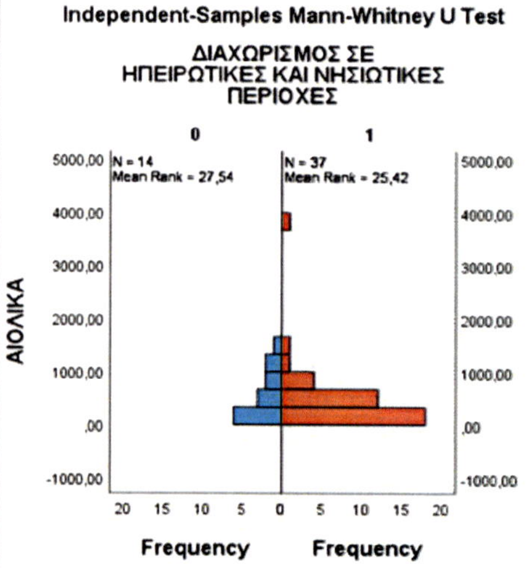

- The null hypothesis is that the distribution for biomass is the same in all categories: Separation into mainland and island areas with the Mann–Whitney U test, and it was concluded that the null hypothesis is rejected ($p = 0.000$).
- The null hypothesis is that the distribution for small hydro is the same in all categories: Separation into mainland and island regions with the Mann–Whitney U test, and it was concluded that the null hypothesis is rejected ($p = 0.000$).

The natural interpretation of the differentiation reported regarding biomass and small hydropower projects (SHP) is due to the fact that the primary energy sources: biomass produced by specific plant species, as well as the capacity scale of SHP, varies both between the mainland-island scale of analysis, as well as between different areas of the continent and the island regions. Indeed, mainland sustains a greater potential for arable land or natural biomass growth, as well as (due to climatic conditions) waterfalls compared to relatively anhydrous island areas. Contrarily, the RES types of wind energy and biogas sustain specific and stable technologies of electricity generation depending on the desired capacity (rated power of the energy plants) setting by the energy manufacturer but not depending on the geomorphological configuration, which is equally capable to be installed (wind farms and bioenergy units) at both mainland and island Greece. Contrarily, in the case of biomass and SHP, the geomorphology of the installation area depends on both the endemic/indigenous energy crops and the stored water supply/rainfall (thus determining the exploitable amount of stored water at the SHP in operation). To this end, the non-statistically significant difference in wind and biogas was observed.

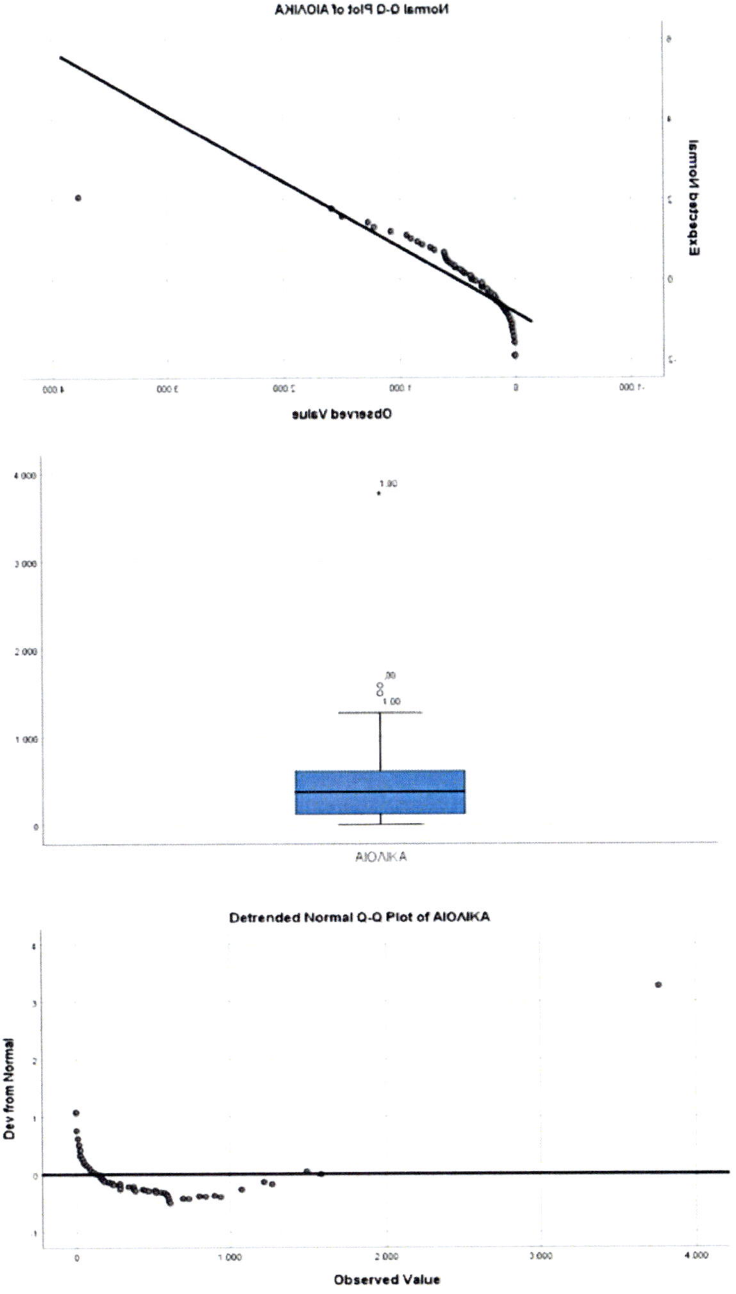

Fig. 1 Graphical normality checks of wind RES

4 Discussion

Based on the existing literature, the use of RES technologies for electricity generation poses the following questions: What technologies will be needed most? In what technologies there is room for further development? What political considerations will affect RES development and penetration in the respective state energy mix? and What are the implementation problems may drive or impede the development of RES technologies, by co-evaluating a wide range of energy policies related to electrochemical, mechanical and thermal energy storage technologies? At the same time, the structure and the applicability of the findings from a global regulatory and development tools; being based on the post-2015 development agenda of the United Nations Conference on Sustainable Development, they should be adapted to electricity production in the framework of the "Sustainable Development Goals" (SDGs) and "Millennium Development Goals" (MDGs) policy initiatives. Moreover, it is critical to address key research issues regarding the operation, applicability and pricing options of energy storage technologies, especially regarding the future prospects offered (Kyriakopoulos & Arabatzis, 2016).

At the scales of research modeling and small-scale planning, the diffusion of RES technologies to the current energy demand plays a determining role, as well as the role of energy storage systems (EES) in the context of smarter and more sustainable energy systems. These two characteristics can be modeled and analyzed according to energy damping and demand response flexibility, EES capacity and efficiency, as well as to ways of demand response and ways of EES affecting both system reliability and electricity generation capability/adequacy (Zhou et al., 2015). Demand response as well as EES they can reduce the frequency and the duration of power outages, which are also considered among the most serious risk events in electricity generation. To this end, the quantification of energy production allows the formulation of energy policies regarding (a) the possibilities of responding to demand and (b) the contribution of EES systems to ensure increased energy capacity in the electricity market (Zhou et al., 2015).

At the large-scale planning and implementation, macroeconomic studies carried out from EU member states in the hydropower sector, and they were focused on the direct and the indirect benefits of RES utilization in Europe and internationally, stating that electricity generation could increase especially in Europe by around 20% before 2030. Such planning can also support policy-makers—both at national and European level—to improve the potential for hydropower development, making a decisive contribution to achieve the low-carbon goals for 2030 and beyond. The current state of hydropower has the potential to increase the electricity share produced by different RES, taking into consideration the ways in which storage and pump technologies of the hydroelectric system evolve, while supporting further the RES integration into a controlled, regulated, and operative European energy system (GlobalData, 2015).

At the societal level of large-scale implementation, it can be signified the cooperation between local communities in the development of electricity generation units, as it provides significant opportunities to develop capacity and skills in a variety of other (non-energy) rural development activities (Greacen, 2004). This collaborative process brings significant benefits to the development of a local energy project that approaches electrification less as an end in itself and more as a necessary by-product of a process in which local residents develop the expertise and confidence to implement significant self-financed development projects. Income generation, education, health care, forestry, and other parameters are factored into an integrated design to maximize social benefits from energy units at a local–regional scale of performance (Greacen, 2004).

5 Conclusions

The utilization of RES technologies for electricity generation is consistent with the fact that some selected technologies sustain increased yields and positive technological expansion in local networks, qualifying them over other (of lower yield) technologies (Greacen, 2004). At the same time, the heterogeneous access to necessary energy resources points out the decisive role of RES technologies under R&D perspectives. Subsequently, under limited/finite/shortage of resources, the "divisibility of an innovation" is crucial in the final selection. In this case, large-scale investments in the energy sector become unfavorable (Greacen, 2004), while the technologies in small hydroelectric projects are by their nature successfully adaptable at a local level of analysis and installation, the implementation of which requires (most often) financial support and construction workload from the budgets and the local communities, respectively (Greacen, 2004). At the same time, coal replacement by natural gas, although initially leads to higher global temperatures comparing to historically recorded carbon prices, in the long run global temperature is expected to decrease if lower emissions of carbon dioxide and black carbon (BC) are achieved (Greacen, 2004).

For Greece, it is noteworthy that, due to the correlation of electrical dynamics with the sun and with the hydrological cycle (evaporation, water precipitation, water runoff), hydroelectricity sustains a significant energy potential, and it is considered one of the most suitable RES technologies that contributes to the increasing coverage of electricity generation in Greece (Kaldellis, 2008). At the same time, according to the statistical results of this study, it was clearly demonstrated the differentiation of the degree and the sources of RES utilization between the island and mainland regions of Greece, mainly due to the fact that while there is technological convergence of RES for each distinct renewable source itself; however, there is an energy pluralism and diversification of distributed RES throughout Greece.

At the same time, the heterogeneous access to necessary energy resources to adopt a RES technology for power generation is vital in the R&D perspective of RES diffusion/spur. Consequently, access to financial investment funds with an income prospect, it is a decisive planning parameter. To the extent that low-carbon technologies are of interest at the local level, this is due to the fact that citizens in rural areas appreciate the importance of electricity in their daily lives, and they are willing to take the necessary initiatives to operate local RES units, liking to the form of local cooperatives for the disposal of their agricultural production in a collective manner. On the contrary, the consequences of an inefficient energy investment burden consumers and taxpayers of both developed (but most intensively in) developing economies (Greacen, 2004). A recent example is the subsidies offered in Greek citizens to purchase electric vehicles thus fostering electrification and to simultaneously support the energy bills paid by the energy end users or offering them flexibility in registering at different electricity providers under an open competitive energy market.

References

Adams, P., Bridgwater, T., Lea-Langton, A., Ross, A., & Watson, I. (2017). Biomass conversion technologies. Report to NNFCC. Greenhouse Gas Balances of Bioenergy Systems (pp. 107–139). https://doi.org/10.1016/B978-0-08-101036-5.00008-2

Arabatzis, G., Galatsidas, S., Intze, C., Chalikias, M. S., Tsiantikoudis, S., & Mamalis, S. (2015). Green Entrepreneurship and Green Products: Consumers' Views and Attitudes in Regional Unit of Evros. In *HAICTA* (pp. 291–297).

Arabatzis, G., Aggelopoulos, S., Tsiantikoudis., S. (2010). Rural development and LEADER + in Greece: Evaluation of local action groups. *International Journal of Food, Agriculture and Environment, 8*(1), 302–307.

Bilgili, M., Bilirgen, H., Ozbek, A., Ekinci, F., & Demirdelen, T. (2018). The role of hydropower installations for sustainable energy development in Turkey and the world. *Renewable Energy, 126*, 755–764.

Castagneto Gissey, G., Dodds, P. E., & Radcliffe, J. (2018). Market and regulatory barriers to electrical energy storage innovation. *Renewable and Sustainable Energy Reviews, 82*, 781–790. https://doi.org/10.1016/j.rser.2017.09.079

Chalikias, M. S., & Kolovos, K. G. (2013). Citizens' views in Southern Greece part II. Contribution of forests to quality of life. *Journal of Environmental Protection and Ecology, 14*(2), 629–637.

Chalikias, M. S., Kyriakopoulos, G., & Kolovos, K. G. (2010). Environmental sustainability and financial feasibility evaluation of woodfuel biomass used for a potential replacement of conventional space heating sources. Part I: A Greek Case Study. *Operational Research, 10*(1), 43–56.

Chalikias, M., Manolesou, A., & Lalou, P. (2015). *Research methodology and introduction to statistical data analysis with IBM SPSS statistics*. Greek Academic Electronic Textbooks and Aids.

Crossland, A., Scoles, K., Wang, A., Groves, C., & Sun, S. (2020). *Assessment of electricity decarbonization scenarios for new Zealand and great Britain using a plant dispatch and electrical energy storage modelling framework energies, 13*(11), 2799. https://doi.org/10.3390/en13112799

del Mar Rubio, M., & Tafunell, X. (2014). Latin American hydropower: A century of uneven evolution. *Renewable and Sustainable Energy Reviews, 38*, 323–334.

GlobalData. (2015). Study identifies benefits of hydro in Europe. Progressive Digital Media Oil & Gas, Mining, Power, CleanTech and Renewable Energy News, 07 July 2015. https://proxy.eap.gr/login?url=https://www-proquest-com.proxy.eap.gr/wire-feeds/study-identifies-benefits-hydro-europe/docview/1694695032/se-2?accountid=16059

Greacen, C. E. (2004). The Marginalization of "Small is Beautiful": Micro-hydroelectricity, common property, and the politics of rural electricity provision in Thailand (pp. 231). Doctor of

Philosophy in Energy and Resources in the Graduate Division of the University of California, Berkeley, USA, UMI Number: 3183808.

Hermerschmidt, F., Pouloupatis, P. D., Partasides, G., Lizides, A., Hadjiyiannakou, S., & Choulis, S. A. (2015). Beyond solar radiation management—The strategic role of low-cost photovoltaics in solar energy production. *International Journal of Sustainable Energy, 34*(3–4), 211–220. https://doi.org/10.1080/14786451.2013.854244

Holmgren, K. M., Berntsson, T., & Lönnqvist, T. (2018). Profitability and greenhouse gas emissions of gasification-based biofuel production—Analysis of sector specific policy instruments and comparison to conventional biomass conversion technologies. *Energy, 165*, 997–1007. https://doi.org/10.1016/j.energy.2018.09.105

Kaldellis, J. K. (2008). Critical evaluation of the hydropower applications in Greece. *Renewable and Sustainable Energy Reviews, 12*, 218–234.

Kolovos, K. G., Kyriakopoulos, G., & Chalikias, M. S. (2011). Co-evaluation of basic woodfuel types used as alternative heating sources to existing energy network. *Journal of Environmental Protection and Ecology, 12*(2), 733–742.

Koutroumanidis, T., Zafeiriou, E., & Arabatzis, G. (2009). Asymmetry in price transmission between the producer and the consumer prices in the wood sector and the role of imports: The case of Greece. *Forest Policy and Economics, 11*(1), 56–64.

Kyriakopoulos, G. L., & Arabatzis, G. (2016). Electrical energy storage systems in electricity generation: Energy policies, innovative technologies, and regulatory regimes. *Renewable and Sustainable Energy Reviews, 56*, 1044–1067. https://doi.org/10.1016/j.rser.2015.12.046

Kyriakopoulos, G., Chalikias, M. S., Kalaitzidou, O., Skordoulis, M., & Drosos, D. (2015). Environmental viewpoint of fuelwood management. In *HAICTA* (pp. 416–425).

Kyriakopoulos, G., Kolovos, K. G., Chalikias, M. S. (2010). Environmental sustainability and financial feasibility evaluation of woodfuel biomass used for a potential replacement of conventional space heating sources. Part II: A combined Greek and the nearby Balkan countries case study. *Operational Research, 10*(1), 57–69.

Lai, C. S., Jia, Y., Xu, Z., Lai, L. L., Li, X., Cao, J., & McCulloch, M. D. (2017). Levelized cost of electricity for photovoltaic/biogas power plant hybrid system with electrical energy storage degradation costs. *Energy Conversion and Management, 153*, 34–47. https://doi.org/10.1016/j.enconman.2017.09.076

Manzano-Agugliaro, F., Taher, M., Zapata-Sierra, A., Juaidi, A., & Montoya, F. G. (2017). An overview of research and energy evolution for small hydropower in Europe. *Renewable and Sustainable Energy Reviews, 75*, 476–489.

Moser, M., Gils, H.-C., & Pivaro, G. (2020). A sensitivity analysis on large-scale electrical energy storage requirements in Europe under consideration of innovative storage technologies. *Journal of Cleaner Production, 269*, 122261. https://doi.org/10.1016/j.jclepro.2020.122261

Ntanos, S., Kyriakopoulos, G., Chalikias, M., Arabatzis, G., Skordoulis, M., Galatsidas, S., & Drosos, D. (2018a). Social assessment of renewable energy sources usage and contribution to life quality: The case of an Attica Urban Area in Greece. *Sustainability., 10*(5), 1414.

Ntanos, S., Kyriakopoulos, G., Skordoulis, M., Chalikias, M., & Arabatzis, G. (2019). An application of the new environmental paradigm (NEP) scale in a Greek context. *Energies, 12*(2), 239.

Ntanos, S., Skordoulis, M., Kyriakopoulos, G., Arabatzis, G., Chalikias, M., Galatsidas, S., ... & Katsarou, A. (2018b). Renewable energy and economic growth: Evidence from European countries. *Sustainability, 10*(8), 2626.

Pacca, S. A. (2003). Global warming effect applied to electricity generation technologies. Doctoral dissertation in Energy and Resources, Graduate Division of the University of California, Berkeley, USA, UMI Number: 3105330, pp. 208.

Papageorgiou, K., Carvalho, G., Papageorgiou, E. I., Bochtis, D., & Stamoulis, G. (2020). Decision-making process for photovoltaic solar energy sector development using fuzzy cognitive map technique. *Energies, 13*(6), 1427. https://doi.org/10.3390/en13061427

Romano, A. A., & Scandurra, G. (2011). The investments in renewable energy sources: Do low carbon economies better invest in Green technologies? *International Journal of Energy Economics and Policy, 1*(4), 107–115.

Roodenrijs, J. C. M., Hegger, D. L. T., Mees, H. L. P., & Driessen, P. (2020). Opening up the black box of group decision-making on solar energy: The case of strata buildings in Amsterdam, the Netherlands. *Sustainability, 12*(5), 2097. https://doi.org/10.3390/su12052097

Sundararagavan, S., & Baker, E. (2012). Evaluating energy storage technologies for wind power integration. *Solar Energy, 86*, 2707–2717.

Sweerts, B., Pfenninger, S., Yang, S., Folini, D., van der Zwaan, B., & Wild, M. (2019). Estimation of losses in solar energy production from air pollution in China since 1960 using surface radiation data. *Nature Energy, 4*(8), 657–663. https://doi.org/10.1038/s41560-019-0412-4

Tavoni, M., de Cian, E., Luderer, G., Steckel, J. C., & Waisman, H. (2012). The value of technology and of its evolution towards a low carbon economy. *Climatic Change, 114*(1), 39–57. https://doi.org/10.1007/s10584-011-0294-3

Timmons, D., Elahee, K., & Lin, M. (2020). Microeconomics of electrical energy storage in a fully renewable electricity system. *Solar Energy, 206*, 171–180. https://doi.org/10.1016/j.solener.2020.05.057

Tortajada, C., & Saklani, U. (2018). Hydropower-based collaboration in South Asia: The case of India and Bhutan. *Energy Policy, 117*, 316–325.

UEMOA (2019). "5. Biomass Conversion Technologies". http://www.globalproblems-globalsolutions-files.org/gpgs_files/pdf/UNF_Bioenergy/UNF_Bioenergy_5.pdf. Πρόσβαση στις 02 Φεβρουαρίου 2019.

Vatalis, K.I., Laaksonen, A., Charalampides, G., & Benetis, N. P. (2012). Intermediate technologies towards low-carbon economy. The Greek zeolite CCS outlook into the EU commitments. *Renewable and Sustainable Energy Reviews, 16*(5), 3391–3400. https://doi.org/10.1016/j.rser.2012.02.031.

Voutsinas, S. G. (2007). *Wind energy—Educational material* (p. 64). National Technical University of Athens.

Willis, D. J., Niezrecki, C., Kuchma, D., Hines, E., Arwade, S. R., Barthelmie, R. J., DiPaola, M., Drane, P. J., Hansen, C. J., Inalpolat, M., Mack, J. H., Myers, A. T., & Rotea, M. (2018). Wind energy research: State-of-the-art and future research directions. *Renewable Energy, 125*, 133–154.

Zhao, X., Li, S., Zhang, S., Yang, R., & Liu, S. (2016). The effectiveness of China's wind power policy: An empirical analysis. *Energy Policy, 95*, 269–279.

Zhou, Y., Mancarella, P., & Mutale, J. (2015). Modelling and assessment of the contribution of demand response and electrical energy storage to adequacy of supply. *Sustainable Energy, Grids and Networks, 3*, 12–23.

Fostering Sustainability in Tourism Destinations Through Residents' Assessment

Pedro Liberato, Dália Liberato, and Ricardo Cerqueira

Abstract The growth of the tourism sector has significant impacts on tourism destinations. Therefore, sustainability is a very current issue and has received special attention in recent years. Considering the problems that the uncontrolled growth of tourism has caused in certain tourism destinations, strategic planning and sustainable development are determinant for their success. Tourism activity has a direct impact on residents' lives; however, the maximization of economic benefits and the attraction of investment continue to emerge as a highlight in most destinations. Despite the resident's critical role in the success of tourism destinations, their perspective still takes a secondary place, which must be reconsidered by destination managers, particularly DMOs. The research focuses on the characteristics of residents in two cities in the North of Portugal, as well as their perception about the tourism impacts at economic, sociocultural, and environmental level. A quantitative methodology was applied, using a questionnaire survey, to 810 residents, in the cities of Porto and Vila Nova de Gaia. By analyzing residents' perceptions about the tourism impacts in the region, the results identify which factors need to be improved to influence residents' opinions over tourism and leverage their intention to support the sector's development, in advance of strategies addressed to DMOs.

Keywords Sustainability · Sustainable tourism · Residents · Porto · Vila Nova de Gaia

JEL Classification Z32

P. Liberato · D. Liberato · R. Cerqueira
School of Hospitality and Tourism, Polytechnic Institute of Porto, Vila Do Conde, Portugal
e-mail: dalialib@esht.ipp.pt

R. Cerqueira
e-mail: ricardo.dp.cerqueira@gmail.com

P. Liberato (✉) · D. Liberato
CiTUR Researcher, (Centre for Tourism Research, Development, and Innovation), Porto, Portugal
e-mail: pedrolib@esht.ipp.pt

© The Author(s), under exclusive license to Springer Nature Switzerland AG 2023

V. Katsoni (ed.), *Tourism, Travel, and Hospitality in a Smart and Sustainable World*, Springer Proceedings in Business and Economics,
https://doi.org/10.1007/978-3-031-26829-8_6

1 Introduction

The impacts of tourism activity can be related to factors that cause stress, which makes it possible that certain tourism impacts can have consequences that go beyond residents' perspectives and attitudes, due to stress causing various health problems and influencing residents' well-being and quality of life (Jordan et al., 2021). According to the research of Munanura et al. (2021), residents' emotional solidarity with tourists can help to understand how they support tourism. Perceptions of tourism impacts can directly affect residents' emotional solidarity through changes in the individual's well-being. However, the results of this study indicate that the positive effects of tourism impacts on residents' emotional solidarity are more important than the negative ones. From this perspective, residents' decision to support tourism may depend on how tourism impacts affect the individual's well-being, and if they are positive, they may trigger a positive attitude toward tourism and tourists. This research confirms that the prospect of changes in the individual's well-being due to increased tourism and emotional solidarity toward tourists play a key role in the relationship between the perception of tourism impacts and residents' support for tourism. Destination managers and policy-makers should promote activities, festivals, and events so that residents can develop greater emotional solidarity with tourists, which will contribute to residents' support of tourism activity in their destinations. Tourism planners should provide opportunities for residents to share their views on tourism, either through surveys or through residents' participation in planning meetings, so that their ideas and concerns can be integrated into the planning strategy, enabling the sustainability of tourism in the destination (Erul & Woosnam, 2021; Su & Swanson, 2019). Zheng et al. (2021) mention that residents may display various emotional responses, such as affection, happiness, or anger, an approach that differs from the traditional level of resident satisfaction with tourism development. The results of this research show that interest and affection are the most common emotions surrounding initiatives called "*tourism permorfing arts*" (TPA), a variant of creative tourism, revealing that these are not only a product for tourists, but also an attractive resource for residents. Emotions such as pride and gratitude indicate that Hangzhou residents place a high value on their culture, with TPA presenting local history, music, art, and traditions to tourists. Residents believe that the development of these initiatives can contribute to the promotion and preservation of their culture. When residents of a tourist destination show dissatisfaction due to the contact of negative impacts generated by the tourism industry, the entire industry is indirectly affected, this is because the overall level of resident satisfaction, can define the success or failure of tourism in the region. Understanding the level of resident satisfaction toward tourism can be a proactive approach to decrease potential negative impacts.

This study aims to understand residents' perceptions regarding the impacts of tourism activity in the municipality of Porto and Vila Nova de Gaia. Two research hypotheses were associated to this objective: H1: Residents' perception of the economic impacts of tourism influences residents' opinion of the sociocultural impacts and H2: Residents' perception of the economic impacts of tourism influences residents' opinion of environmental impacts. To respond to the general objective proposed, a quantitative methodology was applied, using a questionnaire survey. This survey was applied to 810 residents, in the city of Porto and Vila Nova de Gaia, in the Northern region of Portugal. The following section discusses the literature review, and the methodology chapter will address the quantitative research instrument. The two final sections will discuss the results and focus on the main research conclusions.

2 Literature Review

Destination managers should develop protocols to inform and educate residents on a regular basis about the short-and long-term goals of destination tourism planning (Aleshinloye et al., 2021). Adding the residents' perspective is key, as are meetings being held on neutral grounds to reduce inequalities between the various *stakeholders* and potential conflicts of interest. Destinations that do not include the active involvement of residents and their support for the tourism sector will have great difficulty in attracting visitors in the short and long term and will be less competitive given the diversity of supply in today's tourism market (Lawton & Weaver, 2015; Sharpley, 2014). According to Moscardo (2008), the absence of local participation in tourism development and the exclusion of local communities from tourism planning are an issue that requires attention. This exclusion of residents from tourism development may be a consequence of foreign private investment in project implementation, especially in developing countries (Stone & Nyaupane, 2020; Teye et al., 2002). The participation of residents in decision-making in tourism planning can contribute to the development of positive attitudes toward tourism (Vargas-Sánchez et al., 2015). Involving and perceiving residents in matters about tourism activity may lead to the development of a participatory and empowered community, something fundamental in terms of the sustainability and promotion of tourism (Nunkoo & Gursoy, 2012). Additionally, Obradović et al., (2021), conclude that residents are not involved in the decision-making process of tourism development, highlighting the lack of communication and interaction between residents and the entities that govern tourism. The benefits of tourism activity such as employment opportunities and improved quality of life in the community do not occur, something that the authors relate to the lack of involvement of residents in the activities implemented by the entities responsible for tourism. All these factors together may lead residents to develop negative attitudes toward tourists. Considering the research of Kim & Kang (2020) which shows how residents are confronted with uncomfortable situations from the first stage of tourism development to the point of saturation, the accumulation of these negative experiences has led to the development of negative attitudes toward tourists by residents. Residents need to be part of the decision-making process of their municipality. They can be invited to participate in meetings of their municipal/parish councils and regular meetings with community leaders through the creation of youth organizations and other interest groups. DMOs should assure residents that destinations are for residents, in the presence and absence of tourists, and that DMOs are interested in residents for the long term (Aleshinloye et al., 2021). The research results of Nugroho and Numata (2020) show that local community involvement in planning, decision-making, and participation in tourism activities is related to residents' support for tourism development. Residents' satisfaction with tourists at the destination is a major factor in the decision-making process of tourism managers, researchers, and planners because the success of any development is dependent on the active support of the local community (Lee and Jan, 2019). Residents' dissatisfaction can lead to the decline or even elimination of tourism by the receiving community (Gannon et al., 2021).

Residents may support tourism even though they are aware of the negative impacts of the activity. The research of Çelik and Rasoolimanesh (2021) states that residents of *Şırnak* are cognizant of the negative and positive impacts of tourism, and despite the existence of negative impacts, their decision to support the industry arises through cost–benefit analysis. Taking into consideration that this research took place in a province whose tourism sector is not yet developed, the investment related to tourism will be supported by the public. Corporate social responsibility-based practices were the focus of the research of Pereira and Gadotti (2021). The authors refer to how these direct or indirect practices reflect commitment to and support of tourism development, particularly those related to the social and economic dimension. This research demonstrates that residents' perceptions of social responsibility practices, by accommodation establishments, have an influence on their satisfaction, directly or indirectly through perceived benefits. Considering the research results, the authors recommend that companies form agreements with non-governmental organizations implement sociocultural initiatives, invest in training, and promote the quality of life of their employees. For these measures to have a positive effect on the community, tourism managers should inform and reinforce the importance of the tourism sector to the community.

3 Methodology

The municipality of Vila Nova de Gaia is part of the Metropolitan Area of Porto, in the northern region of Portugal. Currently, the municipality of Vila Nova de Gaia is the third most populous municipality in the country with 302,295 inhabitants, and the one with the largest population in the northern region. The municipality is internationally known for its Port and Douro wine companies and for its diverse industries (glass, automotive, and electronic components). The city of Porto is the second largest city in the country and the most representative of the Northern Region of Portugal. Its historic center was considered Cultural Heritage listed by UNESCO in 1996. The city's name is worldwide recognized thanks to the Port wine and its historical heritage (AMP, 2021).[1] The municipality of Porto, currently with 237,591 residents, is the fourth most populous municipality in Portugal.

This study aims to understand residents' perceptions regarding the impacts of tourism activity in the municipality of Porto and Vila Nova de Gaia. Two research hypotheses were associated to this objective, and their validation is subsequently verified in this document.

H1: Residents' perception of the economic impacts of tourism influences residents' opinion of the sociocultural impacts (relates Q.10 to Q.11).

If we consider the argument of Muresan et al., (2016) that sustainable tourism development arises as a need to ensure the efficiency of tourism activity, taking into account the environmental interest, the economic and sociocultural needs of local communities in the region, maintaining a balance between the negative and positive impacts resulting from these components, we can say that there is a relationship between them, and that the economic impacts, as a rule, are those most valued and easily identified by residents. Their perception of the economic impacts may influence their opinion regarding the sociocultural impacts. Antunes et al. (2020) refer that in several cases the increase in rental prices (negative economic impact) may lead to gentrification (negative sociocultural impact), which may cause the abandonment of the destination's natural residents (negative sociocultural impact) and simultaneously increase the number of businesses in the destination (positive economic impact).

H2: Residents' perception of the economic impacts of tourism influences residents' opinion of environmental impacts (relates Q.10 to Q.12).

Several authors argue that the concept of sustainable tourism should be constituted by a dynamic framework, with the ability to have approaches depending on the various circumstances, such as the characteristics of the destination and its residents, which aim at the protection of future tourism development (Hunter, 1997; Janusz and Bajdor, 2013). Through such a framework, it may be possible to balance environmental conservation and community development while simultaneously meeting the needs of residents, tourists, and the local environment (Savage et al., 2004). Residents value the ability of the tourism sector to protect and preserve the environment, as well as the creation of programs aimed at increasing knowledge about the environmental status (Sharma et al., 2016). In addition to protection, preservation, and increased knowledge, tourism is also able to increase the general public's appreciation of the environment, this is because the interaction between people, the environment and nature provided by tourism, has the consequence of increasing the public's knowledge of environmental problems (Stainton, 2019).

Following this framework, 810 questionnaires were applied to residents of Porto and Vila Nova de Gaia. The sample of residents of the municipality of Porto and Vila Nova de Gaia consists of 810 participants. For a sample size consisting of 810 individuals, and using a 95% confidence level, for a tendentially infinite population and a sample of n = 810, the Error (B) = 3.44% (Table 1).

---

[1] Source: AMP, http://portal.amp.pt/pt/4/municipios/porto/

Table 1 Research sample

Age (years)	Frequency	Percentage (%)
18–24	133	16,4
25–34	255	31,5
35–44	182	22,2
45–54	145	17,9
55–64	73	9
+65	22	2,7
Qualifications	Frequency	Percentage (%)
Elementary education	58	7,2
High school	327	40,4
Medium Course	36	4,4
Higher education or equivalent	389	48
Gross monthly income (Individual)	Frequency	Percentage (%)
Up to 1000€	412	50,9
Between $1001 and $2000	220	27,2
Between 2001 and 3000 euros	47	5,8
Greater than or equal to 3001	23	2,8
Don't know/No answer	108	13,3
Total	810	100

Cronbach's alpha value is higher than 0.80 for all three dimensions presented in the scientific hypotheses, so it is possible to construct a variable to measure each dimension adequately. To test the hypotheses defined above, the following statistical methods were used: descriptive statistics, Pearson's correlation, Kruskal–Wallis test, and ANOVA tests.

4 Results

Hypothesis H1: Residents' perception of the economic impacts of tourism influences residents' opinion of the sociocultural impacts.

According to Table 2, which presents the results of Pearson's correlation, we only verify the existence of statistically significant positive relationships, whose results indicate that residents who have a positive perception about the economic impacts of tourism, have a positive opinion about the sociocultural impacts; however, those who have a negative perception about the economic impacts derived from tourism have a negative opinion about the sociocultural impacts.

Table 2 Pearson's correlation: relationship between Q.10 and Q.11

		11.1 Tourism has fostered the creation and enhancement of the region's natural and cultural heritage	11.2 Tourism is one of the reasons why there is a better entertainment offer in the region	11.3 My municipality mainly values cultural heritage as a tourism resource	11.4 An increase in social problems due to tourism	11.5 Tourism contributes to the improvement of essential infrastructure
Q10.1 Can you see a significant improvement in the living conditions of the population, due to tourism	r	0,479[b]	0,463[b]	0,457[b]	0,170[b]	0,466[b]
	p	0,000	0,000	0,000	0,000	0,000
Q10.2 The cost of living has increased significantly (rent/house price inflation)	r	0,427[b]	0,396[b]	0,366[b]	0,302[b]	0,367[b]
	p	0,000	0,000	0,000	0,000	0,000
10.3 Tourism has led to increased prices of products and services	r	0,378[b]	0,367[b]	0,334[b]	0,368[b]	0,333[b]
	p	0,000	0,000	0,000	0,000	0,000
10.4 A growth in the number of businesses is observed	r	0,543[b]	0,508[b]	0,479[b]	0,255[b]	0,482[b]
	p	0,000	0,000	0,000	0,000	0,000
10.5 Tourism contributes to increased income for the chamber and parish councils, which is used in the improvement and preservation of infrastructure and heritage	r	0,547[b]	0,512[b]	0,535[b]	0,256[b]	0,538[b]
	p	0,000	0,000	0,000	0,000	0,000

(continued)

Table 2 (continued)

		11.1 Tourism has fostered the creation and enhancement of the region's natural and cultural heritage	11.2 Tourism is one of the reasons why there is a better entertainment offer in the region	11.3 My municipality mainly values cultural heritage as a tourism resource	11.4 An increase in social problems due to tourism	11.5 Tourism contributes to the improvement of essential infrastructure
10.6 The tourist activity has contributed to the increase of employment in the city	r	0,610[b]	0,556[b]	0,538[b]	0,242[b]	0,554[b]
	p	0,000	0,000	0,000	0,000	0,000
10.7 Jobs resulting from tourism activity are seasonal or low salary	r	0,287[b]	0,307[b]	0,276[b]	0,337[b]	0,278[b]
	p	0,000	0,000	0,000	0,000	0,000
10.8 The tourism sector is the most important economic activity for the region, attracting investment and increasing consumption	r	0,492[b]	0,477[b]	0,528[b]	0,313[b]	0,463[b]
	p	0,000	0,000	0,000	0,000	0,000
10.9 Tourism generates more benefits than costs	r	0,564[b]	0,545[b]	0,548[b]	0,236[b]	0,515[b]
	p	0,000	0,000	0,000	0,000	0,000
10.10 The economic benefits of tourism activity for the region and residents are visible	r	0,583[b]	0,570[b]	0,581[b]	0,207[b]	0,564[b]
	p	0,000	0,000	0,000	0,000	0,000
10.11 COVID-19 caused economic stagnation due to decreased tourism activity in the region	r	0,543[b]	0,514[b]	0,483[b]	0,260[b]	0,449[b]
	p	0,000	0,000	0,000	0,000	0,000

(continued)

Table 2 (continued)

		11.6 Tourism leads to congestion on roads, commercial spaces, and leisure parks	11.7 Loss of local authenticity is observed due to tourism	11.8 Tourism allows you to get to know new cultures through contact with tourists	11.9 My degree of involvement with the tourism sector is high	11.10 Tourism maintains my cultural heritage	11.11 COVID-19 made me feel insecure when interacting with tourists
Q10.1 Can you see a significant improvement in the living conditions of the population, due to tourism	r	$0{,}234^b$	$0{,}075^a$	$0{,}405^b$	$0{,}350^b$	$0{,}407^b$	$0{,}132^b$
	p	0,000	0,033	0,000	0,000	0,000	0,000
Q10.2 The cost of living has increased significantly (rent/house price inflation)	r	$0{,}434^b$	$0{,}390^b$	$0{,}434^b$	$0{,}246^b$	$0{,}241^b$	$0{,}274^b$
	p	0,000	0,000	0,000	0,000	0,000	0,000
10.3 Tourism has led to increased prices of products and services	r	$0{,}486^b$	$0{,}446^b$	$0{,}411^b$	$0{,}288^b$	$0{,}256^b$	$0{,}323^b$
	p	0,000	0,000	0,000	0,000	0,000	0,000
10.4 A growth in the number of businesses is observed	r	$0{,}344^b$	$0{,}285^b$	$0{,}508^b$	$0{,}363^b$	$0{,}408^b$	$0{,}159^b$
	p	0,000	0,000	0,000	0,000	0,000	0,000
10.5 Tourism contributes to increased income for the chamber and parish councils, which is used in the improvement and preservation of infrastructure and heritage	r	$0{,}317^b$	$0{,}191^b$	$0{,}495^b$	$0{,}367^b$	$0{,}398^b$	$0{,}200^b$
	p	0,000	0,000	0,000	0,000	0,000	0,000
10.6 The tourist activity has contributed to the increase of employment in the city	r	$0{,}304^b$	$0{,}213^b$	$0{,}556^b$	$0{,}412^b$	$0{,}432^b$	$0{,}206^b$
	p	0,000	0,000	0,000	0,000	0,000	0,000

(continued)

Table 2 (continued)

		11.6 Tourism leads to congestion on roads, commercial spaces, and leisure parks	11.7 Loss of local authenticity is observed due to tourism	11.8 Tourism allows you to get to know new cultures through contact with tourists	11.9 My degree of involvement with the tourism sector is high	11.10 Tourism maintains my cultural heritage	11.11 COVID-19 made me feel insecure when interacting with tourists
10.7 Jobs resulting from tourism activity are seasonal or low salary	r	0,428[b]	0,462[b]	0,331[b]	0,263[b]	0,210[b]	0,292[b]
	p	0,000	0,000	0,000	0,000	0,000	0,000
10.8 The tourism sector is the most important economic activity for the region, attracting investment and increasing consumption	r	0,391[b]	0,280[b]	0,477[b]	0,403[b]	0,404[b]	0,272[b]
	p	0,000	0,000	0,000	0,000	0,000	0,000
10.9 Tourism generates more benefits than costs	r	0,281[b]	0,179[b]	0,490[b]	0,388[b]	0,438[b]	0,178[b]
	p	0,000	0,000	0,000	0,000	0,000	0,000
10.10 The economic benefits of tourism activity for the region and residents are visible	r	0,326[b]	0,140[b]	0,526[b]	0,394[b]	0,477[b]	0,229[b]
	p	0,000	0,000	0,000	0,000	0,000	0,000
10.11 COVID-19 caused economic stagnation due to decreased tourism activity in the region	r	0,388[b]	0,303[b]	0,492[b]	0,316[b]	0,335[b]	0,276[b]
	p	0,000	0,000	0,000	0,000	0,000	0,000

[a] $p < 0.05$
[b] $p < 0.01$ Source: Own elaboration based on SPSS outputs

Considering the existence of significant relationships, we can conclude that the "Hypothesis H1: The perception of residents regarding the economic impacts of tourism influences the opinion of residents regarding the sociocultural impacts" is verified. Those who observe a significant improvement in the living conditions of the population due to tourism have the opinion that "Tourism has fostered the creation and enhancement of the natural and cultural heritage in the region," those who observe an increase in the number of businesses due to tourism have the opinion that "Tourism has fostered the creation and enhancement of the natural and cultural heritage in the region," those who perceive that "Tourism has contributed to an increase in the income of the town and parish councils, which is used for the improvement and preservation of the infrastructure and heritage," have the opinion that "Tourism has fostered the creation and enhancement of the natural and cultural heritage in the region" and that "Tourism contributes to the improvement of essential infrastructure," those who perceive that "Tourism has contributed to the increase of employment in the city" have the opinion that "Tourism has promoted the creation and enhancement of the natural and cultural heritage in the region," "Tourism is one of the reasons why there is a better offer of entertainment in the region," and "Tourism allows to know new cultures through the contact with tourists," those who perceive that "The tourism sector is the most important economic activity for the region, by attracting investment and increase" have the opinion that "My municipality values mainly the cultural heritage as a touristic resource," those who perceive that "Tourism generates more benefits than costs" have the opinion that "Tourism has fostered the creation and enhancement of natural and cultural heritage in the region," "My municipality mainly values the cultural heritage as a tourism resource," "Tourism is one of the reasons why there is a better entertainment offer in the region" and "Tourism contributes to the improvement of essential infrastructures," those who have the perception that "The economic benefits of tourism activity for the region and residents are visible" have the opinion that "Tourism has fostered the creation and enhancement of the natural and cultural heritage in the region," "My municipality mainly values cultural heritage as a tourism resource," "Tourism is one of the reasons why there is a better entertainment offer in the region," "Tourism contributes to the improvement of essential infrastructure" and "Tourism allows to know new cultures through contact with tourists" and those who have the perception that "COVID-19 caused economic stagnation due to the decrease of tourism activity in the region" have the opinion that "Tourism fostered the creation and enhancement of natural and cultural heritage in the region," "Tourism is one of the reasons why there is a better offer of entertainment in the region" and "Tourism allows to know new cultures through contact with tourists." It can be seen in this analysis that residents who have a positive perception of the economic impacts have a positive opinion about the sociocultural impacts of tourism.

For those residents who perceive that "The cost of living has increased significantly (rent/real estate price inflation)" have the opinion that "Tourism leads to congestion on roads, shopping areas and recreational parks" and that "Tourism enables one to get to know new cultures through contact with tourists," those who perceive that "Tourism has led to increased prices of products and services" have the opinion that "Tourism leads to congestion on roads, Those who have the perception that "Tourism leads to increased price of products and services" have the opinion that "Tourism leads to congestion on roads, commercial spaces, and leisure parks" and that "Tourism leads to loss of local authenticity due to tourism" and that "Tourism leads to congestion on roads, commercial spaces, and leisure parks." We can see that those who have a negative perception of the economic impacts have a negative opinion about the sociocultural impacts, except for the question "The cost of living has increased significantly (rent/real estate price inflation)" where there is an equality of values in the two opinions with the highest correlation, one being positive and the other negative.

In summary, statistically, significant relationships are found in all impacts and opinions, and we chose to analyze the most relevant ones. It is concluded that Hypothesis H1 is valid.

Hypothesis H2: Residents' perceptions of the economic impacts of tourism influence residents' opinions of the environmental impacts.

According to Table 3, which presents the results of Pearson's correlation, it is observed only the existence of statistically significant positive relationships, whose results indicate that residents who have a positive perception about the economic impacts of tourism have a positive opinion about the environmental impacts, however those who have a negative perception about the economic impacts of tourism have a negative opinion about the environmental impacts. Considering the existence of significant relationships, it is possible to conclude that the "Hypothesis H2: Residents' perception of the economic impacts of tourism influences residents' opinion toward environmental impacts" is verified. Those who observe a significant improvement in the living conditions of the population due to tourism have the opinion that "Overall, I am satisfied with tourism in the region" and that "There is an increase in incentives by the municipality aimed at environmental protection," those who observe a growth in the number of businesses have the opinion "Overall, I am satisfied with tourism in the region" and that "There is an increase in incentives by the municipality aimed at environmental protection," those who have the perception that "Tourism contributes to an increase in the income of the city council and parish councils, which is used for the improvement and preservation of infrastructure and heritage" have the opinion that "Overall, I am satisfied with tourism in the region," "There is an increase in incentives by the municipality aimed at environmental protection" and that "Tourism has led to the promotion and adoption of environmentally responsible behavior/measures by the municipality," those who perceive that "Tourism activity has contributed to increased employment in the city" have the opinion that "Overall, I am satisfied with tourism in the region," "There is an increase in incentives by the municipality aimed at environmental protection" and that "Tourism contributes to the protection and conservation of the environment and natural habitats in the region," those who have the perception that "The tourism sector is the most important economic activity for the region, attracting investment and increasing consumption" have the opinion that "Overall, I am satisfied with tourism in the region" and that "My municipality mainly values the natural heritage as a tourism resource," those who have the perception that "Tourism generates more benefits than costs" have the opinion that "Overall I am satisfied with

Table 3 Pearson's correlation: relationship between Q.10 and Q.12

		12.1 There is an increase in visual and noise pollution as a result of tourism activity	12.2 Tourism contributes to the protection and conservation of the environment and natural habitats in the region	12.3 Tourism has led to the promotion and adoption of responsible behavior/measures toward the environment by the municipality	12.4 There is an increase in incentives by the municipality aimed at environmental protection	12.5 Tourism contributes to the residents' knowledge of environmental problems	12.6 Tourism does not contribute to degradation/ disfiguring the region's natural landscape	12.7 My municipality mainly values the natural heritage as a tourist resource	12.8 A decrease in the ecological footprint of tourism is observed due to COVID-19	12.9 Overall, I am satisfied with tourism in the region
Q10.1 Is there a significant improvement in the living conditions of the population, due to tourism	r	0,198[a]	0,417[a]	0,407[a]	0,435[a]	0,415[a]	0,351[a]	0,339[a]	0,301[a]	0,520[a]
	p	0,000	0,000	0,000	0,000	0,000	0,000	0,000	0,000	0,000
Q10.2 The cost of living has increased significantly (rent/house price inflation)	r	0,395[a]	0,197[a]	0,242[a]	0,278[a]	0,173[a]	0,175[a]	0,309[a]	0,366[a]	0,350[a]
	p	0,000	0,000	0,000	0,000	0,000	0,000	0,000	0,000	0,000
10.3 Tourism has led to increased prices of products and services	r	0,450[a]	0,203[a]	0,249[a]	0,258[a]	0,184[a]	0,167[a]	0,266[a]	0,317[a]	0,275[a0]
	p	0,000	0,000	0,000	0,000	0,000	0,000	0,000	0,000	0,000
10.4 A growth in the number of businesses is observed	r	0,335[a]	0,373[a]	0,310[a]	0,377[a]	0,302[a]	0,257[a]	0,320[a]	0,317[a]	0,468[a]
	p	0,000	0,000	0,000	0,000	0,000	0,000	0,000	0,000	0,000

(continued)

Table 3 (continued)

		12.1 There is an increase in visual and noise pollution as a result of tourism activity	12.2 Tourism contributes to the protection and conservation of the environment and natural habitats in the region	12.3 Tourism has led to the promotion and adoption of responsible behavior/measures toward the environment by the municipality	12.4 There is an increase in incentives by the municipality aimed at environmental protection	12.5 Tourism contributes to the residents' knowledge of environmental problems	12.6 Tourism does not contribute to degradation/ disfiguring the region's natural landscape	12.7 My municipality mainly values the natural heritage as a tourist resource	12.8 A decrease in the ecological footprint of tourism is observed due to COVID-19	12.9 Overall, I am satisfied with tourism in the region
10.5 Tourism contributes to increased income for the chamber and parish councils, which is used in the improvement and preservation of infrastructure and heritage	r	0,208[a]	0,393[a]	0,440[a]	0,465[a]	0,385[a]	0,358[a]	0,371[a]	0,326[a]	0,518[a]
	P	0,000	0,000	0,000	0,000	0,000	0,000	0,000	0,000	0,000
10.6 The tourist activity has contributed to the increase of employment in the city	R	0,253[a]	0,410[a]	0,400[a]	0,431[a]	0,387[a]	0,321[a]	0,339[a]	0,347[a]	0,556[a]
	P	0,000	0,000	0,000	0,000	0,000	0,000	0,000	0,000	0,000
10.7 Jobs resulting from tourism activity are seasonal or low salary	r	0,389[a]	0,183[a]	0,241[a]	0,233[a]	0,175[a]	0,180[a]	0,252[a]	0,331[a]	0,259[a]
	P	0,000	0,000	0,000	0,000	0,000	0,000	0,000	0,000	0,000

(continued)

Table 3 (continued)

		12.1 There is an increase in visual and noise pollution as a result of tourism activity	12.2 Tourism contributes to the protection and conservation of the environment and natural habitats in the region	12.3 Tourism has led to the promotion and adoption of responsible behavior/measures toward the environment by the municipality	12.4 There is an increase in incentives by the municipality aimed at environmental protection	12.5 Tourism contributes to the residents' knowledge of environmental problems	12.6 Tourism does not contribute to degradation/ disfiguring the region's natural landscape	12.7 My municipality mainly values the natural heritage as a tourist resource	12.8 A decrease in the ecological footprint of tourism is observed due to COVID-19	12.9 Overall, I am satisfied with tourism in the region
10.8 The tourism sector is the most important economic activity for the region, attracting investment, and increasing consumption	r	0,317[a]	0,334[a]	0,375[a]	0,375[a]	0,377[a]	0,257[a]	0,423[a]	0,318[a]	0,446[a]
	p	0,000	0,000	0,000	0,000	0,000	0,000	0,000	0,000	0,000
10.9 Tourism generates more benefits than costs	r	0,180[a]	0,445[a]	0,448[a]	0,455[a]	0,408[a]	0,369[a]	0,369[a]	0,348[a]	0,575[a]
	p	0,000	0,000	0,000	0,000	0,000	0,000	0,000	0,000	0,000
10.10 The economic benefits of tourism activity for the region and residents are visible	r	0,185[a]	0,417[a]	0,445[a]	0,469[a]	0,424[a]	0,385[a]	0,395[a]	0,325[a]	0,585[a]
	p	0,000	0,000	0,000	0,000	0,000	0,000	0,000	0,000	0,000
10.11 COVID-19 caused economic stagnation due to decreased tourism activity in the region	r	0,276[a]	0,303[a]	0,330[a]	0,339[a]	0,249[a]	0,236[a]	0,339[a]	0,390[a]	0,457[a]
	p	0,000	0,000	0,000	0,000	0,000	0,000	0,000	0,000	0,000

< 0.05 $^a p < 0.01$. *Source* Own elaboration based on SPSS outputs tourism in the region," "There is an increase in incentives by the municipality aimed at environmental protection," "Tourism has led to the promotion and adoption of environmentally responsible behavior/measures by the municipality" and "Tourism contributes to the protection and conservation of the environment and natural habitats in the region," those who have the perception that "The economic benefits of tourism activity for the region and residents are visible" have the opinion that "Overall, I am satisfied with tourism in the region," "There is an increase in incentives by the municipality aimed at environmental protection," "Tourism has led to the promotion and adoption of environmentally responsible behavior/measures by the municipality" and "Tourism contributes to residents' knowledge of environmental problems," and those who have the perception that "COVID-19 has caused economic stagnation due to the decrease in tourism activity in the region" have the opinion that "Overall, I am satisfied with tourism in the region" and that "A decrease in the ecological footprint of tourism is observed due to COVID-19." It can be seen from this analysis that those who have a positive perception of the economic impacts have a positive opinion about the environmental impacts of tourism.

For residents who perceive that "The cost of living has increased significantly (rent/real estate price inflation)" have the opinion that "There is an increase in visual and noise pollution as a result of tourism activity" and that "There is a decrease in the ecological footprint of tourism due to COVID-19," those who have the perception that "Tourism has led to increased prices of products and services" have the opinion that "An increase in visual and noise pollution is observed as a result of tourism activity" and that "A decrease in the ecological footprint of tourism is observed due to COVID-19," those who perceive that "Jobs from tourism activity are seasonal or low salary" have the opinion that "There is an increase in visual and noise pollution as a result of tourism activity" and that "There is a decrease in the ecological footprint of tourism due to COVID-19." We can see that those who have a negative perception about the economic impacts have a negative opinion about the environmental impacts. The opinion that "A decrease in the ecological footprint of tourism is observed due to COVID-19" appears with the highest ratio after the opinion. "An increase in visual and noise pollution is observed as a result of tourism activity," which we consider to be two negative opinions, both of which refer to pollution from tourism activity.

In summary, statistically significant relationships are found in all impacts and opinions, and we chose to analyze the most relevant ones. It is possible to conclude that Hypothesis H2 is valid.

5 Conclusions

Collaboration between local communities, government entities, and their DMOs is critical for sustainable tourism development. However, sustainable tourism can only be achieved if the ideas and visions of local community members and other *stakeholders* are recognized, encouraged, and included in tourism strategic planning (Ribeiro et al., 2018).

The general objective of this research is to understand residents' perceptions of the impacts of tourism in the municipality of Porto and Vila Nova de Gaia. To achieve the objective of this research, a quantitative approach methodology was used. Questionnaire surveys were applied to 810 residents of the municipality of Porto and Vila Nova de Gaia, in May, June, and July 2021.

H1 Residents' perceptions of the economic impacts of tourism influence residents' opinions of the sociocultural impacts.

It was found that those who have a positive perception of the economic impacts of tourism also have a positive opinion about the sociocultural impacts of tourism. However, when there is a negative perception regarding the economic impacts of tourism, the opinion about the sociocultural impacts is also negative. Regarding the statement "The cost of living has increased significantly (rent/real estate price inflation)," the results show an equality of values in relation to the opinion about the sociocultural impacts, one being negative and the other positive.

The results of this research coincide with the results of Jaafar et al. (2017) and MacKenzie & Gannon (2019) where it is found that tourism can contribute to the preservation of traditional culture and arts. However, negative effects may arise such as congestion of commercial, public spaces, and increased traffic on roads. Despite the perception of negative economic impacts, which have a negative influence on residents' opinion about sociocultural impacts, the research of Qin et al. (2021) reports that residents are able to tolerate certain effects of negative tourism impacts as long as tourism activity is positive for the destination overall, continuing residents' support for the development of the sector. The research results of Qin et al. (2021) may be a possible explanation for the exception mentioned above, where the perception of a negative economic impact had a positive and negative influence on the opinion about sociocultural impacts. If we consider the research results of Gannon et al. (2021), where residents are more likely to support tourism development if they perceive benefits, such as job creation, infrastructure development, improved quality of life (positive economic impacts), and the promotion and preservation of local communities' culture (positive sociocultural impacts), we can infer that residents of Porto and Vila Nova de Gaia are willing to support tourism development. However, residents of Porto and Vila Nova de Gaia also have negative perceptions about the economic impacts of tourism. Taking into consideration, the study of Çelik e Rasoolimanesh (2021), where residents are aware of the existence of negative tourism impacts, residents support tourism development in the region through cost–benefit analysis, something that can be applied to the case in question.

These results are compatible with the results of the research of Martín et al. (2018), in which positive perceptions of economic and sociocultural impacts positively influence residents' attitudes toward tourism; the opposite is also true, when perceptions of tourism impacts are negative. However, this research, in addition to verifying the negative and positive perceptions of economic and sociocultural impacts, concludes that the perception of economic impacts influences the opinion about sociocultural impacts, with those who have a positive perception of the economic impacts of tourism having a positive opinion about the sociocultural impacts of tourism, and those who have a negative perception of the economic impacts of tourism having a negative opinion about the sociocultural impacts. Thus, H1 was accepted.

H2: Residents' perceptions of the economic impacts of tourism influence residents' opinions of environmental impacts.

It was found that those who have a positive perception of the economic impacts of tourism demonstrate a positive opinion about the environmental impacts of tourism. However, those who have a negative perception of the economic impacts of tourism demonstrate a negative opinion of the environmental impacts.

The results of H1 are similar to those of H2. There is an intention of Porto and Vila Nova de Gaia residents to be willing to support the development of the sector (Çelik & Rasoolimanesh, 2021; Qin et al., 2021). Moreover, the results of this research coincide with the results of Sharma et al. (2016) and Stainton (2019) where residents value the ability of tourism to protect and preserve habitats and natural heritage, as well as the creation of programs aimed at increasing residents' knowledge of the environment and tourists' awareness of natural heritage. The results in which negative perceptions about the economic impacts of tourism have a negative influence on residents' opinions about environmental impacts focus on noise and visual pollution and the ecological footprint of tourism. These results, which coincide with those of the research of Brankov et al. (2019); Gu et al. (2020) and Xue et al. (2017). In addition to the agreement with previous research, the results of this research support the suggestion of Charag et al. (2020) stating that economic incentives should be used to create incentives for the protection and conservation of the environment and habitats of the region and the promotion/adoption of environmentally responsible behavior/measures by the municipality. It is important to note that in addition to the perception of the positive economic impacts of tourism, the way in which the benefits from these impacts are used to improve pre-existing environmental issues, by those responsible for the destination, has an influence on the intention of residents to support the development of the sector, in line with the conclusions of Hunt et al. (2015).

It is found that residents' perception of the economic impacts of tourism influences residents' opinion toward environmental impacts. Thus, H2 was accepted.

This research allows us to infer that the residents of Porto and Vila Nova de Gaia have a positive perception about tourism in the region. The concept of sustainable tourism may not be something achievable in practical terms; however, it is a goal that should be pursued. This concept considers the destinations' residents as the main focus of tourism planning, since without their support and satisfaction, no destination can be successful in the long term. As evidenced throughout this paper, based on other research, a positive perception by residents of tourism impacts increases residents' intention to support the development of the sector in the region. Despite the existence of negative perceptions about tourism impacts by residents, we can verify that they are lower than the positive perceptions, which through the cost–benefit analysis, indicates that residents are willing to support the development of the sector, even though they are aware of the negative impacts generated by tourism. Considering the main results, there is a low level of engagement between residents and the tourism sector. This factor should be given attention by those responsible for the destination.

Considering the studies mentioned throughout this research, it can be seen that the level of involvement of residents with the tourism sector is closely linked with their intention to support the development of the sector. Analyzing this factor from the sustainability perspective, specifically in its three dimensions, resident involvement can be a strategy for the prevention of environmental overload, for increasing destination competitiveness, for attracting visitors in the short and long term, and for protecting and preserving the destination's cultural resources. Involving residents, in addition to contributing to their intention to support tourism development, can also contribute to sustainable development. It is essential that those responsible for destination management develop strategic plans with the active involvement of local communities as a premise. Involvement, however, should not be limited to participation in tourism-related activities, but also in the decision-making process regarding the tourism sector. The ecological footprint of tourism, such as noise and visual pollution proves to be one of the most significant negative perceptions of tourism impacts for residents, suggesting that destination managers should take measures to mitigate these negative impacts, such as waste management and the development of infrastructure capable of controlling and protecting existing ecosystems in destinations. The development of these measures enables a greater intention by residents to support the development of the sector. In terms of the economic dimension, the increase in prices of goods and services and inflation in the real estate market are the predominant negative perceptions among residents, something expected considering the current state of the regions under study, specifically the municipality of Porto which has difficulties in providing housing for the neediest groups and in keeping the natural residents of the historic center. The municipalities of Vila Nova de Gaia and Porto present different realities, factors such as the difference in size of the municipalities and tourism demand between them are examples of how these municipalities diverge. However, Vila Nova de Gaia should anticipate this issue and observe the municipality of Porto to prevent certain situations, namely the housing issue. As far as the sociocultural dimension is concerned, the issue of lack of involvement by residents with the sector has already been addressed; however, the issue of tourism leading to congestion of roads and leisure places, as well as the loss of local authenticity, should also be targeted. The problem of congestion is recurrent in both municipalities; as many of the residents may live in one municipality and work in the other, traffic can be heavy at certain times of the day; with the addition of tourism activity, it would be expected that residents would have this perception. Those responsible for the destinations are aware of these difficulties, so recently there are plans to build a new access bridge between Porto and Vila Nova de Gaia, the requalification of the lower deck of the "D. Luís I" bridge and the expansion of the metro line. The issue of loss of authenticity is worrisome. Besides the adversities faced by residents, the loss of authenticity directly affects the attractiveness of the destination and its sustainability. Tourism destinations' management organizations should adopt measures to preserve the cultural heritage of the residents, as well as the incentive to maintain the cultural values and traditions existing in the destination.

Acknowledgements The authors acknowledge the financial support of CiTUR, R&D unit funded by the FCT—Portuguese Foundation for the Development of Science and Technology, Ministry of Science, Technology and Higher Education, under the scope of the project UID/BP/04470/2020.

References

Aleshinloye, K. D., Woosnam, K. M., Tasci, A. D. A., & Ramkissoon, H. (2021). Antecedents and outcomes of resident empowerment through tourism. *Journal of Travel Research*, 004728752199043. https://doi.org/10.1177/0047287521990437

AMP (2021). Caracterização da AMP: Demografia. Área Metropolitana do Porto. http://portal.amp.pt/pt/4/municipios/vngaia/stats/demografia/

Antunes, B., March, H., & Connolly, J. J. T. (2020). Spatializing gentrification in situ: A critical cartography of resident perceptions of neighborhood change in Vallcarca, Barcelona. *Cities, 97*, 102521. https://doi.org/10.1016/j.cities.2019.102521

Brankov, J., Jojic Glavonjic, T., Milanović Pešić, A., Petrović, M., & Tretyakova, T. (2019). Residents' perceptions of tourism impact on community in national parks in serbia. *European Countryside, 11*, 124–142. https://doi.org/10.2478/euco-2019-0008

Çelik, S., & Rasoolimanesh, S. M. (2021). Residents' attitudes towards tourism, cost-benefit attitudes, and support for tourism: A pre-development perspective. *Tourism Planning & Development*, 1–19. https://doi.org/10.1080/21568316.2021.1873836

Charag, A. H., Fazili, A. I., & Bashir, I. (2020). Residents' perception towards tourism impacts in Kashmir. *International Journal of Tourism Cities, ahead-of-print*(ahead-of-print). https://doi.org/10.1108/IJTC-11-2019-0202

Erul, E., & Woosnam, K. M. (2021). Explaining residents' behavioral support for tourism through two theoretical frameworks. *Journal of Travel Research*, 004728752098761. https://doi.org/10.1177/0047287520987619

Gannon, M., Rasoolimanesh, S. M., & Taheri, B. (2021). Assessing the mediating role of residents' perceptions toward tourism development. *Journal of Travel Research, 60*(1), 149–171. https://doi.org/10.1177/0047287519890926

Gu, X., Hunt, C. A., Lengieza, M. L., Niu, L., Wu, H., Wang, Y., & Jia, X. (2020). Evaluating residents' perceptions of nature-based tourism with a factor-cluster approach. *Sustainability, 13*(1), 199. https://doi.org/10.3390/su13010199

Hunt, C. A., Durham, W. H., Driscoll, L., & Honey, M. (2015). Can ecotourism deliver real economic, social, and environmental benefits? a study of the Osa peninsula, costa rica. *Journal of Sustainable Tourism, 23*(3), 339–357. https://doi.org/10.1080/09669582.2014.965176

Hunter, C. (1997). Sustainable tourism as an adaptive paradigm. *Annals of Tourism Research, 24*(4), 850–867. https://doi.org/10.1016/S0160-7383(97)00036-4

Jaafar, M., Rasoolimanesh, S. M., & Ismail, S. (2017). Perceived sociocultural impacts of tourism and community participation: A case study of Langkawi Island. *Tourism and Hospitality Research, 17*(2), 123–134. https://doi.org/10.1177/1467358415610373

Janusz, G. K., & Bajdor, P. (2013). Towards to sustainable tourism–framework, activities and dimensions. *Procedia Economics and Finance, 6*, 523–529. https://doi.org/10.1016/S2212-5671(13)00170-6

Jordan, E. J., Lesar, L., & Spencer, D. M. (2021). Clarifying the interrelations of residents' perceived tourism-related stress, stressors, and impacts. *Journal of Travel Research, 60*(1), 208–219. https://doi.org/10.1177/0047287519888287

Kim, S., & Kang, Y. (2020). Why do residents in an overtourism destination develop anti-tourist attitudes? an exploration of residents' experience through the lens of the community-based tourism. *Asia Pacific Journal of Tourism Research, 25*(8), 858–876. https://doi.org/10.1080/10941665.2020.1768129

Lawton, L. J., & Weaver, D. B. (2015). Using residents' perceptions research to inform planning and management for sustainable tourism: A study of the gold coast schoolies week, a contentious tourism event. *Journal of Sustainable Tourism, 23*(5), 660–682. https://doi.org/10.1080/09669582.2014.991398

Lee, T. H., & Jan, F.-H. (2019). Can community-based tourism contribute to sustainable development? evidence from residents' perceptions of the sustainability. *Tourism Management, 70*, 368–380. https://doi.org/10.1016/j.tourman.2018.09.003

MacKenzie, N., & Gannon, M. J. (2019). Exploring the antecedents of sustainable tourism development. *International Journal of Contemporary Hospitality Management, 31*(6), 2411–2427. https://doi.org/10.1108/IJCHM-05-2018-0384

Martín and de los Salmones Sánchez, M. M. G., & Herrero, Á. , 2018 Martín, H. S, de los Salmones Sánchez, M. M. G., & Herrero, Á. (2018). Residents' attitudes and behavioural support for tourism in host communities. *Journal of Travel & Tourism Marketing, 35*(2), 231–243. https://doi.org/10.1080/10548408.2017.1357518

Moscardo, G. (Ed.). (2008). *Building community capacity for tourism development*. CABI.

Munanura, I. E., Needham, M. D., Lindberg, K., Kooistra, C., & Ghahramani, L. (2021). Support for tourism: The roles of attitudes, subjective wellbeing, and emotional solidarity. *Journal of Sustainable Tourism*, 1–16. https://doi.org/10.1080/09669582.2021.1901104

Muresan, I., Oroian, C., Harun, R., Arion, F., Porutiu, A., Chiciudean, G., Todea, A., & Lile, R. (2016). Local residents' attitude toward sustainable rural tourism development. *Sustainability, 8*(1), 100. https://doi.org/10.3390/su8010100

Nugroho, P., & Numata, S. (2020). Resident support of community-based tourism development: Evidence from Gunung Ciremai national park Indonesia. *Journal of Sustainable Tourism*, 1–16. https://doi.org/10.1080/09669582.2020.1755675

Nunkoo, R., & Gursoy, D. (2012). Residents' support for tourism. *Annals of Tourism Research, 39*(1), 243–268. https://doi.org/10.1016/j.annals.2011.05.006

Obradović, S., Tešin, A., Božović, T., & Milošević, D. (2021). Residents' perceptions of and satisfaction with tourism development: A case study of the Uvac special nature reserve, Serbia. *Tourism and Hospitality Research, 21*(1), 31–43. https://doi.org/10.1177/1467358420946789

Pereira, T., & Gadotti dos Anjos, S. J. (2021). Corporate social responsibility as resource for tourism development support. *Tourism Planning & Development*, 1–21. https://doi.org/10.1080/21568316.2021.1873834

Qin, X., Shen, H., Ye, S., & Zhou, L. (2021). Revisiting residents' support for tourism development: The role of tolerance. *Journal of Hospitality and Tourism Management, 47*, 114–123. https://doi.org/10.1016/j.jhtm.2021.02.010

Ribeiro, M. A., Woosnam, K. M., Pinto, P., & Silva, J. A. (2018). Tourists' destination loyalty through emotional solidarity with residents: An integrative moderated mediation model. *Journal of Travel Research, 57*(3), 279–295. https://doi.org/10.1177/0047287517699089

Savage, V. R., Huang, S., & Chang, T. C. (2004). The singapore river thematic zone: Sustainable tourism in an urban context. *The Geographical Journal, 170*(3), 212–225. https://doi.org/10.1111/j.0016-7398.2004.00121.x

Sharma, R., Chaurasia, S., & Bhattacharya, A. K. (2016). Residents' perception of environmental impacts of tourism recreational activities in protected area. *Journal of Environmental Management and Tourism, 2*, 238–249%V 7.

Sharpley, R. (2014). Host perceptions of tourism: A review of the research. *Tourism Management, 42*, 37–49. https://doi.org/10.1016/j.tourman.2013.10.007

Stainton, H. (2019). *TEFL tourism: Principles*. CABI.

Stone, L. S., & Nyaupane, G. P. (2020). Local residents' pride, tourists' playground: The misrepresentation and exclusion of local residents in tourism. *Current Issues in Tourism, 23*(11), 1426–1442. https://doi.org/10.1080/13683500.2019.1615870

Su, L., & Swanson, S. R. (2019). The effect of personal benefits from, and support of, tourism development: The role of relational quality and quality-of-life. *Journal of Sustainable Tourism, 28*(3), 433–454. https://doi.org/10.1080/09669582.2019.1680681

Teye et al., 2002 Teye, V., Sirakaya, E. & Sönmez S. F. (2002). Residents' attitudes toward tourism development. *Annals of Tourism Research, 29*(3), 668–688. https://doi.org/10.1016/S0160-7383(01)00074-3

Vargas-Sánchez and Oom do Valle, P., da Costa Mendes, J., & Silva, J. A. , 2015 Vargas-Sánchez, A., Oom do Valle, P., da Costa Mendes, J., & Silva, J. A. (2015). Residents' attitude and level of destination development: An international comparison. *Tourism Management*, 48, 199–210. https://doi.org/10.1016/j.tourman.2014.11.005

Xue, L., Kerstetter, D., & Hunt, C. (2017). Tourism development and changing rural identity in China. *Annals of Tourism Research, 66*, 170–182. https://doi.org/10.1016/j.annals.2017.07.016

Zheng, D., Ritchie, B. W., & Benckendorff, P. J. (2021). Beyond cost–benefit analysis: Resident emotions, appraisals and support toward tourism performing arts developments. *Current Issues in Tourism, 24*(5), 668–684. https://doi.org/10.1080/13683500.2020.1732881

Digital Innovation and Sustainability Practices in Tourism: An Overview

Artemis Giourgali

Abstract This article seeks to develop a theoretically oriented reflection on sustainability digital innovations within the context of tourism and the potentials of this relationship. To this end, a literature review analysis was chosen for the collection, selection and critical evaluation of the literature on this topic. The authors assert that there is an ongoing process of evolutionary sustainability practices in tourism, indicating a theoretical approach rather than specific actions. The significant progress on the acceptance of the basic principles of sustainability from businesses and individuals still lacks the application of the appropriate instruments in the tourism industry so as to maximize the efficiency and successful development of sustainable digital innovations.

Keywords Sustainability practices · Tourism sustainability · Digital innovation · Tourism development

JEL Classifications Z30 · Z32 · Q01 · Q56 · O33

1 Introduction

With digital technologies, organizations have adopted innovative approaches to tackling societal challenges. These technologies help address grand challenges like climate change and promote sustainable development. They are generally referred to as digital sustainability activities. Digital sustainability can drive empirical advances in entrepreneurship, innovation and has the potential to positively impact society (George G. et al, 2020).

A. Giourgali (✉)
Department of Business Administration, University of the Aegean, Chios, Greece
e-mail: giourgali.arte@gmail.com

© The Author(s), under exclusive license to Springer Nature Switzerland AG 2023

V. Katsoni (ed.), *Tourism, Travel, and Hospitality in a Smart and Sustainable World*, Springer Proceedings in Business and Economics,
https://doi.org/10.1007/978-3-031-26829-8_7

Many large companies have created corporate social responsibility departments to generate social outputs, but they still separate these activities from their core processes. A smaller subset of companies are trying to integrate pro-social choices into their core strategies, processes and practices. By detecting the emergence of STE in practices, they initiate modifications in goal setting that focus on reducing environmental impact and going "green". A few and more ambitious sustainable entrepreneurs deliberately pursue net positive environmental impacts. As sustainable entrepreneurs reshape capital structures and corporate cultures, they are creating a growing population of organizations for which the pursuit of sustainability has become a core economic opportunity (George G. et al, 2020).

There are many substantial challenges to the application of the principles of sustainable development in tourism. Some of them include the nature of the tourism product, the fragmented way in which critical decisions are made in tourism and the different, often conflicting interests of those involved in tourism development. This article examines the application of sustainable practices in the tourism industry with a theoretical presentation of development strategies and a focus on sustainable approaches in practice.

2 Literature Review

2.1 Digital Innovation in Tourism

Digital technologies have revolutionized the tourism industry by bringing major and remarkable changes to tourism businesses, products, experiences, business ecosystems and destinations. Digitization has not only transformed the producers, but also the consumers of traditional tourism, with the emergence of new roles, relationships and references. The variety and volume of tourism products have increased, and consequently, the functionality, the speed of financial transactions and the feedback have also increased. Digitalization in tourism not only leads to creativity and innovation, but also facilitates an increased customization of visitors' experiences, which enhances their satisfaction. Digitalization contributes to the configuration of new destinations, new business models and business ecosystems which open up new roles for consumers, producers and DMOs (Dredge, et al., 2019). The journey from traditional operators to a smart destination is long, although coordinated efforts to promote an innovative digital culture can ensure the competitiveness of destinations, as shown in detail in the following figure.

"The journey towards digitalization in Tourism". Source: Adapted from Dredge, D, et al. (2019).

Due to the progress in digitalization, the tourism sector has undergone a remarkable change and has affected the economic efficiency of the industry. The presence of factors influencing consumers and economic growth is proven, as are the economic benefits associated with the development and use of digitalization (Filipiak et al, 2020). Some of the implementations that could support the empowerment of tourism operators are the below forms of digital innovation usage:

- Internet of things: A key promoter of tourism has been the development of network capabilities. The increase of broadband infrastructures that allow users to access multimedia content in real time is a determining factor for the quantity and quality of information that can be presented on the Internet. This fact is particularly important in the tourism market, as the choice of a specific destination is directly related to the quality of information available to the potential customer (Filipiak et al, 2020). The existence of images and detailed information on a web site plays a crucial role in the customer's view of the quality of a product. IoT connects various devices online and can have a major impact in cost-effectiveness of daily operations (Ivan, 2020).
- Recognition technology: A second dimension in the field of networks is related to security and personal data protection technologies. The progress that has been made in this area ensures to a very high degree the security of transactions that also play a key role in e-business. One of the technological trends that are being studied for use in the tourism industry to provide smoother biometric authentication is fingerprints or face scans (Ivan, 2020).
  Finally, the integration of systems is an important technological factor that influences the development of tourism entrepreneurship. At the same time, services that create new value by coordinating data across industries and platforms remain a work in progress.
- Big data: In the hospitality industry, big data helps operators offer a more personalized experience, suggest attractive alternatives, evaluate performance and make

data-based analysis for business purposes. Interoperability includes both data and IT and telecommunications' applications, which provides for the creation of an environment, transparent to the technology used and a shared data and process environment. New programming trends that improve the reliability and performance of the software improve the price/performance ratio and its ability to carry out extremely complex applications.

- Artificial intelligence (AI): Through AI analysis and service automation, operators can improve performance, make more accurate predictions about consumer needs create predictions and reduce costs. AI offers advanced data, computing capabilities, ample storage and super-fast speed, enabling analysis like capacity expectations, online and fast customer service and augmentation of the process of collaboration, so it can become more memorable for customers and increase the service-delivery competencies for firms (Solakis et al, 2022). A wide range of applications are available in the market, general and specialized, which solve the issue of internal and external communications of a company with business partners and customers, but also issues of data management and distribution and trade (Filipiak et al, 2020).
- Augmented reality (AR): When promoting is required, AR technology plays an important role. The tourism product is often becoming more credible and satisfactory, with virtual tours and testers offered.

The indicators related to the resources on which the development of tourism is based are many and important. In a first stage of evaluating indicators for sustainable tourism development, it is required to identify these resources, in order to understand their importance both to tourists and residents, businesses and local authorities (Kozirakis, 2014). Sustainability is becoming an increasingly critical requirement for multinational companies that have often been identified as primary contributors to unsustainable practices in environment and society. According to the Carbon Majors Report (CDP, 2017), more than 70% of the world greenhouse gas emissions since 1988 come from just 100 companies. International tourist arrivals are constantly increasing by 25 million worldwide in 1950, to 527 million in 1995 and 1.32 billion in 2017. It is expected to reach 1.8 billion by 2030 (UNWTO, 2016). The number of cars on the streets exceeded 1 billion in 2010. In 2014, there were 1.2 billion and will be 2 billion by 2035 (Green Car Reports, 2014). Investments in sustainable education and infrastructure are now paramount as employers and employees must be trained for environmental responsible choices. Therefore, an identification and assessment are made, for environmental, economic, social and cultural indicators in tourism, to evaluate their positive or negative effects related to biodiversity, air, noise, water, waste generation, energy use and cultural heritage. These impacts create a set of indicators for monitoring situations in tourist areas and drawing up strategic plans for protection, impact control and sustainability practices (Kozirakis, 2014).

2.2 Digital Innovation and Sustainability

This golden age of digital innovation provides an unprecedented opportunity in the field of sustainability. The world's attention is focused on technology and innovation, presenting an opportunity to review and redesign processes and digital technologies that can benefit the planet's well-being. The current pandemic highlighted the importance of digitization and digital transformation and accelerated digital transformation across all sectors. The importance of digital innovation for organizations is great in the context of sustainability. While one main objective of digitalization is to create economic value, sustainable business models empowered by digital technologies also set sights on higher ecological and social value (Bähr & Fliaster, 2022). Without the digital transformation of businesses, the economic and environmental challenges of the future cannot be met in a sustainable way. The role of digital innovation is central to achieving long-term economic growth, reducing unemployment, improving the quality of life, providing wider access to public services and reducing costs in many sectors of the economy. Thus, the challenges of digitalization and sustainability transitions are likely to be strategically relevant and particularly decisive for the vast majority of industry sectors.

Sustainability and business digitalization topics are burgeoning, and recent studies are increasingly addressing digital sustainability. Digital sustainability can be defined as the activities that aim at achieving the goals of sustainable development through a creative use of digital technologies (Yousaf et al, 2021). Sustainable innovation is a process where sustainability considerations, environmental, social and financial, are integrated into the equation. This applies to products, services, technologies, as well as to new business models. The reference to the concept of sustainability as a key element that guides innovation refers to the relevance of these dimensions in the innovation process. The economic dimension of sustainability, for instance, is related to the profit component, the economic growth, the efficient use of resources and the viability of businesses. The environmental dimension focuses on combating pollution and the efficient and prudent use of natural resources. The social dimension concerns issues such as equal opportunities, justice in the distribution and equality. The concern and combination of all the above results to a smart, sustainable and inclusive growth (Ferreira & Serpa, 2018).

In the below figure, the analysis of the relationship between these dimensions is shown, creating a systemic relationship between innovation and sustainable development (Ferreira & Serpa, 2018). This intersection of the elements emphasizes that social sustainability is related with economic sustainability with a socio-innovative lace that includes community equity and economic growth, while the relation between the economy and the environment creates a concept of eco-innovation. The social dimension indicates that the focus of innovation is not the technology in itself but rather the need to find solutions that are more effective, sustainable and ethically adequate.

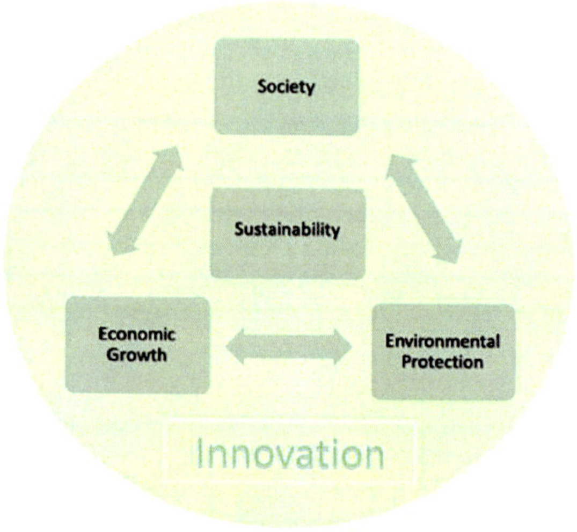

Source: Adapted from Ferreira & Serpa (2018). ***Sustainability tripod within the context of innovation.***

Organizations can play an important role in how the global industrial system manages the great challenges of sustainable innovation (Yousaf et al, 2021). There should be a replacement of fossil energy with renewable energy systems, enabled in particular, by technological innovations. An example of such technologies is virtual power plants (VPPs), who virtually integrate several distributed power-generating, power-storing and power-consuming units to allow aggregation and remote control of individual units with alternative digital technologies (Bähr & Fliaster, 2022). A study made by Bähr & Fliaster in the context of business strategies and the environment in 2022 concluded that the perception of primary sustainable changes in businesses comprises the following three aspects: Firstly, the erosion of the traditional core business can come as a result from changes in the competitive landscape, stemming from the decentralization of energy production. Secondly, the technological advancements are expanding in importance, with the emergence of artificial intelligence, cloud-based services, big data, etc. These digital technologies enable the automation of processes. Similarly, the management of minor decentralized power-generating units necessitates more digital technology implementation. Thus, the existing infrastructure either enables or limits the opportunities. Thirdly, there seems to be a high societal awareness and acceptance of current transitions. At the same time, the transitions are dependent on regulations due to their imperfect and fast-changing nature.

Another strong example of innovative sustainable technologies is the Innovative Optical and Wireless Network (IOWN), an initiative infrastructure, built around photonic technology for high capacity, low latency and low power consumption (NTT, 2022). With core technological infrastructures, they offer an environmental sustainability value to society with the following ways (NTT, 2022):

- Sourcing information from sensors and devices, the manage urban assets creating digital spaces, offices, commercial venues and residential places to optimize usage, based on preferences and predictive movements of people.
- Predicting environmental disasters with real-time understanding of environmental
- activity. The habilitation of more precise weather simulations and predictions facilitates the environmental conservation and advanced preparation to mitigate natural disasters. Using satellites and meteorological sensor devices to acquire detailed data, diversified energy networks, including lightning power generation and space solar generation, a chance for flexibility against environmental change is given to society. As more future disasters loom, preparation and better management of energy consumption are essential.
- Work is being done to manage nuclear fusion and optimize the operation of nuclear fusion reactors, while experimental results show that it allows computers to transmit vast amounts of information in extremely low time.
- With supply and demand control and forecasting technologies, a better energy distribution network is created, with greater precision and use of renewable sources.

Sustainable digital innovation helps to apply the latest digital technology to achieve creative solutions that will address long-term social, economic and environmental issues. It is a combination of digitized technologies and the digitized management system for business process innovation in a sustainable way. The transition process presupposes the confrontation of ambiguous, uncertain and extremely complex challenges. It also involves the introduction of significant changes in the established sociotechnical systems, in order to give priority to a more environmentally friendly production and consumption. These changes depend to a large extent on technology, institutions and the social sphere (Bähr & Fliaster, 2022). To meet the goals of sustainability, there must be a commitment to digital innovation in a more complex way, that is, by developing whole business models focused on this specific goal.

2.3 Sustainable Tourism Practices

Tourism should take into account its current and future economic, social and environmental impacts, while addressing the needs of visitors, industries and host communities. Sustainable tourism development guidelines and practices are applicable to all forms of tourism and all types of destinations. Sustainability principles refer to the environmental, economic and sociocultural balance that must be established between tourism development and actions to guarantee long-term sustainability (Unwto &

Unep, 2005). In order to be able to approach and apply the principles of sustainable tourism in practice, we have to formulate and categorize a set of specific directions—guidelines for practical application. Kozirakis in his research on environmentally sustainable tourism development in 2014 identified the following key factors to indicate sustainable tourism:

- Pressure: number of tourists visiting the location.
- Use intensity: intensity of use during peak periods.
- Social implications: ratio between tourists and residents.
- Development control: existence of official controls for the development of the space and the intensity of its use.
- Waste management: percentage of wastewater in the treated area.
- Programming procedure: existence of an organized regional plan.
- Critical ecosystems: number of rare or endangered species.
- Consumers' satisfaction: levels of visitors' satisfaction (based on research).
- Local residents' satisfaction: levels of locals' satisfaction (based on research).
- Tourism contribution in the local economy: proportion of total economic activity resulting from tourism.
- Bearing capacity: the ability of the area to receive different sizes of tourists.
- Space pressure: impact level in the area's elements due to tourism.
- Attractiveness: quality measure of those elements that make the area attractive in tourism.

As seen above, the indicators are plenty and relate to a variety of tourist destinations. They assess the pressure and the effects that tourism causes on the environment and the society, the limits of the carrying capacity, the attractiveness of the destination, the satisfaction of the tourists and the locals, which clearly show the dimension of sustainable development for which relevant controls should be made (Kozirakis, 2014). The use of indicators' evaluation frameworks helps to understand all the dimensions and sources of relevant problems in the sustainable development processes and serves as a refocus to minimize any negative impacts that may arise as the result of tourism activities.

Some of the sustainable tourism practices include the following actions (Unwto & Unep, 2005):

- The optimal use of environmental resources by maintaining necessary ecological processes and contributing to the preservation of biodiversity, avoiding any form of natural degradation or exploitation.
- Equally important is the respect for the authenticity of the host destinations, preserving the cultural heritage, values and tradition. In this, cross-cultural understanding and tolerance are essential.
- The fair distribution of employment and income opportunities in the host communities contributes to ensuring the viability and competitiveness of tourism destinations and enterprises, sustainable and long-term economic functions, as well as poverty alleviation.

- Achieving sustainable tourism requires continuous monitoring of impacts, introducing the necessary preventive and corrective measures whenever necessary. To ensure broad participation and consensus building, it requires organization and strong political leadership.
- Finally, sustainable tourism should maintain a high level of tourist satisfaction, while increasing their awareness of sustainability issues and promoting sustainable tourism practices.

3 Methodology

The methodology used for the present paper is a narrative literature review, summarizing and synthesizing the literature found with a critical analysis. By presenting a comprehensive background of the literature within the interested topic, theoretical and conceptual frameworks were developed, with the objective of systematizing the relation between digital innovation, sustainability and tourism. All the presented information were gathered from official scientific databases such as Google Scholar or Scopus. The data were collected from credible articles, journals, books and scientific websites of relevance. The criteria in order to select the literature information were the relevance to topic and the publishing date. The review concerns an initial critical reading driven by the main research question: How is digital innovation employed, in addressing environmental sustainability?

The critical analysis of the articles is already discussed in the literature review section.

4 Conclusion

Sustainable tourism is not a particular form of tourism but a philosophy and a set of guidelines that can be used as a framework for the development of all types of tourism, regardless of size or type (Waligo et al., 2013). There is an urgent need to begin describing what actions a responsible tourist must take, increase the levels of awareness and enhance pro-environmental behavior. The recognition of the urgent need to redefine the tourism–environment relationship has led to it being one of the main policy issues nowadays. Sustainable is designed to combine the interests of all the parties involved. The host habitats and communities, the tourists and the industry itself. The aim is both at the protection of the tourism industry and the protection of the cultural and natural resources of an area (Tsartas et al, 2010).

This paper presented a review for the implementation of practical tools and means, their practical application and even beyond that, a concept of a proactive rather than a recovery action plan. Through digitalization, society must see how paths to sustainable development can be made. New technological tools incorporating and facilitating sustainability plans should be developed from governments and tourism industry enterprises. There must be a strong commitment to linking digitalization to these goals, and a common policy must be put in place to support sustainable development (Filipiak et al, 2020). Digitalization offers multiple opportunities to access new markets, develop new roles and configurations and evolve new travel products and services.

Companies frame new digital technologies, promote business digitalization and contribute to sustainability transitions beyond their organizational boundaries. Although promising technologies exist to face environmental sustainability challenges, failing to address most of the challenge makes the application of the technologies unsustainable (Kwame et al, 2022). Particularly, different technology frames lead to different value propositions and contributions to sustainability transitions. The impact of technology on humans and the planet must be positive, encouraging progress in a way that does not harm or imply any kind of injustice. The concept of sustainable development occupies a prominent place in the political agenda, and the aim is to transform it into a functional set of policy goals and measures. With all the above parameters in mind, not only can we largely cover the behaviors of the stakeholders, but it is possible to predict possible reactions, resulting in the reconsideration and adoption of the most appropriate development option regarding the environment and so the implementation of a win–win situation.

References

Bähr, K., Fliaster, A. (2022). The twofold transition: Framing digital innovations and incumbents' value propositions for sustainability. *Business Strategy and the Environment*. Wiley Online Library. [Retrieved on 07/06/2022 from https://doi.org/10.1002/bse.3082]

CDP Carbon Majors Report. (2017). The carbon majors database. Retrieved March 2, 2022, from https://www.cdp.net/en/articles/media.

Dredge, D., Phi, G. T. L., Mahadevan, R., Meehan, E., & Popescu, E. (2019). Digitalisation in tourism: In-depth analysis of challenges and opportunities. Executive Agency for Small and Medium-sized Enterprises (EASME), European Commission.

Ferreira, C., & Serpa, S. (2018). Society 5.0 and social development: Contributions to a discussion. *Management and Organizational Studies*. [Retrieved on 28/06/2022 from: https://doi.org/10.5430/mos.v5n4p26

Filipiak, B., Dylewski, M., & Kalinowski, M. (2020). *Economic development trends in the EU tourism industry*. Springer.

George G., Merrill, R. K., & Schillebeeckx, S. J. D. (2020). Digital sustainability and entrepreneurship: How digital innovations are helping tackle climate change and sustainable development. *Entrepreneurship Theory and Practice*. 2021, *45*(5), 999–1027. https://doi.org/10.1177/1042258719899425

Ivan I. (2020). *Effects of Dynamic Organization and Digital Innovation on the Hotel Tourism Industry during the Coronavirus Pandemic Period.* Available at [Retrieved August 22, 2022, from: https://doi.org/10.2139/ssrn.3617528]

Kozirakis, G. (2014). Environmentally sustainable tourism development. Evaluation Framework, The Case of Crete. Panteion University Of Social And Political Sciences. Department Of Economics & Regional Development, (3), 50–67.

Kwame, Nti. E., Jerry Cobbina, S., Efua Attafuah, E., Opoku, E., & Amoah Gyan, M. (2022). *Environmental sustainability technologies in biodiversity, energy, transportation and water management using artificial intelligence: A systematic review. 4*, 100068. ISSN 2666-1888, [Retrieved August 21, 2022, from: https://doi.org/10.1016/j.sftr.2022.100068]

NTT Corporation. (2022). *IOWN technology in action, offering a new value to society.* Retrieved August 19, 2022, from: https://www.global.ntt/isf/index.html

Nunes, Sérgio., & Cooke, Philip. (2021). New global tourism innovation in a post-coronavirus era. *European Planning Studies, 29*(1), 1–19. https://doi.org/10.1080/09654313.2020.1852534

Solakis, K., Katsoni, V., Mahmoud, A.B. & Grigoriou, N. (2022). Factors affecting value co-creation through artificial intelligence in tourism: A general literature review. *Journal of Tourism Futures.* [Retrieved August 21, 2022, from https://doi.org/10.1108/JTF-06-2021-0157]

Tsartas, P., Stavrinoudis, Th., Zagkotsi, S., Kyriakaki, A., & Vassiliou, M. (2010). *Tourism and Environment*. WWF Greece.

UNWTO (2016). UNWTO Annual Report 2015. Retrieved March 3, 2022, from: https://www.unwto.org/archive/global/publication/unwto-annual-report-2015

UNWTO & UNEP, (2005). Making tourism more sustainable-a guide for policy makers, 11-12.

Waligo, V. M., Clarke, J., & Hawkins, R. (2013). Implementing sustainable tourism: A multi-stakeholder involvement management perspective. *Tourism Management, 36*, 342–353.

Yousaf, Z., Radulescu, M., Sinisi, C. I., Serbanescu, L., & Paunescu, L. M. (2021). Towards sustainable digital innovation of SMEs from the developing countries in the context of the digital economy and frugal environment. *Sustainability MDPI, 13*, 5715.

Transformative Tourism as a Mean of Region's Sustainable Development

Ekaterina Buzulukova

Abstract Transformational tourism opens new opportunities for travel companies and recovering tourism industry in the period of pandemic and restrictions. Despite this constrains, travelers are looking for new ways of traveling and getting impressions. Transformative tourism aims to expand the vision of familiar things and rediscover them, sometimes even making changes in tourist attitudes and beliefs. This article goal is to show possibilities of transformative tourism as a tool to overcome the crises in the tourism industry for sustainable regions' development. Based on transformative tourists experience in Russia, South Africa and Turkey, content analysis of feedbacks, qualitative and quantitative research, we have discovered great potential for the travel industry recovery by means of the transformative tourism: preservation of the region's heritage and nature, support of SMEs, increasing interest to local cultures, improving the place image, create a tolerance to different cultures and nations, shape respect for the environment and nature, foster personal development. The article will be interesting for municipal leaders, travel companies and researchers.

Keywords Transformative tourism · Sustainability · Regional tourism

JEL Classifications Z31 · Z32

1 Introduction

Currently, there are quite a lot of publications on the search for ways to survive travel companies during the pandemic and political crises (Varzaru et al., 2021; Abbas et al., 2021; Sheresheva et al., 2021). One of these opportunities is transformational tourism. The first publications on the topic appeared relatively recently, and transformational tourism is gaining popularity both among researchers (Ateljevic, 2020; Pung et al, 2020) and among the offers of travel companies. Many tourists prefer non-traditional

E. Buzulukova (✉)
Senior researcher of Lomonosov's Moscow State University, Moscow, Russia
e-mail: buzulukova@econ.msu.ru

© The Author(s), under exclusive license to Springer Nature Switzerland AG 2023
V. Katsoni (ed.), *Tourism, Travel, and Hospitality in a Smart and Sustainable World*, Springer Proceedings in Business and Economics,
https://doi.org/10.1007/978-3-031-26829-8_8

travel destinations, and they are looking for new experiences, new knowledge, new travel opportunities (Ateljevic et al., 2016; Kirilova et al., 2017; Sheldon, 2020), such as to take a break from the hustle and bustle of the city, stay in nature, reduce stress, discover something new for themselves. According to the Transformational Travel Council (2021), transformational journey is a way to learn, grow and interact with the world. Travel can change a person's outlook on life and change their knowledge, values, attitudes, behavior, as well as how people spend their time and use resources (Sheldon, 2020).

In this article, author defines transformational tourism as a journey to find new cultural and spiritual experiences, discover the world for yourself, immerse yourself in the local culture and lifestyle of other people. For transformational travel, in addition to a uniqueness of place, people a tourist meets are also important, their willingness to openly talk about their culture and way of life. Other factors like co-travelers (travel partners), personal characteristics of traveler, opportunity to immerse in local culture are also important. The result of such a trip is that each participant should become the best version of himself.

The leading travel agencies and hotels (e.g., Marriot) are working on the possibility of organizing transformational trips, opening organic and vegetable gardens, various culinary studios and art workshops where you can do something with your own hands, personal trainers and yoga instructors. For example, The Ritz-Carlton and St. Regis Hotels and Resorts in 2018 offer their clients who want to experience a transformational travel experience to choose tours to the hidden places of Los Cabos (Mexico), Cairo (Egypt) and Amman (Jordan).

Despite the relevance of transformational tourism, there is still a relatively small amount of research about the possibility of transformational tourism to influence the sustainable development of the world and regions, thus, the purpose of the article is to show the possibilities of transformative tourism as a mean for sustainable regions' development through the analysis of the travelers' experience in Turkey, South Africa and Russia.

The main research question is whether transformational tourism can become a reliable support and inexhaustible source for the development of regions, travel companies, that is able to overcome the crisis of the tourism industry and meet the growing demand for transformational travel services from tourists.

2 Literature Review

2.1 Definition, Effects and Types of Transformative Tourism

The term first appeared in the academic discourse by Kottler (1997), who stressed that traveling gives a lot of opportunities for personal change and mention its transformative effect on tourists. UNWTO (2016) mentions that transformative tourism gives opportunity to travelers "to re-invent themselves in the world". Nowadays transformative tourism is a widely spread concept and its definition has been deepened and expanded. Transformational Travel Council (2021) defines transformation travel as "a deliberate journey to expand, learn and develop new ways of existing and interacting with the world". This definition emphasizes the distinctive feature of this type of travel—the search for new opportunities for self-knowledge and the world. Kirillova, Letho & Cai (2017) mention that transformative experiences are "those special extraordinary events that do not only trigger highly emotional responses but also lead to self-exploration, serve as a vehicle for profound intrapersonal changes, and are conductive to optimal human functioning" (2017, p. 498). Another definition was given by Tomljenovic and Ateljevic (2015, p.14) "transformative tourism is a type of tourism where tourists participate in travel that offers a high level of immersion with the host culture". So, the transformation can only happen if there is interconnection and openness between traveler and local people. Noy (2004) stressed that transformation is positive self-change of a person because of intentional and unintentional interactions between traveler and destination.

Morgan (2010) reveals factors that important to transformation: nature, culture, characteristics of the destination, traveler's personal motivation, intentional or unintentional forms of experience, duration of travel (long or short term) and distance (both physical and cultural) also affect the intensity of transformation. Pung & Del Chiappa (2020) also identified a set of essential characteristics for a deep transformation effect: interacting with locals and other tourists, facing challenges, experience the sense of place, long-stays and post-travel reflection. Pine & Gilmore (2011) mentioned individuals' personal backgrounds, events, perceptions and interactions at the destination as crucial factors for transformation tourism. However, Kirillova et al. (2017) proposed that the transformative effect does not lie in the nature of the tour or in the environmental factors, but in the way the tourist perceives those experiences and gives them meaning. Tasci and Godovykh (2021) revealed that travelers change more when they have contact with nature and people and participate in cultural activities. Cultural background and personal characteristics of the tourist might also affect the intensity of transformation (Cetin & Bilgihan, 2016).

Transformational tourism attracts the attention of not only researchers, but also practitioners. Melo et al (2021) made content analysis of websites of travel agencies published as partners in the Transformational Travel Council webpage (www.transformational.travel). The most common types of companies that were mentioned are travel experiences, followed by the life and travel coaches and retreat companies. Most frequent categories that were detected in tours description are: process (transformation), experience, traveler characteristics and context (place or environment).

Different types of transformative trips also were identified from theoretical and field research: backpacking tourism, education, tourism, ecotourism, extreme sports, yoga and other spiritual practices tourism, creative tourism, cultural tourism, spiritual, religious tourism, volunteer tourism, WWOOF-ing (agricultural tourism), health (wellness) tourism , gastronomic, sightseeing and immersion in culture, retreat tourism (Melo et al, 2021).

2.2 *Transformational Tourism as an Opportunity for Regions' Sustainable Development*

The benefits of transformative experiences for the tourism industry, travel agencies and individuals were mentioned in the UNWTO report (UNWTO, 2016). The report notes that with current environmental problems, tourism can become an important aspect of preserving cultural heritage, the unique nature of the regions, thanks to a new type of travelers who lead a conscious lifestyle.

The previous development of tourism industry has led to over-tourism, over-pollution and over-consumption (Sheldon, 2020) in the most popular travel destination. In several regions, it has led to the destruction of a unique ecology (lake Baikal), causing significant damage to cultural heritage (one can easily recall the Great Wall of China), discontent of the local population (Easter Island, etc.).

Transformative tourism, according to a number of researchers, has the potential to make the world a better place (Ateljevic, 2020). Transformational tourism can help overcome the emerging crises in the tourism industry thanks to the following characteristics:

- Transformative tourism experiences facilitate understanding of different cultures, developing a more tolerant and universal worldview. Travelers accept easily other countries representatives and culture (Brown, 2009). For example, Pung and Del Chiappa (2020) found that Italian travelers learned to be non-judgmental toward others (people or situations) through transformative experiences during their trips.
- Being in such travel stimulates self-reflection, critical thinking, emotional connections, that leads to personal development (Soulard et al., 2020).
- Transformative tourism showing uniqueness of nature, foster responsibility, and a more careful attitude to the environment. Tasci & Godovykh (2021) proved that transformative travelers become more environmentally conscious after tour.
- Transformative tourism can even change the tolerance to warring nations. For example, Pizam et al. (2002) found more positive perception of Jordanian people by Israelis after their eco-traveling tour to Jordan. These attitudinal changes are claimed to serve world peace (Bruner, 1991).
- Transformative model of tourism can preserve the local heritage and nature (UNWTO, 2016).
- Small- and medium-sized enterprises are the main actors in transformative tourism because only they can create unique tours with a meeting with original people for the small groups of tourists. Transformative tourism opens wide opportunities for them, that can increasing workplaces and tax revenues.
- Creation of new transformative tours can lead to improving the city or place image and attracts new tourists to the territory.

Thus, transformational tourism has a number of features that can lead to a more sustainable development of regions: the preservation of their natural resources, the maintenance of cultural heritage, attracting attention to unique entrepreneurs and craftsmen working on the territory, attracting additional funds, stimulating small-and medium-sized enterprises.

Of course, the capabilities of the world's leading countries are extensive in preserving their natural resources and cultural heritage. We are also interested in whether transformational tourism can increase opportunities for sustainable development in developing economies. To do this, we will analyze the experience of Turkey and Africa, and then explore the experience of Russia.

2.3 *Transformative Tourism Experience in Turkey*

Pale and Settings (2021) based on the experience of Turkish travelers investigated the factors influencing the transformational experience. Using semi-structured interviews, they studied travelers who had been abroad more than three times in the past two years. By studying the changes that travelers noticed in themselves, they identified factors that influence the beliefs and behaviors of tourists and the change in their daily lives.

The author investigated tripographic factors (e.g., length of travel), personal characteristics (e.g., cultural background) and destination features (e.g., cultural distance). Types of transformations were also categorized under three groups: behavioral, attitudinal and personality changes. Motivation, duration of the trip, co-travelers, organization of the trip, type of planning, cost of the trip and type of accommodation were identified as the main topics within the tripographic factors. Personal characteristics: demographic factors, personality traits, previous tourist experience, place of residence and nationality were marked by factors affecting the depth of transformational experience. The distance to place of travel, the language, the level of interaction and the characteristics of the destination (history, heritage, customs and people) also influenced the transformations. This effect can be caused by material factors such as landscape, nature, infrastructure and architecture, or by intangible factors such as culture, values, lifestyle, relationships and cost of living.

Pala and Cetin (2021) found that nature-based travel experiences lead to environmental awareness, behavioral change, self-actualization, personal development, sociocultural alterations and improved health and well-being in a nature-based tourism setting.

2.4 *Transformative Tourism in South Africa*

The study by Mkhize and Ivanovic (2020) was also based on a qualitative method—interviewing international tourists who visited Soweto in the period 3 January to 30 March 2018. The authors asked about the experience of tourists when walking through the township; living as the locals do; and dining with the locals. The authors identified the following factors that have a greater impact on the transformational experience of travelers: cultural and historical interest, education, conviction in environmental and social justice, the values of the traveler, the desire for authenticity. Guided by their conscience, some tourists demonstrate self-limiting and responsible tourist behavior that reflects their social and moral principles, as well as life values.

The authors also noted that a new type of tourists is looking for highly personalized, independent type of travel and avoid pre-planned popular commercial tours. They are looking for opportunities to meet and talk with residents, to get to know the city, the possibility to experience deep immersion in the environment and gain life experience.

The interviewed respondents in this study expressed a need for a "live like a local" experience. There is a need to eat what the locals and do what a local would do rather than visit the overwhelmed restaurant or to stay in a hotel.

Authors came to the conclusion that the more such travelers will appear, the more locals will be encouraged to save and promote their local heritage and extend personal knowledge about native culture in order to meet the demand of foreigners. However, development of the tourism should not damage the local culture, and government should establish an open dialog with the locals aligned with the principles of responsible tourism. Transformative tourism by the view of Mkhize and Ivanovic (2020) can set a new trajectory toward sustainable post pandemic world and make this type of tourism a dominant type of travel.

Now let us move to the experience of Russian travelers.

3 Methodology

The study was conducted in the period 2021–2022. The object of the study was the tourists of a transformational travel company. In order to make sure that the selected company really provides services in the field of transformational tourism, and not just declares about it, we conducted an interview with the owner of the company, a leading guide, three travelers and also got acquainted with travelers' feedback on the company's website. Based on the information received, we came to the conclusion that the chosen company Glukhoman-Go (possible translation backwoods-go) corresponds to the transformational model of tourism. In the period December 2021, 7 interviews were conducted: with the owner, with the leading guide, with three clients, as well as with two destination-based agents (cheese factory, Gzhel crafts). In the company's communication channels (Instagram, Telegram, WhatsApp) for the period from 2020 to 2022 years 32 reviews were selected, where travelers shared their experiences and the transformational effect that travel had on them, and not just gratitude for the tour. Since the information received was insufficient from the point of view of studying the intensity of the transformational experience and the factors that influenced it, we launched a questionnaire among the participants of the tours in June 2022. Although there is not much data at the moment (there are only 20 answers), it is possible to make only general conclusions in the form of assumptions. The questionnaire was limited to 15 questions, because that was the request of the travel company. The CEO mentioned that questionnaire should not be too long and difficult, as it can affect the attitude of tourists.

Thematic content analysis was applied to analyze the data from reviews and interviews. The author shared transcription with colleagues who independently categorize the variables. Based on the thematic content analysis, several keywords that affect transformative travel experiences were revealed: personal characteristics, co-travelers, local people, places of visits and destination-based agents, who are ready to share their experience in different fields such as making cheese, sheep rearing, preservation of cultural heritage (collection of platbands), developing of crafts (Gzhel), people who take care of nature and know unique places to visit (Dagestan). Feedback overview was also made with the help of content analysis and frequency analysis. The amount of data obtained during the questionnaire survey is still too small to use complex methods of data analysis, that is why received outcomes require further verification.

4 Results

The results of the analysis of reviews showed that the most common topics are the place of travel, admiration for local agents, deep emotions experienced by tourists (some tourists said that were in fairy tale or a dream), gratitude to the organizers for unique tours and transformational effects that the trip had. Co-travelers, the novelty of knowledge, the expansion of horizons that travelers felt were also often noted. The number of mentions can be seen on Fig. 1.

According to earlier researchers, we can also see such factors as deep emotions, transformational effect, description of the place of travel, expansion of horizons and novelty of the acquired knowledge. But there is a high frequency of answers about importance of co-travelers and gratitude to them. In our study, an important factor is

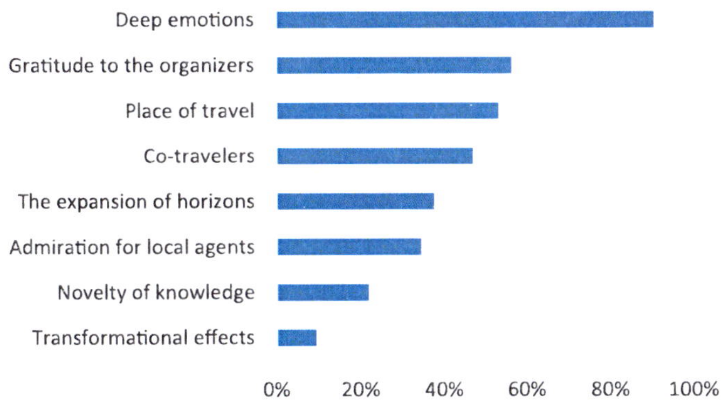

Fig. 1 Frequency analysis of the mentioned words in the feedbacks to tours

also the fascination of local agents. For example, phrases such as "they fall in love with what they are doing", <they> "leave a bright mark in everyone's heart".

In response to the question how much the trip affected you and your initial view of the place of travel, 85% of tourists replied that the effect is significant. At the same time, 65% replied that they were energized by the tour participants, guides, and 10% that they completely changed their perception of the place of travel.

In interview with tourists also the deep feelings that they received from excursions to local agents and people were mentioned: "The people who tell you, they reveal some secret to you, they are not just guides, they are those who want to convey to you all their love that they have for this place, for this business. You may not know some dates, etc., but you become one of them" (female, 41).

As for the co-travelers, people mention that they are not just individual participants in the tour, but "from the very beginning of the journey, you immediately plunge into some kind of unifying process" (female, 41). "There is an element of networking in trips, when we all participate in a challenge, everyone pollinates each other, and these people are usually very interesting. They are non-serial travelers" (male, 40).

The reasons why the participants chose the company for transformational tourism: unusual travel, non-standard tours, unique places, immersion in the lives of other people, a different reality, the opportunity to touch everything with their hands, affordable prices, spontaneity of decisions that are made, impressions that cannot be obtained independently, responsiveness of the organizers, a great travel partners, the opportunity to feel personal uniqueness and get new experiences, a small group of travelers.

For the revealing possibilities of transformational tourism for the sustainable development of regions, we have asked local agents (the head of the unique Gzhel Sea project—dishes with hand-painted authentic paintings and the head of the cheese factory) about their view on the topic. Each of them presents a unique project that is the love of their whole life, and they are ready to talk about it for very long and share their experience with others. They see the preservation of cultural heritage as their personal goal, e.g., "our goal is popularization of the development of Gzhel majolica and preservation of historical heritage. We carry out an educational function and promote projects on the territory of Gzhel, foster the popularization of Majolica" (male, 48). The main motivation for their participation in travel project is intangible benefits—to interest people in what they are doing, share their passion, find the supporters. At the same time, they say that such trips are not a mass product, and it does not bring much income but requires a lot of time and effort: "Physically, it is very difficult, I would not conduct author tours often. The places I show are unconventional, e.g., we organize events with artists who are not always ready to talk to tourists" (male, 48).

Summarizing the obtained results, we can see common features in the behavior patterns of transformative tourism travelers in Turkey, Africa and Russia. These tourists are looking for a unique travel experience that will allow them to see common things in a new way, to become netter version of themselves, they try to feel deep immersion in the local culture, to talk to interesting people in places of travel. Our research additionally revealed the high importance of such a factor as co-travelers, who make trips even more unforgettable and with whom they can share received experience. In addition, interviews with local agents showed that they are primarily motivated by intangible benefits such as recognition and preservation of cultural heritage. Each tour is unique for them, and they believe that this type of travel product cannot be mass produced, because in this case, authenticity will be lost.

5 Conclusion

In this study, we have studied the possibilities of transformational tourism to achieve the goals of sustainable development of territories. The following characteristics of transformational tourism contribute to sustainable development: tolerance to different cultures and nations, preservation of the local heritage and nature, extend the worldview, personal development through self-reflection, critical thinking, emotional connections, foster responsibility and a more careful attitude to the environment and nature, contribute to the small- and medium-sized enterprises development, can improve the place image.

The experience of Turkey, South Africa and Russia was analyzed, which showed many similarities among travelers choosing this type of tourism: cultural and historical interest, education, enthusiasm, middle and high income, having hobbies, convicted in environmental and social justice, with the desire for authenticity".

The following results are of particular interest: the importance of travel partners and the possibility of direct contact with local agents, which allows to dive deeply into the local culture, share views and way of thinking, imbued with ideas, which ultimately leads to a long-lasting transformational effect of the trip.

Interviews with local agents, carriers of unique cultural heritage and knowledge, showed that the main motive for participating in the project for them is the intangible component and they doubt the possibility of scaling such trips to large groups of people. In their opinion, then the journey will lose its transformational effect and will not be so exciting. Thus, we come to the conclusion that although the possibilities of transformational tourism for sustainable development are wide, nevertheless, the uniqueness of trips and need of deep immersion in culture makes it difficult to widely spread such trips and the opportunity to significantly influence the tourism industry. Of course, it is necessary to consider the serious limitations of this study: a small sample of respondents, the qualitative nature and focus on the case of only one company in the Russian market.

It should not be forgotten that other studies have emphasized the need for local government agencies to support such transformative tourism projects, stressing that they will help to preserve an original culture, facilitate nature-based tours, promote environmentally friendly destinations, foster the development of small- and medium-sized enterprises, increase residents' interest in their own culture and bring additional income to the budget.

Acknowledgements This research was funded by the Lomonosov Moscow State University's Faculty of Economics under the research project "Entrepreneurial networking as a self-organization tool for the sustainable development of municipalities".

References

Abbas, J., Mubeen, R., Iorember, P. T., Raza, S., & Mamirkulova, G. (2021). Exploring the impact of COVID-19 on tourism: Transformational potential and implications for a sustainable recovery of the travel and leisure industry. *Current Research in Behavioral Sciences, 2*, 100033.

Ateljevic, I. (2020). Transforming the (tourism) world for good and (re) generating the potential 'new normal. *Tourism Geographies, 22*(3), 467–475. https://doi.org/10.1080/14616688.2020.1759134

Ateljevic, I., Sheldon, P., & Tomljenovic, R. (2016). *Global report on the transformative power of tourism.* UNWTO.

Brown, L. (2009). The transformative power of the international sojourn: An ethnographic study of the international student experience. *Annals of Tourism Research, 36*(3), 502–521.

Bruner, E. (1991). Transformation of self in tourism. *Annals of Tourism Research, 18*, 238–250.

Cetin, G., & Bilgihan, A. (2016). Components of cultural tourists' experiences in destinations. *Current Issues in Tourism, 19*(2), 137–154.

Kirillova et al., 2017 Kirillova, K., Lehto, X., & Cai, L. (2017) What triggers transformative tourism experiences?. *Tourism Recreation Research, 42*(4), 498–511

Kottler, J. (1997). *Travel that can change your life - how to create a transformative experience.* Jossey-Bass Publishers.

Martins, M., & Santos, L. (2022). "Transformational marketing and transformational travel". *Journal of Tourism Futures*, Vol. ahead-of-print No. ahead-of-print. https://doi.org/10.1108/JTF-01-2022-0025

Melo, C., Richards, G., & Smith, M. K. (2021). Transformational tourism experiences: A new communication path for destinations and service providers. In M. G. Dinis, L. Bonixe, S. Lamy, & Z. Breda, (Eds) *Impact of New Media in Tourism* (pp. 210-233), Hershey, PA: IGI Global. doi: https://doi.org/10.4018/978-1-7998-7095-1.ch013

Mkhize, D.S.L., & Ivanovic, M. (2020). Building the case for transformative tourism in South Africa. *African Journal of Hospitality, Tourism and Leisure, 9*(4), 717–731. doi: https://doi.org/10.46222/ajhtl.19770720-47

Morgan, A. (2010). Journeys into transformation: Travel to an 'other' place as a vehicle for transformative learning. *Journal of Transformative Education, 8*(4), 246–268.

Noy, C. (2004). This trip really changed me–Backpackers' narratives of self-change. *Annals of Tourism Research, 31*(1), 78–102.

Pala, T., & Cetin, G. (2021). *Tourism & Management Studies, 18*(2), 7–17. https://doi.org/10.18089/tms.2022.180201

Pine, B. J., & Gilmore, J. H. (2011). *The experience economy.* Harvard Business Review Press.

Pizam, A., Fleischer, A., & Mansfeld, Y. (2002). Tourism and social change: The case of Israeli ecotourists visiting Jordan. *Journal of Travel Research, 41*, 177–184.

Pung, J. M., & Del Chiappa, G. (2020). An exploratory and qualitative study on the meaning of transformative tourism and its facilitators and inhibitors. *European Journal of Tourism Research, 24*, 2404–2425.

Pung, J. M., Gnoth, J., & Del Chiappa, G. (2020). Tourist transformation: Towards a conceptual model. *Annals of Tourism Research, 81*, 102885.

Sheldon, P. J. (2020). Designing tourism experiences for inner transformation. *Annals of Tourism Research, 83*, 1–12.

Sheresheva, M., Efremova, M., Valitova, L., Polukhina, A., & Laptev, G. (2021). Russian tourism enterprises' marketing innovations to meet the COVID-19 challenges. *Sustainability, 13*, 3756. https://doi.org/10.3390/su13073756

Soulard, J., McGehee, N., & Knollenberg, W. (2020). Developing and testing the transformative travel experience scale (TTES). *Journal of Travel Research, 60*(5), 923–946.

Tasci, A. D., & Godovykh, M. (2021). An empirical modeling of transformation process through trip experiences. *Tourism Management, 86*, 104332.

Tomljenovic, R., & Ateljevic, I. (2015). "Transtourism–operational definition and typology of transformative travel experiences–working paper No 2", doi: https://doi.org/10.13140/RG.2.1.1758.6325. Retrieved August 29, 2022, from https://www.researchgate.net/publication/299489365_Transformative_tourism_WORKING_DEFINITION_AND_TYPOLOGY_OF_TRANSFORMATIVE_TRAVEL.

Transformational Travel Council. (2021). A more mindful approach. Retrieved August 29, 2022, from https://www.transformational.travel.

United Nations World Tourism Organization [UNWTO]. (2016). The transformative power of tourism: A paradigm shift towards a more responsible traveller. Affiliate members global reports, volume 14. Retrieved August 29, 2022, from https://doi.org/10.18111/9789284417834.

Vărzaru, A. A., Bocean, C. G., & Cazacu, M. (2021). Rethinking tourism industry in pandemic COVID-19 period. *Sustainability, 13*, 6956. https://doi.org/10.3390/su13126956

Wolf, I. D., Ainsworth, G. B., & Crowley, J. (2017). Transformative travel as a sustainable market niche for protected areas: a new development, marketing and conservation model. *Journal of Sustainable Tourism, 25*(11), 1650–1673. https://doi.org/10.1080/09669582.2017.1302454

Sustainability Practices in Events' Organization in Lisbon. Empirical Study of the Rock in Rio Music Festival

Dália Liberato, Elga Costa, Cláudia Moreira, Pedro Liberato, and Joaquim Ribeiro

Abstract In recent years, the number of music festivals held throughout the world has increased significantly. These festivals are diverse and feature a variety of musical genres, and some aim to celebrate local culture and traditions. Therefore, several reasons drive a visitor to participate in the event, and the impacts that such a large event produces on the destination where it takes place are unquestionable. This research aims to study music events, in particular the Rock in Rio festival, and to understand the motivations that lead visitors to participate in this event, as well as the impacts at the destination where it is held. From this perspective, the main objective of this study is to identify main sustainable practices in the Rock in Rio event and their impact on the destination. A qualitative methodology was used engaging members of the Turismo de Portugal (national DMO), VisitLisboa (regional DMO), and hotel units in the city of Lisbon. Considering that social sustainability is a topic still underdeveloped in scientific studies, particularly when related to music events, this study will bring new contributions. The main results suggest that the Rock in Rio event impacts Lisbon, positively and negatively, as a destination, with the most evident positive impacts related to the city destination's promotion, supply development, and the image and reputation creation. Even though the sustainability

D. Liberato (✉) · E. Costa · C. Moreira · P. Liberato · J. Ribeiro
School of Hospitality and Tourism, Polytechnic Institute of Porto, Vila do Conde, Portugal
e-mail: dalialib@esht.ipp.pt

E. Costa
e-mail: elgacosta@esht.ipp.pt

C. Moreira
e-mail: claudiasgm98@hotmail.com

P. Liberato
e-mail: pedrolib@esht.ipp.pt

J. Ribeiro
e-mail: jribeiro@esht.ipp.pt

D. Liberato · E. Costa · P. Liberato · J. Ribeiro
CiTUR researcher (Centre for Tourism Research, Development, and Innovation, Vila do Conde, Portugal

© The Author(s), under exclusive license to Springer Nature Switzerland AG 2023

V. Katsoni (ed.), *Tourism, Travel, and Hospitality in a Smart and Sustainable World*, Springer Proceedings in Business and Economics,
https://doi.org/10.1007/978-3-031-26829-8_9

approach is already in everyday life, sustainable practices are not as evident in events as in destinations, particularly in the hotel sector. However, they are already defined as a valued element for choosing destinations and events.

Keywords Events · Events tourism · Rock in Rio · Music festivals · Portugal

JEL Classification Z32

1 Introduction

Events play an important role in encouraging tourism and developing a destination and its marketing plan, which results in increasing its competitiveness. They are highly valued as attractions, catalysts, and local marketers, helping community building and urban renewal, and developing local culture, thus building an image of the destination (Getz, 2008). The attractiveness of events is intrinsically linked to the fact that they are unique experiences. This uniqueness is due to several factors, such as the place where the event is held, the participants and organizers involved, and the event management (Getz, 2008). Several businesses operate year-round, which creates a need for adaptation and flexibility on the part of attraction managers (By & Dale, 2008), by developing products and marketing strategies to make the offer increasingly varied and appealing, in order to attract potential visitors in low seasons (Connell et al., 2015). Cultural events work as economic and urban revitalizers (Finkel & Platt, 2020), by enhancing its image and repositioning it in world tourism markets. These events also result in the generation of profits for the tourism sector and local investment (Liu, 2014). Festivals, fairs, and carnivals are some examples of cultural events that provide new sensations and arouse emotions in their participants. As such, they are a highly important tool in urban development and in changing the negative image of the destination among visitors to the destination (Finkel & Platt, 2020).

Festivals, associated with a cultural aspect, also encompass social, economic, and political aspects (Mair, 2018). They act as a link between the community and the destination (Finkel & Platt, 2020) highlighting the heritage and local cultural traditions (Derrett, 2004). The attractiveness of these events is perceived by the community as a way to solve problems such as community spirit and pride, heritage, investment and economic development, employability, and urban renewal (Derrett, 2004), so a large part of the festivals are currently included in marketing plans, promoted in social networks, which will increase the number of participants and their consumption, benefiting both the destination and the events (Finkel & Platt, 2020). Entertainment events, including music events such as concerts and festivals, attract large numbers of participants from different demographic and socio-economic profiles (Llopis-Amoró et al., 2019). The use of social networks as a tool for dissemination by the destination and the event organizers allows the public to connect more easily and share opinions and experiences about it (Brown et al., 2019; Monteiro et al., 2019).

The main objective of this study is to identify sustainable practices in the Rock in Rio music festival and their impact on the destination, through a qualitative methodology research. The following section discusses the literature review, and the methodology chapter will address the qualitative research instrument. The two final sections will discuss the results and focus on the main research conclusions.

2 Literature Review

The exponential growth experienced in recent years by music festivals in the tourism sector is due to the contribution of this event typology to destination development (Liberato et al, 2020; Pavluković et al., 2017), such as urban revitalization, repositioning of the destination in tourism markets, its economic restructuring and the construction of a new tourism image (Cardoso et al., 2019; Che, 2008; Liu & Chen, 2007; Pavluković et al., 2017), and also an increase in demand for tourism services when they occur (Lee et al., 2015).

Considering the size of this event, the number of people involved is vast and distinct, ranging from stakeholders, namely, government officials, private companies, non-profit organizations, artists, technical staff, managers, equipment suppliers, volunteers, sponsors, promoters (Brown et al., 2019; Lee et al., 2015) and residents and local business owners (Lee et al., 2015). Thus, participation in an entertainment event can be direct or indirect, since residents and local business owners may not attend the festival but are part of its environment (Lee et al., 2015). Participation in a festival encompasses the creation of cultural, human, and social relationships (Forga, 2014). It is an overall experience in which social experience (Borges et al., 2020) and socialization are one of the most important factors due to the opportunity to share beliefs, values, identities, experiences, and traditions, thus creating a temporary sense of community (Brown et al., 2019), while experiencing diverse cultural events. To Pérez Gálvez et al. (2021), the segmentation of participants in an event is essential to adapt the offer. To ensure that the offer is varied and meets the consumption practices of the various participants of the event, the processes of collaboration and co-creation are of high importance, and they do not interfere with the experience of the participants. The addition of the festival location to the participants' perceived experience, as well as the participants' contribution to the dynamic, performative, and communicative festival environment, results in the creation of socio-spatial authenticity (Szmigin et al., 2017).

2.1 Economic Impacts

According to Yürük-Kayapınarn et al. (2017), events can impact destinations economically, environmentally, politically, socially, and culturally and can be negative or positive (Getz e Page, 2016). The most evident is the positive ones, generally associated with the economic and social aspects (Moisescu et al., 2019; Yürük-Kayapınar et al., 2017), by attracting resident and non-resident visitors who induce investment and financial resources to the community, resulting in economic benefits to the destination (Popescu & Corboş, 2012; Reid, 2004). Mega-events are considered increasingly important for improving the performance and competitiveness of a destination, being a fundamental part of a destination's development and marketing strategies (Mirzayeva, et al, 2020). Urban revitalization is one of the most significant impacts for a destination as the host of an event, due to the stimulation in urban infrastructure investment (Mirzayeva et al., 2020); the construction of event-specific infrastructure is perceived by a portion of residents as an unnecessary financial investment due to the use of public funds for the benefit of event organizers and managers, usually associated with elites (Liu & Chen, 2007; Moisescu et al., 2019; Yolal et al, 2016). Holding an event results in increased employment. From the preparation phase to the post-event phase, there is a high demand for labor. The construction of infrastructures, the assistance during the event of technical teams, security guards, cleaning, and the sale of food and souvenirs are some examples of activities that contribute to the increase of employability (Mirzayeva et al., 2020; Yolal et al., 2016). Events can also economically impact a destination in terms of business opportunities, tax revenues for the local government, and revenues for the local community and businesses (Gursoy et al, 2004; Yolal et al., 2016).

2.2 *Social, Cultural, and Political Impacts*

The hosting of an event results in impacts, both positive and negative, on the quality of life of the residents of the destination where it takes place (Yürük-Kayapınar et al., 2017). The development of infrastructure, which after the event can be used for other community activities, and the employment growth are two examples of impacts that contribute to the community at the economic and social level (Mirzayeva et al., 2020; Moisescu et al., 2019; Pavluković et al., 2017; Wilson et al., 2017; Yolal et al., 2016). The improvement of infrastructures such as accessibility and water and sanitation systems are some benefits associated with the development of infrastructures that not only improve the quality of life of the local community but also enhance the experience of tourists and visitors (Reid, 2004).

Events result in benefits for the local community. The creation of a sense of community pride (Gursoy et al., 2004; Mirzayeva et al., 2020; Yürük-Kayapınar et al., 2017), which results in residents interacting, and the strengthening of cultural and social identity lead to the development of social cohesion (Chi & Qu, 2008; Gursoy et al., 2004; Reid, 2004), sense of community improving the quality of life for local residents (Yürük-Kayapınar et al., 2017), and also the revitalization of local traditions (Mirzayeva et al., 2020). Socialization as well as the opportunity to participate in recreational activities and the creation of an atmosphere of celebration are other benefits, which come from holding events, for the local community (Reid, 2004). Participation in organizing events by local residents results in the acquisition of new skills, which contributes to a more positive perception of the event (Acordia & Whitford, 2006; Reid, 2003). According to Getz (2010), events, particularly festivals, can improve the image of the destination, contribute to the conservation of culture, art, heritage, and nature, develop the community, and lengthen the tourist season. In this way, it is possible to maintain tourist activities and tourist flow during low seasons (Connell et al., 2015).

On the other hand, events can negatively impact a destination and the social life of the local community. Increased traffic, noise, litter, and crowding are some examples of negative impacts, as well as loss of amenities and increased prices at the level of services and food (Acordia & Whitford, 2006; Pavluković et al., 2017; Reid, 2004). Increased crime, alcoholism, sexual exploitation, drug and other substance abuse, xenophobia, and abusive behaviors are other negative aspects resulting from hosting an event that can affect the public's perception (Acordia & Whitford, 2006; Getz & Page, 2016; Moisescu et al., 2019; Pavluković et al., 2017; Reid, 2004), affecting the image of the destination and the event, resulting in a decrease in community support (Pavluković et al., 2017). Politically, events can be supported by local authorities through investments for the restoration of infrastructure, historical sites, preservation of natural resources and local culture, and the creation of specific laws and regulations (Chen, 2011; Moisescu et al., 2019). A festival that runs smoothly results in satisfied consumers with the intention of repeating the event and revisiting the destination and in a greater dissemination of the event.

2.3 *Environmental Impacts*

The high concentration of individuals in a destination, as a consequence of holding an event, creates pressure in infrastructures and local environment (Wu et al, 2016). Over the years, environmental strategies have become a factor of high importance for event sustainability management plans (Mirzayeva et al., 2020). Defining actions will minimize negative environmental impacts resulting from an event (Parkes et al., 2016).

Considering the preparing processes of an event, the course of the event, and the post-event venue renovation, noise and air pollution, the use of resources and energy, the high portions of construction materials, and the waste generated are some of the resulting environmental impacts (Getz, 2007; Mirzayeva et al., 2020). Getz (2007) presents a diversity of responses that destinations and events can adopt in order to minimize the negative environmental impacts of an event: Avoid sensitive areas when selecting the space for the event; use part of the investment to clean up affected areas; regulate the development of environmental sustainability; impose design standards; impose an accounting life cycle; concentrate events; reinforce smaller scale events; favor/require public transportation; adoption of sustainable practices (Reduce, Reuse, Recycle); cleaning of spaces after events; educate the visitors; community-based planning for the accommodation and regulation of events; improved support and pressure for environmental management; pressure from special interest groups; contribution from various stakeholders; shaping politics and public opinion, through the press.

Festival attendees, unlike the destination community, are more concerned about the environment inside/intrinsic to the event, while residents show more concern regarding the physical environment outside the event. Since the environment intrinsic to the event directly influences the experience and satisfaction of attendees, they show more concern about the layout of the venue, queuing efficiency, security measures, and other environmental aspects such as litter, noise, and traffic (Acordia & Whitford, 2006; Gursoy et al., 2004; Lee, 2016).

Active government support for the creation of a desirable destination and event environment is one of the key factors in achieving a satisfactory experience for tourists. Such support is needed to control and monitor the negative aspects that may arise from a festival, such as noise, increased traffic, environmental problems, and waste (Acordia & Whitford, 2006; Lee, 2016; Moisescu et al., 2019). Green practices and standards for events and event venues are increasingly implemented by organizations and event management (Getz & Page, 2016).

Sustainable development is defined as development that involves the protection of the environment and natural resources, the well-being of the local community and respect for its culture, and the creation of economic benefits for the destination's *stakeholders,* thus ensuring a positive outlook for future generations (Fairer-Wessels & Malherbe, 2012; Wise et al, 2019; Wise & Harris, 2017).

The key concepts that constitute the Triple Bottom Line are three: economic responsibility, social responsibility, and environmental responsibility. These are considered to be the basic pillars of sustainability, playing a key role in improving the image, competitiveness, and sustainability of a tourism event and destination. The three concepts (economic, social, and environmental responsibility) are also called the three P's of sustainability (*People, Planet,* and *Profit*) (Fairley et al, 2011; Wise, 2020).

2.4 Economic Sustainability

The main purpose of holding an event relates to the goal of gaining profits and other economic benefits (Getz, 2008). A successful event results in motivating tourists and, consequently, increasing their stay. Currently, sustainable development is at the core of the structural reorganization of the economic system. A balanced socio-economic system implies overcoming inequality and promoting economic development, through the efficient use of available resources, based on the interests of the local community. Thus, sustainable development, at the socio-economic level, implies increasing the quality of life and well-being of international entities with the influence of the resources necessary for economic growth (Dalevska et al., 2019).

2.5 Social Sustainability

A social sustainable event builds trust and creates a sense of belonging among the community, preserving local traditions and heritage (Gursoy et al., 2004; Mirzayeva et al., 2020). Festivals are increasingly an example of social sustainable events due to the development of a sense of community, identity and local pride (Bres & Davis, 2001) a shared sense of belonging and identity, and social cohesion (Derrett, 2004; Duffy & Waitt, 2011; Getz, 2008). According to Dempsey et al (2011), there are several factors that contribute to social sustainability, among them: education, health, quality of life and well-being, employment, fair distribution of earnings, social inclusion, social interaction and cohesion, local democracy, and social justice. According to Vallance et al (2011), social sustainability can be considered in three different ways: *Development sustainability*, associated with basic needs, capital creation, and social justice, among others; *Bridge sustainability*, with respect to the behaviors adopted to achieve environmental goals; and *Maintenance sustainability*, with respect to the preservation of socio-cultural characteristics.

Events and festivals contribute to the preservation and celebration of culture (Derrett, 2004) and values (Kim et al, 2006). The adoption of values intrinsic to festival participants, by the event organization, results in increased support from consumers (Grappi & Montanari, 2011; Thrane, 1997) and other stakeholders thus legitimizing the festival organization. Voss et al (2006) and Chaney & Martin (2016) argue that the core values of festivals are important for those who participate since they help to build and maintain a relationship between the festival and those who participate in it. Shared values are ideas, beliefs, and meanings that are accepted by both festival participants and festival organizations (Chaney & Martin, 2016). Several authors argue that there is a direct relationship between a visitor's loyalty to the event and the creation of shared value by the visitor. Due to the constant change and renewal of music festivals, obtaining attendee loyalty proves to be a difficult process (Chaney & Martin, 2016).

2.6 Environmental Sustainability

Sustainability includes concerns regarding global warming and natural resource degradation. Governments, communities, and organizations are increasingly pressured to decrease negative environmental impacts by increasing the environmental protection of destinations through sustainable tourism development strategies and principles (Dickson & Arcodia, 2010). Over the years, the events sector, as part of the tourism industry, has generated substantial amounts of waste, negatively impacting the environment. In order to minimize negative impacts on natural and sociocultural environments, organizations and sponsors need to understand the importance of sustainable practices in events in order to achieve long-term success (Fairer-Wessels & Malherbe, 2012).

Events can act as sources of awareness (Dugalić, 2020), environmental responsibility, and understanding among tourists, local community, and tourism sector. Sustainable practices at events influence the development of environmental-friendly transport systems and infrastructure, waste management, recycling, *ecofriendly* energy sources, and environmental improvement (Collins et al, 2009).

Considering the demand attending festivals that crowd into a restricted space, there are several resulting impacts, among them: pollution, waste of water and resources, noise, and congestion (Collins et al., 2009). Dugalić (2020) argues that the largest amounts of waste associated with tourist events relate to organic waste (paper), plastic, metal, and glass (consumption of food and beverages), all of which can be recycled. The use of non-renewable natural resources and carbon dioxide emissions is other examples of negative impacts on the ecosystem when events such as festivals are held, which, over the years, induces climate change (Collins et al., 2009). In order to minor all these impacts, the tourism events' sector should adopt quality management strategies associated with sustainability factors, considering the number of tourists at the destination compared to the number of local residents, water supply, waste treatment through recycling, and raising awareness about waste (Dugalić, 2020).

3 Methodology

Qualitative methodology comprises the research of human experiences in social and personal contexts, allowing researchers to understand the factors that influence these experiences (Bricki & Green, 2007; Gelling, 2015). This methodology not only seeks a better understanding of human experiences, but also aims to understand human beliefs, attitudes, behaviors, and interactions. To carry out this study, we chose to conduct semi-structured interviews. All interviewees were selected based on the relevance of their functions, as well as of the institutions/organizations to which they belong, regarding the possibility of providing a valid contribution in terms of knowledge of the reality under study (Table 1).

To conduct this study, several objectives were set based on the several topics addressed in the literature review.

The general objective of this study is to understand the relationship between the existence of sustainable practices at the Rock in Rio event and the motivations to participate in the festival. To complement this objective, four specific objectives were defined (Table 2). The information collection for this research was based on a semi-structured interview with 14 open-ended questions (Table 3).

Table 1 Interviewees

Interviews	Entity
E1	Directorate of investment support (Tourism of Portugal)
E2	Hotel unit #1
E3	Directorate of supply enhancement (Tourism of Portugal)
E4	Operations director (E. R. Turismo do Porto e Norte de Portugal)
E5	Hotel unit #2
E6	Hotel unit #3
E7	Brand activation director (Ágora)
E8	Hotel unit #4
E9	Hotel unit #5
E10	Rock in rio organization

Table 2 General objective and specific objectives

General objective	*– To understand the relationship between the existence of sustainable practices at the Rock in Rio event and the motivations for attending the festival*
Specific goals	– Identify the profile of tourists who participate in the Rock in Rio event – Evaluate the sustainable practices at the Rock in Rio event – Identify the impacts of the Rock in Rio event on the tourist destination – To ascertain the possible relationships between the sustainable practices applied at the festival and the attendees' behavioral intentions toward the event and the destination

4 Results

4.1 Event Planning and Organization

The questions related to the planning and organization of events ("What is the intervention of the regional and national DMOs in the planning, organization, dynamization, and promotion of music events at the destination?"; "What characteristics differentiate the City of Lisbon from other places where the Rock in Rio event takes place?") at the Lisbon destination were aimed not only at the DMO but also at the organizers of the Rock in Rio event in order to understand what connection these two entities have with each other and how the whole planning process is carried out, as well as how the City of Lisbon differentiates itself as a destination as host of this event.

The DMOs help to promote and market the events by consistently presenting the different hotel or other accommodation options, transportation services for example, restaurants, among others. (E1), ... *guaranteed accessibility and adjacent offers*

Table 3 Theoretical basis of the interview guide

Interview questions	Theoretical framework
Objective 1: Identify the impacts of the Rock in Rio event in tourism destination	
1. What is the intervention of the regional and national DMOs in planning, organizing, energizing, and promoting music events at the destination?	Che (2008); Connell et al. (2015); Getz (2008); Getz and Page (2016)
2. What are the characteristics that differentiate the city of lisbon from other places where the rock in rio event takes place?	
3. What do you think are the positive impacts of rock in rio for the lisbon destination? and the negative impacts?	Acordia and Whitford (2006); Getz (2007); Getz (2010); Getz e Page (2016); Gursoy et al. (2004); Liu and Chen (2007); Mirzayeva et al. (2020); Moisescu et al. (2019); Parkes et al. (2016); Popescu and Corboş (2012); Reid (2004); Wilson et al. (2017); Wu et al. (2016); Yolal et al. (2016); Yürük-Kayapınar et al. (2017)
4. How has lisbon's destination changed due to the successive editions of rock in rio?	
5. In your opinion, what are the contributions of the festival to the city/destination?	
6. How does tourist demand evolves in rock in rio years, compared to years when the event does not take place? and in the following years?	Connell et al. (2015); Lee et al. (2015)
7. What activities emerged at the destination because of the event, in terms of differentiated offer?	Crompton and McKay (1997); Gursoy et al. (2004); Mirzayeva et al. (2020); Popescu and Corboş (2012); Reid (2004)
8. What are the objectives that the destination intends to achieve through the realization of rock in rio? (disclosure, destination growth in the tourism market, increased demand in specific issuing markets, improvement of the local economy, urban revitalization, infrastructure improvement, resident community satisfaction, and greater involvement and empathy with the tourism and events' sector, among others)	Acordia and Whitford (2006); Getz (2010); Gursoy et al. (2004); Liu and Chen (2007); Mirzayeva et al. (2020); Moisescu et al. (2019); Pavluković et al. (2017); Reid (2003); Yolal et al. (2016); Yürük-Kayapınar et al. (2017)
Objective 2: To identify the profile of tourists who participate in the rock in rio event	
9. In your perspective, do the tourists who come to the destination for the event repeat the destination? do they repeat the destination or the event?	Christou et al. (2018); Grappi & Montanari (2011); Yoon et al. (2010)
10. Are questionnaires administered to tourists? are the results/opinions expressed discussed?	Alegre and Garau (2010); Pivac et al. (2019); Wong & Law (2003)
11. Do they have access to data about the degree of satisfaction with the event and the destination?	Pivac et al. (2019); Wang & Leou (2015)
Objective 3: Evaluate the sustainable practices in the rock in rio event	

(continued)

Table 3 (continued)

Interview questions	Theoretical framework
12. What sustainable practices have been implemented in the various editions of the event that contribute to the sustainability of the destination?	Bres e Davis (2001); Collins et al. (2009); Dickson and Arcodia (2010); Dugalić (2020); Fairer-Wessels and Malherbe (2012); Fairley et al. (2011); Vallance et al. (2011); Wise (2020)
13. What are the sustainable practices implemented by the destination?	
Objective 4: To evaluate the possible relationship between the sustainable practices applied at the festival and the participants' behavioral intentions toward the event and the destination	
14. What is your opinion regarding the existence of a relationship between the motivations to visit rock in rio and the sustainability practices carried out in the event?	Bres e Davis (2001); Chaney e Martin (2016); Collins et al. (2009); Derrett (2004); Duffy e Waitt (2011); Dugalić (2020); Fairer-Wessels e Malherbe (2012); Getz (2008); Gursoy et al. (2004); Mirzayeva et al. (2020).

(accommodation, means of communication...) (E1), (…) *more music events are used as instruments of tourism promotion, and DMOs play an important role in several moments of its realization.* (E3). *In my opinion the main factors that differentiate Lisbon are: safety, genuine/inclusive people who know and like to receive, good accessibility.* (E2)

Lisbon has its own characteristics that make everything that happens inside and outside the Rock in Rio space unique (…) Lisbon is one of the main destinations for events. (E3)

A greater articulation with the event organizers, in the logic of elaborating packages where, besides lodging and transport, other activities are included, as well as entrance to the event. (E3)

DMOs contribute not only to the promotion of the event but also to the presentation and promotion of other tourist services complementary to the stay of a participant of the event under study, which makes this a leading destination in the European Union. According to interviewees E3 and E4, it is necessary to develop a higher level of partnerships between DMOs, hotel units, and event organizers in order to attract more visitors to both the event and the destination, thus developing the region's tourism sector.

4.2 *Impacts of the Event at the Destination*

It was intended to find out which is the predominant typology of impacts (positive or negative) and in what way they occur in the City of Lisbon and in the hotel sector (Getz e Page, 2016). Thus, concerning the questions "What do you think are the positive impacts of the Rock in Rio festival for the Lisbon destination? And the negative impacts?", the following answers were obtained from the DMO and hotel units, respectively:

National and international projection and notoriety, as a destination city, will be positive; negative will be the tourist pressure and the volume of demand for the event. (...) expands knowledge and projection of the city as a strong capital of a European Union country, with the image it carries of a leading destination, with vast accessibility and adjacent offers for tourism. The event enhances the destination as an aggregator of cultural supply and attractiveness, (...) transportation and accommodation, among other activities. The growth has been constant in the city of Lisbon, not only as a result of the event. (E1)

When we make an analysis of all events of this nature and scale, they have contributed to a greater awareness of the destination in the music scene and helped build the image of a "cool" destination and attractive to a foreign audience, especially younger age groups, which contribute to the sustainability of demand for the destination. (E2)

I think that the destination has changed over the last few years, as a result of better external communication, which goes "hand in hand" with the valorization of the destination by all the stakeholders in the sector. (E2)

The festival contributes to the image of a city capable of organizing events of international dimension and quality and that helped Lisbon to reposition itself as a "trendy" destination among various publics/age groups and in relation to quite consolidated destinations (especially European ones). (E2) *It is an important tourist promotion factor that takes the name of Lisbon to many destinations and helps to enhance the visit to the city either to participate in music festivals or to perform other activities.* (E3)

Lisbon has not changed due to the event, but over the various editions of the event it has asserted itself as a cosmopolitan city, but with a "human/visible" scale for those who seek it. Today Lisbon is known for several factors, but turistically for being a city/destination that has a significant variety of offers, spread over an increasingly decentralized geographical area (Arrábida, Sintra, South Bank of the Tagus, among others). (E3)

(...) it has a huge impact, especially when we are talking about an urban center that hosts many events, from sports, music, art, congresses, conferences, professionals, events closed to companies... There is a giant myriad of types of events that take place every year (...) that affect us directly. (E5) *(...) the festivals, sporting events, cultural events too (...) Congresses too, of course, MICE are also events that often happen in the city and that end up affecting us positively, once that they bring us occupation.* (E6) *We, hotel units, are winning (...) festivals are no longer specifically for an ultra-young public, the public is very varied.* (E5)

The Rock in Rio event contributes positively and negatively to the destination, once it leads to its national and international dissemination, as well as to the development of the supply at the cultural level, transportation, among other tourism services (Mirzayeva et al., 2020; Moisescu et al., 2019; Pavluković et al., 2017; Wilson et al., 2017). In addition to the contribution of Rock in Rio as a destination attraction, other events of equal size attract large numbers of visitors to the destination, such as the

Web Summit (E8), other festivals like NOS Alive and Super Bock Super Rock (E5). One of the most negative aspects is related to the *tourist pressure and volume of demand* (E1), evident at the destination.

4.3 Sustainable Practices

Sustainable development involves the protection of the environment and natural resources, the well-being of the local community and respect for its culture, and the creation of economic benefits for the destination *stakeholders,* thus ensuring a positive outlook for future generations (Fairer-Wessels & Malherbe, 2012; Wise, et al, 2019; Wise & Harris, 2017). Thus, sustainable practices are an important topic that still lacks studies in the events sector, especially in terms of social sustainability. The following questions were presented: "What sustainable practices have been implemented in the various editions of the event that contribute to the sustainability of the destination? And what are the sustainable practices implemented by the destination?".

Gender equality, circular economy, rationalization in water consumption. Although the negative impacts in environmental terms outweigh the notion of sustainability that we try to promote. (E1) *Encouraging the use of bicycles, less polluting public transportation or with zero pollution, stimulating the circular economy, accommodation and other tourist equipment with good sustainable practices in terms of water consumption, energy saving, among others* (E3) *Rock in Rio is an event that was born with the purpose of "a better world". It was like that in '85, when the event was born by the hand of a dreamer, passionate about the city he lived in, and who idealized a project capable of attracting tourism to, in this way, generate revenues and positive impact for the city. This is how Rock in Rio continued to be until 2001, when the Social Project was born, as a way for the festival to affirm its performance in sustainability. Since then, it has established itself as the first major music event in the world to offset its Carbon Footprint (15 years ago) and to be ISO 20121-Sustainable Events certified. The festival's efforts towards sustainability have not stopped here and its legacy already includes milestones such as having reduced fuel consumption by 62%-even having doubled its size-, being Zero Waste in the Lisbon editions and having reduced 10 tons of waste in 2018 with the adoption of the reusable cup, having facilitated access to health care for more than 500.000 people or having invested 35 million Euros in social and environmental projects benefiting 1 million people, supported more than 200 entities, funded 28 UNESCO projects and assist 56,000 beneficiaries annually, all over the world. We promote a safe event, considering health, safety, comfort, accessibility, equality, inclusion and well-being of all, complying in addition with all applicable Human Rights, Labor Legislation and other relevant Regulations, aiming at best practices. We promote through example and dissemination the construction of sustainable practices that generate economic, social and environmental value, directly or indirectly, to all stakeholders in order to achieve our purpose of a "Rock in Rio for a Better World". We guarantee the design and implementation of engagement processes, content programming, entertainment and information provision in a way that is inclusive and accessible to all stakeholders in the Rock in Rio experience. We guarantee transparency and integrity in our business transactions, institutional and labor relations. We are committed over the long term to promoting and maintaining management based on continuous improvement, providing the resources and knowledge necessary to meet Rock in Rio's sustainability goals.* (E10)

Thus, in all editions of RiR there are campaigns about sustainable mobility, a mobility and accessibility plan for the public is developed, as well as a sustainability plan for the organization, sponsors and suppliers, as well as awareness campaigns about good sustainability practices developed for artists, sponsors, suppliers, public and community (...). In addition to all this there is also a demanding waste management plan with materials' donation at the end of the event and the donation of food remains in good condition (Lisbon and Las Vegas). All artists receive a Zero Carbon certificate as well as an award for sponsors and suppliers with best practices of sustainability in the City of Rock. (E10)

In each edition of the festival several sustainability goals are outlined. Already for the 2022 edition the goals are: leave a positive social impact on the local community; improve the look of plurality and accessibility; reduce waste generation in the City of Rock; increase recycling and recovery; combat food waste. (E10)

Despite the various practices adopted based on sustainability, the negative impacts still outweigh these: *Although the negative impacts in environmental terms outweigh the notion of sustainability that one tries to promote* (E1). Another perspective refers to the adoption of other means of transportation for commuting, such as the bicycle and *less polluting public transportation or with zero pollution* (E3). Saving water and energy is sustainable practices also adopted by the destination and its visitors in order to reduce the ecological footprint of each individual (Collins et al., 2009; Dickson & Arcodia, 2010; Dugalić, 2020). Throughout the Rock in Rio event editions, several social, economic and environmental sustainability practices have been adopted based on its main logo "*for a better world*", being also an ISO 20121—Sustainable Event certified event (E10). In addition to the sustainability programs created (For a Better World; Social Project; Zero Waste), Rock in Rio establishes objectives and targets for sustainability. In the next edition to be held in Portugal (2022), there are already several goals set by the organization: *leave a positive social impact on the local community; improve the look of plurality and accessibility; reduce waste generation in the City of Rock; increase recycling and recovery; combat food waste* (E10).

4.4 Motivations Versus Sustainable Practices

There are several participants who visit a destination and, more specifically, the Rock in Rio event. To find out if sustainable practices are a factor that contributes to the election of the event and the destination, the following questions were developed:

"What is your opinion about the existence of a relationship between the motivations to visit Rock in Rio and the sustainability practices carried out in the event?".

The word sustainability has gone from being a "buzz word" to something that is increasingly present in our daily lives (...) it is already a very valued element at the moment of decision of participating or not in an event and that in the future it will be a factor "required" by everyone. (E2)

4.5 Behavioral Intentions Relative to the Event and/or Destination

Satisfaction is one of the factors that influence consumer loyalty and loyalty. To confirm if satisfaction influences the repetition of a destination or event, the following question was applied: "In your perspective, do tourists who come to the destination for the event repeat the destination? Do they repeat the destination or the event?".

Both are repeated, because Lisbon can show that there is much to know beyond the event to those who only go to the event. As the event has a strong commercial component, it is a catalyst for the repetition of the same event, and there are many people who repeat it every time it happens. (E1) *There are many probabilities of both dynamics happening, that is, those who come to the event will probably want to repeat it and find out more about the destination and those who come to the destination and end up going to the event will also want to come back.* (E2) *Nowadays it is more difficult to retain the loyalty of tourists to destinations, however this type of events can combat this issue in two aspects: a) young public who may want to return later; b) tourists who come for the event but are curious to know what they could not see during their stay.* (E3) *Customers come to attend a certain event, or come to participate in a certain event, and because of the good experience they had as participants of the event, the perception they have of the destination ends up being good, it increases its notoriety, which will really attract, or will make the customer have the desire to return to the destination and enjoy the destination now in a more relaxed way.* (E9) *Lisbon and Portugal have that advantage. It's the climate, (...) the proximity between things that, normally, opens the appetite for people to come back after the events.* (E8) *In general, Rock in Rio spectators repeat their presence in more than one edition.* (E10)

Through this question it was possible to conclude that both options are valid (E2) and that *Lisbon has the ability to show that there is much to know beyond the event to those who only go to the event* (E1). The event, in turn, is *a catalyst for its repetition* (E1). A good experience regarding the event is reflected in a good impression of the destination, which results in an intention to revisit the destination by the destination (E9). Regarding the organization of the event, most event participants repeat the event (E10). *Yes, in all editions we conduct satisfaction surveys in the festival grounds, which allow us to assess how the public perceives and lives the entire festival experience, from the moment they leave home until the moment they return, and it is also a useful tool for us to implement improvements and innovations from year to year.* (E10)

E10 states that satisfaction questionnaires are applied to the participants of the event in order to evaluate their experience *from the moment they leave home until the moment they return and* that these are an important tool in the development of improvements in future editions of the festival.

5 Conclusion

DMOs help in the promotion of the Rock in Rio festival, presenting and promoting products and services that complement the experience of its participants, making Lisbon a leading destination in the European Union. Thus, DMOs are key partners for the planning and organization of events at the destination (Connell et al., 2015; Getz, 2008; Getz & Page, 2016).

Regarding the impacts and at a positive level is observed the contribution of the event to the dissemination of the destination, both nationally and internationally, the development of the cultural offer and tourist services such as transportation. Rock in Rio festival also contributes to the creation of a positive destination image (Liu & Chen, 2007; Mirzayeva et al., 2020; Moisescu et al., 2019; Yolal et al., 2016).

Considering the importance of the sustainability theme and the limited number of studies, especially in terms of social sustainability, throughout the various editions of the festival, several sustainability practices have been adopted at the social, economic, and environmental level through a variety of projects (For a Better World, Social Project, Zero Waste). Thus, Getz's theory (2009) emphasized by Fairley et al. (2011) and Wise (2020) considered the basic pillars of sustainability, playing a key role in improving the image, competitiveness, and sustainability of a tourist event and destination: campaigns on sustainable mobility, a mobility and public accessibility plan, a sustainability plan for the organization, sponsors and suppliers, awareness campaigns on good sustainability practices developed for artists, sponsors, suppliers, public and community, a waste management plan, donation of materials at the end of the event, donation of food remains in good conditions (Lisbon and Las Vegas), the attribution of a Zero Carbon certificate for all artists, and the attribution of an award for sponsors and suppliers with best sustainability practices during the festival (Collins et al., 2009; Dickson & Arcodia, 2010; Dugalić, 2020; Fairer-Wessels & Malherbe, 2012).

Considering behavioral intentions of festival and destination visitors, the intention to repeat the destination and event is due to the fact that they are catalysts for repetition and spark interest in those who visit. Satisfied participants develop an emotional connection with the event and the destination, thus becoming loyal to the destination (Christou et al., 2018; Lee et al., 2012). The intention to revisit is influenced by several aspects associated with tourists' experiences, satisfaction, and perceived value (Xu, et al., 2016). Satisfaction surveys are an important tool for obtaining information regarding tourists' opinions about the destination and evaluating a festival's performance (Alegre & Garau, 2010; Pivac et al., 2019).

Regarding the possible relationship between the sustainable practices applied at the festival and the participants' behavioral intentions toward the event and the destination, there may exist a relationship between these two variables once the Portuguese Tourism Supply Enhancement Directorate argues that sustainable practices are increasingly encompassed in events and are an element of high importance in the decision-making process of attending an event.

Acknowledgements The authors acknowledge the financial support of CiTUR, R&D unit funded by the FCT—Portuguese Foundation for the Development of Science and Technology, Ministry of Science, Technology and Higher Education, under the scope of the project UID/BP/04470/2020.

References

Acordia, C., & Whitford, M. (2006). Festival attendance and the development of social capital. *Journal of Convention & Event Tourism, 8*(2), 1–18. https://doi.org/10.1300/J452v08n02_01

Alegre, J., & Garau, J. (2010). Tourist Satisfaction and Dissatisfaction. *Annals of Tourism Research, 37*(1), 52–73. https://doi.org/10.1016/j.annals.2009.07.001

Borges, AP., Cunha, C., & Lopes, J. (2020). The main factors that determine the intention to revisit a music festival. *Journal of Policy Research in Tourism, Leisure and Events, 0*(0), 1–22 https://doi.org/10.1080/19407963.2020.1817051

Bres, K. D., & Davis, J. (2001). Celebrating group and place identity: A case study of a new regional festival. *Tourism Geographies, 3*(3), 326–337. https://doi.org/10.1080/14616680110055439

Bricki, N., & Green, J. (2007). *A guide to using qualitative research methodology*. https://fieldresearch.msf.org/handle/10144/84230

Brown, A., Donne, K., Fallon, P., & Sharpley, R. (2019). From headliners to hangovers: Digital media communication in the British rock music festival experience. *Tourist Studies, 20*, 146879761988595. https://doi.org/10.1177/1468797619885954

By, R. T., & Dale, C. (2008). The successful management of organisational change in tourism SMEs: Initial findings in UK visitor attractions. *International Journal of Tourism Research, 10*(4), 305–313. https://doi.org/10.1002/jtr.663

Cardoso, L., Araújo Vila, N., Almeida, Â. de, & Fraiz Brea, J. A. (2019). Music festivals as promoters of tourism destination image. The case of the Vodafone Paredes de Coura Festival. *Investigaciones Turísticas, 17*, 149–167

Chaney, D., & Martin, D. (2016). The role of shared values in understanding loyalty over time: A longitudinal study on music festivals. *Journal of Travel Research*. https://doi.org/10.1177/0047287516643411

Che, D. (2008). Sports, music, entertainment and the destination branding of post-fordist detroit. *Tourism Recreation Research, 33*(2), 195–206. https://doi.org/10.1080/02508281.2008.11081305

Chen, S. C. (2011). Residents' perceptions of the impact of major annual tourism events in Macao: Cluster analysis. *Journal of Convention & Event Tourism, 12*(2), 106–128. https://doi.org/10.1080/15470148.2011.569877

Chi, C.G.-Q., & Qu, H. (2008). Examining the structural relationships of destination image, tourist satisfaction and destination loyalty: An integrated approach. *Tourism Management, 29*(4), 624–636. https://doi.org/10.1016/j.tourman.2007.06.007

Christou, P., Sharpley, R. A. john, & Farmaki, A. (2018). *Exploring the emotional dimension of visitors' satisfaction at cultural events*. https://doi.org/10.3727/152599518X15173355843389
Collins, A., Jones, C., & Munday, M. (2009). Assessing the environmental impacts of mega sporting events: Two options? *Tourism Management, 30*(6), 828–837. https://doi.org/10.1016/j.tourman.2008.12.006
Connell, J., Page, S. J., & Meyer, D. (2015). Visitor attractions and events: Responding to seasonality. *Tourism Management, 46*, 283–298. https://doi.org/10.1016/j.tourman.2014.06.013
Crompton, J. L., & McKay, S. L. (1997). Motives of visitors attending festival events. *Annals of Tourism Research, 24*(2), 425–439.https://doi.org/10.1016/S0160-7383(97)80010-2
Dalevska, N., Dalevska, N., Khobta, V., Khobta, V., Kwilinski, A., Kwilinski, A., Kravchenko, S., & Kravchenko, S. (2019). A model for estimating social and economic indicators of sustainable development. *Entrepreneurship and Sustainability Issues, 6*(4), 1839–1860.
Dempsey, N., Bramley, G., Power, S., & Brown, C. (2011). The social dimension of sustainable development: Defining urban social sustainability. *Sustainable Development, 19*, 289–300. https://doi.org/10.1002/sd.417
Derrett, R. (2004). Chapter 3—Festivals, events and the destination. In Em I. Yeoman, M. Robertson, J. Ali-Knight, S. Drummond, & U. McMahon-Beattie (Eds.), *Festival and Events Management* (pp. 32–50). Butterworth-Heinemann. https://doi.org/10.1016/B978-0-7506-5872-0.50008-9
Dickson, C., & Arcodia, C. (2010). Promoting sustainable event practice: The role of professional associations. *International Journal of Hospitality Management, 29*(2), 236–244. https://doi.org/10.1016/j.ijhm.2009.10.013
Duffy, M., & Waitt, G. (2011). Rural festivals and processes of belonging. *Festival Places: Revitalising Rural Australia*, 44–57.
Dugalić, S. (2020). *Event Tourism and Sustainable Development, 10*, 87–98.
Fairer-Wessels, F., & Malherbe, N. (2012). Sustainable urban event practice: The role of corporate sponsors within gauteng. *Urban Forum, 23*. https://doi.org/10.1007/s12132-011-9128-7
Fairley, S., Tyler, B. D., Kellett, P., & D'Elia, K. (2011). The formula one australian grand prix: Exploring the triple bottom line. *Sport Management Review, 14*(2), 141–152.
Finkel, R., & Platt, L. (2020). Cultural festivals and the city. *Geography Compass, 14*(9), e12498. https://doi.org/10.1111/gec3.12498
Forga, J. (2014). Las relaciones sociales y las motivaciones para asistir al Festival de Música de Peralada. *Boletin de la Asociacion de Geografos Espanoles*, 207–221. https://doi.org/10.21138/bage.1787
Gelling, L. (2015). Qualitative research. *Nursing Standard, 29*(30), 43.
Getz, D. (2007). *Event studies: Theory*. Routledge.
Getz, D. (2008). Event tourism: Definition, evolution, and research. *Tourism Management, 29*(3), 403–428.
Getz, D. (2010). The nature and scope of festival studies. *International Journal of Management Research, 5*(1), 1–47
Getz, D., & Page, S. J. (2016). Progress and prospects for event tourism research. *Tourism Management, 52*, 593–631. https://doi.org/10.1016/j.tourman.2015.03.007
Grappi, S., & Montanari, F. (2011). The role of social identification and hedonism in affecting tourist re-patronizing behaviours: The case of an Italian festival. *Tourism Management, 32*, 1128–1140. https://doi.org/10.1016/j.tourman.2010.10.001
Gursoy, D., Kim, K., & Uysal, M. (2004). Perceived impacts of festivals and special events by organizers: An extension and validation. *Tourism Management, 25*(2), 171–181.
Kim, H., Borges, M. C., & Chon, J. (2006). Impacts of environmental values on tourism motivation: The case of FICA, Brazil. *Tourism Management, 27*(5), 957–967. https://doi.org/10.1016/j.tourman.2005.09.007
Lee, Y.-K. (2016). Impact of government policy and environment quality on visitor loyalty to Taiwan music festivals: Moderating effects of revisit reason and occupation type. *Tourism Management, 53*, 187–196. https://doi.org/10.1016/j.tourman.2015.10.004

Lee, Y.-K., Liu, T.-L., Chung, F.-T., & Ho, H.-H. (2015). Investigating the role of government policy and the environment on locals' loyalty to spring music festivals. *Contemporary Management Research, 11*(1), 33–51. https://doi.org/10.7903/cmr.13791

Lee, J. J., Kyle, G., & Scott, D. (2012). The mediating effect of place attachment on the relationship between festival satisfaction and loyalty to the festival hosting destination. *Journal of Travel Research, 51*, 754–767. https://doi.org/10.1177/0047287512437859

Liberato, D., Costa, E., Liberato, P., Ribeiro, J. (2020). The role of events and music festivals in urban tourism: Case study. In Á. Rocha, A. Abreu, J. de Carvalho, D. Liberato, E. González, P. Liberato (Eds.), *Advances in Tourism, Technology and Smart Systems. Smart Innovation, Systems and Technologies* (vol. 171, pp. 537-549). Singapore: Springer. https://doi.org/10.1007/978-981-15-2024-2_47

Liu, Y., & Chen, C. (2007). The effects of festivals and special events on city image design. *Frontiers of Architecture and Civil Engineering in China, 1*(2), 255–259. https://doi.org/10.1007/s11709-007-0032-0

Liu, Y.-D. (2014). Cultural events and cultural tourism development: Lessons from the European capitals of culture. *European Planning Studies, 22*(3), 498–514. https://doi.org/10.1080/09654313.2012.752442

Llopis-Amorós, M-P., Gil-Saura, I., Ruiz-Molina, M. E., & Fuentes-Blasco, M. (2019). Social media communications and festival brand equity: Millennials versus centennials. *Journal of Hospitality and Tourism Management, 40,* 134–144 https://doi.org/10.1016/j.jhtm.2019.08.002

Mair, J. (2018). *The routledge handbook of festivals*. Routledge.

Mirzayeva, G., Turkay, O., Akbulaev, N., & Ahmadov, F. (2020). The impact of mega-events on urban sustainable development. *7*, 1653–1666. https://doi.org/10.9770/jesi.2020.7.3(15)

Moisescu, O., Gica, O., Coros, M., & Yallop, A. (2019). The UNTOLD story: Event tourism's negative impact on residents' community life and well-being. *Worldwide Hospitality and Tourism Themes, 11*, 492–505. https://doi.org/10.1108/WHATT-06-2019-0036

Monteiro, V., Macdonald, C., Ounis, I., Perego, R., Renso, C., & Times, V. (2019). Event attendance classification in social media. *Information Processing and Management, 56*, 687–703. https://doi.org/10.1016/j.ipm.2018.11.001

Parkes, O., Lettieri, P., & Bogle, I. D. L. (2016). Defining a quantitative framework for evaluation and optimisation of the environmental impacts of mega-event projects. *Journal of Environmental Management, 167*, 236–245. https://doi.org/10.1016/j.jenvman.2015.11.009

Pavluković, V., Armenski, T., & Alcántara-Pilar. J. (2017). Social impacts of music festivals: Does culture impact locals' attitude toward events in serbia and hungary?. *Tourism Management*. https://doi.org/10.1016/j.tourman.2017.06.006. In press

Pérez Gálvez, J. C., López-Guzmán, T., Gómez-Casero, G., & Jesús Medina-Viruel, M. (2021). Análisis de la satisfacción y la lealtad en festivales musicales. Un estudio de caso en base al grado de conocimiento del jazz. *Satisfaction and loyalty in musical festivals. Study based on the level of jazz musical knowledge, 21*(1), 115–124. https://doi.org/10.5295/cdg.180901mm

Popescu, R-I., & Corboş, R-A. (2012). The role of festivals and cultural events in the strategic development of cities. Recommendations for Urban Areas in Romania. *Undefined*. https://www.semanticscholar.org/paper/The-Role-of-Festivals-and-Cultural-Events-in-the-of-Popescu-Corbo%C8%99/8de48d4c1c84258e741988dd9f5639acc4518def

Pivac, T., Blesic, I., Bozic, S., Besermenji, S., & Lesjak, M. (2019). Visitors' satisfaction, perceived quality, and behavioral intentions: The case study of exit festival. *Journal of the Geographical Institute Jovan Cvijic, SASA, 69*, 123–134. https://doi.org/10.2298/IJGI1902123P

Reid, S. (2003). Community participation in rural events: The potential to develop and utilize social capital. In *Advances in Convention, Exhibition and Event Research*. Hong Kong: Convention and Expo Summit. https://www.semanticscholar.org/paper/Community-Participation-in-Rural-Events%3A-The-to-and-Reid/cbb86d9db4f5666f5c3bca1f12af400a53b19c9b

Reid, S. (2004). *The social consequences of rural events: The inglewood olive festival*. https://research-repository.griffith.edu.au/bitstream/handle/10072/47967/55931_1.pdf?sequence=1&isAllowed=y

Szmigin, I., Bengry-Howell, A., Morey, Y., Griffin, C., & Riley, S. (2017). Socio-spatial authenticity at co-created music festivals. *Annals of Tourism Research, 63*, 1–11. https://doi.org/10.1016/j.annals.2016.12.007

Thrane, C. (1997). Values as segmentation criteria in tourism research: The Norwegian monitor approach. *Tourism Management, 18*(2), 111–113. https://doi.org/10.1016/S0261-5177(97)80777-7

Vallance, S., Perkins, H. C., & Dixon, J. E. (2011). What is social sustainability? A clarification of concepts. *Geoforum, 42*(3), 342–348. https://doi.org/10.1016/j.geoforum.2011.01.002

Voss, Z., Cable, D., & Voss, G. (2006). Organizational identity and firm performance: What happens when leaders disagree about? who we are? *Organization Science, 17*, 741–755. https://doi.org/10.1287/orsc.1060.0218

Wang, X., & Leou, C. H. E. (2015). A study of tourism motivation, perceived value and destination loyalty for macao cultural and heritage tourists. https://doi.org/10.5539/IJMS.V7N6P83

Wise, N. (2020). Urban and rural event tourism and sustainability: Exploring economic, social and environmental impacts. *Sustainability, 12*(14). https://doi.org/10.3390/su12145712

Wise, N., & Harris, J. (2017). Introduction: Framing sport, events, tourism and regeneration. In *Sport, Events, Tourism and Regeneration*. 1–8. https://doi.org/10.4324/9781315629728

Wise, N., Peric, M., & Djurkin, J. (2019). Benchmarking service delivery for sports tourism and events: Lessons for gorski kotar, croatia from pokljuka, slovenia. *European Journal of Tourism Research, 22*, 107–128.

Wilson, J., Arshed, N., Shaw, E., & Pret, T. (2017). Expanding the domain of festival research: A review and research agenda. *International Journal of Management Reviews, 19*, 1950213. https://doi.org/10.1111/ijmr.12093

Wong, J., & Law, R. (2003). Difference in shopping satisfaction levels: A study of tourists in Hong Kong. *Tourism Management, 24*(4), 401–410. https://doi.org/10.1016/S0261-5177(02)00114-0

Wu, Y., Li, X., & Lin, G. (2016). Reproducing the city of the spectacle: Mega-events, local debts, and infrastructure-led urbanization in China. *Cities, 53*, 51–60. https://doi.org/10.1016/j.cities.2016.01.004

Xu, Y. H., Wong, I. A., & Tan, X. S. (2016). Exploring event bundling: The strategy and its impacts. *Tourism Management, 52*, 455–467. https://doi.org/10.1016/j.tourman.2015.07.014

Yolal, M., Gursoy, D., Uysal, M., & Kim, H. (Lina), & Karacaoğlu, S. (2016). Impacts of festivals and events on residents' well-being. *Annals of Tourism Research, 61*, 1–18. https://doi.org/10.1016/j.annals.2016.07.008

Yoon, Y.-S., Lee, J.-S., & Lee, C.-K. (2010). Measuring festival quality and value affecting visitors' satisfaction and loyalty using a structural approach. *International Journal of Hospitality Management, 29*(2), 335–342.

Yürük-Kayapınar, P., Akyol, A., & Şimşek, G. (2017). Analyzing the effects of social impacts of events on satisfaction and loyalty. *Tourism Management, 60*, 367–378. https://doi.org/10.1016/j.tourman.2016.12.016

Sustainable Tourism and Degrowth: Searching for a Path to Societal Well-Being

Zuzana Gajdošíková

Abstract Tourism has long been seen as a positive phenomenon that meets the needs of visitors, provides income, creates jobs, raises the living standards of local people, and develops destinations. This one-sided perception of the positive effects of tourism and highlighting just its ability to generate revenues and create jobs has led to the support of its quantitative growth, increasing the number of visitors and efforts to maximize its economic benefits. It is more than obvious that the mass and uncontrolled growth of tourism will hit the limits of the destination and negative effects begin to occur, although the principles of sustainable development have been already known for more than three decades. There is a need to examine the truly essence of sustainable tourism, its real application into praxis, and its ability to ensure a more sustainable tourism sector. Moreover, it is desirable to more explore the issue of the degrowth concept and to try to implement its principles into tourism development. The aim of the paper is to analyze the shift in the tourism paradigm from the needs of visitors to the living space of local inhabitants. According to the literature review focused on sustainable development, tourism degrowth, and local inhabitants in tourism destinations, the article analyzes the crossing needs and opinions of local inhabitants and visitors in a mature destination—the High Tatras in Slovakia with the intention to find out how tourism development should change in order to create a better living space for local inhabitants and be aware of the principles of sustainable development.

Keywords Degrowth · Living space · Local inhabitants · Sustainable development

JEL Classifications L83 · Q1 · Z32

Z. Gajdošíková (✉)
Faculty of Economics, Department of Tourism, Matej Bel University, Banská Bystrica, Slovakia
e-mail: zuzana.gajdosikova@umb.sk

© The Author(s), under exclusive license to Springer Nature Switzerland AG 2023

V. Katsoni (ed.), *Tourism, Travel, and Hospitality in a Smart and Sustainable World*, Springer Proceedings in Business and Economics,
https://doi.org/10.1007/978-3-031-26829-8_10

1 Introduction

The Covid-19 pandemic has mirrored the way tourism has worked so far. Many people from theory, praxis, and the international organizations perceive the break brought by the pandemic as an important chance to rethink the way, essence, circumstances, and impacts of previous tourism development. One of the most important findings is that tourism development was unfair to the environment and local communities. Considering the needs and rights of the local people, something went wrong in the tourism development. Local communities in many destinations around the world experienced the negative effects of tourism that affected their lives. Some of them were dependent on revenues only from tourism, and others were forced to change the way of their lives due to the mass tourism development and the consequences it has brought. Now, after a two-year break, enriched with a new experience, we have a more open mind and opinions to change the way how tourism influences the environment and local communities. Higgins-Desbiolles et al. (2019) argue for defining tourism by the local community instead of the conventional tourism definition based on the tourism industry that supplies products and services to meet the demand of tourists for their experiences (Higgins-Desbiolles & Bigby, 2022). Higgins-Desbiolles (2020) demand to make tourism responsive and accountable to the society in which it occurs. However, is this possible? What needs to be changed in tourism development, as we know now led by numbers of tourism revenues and number of visitors (ideally foreign ones) to stop transforming home places to tourism destinations and the local inhabitants to hosts communities? Is sustainable tourism with its principles a sufficient solution to this problem or do we need to take action and make another intervention in tourism development? The findings so far (Liu, 2003; Mika, 2015; Sharpley, 2020) state that the concept of sustainable tourism is just a small correction of constant tourism growth with all its impacts.

2 Literature Review

Several studies, theories, and papers have been used to explain the concept of sustainable tourism development, tourism degrowth, and the role of local inhabitants in tourism. They will be used as a starting point for the present study, which focuses on the importance of local communities in the tourism development according to the principles of sustainable development.

2.1 Sustainable Tourism Development and its Principles

Tourism has long been perceived as a positive phenomenon that satisfies the needs of visitors, provides economic income, creates jobs, increases the living standard of local inhabitants, and develops destinations. This one-sided perception of the positive effects of tourism and its emphasis on its ability to generate income and create jobs led to the promotion of its quantitative growth, increasing the number of visitors and the effort to maximize economic effects. However, it is important to note that not all cases involved massive and uncontrolled tourism growth support. Tourism studies have a tradition of seeking alternative pathways to economic development that minimize negative externalities to destinations (Hall, 2009).

In the 1980s, we observe the first official initiative aimed at analyzing how the development of tourism has affected the territory of the Alps during the last 100 years. The initiative was carried out by the international organization UNESCO as part of the "Man and the Biosphere" program. The study emphasizes the need to strengthen nature conservation, limit construction activity, control land development plans, carefully build infrastructure, protect nature and the landscape, strengthen agriculture and forestry, improve the quality of jobs, maintain local culture, and promote the marketing of "soft" tourism. Even then, the need to move away from mass to sustainable development of tourism was being promoted. Based on the results of this study, Krippendorf (1987) published a pioneering publication entitled Alpine Blessing? Bad Dream (Alpsegen? Alptraum), in which he states and explains the requirements for tourism in the Alps in accordance with the principles of sustainable development. This act attracted the attention of both theory and practice and stirred the investigation of the relationship between tourism and sustainable development.

Considering the development and state of the world until then, the United Nations Commission on Environment and Development published a document called Our Common Future (Commission on Environment, 1987), also known as the Brundtland Report. The United Nations Commission on Environment and Development understands sustainable development as "development that meets the needs of the present without compromising the ability of future generations to meet their own needs". The Swiss experience and the document Our Common Future can be considered a precursor to Agenda 21 as a result of the Declaration adopted at the UN Conference on Environment and Development held in 1992 in Rio de Janeiro. Agenda 21 provides a program to ensure sustainable development on Earth.

The initiator of the implementation of the concept of sustainable development to the tourism industry was the World Tourism Organization (UNWTO), which in 1995 expanded the basic definition of sustainable development for the needs of the tourism industry: "Tourism that takes full account of its current and future economic, social and environmental impacts, addressing the needs of visitors, the industry, the environment, and host communities". Since 1999, tourism as a factor of sustainable development has been enshrined in the third and fourth articles of the Global Tourism Code of Ethics (UNWTO, 1999).

A milestone for global development was the year 2015, when governments adopted the 2030 Agenda for Sustainable Development along with the Sustainable Development Goals. The 2030 Agenda sets out a global framework to end extreme poverty, fight inequality and injustice, and address climate change by 2030. It represents a set of 17 Sustainable Development Goals. In this sense, the 70th General

Assembly of the United Nations designated 2017 as the International Year of Sustainable Tourism for Development. Sustainable development represents a response to the acknowledged inseparability of the environment and human existence, which means to enact a positive vision of a world in which basic human needs are met without destroying the natural systems on which we all depend.

In addition to the initiatives of international organizations, attention is also paid to sustainable tourism in theory, while many experts who take into account various points of view constantly investigate its essence, principles, provision in practice, as well as criticism and weaknesses. At the beginning of sustainable tourism research (1990s), authors paid attention mainly to the general strategy and policy of sustainable tourism (e.g., Cronin, 1990; Pigram, 1990; Bramwell & Lane, 1993; Forsyth, 1996; Clarke, 1997; Hunter, 1997; Middleton & Hawkins, 1998; Place & Hall, 1998; Butler, 1991; Swarbrooke, 1999). Over time, the scope of research is narrowed down to specific selected problems related to the sustainable development of tourism, especially tourism indicators, the competitiveness of destinations in terms of sustainable development, the impact of tourism on the environment, climate change, the stakeholders involvement, the importance of local residents, etc. (Sharpley, 2000; Miller, 2001; Hall & Richards, 2003; Ritchie & Crouch, 2003; Robinson & Picard, 2006; Mowforth & Munt, 2008; Edgell, 2016; Kuščer, et al., 2017; Gajdošík, et al., 2018; Klimek & Doctor, 2019; Gajdošíková, 2020) to criticism of the essence and principles of the concept of sustainable tourism (Buckley, 2012; Liu, 2003; Mika, 2015; Sharpley, 2000, 2020).

Some authors agree (Galvani et al., 2020; Sharpley, 2020) that the concept of sustainable tourism is only a band-aid to the real problems caused by constant growth. Consequently, questioning growth itself as the basis of the sustainable tourism has now become part of sustainable tourism discourse (Büscher & Fletcher, 2017). They considered sustainable development only as a correction of economic growth, not as a tool for systemic changes. Sustainable tourism is perfectly explained in theory, but, except for a few small case studies, it is more difficult to extend it in practice. Despite continuing alignment between tourism and sustainable development, there is little evidence of a more sustainable tourism sector (Sharpley, 2020).

2.2 *The Essence of Tourism Degrowth*

In the beginning phase of shaping the sustainable tourism theory, another approach began to appear which criticized quantitative growth in a limited space, namely the so-called theory of degrowth. The degrowth theory arose in the 1970s, when the negative aspects of technological progress began to be known, and the world began to reach the limits of economic growth (Fletcher et al., 2021). Meadows's report The Limits to Growth (1972) is considered the initiator of the degrowth theory, in which he draws attention to the excessive growth of the economy, which attracted attention in both academic and political circles and stimulated the emergence of environmental movements. The term degrowth (from French décroissance) was first used by the French journalist André Gorz in connection with the need to reduce consumption. The essence of the degrowth theory is "a socially reduced production and consumption that increases human satisfaction and improves ecological conditions" (Schneider et al., 2010). More simply, degrowth signifies a critique of growth (D'Alisa et al., 2014). Explicit discussion of degrowth has grown quickly since the turn of the 21th century in particular, to include a number of interconnected threads (Fletcher et al., 2021). Degrowth represents a direction in which societies will use fewer natural resources and will organize and live differently, better. Sharing, simplicity, conviviality, care, and the commons are primary significations of what society in degrowth might look like (D'Alisa et al., 2014). The degrowth economy is considered radical in the ways in which it wants to solve the ecological crisis "from the bottom" unlike sustainable development which is satisfied with solving the problem through "green, cosmetic changes" (Latouche, 2009). Some degrowth characteristics are already known (Fletcher et al., 2021):

- emphasis on quality of life, not consumption,
- satisfying the needs of all people on Earth,
- constant reduction of dependence on economic activities, increase in free time,
- volunteering, hospitality, community, individual and collective health,
- encouraging self-reflection, balance, creativity, flexibility, good citizenship, generosity, and immaterialism,
- based on the principles of equality, respect for human rights, and cultural diversity.

The potential contribution of degrowth theory to tourism was offered by Bourdeau and Berthelot (2008) at the first international degrowth conference in Paris. This idea was then elaborated by Hall (2009), who stated "the contribution of tourism to sustainable development should be understood in the context of degrowth processes that offer an alternative discourse to the economism paradigm that reifies economic growth in terms of GDP". Explicit discussion of degrowth in tourism has grown rapidly since the Covid-19 pandemic sustained the ordinary tourism development (Butcher, 2021; Seyfi & Hall, 2021; Vogler, 2022).

As is already known, tourism is affected by the negative effects of its growth. Concerning the degrowth principles, we should probably reconsider the "fastest growing sector", "one of the largest industries" or "major exporter" and rather seek to make the economic, environmental, and social effects of tourism more balanced, systematic, integrated, and sophisticated (Higgins-Desbiolles et al., 2020). The degrowth theory criticizes tourism for several reasons (Sekulova et al., 2017):

- tourism has become a benchmark for a good, quality life,
- tourism means consumption, and stimulating further consumption, unnecessary consumption, vain consumption,
- tourism often means destruction (mountain slopes for skiing, wild beaches for resorts),
- tourism creates an illusionary paradise, thus hiding dirty and sad realities, excluding those who are there to serve, in a semi-slavery mode of operation, for those who ingratiate with the idea of buying an experience of relaxation,

- tourism is grounded in the illusion that the best experience can and must be bought with money,
- tourism is like a drug that helps one escape and recharge.

Latouche (2003) contributed to degrowth discourse by defining it as a necessity, not a principle. He recommends to achieve social prosperity without the need for infinite growth. The essence of degrowth in tourism lies in this form of tourism, which is not aimed at economic benefits, but saves resource consumption and minimizes the negative effects of tourism.

2.3 *Changing the Tourism Paradigm from the Needs of Visitors to the Living Space of Local Inhabitants*

In the past, it was a frequent and repeated claim that residents' receptiveness to both visitors and tourism development plays an important role in attracting and pleasing visitors (Davis, et al., 1988; Cooke, 1982). A group of tourism authors has argued that residents have been marginalized and subordinated to the edge of the tourism development process and that residents should be included as major stakeholders in the tourism planning and development process (Jamal & Getz, 1995; Joppe, 1996; Shmelev, 2001).

Current situations (environmental crisis, pandemic, war, etc.) in the world have a major impact on the people living on the Earth and their lives. Thanks to the expansion in information technologies, governments, companies, researchers and citizen groups have access to data and information, which are bigger, faster, and more detailed than ever before (UN, 2014). This allows us to better examine the true tourism development with the majority of its impacts on the environment. We are witnessing the effects that the uncontrolled growth of tourism has on the country and local communities, which forces us to change our perception from what visitors need and how to best meet their needs to what benefits tourism can bring to the country and local people. Instead of the conventional definition of tourism as "the industry that supplies products and services to tourists to meet their demand for tourism experiences", the tourism should be defined by the local community (Higgins-Desbiolles & Bigby, 2022). The question: "What can we do for visitors and how can we as best as possible meet their needs?" should change to "What can tourism do and bring for the local inhabitants" How can tourism make their life and living space richer not poorer?". Higgins-Desbiolles (2020) call for "socialization of tourism". We have observed mainly from the Covid-19 pandemic changes in the perception of tourism toward the environment and local communities in which it develops. Especially in research (Tomassini & Cavagnaro, 2020, Mika & Scheyvens, 2022; Schweinsberg et al., 2021) and statements of international organization (UNWTO, WTTC and others), there is much more attention aimed at the responsible tourism development.

Higgins-Desbiolles and Bigby (2022) introduce the terminology of local community and explain the understanding of local community as local ecology with the living air, land, and waterscapes, more than human beings and all generations pertaining to the place. The local turn in tourism studies, describe Higgins-Desbiolles and Bigby (2022) as an important catalyst to change our consciousness, relationships and activities to prevent and mitigate the negative tourism outcomes. All efforts must be made to respect traditional lifestyles and cultures. Tourism must become a vehicle by which community tourism destinations can ensure their long-term viability (Choi & Sirakaya, 2005). It is important to change the perception from the tourist destination to the living space of local communities (Pechlaner, 2019) who lived there before the place was discovered by tourists. There is already existing life in the place with all its customs, norms, and traditions, and tourism development has to be integrated to the living space in such a way not to disturb the people, their usual environment, and lives.

3 Methodology

The aim of the paper is to analyze the shift in the tourism paradigm from the needs of visitors to the living space of local inhabitants. Our investigation based on the research question: "How should tourism development change in order to create a better living space for local inhabitants and be aware of the principles of sustainable development?" is an answer to the call for multidisciplinary approaches to develop the local turn in tourism (Higgins-Desbiolles & Bigby, 2022), and at the same time, we follow up the survey from 2017 on the dynamics of the destination structure revisited in view of the community and corporate model (Gajdošík et al., 2017) that aimed to find the model of the destination organization structure (Fig. 1).

We believe that to be able to say that a local turn has taken place in tourism development, residents would have to feel that they are a priority in tourism development planning. Tourism must develop in such a way as to bring as many benefits as possible to local residents and their lives. To provide stronger evidence for changing paradigm in tourism development toward the local communities and their living space, High Tatras, the developed tourist destination with history of tourism development since

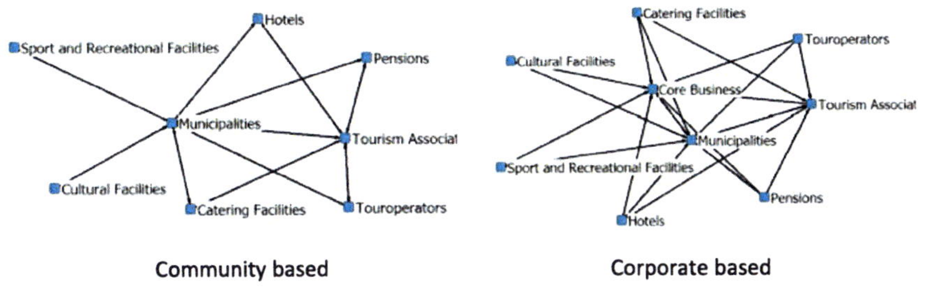

Fig. 1 Graphical interpretation of the model of the destination organization structure. *Source* Gajdošík et al., 2017

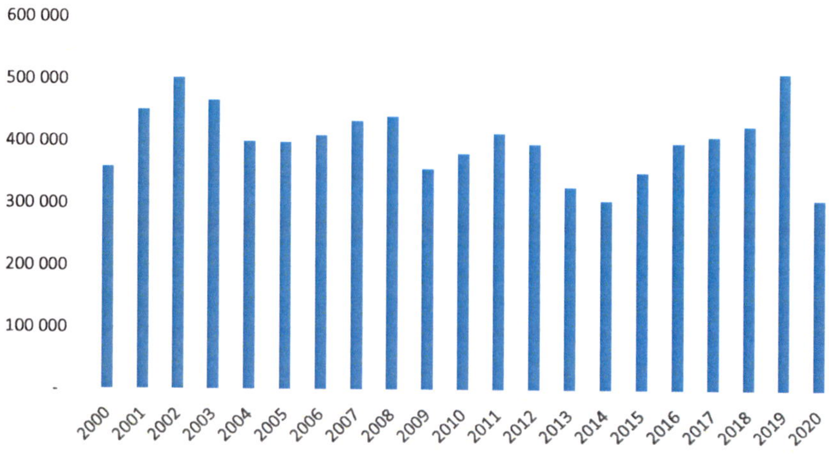

Fig. 2 Number of visitors in accommodation facilities in the High Tatras. *Source* slovak.statistics.sk, 2022

the end of the nineteenth century, was chosen. This destination is a representative of a typical European destination, as it is a destination in the maturity stage of the destination life cycle and is in the process of changing the destination structure model (Gajdošík et al., 2017). The High Tatras (Fig. 2) is a typical winter destination with a history first as a climatic spa destination, then a winter sport destination and is currently trying to profile itself as a year-round destination.

We want to demonstrate the role of local communities in the tourism development in the High Tatras, and we are interested in the current phase of tourism development according to the principles of local turn and changing the tourism paradigm from the needs of visitors to the living space of local inhabitants on the example of High Tatras destination. With the intention of finding this, a qualitative research approach was realized: interviews with local inhabitants. The questions were aimed at finding out whether the inhabitants feel that they are a part of the planning and development of tourism. Interviews were conducted in the spring of 2022 with the local inhabitants, who showed willingness to express their opinion on the tourism development in High Tatras, while we focused on the residents with higher provenance. We conducted 47 interviews (High Tatras currently have 3 851 inhabitants). The average age of the responded local inhabitants is 45 years, and the average provenance of living in the High Tatras is 33 years. Participants were heterogeneous with respect to gender, age (from 20 to 78), place of residence in High Tatras, employment in tourism, and the status of respondents. The interviews lasted between 30 and 45 min. The sample size was not set in advance; the approach of seeking the data saturation point (Glaser & Strauss, 1967) was used. The point of data saturation is reached when additional interviews do not lead to additional insights.

4 Results

The research follows on findings about the organization structure model from 2017 (Gajdošík et al.), where we found that in the High Tatras, the community model (Flagestad & Hope, 2001) was dominant until roughly 2015, when the monopoly dominance of one private stakeholder is strengthening and thus changing to the corporate-based model of destination, where one stakeholder controls dominant destination resources and has a significant impact on strategy and policy with a further impact on development of a destination (Fig. 3).

The results of the study indicated that the corporate model may perform better in creating customer satisfaction. The corporate model is in a position to manage more professionally a customer-oriented destination development by controlling a critical mass of service providers. This is in contrast to the community model where the stronger stakeholder involvement may perform better in environmental and social dimensions of sustainability.

Several years have passed since the initial research, and nowadays, we want to research the role of local inhabitants in the tourism development and whether there are any indications of the announced shift to the local turn in High Tatras (Fig. 4).

In fact, in the High Tatras, there is no systematic tool to involve local inhabitants in decision-making to express their opinions about tourism development. The results indicated that almost two-thirds of the study participants stated that they do not know the possibility of expressing their view on the tourism development in High Tatras. The remaining one-third of respondents can do so because they are either directly active in organizations or councils responsible for tourism development or are in familiar with somebody who is in such a body. Moreover, all the participants expressed agreement that they would like to be involved in tourism policy and planning, knowing the plan (strategy) for the development of tourism in the High Tatras and being able to express their opinion on it.

The majority of the respondents are satisfied with the current state of tourism development in the High Tatras, some of them said: *I support tourism development;*

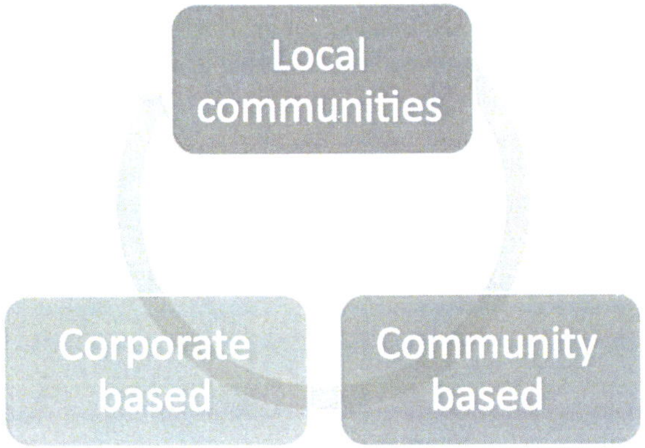

Fig. 3 Graphical interpretation of the turn from corporate-based model to the local communities. *Source* Own elaboration, 2022

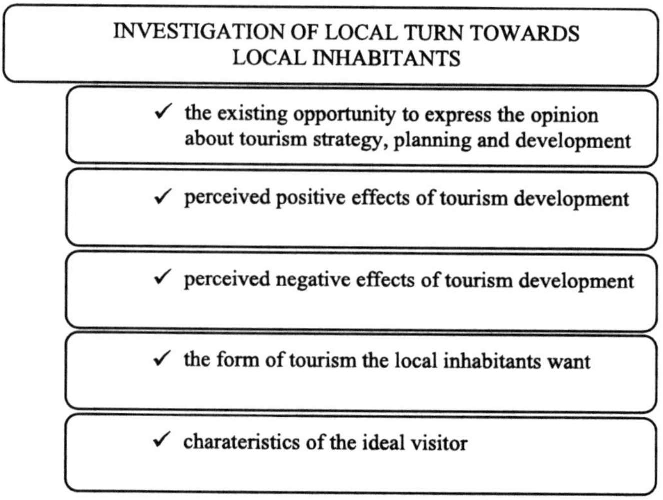

Fig. 4 Investigated factors of local turn. *Source* Own elaboration, 2022

I want tourism to continue to develop in the High Tatras; I think the High Tatras should continue to remain a popular tourist destination; Due to the development of tourism, the High Tatras look better; Thanks to tourism, there are more options to shop and spend free time in the High Tatras; Thanks to tourism, there are more business opportunities and opportunities to sell regional products in the High Tatras. On the other hand, local inhabitants perceive sensitively the impacts that tourism leaves on their living space. Some of the respondents state that: *There are more negative effects that profits from tourism.* Eighteen respondents indicated that *they do not want the additional growth in the number of tourists in the High Tatras.* They answered the question why with different answers: *Attractions are crowded with tourists, Due to tourism development, life in the High Tatras is more expensive, Tourism causes traffic and parking problems, Tourism contributes to the destruction of nature, pollutes water, soil and causes excessive noise.* A total of 35 respondents felt that there are already enough tourists in the Tatras.

Based on the results of the analysis, we can conclude that local residents are sensitive to the development of tourism and could contribute to its better development with many opinions and experiences if there were a space where they could express their opinions and where they would be accepted. We also observe that there is currently no official and systematic platform in the High Tatras for the involvement of local residents and their needs in the development of tourism. However, while ascertaining their opinions, we also learned the following facts, which have an impact on further development and direction: *I want the kind of tourism development that will not negatively affect life in my region; which will not take my parking space; which will not pollute the environment; which will not raise the prices; which will not cause traffic collapse; which will not cause overcrowding of protected areas.*

The locals also have an opinion on the characteristics of visitors they want to have in their living space. This gives us valuable findings for shaping destination marketing to aim the marketing communication on such segments that are suitable for the local people and respect life existing in the place. The respondents prefer to attract visitors *whose priority is respect for nature, who respect the local inhabitants, who are considerate, polite, thinking of others and not only on themselves, who follow the rules*. Nobody from the respondents said that he wanted still more visitors from abroad who would bring a lot of money, as we often hear it from the tourism representatives.

5 Conclusions

Residents are major actors in the tourism development process since they are directly affected by it (Ap, 1992; Murphy, 1985; Gunn, 1994). The results of the analysis carried out in the High Tatras indicate that the local inhabitants have a huge potential to be a part of decision-making and tourism planning. But, to our knowledge, so far the local turn toward the needs and opinions of local inhabitants has not taken place in the High Tatras. The corporate model with one dominant stakeholder, who owns the most important infrastructure points, thus influences the leadership in the destination, still prevails. However, the quantitative expansion of the economic subsystem increases environmental and social costs faster than production benefits, making us poorer not richer (Hall, 2009). As Higgins-Desbiolles argues (2010): achieving a truly "sustainable tourism necessitates a clear-eyed engagement with notions of limits that the current culture of consumerism and pro-growth ideology precludes".

There are known important signs from the literature review and analyzed destination about sustainable development and degrowth principles, which we can start changing toward a tourism development better for local inhabitants and their living space. Hall (2009) advocated a "steady-state tourism that encourages qualitative development but not aggregate quantitative growth to the detriment of natural capital". Degrowth means that being smaller can also be beautiful. The Berlin Declaration (1997) made a strong normative point by suggesting that tourism should be developed in a way that benefits the local communities, strengthens the local economy, employs the local workforce, and wherever ecologically sustainable, uses local materials, local agricultural products, and traditional skills. Mechanisms, including policies and legislation, should be introduced to ensure the flow of benefits to local communities.

We believe that tourism should develop according to the following principles:

- less consumed resources,
- creating a pleasant environment for local inhabitants and visitors,
- fewer, but better quality visitors,
- truly sustainable-slow but high quality tourism development.

In terms of sustainable tourism pillars (environmental, economic, social) and degrowth principles, we expect one of the following scenarios in tourism development in destinations: (1) economic development: development of tourism, where the economic profits still prevail, (2) rhetorically sustainable: development of tourism, where the principles of sustainable development are known, but not used in the practice, (3) sustainable development: development of tourism, where the principles of sustainability are a part of all activities, (4) degrowth: development of tourism according to the principles of degrowth theory with the aim to ensure quality, stability, and responsibility. The ideal solution would be to create such an environment for local communities that they can relax (regenerate, have experiences, gain energy) also in their surroundings and not live in a stressful environment and then need to travel halfway around the globe to recover. The goal of the tourism development should be well-being of individuals and societies, who are living in such a space, where they are able to enjoy meaningful lives and satisfy their needs.

This debate on sustainable tourism, degrowth, and the importance of local communities in tourism development should contribute to its transformation. We should develop forms of tourism that are possible within planetary boundaries. There is a need for fundamental rethink about how tourism is produced, managed, and consumed (Sharpley, 2020) in order to create a different system where expansion will no longer be a necessity.

The paper presents our view toward the research of the local turn in tourism. The present study is limited in that it focuses on the research of local inhabitants only. To support the outputs, detailed research on the stakeholders' opinion is lacking. It is needed to develop research on the imagination of stakeholders about future tourism development and their attitude to the local turn in tourism development.

Acknowledgements This research was funded by the Scientific Grant Agency of the Ministry of Education, Science, Research and Sport of Slovak Republic VEGA, grant number 1/0237/20 "Tourism 4.0: Intelligent and sustainable development of tourism in a competitive environment".

References

Ap, J. (1992). Residents' perceptions on tourism impacts. *Annals of Tourism Research, 19*, 665–690.

Bourdeau, P., & Berhelot, L. (2008). *Tourisme et Décroissance: de la critique à l'utopie?* First international conference on economic de-growth for ecological sustainability and social equity, Paris, April 18–19th, 2008.

Bramwell, B., & Lane, B. (1993). Sustainable tourism: An evolving global approach. *Journal of Sustainable Tourism, 1*(1). https://doi.org/10.1080/09669589309450696.

Buckley, R. (2012). Sustainable tourism: Research and reality. *Annals of Tourism Research, 39*(2), 528–546. https://doi.org/10.1016/j.annals.2012.02.003.

Büscher, B., & Fletcher, R. (2017). Destructive creation: Capital accumulation and the structural violence of tourism. *Journal of Sustainable Tourism, 25*(5), 651–667. https://doi.org/10.1080/09669582.2016.1159214.

Butcher, J. (2021). Covid-19, tourism and the advocacy of degrowth. *Tourism Recreation Research*. https://doi.org/10.1080/02508281.2021.1953306.

Butler, R. W. (1991). Tourism, environment and sustainable development. *Environmental Conservation, 18*(3), 201–209. https://doi.org/10.1017/S0376892900022104.

Choi, H. S. C., & Sirakaya, E. (2005). Measuring residents' attitude toward sustainable tourism: Development of sustainable tourism attitude scale. *Journal of Travel Research, 43*(4), 380–394. SAGE Publications Ltd. https://doi.org/10.1177/0047287505274651.

Clarke, J. (1997). A framework of approaches to sustainable tourism. *Journal of Sustainable Tourism, 5*(3), 224–233. https://doi.org/10.1080/09669589708667287.

Cooke, K. (1982). Guidelines for socially appropriate tourism development in British Columbia. *Journal of Travel Research, 21*(1). https://doi.org/10.1177/004728758202100106.

Commission on Environment, W. (1987). *Report of the world commission on environment and development: Our common future towards sustainable development 2. Part II. Common challenges population and human resources, 4.*

Cronin, L. (1990). A strategy for tourism and sustainable development. *World Leisure and Recreation, 32*(3), 12–18. https://doi.org/10.1080/10261133.1990.10559117.

D'Alisa, G., Demaria, F., & Kallis, G. (2014). Degrowth: A vocabulary for a new era. *Teknokultura. Revista de Cultura Digital y Movimientos Sociales, 14*(1), 185–186. https://doi.org/10.5209/tekn.55414.

Davis, A. et al. (2009). Avatars, people, and virtual worlds: Foundations for research in metaverses. *Journal of the Association for Information Systems, 10*(2), 90–117. https://doi.org/10.17705/1jais.00183

Edgell, D. L. (2016). *Managing sustainable tourism: A legacy for the future.* Binghamton: Haworth Press

Flagestad, A., & Hope, C. A. (2001). Strategic success in winter sports destinations: A sustainable value creation perspective. *Tourism Management, 22*(5), 445–461. https://doi.org/10.1016/S0261-5177(01)00010-3

Fletcher, R. et al. (Eds). (2021). *Tourism and degrowth. Towards a truly sustainable tourism.* Oxon: Routledge.

Forsyth, T. (1996). *Sustainable tourism: Moving from theory to practice.* Godalming: World Wildlife Fund.

Gajdošík, T. et al. (2018). Residents' perception of sustainable tourism destination development—A destination governance issue. *Global Business & Finance Review, 23*(1), 24–35.

Gajdošíková, Z. (2020). *Udržateľný rozvoj cestovného ruchu v horských strediskách.* Banská Bystrica: Belianum.

Gajdošík, T., Gajdosikova, Z., Maráková, V., & Flagestad, A. (2017). Destination structure revisited in view of the community and corporate model. *Tourism Management Perspectives, 24*, 54–63. https://doi.org/10.1016/j.tmp.2017.07.009.

Galvani, A., Lew, A. A., & Perez, M. S. (2020). COVID-19 is expanding global consciousness and the sustainability of travel and tourism. *Tourism Geographies*, 567–576. https://doi.org/10.1080/14616688.2020.1760924.

Glaser, B. G., & Strauss, A. (1967). *The discovery of grounded theory: Strategies for qualitative research.* Chicago, IL: Aldine.

Gunn, C. (1994). *Tourism planning: Basics, concepts, cases.* NewYork: CBS College Publishing.

Hall, C. M. (2009). Degrowing tourism: Décroissance, sustainable consumption and steady-state tourism. *Anatolia, 20*(1), 46–61. https://doi.org/10.1080/13032917.2009.10518894.

Hall, D., & Richards, G. (2003). *Tourism and sustainable community development.* London: Routledge.

Higgins-Desbiolles, F. (2020). The "war over tourism": Challenges to sustainable tourism in the tourism academy after COVID-19. *Journal of Sustainable Tourism*, 29(4), 551–569. https://doi.org/10.1080/09669582.2020.1803334.

Higgins-Desbiolles, F., & Bigby, B. C. (2022). A local turn in tourism studies. *Annals of Tourism Research, 92*. https://doi.org/10.1016/j.annals.2021.103291.

Higgins-Desbiolles, F., Carnicelli, S., Krolikowski, C., Wijesinghe, G., & Boluk, K. (2019). Degrowing tourism: Rethinking tourism. Journal of Sustainable Tourism, 27(12), 1926–1944. https://doi.org/10.1080/09669582.2019.1601732.

Higgins-Desbiolles, F. (2010). The elusiveness of sustainability in tourism: The culture-ideology of consumerism and its implications. *Tourism and Hospitality Research*, 10(2), 116–129.

Hunter, C. (1997). Sustainable tourism as an adaptive paradigm. *Annals of Tourism Research*, 24(4), 850–867.

Jamal, T. B., & Getz, D. (1995). Collaboration theory and community tourism planning. *AnnMs ofTourismResearch, 22*(1).

Joppe, M. (1996). Sustainable community tourism development revisited. *l~mrism Management, 17*(7).

Klimek, K., & Doctor, M. (2019). Are alpine destination management organizations (DMOs) appropriate entities for the commercialization of summer tourism products? *Journal of Destination Marketing & Management, 10*(6), 181–190. https://doi.org/10.1016/j.jdmm.2018.09.005.

Krippendorf, J. (1987). *Alpsegen, Alptraum. Für eine Tourismus-Entwicklung im Enklang mit Mensch und Natur*. Bern: Kűmmerli.

Kuščer, K., Mihalič, T., & Pechlaner, R, H. (2017). Innovation, sustainable tourism and environments in mountain destination development: A comparative analysis of Austria, Slovenia and Switzerland. *Journal of Sustainable Tourism, 25*(4), 489–504. https://doi.org/10.1080/09669582.2016.1223086.

Latouche, S. (2009) *Farewell to growth*. Cambridge: Polity Press.

Liu, Z. (2003). *Sustainable Tourism Development: A Critique*.

Middleton, V. T. C., & Hawkins, R. (1998). *Sustainable tourism: A marketing perspective*. Oxford: Butterworth, Heinemann.

Miller, G. (2001). The development of indicators for sustainable tourism: Results of a delphi survey of tourism researchers. *Tourism Management, 22*(4), 351–362. https://doi.org/10.1016/S0261-5177(00)00067-4.

Mika, J. P., & Scheyvens, R. A. (2022). Te Awa Tupua: peace, justice and sustainability through Indigenous tourism. *Journal of Sustainable Tourism, 30*(2–3), 637–657. https://doi.org/10.1080/09669582.2021.1912056.

Mika, M. (2015). Sustainable tourism: A critique of the academic feasibility of the concept. *Tourism(Poland), 25*(1), 9–17. https://doi.org/10.2478/tour-2014-0015.

Mowforth, M., & Munt, I. (2008). *Tourism and sustainability: Development, globalisation and new tourism in the third world*. London: Routledge.

Murphy, P. (1985). *Tourism: A community approach (RLE Tourism)* (1st Ed.). Routledge. https://doi.org/10.4324/9780203068533

Pigram, J. (1990). Sustainable tourism: Policy considerations. *Journal of Tourism Studies, 1*(2), 2–9.

Place, S. E., & Hall, C. M. (1998). *Sustainable tourism: A geographical perspective*. Editorial Harlow

Ritchie, J. R. B., & Crouch, G. I. (2003). *The competitive destination: A sustainability perspective*. Wallingford: CABI International.

Robinson, M., & Picard, D. (2006). *Tourism, culture and sustainable development*. Division of Cultural Policies and Intercultural Dialogue, Culture Sector, UNESCO.

Schneider, F., Kallis, G., & Martinez-Alier, J. (2010). Crisis or opportunity? economic degrowth for social equity and ecological sustainability. Introduction to this special issue. *Journal of Cleaner Production, 18*(6), 511–518. https://doi.org/10.1016/j.jclepro.2010.01.014.

Schweinsberg et al., 2021 Schweinsberg, S., Fennell, D., & Hassanli N. (2021). Academic dissent in a post COVID-19 world. Annals of Tourism Research, *91*. https://doi.org/10.1016/j.annals.2021.103289.

Seyfi, S., & Hall, C. M. (2021). *COVID-19 pandemic, tourism and degrowth*. Retrieved, from https://www.researchgate.net/publication/344215235.

Sharpley, R. (2000). Tourism and sustainable development: Exploring the theoretical divide. *Journal of Sustainable Tourism, 8*(1), 1–19. https://doi.org/10.1080/09669580008667346.

Sharpley, R. (2020). Tourism, sustainable development and the theoretical divide: 20 years on. *Journal of Sustainable Tourism, 28*(11), 1932–1946. https://doi.org/10.1080/09669582.2020.1779732.

Shmelev, S. (2001). Voices of a natural prison: Tourism development and fisheries management among the political ghosts of pisagua, chile. *Journal of Political Ecology, 8*(1). https://doi.org/10.2458/v8i1.21580.

Sekulova, F. et al. (2017). A 'fertile soil' for community initiatives: A new analytical framework. *Environment and Planning, 49*(10), 2362–2382.

Swarbrooke, J. (1999). *Sustainable tourism management*. Oxon: CABI Publishing.

Tomassini, L., & Cavagnaro, E. (2020). The novel spaces and power-geometries in tourism and hospitality after 2020 will belong to the 'local'. *Tourism Geographies, 22*(3), 713–719.

UN. (2014). A World that counts. Mobilising the Data Revolution for Sustainable Development. 32 p.

UNWTO. (1999). *Global code of ethics for tourism.*

Vogler, R. (2022). Revenge and catch-up travel or degrowth? debating tourism Post COVID-19. *Annals of Tourism Research, 93*. Elsevier Ltd. https://doi.org/10.1016/j.annals.2021.103272.

The Sustainable Strategic Innovation Adopted by the Archaeological Museum of Taranto "MArTA" Which Drives the Repositioning of Taranto as a Cultural Tourist Destination

Nicolaia Iaffaldano

Abstract Policymakers' strategic choices are directed to seek the endogenous resources to create sustainable development paths in Taranto, a city known as a symbol of industrial pollution because of its steel mill. The research objective is trying to give a strategic framework in order to start development paths, thus making the city of Taranto rediscover a cultural and tourist vocation, thanks to its historical endowment of local resources. The analysis of the archaeological museum of Taranto, "MArTA", clarifies how a sustainable museum model can help local development and civic innovation thanks to its inclusion potential and its ability to involve various local stakeholders. Therefore, the paper aims to shed light on the path of sustainable strategic innovation undertaken by museum organizations, in particular by the MArTA museum organization. Faced with the social, economic and technological changes that have characterized the last few decades, museums have gradually initiated transformative policies of their *core concept*. In this way, they have rediscovered themselves as an active part in the process of social innovation and territorial development, following the *Sustainable Development Goal* 11 of the UN 2030 Agenda. In-depth interviews with the museum management, the local community and the local public and private stakeholders involved in the socio-economic development of the city will be adopted. The paper provides policy indications to public decision-makers operating in the cultural sector and in the local administration in order to design and develop a management model that responds to a systemic perspective.

Keywords Strategic innovation · Management of cultural heritage in a systemic perspective · Ecomuseum · Landscape · Territorial planning · Social sustainability

JEL Classification Z32

N. Iaffaldano (✉)
University of Bari "Aldo Moro", Bari, Italy
e-mail: nicolaia.iaffaldano@uniba.it

© The Author(s), under exclusive license to Springer Nature Switzerland AG 2023
V. Katsoni (ed.), *Tourism, Travel, and Hospitality in a Smart and Sustainable World*,
Springer Proceedings in Business and Economics,
https://doi.org/10.1007/978-3-031-26829-8_11

1 Introduction

The paper aims to strengthen the relational capacities of museum institutions and to improve the attractiveness of the cultural and landscape heritage of Italian southern regions to support their development and competitiveness.

This research wants to provide indications to define the strategic lines which can create a coherent and integrated path of valorization for cultural and tourist purposes in the area of Taranto and can go beyond current environmental and social emergencies, thus contributing to the wider design of urban redevelopment and regeneration.

Moreover, the peculiarity of Taranto in being a perfect "one company town" has determined in the city inhabitants a sort of suspension of any vision on its future.

Today, this scenario has completely changed, because of the well-known environmental and employment problems due to the crisis of the steel plant, and challenges the institutional, professional and entrepreneurial forces of the city in order to awaken a shared identity on the future and engage these forces in the definition of a strategy that can launch urban conversion paths able to give new development prospects to the city (Law, 1992; Valdani & Ancarani, 2000).

Thanks to their strategic choices, it seems that local public policymakers are willing to mend the relationship between inhabitants, economy and territory that has been cut by both the State industry imposed from above and the top-down intervention logic that fall on the city without considering its endogenous potential. Therefore, the choices are aimed at researching in history, in cultural–archaeological and natural values, the endogenous resources on which to recompose and integrate the different identities of the city, which are the preconditions to build sustainable development paths in Taranto (Elkington, 1997).

This research aims to give a strategic framework to start development paths, making the city rediscover a clear cultural and tourist vocation, starting from its unique and inimitable historical endowment of local resources and specificities (Caroli, 2013; Cerquetti, 2014).

The analysis of MArTA will demonstrate how a sustainable museum model can be the driving force of local development and civic innovation thanks to its potential for inclusion.

2 The Open Organization as Model of Sustainable Strategic Innovation

As the last decades have been characterized by social, economic and technological changes, museums have gradually launched transformative policies of their core concept, thus being again an active part in the process of social innovation and territorial development, a role that meets the Goal 11 requirements of the 2030 Agenda of the United Nations «sustainable cities and communities; making cities and human settlements inclusive, safe, sustainable and long-lasting» and those of Goal 12 «ensuring sustainable patterns of production and consumption».

The paper therefore provides a theoretical framework for the concept of sustainable strategic innovation, highlighting the link between sustainability and organizational change. This requires a greater level of involvement that can reach all local stakeholders. In this regard, the discussion will then describe the open organization as a model of sustainable strategic innovation. This museum model can potentially be the driving force of local development and civic innovation, thanks to its inclusion potential, but it can also be a strategic tool for the analysis of decision-making processes.

By stressing the specific response to Goals 11.3, 11.4, 12.b of the UN 2030 Agenda, the paper identifies and provides an analysis perspective that highlights the impacts of organizational (as well as technological) innovation policies. These polices tend towards an inclusive museum model based on the active participation of visitors/citizens and on the link of identity between the museum offer and the archaeological , historical and cultural heritage of the territory, thus revealing also a strategic territorial approach that can bridge the gap between the identity and image of the territory itself.

The imperative that the current extreme socio-economic uncertainty increasingly imposes on organizations is to renew their value, not just create it. This is the change paradigm that companies are required when they want to renew their growth conditions in order to sustain their competitive advantage. It is a path that requires the ability to innovate strategically and that goes beyond the idea of an enterprise that is stubborn in putting the opportunistic logic of financial return first. This requires a shift from the shareholders-oriented model to a broader stakeholders-oriented vision, in a synergistic and engagement perspective, capable of creating shared value—or CSV, "Creating Shared Value" (Porter & Kramer, 2011)—without neglecting strictly financial logic and performance.

After all, these are the rationales behind CSR, "Corporate Social Responsibility" (Carrol, 1979; Carrol, 1991; Perrini & Tencati, 2008a, b; Schwartz, 2011): it is what management literature defines as «the voluntary integration of social and environmental objectives and practices in business processes and relations with stakeholders (...) the result of the company's progressive recognition of the need to open up to dialogue, cooperation, with its own reference context in order to create sustainable value and focus on the long term» (Perrini, 2016, 7). The combination of CSR principles and shared value creation perspectives inevitably results in a sustainable strategic innovation model, based on the ability of organizations to meet the needs and expectations of different stakeholders, following a collaborative and sustainable relationship approach (Freeman et al., 2007).

Therefore, due to the multiplicity of resources offered, it is necessary to consider the dependence of people and organizations on the environment in which they live. From this point of view, the worsening nature of the complexity and structural dynamism of the reference environments leads to the change of balances and relationships, marked by the transition from competitive logics to increasingly collaborative approaches, from a narrower intraorganizational perspective—which considers organizations as «closed mechanical systems» (Morgan, 2002, 62 et seq.)—to an interorganizational one, according to what Von Bertalanffy (1968) defined as open systemic approach or open systems' theory, drawing inspiration from living biological systems whose characteristics are also valid for organizational systems. Their continuous interaction with the macro-environment is no longer limited to being a mere solicitation for organizations, aimed exclusively at bridging a strategic gap, but it becomes, in perfect harmony with the principles of the "Economics of Collaboration" (Maiolini, 2016; Manzo & Ramella, 2019; Tencati & Zsolnai, 2009), a real opportunity for sustainable development both for the company and for the community of which it is an integral part.

It is essential to note that corporate models and practices are also perfectly associated with museums when they define strategic tools, organizational structures and management models. In fact, although public museums have eminently cultural institutional purposes that come from an appreciation of merit shared by the community (Montella, 2003, 270–271; Musgrave, 1995, 184–191), it is precisely in the purposes' management that we identify the need to define the museum enterprise as the «economic dimension of the museum-institute» (Bagdadli, 2000; Dubini, 2017, 107 et seq.; Grandori, 1995).

3 The Network Between MArTA and the Mar Piccolo Ecomuseum

A series of objectives that the Direction of the MArTA intends to pursue have been identified within the strategic plan of the museum. Focus will be placed above all on the objective of encouraging a "territory" project trying to transform the museum not only into a center of research, education, pedagogy and preservation of heritage assets, but also into driving force of local growth that is able to embrace the socio-cultural, tourist and economic dimensions of the territory. Therefore, in addition to the cultural heritage, also the identity of the city should be fostered, thus making the museum the main actor of this relaunch.

The Italian experience places ecomuseums in a position of complementarity and not of antagonism with museums (Montella, 2003). In the territory of the "widespread museum" (Iaffaldano, 2013), there are many examples of positive collaboration between ecomuseums and museum networks. One of them exists between the ecomuseum of the Mar Piccolo and the MArTA.

The experiences realized in Apulia offer an original point of view on the perception that citizens have of the cultural heritage protection of their territory, creating a bottom-up cartography. The project on "Ecomuseums" is committed to encouraging the perceptive interpretation of places by the local population, as it is the main actor of the transformation dynamics of the territory in all its complexity.

At the organizational level, the project is structured around the ecomuseum laboratories, as they are spaces where citizens meet to develop a mechanism of recognition, value attribution, sharing, protection and enhancement of the tangible and intangible heritage (traditions, ways of life, languages, music, etc.) that individual communities have produced over time.

The research path is therefore based on the participatory construction of landscapes, identifying the ecomuseum as the most suitable formula for the realization of the network process for local and sustainable territorial development.

For this reason, one of the aspects that distinguishes the ecomuseum design is the high rate of relational activity: much of the work is carried out in a collective dimension, in which the community, the facilitators, the technicians and the actors of the territory interact. The ecomuseum project is strongly influenced by the quality of such relationships and interactions. They take place in different situations which may be public meetings, consultation tables, workshops, working groups, etc., more or less suitable to stimulate creativity, collective intelligence and the so-called "community visioning" (Baratti, 2012; Gavinelli, 2012).

It is particularly interesting that ecomuseums are associated with local networks. Ecomuseums depend on the internal networks of the community and on the territory of reference, but they cannot develop, sustain themselves and open up to the outside world without also appealing to partners, supporters and external financiers.

It is a matter of ensuring the cooperation and collaboration of those we could call the stakeholders of ecomuseums: all those who have an interest in the existence and activities of ecomuseums and can take a cultural, social, educational or economic advantage of them.

To give a few examples, researchers and university laboratories are interested in the help that an ecomuseum can bring to their research work in the field of natural sciences and humanities; tourist agencies want to offer their customers a wide range of cultural tourism, ecotourism, family stays and promote local hospitality; enterprises exploit local resources, use the local labor force or wish to invest and settle in a territory, whose heritage and cultural activities are attractive factors.

These subjects are carriers of different interests that an ecomuseum will have to organize in order to offer them the services required and thus obtain the resources that will enable that museum to keep existing and grow for the good of the community. It will also be necessary to compensate the potential negative impact on the territory for the population (e.g., pollution, threats to the landscape, etc.) and possibly increase a resistance able to counteract unsustainable impacts. All these presuppose the creation of many new networks and are working together without being dominated or exploited (de Varine, 2014, 7–19). For instance, too many scientific research programs in the area, with the help of an ecomuseum, can make a community feel disaffected, as they realize that "their museum" serves external researchers at the expense of the real interests of the community.

Ongoing experiences in various countries, both European and non-European, have shown that it is possible to promote innovative cultural projects such as ecomuseums, when they aim to exalt the territory as a privileged area of relationships between humans and nature, as a place of knowledge of local communities, as a testimony to environmental values, as a space that puts in synergy the relational skills of the inhabitants in order to set up long-lasting, endogenous and eco-sustainable development processes.

An ecomuseum can be defined as a cultural practice of identification and protection of the local system that involves the local community for the enhancement of the "territorial heritage" in the perspective of sustainable development. It is based on a pact with the local community that obliges the inhabitants to take care of the landscape. However, it is not a set of rules that oblige or prohibit something, but an unwritten and generally shared agreement.

The most important activity of ecomuseum workshops was to promote the construction of so-called landscape community maps (Baratti, 2012, 76; Leslie, 2006), a participatory method that has allowed the involvement of inhabitants in territorial planning.

Community maps—essential tools to understand if it is possible to create an ecomuseum in a territory—constitute a participatory process that involves inhabitants in an exercise of identity self-representation and recognition of the typical values of the place they live in. Maps are actually built to ensure that the diversity of a territory continues to exist (Clifford et al., 2006; Clifford & King, 1996; Magnaghi, 2010; Reina, 2014).

They are very different from topographic maps, as they are sentimental maps that transmit the consistency of the places value.

4 Research Aim and Methodology

The research objective is trying to give a strategic framework in order to start development paths, thus making the city of Taranto rediscover a cultural and tourist vocation, thanks to its historical endowment of local resources.

The analysis of the Archaeological Museum of Taranto, "MArTA", clarifies how a sustainable museum model can help local development and civic innovation thanks to its inclusion potential and its ability to involve various local stakeholders. Therefore, the paper aims to shed light on the path of sustainable strategic innovation undertaken by museum organizations, in particular by the MArTA museum organization.

Faced with the social, economic and technological changes that have characterized the last few decades, museums have gradually initiated transformative policies of their core concept. In this way, they have rediscovered themselves as an active part in the process of social innovation and territorial development, following the Sustainable Development Goal 11 of the UN 2030 Agenda.

As regards the methodology, a qualitative approach was adopted, as it provides flexibility and adaptability to the research aims and is more interactive, in-depth and sensitive when studying human behaviors, attitudes, ways of thinking and motivations (Carson et al., 2001; Silverman, 1998). For this reason, 25 in-depth, face-to-face, semi-structured interviews were conducted, each lasting approximately 40–50 min.

The respondents belong to various categories of stakeholders: city and regional policymakers; director and operators of the "MArTA" museum; representatives of the local community; representatives of cultural, sports, environmental and social promotion associations; people belonging to entrepreneurial and commercial associations as well as to universities and other research centers; operators of local production sectors; port authority. The sample was composed as follows: 7% of the sample interviewees are city and regional policymakers; 17% representatives of local community; 6% management of the "MArTA" museum; 24% representatives of cultural, sports, environmental and social promotion associations; 7% entrepreneurial and commercial associations; 18% universities and other research centers; 19% operators of local production sectors and 2% port authority.

All interviews were recorded and transcribed on the same day to ensure minimal information loss. A subsequent step was to structure meanings by using narrative and leaving participants' stories as they were told. The narrative provides more in-depth information about participants' ideas and attitudes (Saunders et al., 2009).

Data analysis followed a deductive coding approach. The data were managed by hand through a thematic analysis. The three main themes identified are as follows:

- The wealth of knowledge and distinctive expertise matured and sedimented in a specific context and fed into the production processes of the "made in Italy" companies enrich the output and are a factor of competitive advantage.
- The presence of collaborations between institutions and citizens, between businesses and citizens, companies' mutual collaboration to measure the relational skills of the territory and how all that benefits the local community.
- The presence of tourist-cultural itineraries or routes involving different actors (public and private) through actions aimed at the composition of an integrated offer for the measurement of the relational assets of the territory and the measurement of how all that benefits local public and private actors.

5 Results

The wealth of knowledge and distinctive expertise matured and sedimented in a specific territorial context are a factor of competitive advantage of the "made in Italy" companies.

As regards the valorization and promotion of the territory as source of vitality for old and brand-new enterprises, the director of the MArTA museum asserted that: "*As part of the project, the museum has contributed to the realization of experimental archaeology laboratories in various sectors: the naval carpentry, the production of purple dye, the construction of fishing nets and lobster pots, salt production and use, and the re-creation of fish preserves drawn from ancient recipes*". When he/she talked about the laboratory on the fishing techniques that have ancient origins, he/she described the use of the lobster pots that are depicted in some North African mosaics of the third and sixth centuries AD, emphasizing that they are still used by Taranto fishermen. Moreover, he/she illustrated the ancient and still used "seine" technique where fishing nets were dragged on boats and/or ashore on foot from their flaps.

A representative of a social promotion association expressed a similar opinion: "*Our schools decided to preserve the ancient craft of naval carpentry that was disappearing. In our opinion, this can be a good way to revitalise the city. At present we are the first and only ones to do this type of activity, and many operators have contacted us, and the city mayor told us that our association, "Officina Maremosso", will be the fulcrum of this sector, aiming at recreating the image of Taranto as a seaside town. For this reason, we try to use the ancient techniques for the maintenance, the construction and the restoration of wooden boats and to involve young people*".

A representative of a cultural association, talking about the laboratory that deals with the production of purple dye, said that the pigment was extracted from Murex shells. He/she also explained that their yellowish glands were extracted after breaking the shells. Then, they were immediately left in the sun to make them become bluish or purple in color. Purple dyed products were very valuable, because it was calculated that it took about 8,000 Murex shells to get a gram of purple to dye a tunic. He/she added that, for this reason, the cost of a purple tunic was very high, even because there was a long color fixing process of the fabrics that involved many stages. In ancient times, this manufacture became an expression of refined outfit.

He/she said that: "*The purple of Taranto was defined "rubra tarentina" by Pliny the Elder and was very appreciated for its chromatic hue, which is the same shade of the clothes of some statuettes that can be seen in the museum*". He/she then added that quite certainly the production laboratories of purple dye were located along the shores of Mar Piccolo, because the remains of shredded Murex shells were found there. That heap of broken shells was later called "Monte dei Coccioli", that is Mount of Shards.

A marine biologist of Taranto University stated that salt was undoubtedly an important resource of the sea for the city. Ancient scholars exalted its excellent quality, refinedness and whiteness. Moreover, it was also very appreciated in the medical field, because it was the so-called "pond salt" present both in the "salina piccola" (small saltworks) and "salina grande" (large saltworks), which are located in the southeastern area of the current urban center. The availability of salt led to the development of several related activities, from the manufacture of wool—very renowned for its quality—to the production of fish salting and the making of purple dye. An important historical record of a well-known Athenian physician of the second century AD mentioned salt-cured tuna from Taranto, which was transported by sea. Scholars think that Taranto played an important role in the export of tuna fish both locally and all over Southern Italy.

The presence of collaborations between institutions and citizens, between businesses and citizens, companies' mutual collaboration for the measurement of the relational skills of the territory.

A city councilor said that the previous municipal administration had triggered a synergistic process where they had set the objective of becoming the leader of a project involving other political bodies, such as the Region, the national Government and the European Union. He/she pointed out: "*The municipal administration has defined the strategic plan "Ecosistema Taranto". It started from a cultural approach that is different from what the history of this city has produced, namely the great disasters of the local economic, social and also cultural system. The vision is to move from an "egocentric system" to an "eco-centric system", i.e., a system where the main actors of the territory come together, work in synergy so that we can work on the construction of a sustainable and circular economy based on what were and are the peculiarities of the territory. For this reason, we have defined a broad strategic plan ranging from urban regeneration to local community participation to the urban regeneration process through an integrated, cross-sectoral, economic system".*

A representative of a commercial association expressed an opposite opinion. He/she said: "*In my opinion, the first thing that must be done is to revitalise the city, because it is a dead city as it has been brought on its knees. Taranto people just can't understand how the dilemma of shutting down or not shutting down the Ilva steel plant will be solved. Today the Ilva is a "hanging sword over our heads" that does not allow us to do and say anything else, because you know it is there but cannot get what its destiny will be. Moreover, to my mind, this city has not got ready for the so-called urban renovation process. Indeed, it's true that we complain about shutting down the Ilva plant, but if we assume to shut it down for a moment, can we say what we have built as an alternative? Nothing! What is the system we want to create in order to grow? In my opinion, it is the tourism sector, where all the actors must think in the same way, that is in terms of system, to create an efficient system of hospitality, reception services, hotels, restaurants, transport*".

The director of the archaeological museum said: "*I felt that the ecomuseum experience was the most suitable for the context of Taranto because of the industrial and socio-economic crisis. In fact, it is necessary to encourage the active participation of the local community, who are the holders of the stratified knowledge of the history of their territory, to rediscover forgotten pieces of the landscape, and to re-establish an active relationship with the sea and its resources*".

A representative of a cultural association stated: "*To implement this project aiming to establish an Ecomuseum of Mar Piccolo, we worked in the neighbourhoods of the coastal strip of the city, met and interviewed students, inhabitants, fishermen and seafarers, mapped the significant places of the surrounding territory, collected stories, photographs, videos and typical recipes, organized activities and meetings in public spaces, in aggregation places and in the MArTA museum*".

The same narrative was provided by a university archaeologist: "*In recent months we climbed on fishermen's boats, lived next to them, recorded the following data: the individual actions which accompany their use of tools; their experiences and their stories of the sea crafts history; the contribution of fishermen, shipwrights, and sailors to the construction of the cultural identity of Taranto and Apulia*".

The director of the MArTA museum expressed a similar opinion: "*The central element of the project is the participation of the citizens, which is significantly exemplified in the Mar Piccolo Fishermen Community Maps, as they represent the operational tools of the participatory process involving the inhabitants in an exercise of identity self-representation and recognition of the typical values of the place they live in*". He/she added that these initiatives are intended to encourage the discovery of new functions and new social roles, as well as to create new professions and offer job opportunities for young people, thus strengthening social awareness through the knowledge, valorization, communication and fruition of the territorial assets.

The presence of tourist-cultural itineraries or routes involving different actors (public and private) for the measurement of the relational assets of the territory.

The Director of the MArTA museum pointed out: "*Through the ecomuseum laboratory three thematic itineraries within the Mar Piccolo have been identified: the mussel cultivation itinerary, the history, archaeology and industrial archaeology itinerary, the environment and nature itinerary that enhances the richness of marine and terrestrial biodiversity*". He/she continued: "*The itineraries identified are covered with a "slow" mobility of canoes and light boats. Mobility is on traditional boats along the routes with the guidance of archaeologists, and/or tour guides and fishermen who tell about traditional fishing techniques and the tools they use*". He/she added that there are also visits to mussel farms and to a wooden "Pagliaro", a typical hut linked to the life of mussel farmers. Moreover, there can be various events and theatrical performances along the coast of the Mar Piccolo thanks to some theatrical companies able to enhance the coastal heritage and the resources of the sea. He/she specified: "*At the MArTA museum there were two events: an exhibition whose name was "Taranto e il mare", Taranto and the sea, and a virtual tour of the Mar Piccolo Ecomuseum by means of projections on a giant-immersive screen. We have created a multimedia app with animations dealing with "Taranto and the sea" that can be used while visiting the museum. It illustrates the artifacts of the sea displayed in the museum and the places where they were found, that is along the shores of the two inlets of the Mar Piccolo*".

A representative of an environmental association expressed a similar opinion. He/she said: "*If I had to design a sustainable development strategic plan, I would focus on some tourism segments: nature tourism in the Mar Piccolo, tourism linked to mussel farming and tourism related to cetaceans, namely, the dolphins in the Gulf of Taranto. These are three realities that already exist, and they only need to be better organised*". He/she said that the region authorities have recently established the Regional Natural Park in the Second Inlet of Mar Piccolo, while in the First Inlet, reclamations are ongoing. Therefore, tourism is undoubtedly linked to nature tours. In the first place, it is necessary to rebuild habits and provide both the reclamation and the renovation of natural environments. In the second place, it will be necessary to trace the naturalistic routes that WWF has been organizing for many years, canoeing or kayaking, and finally promoting birdwatching.

The same narrative is reported by an operator of the local mussel farming production sector. He/she said: "*I've always accompanied tourists to show them the mussel farms in Mar Piccolo and to make them taste our mussels. Moreover, through a story I created, I use to tell them about the previous scenario of the place and the types of mussel and oyster farms that were made there. I also talk to them about older businesses existing here: the production of sea-silk from a protected species of shell called "pinna nobilis" used to make fabrics and the production of purple dye from the murex shells used to colour fabrics. All that was very linked to our local identity. This ongoing process of urban renovation and repositioning of the city will be important for us, mussel farmers, because the activities related to what we do will be enhanced, namely, to make mussel farming a form of experiential tourism*".

An operator of the local fishery production sector expressed an opposite opinion. He/she said: "*The real problem is the lack of logistics because we fishermen have no available areas to use. What I mean is that we have not got the area logistics that can offer a touristic route to those who come to Taranto and make a stop at its historical and picturesque mussel farms. For this reason, if this logistical surface isn't created, it will be difficult to think of mussel farmers welcoming tourists, maybe offering them the experience of tasting the typical mussel "impepata" (fresh mussels opened in a hot pan and seasoned with oil, garlic and pepper) that they can do on the spot*".

6 Conclusions

As MArTA is a site museum and not a collection one, everything that is exposed in it comes from the territory. This feature assigns to the archaeological museum of Taranto a role of "physical, organizational and intellectual gateway to the Mediterranean" for an inclusive museum model participated by the community. It is a museum model that opens its display cases to the territory, thus overcoming the paradigm of "museum-boredom" and becoming a driving force for a complementary development alternative to the large-scale industry and the environmental pollution that has been the image of this Apulian provincial capital for several decades. This concept is not addressed only to visitors–tourists. Actually, it gives visitors–citizens a central position, as they are part of the territory and the community: they become the protagonists and co-owners of the archaeological and cultural heritage enhanced by the MArTA and the ecomuseum. The Italian experience places ecomuseums in a complementary and not antagonistic position to museums, as it luckily happens in Taranto.

Moreover, the synergistic cooperation with students, inhabitants, fishermen and seafarers has allowed to create community maps of Taranto territory. Now, thanks to these maps, in a co-planning perspective, students, as younger citizens, are able to detect future of the city and know their territory better. The transformative approach adopted by the MArTA helps to understand how the logic of sustainability is able to converge the three declinations of innovation: technological, strategic organizational and civic social. The collaborative wealth of the MArTA is beneficial for its organization as it allows the museum to establish itself as a solid and authoritative territorial reference, able to provide new and more sustainable scenarios of social and urban development. Therefore, from this consideration, some elements become clear: not only the particular applicability of tools and models of strategic innovation in museum organizations but, even more obviously, the indispensable correlation between them and the sustainability objectives included in the 2030 agenda of the United Nations.

The limit of this research lies in the awareness that the adoption of the above-mentioned innovative and inclusive policies meets a need: a broader and systemic tourist orientation that concretely involves all economic and institutional actors. To this end, future research perspectives could focus on the analysis of the consequences of this new strategic organizational approach on museum marketing and territorial marketing.

References

Bagdadli, S. (2000). *Il museo come azienda. Management e organizzazione al servizio della cultura.* Milano: Etas.

Baratti, F. (2012). *Ecomusei, paesaggi e comunità. Esperienze, progetti e ricerche nel Salento.* Milano: Franco Angeli.

Caroli, M. G. (2013). *Il marketing territoriale. Strategie per la competitività sostenibile del territorio.* Milano: Franco Angeli.

Carrol, A. B. (1979). A three-dimensional conceptual model of corporate performance. *Academy of Management Review, 4*(4), 497–505. https://doi.org/10.5465/amr.1979.4498296

Carrol, A. B. (1991). The pyramid of corporate social responsibility: Toward the moral management of organizational stakeholders. *Business Horizons, 34*(4), 39–48. https://doi.org/10.1016/0007-6813(91)90005-G

Carson, D., Gilmore, A., Perry, C., & Gronhaug, K. (2001). *Qualitative marketing research.* London: Sage Publications. https://doi.org/10.4135/9781849209625

Cerquetti, M. (2014). *Marketing museale e creazione di valore: strategie per l'innovazione dei musei italiani.* Milano: Franco Angeli.

Clifford, S., & King, A. (1996). *From place to PLACE: Maps and parish Maps.* London: Common Ground.

Clifford, S., Maggi, M., & Murtas, D. (2006). *Genius Loci. Perché, quando e come realizzare una mappa di comunità.* Torino: Ires Piemonte.

de Varine, H. (2014). Gli ecomusei. Una risorsa per il futuro. In G. Reina (a cura di), *Gli ecomusei. Una risorsa per il futuro* (pp. 7-19). Venezia: Marsilio Editori.

Dubini, P. (2017). Le aziende che operano nelle arti e nella cultura e i loro assetti istituzionali. In P. Dubini, F. Montanari, & A. Cirrincione (a cura di), *Management delle aziende culturali* (pp. 97–134). Milano: Egea.

Elkington, J. (1997). *Cannibals with forks. The triple bottom line of 21st century business.* Oxford: Capstone.

Freeman, R. E., Harrison, J. S., & Wicks, A. C. (2007). *Managing for stakeholders: Survival, reputation and success.* Yale University Press.

Gavinelli, L. (2012). *Territorio, networking a management come dimensioni di analisi per le decisioni degli ecomusei italiani.* Lavis, TN: Cedam.

Grandori, A. (1995). *L'organizzazione delle attività economiche.* Bologna: Il Mulino.

Iaffaldano, N. (2013). Il 'museo diffuso' e le risorse immateriali quali determinanti della competitività di una destinazione turistica culturale. *Esperienze D'impresa, 2*, 43–67.

Law, C. M. (1992). Urban tourism and its contribution to economic regeneration. *Urban Studies, 29* (3–4), 599–618.

Leslie, K. (2006). *A sense of place: West Sussex Parish Maps.* Chichester, UK: West Sussex County Council.

Magnaghi, A. (2010). *Il progetto locale. Verso una coscienza di luogo.* Torino: Bollati Boringhieri.

Maiolini, R. (2016). L'evoluzione delle relazioni con gli stakeholders. In M. G. Caroli (a cura di), *L'innovazione delle imprese leader per creare valore sociale. Terzo rapporto CERIIS sull'innovazione sociale* (pp. 89–107). Milano: Franco Angeli.

Manzo, C., & Ramella, F. (2019). *L'economia della collaborazione.* Bologna: Il Mulino.

Montella, M. (2003). *Musei e beni culturali. Verso un modello di governance*. Milano: Electa Mondadori.
Morgan, G. (2002). *Images. Le metafore dell'organizzazione*. Milano: Franco Angeli.
Musgrave, A. R. (1995). *Finanza pubblica, equità, democrazia*. Bologna: Il Mulino.
Perrini, F. (2016). *Management. Economia e Gestione delle Imprese*. Milano: Egea.
Perrini, F., & Tencati, A. (2008a). *Corporate social responsibility. Un nuovo approccio strategico alla gestione d'impresa*. Milano: Egea.
Perrini, F., & Tencati, A. (2008b). La responsabilità sociale d'impresa: Strategia per l'impresa relazionale e innovazione per la sostenibilità. *Sinergie, 26*(77), 23–43.
Porter, M. E., & Kramer, M. R. (2011). Creating Shared Value. *Harvard Business Review, 89*(1–2), 62–77.
Reina, G. (a cura di). (2014). *Gli ecomusei: Una risorsa per il futuro*. Venezia: Marsilio Editori.
Saunders, M., Lewis, P., & Thornhill, A. (2009). *Research methods for business student*. New York, USA: Pearson Education.
Schwartz, M. S. (2011). *Corporate social responsibility: An ethical approach*. New York, USA: Broadview Press.
Silverman, D. (1998). Qualitative research: Meaning or practices? *Information Systems Journal, 8*(1), 3–20. https://doi.org/10.1046/j.1365-2575.1998.00002.x
Tencati, A., & Zsolnai, L. (2009). The collaborative enterprise. *Journal of Business Ethics, 85*(3), 367–376. https://doi.org/10.1007/s10551-008-9775-3
Valdani, E., & Ancarani, F. (a cura di). (2000). *Strategie di marketing del territorio. Generare valore per le imprese e i territori nell'economia della conoscenza*. Milano: Egea.
Von Bertalanffy, L. (1968). *Teoria generale dei sistemi. Fondamenti, sviluppi, applicazioni, tr.it.* Milano: ILI.

Experience Tourism as a Smart and Sustainable Form of Tourism in the Twenty-First Century

Lubica Sebova, Radka Marcekova, Ivana Simockova, and Kristina Pompurova

Abstract Experience tourism is a form of tourism in which participants are motivated to seek experiences that are different from everyday life. It has two extreme approaches. The first is seeking and experiencing social contacts and contact with nature and cultural environments, which supports sustainable development of the destination, and the second is seeking experiences within attractions. Experiential tourism is associated with a personal and subjective experience in which main attribute is interactivity. Experience tourism can be distinguished from traditional tourism by the fact that it engages all human senses. In experience tourism, it is important to place greater emphasis on reconciling the economic, social and environmental objectives of the destination in order to ensure its sustainable development. It is essential that visitor gains unforgettable experiences, i.e., different from everyday life. The aim of this paper is to examine the supply and demand of experiential tourism products in a selected region with respect to smart and sustainable tourism development. Promoted visitor experience can be an important tool for destination marketing communication in the twenty-first century. If the destinations want to be competitive in a dynamically developing tourism market, the must adapt to this situation and offer the visitors an unforgettable authentic experience in smart and sustainable way. The goal is to create an offer that allows the visitors to actively engage in tourism activities and thus to intensify their emotions.

Keywords Destination · Experience · Sustainability · Tourism

JEL Classification L83 · Z32 · M31

1 Introduction

Participation in experience tourism is a personal and highly subjective experience for everyone. Visitors are primarily motivated by seeking out experience they do not

L. Sebova (✉) · R. Marcekova · I. Simockova · K. Pompurova
Matej Bel University, Banska Bystrica, Slovakia
e-mail: lubica.sebova@umb.sk

© The Author(s), under exclusive license to Springer Nature Switzerland AG 2023

V. Katsoni (ed.), *Tourism, Travel, and Hospitality in a Smart and Sustainable World*, Springer Proceedings in Business and Economics,
https://doi.org/10.1007/978-3-031-26829-8_12

yet know and are interested in experiencing new situations. The aim of experience tourism is to offer visitors different types of attractions, ranging from adrenaline to sightseeing. Tourism businesses that offer such activities do not only make money from entrance fees or other activities, but also raise awareness of the destination and their brand. This motivates the visitor to stay longer in the destination, leading to increased tourism revenue and a longer season in the area. Experience tourism is a smart and sustainable type of tourism, it differences from mass tourism, and it offers something special and personal to every visitor.

Experience tourism and its attendance are currently a trend that greatly affects the supply of tourism products and provides the transition from mass tourism to sustainable and smart tourism, from offering products to everyone to offering products tailored to individual customer requirements, from consumption to experience (Adillon, 2019). The desire to experience something new, unrepeatable, exceptional stems from today's fast and often stereotypical way of life. The experience creates in each person personal feelings and emotional experiences that cannot be transferred to other people and are a guarantee of the satisfaction of the visitor. Getting a complex experience is one of the main conditions for tourism realization.

2 Literature Review

Experience tourism is a subject of interest of many domestic and international authors, especially from Kotíková (2000, 2013), Seaton and Bennett (2001), Gúčik (2006), Tarssanen (2009), Sharpley and Stone (2012), Tangeland and Øystein (2011), Dubská (2012), Zelenka and Pásková (2012), Nowacki (2013), Goeldner and Ritchie (2014), Voříšková and Indrová (2015), Pomfret (2018), Bergstrom (2021) and others.

Experience tourism is generally based on a number of basic principles: the added value of activities, experiences, catering services and accommodation or infrastructure (Tangeland & Øystein, 2011; Seaton & Bennett, 2001). Authors Sharpley and Stone (2012) claim that tourism in general is about experience. According to them, the experience consists of four components—emotions that relate to the choice of destination, excitement from planning and route preparation, sense of adventure and a set of emotions that visitors get after experiencing the tourism product. An experience can represent the inner state of an individual resulting from something he or she personally met or survived (Nowacki, 2013). Every tourist expects to experience something extraordinary, exciting and unrepeatable. Therefore, experience is a prerequisite for his satisfaction (Gúčik et al., 2006, p. 209, password Experience). The experience can be defined by several attributes. It is a strong emotional engagement (excitement), it only takes place at a certain time and under certain conditions (finiteness), it cannot be replaced by another experience, and it is always unrepeatable (uniqueness) and cannot be transferred to another person (Kotíková, 2000). According to Gúčik (2006), experience can be an exploratory (something new, unusual), somatic and physiological, social (social contact, society) and additional (to be successful as a person), and like Kotíková (2013), he claims that experience is associated with emotional engagement of individual, which is unrepeatable and passable.

In tourism, the "experience environment" is made up of all the people and things surrounding the tourists. It does not only include the time when people travel, but also the period before they decide to travel and after they return home. Consequence of this is that every person is surrounded by the unique "tourist experience network" of all the participants included in his or her experiences, whether real, virtual or even fictitious. Various actors contribute in gaining tourist experiences. Tourist experience networks are enormous and connect human beings with the people they travel with (friends, partners, family, interest groups, colleagues), with the internet, with a virtual travel community, travel agencies, tour operators and travel service providers, hoteliers, guides, local businesses offering activities in a given destination, with local residents, monuments, activities and various attractions, typical landmarks, museums, places of cultural heritage, events, natural scenery, techniques, and so on (Bergstrom, 2021; Valeri, 2021).

To the theoretical definition of experience tourism is paid more and more attention. To the experts in the field of experience tourism belongs Tarssanen (2009), who considers experience tourism for a phenomenon of the twenty-first century. She characterizes it as a positive, extensive and multi-sensory perceived emotional experience that evokes in every visitor various personal feelings. According to the author, individuality, authenticity, story, perceptions of multiple senses, interaction and contrast are basic elements of the experience tourism.

Czech authors define experience tourism as a form of tourism, in which visitors are motivated to find experiences different from their normal life. We can divide it according to two approaches. The first approach is to find and experience social contacts, contacts with nature and the cultural environment in the destination. The second approach is to find experiences in the different types of attractions. Among the categories of experience tourism, they include gastronomic, adventurous, gambler, meditation, sport, virtual or cosmic tourism (Zelenka & Pásková, 2012).

The tourism product is defined by UNWTO (2022a, b). The product of experience tourism and its categorization is also discussed by Pine and Gilmore (1999). They divide experience tourism products into four categories, based on the identification of two basic dimensions. One dimension is activity–passivity in the sense of whether the consumer of an experience tourism product is actively engaged in an activity or is merely a passive spectator. The second dimension distinguishes between mental and physical participations, namely whether the consumer of the experiential tourism product is mentally immersed in the consumption of the product or whether he or she is also physically engaged in participating in experiential tourism.

The biggest difference between traditional tourism and experience tourism is in human perception. In the case of traditional tourism, people look in particular at a pleasant and stimulating environment, and the experience is only the result of their traveling. Conversely, in the context of experience tourism, emphasis is placed on personal subjective experiences, where the main attribute is interactivity (Voříšková & Indrová, 2015). The source of the tourist experience can be something else for different people. It can be nature and the natural environment, the social environment and contact with the local population, the quality of services or the level of programs—cultural, sport, cognitive, social or educational. In order for all these components to become an experience, they must be new and unusual for the visitor or at least perceived in a new environment or situation. In this case, for somebody, a common phenomenon can be an experience, such as sunset or service, the first flight by airplane, accommodation in a first-class hotel (Kotíková, 2000). Based on their research, Voříšková and Indrová (2015) identified four main groups of the experiences—gastronomic experiences, adrenaline adventure experiences, unusual visits of historic and natural places and wellness and health.

3 Methodology

The aim of the paper is to characterize experience tourism on the example of selected region, as well as to analyze the supply and demand for experience tourism products in the region. The aim of experience tourism is to evoke in the visitor the most intense authentic experience combined with strong emotions. We understand the experience tourism as a form of smart and sustainable tourism.

We evaluate the potential for the development of experience tourism in the Liptov region, based on the market research, demand and supply analysis. The subject of exploration is experience tourism. The object is the destination of Liptov region in Slovakia. Liptov region (2022) belongs to the three most tourism developed areas in Slovakia, whether in terms of number of overnights, visitors or tourism attractions. It has wide primary (natural attractions, cultural and historical monuments, organized events) as well as a secondary offers (accommodation, catering, sports and recreation facilities, public and private institutions). It has borderline with Poland in the north (Fig. 1).

We are asking following research questions: What is the offer of experience tourism products in Liptov region? What is the demand for experience tourism products in the Liptov region?

We get an answer to the first question by analyzing the secondary resources which we obtain by reviewing the websites of the experience tourism product providers. By supply analyzing, we focus on the tourism businesses and institutions operating in the region and the offer of their activities. Individual activities of adventure tourism will be divided according to the themes (adventure, sport, cognition, culture, health, recreation and gastronomy), on the basis of which selected businesses provide them to the visitors to ensure their complex experience.

We obtain the answer to the second question by analyzing the demand for experience tourism in the Liptov region through the sociological survey method. The data collection technique was a questionnaire survey. We have distributed the questionnaires electronically. The survey was realized in the months of March–May 2022. We were able to get 154 responses. The questionnaire consisted of 15 questions and was divided into two parts. The first part was focused on finding out whether visitors

Fig. 1 Liptov region localization. *Source Liptov region map*, 2022

and residents of Liptov region use the offer of experience tourism products. We used closed and semi-closed questions with the choice of one or more answers. In the questions, we have found out from where the visitors have received the necessary information about the region, how often and with whom they visit the region, what are their main motives and reasons for selecting this destination or what factors affect them in their selection. We have also found out which services and specific activities were performed during their stay in the Liptov region. In the second section, we have aimed on the demographic and other visitors' data. At the conclusion of the questionnaire, we have left the space for the opinions of the respondents who visited Liptov region and have experienced the offer of the products of experience tourism.

Based on the answers to the research questions, we evaluate whether there is potential for the development of experience tourism in Liptov region.

4 Results

The Liptov Region is located in the northwestern part of the Slovak Republic, on the north it borders with Poland, thanks to the excellent natural conditions, cultural and historical monuments and preserved folk architecture; Liptov is one of the most beautiful and most sought-after regions in Slovakia. The combination of natural attractions, cultural and historical richness, opportunities for sporting activities and active leisure time together with the possibilities of relaxation and experiences make the Liptov region an attractive destination not only for international but also for domestic visitors. Experience tourism could not be developed if the region itself did not have suitable potential for its development.

Experience tourism supply in the Liptov region

The region has excellent natural and cultural conditions for the development of experience tourism, even in winter or summer months. The Liptov region offers mountaineering, alpine hiking, trekking, cycling, yachting, water skiing, wakeboarding, swimming, fishing, caving, bungee jumping, horse riding, hunting, gyms, tennis, squash, badminton, ballooning, sport flying, sightseeing flights helicopters, quad biking, archery, downhill skiing, cross-country skiing, ski mountaineering, snowboarding, mountain scooters, go-karting, bowling and many other activities. The typical adventure or adrenaline activities include paintball, wakeboarding, paragliding, rafting, wakesurfing and others. For canoeing and rafting, there is also a modern water slalom area with artificial canal and major international events are held here every year.

Thanks to the rich primary offer, the region is also visited for hiking or cycling. In the Liptov region, there are over 50 hiking trails and marked trails with a total length of up to 1 185 km. Liptov is surrounded on all sides by mountains, but hiking in them is more focused on medium and less physically fit hikers and thanks to the cable cars to Chopok and Malinô Brdo, even the youngest children can manage the trips. In order to not to forget the youngest visitors, the Liptov region has created the experience tourism product called Liptov with children, which consists of a list of selected places in Liptov where families can spend a pleasant time.

Adventure activities are often combined with sports activities. These are provided mainly by operators of ski resorts. Although in this case it is mainly the winter season, they try to extend their offer to summer sports and thus ensure that the resorts are used all year round. Ski resorts offer skiers excellent conditions for downhill and cross-country skiing. Jasná connects both the north and south sides of Chopok, creating an unrivaled number of ski slopes in Slovakia. The resort also includes freeride zones for skiers. In the region, there are also valleys that meet the conditions for ski mountaineering. Within the ski resort, there are a number of hotels in in the immediate vicinity of the ski slopes, which offer a large capacity accommodation beds at a high level. After-skiing activities are also catered for through the after-ski bars and the recently built Music Club, which hosts concerts and parties of all kinds all year round. There are also shops with souvenirs or sporting goods for skiing or hiking in summer. The resort offers many leisure activities even in summer, such as a bike park with bike rental or a rope course, archery and various other activities (www.visitliptov.sk, 2022).

Experience tourism also includes activities related to exploration. These activities include tours of castles, chateaus, open-air museums, manor houses, etc. Visits to these monuments are often associated with various events that are organized in their vicinity. For example, there may be demonstrations of traditional crafts or folk customs, in which case visitors can actively participate and try out the crafts or customs themselves. Traditions are also followed by many events that are organized throughout the year, from folklore festivals to concerts, sports and adrenaline events. Every year the most famous folklore festival is held in the village of Východná. It is the largest folklore event in Slovakia attracting thousands of visitors from all over Europe. Vlkolínec was listed as a world heritage site by the UNESCO, as a monumental reservation of folk architecture. Liptov is also known for its gastronomic specialties, which are worth tasting in typical cottages or modern restaurants. The modern gastronomic facilities are focused on the experience gastronomy, which includes interconnection between Slovak and international cuisine, serving of quality wines and seating in a pleasant environment.

By analyzing the offer of adventure tourism in the Liptov region, we have focused on the tourism enterprises and institutions operating in the region and their offer of activities. We have divided the various activities of experience tourism according to the motives on the basis of which they are provided to the visitors to secure their complex experience. We have identified several main themes—adventure, sport, cognition, culture, health, recreation and gastronomy. The motive of adventure mainly includes activities taking place in the nature (water areas, rivers, mountains) and in artificially created recreational facilities. These activities are provided by interest associations or private companies.

Conditions for the tourism development and the supply of experience tourism activities at the level of the Liptov region are represented by destination management organization Region Liptov.

Demand analysis of experience tourism in the Liptov region

The analysis of the demand for experience tourism in the Liptov region was obtained through the method of sociological inquiry. We have found out that the most common purpose of participation in experience tourism was the experience connected with sport activities (53.2%). Under this experience belong various adventure sports or participation in a sporting event. The second most common reason was cultural experience (41.6%), where it was an unusual visit to a museum, gallery, theater or castle. Respondents in the region also sought adrenaline experiences (17.6%), gastronomic experiences (13.6%). The purpose of their visit was also wellness stays, family visits, recreational stays or tourism itself (26.6%). After identifying the main purpose of the visit of the respondents, we were interested to know that on the basis of what they decided to participate in the experience tourism in the Liptov region. Almost half of the respondents stated that based on their own experience (43.5%), they either come from the region or visit the region frequently. As many as 33.8% of respondents decided to participate on the basis of the information found on Internet and social networks, a quarter of respondents chose to participate in an experience tourism based on recommendations from family or friends (16.2%). Respondents were least influenced by the other media (4.5%), the poster (1.3%) and also by the travel offer agency (0.6%).

More than a half of the respondents are tourists on a short stay (51.9%), who spent at least three nights in the Liptov region. Approximately 27.8% of the respondents stayed in the region for more than four nights. Another group of respondents consists of day-trippers (14.3%) who spent less than 24 h in the region without overnight stay. The last group is respondents who come directly from the Liptov region (5.8%), i.e., they are residents. With the length of visit, we were interested in the survey to know how often the respondents visit the Liptov region or whether they return to a given destination. The majority of respondents visit the region occasionally (37%), i.e., 1 to 2 times a year. Fewer respondents (29.2%) visit the region more than once a year. Approximately 27.8% of respondents rarely visit the Liptov region. In the survey, we also received the responses from the above-mentioned residents (5.8%).

Furthermore, in the survey, we were interested in how the respondents got to the region. More than 88% of respondents had traveled to the region by car, which is not the most sustainable way of transportation, but it is the most efficient when traveling or visiting several places at once. The second mode of transport was the train (4.5%), and this mode of transport is particularly worthwhile for students, as they can travel by train for free. Some respondents used the bus (1.3%). The rest of the respondents answered that they were directly from the region (5.8%).

Naturally, the respondents used a variety of services during their visit to the region. Therefore, in the survey, we were interested to know what services they used, as everyone's needs during travel are different. The services mostly used by respondents included food services (91.6%) and accommodation services (76.6%). As mentioned above, almost half of the respondents visited the region for a cultural experience, i.e., approximately 52.6% of the respondents used services related to exploring the natural and cultural heritage in the region. In the case of holidays, these were services related to rest and recreation (45.5%) and related wellness services (48.7%). Transport services were used by 16.2% of respondents. Only 26% of respondents in the region used services that related to sports activities. Less used services included information services (10.4%), guide services (7.1%) and health tourism services (3.2%). Based on findings on the services used, we were further interested to know whether the respondents were satisfied with these services. As many as 96.1% of them were satisfied with the services. Only four respondents were unable to judge, which leads us to conclude that they did not use as many services to be able to rate them sufficiently. Two respondents were not satisfied with the services (1.3%) and gave us the reason as too high prices and poor sanitary facilities.

A positive finding of our survey was that more than half of the respondents (78.6%) are looking for tourism activities that offer some form of experience. Only 8.4% of respondents do not seek these activities and 13% of respondents could not judge, whether they did or not. Having established the answers to the above question, we next investigated what experience tourism products they consumed in the region. Most respondents (96.8%) were looking for hiking experience in the Liptov region, as thanks to the excellent natural conditions, this region is the most sought after by tourists. Among the most popular products of experience tourism belong also visits to caves and other natural attractions (73.4%), castle or museums tours (59.1%). In winter, respondents visit the region for wellness (50%), skiing or snowboarding (37%). Organized sports (14.9%) and cultural events (18.2%) are also a draw. Horseback riding is a less popular product of experience tourism in the region (3.9%). Respondents also visit the region for experience gastronomy (16.2%). The longest river in Slovakia, the Váh, which flows through the region, also has great potential and is used for sightseeing cruises, which respondents identified (19.5%) as a popular product of experience tourism. Almost 30% of respondents visit the region because of the cycling routes located there. Other products of experiential tourism that respondents mentioned in the open question were mountaineering or rafting (3.9%).

At the end of the questionnaire, we were interested in respondents' insights and suggestions for improvement experience tourism offer in the region. The most frequently mentioned suggestions included agro-tourism, rural tourism, building more cycle paths (better signposted and landscaped), more sport activities (bungee jumping, water skiing). There should also be focus on adventure activities (snow kiting adventure cave tours) or organize adventurous stays in nature (surviving alone in the forest), for example. In the case of hiking trails, more hiking maps or better signage should be developed. As we have already mentioned above, there is a well-known water reservoir in the region, the Liptovská Mara, and better access to it or its use could attract more visitors not only domestic but also foreign visitors. In the case of better use of the reservoir, the boat trips could be organized. The Liptov region has not only natural but also cultural heritage. It is therefore important to preserve this heritage and folk traditions of the region and to promote it. To show visitors how life used to be in the region, organize more cultural events or invite people to visit farms, ranches, or shepherds' huts, that are self-sufficient and produce their own products, and thus teach visitors make these products.

About 61.7% of women and 38.3% of men were participated in the questionnaire survey. From the perspective of age, most people were in the age group from 36 to 45 (28.6%) and 26 to 35 years (26.6%). People aged 46–55 created 18.8% of all respondents, people from 18 to 25 years 14.3%, people over 56 years 11.7%. The most responses were from people with university education (48.7%), people with secondary school education (43.5%), respondents with university third degree (PhD.) education (2.6%) and primary education (0.6%). According to the social status, the most numerous group of respondents is employed (71.4%), and then, we have students (10.4%), entrepreneurs (8.4%), retires (5.2%), respondents on parental leave (3.9%). More than 37% of the respondents have monthly incomes between 500 and 999 €. Right behind were the respondents that earn between 1,000 and 1,500 €. More than €1,500 is earned by approximately 18.2% of respondents and less than €500 by approximately 13%.

5 Conclusion

On the basis of our analysis, we can conclude that the Liptov region has a great potential for the development of experience tourism. Suitable conditions are created in particular by natural, cultural and historical assumptions. In this paper, we asked two research questions. On the basis of our research, we found that the offer in the field of experience tourism in the Liptov region is attractive not only for domestic and foreign visitors but also for the local population. The region has not only an attractive offer but also a suitable potential for the development of experiential tourism. In terms of products that would stimulate the interest of visitors, these are mainly cycling, hiking and cultural and historical monuments, with the intention of adapting products to individual needs.

Liptov region belongs to the most visited parts of Slovakia, and it has many tourist attractions which are exploited in experience tourism. The task is to develop the tourism in smart and sustainable way—as the experience tourism is based on individual expectations, we assume that with the coordination of subject interested in the development of tourism in the region, this could be the appropriate way. The trend of short-term stays has also been reflected in our questionnaire survey, we have found out that visitors prefer shorter stays. Most visitors visit the region with their family, partner or friend. We have also examined the reasons for selecting the destination, where we have found out that to the main ones belong a previous visit and positive recommendations. This fact results from the satisfaction of the visitors and the subsequent spread of the positive image about the region. Among the motives of the visit clearly dominated the theme of recreation. According to our findings, the most important factors on the basis of which the visitors have chosen the destination were the natural and cultural conditions of the region. Visitors consider the quality of services, health and wellness services, sports services or adventure activities for less important factors. The most popular services from the visitor's point of view are recreation and leisure services and natural heritage recognition services. Visitors prefer visits of caves and natural beauties, tours of castles, museums, galleries and open-air museums.

The visitors consider the offer of the experience tourism products as sufficient, and they would use this offer again. According to the respondents, prices correspond to the quality of the services provided. From respondents' opinions, we have found out that there is a need to improve the infrastructure, especially for cyclists, to make cultural–historical sights more attractive and to improve the promotion of individual attractions in the region. According to the views of visitors and exploring the supply of experience tourism in the Liptov region, we think that the region has great potential for the further development of experience tourism.

Liptov is a destination offering many tourism opportunities. Destination management organization tries to bring the tourism subjects to cooperate. The aim of the collaboration is to jointly promote the region as a "green" region for active and adventurous leisure time. Such cooperation is essential for the region, as they can create joint smart and sustainable experience tourism products. In this case, we would recommend focusing mainly on the needs of the visitors by adapting new trends in the development of experience tourism.

Acknowledgements This research was funded by the Scientific Grant Agency of the Ministry of Education, Science, Research and Sport of Slovak Republic VEGA, grant number 1/0368/20 "Sharing economy as an opportunity for sustainable and competitive development of tourist destinations in Slovakia".

References

Adillon, R. (2019). Diamond model: A theoretical framework for the sustainable development of tourism. *Arts and Humanities Open Access Journal, 3*(1), 11–23. https://doi.org/10.15406/ahoaj.2019.03.00099

Bergstrom, N. (2021). How the rise of experience tourism is changing travel. Retrieved from: https://www.traveloffpath.com/how-the-rise-of-experience-tourism-is-changing-travel/. Accessed the 26th of June 2022.

Dubská, M. (2012). Emotions and experience creation in culture and tourism. *Culture and Tourism Context, 5*, 26–30.

Goeldner, C. H., & Ritchie, B. (2014). *Tourism—principles, examples and trends*. Brno: Albatros Media.

Gúčik, M., et al. (2006). *Dictionary—Tourism, hospitality*. Bratislava: Slovenské pedagogické nakladateľstvo.

Kotíková, H. (2000). Experience and adventure as a basis for the satisfaction of the customer with the tourism product. *Economic Review of Tourism, 33*, 13–15.

Kotíková, H. (2013). *New trends in tourism offer*. Grada Publishing.

Liptov region. (2022). *Liptov information*. Retrieved from http://visitliptov.sk. Accessed the 26th of June 2022.

Liptov region map. (2022). *Map*. Retrieved from https://liptovsky-region.webnode.sk/o-nas. Accessed the 26th of June 2022.

Nowacki, M. (2013). *The determinants of satisfaction of tourist attractions visitors*. Poznan: ACTIVE.

Pomfret, G. (2018). Conceptualising family adventure tourist motives, experiences and benefits. *Journal of Outdoor Recreation and Tourism, 27–28*, 463–465.

Seaton, A. V., & Bennett, M. M. (2001). *The marketing of tourism products: Concepts, issues and cases*. Thomson Learning.

Sharpley, R., & Stone, P. (2012). Tourist experience: Conterporary perspectives. *Journal of Heritage Tourism, 7*, 187–188. https://doi.org/10.1080/1743873X.2011.638459

Tangeland, T., & Øystein, A. (2011). Household composition and the importance of experience attributes of nature based tourism activity products—A Norwegian case study of outdoor recreationists. *Tourism Management, 32*(4), 822–832. https://doi.org/10.1016/j.tourman.2010.07.005

Tarssanen, S. (2009). *Handbook for experience stagers*. Rovaniemi: Lapland Center of Expertise for the Experience Industry.

UNWTO. (2022). *Product development*. Retrieved from https://www.unwto.org/tourism-development-products. Accessed the 8th of January 2022.

UNWTO. (2022). *Sustainable development*. Retrieved from https://www.unwto.org/sustainable-development. Accessed the 8th of January 2022.

Valeri, M. (2021). *New governance and management in touristic destinations*. Pennsylvania: IGI Global.

Voříšková, B., & Indrová, J. (2015). Consumption of the experience tourism products by Czech tourists. *Economic Review of Tourism, 48*, 142–149.

Zelenka, J. & Pásková, M. (2012). *Tourism—Dictionary*. Praha: Linde Praha.

The Needs of the Hospitality Industry in Its Transition to the Circular Economy

Marival Segarra-Oña, Blanca De-Miguel-Molina, Angel Peiró-Signes, and María De-Miguel-Molina

Abstract Tourism is one of the largest industries in Spain. Although it generates a wide range of economic benefits, it also produces a negative environmental impact. The circular economy, which includes eliminating waste and contamination, circulating products and materials and regenerating nature, is clearly the path to follow to overcome these negative impacts. The objective of this study is to analyze the needs of companies in the tourism industry regarding its transition to a circular economy. We obtained the information from three focus groups with 18 experts from hotels and restaurants as well as their suppliers and analyzed the content using QDAMiner 5 software. Three main conclusions are drawn from the analysis. First, size, management and financial resources are the main barriers to the circular economy. Second, companies do not receive real support from the public administration and they feel penalized in relation to those that do not apply these measures or to suppliers in the value chain that transfer their own waste. Third, there is a lack of a circular economy culture, which indicates the need for training at every level, so that the transition can be shared by all.

Keywords Circular economy · Tourism · Hospitality · QDAMiner

1 Introduction

Tourism moves billions of travelers every year. This is a figure that will continue to grow, according to the forecasts of the World Tourism Organization (WTO), at an average of 3.3% per year until 2030, when it will reach 1,800 million. According to the latest data (WTO), the Spanish tourism industry is currently the second most powerful in the world by income and the third in terms of tourist arrivals, with a consequent contribution to the country's gross domestic product (GDP) and its economic development. Indeed, according to Spanish Institute of Statistics (INE)

M. Segarra-Oña (✉) · B. De-Miguel-Molina · A. Peiró-Signes · M. De-Miguel-Molina
Universitat Politècnica de València, Valencia, Spain
e-mail: maseo@omp.upv.es

© The Author(s), under exclusive license to Springer Nature Switzerland AG 2023

V. Katsoni (ed.), *Tourism, Travel, and Hospitality in a Smart and Sustainable World*, Springer Proceedings in Business and Economics,
https://doi.org/10.1007/978-3-031-26829-8_13

data for 2017, Spain received 82 million visitors, who generated an income of more than 87,000 million euros, which increased the contribution of tourism to the national economy up to 14.9% of GDP. Figures from the Ministry of Employment and Social Security show that in 2017 this sector generated 2,297,331 jobs in Spain (77,501 of which were new positions). In the Valencian community specifically, the tourism sector is considered to be the main driving force of the economy and its specific weight has increased continuously since 2011, reaching 15.1% of the total employment and 14.6% of the total GDP of the area in 2017 (IMPACTUR Comunitat Valenciana Study, 2017). In this context, the tourism industry is a strategic sector for the Valencian community.

However, parallel to this growth, the environmental impacts of tourism have also been increasing. Tourism has not only physically transformed tourist areas, but has also generated significant environmental impacts, such as the destruction of ecosystems, reduction in the quantity and quality of water, impoverishment and contamination of soils, and generation of waste, as well as environmental contamination, in addition to terrestrial and maritime effects, etc. Some studies (Gómez-Martín, 2017) estimate that every million tourists that Spain receives generate 25 million kilograms of carbon dioxide, 1.5 million kilograms of waste and 300 million liters of wastewater. In addition, 11 million liters of fuel, 300 million liters of water and 2 million kilograms of food are consumed. In some locations, such as the Valencian community, the consequences of these impacts are even greater due to the regional situation, the scarcity of water resources, the erosion and the loss of biodiversity, without counting the impacts due to pollution, the consumption of energy resources or the generation of waste. It is evident that in this situation, it is necessary to promote a growth model based on sustainable development. By implementing the results of eco-innovative research, development and innovation (R&D&i), concepts and methodologies in favor of the circular economy in tourism, apart from improving the efficiency and sustainability of this industry, can contribute to raising awareness in society about the sustainable use of resources. Because of the importance of these economic, social and environmental reasons, the Spanish Tourism Council agreed to carry out a broad strategic review of the sector to assess the challenges and trends that lie ahead, which has materialized in the Horizon 2020 Spanish Tourism Plan (Secretary of State for Tourism (SETUR)).

The circular economy opens up a scenario of key business opportunities for the present and future competitiveness of companies. In this context, the InnoEcoTur project aims to promote models and strategies based on the circular economy in the Valencian community. The focus of the project is the tourism sector, for which it proposes the creation of a network involving hotels and restaurants, the result of which will be the proposal for the implementation of a "Circular Economy Strategy in the Tourism Sector of the Valencian Community". The project also intends to transfer the results to companies in the sector. Among the different activities that are planned with key agents of the Valencian community, we can highlight:

- Carrying out an audit of the needs of the sector and detecting the potential for improvement in favor of sustainability, through the participation of the different stakeholders involved
- Analyzing the results to connect them with the R&D&i sector, which will allow the development and incorporation of eco-innovative technologies and methodologies in these fields of improvement.

The objective of this paper is to present the design and results of the first stage using focus groups in relation to the detection of barriers that hotels and restaurants encounter in their transition to the circular economy and the good practices that they have adopted in this area.

This paper is organized as follows. Section 2 is centered on the main theoretical framework. Section 3 presents the methodology of this study and includes the explanation of the profile of the focus group members and the structure of the focus group sessions. Section 4 examines the empirical evidence extracted from the content analysis of the sessions. The final section including the conclusions and policy implications completes this work.

2 Literature Review

The year 2017 was declared the International Year of Sustainable Tourism for Development by the United Nations General Assembly. According to the UNWTO, sustainable tourism takes into account all these impacts when meeting the needs of the agents that constitute it. To achieve sustainability, natural resources must be optimized and conserved, sociocultural authenticity must be respected and poverty must be reduced by creating stable employment opportunities. The state of the environment and climate change requires the adoption of clean and environmental-friendly innovations on a large scale and the proposing of a series of priorities and actions that increase the demand for environmental technologies and eco-innovations.

However, as in other technological and social transformations, there is currently a distance between the existing economic and business structures and institutions and what society needs (Spiegel-Rösing & de la Solla Price, 1977). The importance of sustainability, on the one hand, and innovation, on the other, in global development is beyond question. Although the phenomenon of eco-innovation has been studied from a strategic and economic point of view in relation to companies (Bitencourt et al., 2020), its integration is lacking at a horizontal level in an environment of inter- and intra-industrial cooperation.

The transition to a circular economy represents a systemic change, in which research and innovation, both technological and non-technological, play a key role. For the redefinition of the dominant production and consumption model, the implementation of new processes that allow for transforming waste into new inputs and products that become part of the new value chain or the recovery of raw materials to be reintroduced in the productive processes, it is necessary to improve the basis of our scientific and technical knowledge, develop new technologies and redesign production processes as well as business and consumption models that will shape a new economy and society. Therefore, the promotion of R&D&i in these areas is a crucial element in this transition process, making, at the same time, a contribution to business modernization, development, growth and competitiveness which will also assist in advancing the competitiveness and modernization of the EU industry. Indeed, the Spanish Circular Economy Strategy highlights the importance of the awareness and participation of citizens and stakeholders, as well as the promotion of R&D&i in favor of the competitiveness of the sectors and the generation of employment and the training required for it.

2.1 The Circular Economy Concept

According to the Organization for Economic Co-operation and Development (2020), there are more than a hundred definitions of the circular economy, but there is a common thread between them. They summarize design without generating waste and pollution. Compared to the linear manufacturing pattern based on the extract–make–use—dispose model (European Commission, 2020), the circular economy allows companies to improve their economic results while generating positive impacts on the environment and society (Pacto Mundial de Naciones Unidas España, 2021). Current scenarios and projections indicate that by 2030 the circular economy could generate a benefit of 1.8 billion euros in the European Union as a whole, 0.9 billion more than the current linear economy model (Circular Spain 2030; Spanish Circular Economy Strategy).

"The circular economy is in fashion" affirms one of the participants in the focus groups that we present in Section 4. Indeed, the European Commission recently published a report (2018) on the advisability of migrating from a linear economy to a circular one, where "the value of products, materials and resources is in the economy for as long as possible, and in which the generation of waste is minimized".

The Ellen MacArthur Foundation, one of the most prominent organizations in the dissemination of the circular economy, uses three principles and seven "Rs" to support the transition to the circular economy (Ellen MacArthur Foundation, 2021). The three principles on which circularity is based are:

- Eliminate waste and pollution.
- Circulate products and materials.
- Regenerate nature.

The seven "Rs" (Table 1) reflect some of the options that can be used to promote the circular economy.

It should be noted that eco-innovation and the circular economy are concepts that feed each other. Both have the same basic objective: to promote a sustainable future with a balance of the social, economic and the environmental dimensions. In fact, it is now beginning to be believed that both can favor more radical transformations

Table 1 Seven "Rs" of the circular economy for companies

Rs	Description
Reduce	Reduce the use of raw materials and design products to be more durable, repairable and recyclable
Reuse	Share, rent, lease products and buy second-hand products, to encourage their reuse and keep them in use for longer
Redesign	Design products in such a way that they can be remanufactured
Repair/Rehabilitate	Partially repair/renovate goods instead of scrapping them. Companies can create products that are repairable
Recondition/Rework	Restore products to their original functionality
Return/Recover	Return/recovery of products, materials and packaging
Recycle	Recycle materials or return them back to the production cycle

Source Ellen MacArthur Foundation (2013), Ellen MacArthur Foundation (2021) and Capgemini Research Institute (2021)

than traditional innovations, with more forceful effects on the efficiency of processes and on economic and social results. The circular economy can contribute to the establishment of networks, the formation of social capital, business cooperation and the creation of new relationship models between the public and private sectors.

2.2 The Circular Economy, SDGs and Climate Neutrality

The importance of the circular economy can be understood from its relationship with the Sustainable Development Goals (SDGs) and the 2030 and 2050 Agendas that seek climate neutrality. The circular economy is seen as the solution to current problems such as increasing waste and greenhouse gases (Pacto Mundial de Naciones Unidas España, 2021).

The SDGs are considered a reference by companies when defining objectives related to responsibility. These objectives mark what must be achieved according to the 2030 Agenda. Of the 17 SDGs, the SDGs 6, 7, 11, 12, 13, 14 and 15 are directly related to the environment and the circular economy. In particular, the circular economy is highly linked to the United Nations SDG 12, consumption and production, which aims to reduce the use of resources and pollution and analyzes the use of goods throughout their entire life cycle. For this objective to be successfully achieved, a broad cooperation between all interested parties (stakeholders), companies, final consumers, researchers, and public managers (policy-makers) is required.

The EU, through programs such as H2020, with the specific focus "Industry 2020 in Circular Economy", has allocated 650 million euros to demonstrate the economic, environmental and social viability of specific areas of the circular economy.

In 2020, the European Commission adopted the New Action Plan for the Circular Economy (European Commission, 2020), a prerequisite for reaching the goal of climate neutrality in the year 2050. In this plan, the commission focuses on the sectors that it estimates use the most resources, so that the measures taken in relation to them will generate a greater impact on circularity. Table 2 summarizes the sectors which the New Plan most affects, with the objectives for each one and some examples.

Another aspect to take into account at a European level is the inclusion of the circular economy among the highlighted priorities. Examples include the different research strategies for intelligent specialization (RIS3) or the European programs LIFE and COSME.

The Circular Economy Strategy in Spain (MITECO, 2020) takes as a reference the sectors included in Table 3, among which is the tourism sector, due to its importance in the Spanish economy.

Table 2 Sectors in which actions can mostly influence circularity

Sector	Goal	Examples
Electronics and ICT	Extend the life of products	– Right to repair – Common mobile phone charger – Improvement of waste collection and treatment
Batteries and vehicles	Increase the circular potential of batteries	– Improve battery collection and recycling – More circular business models for vehicles
Containers and packaging	That all packaging is reusable and recyclable	– Reduce waste – Design for reuse and recyclability
plastics	Fight against pollution caused by plastics	– Recycling and waste reduction
Textile products	Sustainable and circular textile products	– Separate collection of textile waste – Classification, reuse and recycling of textile products
Construction and buildings	Comprehensive sustainable built environment and circularity throughout the life cycle of buildings	– Improvements in durability and adaptability in line with circular economy principles applicable to building design – Initiatives to rehabilitate abandoned lots
Food, water, nutrients	Mitigate the negative impacts of resource extraction and use on the environment	– Food waste reduction – Replace single-use containers, crockery and cutlery with reusable products – Water reuse and water efficiency – More sustainable application of nutrients

Source European Commission (2020)

Table 3 Priority sectors in the 2030 Circular Economy Strategy of Spain

Sector	Cause	Circular economy. examples
Building	It generates 40% of waste and emits 35% of greenhouse gases	– Waste management, including separation and classification – Increased recycling and reuse
Agrifood, fishing and forestry	Waste of natural resources such as water and food waste	– Reduce food waste – Recovery of discarded food and by-products for animal feed
Industrial	The manufacturing industry generates 11% of the waste Food waste in the HORECA channel	– Use of technologies that favor a better use of resources, reduce waste
Consumer goods	The useful life of consumer goods has been reduced	– Reduce the waste generated through the reduction of single-use products and counteract premature obsolescence – Reuse of discarded products
Tourism	Intensive use of water resources, high generation of waste	
Textile and clothing	Between 2 and 10% of the environmental impact of consumption. Among the countries that discard the most amount of clothing	– Separate collection of containers – Textile waste management

Source MITECO (2020)

2.3 The Circular Economy in Tourism

The Spanish Circular Economy Strategy considers the tourism sector to be one of the five priority sectors and a target for its action plans. It includes programs, such as Smart Tourist Destinations, oriented to the incorporation of technologies, sustainability and innovations as well as an increase in accessibility. On the other hand, the SETUR in the 2018–2020 strategy incorporates the concept of sustainability transversally to the associated actions. Thus, the Horizon 2020 Spanish Tourism Plan defines progress in circular economy practices among its strategic lines.

To promote the reconversion of tourism in Spain and the introduction of the circular model, the ministry responsible for the environment and the Secretary of Tourism have proposed different initiatives such as the plan to promote the environment. This plan promotes the energy rehabilitation of hotel facilities, a carbon footprint registry and CO_2 absorption projects, in addition to signing an agreement

Table 4 Example actions by Marriott and Wyndham related to the circular economy

Examples	Marriott	Wyndham
Buildings	Reuse existing land and buildings by giving them new life instead of destroying them to build new ones	
Energy	Generation of renewable energy in hotels (solar panels)	Low consumption lighting. Generation of renewable energy in hotels
Water	Low-flow toilets and showers, laundry water reuse, smart irrigation controllers	Low-flow toilets and showers. Reuse sheets and towels. Irrigation control in gardens
Waste	Alternatives to plastic water bottles: refill glass bottles Reduction of food waste and separation for compost. Food donation	Minimize single-use plastic. Recycle materials. Food waste for compost
Supply	Support local farmers	Supply options without cardboard boxes

Source Wyndham (2021) and Marriott (2021)

with the Hotel Technical Institute (ITH) to develop a "Tourism sustainability model and improvement plans" in hotel establishments.

In any case, the adoption of environmental measures by the tourism sector will be determined not only because it is on the political agenda, but above all because good conservation of the environment and efficient use of resources will affect the future of the sector, both from the point of view of competitiveness and due to greater sensitivity to a demand that will increasingly require economic, environmental and social sustainability in its purchasing options.

In this line, frontrunners in the industry, such as large hotel chains like Marriott, can serve as a reference for ideas about the circular economy in the sector. For this reason, in Table 4, examples obtained from the sustainability reports of two large chains, Marriott and Wyndham, have been compiled.

3 Methodology

The method selected to obtain information on the needs of companies in the application of the circular economy in hotels and restaurants is the focus group. Through this methodology, we can involve companies in the tourism industry from the beginning of the project. The focus group is an information collection technique that seeks to know what the participants think about an idea (Krueger, 2014). In particular, the objective of these particular focus group sessions is to identify challenges and needs in relation to the transition to the circular economy by companies in the tourism sector (hotels and restaurants) in the Valencian community.

The participants in a focus group form a small group of people, around six, with a specific profile, who will discuss a topic of interest, guided by a person who moderates the session. These participants will provide valuable information on the main difficulties they encounter when applying the circular economy, as well as on some measures that they have already applied in their companies or that they are currently developing.

3.1 Design of the Focus Groups for the Project

The design of a focus group includes decisions about the number of participants and their profiles, as well as the guide to the session dynamic, in addition to who will moderate the dynamic, how the session will be recorded, if it will be transcribed and how the analysis of the information collected will be carried out.

The selection of the participants is crucial for gathering valuable information that allows the objective of the focus group to be achieved. Groups require a certain amount of intra-group homogeneity to assure the significance of the information that is gathered. This means that focus members have similar professional situations and contexts, socioeconomic and cultural backgrounds and working experience. Members should also be from similar seniority levels to avoid one member dominating the discussion because of his/her position. In this selection, we should avoid excessive homogeneity to overcome redundancy in the discussion.

Additionally, we will ensure a certain level of intra-group heterogeneity to assure a wide variety of points of view in the information. It is necessary that members have differences and contradictions. For example, we will look for parity in the groups, so the different points of view and sensibilities of men and women can be accounted for. We will search for participants from different activities in the industry, combining profiles from different business stakeholders.

In the first stage of the project, the participants were required to be experts in the tourism sector with experience in sustainability aspects and with the capacity to make decisions about the transition of their companies to the circular economy. It was also taken into account that the age range of the participants ensured the regular use of new technologies.

Thus, the selected profile for the focus members included professionals between 30 and 50 years old, with technological skills, who belonged to a managerial level. A role such as quality, operations or commercial director with decision-making capacity would be adequate to guarantee the provision of valuable information. Finally, we selected participants from different hospitality actors: hotels with a minimum of four stars, restaurants with verifiable recognition, such as being included in a ranking or having received a gastronomy award, and hotel and restaurant suppliers.

Finally, we wanted to gather different views from the various types of tourism present in the Valencian community. For instance, Alicante area is typically focused on sun and beach tourism, with a combination of foreign and national tourists, while the Valencia area is more oriented to the urban type and the Castellon area to more rural and national and local tourism. Therefore, using the profiles, the partner universities looked for participants in their region.

Three focus groups were created, one per region, with six participants in each group, including two hoteliers, two restaurant managers, one hotel supplier and one restaurant supplier. The latter included providers of plastic products and frozen foods, as well as institutional hygiene, laundry, marketing and communication and tourism services.

3.2 *Group Dynamics*

The three focus groups conducted their sessions during a period of one week. Each focus group session was held in a controlled environment in a quiet and comfortable room at the partner university with a table and appropriate number of chairs for the number of participants and the moderator and with sufficient space for a group of two to three observers who would be recording and observing the session. Each focus session was expected to last no more than one and a half hours.

Each partner university selected a moderator for the session to assure the heterogeneity of the direction of the focus group. However, all the moderators were instructed and given the same session script, so they could redirect the session when needed. Having a guide for the dynamic is crucial, since it will facilitate the subsequent analysis of the transcripts. The research group prepared a guide listing the themes or thematic aspects and an explanation of the different steps to follow when moderating the interviews. Although the script does not have a specific sequential order, it supports the moderator in controlling the session and adapting the order of the topics to the dynamics highlighted by the participants in each focus group. The guide then serves as an orientation in the conversations. The guide included the following topics of interest:

1. BARRIERS and SUPPORT in the transition to the circular economy.
2. PRACTICES in the circular economy and industrial symbiosis.
3. DIGITAL TOOLS to measure the circularity of a company.
4. Eco-innovative RESULTS.
5. CHANGE INITIATIVES involving caring for the environment.
6. USERS raising awareness in relation to knowing about and valuing initiatives.

The UPV was in charge of the audiovisual recording, while the other universities (University of Alicante, Jaume I University and University of Valencia) that had moderated and selected the participants created the transcriptions of the focus group sessions. In addition, the UPV prepared the informed consent document, which was signed by the participants, in order to be able to work with the material generated during the course of the sessions and use it in the activities linked to the project. These authorizations by the participants are in accordance with the legislation on data protection.

Finally, the content analysis of the transcripts was carried out, by the UPV, with the qualitative analysis software QDAMiner 5. The opinions of the participants on each of the six topics were coded and grouped by related topics. However, the number of topics could exceed the six in the guide as new topics emerged in the sessions.

3.3 Stages in the Session

The session was structured in different stages:

1. Preparation: The session started with greeting the participants. The participants were identified and given a sticker with his/her name, so they could be identified by the other group members. There was a coffee break at the beginning, so the participants could meet and destress before the session. The participants then sat down and the supporting team began the recording. Each participant had a pen, paper, water and a copy of the authorization.
2. Presentation: The first stage in the session included the member presentations and the explanation of the objective of the focus group. The members were informed about how they had been selected, in addition to what the dynamics of the session and their role would be, as well as the duration and how the session would be recorded. They were also informed about the anonymity of their opinions and the authorizations.
3. Warm up: After the presentation, the moderator encouraged participants to talk about the theme, letting the discussion flow. In this part, it was important for the moderator to encourage the integration of all the participants in the discussion and to prevent certain members from monopolizing the discussion.
4. Development: There was no need to follow a specific order or sequence of the themes. The discussion flow was determined by the group. Using the guide, the moderator directed the participants to cover all the thematic areas.
5. Relaunch: During the sessions, it was expected that there would be some moments in which the moderator would have to participate to center or direct participants to consider a certain theme included in the guide, to encourage discussion on a theme that had emerged in the session that was not considered in the guide or to summarize or clarify the main statements that the group had discussed.
6. Closing: The session was closed by thanking the participants and indicating that they would be provided with the conclusions of this study.

4 Results

In this section, we include the summary of the most important ideas that have been extracted from the three focus groups. During the focus groups, 174 codes were generated, including the main barriers that the participants encounter when applying the circular economy approach or the practices that they support in their companies and that can serve as an example to other hotels, restaurants and their suppliers. The focus groups also produced opinions on the incentives for the circular economy, how the circular economy influences supplier selection, on the importance of training and communication and, finally, about aspects related to waste.

4.1 Barriers to the Implementation of the Circular Economy in the Tourism Sector

The first topic that was raised with the participants concerned the barriers they encounter to the implementation of the circular economy in their companies. The barriers have been grouped into six main themes:

- Economic.
- Business.
- Knowledge.
- Cultural.
- Social.
- Administrative and institutional.

The *economic barriers* topic groups opinions on the cost of implementing measures that help businesses move toward the circular economy. The companies indicate that the necessary investments sometimes involve such high costs that they make their approval difficult. Codes for these barriers include examples such as the following:

- 20G—It implies a series of things that even when you want to go further you cannot do them because the cost of that improvement you cannot afford.
- 20H—We have a fleet with a series of vehicles, and we have done studies to change them to electric or gas, but the refueling, autonomy and infrastructure plans limit you a lot.
- 20I—Looking for a kind of return circuit, in which you have to go back for empty containers, involves a cost for which you can be penalized if your price is higher.

Business barriers refer to the fact that the higher costs for implementing improvements in circularity translate into disadvantages compared to competitors that do not apply them. Among the codes defined, the following would be an example of the concerns expressed by the participants in the focus groups:

- 20S—There is unfair competition in the sector of companies that do not meet the legal minimum, but we are in the next thing, sustainability, and the margins are reduced tremendously.

Knowledge barriers refer to the lack of knowledge about what the circular economy is. This barrier is found both in companies and in the public administration, suppliers and customers. Sometimes, those who transmit the information do not express it in terms that can be understood by their interlocutor, who expects to be warned about the euros that can be saved if they incorporate a specific measure. Examples of the codes defined for this type of barrier are:

- 20D—There is a lot of ignorance about the circular economy, nobody really knows what we are talking about.
- 20F—Sometimes you have to fight against the element of someone who does not understand or who sees it as an extra cost.

Cultural barriers reflect the situations that make it difficult to change perspectives, so it will take time to accept the transition to the circular economy. These include, for example, companies that are not willing to change the way they work and customers that do not evolve toward more circular products. Codes defined for these barriers may include the following:

- 20T—Cultural barrier of risk aversion. Working in a different way than we are used to is difficult.
- 20Q—If the client does not evolve toward more sustainable products.

Social barriers are what hinder the integration of population groups that are more aware of sustainability issues. For example:

- 20G—Take women into account at work, with equal conditions and equal opportunities, who are highly aware of the issue of sustainability.

Institutional and administrative barriers refer to the limits that companies find in the administration's regulations that do not allow them to advance in the transition to the circular economy. Among the examples provided by the participants are barriers to applying solutions that avoid food waste:

- 20K—They want to introduce Too Good To Go in Benidorm hotels, but we have a lot of problems with the governmental health department and we cannot sell it to customers.
- 20P—The requirements for public subsidies make it more profitable not to ask for a subsidy.

4.2 Measures that Have Already Been Applied in Relation to the Circular Economy

Despite the barriers that the participants in the focus groups have pointed out, in this section, we show that their companies are already applying measures for their transition toward the circular economy. The codes shown in Table 5 that were obtained through the analysis of the information offered by the participants have been organized based on the seven "Rs" mentioned in Table 1.

It is important to note that the companies that participated in the focus groups take the SDGs as a guide for their future direction. Therefore, the fact that companies have

Table 5 Code examples of application of the circular economy in companies obtained in the focus groups

Good practice	Code
Measure to reduce	21A—Carbon footprint audit and follow the carbon footprint methodology
	21Z—Study to make truck routes more efficient, so that they consume much less fuel
Reduce	21G—The change of towels in the hotel, advising the client not to ask for it to be changed
	21O—Solar panels
	21W—Change all single-use containers
Reduce by promoting local products	21R—Proximity and ecological product
Reduce + return	21J—Minimize water waste or treat water waste that truly comply with the regulations
Reduce + recover + reuse	21ZI—We treat the water from the cisterns and showers in our treatment plant. One hundred percent of our water is purified and it is with that we irrigate the golf course along with the river water
Repair/rehabilitate	21C—Rehabilitate a building instead of building a new one

the SDGs as a guide for their day-to-day activities will make it easier for them to be open toward the circular economy and for their decisions to be aligned with the 2030 and 2050 Agendas.

The code examples in Table 6 relate to how the companies measure their impact in the environment and the actions that they take to reduce it. Among the alternatives, the calculation of the carbon footprint as well as environmental certifications and seals are among the more popular ones for proving companies' commitment to the environment. Among the actions to reduce the negative impact of their activities, companies try to reduce the consumption of natural resources such as water, substitute energy sources for other renewable ones or limit the waste they generate. Other reduction measures are related to zero-miles food, which reduces the transport of products and favors seasonal ones, supporting local products.

Other groups of codes have been defined to show measures that go beyond reduction and encompass other "Rs" of the circular economy cited in Table 1. It is important to bear in mind that what may be redesign for one company may be recovering and reusing for another. For example, a company that develops technology for hotels and redesigns a system that allows for recovering the water that would be wasted would be applying redesign. However, the hotel would be employing recovering and reusing. By working together, the technology provider and the hotel can incorporate more "Rs" from the circular economy principles.

4.3 Incentives for the Circular Economy

Among the comments of the participants, some refer to incentives for the circular economy. Some comments indicate that there are no such incentives, while others say the opposite:

- 23A—There is a lack of incentives that reward those who do well.
- 23E—Yes, there are many lines of incentives and many lines of subsidies for environmental aspects.

There are also comments about what they expect from such incentives. The incentives should reward companies and do not penalize them:

- 23B—That they do not penalize companies for trying to be more consistent with beliefs and corporate social responsibility.

Lastly, there are comments that stress the lack of information on such incentives:

- 23L—Communication by the public administration to the companies of the public aids and subsidies because the information does not arrive.

4.4 Selection of Suppliers According to Their Effort in This Transition to the Circular Economy

This group of codes includes the comments that affirm that there is a selection of suppliers, based on whether they make the effort to incorporate circularity. In addition, some opinions are added on the reaction of suppliers when requirements are increased.

- 25A—The certification requires it.
- 25F—We are on "Ruta de Sabor" and we select suppliers from the province, rewarding the zero-miles food.
- 25E—I miss having a network of suppliers, an association or group of suppliers so you do not spend the day receiving goods from many different small local suppliers, which takes a lot of time.
- 25B—Some providers, when a certification is required, tell you that they would rather quit being your supplier.

4.5 The Importance of Training and Communication

Training and communication were cited by participants as essential to encourage the transition to the circular economy. Training is considered necessary for small businesses, workers and customers. Communication is perceived as crucial for spreading the message of what the circular economy is and its importance, but they also indicate that selecting the appropriate communication channels will help in the transition to the circular economy.

- 29D—Training for the entire staff, not just the technical services department, on energy saving issues.
- 29G—Customers must also be educated, explaining why we do things.
- 29E—You have to find the right dissemination channels so that it reaches everyone. Reach not only those who already do it, but also the small businessman who has a restaurant or hostel, as the information does not reach them.
- 29M—The fundamental thing is to inform the components of the circle about what the circular economy is and how to make it work. It is fundamental.

4.6 *Waste Management*

The generation of waste is a problem that the circular economy tries to reduce. The codes created in the analysis refer to what companies do to reduce waste and the limitations they find in doing so or the costs due to waste generated from their suppliers. Among the opinions expressed, we can highlight the failure to properly control which companies in the value chain actually create such waste:

- 31D—I have managed to have my own boxes so that the suppliers can transfer them to me without leaving theirs.
- 31G—Almost all of us try to have our containers differentiated, separate glass, separate plastic, separate cardboard, organic waste, oil collection, etc.
- 31B—Segregating and recycling in our hotel costs us money. I call a company to take my cardboard and they are charging me a lot of money.
- 31A—They force the use of compostable plastic, when compostable plastic in the medium and long term is much more polluting.
- 31C—I analyze where all my waste comes from and I can tell you that I think 80 percent comes from my suppliers. But, the city council does not go to the origin to control who is really generating it. It seems totally unfair to me.

5 Conclusions

This work has collected the main results of the analysis of the focus groups in which experts from the tourism sector of the Valencian community with experience in the application of circular economy measures in their companies participated. In total, 18 experts from the three provinces shared their knowledge about the main barriers in the transition to the circular economy, the measures they have already applied in their businesses, the selection of suppliers or the importance of communication and training, among others.

As an introduction to these results, a summary of some important concepts related to the circular economy has been offered, which help in understanding the subsequent definition of codes and the grouping by themes. This study was carried out through content analysis supported by the QDAMiner 5 software.

Three main conclusions have been reached. The first is related to the difficulties that companies encounter in their transition to the circular economy. Taking into account that a significant number of companies in the sector are small- and medium-sized enterprises (SMEs), the financial and management resources required to make this transition hinder the capability of companies to measure and introduce circular practices in their businesses. Second, the businesses do not perceive that they have real support from the administration, but instead feel penalized in relation to those who do not apply measures that reduce their environmental impact, as well as to suppliers in the value chain that transfer their own waste to them. Third, the lack of a circular economy culture indicates that there is a need for training at all levels, so that the transition is shared by all the members of the company and the supply chain.

The results of the analysis highlight that sustainability policies are not being entirely efficient in promoting the circular economy. First, the lack of capabilities and knowledge about the circular economy itself and the availability of technical solutions constrains the ability of the companies to engage in the circular economy. The circular economy is a complex concept that requires multiple actors to be involved. The conjunction of an environment with a lot of SMEs and the lack of efficient and economic ways to implement circular solutions and of a network of actors in the supply chain engaged in circularity hinder the capability of these companies to engage in circular models. In addition, public funding that should be intended to overcome the financial constraints has a lot of requirements and bureaucracy that prevents companies from applying for it. Thus, public funding should aim to create a proper ecosystem to promote economies of scale in circular solutions that can be implemented by the tourism companies and a network of solutions suppliers that can connect different industries and companies in circular models.

From the suppliers' side, companies have to deal with the waste that is generated in other phases upstream. The legislation is mainly oriented at controlling the disposal of the waste rather than its generation. For example, the way a product is packaged influences the capability of this packaging to be reused or recycled. Then, if the supplier is not efficient in providing the product, the client will have to deal with more waste that is not being generated in-house. Thus, companies are being evaluated by the non-sustainable ways in which their suppliers are supplying their products and not by how sustainable they are in providing their services. It is clear that there is a managerial role in this matter that can be used to select more sustainable suppliers. However, this is not always possible. In other sectors, such as food retail, we can see that big companies are able to force suppliers to adapt their supplies to the client's requests. However, the bargaining power of SMEs in the tourism industry is limited. When developing new regulations to promote sustainable solutions and, therefore, motivate suppliers to take action accordingly, the impact of their products downstream in the supply chain should be considered.

Finally, it appears that many customers, managers and employees do not really understand the circular economy concept and how it relates to different stakeholders. There is a need to create a common understanding of the circular economy and its implications in the tourism industry, as well as transferring this knowledge across the industry's value chain to encourage the demand and the offer of circular tourism business models.

Acknowledgements This work was supported by the Valencian Innovation Agency (AVI) Complementary Actions of Promotion and Strengthening of the Innovation 2021. Research Project INNACC/2021/49, "Creation of an Innovation Platform for the Promotion and Implementation of a Circular Economy Strategy in the Tourism Sector of the Valencian Community (InnoEcoTur)".

References

Bitencourt, C. C., de Oliveira Santini, F., Zanandrea, G., Froehlich, C., & Ladeira, W. J. (2020). Empirical generalizations in eco-innovation: A meta-analytic approach. *Journal of Cleaner Production, 245*, 118721.

Capgemini Research Institute. (2021). Circular Economy for a sustainable future. How organizations can empower consumers and transition to a circular economy. Paris: Capgemini. Retrieved from https://www.capgemini.com/wp-content/uploads/2021/12/CircularEconomy_22122021_v12_web.pdf. Accessed the 26 th of January 2022.

European Commission (2018). Communication from the Commission to the European Parliament, the Council, the European Economic and Social Committee and The Committee of the regions on a monitoring framework for the circular economy. Retrieved from https://eurlex.europa.eu/legal-content/EN/TXT/?qid=1516265440535&uri=COM:2018:29:FIN.

European Comission. (2020). New Action Plan for the circular economy for a cleaner and more competitive Europe. Brussels: European Commission. Retrieved from https://eur-lex.europa.eu/resource.html?uri=cellar:9903b325-6388-11ea-b735-01aa75ed71a1.0018.02/DOC_1&format=PDF. Accessed the 6 th of February 2022.

Ellen MacArthur Foundation. (2013). Towards the Circular Economy. San Francisco: Ellen MacArthur Foundation. Retrieved 26, 2022 from https://emf.thirdlight.com/link/x8ay372a3r11-k6775n/@/preview/1?. Accessed the 26 th of January 2022.

Ellen MacArthur Foundation. (2021). CIRCULYTICS. Definitions. San Francisco: MacArthur Foundation. Retrieved from https://ellenmacarthurfoundation.org/resources/circulytics/overview. Accessed the 26 th of January 2022.

Gómez-Martín, M. B. (2017). Retos del turismo español ante el cambio climático. *Investigaciones Geográficas (españa), 67*, 31–47.

Krueger, R. A. (2014). Focus groups: A practical guide for applied research. Sage publications.

Marriott International. (2021). 2021 SERVE 360 Report. Bethesda, MD: Marriott International. Retrieved from http://serve360.marriott.com/wp-content/uploads/2021/09/2021_Serve_360_Report.pdf. Accessed the 22 nd of February 2022.

MITECO. (2020). Circular Spain 2030. Spanish Circular Economy Strategy. Madrid: MITECO, Ministry for the Ecological Transition and the Demographic Challenge. Retrieved from https://www.miteco.gob.es/es/calidad-y-evaluacion-ambiental/temas/economia-circular/espanacircular2030_def1_tcm30-509532_mod_tcm30-509532.pdf. Accessed the 22 nd of February 2022.

OECD. (2018). *OECD Tourism Trends and Policies 2018*. OECD Publishing. https://doi.org/10.1787/tour-2018-en

OECD. (2020). *The Circular Economy in Cities and Regions: Synthesis Report*. OECD Publishing. https://doi.org/10.1787/10ac6ae4-en

Pacto Mundial de Naciones Unidas España. (2021). Modelos de negocio basados en la economía circular: cómo pueden las empresas pasar de la linealidad a la circularidad. Madrid: Pacto

Mundial de Naciones Unidas España. Retrieved from https://www.pactomundial.org/biblioteca/whitepaper-economia-circular/. Accessed the 6th of February 2022.

Spiegel-Rösing, I., & de Solla Price, D. (1977). Science, technology, and society: a cross-disciplinary perspective.

United Nations Global Compact Spain. (2021). Business models based on the circular economy: how companies can move from linearity to circularity. Madrid: United Nations Global Compact Spain. Retrieved from https://www.pactomundial.org/biblioteca/whitepaper-economia-circular/. Accessed the 6 th of February 2022.

Wyndham. (2021). 2021 ESGReport. Parsippany, NJ: Wyndham Hotels & Resorts, Inc.. Retrieved from https://corporate.wyndhamhotels.com/wp-content/uploads/2021/04/2021-WH-ESG-Report.pdf. Accessed the 22 nd of February 2022.

Internal Sustainability Reporting in the Hotel Industry

Lahorka Halmi and Katarina Poldrugovac

Abstract Sustainability reporting is becoming an increasingly widespread form of reporting by various business enterprises. European Commission's Proposal for a Directive on corporate sustainability reporting amending Accounting directive (2013/34/EU), the Non-Financial Reporting Directive (2014/95/EU), the Transparency Directive (2004/109/EC), the Statutory Audit Directive (2006/43/EC) and the Statutory Audit Regulation (No. 5357/2014) moves sustainability reporting to the margins of the norm for regular corporate reporting. Against this background, the objective of this paper is to investigate the de facto level of harmonization of hotel companies' internal sustainability reporting with respect to (1) the measurement and disclosure of environmental and social key performance indicators (KPIs) and (2) the assessment of the importance of environmental and social KPIs. Data were collected using an online questionnaire and examined using qualitative methods, where measurement and disclosure of KPIs were measured with the C index and assessing the importance of KPIs with a nonparametric Kendall's W test of concordance. The research results show that there is a low degree of harmonization in internal sustainability reporting with respect to the two selected topics. The novelty of the study lies in the application of qualitative methods in accounting research and in the discovery of a way to link internal sustainability reporting processes with external sustainability disclosures.

Keywords Sustainability accounting · Sustainability reporting · Internal sustainability reporting harmonization · Hotel industry

JEL Classification Q56 · Q 01 · M56

L. Halmi (✉) · K. Poldrugovac
Karlovac University of Applied Sciences, Karlovac, Croatia
e-mail: lahorka.halmi@vuka.hr

Faculty of Tourism and Hospitality Management, University of Rijeka, Opatija, Croatia

© The Author(s), under exclusive license to Springer Nature Switzerland AG 2023

V. Katsoni (ed.), *Tourism, Travel, and Hospitality in a Smart and Sustainable World*, Springer Proceedings in Business and Economics,
https://doi.org/10.1007/978-3-031-26829-8_14

1 Introduction

Tourism in itself is an epitome of an oxymoron, as on the one hand it requires social and environmental resources to create tourist attraction and interest, and on the other hand it requires the same resources to provide satisfactory service to tourists coming to the area. Historically, hotels have had a major impact on the tourism industry, which has evolved into an industry of its own. Today, the hotel industry has a significant share in the tourism industry that affects, but is not limited to, both environmental and social resources in the immediate area of the site. Therefore, nowadays hotels are turning to a sustainable economy in a more or less agile way. They focus their efforts and business resources on reducing their impact on the environment (reducing water usage and waste, energy saving, GHG emission tracking, etc.) and improving the quality of life, at least within and immediate surroundings (employees work satisfaction, rights of indigenous people, suppliers effect on the society, supporting local businesses, providing good labor conditions, etc.). Therefore, it is not surprising that the world's leading hoteliers readily disclose their environmental and social information in sustainability reports (Jones & Comfort, 2019; Jones et al., 2006, 2014; Ortmann et al., 2020). Nevertheless, sustainability reporting comes with its own problems, which are discussed below.

2 Sustainability Reporting

Many companies are turning to sustainability reporting to communicate their sustainable practices to their customers, business partners, and other stakeholders. The hotel industry is no exception. Yet, whether referring to sustainability reporting in general or in terms of a specific industry, there is a certain amount of confusion and uncertainty about what sustainability topics to disclose and how to disclose them. Sustainability reporting has been researched for over half a century. It has evolved through several stages (Fifka, 2012, p. 62), starting with social reports that presented the impact of corporate activities on health in the 1960s and on energy sources in the 1970s, to environmental protection in the 1980s and environmental reporting dominating 1990s resulting in the emergence of environmental reporting (Bebbington et al., 2014, p. 53, 54; EPA, 1995, p. 5; Ortas & Moneva, 2011, p. 20). Despite all this, it still has not reached the level of quality and accuracy of financial statements and reports. The thinking behind environmental accounting was limited to the historically already established financial reporting and accounting, to which non-financial inputs related to the environment were added (Bartolomeo et al., 2000, p. 33). Since then, the idea of disclosing a company's impact on the environment and society has attempted to evolve to the level of financial reporting in terms of measurement, accuracy, simplicity of form, and yet complexity of information provided. There is extensive research on sustainability reporting and sustainability accounting that provides important insights on this topic, but there is no consensus on what sustainability accounting actually is and how it is should be structured so that its output is sustainability reporting. Until then, this gap will lead to quite a bit of uncertainty about both the form and content of sustainability reports.

Many papers examine this topic from different angles, increasing the number of papers on sustainability reporting and accounting (Bebbington et al., 2014, p. 57; Hahn & Kühnen, 2013, p. 9; Ortas & Moneva, 2011, p. 24) which adds valuable knowledge but does not provide a comprehensive answer on what and how an industry should produce sustainability reports. There are increasingly many sustainability reporting frameworks with completely different approaches to sustainability issues, among the best known being the GRI Standards (GRI, 2016), SASB Standards (SASB, 2013) and Integrated Reporting (IIRC, 2013). As if to add fuel to the fire, in 2021 European Commission (EC) announced a Proposal for a Directive amending Accounting Directive (2013/34/EU), the Transparency directive (2004/109/EZ), the Statutory Audit Directive (2006/43/EZ), and the Statutory Audit Regulation (No. 537/2014), and announced the development of EU standards for non-financial, i.e., EU sustainability reporting (EC, 2021). In this Proposal, EC has identified most of the challenges, so many researchers have been pointed out that "Some companies from which users want sustainability information do not report all the information, while many that do report sustainability information do not report all the information that is relevant for users. When information is reported, it is often neither sufficiently reliable, nor sufficiently comparable between companies. The information is often difficult for users to find and is rarely available in a machine-readable digital form" (EC, 2021, p. 2). This brings us to the next hurdle of sustainability reporting—its harmonization.

Harmonization in accounting in general is "the extent of concentration around a particular accounting policy choice" where "the degree of harmonization increases as the number of companies selecting the same accounting policy increases" (Herrmann & Thomas, 1995, p. 254). **Harmonization in accountancy** in itself requires understanding and explaining the reasons for differences in accounting systems, which are even more significant in sustainability reporting and accounting systems, as sustainability reporting requires all seemingly conflicting values: professionalism and statutory control, uniformity and flexibility, conservatism and optimism, and secrecy and transparency (Roberts et al., 2005, p. 185). Harmonization is divided into de facto harmonization (harmonization of accountancy regulations) and de jure harmonization (harmonization of companies' practices). Various authors have measured the degree of de facto and de jure harmonization using different methods, e.g. with indices (H index, C index, I index), descriptive statistics, nonparametric tests (ordinal data, ANOVA, ponders), parametric statistics (Chi test, Cramer V test, contingency coefficient, linear regression model) and coefficients (Jaccards, Roger-Tanimotov, Lance-Williams; Baker & Barbu, 2007; Bowrin, 2002; Mcleay et al., 1999; Morris & Parker, 1998; Tay & Parker, 1990). Finally, all of these works examined financial harmonization.

Much research on sustainability reporting (Dropulić & Čuler, 2019; Foris et al., 2020; Hladika & Valenta, 2017; Janković & Krivačić, 2014; Markota et al., 2017;

Milanes-Montero & Perez-Calderon, 2014; Soria Leyva & Parra Parra, 2021; Vuković et al., 2020; Zrnić et al., 2020) and sustainability harmonization (Afolabi et al., 2022; Fortanier et al., 2011; Jose, 2017) implicitly point to the above conclusion of EC in their guideline Proposal. Nevetheless, we could not find any authors approaching this topic from an already established and well-known paradigm, i.e., financial reporting, and further looking at this topic through this prism. Most research focuses on de jure harmonization (Aureli et al., 2019; Aureli et al., 2020; Caputo et al., 2020; Jose, 2017; Kinderman, 2019; La Torre et al., 2018; Manes-Rossi et al., 2018; Tschopp & Nastanski, 2013), which is a particularly difficult endeavor in sustainability reporting, because there is no fixed structure as is in financial reporting. Few authors studied de facto harmonization (Manes-Rossi et al., 2018), but did not clearly distinguish these two types of harmonization in their work, so it was up to us to make this distinction. Finally, we came to the concluded that no one explicitly addressed the issue of de facto harmonization of sustainability reporting, certainly not in the hotel industry. This led to our assumption that it is necessary to learn what hotel companies do in their operational and strategic practices in order to produce sustainability reports, i.e., we felt it was necessary to investigate the degree of de facto harmonization in hotel companies.

We approached de facto harmonization as:

- internal de facto harmonization of sustainability reporting
- external de facto harmonization of sustainability reporting harmonization.

At least to our knowledge, no one in sustainability has taken this approach. The roots of this classification of de facto harmonization lie in the structure of financial accounting and reporting, where there are financial reports intended for internal stakeholders (management and non-management employees) and external financial reports intended for external stakeholders (investors, community, legal institutions, etc.). The authors who have studied sustainability reporting harmonization have focused primarily on external de facto harmonization, i.e. the study of available sustainability reports. We therefore chose a less traveled road and decided to look behind the scenes that should be behind all external financial reporting and, by analogy, external sustainability reporting. The goal was to find out what it looks like behind the scenes of external sustainability reporting. Our decision was reinforced by European Commission's Proposal for a Directive on Corporate Sustainability Reporting (EC, 2021) amending the Accounting Directive (2013/34/EU), the Non-Financial Reporting Directive (2014/95/EU), the Transparency Directive (2004/109/EC), the Statutory Audit Directive (2006/43/EC) and the Statutory Audit regulation (No 5357/2014), pushes sustainability reporting to the margins of the norm for regular corporate reporting. If followed, this directive will represent a great leap forward in sustainability reporting by raising the harmonization of sustainability reporting to a de jure level. We believe this EC Proposal will have some impact on the actual practice of sustainability reporting by some companies, i.e. a de facto harmonization of sustainability reporting. However, we also believe that companies should be considered and treated within their respective industry segment in such approaches. Rather than leaving them in the dark, they should be offered concepts and practices to measure and report their environmental and social impact in their internal and external sustainability reports. This was the rationale for our research, which is explained in more detail in this paper.

3 Methodology

The questionnaire was sent to 68 hotel companies in Croatia, of which only 14 completed the online questionnaire either fully of partially. The data collected from those 14 hotel companies is considered relevant and representative as it reflects the number of establishment, average number of employees and income earned. The data were collected in 2019 and refer to 2018, and one segment of the questionnaire focused on (1) measuring and disclosing environmental and social key performance indicators (KPIs) and (2) the assessment the importance of environmental and social KPIs. **Harmonization** of the measurement and disclosure of environmental and social KPIs was measured using the **C index**, while assessment of the importance of environmental and social KPIs was measured using **Kendall's *W***.

For this study, we decided to test the level of a segment (measuring and disclosing environmental and social key performance indicators (KPIs)) harmonization of internal sustainability reporting with **C index** for several primitive reasons. First and foremost, C index "looks at the number of financial statements that are compatible with each other" easily coping with multiple reports within a single country (Roberts et al., 2005, p. 240, 241). The filled-out questionnaires were considered as financial statements, and the answers therein were compared and measured by C index. The paradigm of a single country was replaced with a paradigm of an industry, meaning that instead of measuring the level of (de facto) accountancy harmonization within geography of a country, we measured the level of de facto sustainability accountancy harmonization within the hotel industry. The most compelling reason for choosing this method for measuring de facto harmonization was just 14 observations (or filled-out questionnaires), because more complex statistical methods such as Jaccards or Roger-Tanimotov coefficient cannot be computed with such a low number of observations as we received. C index is calculated as follows (Van der Tas, 1988):

$$\frac{\sum[n_i * (n_i - 1)]}{[N_i * (N_i - 1)]}$$

where: n_i is the number of companies using method i, N is the total number of companies and ranges <0, 1>, with 0 meaning no harmonization and 1 meaning an absolute harmonization.

Nonparametric statistics of Kendall's W test of concordance was applied in measuring harmonization of the importance of environmental and social KPIs. In cotrast to the measurement and disclosure of environmental and social KPIs, the importance of environmental and social KPIs was assessed on an ordinal Likert scale of importance, and the degree of harmonization was measured using the Kendall's W test of concordance. Kendall's W test of concordance is a general measure of association in multiple classifications and measures "the association between the ranks of the k variates of the k sets of rankings of n objects, which in turn estimates some measure of the relationship between the k variates in the population [and] if the variates are independent, there is no association and W is zero, and where is complete dependence, there is perfect agreement and W equals to 1" (Gibbons & Chakraborti, 2003, p. 450). That is, W ranges <0, 1>, where W zero means there is no association, and when W is one, there is absolute association, i.e., harmonization.

4 Research Results

Harmonization of measurement and disclosure of environmental and social key performance indicators (KPIs)

Since sustainability reporting is still a voluntary disclosure in terms of content and form, we felt it was important to investigate the approach hotel companies take to measuring and subsequently disclosing environmental and social KPIs. Table 1 and Table 2 show the results of this investigation. The GRI Standards indicate that environmental and social KPIs can be reported in natural terms as well as in monetary or any type of relative values; we added the option to not measure if this was the case in some hotel companies, but this was not part of the calculation equation.

The most commonly used environmental KPIs are water consumption, measured in 100% of hotels, and energy consumption by source in 93.33% of hotels, which means that almost all of the hotel companies studied measure and track these two KPIs in some way, and the de facto harmonization indicates their importance in the hotel industry. Material consumption by types is slightly lower on the harmonization scale (80% of the hotels), but still quite strong in presence among the hotel companies. Other KPIs that are quite common in the hotel companies are measures to reduce GHG emissions, consumption of detergents and cleaning agents, and waste recycling (in 66.67% of the hotels), which means that the hotel companies consider the information on these KPIs to be quite important. It is interesting to note that most of the proposed KPIs are measured and disclosed with almost equal frequency in hotel companies, ranging from 46.67% to 66,67% of hotels overall, indicating that most of these hotel companies consider most of the proposed KPIs to be of at least somewhat important for their business. Most of these indicators are measured and tracked in monetary terms (in 57.89% cases out of 133 of total possibilities) or not measured/tracked at all (in 42.11% of cases out of 133 total). The index for harmonization of environmental indicators is 0.47, indicating rather low value.

The results for the social KPIs are slightly lower than those for the environmental ones (Table 2).

The social KPIs that all hotels measure are the impact of products/services on customer health and safety, support of cultural events at the destination, employee

Table 1 Measuring and disclosing **environmental** KPIs

Measuring environmental indicators (KPIs)	Not measuring	Natural	Monetary	Relative	Total
Water consumption	1	6	8	1	**15**
Energy consumption by sources	2	5	8	1	**14**
Material consumption by types	3	4	7	1	12
Measures for reducing GHG emissions	3	5	4	1	10
Detergents and cleaning agents' consumption	2	3	7	0	10
Waste recycling	3	4	5	1	10
Destination investments	3	2	7	0	9
Non-compliance with regulation	4	3	5	1	9
Wastewater recycling	3	4	5	0	9
Investing in acquiring eco-certificates and eco-labels of quality	3	4	5	0	9
Solid waste treatment	3	4	4	0	8
Harmful waste treatment	4	3	4	0	7
Impact on biodiversity	7	1	3	0	4
Traffic impact on the environment	7	1	3	0	4
Products/services impact on the environment	7	1	2	0	3
Total	**56**	**50**	**77**	**6**	**133**
C index		**0.47459558**			

loyalty, employee turnover, employee education and development, work, personal and moral employees' rights. It is interesting to note that all other social KPIs seem to be almost equally important to hotel companies, as all other social KPIs are present in 80% and 93.33% of hotel companies.

It is also interesting to note that most of the proposed social KPIs are measured and tracked in monetary or natural (absolute) values; some hotel companies track these KPIs in relative values, while very few are not measure them at all. The C-index for harmonization is 0.36, which is even lower than the C-index for environmental KPIs. Although hotel management recognizes the importance of social KPIs, the way they view their performance measurement and management seems to fall somewhere in between.

Harmonization of the assessment of the importance of environmental and social KPIs

Table 2 Measuring and disclosing **social** KPIs

Measuring social indicators (KPIs)	Not measuring	Natural	Monetary	Relative	Total
Products/services impact on customers' health and safety	0	5	8	2	**15**
Supporting cultural destination events	0	5	7	3	**15**
Employee loyalty	0	5	8	2	**15**
Employee turnover	0	5	6	4	**15**
Employees education and development	0	6	5	4	**15**
Work, personal and moral employees' rights	0	6	4	5	**15**
Regulation and ethical principles application in business	0	6	6	2	14
Working conditions, health and safety of employees	0	5	5	4	14
Investing in projects of improvement of the community quality life	1	3	9	1	13
Ethics and quality principles in communication	1	5	5	3	13
Stakeholder involvement at the destination level	1	5	6	2	13
Promotion and rewarding of efficiency and achievement	0	4	7	2	13
Anti-corruption market behavior	0	7	3	3	13
Incident prevention	2	5	4	3	12
Total	**5**	**72**	**83**	**40**	**195**
C index		**0.356278086**			

We felt it was important to assess the perceptions of hotel company' management regarding the importance of the proposed environmental and social KPIs (Table 3 through Table 6). The importance of the KPIs was rated on a Likert scale from 1 to 5.

As shown in Table 3, the environmental KPIs with the highest rank are those related to energy consumption by source, water consumption, and detergents and cleaning agents' consumption (9.54). The lowest ranked KPIs relate to biodiversity and traffic impact on the environment (6.54). These results indicate that, in

Table 3 Ranking of the assessment of the importance of **environmental** KPIs

Ranks	Mean rank
Energy consumption by sources	**9.54**
Water consumption	**9.54**
Detergents and cleaning agents' consumption	**9.54**
Waste recycling	8.69
Wastewater recycling	8.31
Measures for reducing GHG emissions	8.31
Solid waste treatment	8.12
Harmful waste treatment	7.73
Investing in acquiring eco-certificates and eco-labels of quality	7.73
Non-compliance with regulation	7.73
Destination investments	7.35
Material consumption by types	7.23
Products/services impact on the environment	7.12
Impact on biodiversity	6.54
Traffic impact on the environment	6.54

this respect, hotel management focuses on the operational KPIs, while KPIs that are significant for sustainable strategic positioning and management of the hotels were rated significantly lower.

The importance of environmental KPIs was measured using Kendall's *W* (Table 4) and was 0.160 ($p = 0.010$). This indicates that even the choice of importance of environmental KPIs among hotels in Croatia is harmonized to a very low degree.

The social KPIs with the highest score (Table 5) are employee education and development (8.65), working conditions, health and safety of employees (8.54) and work, personal and moral employees' rights (8.54). The lowest rated social KPIs are stakeholder involvement at the destination level (6.08) and incident prevention (5.81). Just as with the environmental KPIs, hotel managers place greater importance on operational social KPIs than on strategic KPIs.

Kendall's *W* (Table 6) measure for harmonization of the social KPIs is 0.112, which is slightly lower than for the environmental KPIs. This leads to the conclusion that there is no harmonization of importance of neither for environmental nor

Table 4 Kendall's *W* for environmental KPIs

Test statistics	
N	13
Kendall's W[a]	**0.160**
Chi-square	29.069
df	14
Asymp. sig	0.010

Table 5 Ranking of the assessment of the importance of **social** KPIs

Ranks	Mean rank
Employees education and development	**8.65**
Working conditions, health and safety of employees	**8.54**
Work, personal and moral employees' rights	**8.54**
Employee loyalty	8.12
Promotion and rewarding of efficiency and achievement	8.08
Investing in projects of improvement of the community quality life	7.85
Employee turnover	7.62
Regulation and ethical principles application in business	7.58
Supporting cultural destination events	7.08
Ethics and quality principles in communication	7.04
Products/services impact on customers' health and safety	7.04
Anti-corruption market behavior	7.00
Stakeholder involvement at the destination level	6.08
Incident prevention	5.81

social KPIs. Since this research tested the degree of de facto harmonization, that is, the harmonization of existing practices and perception of the environmental and social KPIs, these results imply that there is almost no harmonization between hotels even in the selection of the KPIs. These results of the level of harmonization for environmental and social KPIs suggest that while hotel management recognizes the importance of measuring and disclosing these KPIs, sustainability reporting has not yet advanced to the point where a single industry, at least the hotel industry, could create a list of KPIs that are clearly important to their business. This in turn leads to the issue of de jure harmonization of sustainability, or the harmonization of the laws and regulations related to the sustainability reporting that impose higher standards on companies (de jure) in terms of the sustainability practices they must disclose and report on.

Table 6 Kendall's *W* for social KPIs

Test statistics	
N	13
Kendall's W[a]	**0.112**
Chi-square	18.990
df	13
Asymp. sig	0.123

[a]Kendall's coefficient of concordance

5 Discussion and Conclusion

Sustainability in the hotel industry is becoming increasingly important, and more and more companies are implementing it in their business practices. There are several reasons for it. They have significant implications for economic, environmental and social issues. These issues affect a wide range of hotel performance. Hotel companies have high energy consumption (Arenhart et al., 2022), water consumption (Cruz-pérez et al., 2021) and waste generation (Singh et al., 2015). The hotel industry not only impacts the environment but also social issues. Although the GRI Standards as a prominent frameworok for sustainability reporting, recognize a variety of social issues to be addressed internally and reported in sustainability reports, ranging from human resource issues (e.g., GRI Standards on employment, labor and management relations, occupational health and safety, training and education, etc.), to their impact outside the company (e.g., right of indigenous people, local communities, publica assessment, etc.), most recent research has focused on social sustainability issues that address the interrelations between corporate social responsibility and human resource management (Salama et al., 2022; Youn & Kim, 2022), even being referred to it as "green human resource management" (Haldorai et al., 2022). Other issues that authors have addressed include employee turnover (Vetráková et al., 2019) and their creativity (Guo et al., 2021). It is significant to emphasize that researches on these topics are of particular importance. As mentioned earlier, sustainability accounting at this stage of development does not have internal protocols, systems or forms necessary for clear and harmonized internal and external sustainability reporting, and such research provides deeper insight into these topics and facilitates their measurement, recording, and disclosure in sustainability reports. Because social topics defined in the GRI Standard are so complex, it is understandable that researchers are addressing one issue at a time. However, the impacts that companies have on their external environment (e.g., local communities, public policy, indigenous rights) should not be ignored or left outside the scope of research for too long, because doing business sustainably business means measuring and managing impacts on the external environment as well.

It is important to draw attention that corporate social responsibility has a positive impact on financial performance of hotel companies (Babajee et al., 2022; Ghaderi et al., 2019; Inoue & Lee, 2011), but also on customer satisfaction and loyalty (Mohammed & Rashid, 2018; Olya et al., 2021), as well as the pressure from laws and regulations (Kinderman, 2019), to name a few.

To successfully manage their positive and negative impacts, hotel companies need to measure all aspects of sustainability. Differences in the amount and detail of information disclosed and the lack of common reporting protocols in sustainability reporting contribute to the lack of comparability among sustainability reports (Jones et al., 2014) making it difficult for stakeholders and shareholders to evaluate their performance.

Following the parallels with financial accounting, harmonized protocols, systems, and forms for internal sustainability reporting would be an excellent starting point for achieving a de facto level of harmonization in sustainability reporting from which external sustainability reporting processes can derive and disclose data in external sustainability reports.

Acknowledgments This paper is a result of doctoral thesis research of Ms. Lahorka Halmi, titled "Harmonisation of Sustainability Reporting System in the Hotel Industry", defended at the Faculty of Tourism and Hospitality Management, University of Rijeka.

References

Afolabi, H., Ram, R., & Rimmel, G. (2022). Harmonization of sustainability reporting regulation: Analysis of a contested arena. *Sustainability, 14*, 1–28. https://doi.org/10.3390/su14095517

Arenhart, R. S., Souza, A. M., & Zanini, R. R. (2022). Energy use and its key factors in hotel chains. *Sustainability, 14*, 1–14. https://doi.org/10.3390/su14148239

Aureli, S., Magnaghi, E., & Salvatori, F. (2019). The role of existing regulation and discretion in harmonising non-financial disclosure. *Accounting in Europe, 16*(3), 290–312. https://doi.org/10.1080/17449480.2019.1637529

Aureli, S., Salvatori, F., & Magnaghi, E. (2020). A country-comparative analysis of the transposition of the EU non-financial directive: An institutional approach. *Accounting, Economics and Law: A Convivium.* https://doi.org/10.1515/ael-2018-0047

Babajee, R. B., Seetanah, B., Nunkoo, R., & Gopy-Ramdhany, N. (2022). Corporate social responsibility and hotel financial performance. *Journal of Hospitality Marketing and Management, 31*(2), 226–246. https://doi.org/10.1080/19368623.2021.1937433

Baker, C. R., & Barbu, E. (2007). Evolution of research on international accounting harmonization: A historical and institutional perspective. *Socio-Economic Review, 5*(4), 603–632. https://doi.org/10.1093/ser/mwm014

Bartolomeo, M., Bennett, M., Bouma, J. J., Heydkamp, P., James, P., & Wolters, T. (2000). Environmental management accounting in Europe: Current practice and future potential. *The European Accounting Review, 9*(1), 31–52.

Bebbington, J., Unerman, J., & O'Dwyer, B. (2014). *Sustainability accounting and accountability.* (J. Bebbington, J. Unerman, & B. O'Dwyer, Eds.) (1st ed.). Routledge Taylor & Francis Group. Retrieved from http://books.google.com/books?hl=fr&lr=&id=9LmmgNM3DmMC&pgis=1.

Bowrin, A. R. (2002). Does international financial reporting harmonization equal domestic financial reporting diversity for developing states: the case of Trinidad & Tobago. Retrieved from https://sta.uwi.edu/conferences/financeconference/Conference Papers/Session 29/Does International Financial Reporting Harmonization Equal Domestic Financial Reporting Diversity for Developing States - T&T.pdf.

Caputo, F., Leopizzi, R., & Pizzi, S. (2020). The non-financial reporting harmonization in Europe: evolutionary pathways related to the transposition of the directive 95/2014/EU within the Italian context. *Sustainability, 12*, 1–13.

Cruz-pérez, N., Rodríguez-martín, J., Martín, J. F. A., Ruiz-rosa, I., & Santamarta, J. C. (2021). Improvements in hotel water consumption: case study of a five-star hotel (Canary Islands, Spain).*Urban Water Journal*, 1–8. https://doi.org/10.1080/1573062X.2021.1949480.

Dropulić, I., & Čuler, M. (2019). The Effect of corporate social disclosure practice on reporting quality: Evidence from the insurance sector in Croatia. *Journal of Contemporary Management Issues*, *24*(2), 23–38. https://doi.org/10.30924/mjcmi.24.2.3.

EC. (2021). Proposal for a Directive of the European parliament and of the council amending Directive 2013/34/EU, Directive 2004/109/EC, Directive 2006/43/EC and Regulation (EU) No

537/2014, as regards corporate sustainability reporting. Retrieved from https://eur-lex.europa.eu/legal-content/EN/TXT/?uri=CELEX%3A52021PC0189.

EPA. (1995). An introduction to environmental accounting as a business management tool: Key concepts and terms. *United States Environmental Protection Agency*. https://doi.org/10.1128/AAC.03728-14

Fifka, M. S. (2012). The development and state of research on social and environmental reporting in global comparison. *Journal Für Betriebswirtschaft, 62*(1), 45–84. https://doi.org/10.1007/s11301-012-0083-8

Foris, D., Crihalmean, N., & Foris, T. (2020). Exploring the environmental practices in hospitality through booking websites and online tourist reviews. *Sustainability, 12*(24), 1–18. https://doi.org/10.3390/su122410282.

Fortanier, F., Kolk, A., & Pinkse, J. (2011). Harmonization in CSR reporting: MNEs and global CSR standards. *Management International Review, 51*(5), 665–696. https://doi.org/10.1007/s11575-011-0089-9

Ghaderi, Z., Mirzapour, M., Henderson, J. C., & Richardson, S. (2019). Corporate social responsibility and hotel performance: A view from Tehran, Iran. *Tourism Management Perspectives, 29*(May 2018), 41–47. https://doi.org/10.1016/j.tmp.2018.10.007.

Gibbons, J. D., & Chakraborti, S. (2003). *Nonparametric statistical inference* (4th ed.). New York, Basel: Marcel Dekker, Inc. Retrieved from https://books.google.com/books?id=kJbVO2G6VicC&pgis=1.

GRI. (2016). GRI 101: Foundation 101. *GRI Global, GRI101*(1), 29. Retrieved from https://www.globalreporting.org/standards/gri-standards-download-center/.

Guo, M., Ahmad, N., Adnan, M., Scholz, M., Khalil-Ur-rehman, & Naveed, R. T. (2021). The relationship of csr and employee creativity in the hotel sector: The mediating role of job autonomy.*Sustainability (Switzerland), 13*(18), 1–18. https://doi.org/10.3390/su131810032.

Hahn, R., & Kühnen, M. (2013). Determinants of sustainability reporting: A review of results, trends, theory, and opportunities in an expanding field of research. *Journal of Cleaner Production, 59*, 5–21. https://doi.org/10.1016/j.jclepro.2013.07.005

Haldorai, K., Gon, W., & Garcia, R. L. F. (2022). Top management green commitment and green intellectual capital as enablers of hotel environmental performance: The mediating role of green human resource management. *Tourism Management, 88*(January 2021), 1–11. https://doi.org/10.1016/j.tourman.2021.104431.

Herrmann, D., & Thomas, W. (1995). Harmonisation of accounting measurement practices in the European community. *Accounting and Business Research, 25*(100), 253–265. https://doi.org/10.1080/00014788.1995.9729914

Hladika, M., & Valenta, I. (2017). Integrirano izvještavanje – nova platforma korporativnog izvještavanja. *Zbornik Radova Ekonomskog Fakulteta Sveučilišta u Mostaru, 23*, 166–186.

IIRC. (2013). IIRC. *The International <IR> Framework*. IIRC. https://doi.org/10.2139/ssrn.2378899.

Inoue, Y., & Lee, S. (2011). Effects of different dimensions of corporate social responsibility on corporate financial performance in tourism-related industries. *Tourism Management, 32*(4), 790–804. https://doi.org/10.1016/j.tourman.2010.06.019

Janković, S., & Krivačić, D. (2014). Environmental accounting as perspective for hotel sustainability: Literature review. *Tourism and Hospitality Management, 20*(April), 103–120.

Jones, P., & Comfort, D. (2019). Sustainable development goals and the world's leading hotel groups. *Athens Journal of Tourism, 6*(1), 1–14. Retrieved from https://www.athensjournals.gr/tourism/2019-6-1-1-Jones.pdf

Jones, P., Comfort, D., & Hillier, D. (2006). Reporting and reflecting on corporate social responsibility in the hospitality industry: A case study of pub operators in the UK. *International Journal of Contemporary Hospitality Management, 18*(4), 329–340. https://doi.org/10.1108/09596110610665339

Jones, P., Hillier, D., & Comfort, D. (2014). Sustainability in the global hotel industry. *International Journal of Contemporary Hospitality Management, 26*(1), 5–17. https://doi.org/10.1108/ijchm-10-2012-0180

Jose, T. (2017). Need for harmonisation of sustainability reporting standards. *Journal of Finance and Economics, 5*(6), 253–258. https://doi.org/10.12691/jfe-5-6-1.

Kinderman, D. (2019). The challenges of upward regulatory harmonization: The case of sustainability reporting in the European Union. *Regulation & Governance*, (January), 1–24. https://doi.org/10.1111/rego.12240.

La Torre, M., Sabelfeld, S., Blomkvist, M., Tarquinio, L., & Dumay, J. (2018). Harmonising non-financial reporting regulation in Europe: Practical forces and projections for future research. *Meditari Accountancy Research, 26*(4), 598–621. https://doi.org/10.1108/MEDAR-02-2018-0290

Manes-Rossi, F., Tiron-Tudor, A., Nicolò, G., & Zanellato, G. (2018). Ensuring more sustainable reporting in europe using non-financial disclosure-de facto and de jure evidence.*Sustainability, 10*(4), 2–20. https://doi.org/10.3390/su10041162.

Markota, N. V., Vuković, R., & Calace, D. (2017). Non-financial reporting as a new trend in sustainability accounting. *Journal of Accounting and Management*, *7*(2), 13–26. Retrieved from https://hrcak.srce.hr/194750.

Mcleay, S., Neal, D., & Tollington, T. (1999). International standardisation and harmonisation: A new measurement technique. *Journal of International Financial Management and Accounting, 10*(1), 42–70.

Milanes-Montero, P., & Perez-Calderon, E. (2014). Environmental reporting in the hospitality industry: An international analysis. *Environmental Engineering and Management Journal, 13*(10), 2531–2540. https://doi.org/10.30638/eemj.2014.283.

Mohammed, A., & Rashid, B. (2018). A conceptual model of corporate social responsibility dimensions, brand image, and customer satisfaction in Malaysian hotel industry. *Kasetsart Journal of Social Sciences, 39*(2), 358–364. https://doi.org/10.1016/j.kjss.2018.04.001

Morris, R. D., & Parker, R. H. (1998). International harmony measures of accounting policy: Comparative statistical properties. *Accounting and Business Research, 29*(1), 73–86. https://doi.org/10.1080/00014788.1998.9729567

Olya, H., Altinay, L., Farmaki, A., Kenebayeva, A., & Gursoy, D. (2021). Hotels' sustainability practices and guests' familiarity, attitudes and behaviours. *Journal of Sustainable Tourism, 29*(7), 1063–1081. https://doi.org/10.1080/09669582.2020.1775622

Ortas, E., & Moneva, J. M. (2011). Origins and development of sustainability reporting: Analysis of the Latin American context. *GCG: Revista de Globalización, Competitividad & Gobernabilidad, 5*(2), 16–37. https://doi.org/10.3232/GCG.2011.V5.N2.01.

Ortmann, R., Pelster, M., & Wengerek, S. T. (2020). COVID-19 and investor behavior. *Finance Research Letters, 37*(May), 1–9. https://doi.org/10.1016/j.frl.2020.101717

Roberts, C., Weetman, P., & Gordon, P. (2005). *International financial reporting: A comparative approach* (3rd ed.). Prentice Hall. Retrieved from https://www.books.mec.biz/tmp/books/XKFQ41HIIS5E4HZNX3HL.pdf.

Salama, W., Nor, M., Deen, E., Albakhit, A., & Zaki, K. (2022). Understanding the connection between sustainable human resource management and the hotel business outcomes: Evidence from the green-certified hotels of Egypt. *Sustainability, 14*(5647). https://doi.org/10.3390/su14095647.

SASB. (2013). Conceptual Framework of the Sustainability Accounting Standards Board, (October), 26. Retrieved from sasb.org.

Singh, N., Sundari, S. K., & Nath, A. (2015). Contribution of hotel and restaurant industry to solid waste: An assessment of its impact on environment. *I-Manager's Journal on Management, 10*(1), 22–32. https://doi.org/10.26634/jmgt.10.1.3509.

Soria Leyva, E., & Parra Parra, D. (2021). Environmental approach in the hotel industry: Riding the wave of change.*Sustainable Futures, 3*(April), 1–14. https://doi.org/10.1016/j.sftr.2021.100050.

Tay, J. S. W., & Parker, R. H. (1990). Measuring international harmonization and standardization. *Abacus, 26*(1), 71–88. https://doi.org/10.1111/j.1467-6281.1990.tb00233.x

Tschopp, D., & Nastanski, M. (2013). The harmonization and convergence of corporate social responsibility reporting standards. *Journal of Business Ethics, 125*, 147–162. https://doi.org/10.1007/s10551-013-1906-9

Van der Tas, L. G. (1988). Measuring harmonisation of financial reporting practice. *Accounting and Business Research, 18*(70), 157–169. https://doi.org/10.1080/00014788.1988.9729361

Vetráková, M., Šimočková, I., & Pompurová, K. (2019). Age and educational diversification of hotel employees and its impact on turnover. *Sustainability (switzerland), 11*(19), 1–14. https://doi.org/10.3390/su11195434

Vuković, R., Vukić, N. M., & Sesar, D. (2020). Non-financial reporting as part of sustainability accounting with the examples of good practice. *Journal of Accounting and Management, 10*(1), 41–58.

Youn, H., & Kim, J. (2022). Corporate social responsibility and hotel employees' organizational citizenship behavior: The roles of organizational pride and meaningfulness of work. *Sustainability, 14*(2428). https://doi.org/10.3390/su14042428.

Zrnić, A., Pekanov Starčević, D., & Mijoč, I. (2020). Evaluating environmental accounting and reporting: The case of Croatian listed manufacturing companies. *Pravni Vijesnik, 36*(1), 47–63. https://doi.org/10.25234/pv/8089.

Food Tourism Sustainability in Portugal: A Systematic Literature Review

Teresa Mendes, Pedro Liberato, Dália Liberato, Teresa Dieguez, and Hugo Barreira

Abstract Food is one of the main tourism motivations to visit a destination. Food-related activities and experiences, and food consumption places, have, therefore, been used as unique selling propositions by tourism management destinations employing multiple marketing strategies, which are nowadays strengthened by digital branding. On the other hand, global food supply, and its supply chains sustainability, also became one of the most urgent agendas for all nations, worldwide. Concurringly, United Nations launched in 2016 the Sustainable Development Goals program that the United Nations World Travel Organization officially adopted in 2017. Although international tourism demand has been consistently increasing toward Portugal throughout this decade, being the most searched for European country in 2022, one of the main conclusions of the present research is that both national and international academy has not yet produced significant studies on this country's food sustainability, concerning the tourism market. This paper delivers a systematic literature review on food tourism sustainability concerning tourism destination Portugal, based on a specific articles' selection protocol. The selected articles, analyzed through a visual bibliometric networks' software, produced three main clusters, which presented a

T. Mendes (✉) · P. Liberato · D. Liberato
School of Hospitality and Tourism, Polytechnic Institute of Porto, and CiTUR Researcher (Centre for Tourism Research, Development, and Innovation), Vila do Conde, Portugal
e-mail: teresamendes@esht.ipp.pt

P. Liberato
e-mail: pedrolib@esht.ipp.pt

D. Liberato
e-mail: dalialib@esht.ipp.pt

T. Dieguez
Polytechnic Institute of Cávado and Ave, and CiTUR Researcher (Center for Tourism Research, Development, and Innovation), Braga, Portugal
e-mail: tdieguez@ipca.pt

H. Barreira
Faculty of Arts and Humanities of University of Porto, and CITCEM Researcher (Transdisciplinary Research Center «Culture, Space and Memory»), Porto, Portugal
e-mail: hbarreira@letras.up.pt

© The Author(s), under exclusive license to Springer Nature Switzerland AG 2023

V. Katsoni (ed.), *Tourism, Travel, and Hospitality in a Smart and Sustainable World*, Springer Proceedings in Business and Economics,
https://doi.org/10.1007/978-3-031-26829-8_15

research trend, stressing critical aspects concerning the role of regional governance in food tourism, and the still scarce academic production to this matter.

Keywords Food tourism in Portugal · Food sustainability · Governance in food tourism · Systematic literature review

JEL Classification Z32

1 Introduction

Almost half of tourists' budgets is spent while on vacation on food and drinks (Okumus et al., 2018). Food is, indeed, one of the main motivations to visit a tourism destination (TD) (Okumus, 2021).

This way, Serra et al. (2021) state that food is a cultural symbol, contributing to the international projection of certain regions. It is also a very profitable hospitality strategy because it will increase tourists' expenditure in the TD, and food products can be used as economic drivers in the international tourism market. In fact, according to Figueiredo et al. (2021) "Food does not only convey important cultural functions and values but can also foster social, environmental, territorial and economic cohesion", also contributing to sustaining cultural heritage and food diversity. Food tourism can, therefore, be a successful formula for regional development, and a premium vehicle for local sustainability (Lopes et al., 2021, 2022).

On the other side, Cooke and Nunes (2021) underline that tourism is one of the most polluting industries, reporting that carbon emissions from international and domestic tourism would rise from 5%, in 2018, to 5.3%, by 2030. To this concern, Lukoseviciute and Pereira (2021) point out that United Nations World Tourism Organization (UNWTO) identified coastal areas as the fastest growing tourism demand regions over the last decades, in particular beaches, causing a growing coastal erosion and a sea level rise. In fact, this exploitation of coastal areas and islands has contributed to mass tourism, which increased unsustainability worldwide, like intensive food monoculture, with high seasonality and tourism supply standardization (Valls et al., 2019).

Sustainability is currently one of the most worldwide pressing topics, and it is accepted that food is one of the major concerns in several of the United Nations (UN) Sustainable Development Goals (SDGs) (Viegas & Lins, 2019). Concurrently, Vaz et al. (2014) concluded there is an undeniable conflict between excessive economic growth and the loss of sustainable areas, in Portuguese TDs.

On balance, food tourism academic research has increased in the last decade (Ellis et al., 2018; Okumus, 2021), and its study connected to sustainable tourism has been the most prolific (Okumus et al., 2018). Nonetheless, there is still much scarcity in academic research related to food tourism and sustainability in Portugal (Carvalho et al., 2021b; Santos et al., 2020). In line with Farinha et al. (2021), Rachão et al. (2019), and Real and Graça (2019), governance faces significant obstacles in designing adequate measures to promote food and sustainable tourism between the public and private tourism sector stakeholders.

This paper intends to identify the progress on sustainable food tourism academic studies in Portugal and highlight identified research gaps for further investigation. The methodology is based on an exploratory literature review, using a bibliometric networks software program (BNSP) to enrich intellectual associations on this academic data.

The following section concerns methodology, and will describe the paper selection protocol, as well as the procedure to a correlation map creation. The literature review and results in this chapter will present: (1) papers analysis table per data, (2) keywords' correlation map visualization, and (3) its three clusters concurrent analysis. The two final sections will discuss the results and focus on the main conclusions, evidencing the critical scarce production on this subject, to this date.

2 Methodology

Given the research theme, it should be first recognized that one of the analysis core, the food tourism construct, still presents no academic consensus (Ellis et al., 2018; Mendes et al., 2021; Okumus, 2021; Rachão et al., 2019; Serra et al., 2021). In their systematic analysis on food tourism, Ellis et al. (2018) identified three most common terms. Also, Okumus (2021), applying the same research objective, acknowledged using several keywords to reflect the numerous terms.

Consequently, three sets of research keywords were used to incorporate this multiplicity. Like Martinho (2021), the chosen platform was Web of Science. The used sets were (1) food tourism, sustainability, Portugal; (2) gastronomy tourism, sustainability, Portugal; and (3) culinary tourism, sustainability, Portugal. The two last sets delivered few results, repeated in the largest results set: food tourism, with 44 results. Therefore, it should be noted that, like Mendes et al. (2021) and Rachão et al. (2019) propose, the accepted construct for this paper is food tourism.

The research documents selection was based on the process used by Figueiredo et al. (2021). A three-step protocol was applied, one day in July 2022, and it was used to identify, screen, and elect the 16 final papers, as seen on Fig. 1. A selection criterion was applied: only indexed papers were chosen; the titles, keywords, and abstracts were analyzed, according to this research's relevance. Seventeen papers were chosen, due to their explicit connection to food tourism sustainability in Portugal. Nevertheless, when applying the 3rd protocol step (eligibility), full paper assessment, one paper was excluded, because its data methodology concluded there were no relation to the Portuguese territory.

Contrary to Figueiredo et al. (2021), no time span was used in the selection because of the scarce scientific production. The articles' languages presented no difficulty to these research authors and were all included. Table 1 shows several data identified in the selected papers.

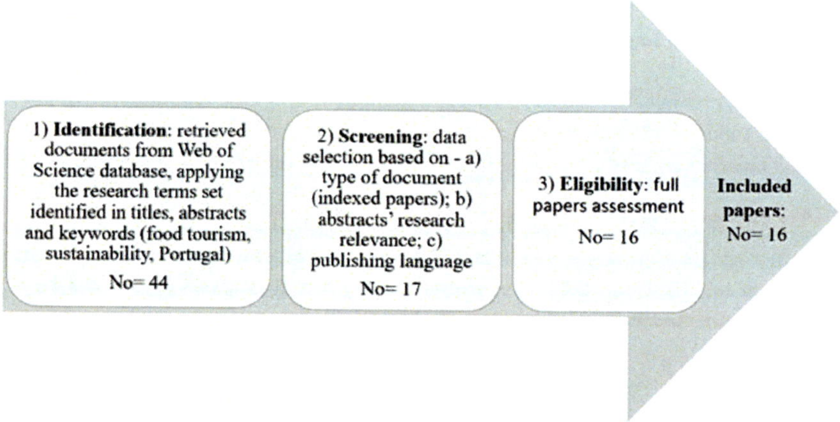

Fig. 1 Paper selection protocol. Adapted from Figueiredo et al. (2021)

To enrich intellectual associations on this academic data, VOSviewer BNSP was used, as in Martinho (2021) and Okumus (2021), producng a correlation map, with three clusters (Fig. 2).

To prevent researcher subjectivity, a double blinded peer coding was used in the clusters' map labeling, to ensure data reliability and trustworthiness (Korstjens & Moser, 2018; Okumus, 2021). Nevertheless, like Okumus (2021) concluded upon the BNSP usage regarding specific clusters viewpoint, this exploratory study did not produce perfect correlations between all documents, and neither did all of them have the same focus.

3 Literature Analysis and Results

The included papers on this study are data categorized as shown on the following table (Table 1). To enrich the intellectual connections and to highlight potential research gaps, a keywords' correlation map was produced, presenting three clusters (Fig. 2).

After the double blinded peer labeling discussion, they were organized in a spread sheet, to reflect the correlation between each clusters' keywords. Each cluster is classified in the following way:

- Cluster 1—Sustainable development in gastronomy-related tourism (9 correlated keywords)
- Cluster 2—Tourists behavior in food and wine regions (7 correlated keywords)
- Cluster 3—Food-related experiences in tourism destinations (5 correlated keywords).

Table 1 Papers analysis table per data

Data	Papers
Publishing journal	Advances in Hospitality and Tourism Research: Rachão et al. (2019)
	Agroecology and Sustainable Food Systems: Vaz et al. (2014)
	Bulletin of the Spanish Geographers Association: Lopes et al. (2021)
	European Planning Studies: Cooke and Nunes (2021)
	European Journal of Tourism, Hospitality and Recreation: Viegas and Lins (2019)
	International Journal of Contemporary Hospitality Management: Santos et al. (2020)
	Slovenian Geographers Minutes' Meetings: Fernandes and Richards (2021)
	Spanish Journal of Health Communication: Real and Graça (2019)
	Sustainability: Carvalho et al. (2021a); Carvalho et al. (2021b); Lopes et al. (2022); Farinha et al. (2021); Serra et al. (2021); Valls et al. (2019)
	Tourism and Management Studies: Lukoseviciute and Pereira (2021)
	Tourism Review: Rachão et al. (2021)
Publishing date	2014: Vaz et al. (2014)
	2019: Rachão et al. (2019); Real and Graça (2019); Valls et al. (2019); Viegas and Lins (2019)
	2020: Santos et al. (2020);
	2021: Carvalho et al. (2021a); Carvalho et al. (2021b); Cooke and Nunes (2021); Farinha et al. (2021); Fernandes and Richards (2021); Lopes et al. (2021); Lukoseviciute and Pereira (2021); Rachão et al. (2021); Serra et al. (2021)
	2022: Lopes et al. (2022)
Publishing language	English: Carvalho et al. (2021a); Carvalho et al. (2021b); Lopes et al. (2022); Cooke and Nunes (2021); Farinha et al. (2021); Fernandes and Richards (2021); Lukoseviciute and Pereira (2021); Rachão et al. (2019); Rachão et al. (2021); Santos et al. (2020); Serra et al. (2021); Valls et al. (2019); Vaz et al. (2014); Viegas and Lins (2019)
	Spanish: Lopes et al. (2021)

(continued)

Table 1 (continued)

Data	Papers
	Portuguese: Real and Graça (2019)
Research centers (1st corresponding author)	Higher Institute of Administration and Management, Porto, Portugal: Rachão et al. (2021)
	Higher School of Hospitality and Tourism of Estoril, Portugal: Viegas and Lins (2019)
	Polytechnic Institute of Viana do Castelo, Portugal: Fernandes and Richards (2021)
	Ryerson University, Canada: Vaz et al. (2014)
	Portuguese Nutrition Association: Real and Graça (2019)
	University of Algarve, Portugal: Lukoseviciute and Pereira (2021); Farinha et al. (2021); Santos et al. (2020); Serra et al. (2021)
	University of Aveiro, Portugal: Carvalho et al. (2021a); Carvalho et al. (2021b); Rachão et al. (2019)
	University of Extremadura, Spain: Lopes et al. (2021, 2022)
	University of Madeira, Portugal: Valls et al. (2019)
	Western Norway University of Applied Sciences: Cooke and Nunes (2021)
Research methodology	Qualitative: Carvalho et al. (2021a); Carvalho et al. (2021b); Cooke and Nunes (2021); Fernandes and Richards (2021); Rachão et al. (2019, 2021); Real and Graça (2019); Viegas and Lins (2019)
	Quantitative: Lopes et al. (2021, 2022); Farinha et al. (2021); Lukoseviciute and Pereira (2021); Santos et al. (2020); Valls et al. (2019); Vaz et al. (2014);
	Mixed: Serra et al. (2021)
Researched region(s)	Spain and Portugal: Lopes et al. (2021)
	Algarve: Farinha et al. (2021); Lukoseviciute and Pereira (2021); Real and Graça (2019); Santos et al. (2020); Serra et al. (2021); Vaz et al. (2014)
	Bairrada wine region: Carvalho et al. (2021a)
	Center of Portugal: Lopes et al. (2022)
	Dão: Carvalho et al. (2021b)
	Douro, Vinho Verde and Trás-os-Montes wine regions: Rachão et al. (2019)

(continued)

Table 1 (continued)

Data	Papers
	Estoril: Viegas and Lins (2019)
	Madeira: Valls et al. (2019)
	Minho: Fernandes and Richards (2021)
	North of Portugal: Rachão et al. (2021)
	Tomar: Cooke and Nunes (2021)

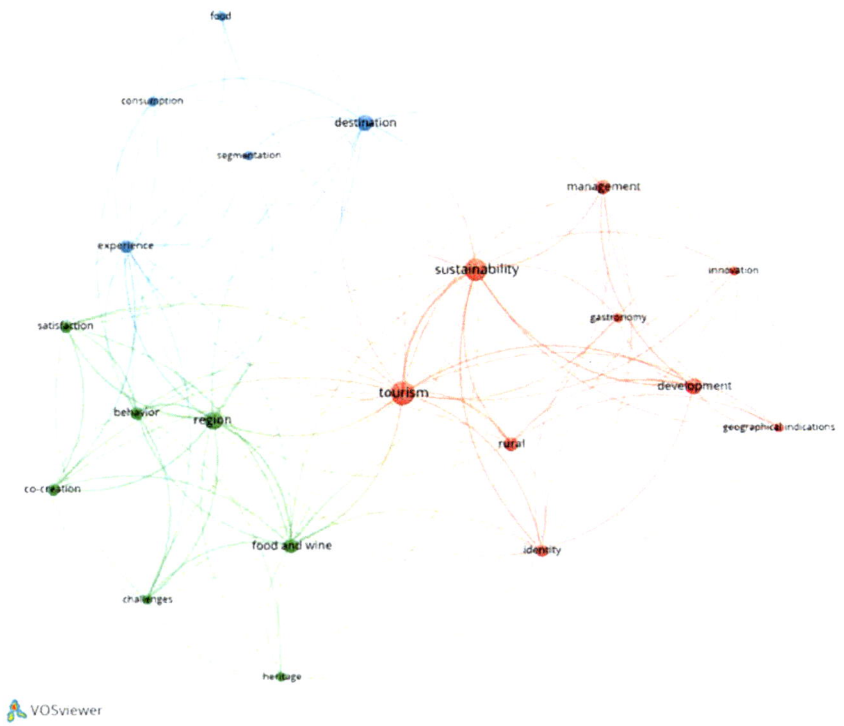

Fig. 2 Keywords correlation map visualization

Cluster 1—Sustainable development in gastronomy related tourism.

Searching sustainability in strategic TDs' management is essential to understand the underlying structure and feedback mechanisms that influence the tourism sector (Farinha et al., 2021). Although the economic benefits are clear, it is vital to emphasize the repercussions on environmental, sociocultural, and institutional dimensions of tourism. The stakeholders' viewpoint should be that all players must be evolved in the creation of tourism strategies.

On the other hand, cooperative governance must ensure successful steps toward integrated sustainable policies (Farinha et al., 2021; Rachão et al., 2019; Romero & Tejada, 2020). Thus, a shared vision is mandatory (Rachão et al., 2019; Romero & Tejada, 2020), beginning with the supply services tourism sector, that must continue adopting innovative sustainable practices (Farinha et al., 2021). This way, the Observatory of Sustainability of the Algarve Region for Tourism Governance is developing an important strategy to raise awareness through the sustainable competitiveness goals (Farinha et al., 2021).

In fact, as observed in Table 1, Algarve was the most researched Portuguese region, and Vaz et al. (2014) shown that landscape metrics can be used to understand and prevent the loss of agricultural land in Algarve due to mass tourism, from an economic perspective.

Also, European Union rural territories are defining different models to enhance their development: promoting their endogenous potential, using combined territorial identity and brands creation. The presence of agri-food products, exclusive to a given territory, used in gastronomy heritage-based tourism, largely benefit both local producers and local economy (Lopes et al., 2021).

Geographical Indication food products contribute to quality gastronomy and differentiating local products, so becoming a suitable formula for tourism attractiveness and regional sustainable development. This development also encourages sustainable agriculture and livestock farming, and these assets are emphasized through food relates events, such as food tasting and educational activities to learn about agri-food production processes (Fernandes & Richards, 2021; Lopes et al., 2021).

In line with these social, environmental, and economical concerns, Serra et al. (2021) created a pairing model for food and wine in the Algarve. In this study, the authors expected to promote an efficient, socially fair, and ecologically sustainable local economy in this coastal destination, using only local products. The same way, Viegas and Lins (2019) intend to increase awareness on health and sustainability food tourism-related concerns in tourism and hospitality students, thus developing plant-based meals projects.

To this point, Lopes et al. (2022), Real and Graça (2019), and Fernandes and Richards (2021) consider that restaurant owners are essential promotion agents, when addressing gastronomy heritage-based tourism, and sustainable identity for TDs.

Still observing this clusters' main keyword, sustainability, although Cooke and Nunes (2021) don't talk about food tourism, they express their concerns toward urban and rural sustainable development, including food supply, partly due to the worldwide massive tourism. Finally, TDs sustainability is a growing tourists' concern, and these consumers demand tailor made activities with this scope (Carvalho et al., 2021b).

Cluster 2—Tourists behavior in food and wine regions

Tourist's behavior in food and wine regions is an interesting research perspective, representing the most researched theme (Carvalho et al., 2021a, b; Rachão et al., 2019, 2021; Serra et al., 2021). This dynamic emphasizes wine tourism experiences, promoting not only rural territories' distinctive wine and food products, but also their local identity, natural and cultural resources, and historical features (Carvalho et al., 2021a).

On the other hand, added value and unique tourism experiences (Carvalho et al., 2021a) deliver high satisfaction scores on these TDs, through interaction and engagement as components for co-creation (Carvalho et al., 2021b; Rachão et al., 2021; Santos et al., 2020). This way, visitors play a central role by sharing creation values with other destination stakeholders, in recreational and holistic experiences. These experiences can occur in wine tourism context (Carmichael, 2005; Cohen & Ben-Nun, 2009), focused on wine cellars and other wine related activities, like: 'winescape' enjoyment, wine-producing process, wine tasting, wine and food pairing, and learning about harvesting techniques.

Simultaneously, the most significant challenges related to wine regions are wine quality promotion, employment, and investment improvement, as well as the regions' tourism development. From the tourists' perspective, an integrative tourism experience, including the abovementioned local natural and cultural resources, are crucial components that contribute to unique, sensory-rich, and emotion-generating experiences (Carvalho et al., 2022; Kastenholz et al., 2021).

In the Bairrada wine region, Carvalho et al. (2021a) suggest a network approach between wineries and local restaurants, enhancing the region's attractiveness. This aspect was observed as a highly appreciated cultural knowledge-transfer strategy when designing wine tourism experiences, since co-creative wine experiences give visitors the opportunity to develop their skills and deepen their knowledge (Carvalho et al., 2021a, b; Kastenholz et al., 2021). Also, Rachão et al. (2021), results' highlight co-creation perceived by tourists as a combination of seven categories: social interaction, novelty, creativity, social sustainability, environmental awareness, enjoyment, and memorable experiences. These authors created a theoretical co-creation model that can be applied on food-and-wine tourism contexts.

Also, in this cluster it should be noted that heritage emerged as one of the keywords. As mentioned by Lopes et al. (2021) Real and Graça (2019), the Mediterranean Diet joint UNESCO classification plays a very important role in the Portuguese food culture identity, and many tourists expect a connection between Portuguese food offer and its gastronomic regions, based on tradition. The memorability component of food and wine tourism is an acknowledged attribute by tourists. Both cultural tangible and intangible components for this type of tourism are used to enhance creative and immersive experiences, whether talking about Mediterranean diet, or Douro's UNESCO cultural landscape (Carvalho et al., 2021b; Rachão et al., 2019).

Cluster 3—Food-related experiences in tourism

It is important to highlight Kotler and Armstrong (2018) study which based their understanding on tourism market segmentation variables on geographic, demographic, psychographics, and behavioral aspects. Recent literature emphasizes some studies considering motivation (Correia et al., 2009; Levitt et al., 2019; Yun et al., 2011) and satisfaction (Correia et al., 2009) to identify tourists' segments regarding food experiences involvement (Santos et al., 2020).

According to Santos et al. (2020), tourists change their food consumption patterns while on holiday, based on factors such accommodation typology and meal plan, as well as tourist origin. To this point, Real and Graça (2019) underline that food and food-related activities are, after all, a dynamic construct and, therefore, a constant research scope.

In tourism research, understanding consumer behavior is crucial, and it was one of the identified keywords for this cluster. Segmentation extracts tourist identities, focus on customer preferences and motivations as key clustering variables, and it can be applied in several tourism destinations (Albayrak & Caber, 2018; Fung & Jim, 2015; Guttentag et al., 2018; Loker-Murphy, 1997; Ramires et al., 2018), to analyze both tourist decisions and needs. Also, clustering is based on another satisfaction evaluation of destination attributes, through the combination "Importance-Performance technique" (Bruyere et al., 2002; Vaske et al., 2009; Wade & Eagles, 2003).

To this purpose, Lukoseviciute and Pereira (2021) identified off-peak season beach users' profiles in the Algarve region and segmented them according to their priorities for sustainable beach management and enjoyment. Most of the interviewed tourists were segmented as comfort and food tourists. Valls et al. (2019) interviewed tourism entrepreneurs in Madeira to assess if they are in line with slow tourism practices. They concluded that this TD's culture and gastronomy experiences should became drivers to visit the region, due to emerging tourism demand. These products should too be worked on through slow and sustainable tourism.

On the other side, when talking about obtainable tourism experiences in destinations, current market trends emphasize regionality, authenticity, uniqueness, sensations, creating one's own journey, vacation rentals, commitment to local culture and history, quiet, de-commoditization, reinforcement on local and indigenous culture, and interactions with residents, with their roots, and with their language and their lifestyle (Valls et al., 2019).

Other aspects valued by tourists, in Rachão et al. (2021), are health concerns, enjoying the scenery and food, dining, and socializing, appreciate experiences that incorporate something from the SDG, like reducing business waste. The social sustainability category was also referred concerning new experiences: social entrepreneurship practices of the food industry, e.g., attending local farmers' markets and having local people teaching how to cook traditional dishes.

4 Discussion and Proceeding Research

Although all articles focus on sustainability, not all really concentrate their scope in engaging this theme to food tourism. Those who did, Carvalho et al. (2021a); Carvalho et al. (2021b); Cooke and Nunes (2021); Farinha et al. (2021); Fernandes and Richards (2021); Lopes et al. (2021, 2022); Rachão et al. (2019, 2021); Serra et al. (2021); and Viegas and Lins (2019), exhibited various angles.

Even though it is undeniable that the most researched region is Algarve, another interesting aspect is that several studies focused on wine tourism-related activities

(Carvalho et al., 2021b; Rachão et al., 2019, 2021; Serra et al., 2021). Conversely, it stood out the growing interest in using the qualitative methodology (Table 1). This corroborates what Okumus et al. (2018) had already concluded: there is in food tourism research "a notably increasing trend in using qualitative and hybrid methods", and this has helped to accelerate the development of the food and gastronomy in the hospitality and tourism area (Okumus et al., 2018, p. 69). Analyzing the discussion and proceeding research sections of the elected papers, some concerns are now highlighted.

Ultra-urban tourism travel can be characterized by the frequent use of social networks, based on the 'attention economy' concept (Cooke & Nunes, 2021), which were celebrities' posts of what they have eaten in a certain TD. This food consumption is frequently healthy and specific to that persons' food habits, with no connection to local gastronomy heritage, thus compromising local food tourism sustainability.

Concurrently, according to Valls et al. (2019), food and wine pairing research still has much to evolve from the common formula of wine and cheese pairing. The destination's food uniqueness is presented as its main asset, related to constructs such as 'foodscape' and 'touristscape', that should be balanced not to grow into commodification and food fictional narratives.

It is paramount, when creating salvage strategies in tourism activities and travel behaviors, to engage with tourists and make them active players in governance decision processes. In the Lukoseviciute and Pereira (2021) research, all the interviewed tourists showed environmental awareness, and all paid much interest in beach cleanness in the Algarve. The consciousness of beach visitors concerning Algarve's current issues will facilitate local governance decision-making processes. Nevertheless, it should be noted that two of the three segmented tourists' profiles, 'comfort and food lovers', as well as 'sustainable minimalists' considered overcrowding an unimportant attribute in beach qualities in this region.

On the other side, Real and Graça (2019) underline the importance of acknowledging food habitats that are constantly changing; thus, food heritage should be observed as a dynamic construct. Mediterranean diet plays an important role in the strengthening of local communities' identity, in culture and nature tourism, in enhancing local economies, preserving tradition lifestyles and even species. This way, a salvage plan ought to be created, focused on its territory. According to these authors, this identity operates best at a regional level, and European joint classification can undermine its authenticity, derived from multiple Mediterranean food regions. Local stakeholders, including local governance, should be engaged in these concerns.

Lopes et al. (2022) acknowledge that restaurants are vital mediators between tourists and local gastronomy, but only a few, in the center region of Portugal, use several agri-food products on their menus. The authors also refer that studies concerning the effects of regional agri-food products usage on restaurants' financial performance and their business sustainability are still scarce. On the other hand, they consider that differentiated quality food products, related to specific territories, should be a premium choice for the regional governance strategies, and restaurant owners.

Also, future studies should analyze further Portuguese wine regions, other than those here presented (Algarve, Bairrada, Dão, Douro, Vinho Verde and Trás-os-Montes wine regions), and establish a comparison between them, and even with foreign regions (Carvalho et al., 2021b; Rachão et al., 2021).

. Still concerning the tourists' perspective, Santos et al. (2020) say that research should analyze tourist budgets, and its relation to food consumption and tourists' desire for local gastronomy experiences, and also whether changes in eating habits on holiday lead to healthy diets. Concerning the Bairrada region attributes, Carvalho et al. (2021a) list in their findings the importance of exploring other tourism attractions than wine tasting, engaging tourists in this research.

According to Farinha et al. (2021), proceeding studies should be addressed, comparing hotels and restaurants from different categories, in the Algarve region. They believe the results would lead to even more specific recommendation to TDs. Other governance stakeholders (e.g., local associations) could be the scope to further research.

A question was left unanswered in Valls et al. (2019) study, that is if tourists are willing to pay more to experience sustainable activities in Madeira, an investigation question is in fact suggested by the same authors, for further studies.

Nonetheless, like Okumus et al. (2018), it is our belief that proceeding research on this subject will benefit from a systematic literature review of the academic production, namely masters dissertations and PhDs thesis. Another interesting perspective, concerning additional studies, would be to engage further, using the same parameters, in coming years.

5 Conclusion

Like Figueiredo et al. (2021), this research concludes the chosen papers show different samples' usage, different techniques and methods, different agendas, different regions, and investigation objectives, as presented in Table 1.

However, to summarize an overall theme for this academic production, a tourism research trend stood out, as previously identified by Ellis et al. (2018): that related to sustainable development and management. These two keywords were present, by order, 5 and 4 times in the analyzed papers. These authors also suggested a wider range of stakeholders and local government involvement, as both part of the answer to sustainable tourism problems, as also further research opportunities.

To this point, it should be mentioned that even though this concept was not analyzed in each cluster, all three reflected findings to join government organizations and other tourism stakeholders, as so concluded in Okumus (2021).

It was clear to Rachão et al. (2019) that there was an evident lack of cooperation between food-and-wine and tourism in the three wine regions under study, which tended to hinder the construction of land identity for development purposes. These authors identified the unquestionable role of public organizations in promoting a shared vision for endogenous products, alongside with local stakeholders.

Lopes et al. (2022) concluded that too local governance should be effective in contributing to implement strategic options and measures, with responsible social practices, concerning food sustainability and the usage of local qualified agri-food products in local restaurants, in the Center region of Portugal.

Real and Graça (2019) concentrated their findings on good practices from another European government that implemented toward the salvage of Mediterranean Diet identity: in Spain a quality label was created for restaurants that use four specific standards, observing this diet maintenance, and helping to preserve its foodways.

Also, Lukoseviciute and Pereira (2021) centered on the importance of sustainable beaches' attributes in Algarve and how they are perceived by tourists, involving regional governance, and advising them to extend this research's methodology to other TD's attributes.

Finally, Farinha et al. (2021) state governance faces significant obstacles in implementing sustainable measures, and there should be greater consensus between government and tourism professional stakeholders to create sustainable integrated strategies in the tourism sector, in this case study, the Algarve.

Even though Figueiredo et al. (2021) used food tourism systematic literature review on rural areas, they underlined that further implications in power relations between TDs' stakeholders should undergo in proceeding studies.

In conclusion, three critical aspects can be drawn from the present research: (1) the state of the art in academic production, relating food tourism and sustainability in Portugal, is too scarce for this country, since in the past years its international demand rose significantly; (2) it is mandatory for TDs' governance organizations to engage more efficiently in food sustainability and tourism issues; and (3) academia should deepen its help in providing more effective solutions to these matters.

Acknowledgements The authors acknowledge the financial support of CiTUR, R&D unit funded by the FCT—Portuguese Foundation for the Development of Science and Technology, Ministry of Science, Technology and Higher Education, under the scope of the project UID/BP/04470/2020.

References

Albayrak, T., & Caber, M. (2018). A motivation-based segmentation of holiday tourists participating in white-water rafting. *Journal of Destination Marketing and Management, 9*, 64–71.

Bruyere, B. L., Rodriguez, D. A., & Vaske, J. J. (2002). Enhancing Importance-Performance analysis through segmentation. *Journal of Travel and Tourism Marketing, 12*(1), 81–95.

Carmichael, B. (2005). Understanding the wine tourism experience for winery visitors in the Niagara region. *Tourism Geographies, 7*(2), 185–204. https://doi.org/10.1080/14616680500072414

Carvalho, M., Kastenholz, E., & Carneiro, M. (2021a). Interaction as a Central Element of Co-Creative Wine Tourism Experiences-Evidence from Bairrada, a Portuguese Wine-Producing Region. *Sustainability, 13*(16). https://doi.org/10.3390/su13169374

Carvalho, M., Kastenholz, E., & Carneiro, M. (2021b). Pairing Co-Creation with Food and Wine Experiences-A Holistic Perspective of Tourist Experiences in Dao, a Portuguese Wine Region. *Sustainability, 13*(23). https://doi.org/10.3390/su132313416

Carvalho, M., Kastenholz, E., Carneiro, M.J. (2022). *Co-creation Experiences and Experiencescape. In* Encyclopedia of Tourism Management and Marketing; Buhalis, D. (Ed.). Edward Elgar Publishing Limited. https://doi.org/10.4337/9781800377486.co-reation.experiences.experiencescape

Cohen, E., Ben-Nun, L. (2009). The important dimensions of wine tourism experience from potential visitors' perception. *Tourism and Hospitality Research, 9*, 20–31.

Cooke, P., & Nunes, S. (2021). Post-Coronavirus regional innovation policies: From mega to giga and beyond through sustainable spatial planning of global tourism.*European Planning Studies*, 1–19. https://doi.org/10.1080/09654313.2021.1936463

Correia, A., Moital, M., Oliveira, N., & Costa, C. F. (2009). Multidimensional segmentation of gastronomic tourists based on motivation and satisfaction. *International Journal of Tourism Policy, 2*(1/2), 37–57.

Ellis, A., Park, E., Kim, S., & Yeoman, I. (2018). What is food tourism? *Tourism Management, 68*, 250–263. https://doi.org/10.1016/j.tourman.2018.03.025

Farinha, F., Bienvenido-Huertas, D., Pinheiro, M., Silva, E., Lanca, R., Oliveira, M., & Batista, R. (2021). Sustainable Competitiveness of Tourism in the Algarve Region. Critical Stakeholders' Perception of the Supply Sector. *Sustainability, 13*(11). https://doi.org/10.3390/su13116072

Fernandes, C., & Richards, G. (2021). Developing Gastronomic Practices in the Minho Region of Portugal. *Acta Geographica Slovenica-Geografski Zbornik, 61*(1), 141–152. https://doi.org/10.3986/AGS.9370

Figueiredo, E., Forte, T., & Eusébio, C. (2021). Rural Territories and Food Tourism – Exploring the Virtuous Bonds Through a Systematic Literature Review. *European Countryside, 13*(3), 622–643. https://doi.org/10.2478/euco-2021-0035

Fung, C. K. W., & Jim, C. Y. (2015). Segmentation by motivation of Hong Kong Global Geopark visitors in relation to sustainable nature based tourism. *International Journal of Sustainable Development and World Ecology, 22*(1), 76–88.

Guttentag, D., Smith, S., Potwarka, L., & Havitz, M. (2018). Why tourists choose Airbnb: A motivation-based segmentation study. *Journal of Travel Research, 57*(3), 342–359.

Kastenholz, E., Cunha, D., Eletxigerra, A., Carvalho, M., Silva, I. (2021). *Exploring Wine Terroir Experiences: A Social Media Analysis*. In Advances in Tourism, Technology and Systems; ICOTTS 2020. Smart Innovation, Systems and Technologies; Abreu, D., Liberato, E.A., González, J.C., Ojeda, G., Eds.; Springer: Singapore, 2021; Volume 209, pp. 401–420.

Korstjens, I., & Moser, A. (2018). Series: Practical guidance to qualitative research. Part 4: Trustworthiness and publishing. *European Journal of General Practice, 24*(1), 120–124. https://doi.org/10.1080/13814788.2017.1375092

Kotler, P. & Armstrong, G. (2018). *Principles of Marketing*. Pearson.

Levitt, J. A., Zhang, P., Di Pietro, R. B., & Meng, F. (2019). Food tourist segmentation: Attitude, behavioral intentions and travel planning behavior based on food involvement and motivation. *International Journal of Hospitality and Tourism Administration, 20*(2), 129–155. https://doi.org/10.1080/15256480.2017.1359731

Loker-Murphy, L. (1997). Backpackers in Australia: a motivation-based segmentation study. *Journal of Travel and Tourism Marketing*, 5(4), 23-45.

Lopes, C., Gallego, J., & Leitão, J. (2021). The importance of quality agro-food products for the promotion of tourism in Spain and Portugal. *Boletin De La Asociacion De Geografos Espanoles, 89*. https://doi.org/10.21138/bage.3020

Lopes, C., Leitão, J., & Rengifo-Gallego, J. (2022). Place-Branded Foods with Responsible and Sustainable Management: A La Carte Serving in Regional Restaurants. *Sustainability, 14*(11). https://doi.org/10.3390/su14116615

Lukoseviciute, G., & Pereira, L. (2021). Tourists' perceptions of beach quality improvement during the off-peak season: A segmentation approach. *Tourism & Management Studies, 17*(2), 17–28. https://doi.org/10.18089/tms.2021.170202

Martinho, V. (2021). Contributions from Literature for Understanding Wine Marketing. *Sustainability, 13*(13). https://doi.org/10.3390/su13137468

Mendes, T., Liberato, P., Liberato, D. F., & Barreira, H. (2021). Food Tourism, an exploratory study in the Portuguese context. *Journal of Tourism & Development*, 33–50. https://doi.org/10.34624/RTD.V1I36.24544

Okumus, B., Koseoglu, M. A., & Ma, F. (2018). Food and gastronomy research in tourism and hospitality: A bibliometric analysis. *International Journal of Hospitality Management, 73*, 64–74. https://doi.org/10.1016/j.ijhm.2018.01.020

Okumus, B. (2021). Food tourism research: A perspective article. *Tourism Review, 76*(1), 38–42. https://doi.org/10.1108/TR-11-2019-0450

Rachão, S., Breda, Z., Fernandes, C., & Joukes, V. (2019). Enogastronomy In Northern Portugal: Destination Cooperation And Regional Identity. *Advances In Hospitality and Tourism Research-Ahtr, 7*(2), 216–237. https://doi.org/10.30519/ahtr.573163

Rachão, S., Breda, Z., Fernandes, C., & Joukes, V. (2021). Food-and-wine experiences towards co-creation in tourism. *Tourism Review, 76*(5), 1050–1066. https://doi.org/10.1108/TR-01-2019-0026

Ramires, A., Brandão, F., & Sousa, A. C. (2018). Motivation-based cluster analysis for international tourists visiting a World Heritage City: The case of Porto, Portugal. *Journal of Destination Marketing and Management, 8*, 49–60.

Real, H., & Graça, P. (2019). Perceptions of use of the concept of Mediterranean diet, potential misuse and prospects to explore. *Revista Espanola de Comunicacion En Salud, 10*(2), 147–159. https://doi.org/10.20318/recs.2019.4824

Romero, I., & Tejada, P. (2020). Tourism intermediaries and innovation in the hotel industry. *Current Issues in Tourism, 23*(5), 641–653. https://doi.org/10.1080/13683500.2019.1572717

Santos, J. A. C., Santos, M. C., Pereira, L. N., Richards, G., & Caiado, L. (2020). Local food and changes in tourist eating habits in a sun-and-sea destination: A segmentation approach. *International Journal of Contemporary Hospitality Management, 32*(11), 3501–3521. https://doi.org/10.1108/IJCHM-04-2020-0302

Serra, M., Antonio, N., Henriques, C., & Afonso, C. (2021). Promoting Sustainability through Regional Food and Wine Pairing. *Sustainability, 13*(24). https://doi.org/10.3390/su132413759

Valls, J., Mota, L., Vieira, S., & Santos, R. (2019). Opportunities for Slow Tourism in Madeira. *Sustainability, 11*(17). https://doi.org/10.3390/su11174534

Vaske, J. J., Kiriakos, R., Cottrell, S. P., & Khuong, M. N. (2009). Importance-performance and segmentation: An application at a biosphere reserve in Vietnam. *Journal of Travel and Tourism Marketing, 26*(1), 30–41.

Vaz, E., De Noronha, T., & Nijkamp, P. (2014). Exploratory landscape metrics for agricultural sustainability. *Agroecology and Sustainable Food Systems, 38*(1), 92–108. https://doi.org/10.1080/21683565.2013.825829

Viegas, C., & Lins, A. (2019). Changing the food for the future: Food and sustainability. *European Journal of Tourism Hospitality and Recreation, 9*(2), 52–57. https://doi.org/10.2478/ejthr-2019-0012

Wade, D. J., & Eagles, P. F. G. (2003). The use of Importance-Performance analysis and market segmentation for tourism management in parks and protected areas: An application to Tanzania's national parks. *Journal of Ecotourism, 2*(3), 196–212.

Yun, D., Hennessey, S.M. & MacDonald, R. (2011). *Understanding culinary tourists: segmentations based on past culinary experiences and attitudes toward food-related behaviour*. International CHRIE Conference Refereed-Track, University of MA, Amherst, pp. 1–13.

Scientific Narratives on Creative Tourism: A Theoretical Framework

Cristina Rodrigues, Dália Liberato, Elsa Esteves, and Pedro Liberato

Abstract In the last decade, the literature on creative tourism has grown tremendously, in effective number of studies and consequently citations, becoming an emerging, relevant, and increasingly complex topic in contemporary tourism literature. This scientific advancement generates a large volume of information, opening space for different perspectives, multiple related issues, but also many consensual ideas. It is therefore essential to synthesize all this knowledge and bring clarity to the state of the art on creative tourism. Following this line of thought, the main objective of this paper is to provide a current scientific framework in this research area, from a conceptual point of view, presenting the most relevant issues, with focus on sustainability. To this end, a large body of scientific literature was analyzed through a combination of two techniques, namely content analysis and systematic literature review. These techniques were applied as a research methodology to identify keyword trends understanding its meaning and analyze the main links between creative tourism and tourism sustainability, respectively. Main results provide a critical analysis of its characteristics as well as the correlations between creative tourism and the sustainable development of tourism activity, highlighting the sustainable matrix of this new model of tourism.

C. Rodrigues
School of Hospitality and Tourism, Polytechnic Institute of Porto, Vila do Conde, Porto, Portugal.
C. Rodrigues, Portugal
e-mail: cir@esht.ipp.pt

D. Liberato (✉) · P. Liberato
School of Hospitality and Tourism, Polytechnic Institute of Porto, Vila do Conde, Portugal, and CiTUR researcher (Centre for Tourism Research, Development, and Innovation), Porto, Portugal
e-mail: dalialib@esht.ipp.pt

P. Liberato
e-mail: pedrolib@esht.ipp.pt

E. Esteves
School of Communication, Administration and Tourism, Polytechnic Institute of Bragança, Portugal, CiTUR researcher (Centre for Tourism Research, Development, and Innovation), Bragança, Portugal
e-mail: elsaesteves@ipb.pt

© The Author(s), under exclusive license to Springer Nature Switzerland AG 2023
V. Katsoni (ed.), *Tourism, Travel, and Hospitality in a Smart and Sustainable World*,
Springer Proceedings in Business and Economics,
https://doi.org/10.1007/978-3-031-26829-8_16

Keywords Creative tourism · Sustainable tourism · Sustainable creative tourism
JEL Z32

1 Introduction

Literature review is an essential feature for advancing scientific knowledge that must be built upon existing studies (Xiao & Watson, 2019). This research method is more relevant than ever, as it allows to keep up with the state of the art, at the forefront of research and evaluate the collective evidence in a particular area (Paul & Criado, 2020; Snyder, 2019). Systematic and narrative are the two main types of review articles and should be combined to provide the best information to advance scientific knowledge (Henry et al., 2018). Particularly in creative tourism, these research methods are beginning to be explored (Baixinho et al., 2020; Carvalho et al., 2019; Galvagno & Giaccone, 2019). In this sense, two techniques were considered: (1) content analysis and (2) systematic literature review.

Some studies addressing creative tourism have already adopted content analysis but for different purposes and other techniques (Cheng & Chen, 2021; Guerreiro et al., 2019; Mohammadi et al., 2019), because content analysis may be as simple or as complex as the research determines it to be (Neuendorf, 2017). Also, there are many different concepts and understandings of content analysis and its various applications in both quantitative and qualitative approaches, but Neuendorf (2017, p. 23) presents a cross-cutting aspect, applied to carry out this paper: "content analysis summarizes rather than reports all details concerning a message set". Only a few studies focused on creative tourism combined content analysis with keyword analysis to identify its main research areas (Galvagno & Giaccone, 2019; Lima Santos et al., 2020). The aim of this research focuses on the collection, review, and analysis of scientific knowledge on creative tourism. Based on the literature that specifically addresses creative tourism, it is hoped to provide a theoretical review and critical analysis of its characteristics as well as identify the correlations between creative tourism and the sustainable development of tourism activity, highlighting the sustainable matrix of this new model of tourism.

The following section discusses the literature review from the selected articles. The Methodology chapter will address each of the techniques, considered at different stages of the research. The two final sections will discuss the results and focus on the main conclusions, emphasizing the critical scientific production on this subject.

2 Literature Review

2.1 *Creative Tourism*

Creative tourism has become popular in the last decade, for that reason its scientific study is relatively recent, and in the last five years, it has grown significantly at different levels. It is a more contemporary form of tourism that promotes endogenous and sustainable development of territories, as well as good relations between residents and tourists. The offer is characterized by small-scale local projects offering varied creative experiences but always characteristic of the destination and usually with a strong focus on cultural heritage and intangible resources. One of the most differentiating aspects is that the experiences are co-created by tourists who can develop their creative potential, knowledge of the destination and of themselves. Creative tourism has been driven in several destinations, at different stages of their life cycle, from the less explored to the more consolidated and in different environments, both urban and rural. It provides authentic tourism experiences and unique learning experiences, diversifies and specializes the destination offer without harming heritage and resources, and brings multiple benefits to the host communities (Baixinho et al., 2021; Dias et al., 2021a, b; Duxbury et al., 2020, 2021; Galvagno & Giaccone, 2019; Marques & Borba, 2017; Richards, 2016, 2019, 2020; Tan et al., 2013, 2014).

2.2 *Experience*

There are several topics that relate the term experience with creative tourism, such as the creative experience, the experience co-creation, the experience economy, the tourist experience, meaningful experiences, authentic experiences, among others. Usually, a mandatory requirement for the construction and enjoyment of a creative experience is the active/immersive participation of the visitor. Despite this association with more engaging and dynamic activities, there are several activities of a more passive character that can be converted or can contribute in some way to develop new creative experiences. This means that the capacity to create creative experiences is very large, and there are many possibilities to boost local attractions, regardless of territorial identity. From another point of view, the creative experience is the combination of personal transformation and development and rich interactions with various elements of the destination. The creative experience provides proximity and interactivity with local actors, namely residents and always reflects the identity of the territory and its community, so each experience is unique and cannot be replicated in other destinations. A creative experience can be related to different heritage and creative resources, for example cooking workshops, artistic expression workshops, and craft activities (Cabeça et al., 2019; Chen & Chou, 2019; Galvagno & Giaccone, 2019; Hung et al., 2016; Pimenta et al., 2021; Remoaldo et al., 2020a, b, c, d; Ross & Saxena, 2019; Tan et al., 2013, 2014; Somnuxpong, 2020).

2.3 Place

Place, as a geographical space (destination), is widely discussed in the literature on creative tourism. It is directly related to this type of tourism, since the identity and characteristics of the place are the main instrument for the creation of creative experiences, defining the type of creative offer provided to the tourist demand. There is a consensual idea in the studies analyzed: This typology has the capacity to develop and boost different geographical, economic, and social contexts. The discussion, among authors, focuses, generally by opposition, the potential of creative tourism in rural and urban places. There is empirical evidence that points to a greater deployment in urban spaces, with more population and tourists, more infrastructure, and a greater diversity of creative resources. On the other hand, the association of creative tourism to enhance the development of more peripheral areas is also perceptible. Portugal has presented, in the last years, some successful cases in low density territories, being the national CREATOUR project frequently mentioned (Bakas & Duxbury, 2018; Dias et al., 2021a, b; Duxbury et al., 2021; Gato et al., 2020; Kostopoulou, 2013; Lavranos et al., 2020; Li & Kovacs, 2022; Nieuwland & Lavanga, 2021; Remoaldo et al., 2020a, b, c, d; Richards, 2019, 2020).

2.4 Involvement

Involvement is a term that cannot be separated from creative tourism. The concept and operationalization of creative tourism require a participative and committed involvement of all stakeholders in the development of creative tourism in destinations. Moreover, the involvement that "creative tourists" assume with experiences, destination, and its resident community is one of the most mentioned aspects among the various studies. Unlike in other types of tourism, where tourists assume a more passive and sometimes disinterested attitude in their tourism consumption, thus, the importance of involvement lies not only in the attitude of tourist demand, which is more committed to sustainability and the positive impacts it generates, but also in the collaborative attitude of operators and "creative entrepreneurs" who build the creative tourism offer (Cabeça et al., 2019; Cruz et al., 2019; Galvagno & Giaccone, 2019; Gato et al., 2020; Marques & Borba, 2017; Marujo et al., 2021; Remoaldo et al., 2020a, b, c, d; Richards, 2016; Tan et al., 2013; Sarantou et al., 2021).

2.5 Motivation

Motivation is a common term in tourism literature, starting with the push and pull factors that lead tourists to choose one destination over another. Particularly in creative tourism, the motivations of tourist demand still have little expression in the literature, but there are clear and relatively consensual points: The motivation of creative tourists is essentially intellectual and relational. Among the different motivations, the following stand out: interacting with cultures different from their own, learning a skill, acquiring knowledge, enjoying unique and authentic experiences, and interacting with local people. These motivations are distinct from other tourism practices, where the tourist mainly seeks to relax or rest in a perspective of escaping from routine. "Creative tourists" are intrinsically creative and sensitive to the creative industries (Chen & Chou, 2019; Choi et al., 2020; Duxbury et al., 2020; Remoaldo et al., 2020a, b, c, d; Tan et al., 2013, 2014).

2.6 *Creativity*

In the literature reviewed, there is a wide range of words linked to creativity (in no specific order): creative destination; creative industries, creative cities; creative development; creative experiences; creative potential and creative economy, among others. Creativity is usually related to "creative experiences" that appeal to tourists' individual creativity or to their self-expression. From another point of view, it is possible to understand that creativity allows developing new tourism products and transforming/innovating existing ones, without the need for major investments, and for that reason creativity can diversify and regenerate the offer of destinations relatively quickly and easily, meeting the interests and expectations of more contemporary tourists and the needs of destinations. Furthermore, several studies link creativity to territorial development and particularly to tourism destination development. All this brings numerous advantages for the tourism industry, mainly in terms of sustainability. (Bastenegar & Hassani, 2018; Cheng & Chen, 2021; Galvagno & Giaccone, 2019; Mareque et al., 2021; Remoaldo et al., 2020a, b, c, d; Richards, 2020; Ross & Saxena, 2019; Tan et al., 2013, 2014; Sarantou et al., 2021).

2.7 *Satisfaction*

Satisfaction, as a pleasant feeling, is mainly understood as a positive consequence or a positive evaluation of tourism consumption, the same happens in creative tourism, for example after the realization of one or more creative experiences in any destination. In other words, satisfaction is linked to the individual perception of the tourist in relation to the destination or to some activity/experience that he/she has experienced. On the other hand, satisfaction can also be linked to the tourists' expectations and intentions considering a particular experience or destination, which means that the tourist can face that experience with a preconceived idea, this can occur for different reasons. Therefore, knowing the motivations of tourism demand is essential to achieve satisfaction. When the tourist gets satisfaction, this leads to positive behavioral intentions, which may translate into a higher frequency of consumption and good feedback, among other aspects that lead to attachment and loyalty to the destination, making it competitive. These indicators translate the results of a positive experience (Chen & Chou, 2019; Cheng & Chen, 2021; Choi et al., 2020; Remoaldo et al., 2020).

2.8 Model

In creative tourism literature, the word "Model" is generally associated with the construction of processes and systems. It can be a way of organizing, developing, implementing, or evaluating something in tourism demand and supply and can be confined to a particular destination, or not, if it is more conceptual. The findings of Tan et al. (2013) suggest that the combination of external interactions and inner reflections builds the model of the creative experience. It was also possible to perceive that the implementation/exploitation of creative tourism in territories must have its own development model and not adapted from another place or another reality. The models of creative tourism, when applied to a particular destination, should reflect its identity and character and should be structured with all actors and stakeholders in that destination, contemplating an appropriate approach to the community, local identity, the business fabric, and resources, considering the capacity and objectives of each destination (Chen & Chou, 2019; Wang et al., 2020; Dias et al., 2021a, b; Duxbury et al., 2021; Henche et al., 2020; Mohammadi et al., 2019; Remoaldo et al., 2020a, b, c, d; Richards, 2019; Tan et al., 2013).

2.9 Co-creation

Co-creation is central to the understanding of creative tourism and is one of the most differentiating elements in relation to other typologies of tourism. Co-creation is mainly related to the sharing of decisions and interactions provided by creative practices that presuppose more meaningful and symbolic tourism experiences. Co-creation is adopted through specific procedures and instruments in creative destinations and is especially desired by creative tourists. As positive impacts or consequences of co-creation, the following aspects stand out: the inclusion and emancipation of the local community (builder and promoter of their territory), the empowerment of visitors to co-create value in their experience (co-creators of the experience), the development of new tourism products, and the regeneration of existing ones, essential for the promotion and competitiveness of destinations (Bertella & Rinaldi, 2021; Dias et al., 2021a, 2021b; Galvagno & Giaccone, 2019; Marques & Borba, 2017; Carvalho et al., 2021; Mohammadi et al., 2019; Richards, 2020; Ross & Saxena, 2019).

2.10 Industries

The word "industries" emerge in many of the studies on creative tourism, mainly in the context of "creative industries", understood as creative classes or clusters that enable the development of creative projects and initiatives, generating innovation, development, and territorial attractiveness. They include artistic and cultural manifestations in the most varied forms. Traditionally, they are found in large cities and metropolitan areas, where there are significant resources and dynamics that help them flourish. However, some studies also show that these creative industries have become more relevant in less globalized places, areas with less social and economic density, but with high creative potential and with technology at their disposal, serving as a strategy to attract investment to those places. In this context, creative tourism takes advantage of all the synergies created by these industries that bring attractiveness and competitiveness to destinations (Bastenegar et al., 2018; Chan et al., 2022; Kostopoulou, 2013; Remoaldo et al., 2020; Richards, 2019, 2020; Chen et al., 2021).

3 Methodology

To expose the methodology followed, Table 1 presents the different stages of the research, the objectives, and the research questions for each of them.

3.1 Content Analysis: Procedures and Results

As the first approach, to provide an overview of the main themes that guide this paper, a content analysis was explicitly applied to analyze keyword trends in the literature on creative tourism. This research technique is one of the most important in the social sciences (Krippendorff, 2018). Keywords can be understood as the

Table 1 Summary of design methodology

Stages of research	Objectives	Questions
1. Content analysis	Identify the keyword trends in the literature and understand its significance in the context of creative tourism	What are the most frequent keywords in the study of creative tourism and their relevance in this field of research?
2. Systematic review	Analyze the main links between creative tourism and tourism sustainability	What observations and findings are evidenced about the sustainable identity of creative tourism? What are the most and least consensual aspects in the literature?

Source Authors

synthesized representation of several ideas from a document (Xiao & Watson, 2019), and this analysis can be instrumental in identifying research themes in a particular area of knowledge (Karimi Takalo et al., 2021; Wang et al., 2021; Zhong et al., 2021). According to Tu et al. (2021), keyword analysis has been used for document categorization, summarization, indexing, and clustering. Many visualization methods are proposed to support this analysis. For McAllister et al. (2021), the VOSviewer maps are best suited for identifying networked keywords. They can help researchers know what to look for in the literature and understand current lines of research and what other lines or topics should be considered in the future. VOSviewer software is also widespread in bibliometric and literature review articles (Cavalcante et al., 2021; McAllister et al., 2021). In this sense, this analysis was conducted through Web of Science (WoS), one of the most reliable databases (Pranckutė, 2021), with the combination of VOSviewer software to create a visual map of the keywords. The following term was identified in WoS—"creative tourism", in line with the methodologies used in similar research that define the best set of terms/keywords to cover a large portion of publications regarding the research topic (Habibi et al., 2021; Rama et al., 2018; Yu et al., 2020). The data collection process took place on June 25, 2022, and allowed the identification of sixty-eight publications (Table 2).

As the objective of this bibliographical research was to collect a comprehensive and representative sample of studies on creative tourism, the filtering process in WoS was limited to a more generic search term "creative tourism", but exactly the focus of the present paper. The title was identified because it presents the whole idea/problem of each research, it is the most representative sentence of its problematic (Pereira et al., 2011; Bahadoran et al., 2019) more than another field. In addition, two filters were applied: theoretical and empirical studies published in articles and only articles written in English, which resulted in a collection of sixty-eight (68) publications. At this point, due to the satisfactory number of publications to achieve this first stage of the research, no additional filtering field was used. Just from the analysis of this bibliographical collection, it is possible to draw some interesting conclusions that highlight the recent interest in this topic and most of the studies that cover the

Table 2 Literature selection and filtering procedures in WoS

Search criteria literature onCreative Tourism	Selection of the (international) database Web of Science
Topic "creative tourism"	(without any refine/filter of results) = = 238 publications
↓Filter Selection ↓	
Title "creative tourism" = 129 publications	Title refers to the title of a journal article, proceedings paper, book, or book chapter. The use of inverted commas allows the search for the union of two words
Document type: only Articles = 76 Languages: only English = 68	Result = 68 articles Export RIS (reference software)

Source Authors

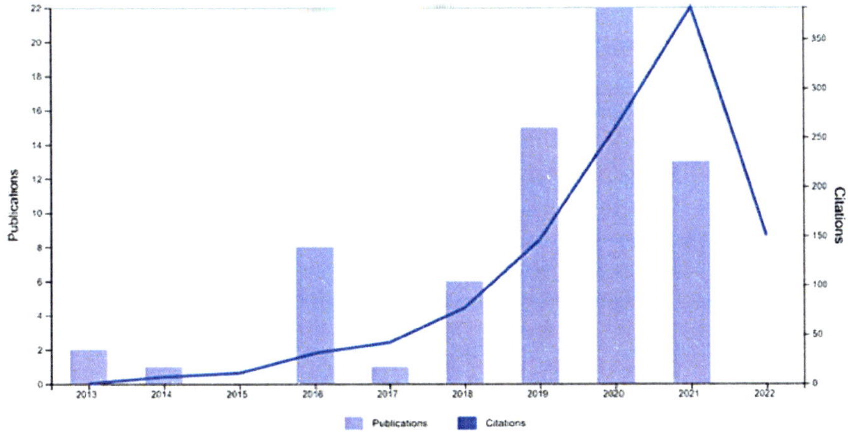

Fig. 1 Times cited and publications over time. *Source* Authors, WoS report

literature on creative tourism highlight this aspect of a certain novelty. International scientific literature about creative tourism still has little expression in quantitative terms, and this is visible when compared with the volume of literature from other types of tourism, such as cultural tourism, which, according to several authors, has driven this new form of tourism. Figure 1 illustrates this growth.

Figure 1 shows the number of articles present in WoS in the time period 2013 to 2021 and the number of citations in each year. The bars identify the number of articles and the line the citations. This corresponds to (2013–2); (2014–1); (2016–8); (2017–1); (2018–6); (2019–15); (2020–22); (2021–13) = N 68. The year with the highest number of citations was 2021, being clear that the number of citations is much higher than the number of articles, with thirteen (13) articles and three hundred and eighty-two (382) citations. Another analysis that WoS allows is the identification of journals, referred to as "publication titles". Figure 2 identifies the top ten of these journals considering the number of articles in the collection, in a logic of the journal relevance/importance. In the total of thirty-eight (38) scientific journals identified in this WoS collection, the ten that stand out for having the highest number of publications are Sustainability; Current Issues in Tourism; Journal of Hospitality Tourism Research; Tourism Management Studies; Annals of Tourism Research; Asia Pacific Journal of Tourism Research; Tourism Geographies; Journal of Travel Tourism Marketing; Journal of Sustainable Tourism; and Journal of Tourism and Cultural Change. Sustainability journal is ranked first, with seven articles. This evidence suggests that many of the studies have a strong link to sustainability—a subject that will be discussed further, given their important connections with creative tourism. After these brief reviews of the selected literature, the analysis of its keywords follows, to specifically answer the first research question: *What are the most frequent keywords in the study of creative tourism and what is their relevance in this field of research?*

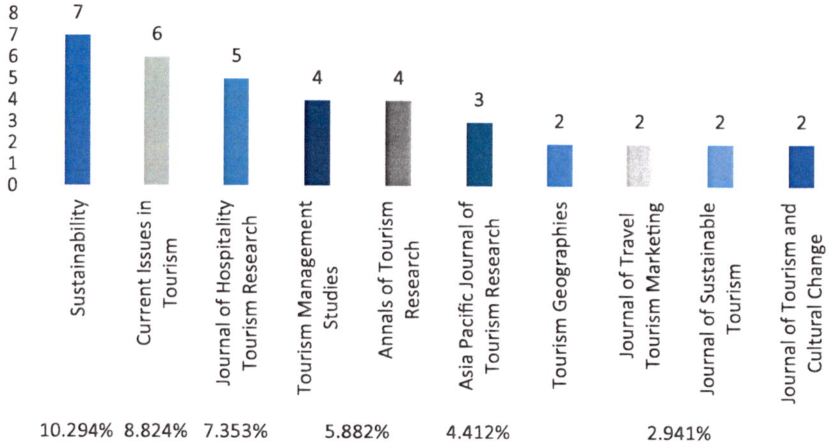

Fig. 2 Ten most relevant journals where the 68 articles were published. *Source* Authros, based on WoS report

For this purpose, the set of 68 publications was loaded into a bibliographic references software (specifically ZOTERO) in order to build a database that could be associated with the VOSviewer software. Figure 3 summarizes the use of WoS and VOSviewer as a combination for content analysis.

Before analyzing the keyword co-occurrence, as described in Figure, a filtering procedure was used in order to extract the most relevant keywords (Faust, 2018; Feng & Law, 2021). This procedure should include (1) converting plurals into a singular form; (2) removing hyphens; (3) converting to lowercase; (4) avoidance of abbreviations; (5) removing redundant words (Feng & Law, 2021). The visual

Fig. 3 Combination of WoS and VOSviewer for keyword analysis. *Source* Authors

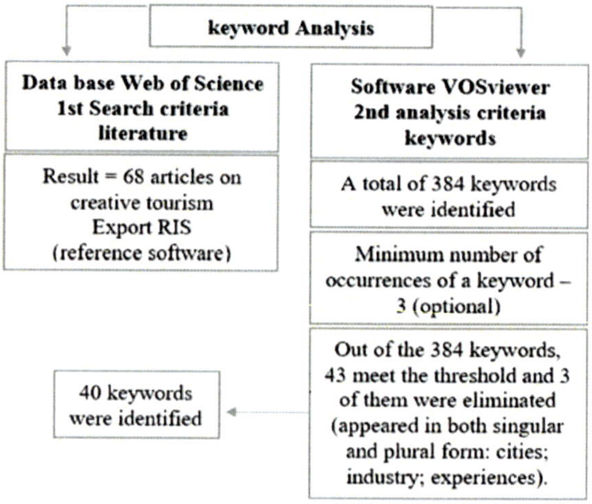

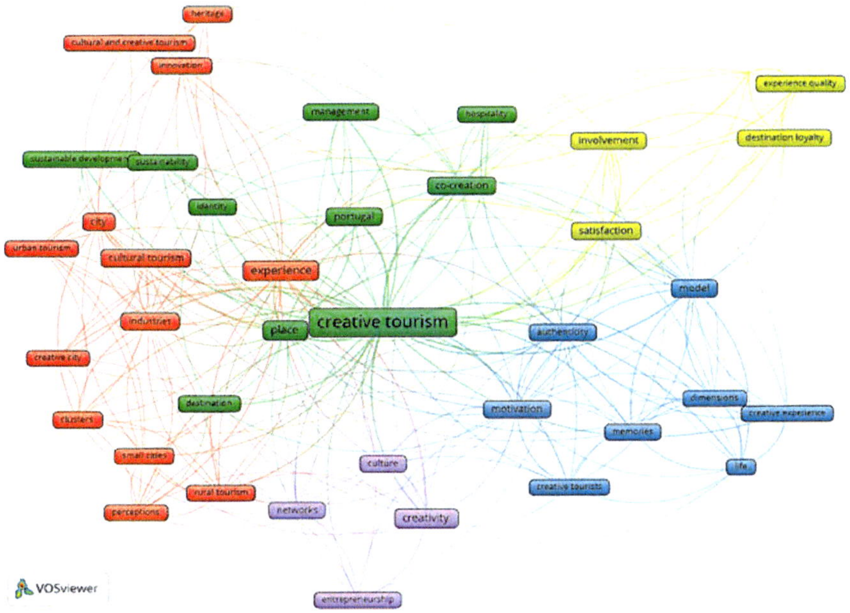

Fig. 4 Visual map concerning the keywords obtained. *Source* Authors, from VOSviewer

map of the keywords can be seen in Fig. 4. The size of the label and an item's box are determined by their weight, which indicates the frequency of keywords' occurrence. The lines between items represent links between keywords; they represent a connection/relationship between them. In this case, the lines represent co-occurrence links, connecting keywords that appear together. Therefore, in general, the closer the keywords are located, the stronger their relatedness is. The color of an item determines the cluster it belongs to. A cluster is a set of closely related nodes. Each node in the network is assigned to exactly one cluster.

According to the software analysis, each keyword appearing on this map has been identified at least three times in the bibliographic collection. It is clearly observed that the label that stands out the most is creative tourism, in the center of the network. In this case, VOSviewer identified five clusters, as identified in Table 3, which simultaneously presents each of the forty keywords. This network of keywords shows that different themes and particular issues have been considered over the last few years regarding this research area.

These ten representative dimensions of the literature on creative tourism, namely experience; place; involvement; motivation; creativity; satisfaction; model; co-creation; and industries, are briefly discussed below, due to their apparent relevance in this field of research. This examination will make it possible to answer the question: *What is the relevance of these terms in this field of research?* To carry out this examination, all Titles, Abstracts, and Keywords of the documents were carefully

Table 3 Keywords identification in each cluster

Clusters	Items	Keywords
1 (red)	13	City; Clusters; Creative city; Cultural and creative tourism; Cultural tourism; Experience; Heritage; Industries; Innovation; Perceptions; Rural tourism; Small cities; Urban tourism
2 (green)	10	Co-creation; Creative tourism; Destination; Hospitality; Identity; Management; Place; Portugal; Sustainability; Sustainable development
3 (blue)	8	Authenticity; Creative experience; Creative tourists; Dimensions; Life; Memories; Model; Motivation
4 (yellow)	5	Destination loyalty; Experience quality; Involvement, Perceived value*; Satisfaction
5 (purple)	4	Creativity; Culture; Entrepreneurship; Networks
N = 40 keywords		

Source Authors, report from VOSviewer
*Perceived value is the only word not identified in the illustration because VOSviewer selects the terms that are considered most relevant on the network, but it is listed among the forty keywords identified, belonging to cluster 4 (yellow). The words sustainability and sustainable development are in bold because they are directly related to the next stage of the research, which aims to analyze the main links between creative tourism and tourism sustainability. Although none of them belong to the top 10, just the fact that they appear in this network of keywords is more than relevant for the purpose of this research

read. Simultaneously, each word was searched in the 68 documents, which proved to be a very useful mechanism to determine the relevant presence or not. Based on this, the documents were selected for full reading in each of the dimensions.

The exploration of these ten terms, the understanding of their meaning and their context within creative tourism, made it possible to understand, in summary form, several important and widely considered aspects among the different authors of the WoS collection (identified in the present paper). An interesting observation in light of these more general reflections is the notion of sustainability among them, which is sometimes explicit. Therefore, to better understand the main links between creative tourism and tourism sustainability, the second stage of the research focuses on the sustainable narratives of creative tourism.

4 Systematic Review: Procedures and Results

Based on these considerations related to the sustainable identity of creative tourism, this second stage of the research aims to answer the question: *What observations and findings are evidenced about the sustainable identity of creative tourism? What are the most and least consensual aspects among the literature?* In order to answer these questions, a systematic review was conducted to identify and understand the literature evidence on links between creative tourism and tourism sustainability. The main feature of this research technique is that it provides a clear explanation of the data collection process, making the research transparent (Henry et al., 2018; Xiao & Watson, 2019). This is achieved through specific mechanisms, such as a clear explanation of inclusion and exclusion criteria and keywords used for searches (Pahlevan-Sharif et al., 2019), as identified for the present analysis, summarizing the best evidence on a specific question (Henry et al., 2018). There are many studies in tourism that, through systematic review, analyze the most relevant fields and aspects of already existing studies and their evidence concerning their research topic (Lima Santos et al., 2020; Pessot et al., 2021; Tölkes, 2018). The present analysis focusing on the sustainable aspects of creative tourism was supported by the visual map retrieved from VOSviewer, presented before (see Fig. 5) which clearly shows the word "sustainable development" and sustainability, grouped in the same cluster. Figure 5 emphasizes more clearly the relationships established from these two words, with others on the map.

Through Fig. 5, it is possible to identify the links that the map established from the keyword "sustainable development", namely creative tourism; city; urban tourism; cultural tourism; sustainability; and identity. Also, it is possible to identify the links

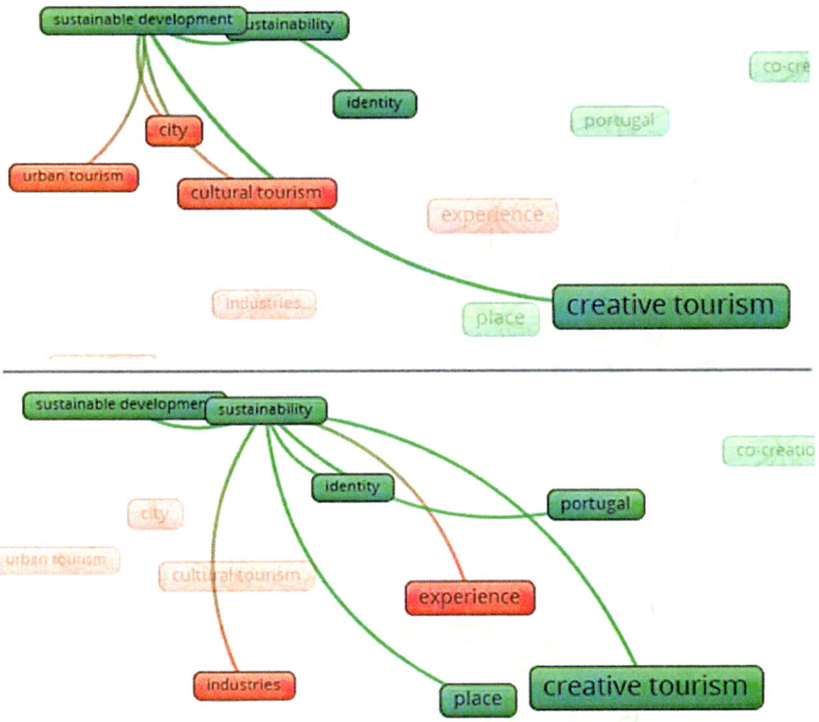

Fig. 5. Connections between sustainable development, sustainability, and creative tourism. *Source* Authors' combination, from VOSviewer

Table 4 Ten most significant keywords of the collection

Keywords (Top Ten)	Ocurrences	Links	Total link strenght
Creative tourism	50	38	140
Experience	15	25	55
Place	9	23	38
Involvement	7	20	35
Motivation	7	21	34
Creativity	10	18	33
Satisfaction	6	20	33
Model	6	17	29
Co-creation	7	20	28
Industries	5	18	26

Source Authors, from VOSviewer

that the map established from the keyword "sustainability", namely creative tourism; experience; place; industries; Portugal; identity; and sustainable development. There are substantial differences, which are expected to be further clarified with the systematic review of some important articles that address several reflections regarding sustainability in creative tourism.

This section was designed to synthesize the available scientific evidence between creative tourism and sustainability, from the same sixty-eight studies previously identified on the Web of Science (first stage of the research). Although all these publications were found with the term "creative tourism" which is the main theme of the present paper, not all of them associate with the same depth the issue of sustainability, focusing on other aspects, also relevant. This is also clear in the identification of the top ten keywords, which do not include sustainability (see Table 4). For this reason, it was necessary to develop a strategy to identify which publications would best answer the second research question, that is, to identify the articles that gathered the most and best information on sustainability in the context of creative tourism. To this end, an inclusion and exclusion criterion was applied to the collection of the sixty-eight WoS publications, as recommended for all systematic reviews (Table 5).

Table 5 Inclusion and exclusion criteria for systematic review

Inclusion criteria	Exclusion criteria
Abstracts and keywords available; Include (somehow) the term sustainable in the keywords; Document with total open access; All kinds of methodology, theoretical, and empirical	Abstracts and keywords unavailable; Do not include (in any way) the term sustainable in the keywords; Document with total close access
Eight (8) publications	Sixty (60) publications

Source Authors

In this follow-up, the term sustainability (in different combinations) was checked in the keywords of the sixty-eight publications, and, in addition, all titles and abstracts were read to confirm the actual relation of the studies to sustainability. After this filtering, eight articles were identified for this review, while the majority, sixty articles, were disregarded for not presenting the defined requirements. As a result of this procedure, one article was identified with the keyword "sustainable governance"; another with "sustainable tourism"; and a third with "sustainable creative tourism". Two articles were also identified with the keyword "sustainability" and three with "sustainable development", resulting in a total of five keywords linking creative tourism and sustainability (Table 6). Once the articles were identified, each was analyzed to understand the focus of their research and find important links between creative tourism and tourism sustainability. To this end, the content of these articles was carefully analyzed, with particular focus on their abstract, results, and conclusions, summarizing each of the eight articles.

All these articles, at different moments of their research, make several considerations about sustainability. These considerations are based on the literature review of each study, but also on the findings and conclusions of their own research. Therefore, different dimensions of sustainability are highlighted, namely social, economic, cultural, and, although less prominent, environmental. Their reading and analysis of these eight articles allowed the identification of some authors' considerations which contribute to the general and specific understanding of mentioned sustainability perspectives.

5 Discussion and Conclusions

Creative tourism is the result of the combination of several aspects: cultural tourism practices, creative potential of places and people, active participation, and economic experience. In general, experience is understood as the tourism product/service of creative tourism, i.e., the activity offered to the visitor during their trip or stay in a particular destination. Furthermore, it is interesting to note that creative tourism is often mentioned as a new (sustainable) model of tourism activity, which involves different dimensions and variables from other tourism typologies. Each place should develop, from its resources and potentialities, its creative offer, which will be exclusive, therefore unique, and competitive against other destinations. As a typology based on co-creation, sharing, and network building, tourists' involvement gains significance for the success of creative destinations. Co-creation can go beyond the relationship between visitors and residents, involving all actors of a visit in a network perspective for more sustainable practices. The values, interests, expectations, and self-knowledge of each individual interact with their motivations, producing different types of experiences. As creative tourism does not necessarily rely on material heritage and other tangible resources (with known limitations), its approaches to local culture and resources bring numerous possibilities for the regeneration of heritage and simultaneously its enhancement, exploitation, and protection (Cabeça et al.,

Table 6 Articles analyzed for systematic literature review

References and journal	Article keywords	Research description and their considerations regarding sustainability
Henche et al., (2020) *Sustainability*	Historic city centers; cultural and creative tourism; urban tourism; **sustainable governance**; social networks; Barrio de las Letras	*"A Sustainable Management Model for Cultural Creative Tourism Ecosystems"* Presents a sustainable management model for urban centers, specifically a historic center of Madrid (Spain), which aims to assist the creation of a competitive cultural/creative/historic tourism ecosystem, able to preserve the sustainability of the local social and commercial fabric; highlights the importance of collaborative networks between the different stakeholders, public and private sector, associations of small businesses, cultural organizations, hotel industry, and commerce. This model comprises a sustainable governance that integrates the inhabitants of these spaces, attending to their permanence and well-being
Gato et al., (2020) *Journal of Hospitality & Tourism Research*	Creative tourism; placemaking strategies; peripheral areas; Portugal; **sustainability**; case study	*"Creative Tourism as Boosting Tool for Placemaking Strategies in Peripheral Areas: Insights from Portugal"* Emphasizes the potential of creative tourism to foster peripheral areas, specifically in Portugal, and discusses how creative tourism experiences can be mobilized by placemaking strategies in these areas to achieve sustainable development, enhancing community resilience, driven by pressing needs for economic development, livelihood diversification, and control over future outcomes
Duxbury et al., (2021) *Sustainability*	Creative tourism; local/regional development; Portugal; **sustainable tourism**; cultural heritage; regenerative tourism	*"Creative Tourism Development Models towards Sustainable and Regenerative Tourism"* Based on the CREATOUR project in Portugal, it understands creative tourism as regenerative tourism that focuses on microscale regeneration initiatives, on "giving back" and contributing to the proactive regeneration of communities, cultures, heritage, and places. Among the five creative tourism models identified, their influences on community dynamics and development can be summarized in three aspects: (1) reinforcing distinctive elements of local identity, (2) instigating flows and connections between the locale and the external, and (3) serving as platforms for local collaboration, exchange, and development

(continued)

Table 6 (continued)

References and journal	Article keywords	Research description and their considerations regarding sustainability
Guerreiro et al., (2019) *Journal of Spatial and Organizational Dynamics*	Cultural tourism; creative tourism; **sustainable development;** celebrations	*"Cultural and creative tourism: the case of 'Celebrations' in the Algarve Region"* Relates culture, creativity, tourism, and sustainable development in a region/destination traditionally linked with "sun and sea" tourism, the Algarve (Portugal). Cultural and creative tourism are identified as having great potential to diversify and add value to its existing tourism products in the region. Three types of local celebrations (festivals; fairs, and events) are identified as interesting for creating cultural products and creative experiences in different municipalities. These celebrations have a high focus on local intangible cultural heritage, playing an important role in diversifying tourism products and enhancing the region's sustainable development
Baixinho et al., (2021) *Land*	**Sustainable creative tourism**; conceptual framework; islands; cultural landscapes; guidelines; policy recommendations	*"Islandscapes and Sustainable Creative Tourism: A Conceptual Framework and Guidelines for Best Practices"* Highlights the importance of creative tourism (sustainable creative tourism—suggested concept) following a set of international principles and policy recommendations (e.g., by the OECD, UNESCO, UNWTO, EU) aimed at promoting sustainable tourism development strategies with more holistic and integrated approaches that consider the economic, social, cultural, and environmental dimensions. Minimizing possible negative impacts that creative tourism can also originate in different destinations, being necessary to apply appropriate indicators and tools for monitoring tourism sustainability in territories with this type of offer

(continued)

Table 6 (continued)

References and journal	Article keywords	Research description and their considerations regarding sustainability
Bertella and Rinaldi (2021) *Journal of Sustainable Tourism*	Collaboration; **sustainable development**; community of practice (CoP); non-governmental development organization (NGDO)	*"Learning communities and co-creative tourism practices in NGDO projects"* This study considers collaboration as a co-creative practice for learning about tourism and sustainable development. Collaborative practices are driven by projects run by non-governmental development organizations (NGDOs), which facilitate the emergence and/or strengthening of local learning communities. The results suggested that NGDO projects can contribute to creating the conditions in which learning communities can emerge and grow as "chains" that evolve over time, promoting constant local innovation. Some factors emerged as particularly important for projects that aim to foster collaborative relationships that extend beyond the limited time horizon of the projects: the identification and involvement of partners/stakeholders that share the same vision; synergies with other projects and sectors; the use of participatory approaches, work methods, and artifacts that promote a sense of empowerment and enhance continuity in the efforts of the local people toward sustainability
Nieuwland and Lavanga (2021) *Journal of Sustainable Tourism*	Creative entrepreneurship; **sustainable development**; creative placemaking; creative tourism; urban tourism	*"The consequences of being 'the Capital of Cool'. Creative entrepreneurs and the sustainable development of creative tourism in the urban context of Rotterdam"* Analyzes creative insertion in tourism and reflects on the role of creative entrepreneurs in sustainable tourism development, as agents of insertion in sustainable urban tourism development in Rotterdam, the Netherlands. The concept of creative placemaking can offer an alternative perspective to help mitigate gentrification processes that displaces local residents and businesses. The study found that creative tourism in an urban context, without paying attention to inclusivity, equality, and gentrification effects, does not lead to more sustainable development of urban tourism

(continued)

Table 6 (continued)

References and journal	Article keywords	Research description and their considerations regarding sustainability
Choi et al. (2020) *Sustainability*	Museums; special events; late opening; visitors' motivations; **sustainability**; creative tourism attractions	"*The Urban Museum as a Creative Tourism Attraction: London Museum Lates Visitor Motivation*" Applied in Museum Lates in London and aims to discover and analyze its visitors' motivations to participate in evening events focused on sustainable behaviors. Contributing to a greater understanding of the relationships between visitor groups and museums' cultural and creative tourism development. Museums as institutions of increasing cultural and creative tourism attractions, which have the capacity to contribute to different dimensions of sustainability, particularly at a cultural level. Museums are an essential attraction for visitors and play a critical role in cultural and creative tourism, providing educational programs and exhibitions highlighting the human–nature relationship and shaping people's values, attitudes, and behaviors, transcending functional displays of artifacts

Source Authors' elaboration

2019; Cruz et al., 2019; Galvagno & Giaccone, 2019; Gato et al., 2020; Marques & Borba, 2017; Marujo et al., 2021; Remoaldo et al., 2020; Richards, 2016, 2019, 2020; Rodrigues et al., 2021, 2022; Tan et al., 2013; Sarantou et al., 2021).

A total of twelve citations, deemed relevant, are presented and grouped into the three most prominent dimensions of sustainability among these eight studies (social, economic, and cultural).

5.1 Sustainable Social Dimension of Creative Tourism

"Supports the coexistence of the private and public sector and sustainable governance models that integrate the inhabitants of city centers with the economic activity generated by urban tourism... the collaboration of cultural organizations, hospitality industry and retail can promote cultural, creative, and sustainable management model of historic urban center" (Henche et al., 2020, p. 1).

"Planned tourism placemaking often occurs as part of a deliberate approach to sustainable development and community resilience, driven by pressing needs for economic development, livelihood diversification, and control over future outcomes." (Gato et al., 2020, p. 14).

"The importance of the community being included and invested in its ongoing success, and the commitment of decision-makers and local agents in the development of a strategy that explicitly considers and prioritizes the holistic sustainability and well-being of the local community" (Duxbury et al., 2021, p. 15).

"Issues such as gentrification and inclusiveness may be addressed early on, to make sure the city is developed for a diverse group of people and creativity is mobilized in service of social justice" (Nieuwland & Lavanga, 2021, p. 939).

5.2 Sustainable Economic Dimension of Creative Tourism

"We talk mainly about economic sustainability for the promoters and chain of partners involved in the experiences, since the other dimensions of sustainability are intrinsic to the concept of creative tourism" (Gato et al., 2020, p. 13).

"The departure point was to present creative tourism experiences as an opportunity to enable actual territorially embedded strategies in such peripheral areas, in order to strengthen sustainable development processes" (Gato et al., 2020, p. 14).

"The pillar supporting this type of tourism is sustainable development, which is understood as transferring human and financial resources from the tourism market to local communities " (Bertella & Rinaldi, 2021, p. 645).

"We use a broad definition of the sustainable development of tourism in which we focus on the community aspect of it. This means we explore the socio-economic and socio-cultural sustainability of the destination, investigating issues such as social

and economic equality, community wellbeing and issues related to overtourism" (Nieuwland & Lavanga, 2021, p. 929).

5.3 Sustainable Cultural Dimension of Creative Tourism

"Creative tourism promises a framework for developing innovative approaches for a sustainable cultural tourism" (Duxbury et al., 2021, p. 1).

"Regenerative tourism goes beyond sustainable tourism to focus on giving back and contributing to the proactive regeneration of communities, cultures, heritage, places, landscapes, and so forth" (Duxbury et al., 2021, p. 3).

"Develop a sustainable creative tourism sector to boost tourism in small cities and rural areas as well as to contribute meaningfully to local cultural vibrancy" (Duxbury et al., 2021, p. 6).

"Distinctive elements and cultural specificities of their locales were the key components from which to build a creative and sustainable tourism for wider community benefit that would safeguard local culture and heritage" (Duxbury et al., 2021, p.15).

The success of creative tourism is largely influenced by its sustainable matrix in the tourism sector.

Acknowledgements The authors acknowledge the financial support of CiTUR, R&D unit funded by the FCT—Portuguese Foundation for the Development of Science and Technology, Ministry of Science, Technology and Higher Education, under the scope of the project UID/BP/04470/2020.

References

Bahadoran, Z., Mirmiran, P., Kashfi, K., & Ghasemi, A. (2019). The principles of biomedical scientific writing. *International Journal of Endocrinology and Metabolism, 17*(4), Article 4. https://doi.org/10.5812/ijem.98326

Baixinho, A., Santos, C., Couto, G., Albergaria, I., Silva, L., Medeiros, P., & Simas, R. (2021). Islandscapes and sustainable creative tourism: A conceptual framework and guidelines for best practices. *Land, 10*(12). https://doi.org/10.3390/land10121302

Baixinho, A., Santos, C., Couto, G., de Albergaria, I., da Silva, L., Medeiros, P., & Simas, R. (2020). Creative tourism on islands: A review of the literature. *Sustainability, 12*(24). https://doi.org/10.3390/su122410313

Bakas, F., & Duxbury, N. (2018). Development of rural areas and small cities through creative tourism: The creatour project. *Anais Brasileiros De Estudos Turisticos-Abet, 8*(3), 74–84.

Bastenegar, M., & Hassani, A. (2018). Tolerance; a spiritual strategy or a strategic spirituality in development of creative tourism. *Bagh-E Nazar, 15*(60), 41–54. https://doi.org/10.22034/bagh.2018.62764

Bastenegar, M., Hassani, A., & Bafruei, M. (2018). Thematic Analysis of creative tourism: Conceptual model design. *Amazonia Investiga, 7*(17), 541–554.

Bertella, G., & Rinaldi, M. (2021). Learning communities and co-creative tourism practices in NGDO projects. *Journal of Sustainable Tourism, 29*(4), 639–657. https://doi.org/10.1080/09669582.2020.1821697

Cabeça, S., Goncalves, A., Marques, J., & Tavares, M. (2019). Mapping intangibilities in creative tourism territories through tangible objects: A methodological approach for developing creative tourism offers. *Tourism & Management Studies, 15*, 42–49. https://doi.org/10.18089/tms.2019.15SI05

Carvalho, M., Kastenholz, E., & Carneiro, M. (2021). Co-creative tourism experiences—A conceptual framework and its application to food & wine tourism. *Tourism Recreation Research*. https://doi.org/10.1080/02508281.2021.1948719

Carvalho, R., da Costa, C., & Ferreira, A. (2019). Review of the theoretical underpinnings in the creative tourism research field. *Tourism & Management Studies, 15*, 11–22. https://doi.org/10.18089/tms.2019.15SI02

Cavalcante, W. Q. D. F., Coelho, A., &. (2021). Sustainability and tourism marketing: A bibliometric analysis of publications between 1997 and 2020 using VOSviewer software. *Sustainability, 13*(9), 4987. https://doi.org/10.3390/su13094987

Chan, C., Chang, T., & Liu, Y. (2022). Investigating creative experiences and environmental perception of creative tourism: The case of PMQ (Police Married Quarters) in Hong Kong. *Journal of China Tourism Research, 18*(2), 223–244. https://doi.org/10.1080/19388160.2020.1812459

Chen, C.-F., & Chou, S.-H. (2019). Antecedents and consequences of perceived coolness for generation Y in the context of creative tourism—a case study of the pier 2 art center in Taiwan. *Tourism Management, 72*, 121–129. https://doi.org/10.1016/j.tourman.2018.11.016

Chen, Z., Chen, X., & Mak, B. (2021). The hybrid discourse on creative tourism: Illuminating the value creation process. *International Journal of Culture Tourism and Hospitality Research, 15*(4), 547–564. https://doi.org/10.1108/IJCTHR-07-2020-0138

Cheng, T., & Chen, M. (2021). Creative atmosphere in creative tourism destinations: conceptualizing and scale development. *Journal of Hospitality & Tourism Research*. https://doi.org/10.1177/10963480211012459

Choi, A., Berridge, G., & Kim, C. (2020). The Urban Museum as a creative tourism attraction: London museum lates visitor motivation. *Sustainability, 12*(22). https://doi.org/10.3390/su12229382

Cruz, A., Perestrelo, M., Gato, M., & Costa, P. (2019). Connecting the dots: Contributions to the creative tourism dynamics in Portugal. *Tourism & Management Studies, 15*, 23–32. https://doi.org/10.18089/tms.2019.15SI03

Dias, A., González-Rodriguez, M., & Patuleia, M. (2021). Creative tourism destination competitiveness: An integrative model and agenda for future research. *Creative Industries Journal*. https://doi.org/10.1080/17510694.2021.1980672

Dias, A., González-Rodriguez, M., & Patuleia, M. (2021). Developing poor communities through creative tourism. *Journal of Tourism and Cultural Change, 19*(4), 509–529. https://doi.org/10.1080/14766825.2020.1775623

Duxbury, N., Bakas, F., & de Carvalho, C. (2020). Why is research-practice collaboration so challenging to achieve? A creative tourism experiment. *Tourism Geographies, 23*(1–2), 318–343. https://doi.org/10.1080/14616688.2019.1630670

Duxbury, N., Bakas, F., de Castro, T., & Silva, S. (2021). Creative tourism development models towards sustainable and regenerative tourism. *Sustainability, 13*(1). https://doi.org/10.3390/su13010002

Faust, O. (2018). Documenting and predicting topic changes in computers in biology and medicine: A bibliometric keyword analysis from 1990 to 2017. *Informatics in Medicine Unlocked, 11*, 15–27. https://doi.org/10.1016/j.imu.2018.03.002

Feng, S., & Law, N. (2021). Mapping artificial intelligence in education research: A network-based keyword analysis. *International Journal of Artificial Intelligence in Education, 31*(2), 277–303. https://doi.org/10.1007/s40593-021-00244-4

Galvagno, M., & Giaccone, S. (2019). Mapping creative tourism research: Reviewing the field and outlining future directions. *Journal of Hospitality & Tourism Research, 43*(8), 1256–1280. https://doi.org/10.1177/1096348019862030

Gato, M., Costa, P., Cruz, A., & Perestrelo, M. (sem data). Creative tourism as boosting tool for placemaking strategies in peripheral areas: Insights from Portugal. *Journal of Hospitality & Tourism Research*. https://doi.org/10.1177/1096348020934045

Gato, M., Costa, P., Cruz, A., & Perestrelo, M. (2020). Creative tourism as boosting tool for place-making strategies in peripheral areas: Insights from Portugal. *Journal of Hospitality & Tourism Research*. https://doi.org/10.1177/1096348020934045

Guerreiro, M., Henriques, C., & Mendes, J. (2019). Cultural and creative tourism: The case of 'Celebrations' in the Algarve Region. *Journal of Spatial and Organizational Dynamics, 7*(4), 320–338.

Habibi, A., Mousavi, M., Jamali, S. M., & Ale Ebrahim, N. (2021). A bibliometric study of medical tourism. *Anatolia*, 1–11. https://doi.org/10.1080/13032917.2021.1954042

Henche, B., Salvaj, E., & Cuesta-Valino, P. (2020). A sustainable management model for cultural creative tourism ecosystems. *Sustainability, 12*(22). https://doi.org/10.3390/su12229554

Henry, B. M., Skinningsrud, B., Vikse, J., Pękala, P. A., Walocha, J. A., Loukas, M., Tubbs, R. S., & Tomaszewski, K. A. (2018). Systematic reviews versus narrative reviews in clinical anatomy: Methodological approaches in the era of evidence-based anatomy. *Clinical Anatomy, 31*(3), 364–367. https://doi.org/10.1002/ca.23042

Hung, W.-L., Lee, Y.-J., & Huang, P.-H. (2016). Creative experiences, memorability and revisit intention in creative tourism. *Current Issues in Tourism, 19*(8), 763–770. https://doi.org/10.1080/13683500.2013.877422

Karimi Takalo, S., SayyadiTooranloo, H., & Shahabaldini Parizi. (2021). Green innovation: A systematic literature review. *Journal of Cleaner Production, 279*, 122474. https://doi.org/10.1016/j.jclepro.2020.122474

Kostopoulou, S. (2013). On the revitalized waterfront: Creative milieu for creative tourism. *Sustainability, 5*(11), 4578–4593. https://doi.org/10.3390/su5114578

Krippendorff, K. (2018). *Content analysis: An introduction to its methodology*. SAGE Publications.

Lavranos, C., Vouvaki-Manousaki, C., & Kostagiolas, P. (2020). Developing creative «spaces» in libraries for creative tourism. *Qualitative & Quantitative Methods in Libraries, 9*(1), 1–7.

Li, P., & Kovacs, J. (2022). Creative tourism and creative spaces in China. *Leisure Studies, 41*(2), 180–197. https://doi.org/10.1080/02614367.2021.1948596

Lima Santos, L., Cardoso, L., Araújo, N., & Brea, J. A. (2020). Sustainability perceptions in tourism and hospitality: A mixed-method bibliometric approach. *Sustainability, 12*, 8852. https://doi.org/10.3390/su12218852

Mareque, M., Creo, E., & Alvarez-Diaz, M. (2021). Exploring creative tourism based on the cultural and creative cities (C3) index and using bootstrap confidence intervals. *Sustainability, 13*(9). https://doi.org/10.3390/su13095145

Marques, L., & Borba, C. (2017). Co-creating the city: Digital technology and creative tourism. *Tourism Management Perspectives, 24*, 86–93. https://doi.org/10.1016/j.tmp.2017.07.007

Marujo, N., Borges, M., Serra, J., & Coelho, R. (2021). Strategies for creative tourism activities in pandemic contexts: The case of the «Saidas de Mestre» Project. *Sustainability, 13*(19). https://doi.org/10.3390/su131910654

McAllister, J. T., Lennertz, L., & Atencio Mojica, Z. (2021). Mapping a discipline: A guide to using vosviewer for bibliometric and visual analysis. *Science & Technology Libraries*, 1–30. https://doi.org/10.1080/0194262X.2021.1991547

Mohammadi, A., Moharrer, M., & Babakhanifard, M. (2019). The business model and balanced scorecard in creative tourism: The ultimate strategy boosters. *Current Issues in Tourism, 22*(17), 2157–2182. https://doi.org/10.1080/13683500.2018.1437714

Neuendorf, K. A. (2017). *The content analysis guidebook*. SAGE.

Nieuwland, S., & Lavanga, M. (2021). The consequences of being «the Capital of Cool»: Creative entrepreneurs and the sustainable development of creative tourism in the urban context of

Rotterdam. *Journal of Sustainable Tourism, 29*(6), 926–943. https://doi.org/10.1080/09669582.2020.1841780

Pahlevan-Sharif, S., Mura, P., & Wijesinghe, S. N. R. (2019). A systematic review of systematic reviews in tourism. *Journal of Hospitality and Tourism Management*, 39, 158–165. https://doi.org/10.1016/j.jhtm.2019.04.001

Paul, J., & Criado, A. R. (2020). The art of writing literature review: What do we know and what do we need to know? *International Business Review, 29*(4), 101717. https://doi.org/10.1016/j.ibusrev.2020.101717

Pereira, H. B. B., Fadigas, I. S., Senna, V., & Moret, M. A. (2011). Semantic networks based on titles of scientific papers. *Physica a: Statistical Mechanics and Its Applications, 390*(6), 1192–1197. https://doi.org/10.1016/j.physa.2010.12.001

Pessot, E., Spoladore, D., Zangiacomi, A., & Sacco, M. (2021). Natural resources in health tourism: A systematic literature review. *Sustainability, 13*(5), 2661. https://doi.org/10.3390/su13052661

Pimenta, C., Ribeiro, J., & Remoaldo, P. (2021). The relationship between creative tourism and local development: A bibliometric approach for the period 2009–2019. *Tourism & Management Studies, 17*(1), 5–18. https://doi.org/10.18089/tms.2021.170101

Pranckutė, R. (2021). Web of Science (WoS) and Scopus: The titans of bibliographic information in today's academic world. *Publications, 9*(1), 12. https://doi.org/10.3390/publications9010012

Rama, D., Maldonado-Erazo, C., & Álvarez García, J. (2018). State of the art of research in the sector of thermalism, thalassotherapy and spa: A bibliometric analysis. *European Journal of Tourism Research, 19*, 56–70.

Remoaldo, P., Matos, O., Freitas, I., Goja, R., Alves, J., Ribeiro, V., Pereira, M., & Xavier, C. (sem data). An international overview of certified practices in creative tourism in rural and urban territories. *Journal Of Hospitality & Tourism Research*. https://doi.org/10.1177/1096348020950792

Remoaldo, P., Ghanian, M., & Alves, J. (2020a). Exploring the experience of creative tourism in the Northern region of portugal-a gender perspective. *Sustainability, 12*(24). https://doi.org/10.3390/su122410408

Remoaldo, P., Matos, O., Freitas, I., Goja, R., Alves, J., Ribeiro, V., Pereira, M., & Xavier, C. (2020b). An international overview of certified practices in creative tourism in rural and urban territories. *Journal of Hospitality & Tourism Research*. https://doi.org/10.1177/1096348020950792

Remoaldo, P., Matos, O., Goja, R., Alves, J., & Duxbury, N. (2020c). Management Practices in Creative Tourism: Narratives by Managers from International Institutions to a More Sustainable Form of Tourism. *Geosciences, 10*(2). https://doi.org/10.3390/geosciences10020046

Remoaldo, P., Serra, J., Marujo, N., Alves, J., Goncalves, A., Cabeça, S., & Duxbury, N. (2020d). Profiling the participants in creative tourism activities: Case studies from small and medium sized cities and rural areas from Continental Portugal. *Tourism Management Perspectives, 36*. https://doi.org/10.1016/j.tmp.2020.100746

Richards, G. (2016). The challenge of creative tourism. *Ethnologies, 38*(1–2), 31–45. https://doi.org/10.7202/1041585ar

Richards, G. (2019). Creative tourism: Opportunities for smaller places? *Tourism & Management Studies, 15*(1SI), 7–10.

Richards, G. (2020). Designing creative places: The role of creative tourism. *Annals of Tourism Research, 85*, 102922. https://doi.org/10.1016/j.annals.2020.102922

Rodrigues, C., Liberato, D., & Melo, C. (2021). Tourism sustainable practices in rural territories: The case of Caretos de Podence. *Journal of Tourism & Development, 36*(1), 205–220. https://doi.org/10.34624/rtd.v1i36.23736

Rodrigues, C., Liberato, D., & Esteves, E. (2022). The intangible cultural heritage as a creative tourism product. The masks of Trás-os-Montes. In A. Abreu, D. Liberato, J. C. Garcia Ojeda (Eds.), *Advances in tourism, technology and systems* (vol 293, pp. 595–609). Smart Innovation, Systems and Technologies. Springer, Singapore. https://doi.org/10.1007/978-981-19-1040-1_50

Ross, D., & Saxena, G. (2019). Participative co-creation of archaeological heritage: Case insights on creative tourism in Alentejo Portugal. *Annals of Tourism Research, 79*. https://doi.org/10.1016/j.annals.2019.102790
Sarantou, M., Kugapi, O., & Huhmarniemi, M. (2021). Context mapping for creative tourism. *Annals of Tourism Research, 86*. https://doi.org/10.1016/j.annals.2020.103064
Snyder, H. (2019). Literature review as a research methodology: An overview and guidelines. *Journal of Business Research, 104*, 333–339. https://doi.org/10.1016/j.jbusres.2019.07.039
Somnuxpong, S. (2020). Chiang Mai: A creative city using creative tourism management. *Journal of Urban Culture Research, 20*, 112–132.
Tan, S. K., Luh, D.-B., & Kung, S.-F. (2014). A taxonomy of creative tourists in creative tourism. *Tourism Management, 42*, 248–259. https://doi.org/10.1016/j.tourman.2013.11.008
Tan, S.-K., Kung, S.-F., & Luh, D.-B. (2013). A model of 'creative experience' in creative tourism. *Annals of Tourism Research, 41*, 153–174. https://doi.org/10.1016/j.annals.2012.12.002
Tölkes, C. (2018). Sustainability communication in tourism—a literature review. *Tourism Management Perspectives, 27*, 10–21. https://doi.org/10.1016/j.tmp.2018.04.002
Tu, Y., Xu, J., & Shen, H.-W. (2021). KeywordMap: Attention-based Visual Exploration for Keyword Analysis. In *2021 IEEE 14th Pacific Visualization Symposium (PacificVis)* (pp. 206–215). https://doi.org/10.1109/PacificVis52677.2021.00034
Wang, C., Liu, J., Wei, L., Zhang, T., & (Christina). (2020). Impact of tourist experience on memorability and authenticity: A study of creative tourism. *Journal of Travel & Tourism Marketing, 37*(1), 48–63. https://doi.org/10.1080/10548408.2020.1711846
Wang, L., Togtokhbuyan, L., & Yadmaa, Z. (2021). Visual analysis of the international wellness tourism WOS literature from 1992 to 2019. *International Journal of Spa and Wellness, 4*(1), 1–16. https://doi.org/10.1080/24721735.2020.1830527
Xiao, Y., & Watson, M. (2019). Guidance on conducting a systematic literature review. *Journal of Planning Education and Research, 39*(1), 93–112. https://doi.org/10.1177/0739456X17723971
Yu, Y., Li, Y., Zhang, Z., Gu, Z., Zhong, H., Zha, Q., Yang, L., Zhu, C., & Chen, E. (2020). A bibliometric analysis using VOSviewer of publications on COVID-19. *Annals of Translational Medicine, 8*(13), 816. https://doi.org/10.21037/atm-20-4235
Zhong, L., Deng, B., Morrison, A. M., Coca-Stefaniak, J. A., & Yang, L. (2021). Medical, health and wellness tourism research—a review of the literature (1970–2020) and research agenda. *International Journal of Environmental Research and Public Health, 18*(20), 10875. https://doi.org/10.3390/ijerph182010875

Tourism Technology and Innovation

Exploring the Role of Technology in Adventure Tourism

Tatiana Chalkidou and George Skourtis

Abstract Adventure tourism has seen a recent surge in popularity, as people are increasingly looking for fear and thrills (Cater, Tour Manage 27:317–325, 2006) and unique and adrenaline-rushing experiences (Pomfret and Bramwell, Current Issues Tour 19:1447–1478, 2014). However, compared to several other specialized tourism studies, the research in this sector has been limited (Buckley, Tour Manage 33:961–970, 2012), while the role of technology in adventure tourism management is still in its infancy, although it is considered one of the most rapidly expanding sectors and a significant part of the tourism industry. Thus, the purpose of this paper is to theoretically discuss how the adventure tourism industry is implementing and managing technology, addressing the changes that technology has brought, and identifying the opportunities that can be exploited. In order to better understand the role of technology in adventure tourism management, we suggest the adoption of service-dominant logic and its ecosystem perspective (Vargo and Lusch, J Acad Market Sci 36:1–10, 2008). More specifically, we propose a transition of adventure tourism management from the perspective of S-D logic, reconsidering the conclusions that have been drawn from traditional perspectives (G-D logic). Importantly, a service-ecosystems view broadens the scope of adventure tourism management and enables researchers and practitioners to zoom out beyond dyadic exchange encounters and to view technology as a necessary resource for value co-creation. To the best of our knowledge, this paper is the first which challenges adventure management literature by adopting an S-D logic perspective.

Keywords Adventure tourism · Technology · Value co-creation

JEL Classification Z32 Tourism and development

T. Chalkidou (✉) · G. Skourtis
Deree–The American College of Greece, Athens, Greece
e-mail: tchalkidou@acg.edu

© The Author(s), under exclusive license to Springer Nature Switzerland AG 2023
V. Katsoni (ed.), *Tourism, Travel, and Hospitality in a Smart and Sustainable World*, Springer Proceedings in Business and Economics,
https://doi.org/10.1007/978-3-031-26829-8_17

1 Introduction

Adventure tourism is a type of tourism that involves taking risks and participating in activities that are outside the realm of normal vacationing. Adventure tourists often seek out destinations that offer opportunities for physical challenges and thrills. Technology has had a major impact on adventure tourism. The popularity of adventure tourism has grown in recent years as more people seek out new and exciting ways to travel. Adventure tourism offers travelers the chance to push themselves physically and mentally, while also experiencing different cultures and landscapes.

2 Origins and Description of Adventure Tourism

Modern adventure tourism originated from alpinism, which is the sport of mountaineering to reach the summit of the Alps. The first mountains were climbed for scientific curiosity and soon after became a competitive sport. In 1786, Montblanc was ascended, and in 1800, Grossglockner was summited. By 1865, all Alpine peaks had been conquered at least once (Standeven & de Knop, 1999). Furthermore, adventure tourism has been around since the late 1800s, when people began to travel to remote locations for hunting and fishing. Skiing and canoeing emerged as popular activities during this time, with France being one of the first countries to really embrace them. In the 1950s and 1960s, adventure tourism became more mainstream, due in part to media coverage of first ascents like climbing Mount Everest.

The boundaries of adventure tourism are debatable, in part because it is difficult to draw a clear distinction between ecotourism and adventure tourism (Fennell & Dowling, 2003; Fennell & Malloy, 1999; Ryan, 1998; Zurick, 1992); however, it is essential to differentiate between adventure tourism and ecotourism, as they are two entirely different things. Adventure tourism includes risk and adrenalin, while ecotourism does not (Beedie, 2005; Buckley, 2010; Hall, 1992; Weber, 2001). When it comes to classifying adventure tourism, strong similarities have been drawn with outdoor recreation and other adventurous pursuits (Pomfret & Bramwell, 2014; Sung et al., 1996). In addition to the above, the term "adventure tourism" itself is not used entirely consistently—other terms such as extreme, trend, active, or outdoor tourism are also employed. Furthermore, it is important to note that what one person perceives as an adventure might not be perceived as such by another individual (ATTA, 2010).

Adventure tourism, as the name implies, is a commercialized adventure activity that takes place outdoors. This activity generally involves a mix of physical exercise and risk and it can be described as a self-initiated recreational activity, typically involving a travel and overnight stay component, that usually involves close interaction with the natural environment, structurally contains elements of perceived or real risk and danger, and has an uncertain outcome that can be influenced by the participant and/or circumstance (Ewert, 2000). This description is very similar to one used by the Adventure Travel Trade Association, which describes adventure travel as "any tourist activity including two of the following three components: a physical activity, a cultural exchange or interaction, and engagement with nature." (ATTA, 2010, p. 2) which is similar to the one produced by UNWTO (2014, p. 10), where adventure tourism is conceptualized as:

[. . .] a trip that includes at least two of the following three elements: physical activity, natural environment, and cultural immersion.

Moreover, the theme of risk and uncertainty is common in most definitions or descriptions used for the specific concept. However, the risk is subjective and dependent on past experiences (Ewert, 1989; Rantala & Rokenses, 2018). It is perceived differently in different situations, and people's risk perceptions are influenced by personality, lifestyle, and level of skill and experience. In addition to the above, the most commonly used models of adventure tourism are founded on these or similar definitions and were built on specific setting-based attributes (Ewert & Hollenhorst, 1989), concepts related to the adventure experience such as fear, eustress, distress, abilities, and attitudes (Priest, 1992) and models of risk recreation (Robinson, 1992).

There are three types of adventure tourism: hard, soft, and nature-based (Cloke & Perkins, 1998; Buckley, 2006, 2010; McKay, 2013). Nature-based adventures are those that include interactions with wild animals, such as elephant walks (Buckley, 2010) whereas some soft adventure activities include backpacking, birdwatching, camping, canoeing, fishing, hiking, horseback riding, kayaking, safaris, sailing, and surfing. Furthermore, hard adventure activities may include caving, rock climbing, and trekking. At this point, it is importance to mention that soft activities performed under manageable conditions (i.e., kayaking on easy rivers) can be classified as hard adventures if performed on more difficult terrain. Thus, in general, hard adventure travel is riskier and requires more skills and experience. In addition, some of the soft adventure activities only seem risky to tourists because they are not accustomed to the activity, even though there is little-to-no objective risk involved (Swarbrooke et al., 2003).

In addition to the above, adventure tourism is usually examined under the goods-dominant (G-D) logic (Vargo & Lusch, 2004) which argues that tourism activities are designed with value and which is based on the soft/hard classification of adventure tourism activities. Janowski, Gardiner, and Kwek (2021) reviewed a number of adventure tourism dimensions that found the following ones to be included: physical activity, natural environment, risk and danger, challenge, socializing and camaraderie, learning and insight, use of skills, novelty, conflicting/intense emotions, thrill and excitement, well-being, cultural experience, involvement and locus of control, accomplishment, fun and enjoyment, flow, exploration, fear, rush, escapism, play, and uncertainty.

It can be argued that in the past two decades, definitions of adventure tourism have been largely based on insight theories, risk paradigms, flow experiences, notions of playfulness, and more recently discovered factors such as "the rush" (Ponte et al., 2021). The term "rush" was recently coined to express the feeling that skilled and experienced adventure tourists get from commercial adventure tours. According to Buckley (2012), "rush" is described as "a particular kind of excitement associated with the physical performance of a specific adventure activity, at the limits of individual capability, under highly favourable circumstances, by a person who is already skilled and trained in the activity concerned" (p. 936). It is important to note that theories about insight, risk, flow, rush, and deep play mostly focus on the demand side of things; however, these don't acknowledge the supply side of adventure tourism products (Swarbrooke et al., 2003). Furthermore, Ponte et al., (2021) examined adventure tourism definitions and found that there is a "shift from the physical aspects of the experience to the psychological aspects, and, more recently, the incorporation of specialized equipment and the role of guided tours" (p. 2), while Rantala et al. (2021) (cited in Ponte et al., 2021) suggested that "the different approaches to these definitions have led some authors to be more inclined to a 'product-oriented' perspective (Ponte et al., 2021, p. 2)".

3 A Shift Toward Adventure Tourism Experience Co-creation

From the goods-dominant perspective, adventure tourism was conceptualized as an activity offered by firms to travelers. Whereas G-D logic views that value is created and delivered by firms as an output that is determined in the exchange, as value-in-exchange, S-D logic acknowledges that value is co-created (FP6) by both firms and customers as resource integrators (FP9) (Vargo & Lusch, 2004, 2008), while it is experienced by beneficiaries in a specific context (Vargo & Lusch, 2008), and therefore understood as value-in-context (Chandler & Vargo, 2011). In an adventure tourism context, this means that firms cannot create and deliver activities i.e., value propositions (embedded with value) in terms of output, because in this manner customers are still considered operand resources (i.e., targets) that accept those activities provided by firms, and consequently value derived from adventure tourism activities is still determined through value-in-exchange. Contrary, S-D logic considers customers as operant resources (Lusch et al., 2007), and thus the fundamental source of strategic benefit (FP4). Moreover, S-D logic embraces a process-driven approach rather than an output approach focus (Vargo & Lusch, 2008) and this contributed to an understanding that adventure tourism requires a resource integration process (co-creation process) rather than the delivery of specific activities. Also, S-D logic posits that value is always determined by the beneficiary FP10, and thus is a context-specific phenomenon. In adventure tourism, acknowledging that value constitutes the travelers' perception of an activity's use value, therefore activity-related value should be determined by the traveler. In addition, because value co-creation is coordinated through institutions and institutional arrangements (FP11), in adventure tourism the role of these institutions must be central in enabling the development of the adventure tourism activities.

In this notion, micro-adventures are a subset of adventure tourism, which is becoming increasingly popular among young urban people. The British Adventurer

Alaistair Humphreys created the concept of micro-adventures to make activities more approachable for a working, mostly urban audience, and as such promote nine-to-five adventures as perfect adventures for people who want to get away from their everyday lives without having to commit to a long vacation (Humphreys, 2014). This trend also presents opportunities for businesses, such as tour operators offering wild camping spots or packaged micro-adventures (Gross & Werner, 2017).

Similar to the benefits accrued from adventure tourism, the benefits of micro-adventures are many and varied, but they can be broadly summed up as being good for one's health and well-being (Kane & Zink, 2004; Triantafillidou & Pet al., 2016), on top of being sustainable tourism experiences that promote the conservation of natural areas. Overall, tourists' adventure experiences are quite complex, and the growth of adventure tourism in recent years is closely linked to the overall increase in nature-based tourist activities.

4 The Role of Technology in Adventure Tourism

Over the years, adventure tourism has gained in popularity, with hotspots in several locations and jurisdictions (ATTA, 2013), as people are increasingly looking for fear and thrills (Cater, 2006) as well as unique and adrenaline-rushing experiences (Pomfret & Bramwell, 2014). The number of participants and adventure travel goods has grown considerably, and in many Western countries, it has emerged as an important element of the tourism industry (e.g., Canada and New Zealand), while in non-western countries adventure tourism is gaining more traction (e.g., China, Brazil, Russia, and India). Indicative of this are the findings of the Adventure Travel Trade Association (2013) that approximately up to 40% of European and South American and almost 15% of North American travelers go on adventure activities these days. The market for international adventure tourism was estimated to be worth USD 683 billion (ATTA, 2018).

Given the diversity of activities involved, it's no surprise that scholarly interest in this sector of the tourism industry has increased as a result of its exponential expansion (Beckman et al., 2017; Cheng et al., 2018). However, compared to several other specialized tourism studies, the research in this sector has been limited (Buckley, 2012). In addition, the literature on technology and adventure tourism management is still in its infancy, although it is considered one of the most rapidly expanding sectors and a significant part of the tourism industry. Understanding the role of technology in adventure tourism is of primary importance since technology not only affects the tourism experience (Neuhofer et al., 2013,) but also remains the central driver of tourism innovations and growth (Croce, 2018).

Overall, the technology sector is in a constant state of flux, with new technologies and innovations disrupting established industries on a regular basis. Technology disruptions can be described as processes whereby a technologically superior product or service replaces an existing technology. One may argue that the technology revolution began in the early 1990s with the advent of the internet and online booking systems. This allowed consumers to shop around for the best deals on travel and accommodation and made it easier to compare prices and find the most affordable options. This increased competition among providers and led to a race to the bottom in terms of prices.

The next major technology disruption came with the development of mobile technology and apps. This allowed consumers to book travel and accommodation on the go and made it even easier to compare prices and find the best deals. This led to further price reductions as providers competed for business. The latest technology disruption to hit the tourism industry is the rise of the sharing economy. This is where people use technology to share resources, such as accommodation or transportation. This has led to a new wave of companies, such as Airbnb and Uber that are shaking up the industry. It is therefore important for companies in the tourism industry to keep up with the latest technology trends and developments, to avoid being left behind.

Since adventure tourism is a subsector of the tourism industry, inevitably it is also affected. In recent years, technology has disrupted the adventure tourism industry in a number of ways, from increasing access to information and booking tools to improving safety and security. The adventure tourism industry has been growing rapidly in recent years, thanks in part to advances in technology which has made adventure travel experiences more accessible and convenient. Technology has made it easier than ever for people to find information about adventure activities and book travel arrangements. In addition, technology has played a role in improving safety and security in the adventure tourism industry. In recent years, safety has become an increasingly important concern for the adventure tourism industry. This is due in part to the increased popularity of extreme sports and other high-risk activities. To address this issue, several companies have developed safety-focused technologies, such as GPS tracking systems and safety monitoring apps.

These tools can help adventurers stay safe while participating in adventure activities. While technology has always been a part of adventure travel (consider GPS units and satellite phones), advances in mobile technologies and social media are making it easier than ever before for travelers to plan and book their adventures as well as to find information about destinations. They are also using technology during their trips to stay connected and share their experiences with others, making the actual experiences more immersive and memorable. There are several apps and websites that specialize in adventure travel, offering everything from trip planning resources to last-minute deals on gear and accommodations. And with more people sharing their adventure travel experiences online, it's easier than ever for tourists to find inspiration for their next trip (Kim & Tussyadiah, 2013; Wang et al., 2002). It has also made it possible for people to share negative experiences with a wider audience, which can hurt a destination's reputation. Furthermore, virtual reality technology, for example, is being used by some tour operators to give potential customers a taste of what they can expect on a trip and drones are becoming increasingly popular for capturing aerial footage of breath-taking landscapes and action-packed activities.

The increased digitalization is one of the primary reasons for the increased demand for adventure tourism. Individuals participating in outdoor adventure activities utilize technological means to travel into remote locations and wilderness areas, which are typically challenging to reach (Berger & Greenspan, 2008). The majority of adventure travelers plan their vacations by browsing online and endorsing adventures and places on social networking sites. Technology, on the other hand, has had an influence not only on traveler behavior at each stage of the purchasing process but also on how adventure tourism businesses and destinations position and market their services. As technology advances, so too does the industry's ability to offer customers a better experience with increased customer service quality at lower costs than ever before (Kim & Ham, 2007) and determines the strategy and competitiveness of tourism organizations and destinations (Buhalis & Law, 2008).

In recent years, there has been much talk about how technological innovation in this sector could help improve services by reducing costs while also streamlining operations (Singh et al., 2006) but what many people don't realize is just as importantly impacts travel safety worldwide. For example, digital maps make navigation easier because they show terrain features like rivers along roadsides which would otherwise be invisible without them. Furthermore, new developments in technology, such as targeted SEO strategies using social media platforms, customer services backed by artificial intelligence (AI), the use of virtual reality (VR) as an effective marketing tool, and the use of Big Data are expected to enhance tourists' (co-creating) adventure experiences as well as marketers' relevant knowledge. However, in spite of the role of technology in the growth of adventure tourism, its role as an operant resource in value co-creation (Akaka & Vargo, 2014) is highly neglected. As Maglio and Spohrer (2008) argued, value is co-created in dynamic configurations of resources (people, technology, information, organizations) connected internally and externally through value propositions. Since technology facilitates (Buhalis, 2019) co-creation experiences, adventure tourism experiences should be reconsidered by considering the role of technology within the adventure tourism service ecosystem.

5 Conclusion

According to a UNWTO study from 2014, worldwide visitor arrivals are predicted to reach 1.8 billion by 2030, and the industry's overall expansion will be aided greatly by established and new adventure tourism markets across the world. Though the latest innovations and dynamic market trends may initially present a challenge for adventure tourism businesses, the sooner they adapt to the new technological advancements, the more competitive they will be in the race to provide personalized and world-class experiences to potential customers, and thus the more growth opportunities will present themselves. We suggest that an S-D logic approach and its service ecosystem perspective (Vargo & Lusch, 2004) could advance adventure tourism research. Looking to the future, it is likely that technology will continue to play a major role in the tourism industry, both in terms of how holidays are booked and in terms of the actual experience of travelling. New transportation options, such as autonomous vehicles, are likely to have a significant impact on the way that we travel, while new platforms and apps will continue to make it easier for travelers to find and book holidays. People will continue to search for unique experiences that go beyond the everyday. This includes new markets such as space and deep sea tourism. In addition to traditional travel, the use of virtual reality and augmented reality looks set to become more widespread, with travelers able to get a realistic sense of their destination before they even leave home. Likewise, the use of artificial intelligence is likely to increase, both in terms of customer service (for example, using chatbots to answer queries) and in terms of things like planning itineraries and making recommendations. All of this is likely to make the adventure tourism industry even more competitive, meaning that those businesses that embrace new technologies and which use these to create value will be the ones who succeed, whereas those which do not, will suffer the consequences. Thus, a service ecosystem approach includes all the aforementioned components and also enables researchers and practitioners to zoom out beyond dyadic exchange encounters and to view technology as a necessary resource for value co-creation.

References

Adventure Travel Trade Association (ATTA). (Ed). (2010). Adventure Tourism Market Report, ATTA, Seattle.

Adventure Travel Trade Association (ATTA). (Ed). (2013). Adventure Tourism Market Study 2013, ATTA, Seattle.

Adventure Travel Trade Association. (2018). 20 Adventure Travel Trends to Watch in 2018. Retrieved from https://cdn.adventuretravel.biz/research/2018-Travel-Trends.pdf.

Akaka, M., & Vargo, S. (2014). Technology as an operant resource in service (Eco)systems. *Information Systems EBusiness Management, 12*, 367–384.

Beckman, E., Whaley, J. E., & Kim, Y.-K. (2017). Motivations and experiences of whitewater rafting tourists on the Ocoee River, USA. *International Journal of Tourism Research, 19*(2), 257–267. https://doi.org/10.1002/jtr.2109

Beedie, P. (2005). The adventure of urban tourism. *Journal of Travel and Tourism Marketing, 18*(3), 37–48.

Berger, I. E., & Greenspan, I. (2008). High (on) technology: Producing tourist identities through technologized adventure. *Journal of Sport & Tourism, 13*(2), 89–114. https://doi.org/10.1080/14775080802170312

Buckley, R. (2006). *Adventure tourism*. CABI.

Buckley, R. (2010). *Adventure tourism management*. Elsevier.

Buckley, R. (2012). Rush as a key motivation in skilled adventure tourism: Resolving the risk recreation paradox. *Tourism Management, 33*(4), 961–970. https://doi.org/10.1016/j.tourman.2011.10.002

Buhalis, D., & Law, R. (2008). Progress in information technology and tourism management 20 years on and 10 years after the internet: The state of etourism research. *Tourism Management, 29*(4), 609–623.

Buhalis, D. (2019). Technology in tourism-from information communication technologies to eTourism and smart tourism towards ambient intelligence tourism: A perspective article. *Tourism Review, 75*(1), 267–272.

Cater, C. I. (2006). Playing with risk? Participant perceptions of risk and management implications in adventure tourism. *Tourism Management, 27*(2), 317–325. https://doi.org/10.1016/j.tourman.2004.10.005

Cheng, M., Edwards, D., Darcy, S., & Redfern, K. (2018). A tri-method approach to a review of adventure tourism literature: Bibliometric analysis, content analysis, and a quantitative systematic literature review. *Journal of Hospitality & Tourism Research, 42*(6), 997–1020. https://doi.org/10.1177/1096348016640588

Chandler, J. D. & Vargo, S. L. (2011). Contextualization and value-in-context: How context frames exchange. *Marketing Theory, 11*(1): 35–49.

Cloke, P., & Perkins, H. C. (1998). Cracking the canyon with the awesome foursome: Representations of adventure tourism in New Zealand. *Environment and Planning D: Society and Space, 16*(2), 185–218.

Croce, V. (2018). With growth comes accountability: Could a leisure activity turn into a driver for sustainable growth? *Journal of Tourism Futures, 4*(3), 218–232. https://doi.org/10.1108/JTF-04-2018-0020

Ewert, A. (1989). *Outdoor adventure pursuits: Foundation, models and theories.* Publishing Horizons.

Ewert, A. W. (2000). Outdoor adventure recreation and public land management: Current status and emerging trends. Paper delivered at the *Tenth Annual World Congress on Adventure Travel and Ecotourism*, Anchorage, AK, September 11–14.

Ewert, A., & Hollenhorst, S. (1989). Testing the adventure model: Empirical support for a model of risk recreation participation. *Journal of Leisure Research, 21*, 124–139.

Fennell, D. A., & Dowling, R. (2003). *Ecotourism policy and planning.* CABI.

Fennell, D. A., & Malloy, D. C. (1999). Measuring the ethical nature of tourism operators. *Annals of Tourism Research, 26*(4), 928–943.

Gross, S., & Werner, K. (2017). Hut-to-Hut-hiking trails—a comparative analysis of popular hiking destinations. In C. M. Hall, Y. Ram, & N. Shoval (Eds.), *The Routledge international handbook of walking* (pp. 159–171). Routledge.

Hall, M. C. (1992). Review: Adventure, sport and health tourism. In M. C. Hall & B. Weiler (Eds.), *Special interest tourism* (pp. 141–158). Belhaven Press.

Humphreys, A. (2014). *Microadventures: Local discoveries for great escapes.* Harper Collins.

Janowski, I., Gardiner, S., & Kwek, A. (2021). Dimensions of adventure tourism. *Tourism Management Perspectives.* Elsevier B.V., 37, 100776. https://doi.org/10.1016/j.tmp.2020.100776.

Kane, M. J., & Zink, R. (2004). Package adventure tours: Markers in serious leisure careers. *Leisure Studies, 23*(4), 329–345.

Kim, W. G., & Ham, S. (2007). The impact of information technology implementation on service quality in the hotel industry. *Information Technology in Hospitality, 4*(4), 143–151.

Kim, J., & Tussyadiah, I. P. (2013). Social networking and social support in tourism experience: The moderating role of online self-presentation strategies. *Journal of Travel & Tourism Marketing, 30*(1–2), 78–92.

Lusch, R. F., Vargo, S. L., & O'Brien, M. (2007). Competing through service: Insights from service-dominant logic. *Journal of Retailing, 83*(1), 5–18.

Maglio, P. P., & Spohrer, J. C. (2008). Fundamentals of service science. *Journal of the Academy of Marketing Science, 36*(1), 18–20.

McKay, T. (2013). Adventure tourism: Opportunities and management challenges for SADC destinations. *Acta Academia, 45*(3), 30–62.

Neuhofer, B., Buhalis, D., & Ladkin, A. (2013). A typology of technology-enhanced tourism experiences. *International Journal of Tourism Research 16*(4). https://doi.org/10.1002/jtr.1958.

Pomfret, G., & Bramwell, B. (2014). The characteristics and motivational decisions of outdoor adventure tourists: A review and analysis. *Current Issues in Tourism, 19*(14), 1447–1478.

Ponte, J., Couto, G., Sousa, A., Pimentel, P., & Oliveira, A. (2021). Idealizing adventure tourism experiences: Tourists' self-assessment and expectations. *Journal of Outdoor Recreation and Tourism, 35*, 100379. https://doi.org/10.1016/j.jort.2021.100379

Priest, S. (1992). Factor exploration and confirmation for the dimensions of an adventure experience. *Journal of Leisure Research, 24*, 127–139.

Rantala, O., & Rokenses, A. (2018). Is adventure tourism a coherent concept? A review of research approaches on adventure tourism. *Annals of Leisure Research, 21*(5), 539–552.

Rantala, O., Rokenes, A., & Valkonen, J. (2021). Is adventure tourism a coherent concept? A review of research approaches on adventure tourism. *Annals of Leisure Research, 21*(5), 539–552.

Robinson, D. (1992). A descriptive model of enduring risk recreation involvement. *Journal of Leisure Research, 24*, 52–63.

Ryan, C. (1998). Saltwater crocodiles as tourism attractions. *Journal of Sustainable Tourism, 6*(4), 314–327.

Singh, A. J., Kim, H., & Huh, C. (2006). Differential impacts of information technology services in the korean hotel industry: A study of management perceptions. *FIU Hospitality Review, 24*(2), 78–89.

Standeven, J., & de Knop, P. (1999). *Sport tourism*. Human Kinetics.

Sung, H. H., Morrison, A. M., & O'Leary, J. T. (1996). Definition of adventure travel: Conceptual framework for empirical application from the providers' perspective. *Asia Pacific Journal of Tourism Research, 1*(2), 47–67. https://doi.org/10.1080/10941669708721975

Swarbrooke, J., Beard, C., Leckie, S., & Pomfret, G. (2003). *Adventure tourism: The new frontier*. Butterworth Heinemann.

Triantafillidou, A., & Petala, Z. (2016). The role of sea-based adventure experiences in tourists' satisfaction and behavioral intentions. *Journal of Travel & Tourism Marketing, 33*(1), 67–87.

Vargo, S. L., & Lusch, R. F. (2004). Evolving to a new dominant logic in marketing. *Journal of Marketing, 68*(January), 1–17.

Vargo, S. L., & Lusch, R. F. (2008). Service dominant logic: Continuing the evolution. *Journal of the Academy of Marketing Science, 36*(1), 1–10.

Wang, Y., Yu, Q., & Fesenmaier, D. R. (2002). Defining the virtual tourist community: Implications for tourism marketing. *Tourism Management, 23*(4), 407–417.

Weber, K. (2001). Outdoor adventure tourism: A review of research approaches. *Annals of Tourism Research, 28*(2), 360–377.

UNWTO. (2014). Global report on adventure tourism. World tourism organization: Madrid. Retrieved from https://www.unwto.org/archive/middle-east/publication/global-report-adventure-tourism

Zurick, D. N. (1992). Adventure travel and sustainable tourism in the peripheral economy of Nepal. *Annals of the Association of American Geographers, 82*(4), 608–628.

Dynamic Areas of Interest Inside an Urban Destination Using Visitors' Geolocation

Ioannis A. Nikas, Athanasios Koutras, and Alkiviadis Panagopoulos

Abstract Nowadays, travel experience tends to become entirely personal. The preparation of modern tourist packages requires a dynamic and personalized adaptation to the needs, capabilities and wishes of travellers. The components of these dynamic packages, consist not only of basic services, but also of other that are driven by the inherent curiosity of travellers to discover unknown destinations and cultures, as well as unprecedented experiences. Hence, the conclusive form of such packages is finalized immediately after the end of the trip, mainly, due to the continuous reformulation of their content by the travellers themselves. But this process is raising several questions: Can we determine these turning points attributed to the travellers' intention to modify or reformulate their tourist package? And if so, can the reasons that led them to such deviation from their original plans be identified and explained? In this work, driven by the above questions and the assumption that travellers' choices have an impact on their footprint, we propose an approximate spatio-temporal representation model to describe their decisions. This model is based on the implicit trace left by them, as it is captured in photo-sharing platforms and social networks. Initially, we use the travellers' digital footprints to map the corresponding routes, and then we transform these into an approximation of the areas they have visited. Finally, a union on these areas is applied to map the actual visited areas inside a tourist destination. Furthermore, the study of these mappings can be used to reveal to policy and decision-makers a dynamic picture of the areas of interest inside destinations, as well as any other areas that can be further developed in the immediate future. Our proposed model is applied and tested in the city of Athens, an urban destination, using data from the photo-sharing platform Flickr.

I. A. Nikas (✉) · A. Panagopoulos
Department of Tourism Management, University of Patras, Patras, Greece
e-mail: nikas@upatras.gr

A. Panagopoulos
e-mail: alkpanagop@upatras.gr

A. Koutras
Department of Electrical and Computer Engineering, University of Peloponnese, Patras, Greece
e-mail: koutras@uop.gr

© The Author(s), under exclusive license to Springer Nature Switzerland AG 2023

V. Katsoni (ed.), *Tourism, Travel, and Hospitality in a Smart and Sustainable World*, Springer Proceedings in Business and Economics,
https://doi.org/10.1007/978-3-031-26829-8_18

Keywords Tourism destination · Dynamic areas of interest · Dynamic packaging · Spatio-temporal analysis · Urban tourism

JEL Classification C23 · C33 · C38 · C55 · L83 · Z32

1 Introduction

Travel plays a vital role in our daily life, whether it concerns simple trips to shops and our workplaces, or longer trips abroad (Niblett & Beuret, 2021). In addition, the desire and need to travel for various reasons (leisure, work, academic, medical, etc.) is so old yet at the same time so current that everything shows that this need will continue to exist in the near and distant future, especially for the younger generations (Xiang et al., 2015; Cavagnaro, Staffieri, & Postma, 2018).

The opportunities that have appeared in the tourism sector through the rapid development and diffusion of new technologies, as well as the transport infrastructures, have created attractive and promising new products, easily accessible and consumable, as well as a new type of destinations, namely the smart tourism destinations (Buhalis & Amaranggana, 2013; Buhalis & Amaranggana, 2015; Boes, Buhalis, & Inversini, 2016; Jasrotia & Gangotia, 2018; Sigala, 2018; Gretzel, Zarezadeh, Li, & Xiang, 2020). Nowadays, the preparation of a trip, as a typical need in combination with the flourish of computer and web technologies in tourism, is a personal affair. It usually starts on the screen of a mobile device, dynamically forms a personal tourist package, and ends up as the initial recipe for another trip, for another traveller, due to web-sharing experiences, using each time the availably accessible digital social networking channels (Kim & Fesenmaier, 2015; Reichenberger, 2017; Liu, Wu, & Li, 2018; Song & Lee, 2020). In this context, travel planning has become part of our living culture.

Personal travel planning can be defined as a series of hierarchical structure of information searches to conclude to a personal final travel decision. Generally, travel planning is currently under the spotlight of the scientific community and is widely studied, discussing mainly either about the reliability and realism of travel planning methods (Bonsall, 2009; Cohen, 2009), or the proposal of recommendation systems for improving tourists' experiences (Chiang & Huang, 2015; Chu & Huang, 2015; Kolahkaj, Harounabadi, Nikravanshalmani, & Chinipardaz, 2020). From the other side, pre-trip travel planning can be considered as a fundamental component of the final trip experience. The components of travel packages are shaped mainly by the professional and social needs, personal desires, and financial potentials of the traveller (Xiang et al., 2015; Huang, Goo, Nam, & Yoo, 2017). The innate curiosity to explore new places and cultures along with the need of numerous services drives the creation of products aimed at living unprecedented and authentic experiences (Garau, 2016). But live experiences offer emotions that most of the time are not captured in their entirety. Thus, travellers having the knowledge and suitable technological assets can reshape their planning constantly and, in this way, the process of planning a trip or creating a tourist package is said to be completed by the end of the personal tourism adventure.

During the co-creation of a trip, events occur, or *tourist opportunities* appear at some time instance, which motivate travellers to modify their trip planning. These spur-of-the-moment decisions are difficult to be detected, as they are only captured or recorded at the end of the journey as part of the overall experience. These momentary choices are the motivation for the present work and questions such as "*Is it possible to determine these spatio-temporal instances attributed to the travellers' intention to modify or reformulate their trip?*" and "*And if possible, can one identify the reasons for these deviations and explain them?*" determined the orientation of this work.

The most common available information for studying travellers' behaviour is the digital footprint they leave on a tourist destination through the social networks they use (Girardin, Blat, Calabrese, Dal Fiore, & Ratti, 2008). This footprint is usually represented by their geolocation. The approach so far has mainly involved studying these geolocations in terms of identifying points of interest, approaching the centres of a set of geolocations as possible centres of their visiting sites, e.g. (Gavric, Culibrk, Lugonja, Mirkovic, & Crnojevic, 2011; Kisilevich, et al., 2013; García-Palomares, Gutiérrez, & Mínguez, 2015; Peng & Huang, 2017; Zhang, Tan, Lei, Guo, & Sun, 2018). In addition, the spatio-temporal study of these geolocations also described the flow of visitors within a destination, e.g. (Girardin et al., 2008; Gavric et al., 2011; Wan, Hong, Huang, Peng, & Li, 2018).

In this paper, under the assumption that the dynamic planning of tourists' trips has an impact on the footprint they leave, an approximate spatio-temporal representation model is proposed to describe these decisions. This model is based on the indirect trace left by a traveller, as captured through social networks and photo-sharing services. Initially, we use the travellers' digital footprints to map the corresponding routes and then we transform these into an approximation of the areas they have visited. Finally, a union on these areas is applied to map the visited areas inside a tourist destination. Furthermore, the study of these mappings can be used to reveal to policy and decision-makers a dynamic picture of the areas of interest inside destinations, as well as any other areas that can be further developed in the immediate future. Our proposed model is applied and tested in the city of Athens, an urban destination, using data from the photo-sharing platform Flickr.

Lastly, it is worth mentioning that reality reveals the difficulty of observing and studying the dynamic nature of this phenomenon, due to the abundance and uniqueness of all these created tourist products. Furthermore, the easy and immediate way to share a constantly shaped experience, as well as the ability of Internet users to directly access it, increases the difficulty of studying the effects of the phenomenon. So, an undesired effect is the creation of outbreaks of overtourism, poor management of available natural resources and a negative impact on the natural environment. In all the above cases, we often lead to either unsustainable tourism development or visitors with reduced empathy and awareness of the fragility of the ecosystems they visit.

2 Literature Review

The evolution of technology and communication devices has allowed the content that exists on the internet to be shaped and created by its users themselves, also known as user-generated content. At the same time, modern devices allow the generation of metadata, such as geodata, which enriched the carrying information with spatio-temporal characteristics. The geographical position of users, namely geolocation, became available mainly using social networks (Batty, 2013; Kitchin, 2013; García-Palomares et al., 2015).

The ability to record a user's geographic location depends on the user himself (Volunteered Geographic Information) and usually accompanies text, hypertext, and multimedia as metadata. It has been ascertained that the value of this data is equivalent to authoritative data (Devkota et al., 2019). In this light, many works have studied the tourism phenomenon using either indirectly (text and multimedia) or directly (geolocation) the geographical information that generated by travellers' interaction with social networks.

Some of the studies concerning travellers were discussing tourism activity analysing social-network data, which embed geotagged information from web photo-sharing services like Flickr and Panoramio (Panoramio was launched in 2005, and closed on November 4, 2016): Girardin et al. (2008) uncover the presence and movements of tourists' in Rome analysing their flows between various points of interest; Popescu, Grefenstette, France, and Moëllic (2009) revealed a large number of trips, proposed new ones, estimated the staying time and provided locations with panoramic views; De Choudhury, et al. (2010) created automatically meaningful travel itineraries for individuals and professionals; Mamei, Rosi, & Zambonelli (2010) developed an intelligent recommendation system using past user experiences; Lu, Wang, Yang, Pang, & Zhang (2010) provided a customized travel plan for tourists in several cities and countries; Gavric et al. (2011) identified attractive locations and some standard routes in the city of Berlin; Kisilevich, et al. (2013) defined several types of spatio-temporal clusters describing travellers' visits and analysed them using various methodologies; Wood, Guerry, Silver, & Lacayo (2013) employ the location metadata embedded in photographs to estimate visitation rates at various recreational sites around the world; Kurashima, Iwata, Irie, and Fujimura (2013) used geotagged photos as a sequence of visited locations and then they recommend travel routes between landmarks; Li (2013) developed a travel planning system to solve the tourists' multi-day and multi-stay travel planning problem; Koerbitz, Önder, & Hubmann-Haidvogel (2013) estimated the actual tourist overnight stays compared with authoritative data; Straumann et al. (2014) study the visiting behaviour of foreign and domestic visitors in Zurich; Sun and Fan (2014) identified social events; García-Palomares et al. (2015) identified and analysed the main tourist attractions in eight major European cities using spatial statistical techniques to analyse location patterns; Memon et al. (2015) proposed a recommendation methodology based on geotagged social media for tourist travel recommendation; Majid, Chen, Mirza, Hussain, and Chen (2015) proposed a recommendation system for interesting tourist locations and tourist travel sequences; Peng & Huang (2017) proposed an integrating spatial clustering and text mining methods for discovering popular tourist attractions; Zhang et al., (2018a, 2018b, 2018c) presented density-based spatial clustering algorithm to detect popular points of interest; Wan et al. (2018) developed a hybrid methodology to make route recommendations; Giglio et al. (2019) determined the attractiveness of various tourism sites in Italy; Höpken, Müller, Fuchs, & Lexhagen (2020) identified tourists' behavioural patterns through the visited POIs.

For the problem of identifying Tourism Areas of Interest (TAOIs), several techniques have been proposed: Spyrou, Korakakis, Charalampidis, Psallas, & Mylonas (2017) presented a geoclustering algorithm for the problem of extracting TAOIs from an urban area; Koutras et al. (2019) extracted TAOIs using volunteered geographic information and density-based spatial clustering algorithms; Devkota et al. (2019) identified in near-real-time TAOIs using density-based spatial clustering algorithm combining in addition night-time light remotely sensed data with promising results, especially for areas with low social media penetration; Karayazi, Dane, & de Vries (2021) investigated the attractiveness and representation of heritage sites; Nikas et al. (2022) combined User-Generated Content and Food & Beverage points of interest to shape areas of interest inside an urban destination.

3 Methodology

In this section, we will discuss the framework for conducting the experiments of this work. The methodology that will be used is described as follows: selection of an area of interest, data collection, data processing, data description, extraction of TAOIs and, finally, discussion of the results.

Areas of Interest

Our study will focus on an urban destination, Athens, Greece. Athens is the largest city and capital of Greece, one of the oldest cities in the world (recorded history from 3200 BC) and one of the most important metropolises in Europe. The central municipalities together with Piraeus consist of 3.218.218 permanent residents.

Some of the most important historical monuments in the world are in Athens (Acropolis, Panathenaic Stadium, etc.). Athens is a pole of attraction for many visitors from all over the world either as an urban destination or as a stopover to one of the Aegean islands. Many researchers have extensively study Athens, evidenced by the publication of 65,800 research papers in the last 20 years, 19,600 in the last five years, and 5030 by now in 2022 (results from Google Scholar with keywords: Athens, tourism). However, COVID19 has led researchers to keep on studying these major tourist destinations regarding the topics of recovery and resilience of tourist destinations (Fernández, Martínez, & Martín, 2022; Bulchand-Gidumal, 2022), as well as the changes in tourists' behaviour (Pappas, 2021; Poulaki & Nikas, 2021). Therefore, the study of important tourist destinations, such as Athens, constitutes a timely research issue.

Experimental Setup

The required data were collected from the photo-sharing platform Flickr (flickr.com). The Flickr platform is a popular photo-sharing service, especially among users who use photography as a job, hobby, or just for fun. The latter category, among others, includes many amateur photographers who use the platform to post their vacation photos. For our study, the choice of the Flickr platform was considered ideal since it is not aimed to users who constantly communicate their social life through social media, but to users who simply want to share photos, whether they are amateurs or professionals. Thus, the studied users, ideally, will post photographic material from points and areas of personal and, by extension, tourist interest.

Photos posted on the Flickr platform are semantically annotated, with metadata containing various information, e.g., the user's Flickr id, the exact date and time the photo was taken and uploaded, the photo's geolocation, and the type and camera used. The data collected was limited to a specific area of the city of Athens. The selected area included the historical centre and neighbouring areas. This area is included in a bounding box with bottom-left and top-right corners coordinates defined by (23. 673, 37.93233), (23.78063, 38.01693). The retrieved data were not only spatially limited but also temporally: data was retrieved using a time query from 01/01/2009 to 25/10/2017. The total retrieved data consisted of 201,100 photos.

The collected data were pre-processed driven mostly by the conducted analysis. Firstly, users identified as locals were excluded from our study. García-Palomares et al. (2015) in a similar study excluded from the data users who appeared in a place for more than a month, based on their posted photos. In this paper, a probabilistic exclusion procedure for residents is proposed. Supposing that the random variable describing the average length of stay follows the Poisson distribution and, knowing the annual average length of stay of tourists in the Attica region, we can estimate the maximum length of stay of tourists in Athens by calculating, annually, the inverse probability

$$P(Stays < x) \leq 99.95\%$$

as shown in the following Table 1. In this way, we have excluded locals from our research, but not the frequent visitors. The occurring database contains 194.307 photos, and these photos were taken by 2753 unique users. The unique users per year are, also, shown in Table 1 and Fig. 1.

Table 1 Estimation of maximum length of stay based on average length of stay and Poisson distribution

	2009	2010	2011	2012	2013	2014	2015	2016	2017
Average length of stay	9.5	9.4	9.2	9.1	8.9	8.4	7.8	7.7	7.7
Estimation of maximum length of stay	21	21	21	21	20	19	18	18	18
Unique non-local users	512	404	386	441	369	361	331	324	279

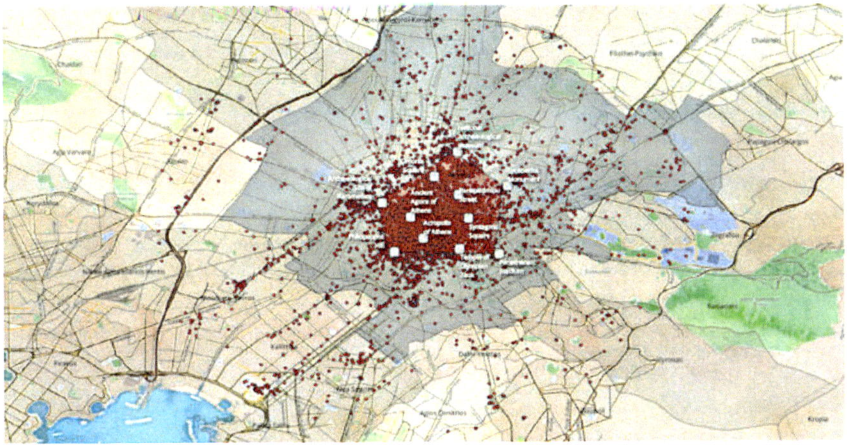

Fig. 1 Collected data inside a bounding box [(23. 673, 37.93233), (23.78063, 38.01693), 01/01/2009–25/10/2017] that includes the historical centre of Athens. The dark red points represent the taken photos, while the grey squares some of the most visited monuments. The darker areas show the city of Athens and around it the neighbouring cities. The data processing has been conducted using R (R Core Team, 2022)

Data Description

A brief description of the processed data is presented below. In Fig. 2, the number of photos taken by year, month, and hour are displayed. We can observe that (a) the total number of photos decrease over time, (b) the most active month is June, and (c) the most active time period is between 9.00 and 15.00.

In Fig. 3, it is shown the number of unique users visited Athens by year, month, and hour. We can observe that (a) the total number of photos decreases over time, (b) the most active month is June, and (c) the most active time period is between 9.00 and 15.00.

The declining number of photos taken in Athens, as well as the declining number of Flickr users visiting Athens could be interpreted as a decaying interest towards Athens. However, correlating the average length of stay with users and the number of photos shows that there is a significant correlation between photos taken per year and average length of stay ($\varrho_{\text{photos}\sim\text{avg stay}} = 0.81$) and unique users and average length of stay ($\varrho_{\text{unique users}\sim\text{avg stay}} = 0.86$), making Flickr platform an important and reliable data source.

Finally, a significant percentage of users (>37.45%) uploaded only one photo to Flickr while over the half of them (>50.67%) uploaded utmost two photos (Fig. 4).

Experiments

The core of our experiments lies in the context of spatial analysis. Spatial analysis tools are used to investigate the behaviour of spatial and sometimes temporal data. The aim of our experiments is to transform the discrete digital footprints left by travellers

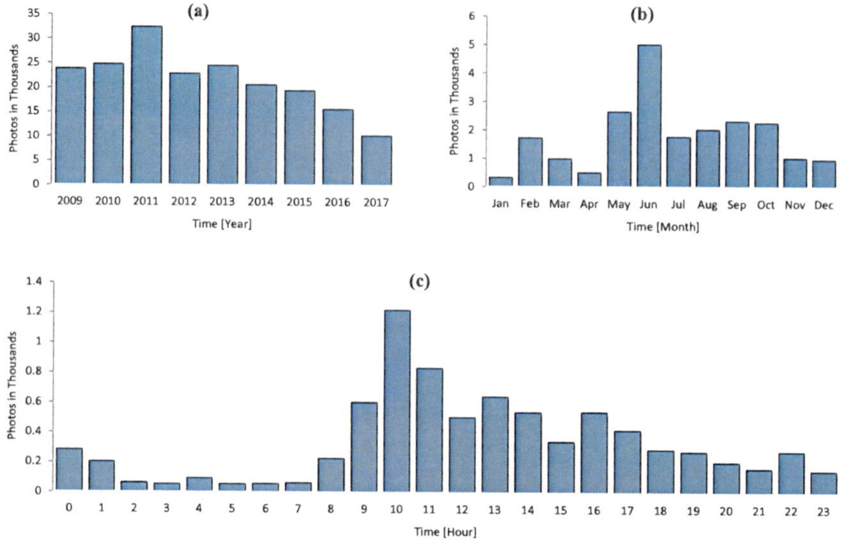

Fig. 2 Frequency bar plots for the number of photos taken annually (**a**), and for the monthly (**b**) and hourly (**c**) average number of photos taken through the period 2009–2017

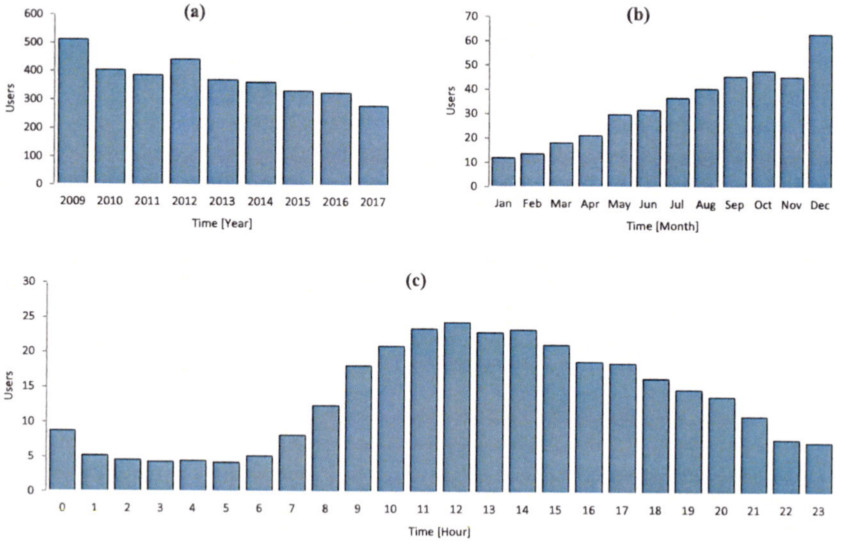

Fig. 3 Frequency bar plots for unique visitors of Athens annually (**a**), monthly (average value) (**b**), and hourly (average) (**c**) through the period 2009–2017

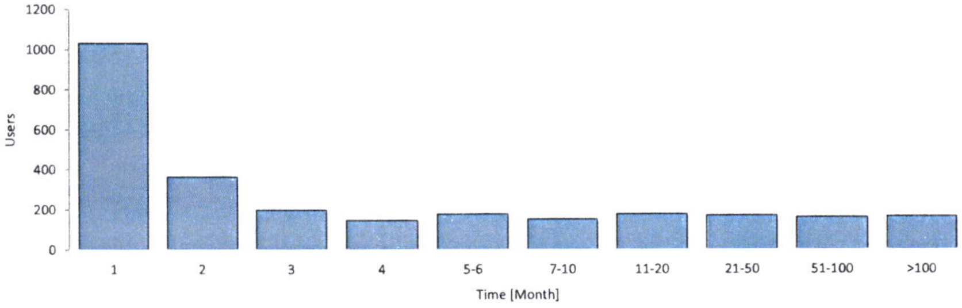

Fig. 4 Frequency bar plots for uploading photos per unique user through the period 2009–2017

using social networks to areas in the form of polygons. These transformations will give us a good approximation of the actual visited places and, hence, potential TAOIs. From the available tools, the convex hull operator will be used. A convex hull is defined as a polygon with the minimum perimeter that encloses a data set. In this work, the Minimum Bounding Geometry toolbox will be utilized as implemented in the QGIS software (QGIS Development Team, 2009).

4 Results

The first task of our experiments is to sketch an approximation of the area of Athens that is visited by travellers and tourists. The TAOI is designed using all the available data, employing the convex hull operator. In Fig. 5, the drawn convex polygon approximates the maximum feasible area that at least a traveller could have visited.

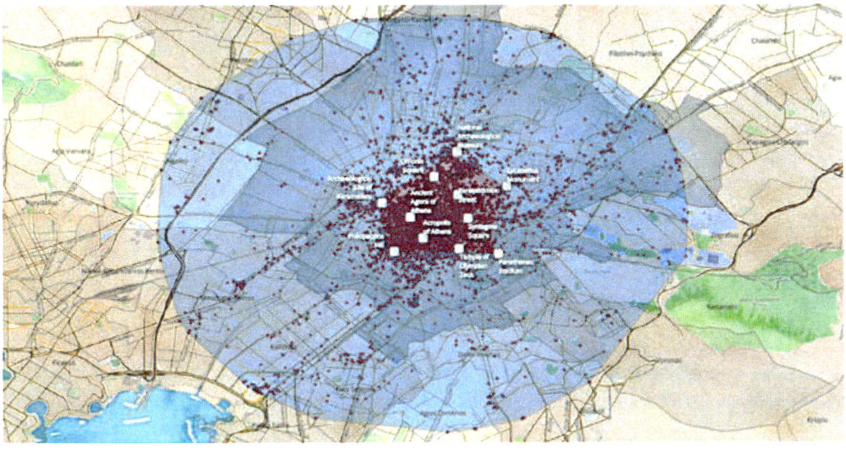

Fig. 5 Convex hull of all data regarding the period 2009–2017

Next, we examine the evolution of TAOIs through the years. In addition, we want to study the seasonal change of TAOIs during a year and during a day. For the first case, the collected data are drawn for each year separately as it is shown in Fig. 6. We notice from the figure that the area of the annual TAOIs is retained over the years, with the expansion towards the whole city of Athens and the neighbouring cities to be evident.

For the later cases, we enclose the photos taken per month and per hour and the results are showing in the Figs. 7 and 8. The monthly analysis of the collected data implies intense activity in the winter months and especially in the month of December. In addition, the areas of interest seem to grow gradually from August to December, the month of major holidays, Christmas, and New Year's Eve. On the other hand, the hourly analysis of the data did not return any significant results, beyond the fact that at night the TAOIs tend to approach the historical centre of Athens.

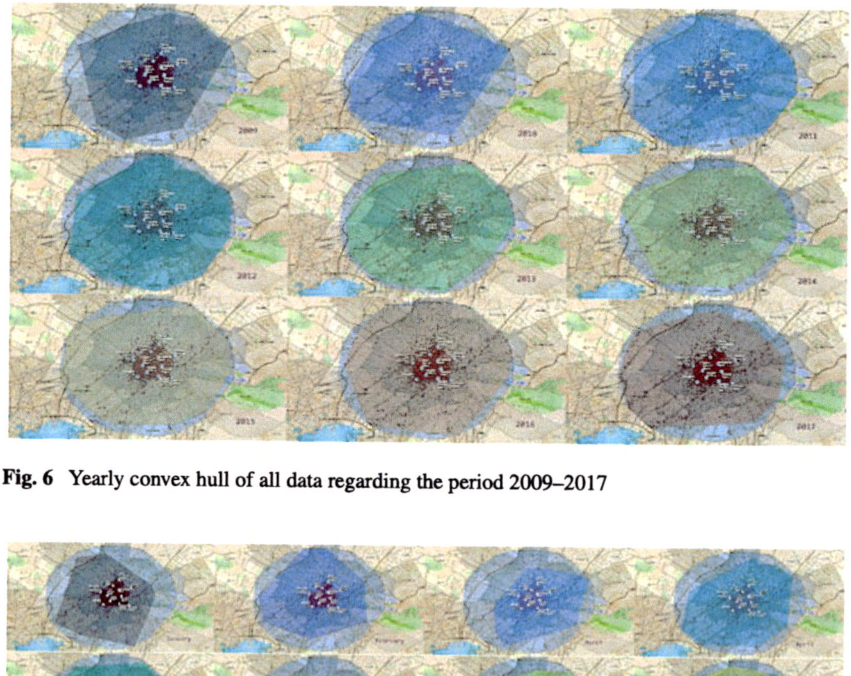

Fig. 6 Yearly convex hull of all data regarding the period 2009–2017

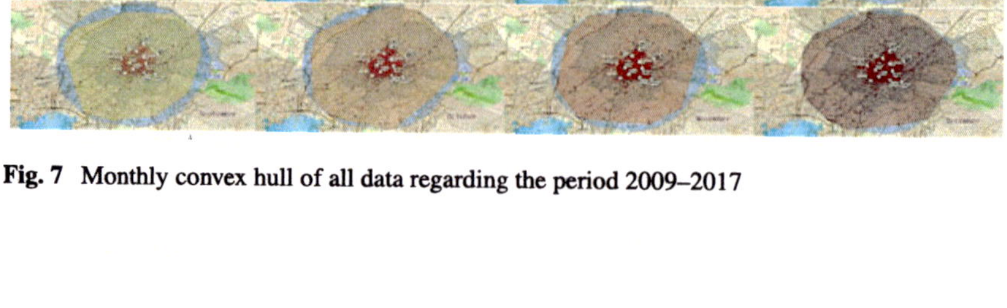

Fig. 7 Monthly convex hull of all data regarding the period 2009–2017

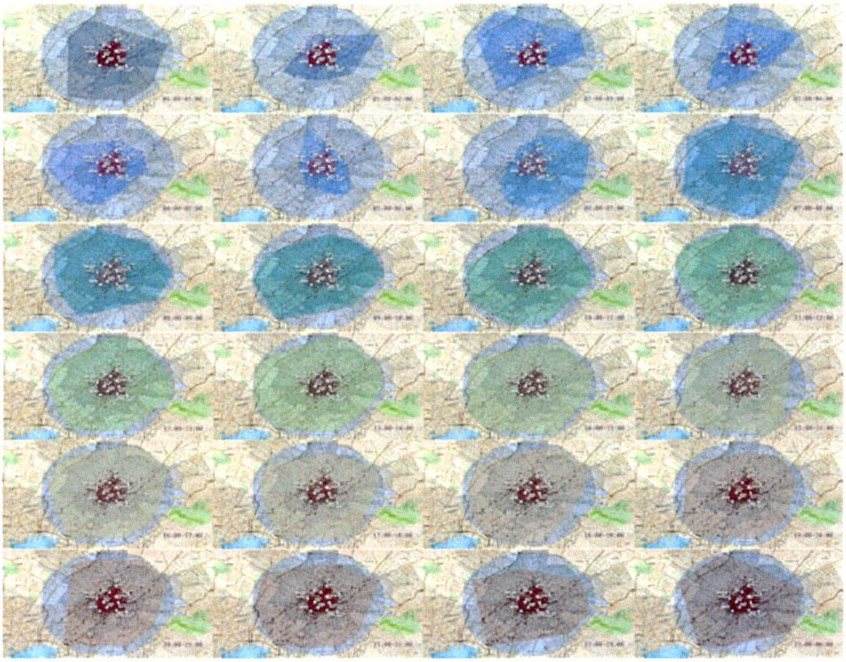

Fig. 8 Hourly convex hull of all data regarding the period 2009–2017

In all cases, the extracted areas of interest contain the well-known urban landmarks of Athens (visitgreece.gr), as well as Food & Beverage points of interest (Koutras et al., 2019; Nikas et al., 2022).

Next, we will approach the TAOIs through the users' photography-area individually. That is, we will isolate regions getting, hopefully, a more detailed picture of the potential areas of interest. From this approach, we will exclude users who uploaded a small number of photos, so that finding the convex hull is meaningful. Under this context, we have produced the following two figures. Figure 9 shows the set of all feasible polygons formulated using the convex hull operator. Beside the fact that the main landmarks are contained in the core of the set of created polygons, no detailed information is revealed about the behaviour of the travellers.

Therefore, we present a statistical approach, drawing the feasible convex hull areas using the corresponding visited areas quantiles. So, in Fig. 10 we can see the users corresponding to the 25, 50, 75, and 95% (a–d) of the visited areas. For the first case the travellers were "wandered", mostly, among the main landmarks of Athens, while in the second case (b) they were covering city's historical centre. For the case (c), we can see that the visiting areas are expanded through the main roads of Athens like Pireos Street and Leoforos Andrea Siggrou, and for the last one (d), the areas tend to cover the total TAOI of Athens.

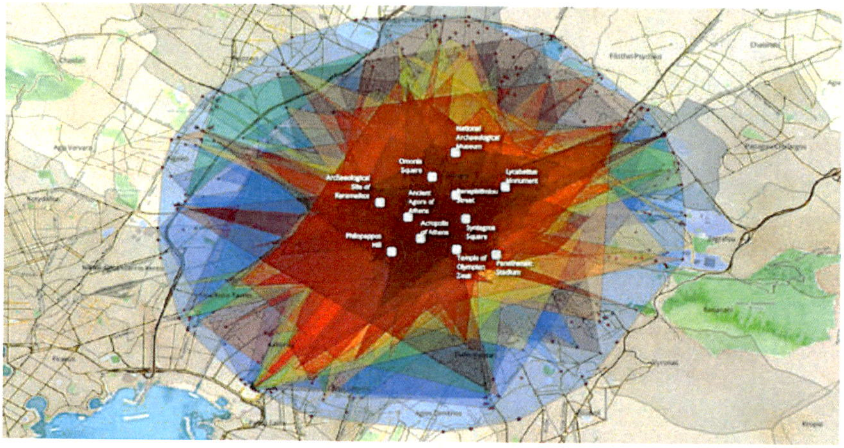

Fig. 9 Individual feasible convex hull of all data regarding the period 2009–2017

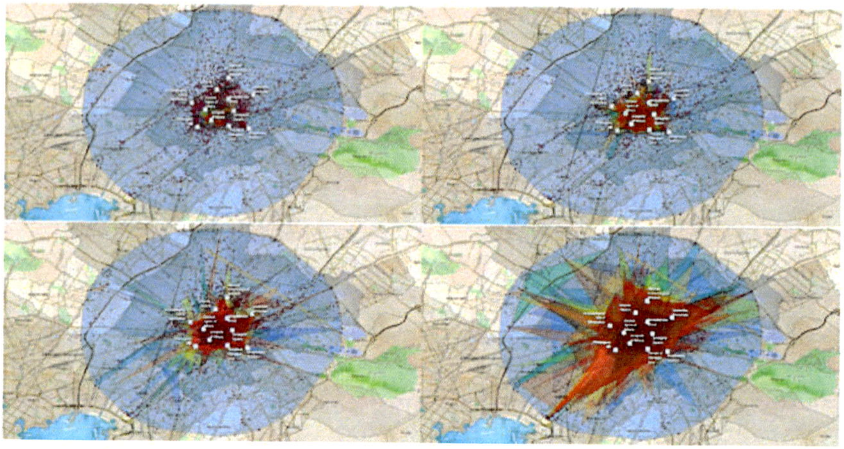

Fig. 10 Individual feasible convex hull of all data regarding the period 2009–2017 based on the quantiles of visited area

Finally, in Fig. 11 we can see the convex hull of users visiting the greater areas (relative to the rest of the visitors), as well as the users corresponding to visited areas between 25 and 75%.

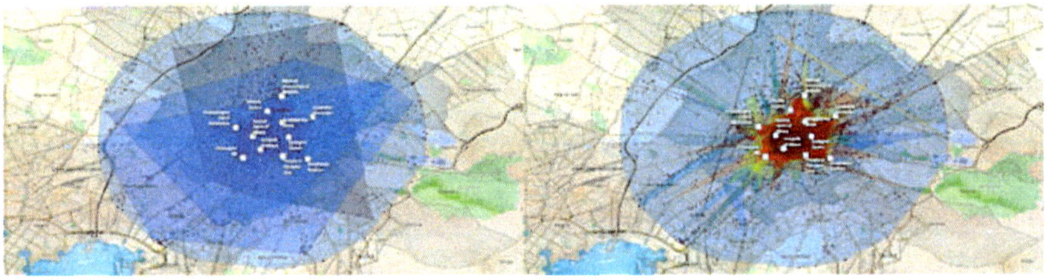

Fig. 11 Individual feasible convex hull of all data regarding the period 2009–2017 of the greater visited areas (relative to the rest of the visitors) and the visited areas between 25 and 75%

5 Conclusion

In this work, a spatio-temporal analysis of geospatial data from an urban destination was conducted to highlight Tourism Areas of Interest. The data used was retrieved from the photo-sharing service Flickr related to the historical city centre of Athens.

These data were processed to separate locals from visitors by assuming that their average length of stay follows a Poisson distribution with mean value the authoritative annual mean. Then by applying the convex hull operator, regions of interest were extracted, all of which contained well-known sights and landmarks of Athens. Furthermore, the study of these mappings can be used to reveal to policy and decision-makers a dynamic picture of the areas of interest inside destinations, as well as any other areas that can be further developed in the immediate future.

The presented analysis highlighted the visitors' intention to explore the city of Athens, apart from visiting only well-known landmarks. In this direction, to feel the real pulse of an urban destination, a combination of many different social networks should be employed, where, in addition to geodata, information indicating their geolocation and travelling intention could be mined. Understanding the tourism phenomenon, especially in the age of personalization of the services and products provided, is imperative. This understanding plays a fundamental role in the recovery process from various crises, such as COVID19, which often brings unsustainable development, overtourism, and unregulated exploitation of natural and tourist resources.

References

Batty, M. (2013, 11). Big data, smart cities and city planning. *Dialogues in Human Geography, 3*(3), 274–279. https://doi.org/10.1177/2043820613513390.

Boes, K., Buhalis, D., & Inversini, A. (2016, 5). Smart tourism destinations: ecosystems for tourism destination competitiveness. *International Journal of Tourism Cities, 2*(2), 108–124.

Bonsall, P. (2009, 11). Do we know whether personal travel planning really works? *Transport Policy, 16*(6), 306–314. https://doi.org/10.1016/J.TRANPOL.2009.10.002.

Buhalis, D., & Amaranggana, A. (2013). Smart tourism destinations. *Information and Communication Technologies in Tourism, 2014*, 553–564. https://doi.org/10.1007/978-3-319-03973-2_40.

Buhalis, D., & Amaranggana, A. (2015). Smart tourism destinations enhancing tourism experience through personalisation of services. *Information and Communication Technologies in Tourism*, 377–389. https://doi.org/10.1007/978-3-319-14343-9_28.

Bulchand-Gidumal, J. (2022, 3). Post-COVID-19 recovery of island tourism using a smart tourism destination framework. *Journal of Destination Marketing & Management, 23*, 100689. https://doi.org/10.1016/J.JDMM.2022.100689.

Cavagnaro, E., Staffieri, S., & Postma, A. (2018, 6). Understanding millennials' tourism experience: values and meaning to travel as a key for identifying target clusters for youth (sustainable) tourism. *Journal of Tourism Futures, 4*(1), 31–42. https://doi.org/10.1108/JTF-12-2017-0058.

Chiang, H., & Huang, T. (2015, 1). User-adapted travel planning system for personalized schedule recommendation. *Information Fusion, 21*(1), 3–17. https://doi.org/10.1016/J.INFFUS.2013.05.011.

Chu, C., & Huang, C. (2015, 9). A platform for travel planning by using Google Maps. In *16th IEEE International Conference on Mobile Data Management* (Vol. 2, pp. 120–125). Institute of Electrical and Electronics Engineers Inc. https://doi.org/10.1109/MDM.2015.13.

Cohen, T. (2009, 11). Evaluating personal travel planning: If it is prohibitively expensive to get a robust answer then what should we do? *Transport Policy, 16*(6), 344–347.

De Choudhury, M., Feldman, M., Amer-Yahia, S., Golbandi, N., Lempel, R., & Yu, C. (2010). Automatic construction of travel itineraries using social breadcrumbs. In *HT'10—Proceedings of the 21st ACM Conference* (pp. 35–44). https://doi.org/10.1145/1810617.1810626.

Devkota, B., Miyazaki, H., Witayangkurn, A., & Kim, S. M. (2019). Using volunteered geographic information and nighttime light remote sensing data to identify tourism areas of interest. *Sustainability, 11*(17), 4718. https://doi.org/10.3390/su11174718.

Fernández, J. A., Martínez, J. M., & Martín, J. M. (2022, 1). An analysis of the competitiveness of the tourism industry in a context of economic recovery following the COVID19 pandemic. *Technological Forecasting and Social Change, 174*, 121301. https://doi.org/10.1016/j.techfore.2021.121301.

Garau, C. (2016, 1). *Emerging technologies and cultural tourism: Opportunities for a cultural urban tourism research agenda* (pp. 67–80). Springer International Publishing. https://doi.org/10.1007/978-3-319-26877-4_4.

Gavric, K., Culibrk, D., Lugonja, P., Mirkovic, M., & Crnojevic, V. (2011). *Detecting attractive locations and tourists' dynamics using geo-referenced images* (pp. 208–211). https://doi.org/10.1109/TELSKS.2011.6112035.

Giglio, S., Bertacchini, F., Bilotta, E., & Pantano, P. (2019, 6). Using social media to identify tourism attractiveness in six Italian cities.*Tourism Management, 72*, 306–312. https://doi.org/10.1016/J.TOURMAN.2018.12.007.

Girardin, F., Blat, J., Calabrese, F., Dal Fiore, F., & Ratti, C. (2008, 10). Digital footprinting: Uncovering tourists with user-generated content, *7*, 36–44. https://doi.org/10.1109/MPRV.2008.71.

García-Palomares, J., Gutiérrez, J., & Mínguez, C. (2015, 9). Identification of tourist hot spots based on social networks: A comparative analysis of European metropolises using photo-sharing services and GIS, *63*, 408–417. Pergamon. https://doi.org/10.1016/J.APGEOG.2015.08.002.

Gretzel, U., Zarezadeh, Z., Li, Y., & Xiang, Z. (2020, 2). The evolution of travel information search research: a perspective article. *Tourism Review, 75*(1), 319–323.

Höpken, W., Müller, M., Fuchs, M., & Lexhagen, M. (2020, 5). Flickr data for analysing tourists' spatial behaviour and movement patterns: A comparison of clustering techniques. *Journal of Hospitality and Tourism Technology, 11*(1), 69–82. https://doi.org/10.1108/JHTT-08-2017-0059.

Huang, C., Goo, J., Nam, K., & Yoo, C. (2017, 9). Smart tourism technologies in travel planning: The role of exploration and exploitation, *54*, 757–770. North-Holland. https://doi.org/10.1016/J.IM.2016.11.010.

Jasrotia, A., & Gangotia, A. (2018). Smart cities to smart tourism destinations: A review paper. *Journal of Tourism Intelligence and Smartness, 1*(1), 47–56.

Karayazi, S., Dane, G., & de Vries, B. (2021, 3). Utilizing urban geospatial data to understand heritage attractiveness in Amsterdam. *ISPRS International Journal of Geo-Information, 10*(4), 198. https://doi.org/10.3390/IJGI10040198.

Kim, J., & Fesenmaier, D. (2015, 12). Sharing tourism experiences: The posttrip experience. *Journal of Travel Research, 56*(1), 28–40. https://doi.org/10.1177/0047287515620491.

Kisilevich, S., Keim, D., Andrienko, N., Andrienko, G., Kisilevich, S., Keim, Á., … Andrienko, G. (2013). *Towards acquisition of semantics of places and events by multi-perspective analysis of geotagged photo collections* (pp. 211–233). Springer. https://doi.org/10.1007/978-3-642-12289-7_10.

Kitchin, R. (2013, 12). Big data and human geography: Opportunities, challenges and risks. *Dialogues in Human Geography, 3*(3), 262–267. https://doi.org/10.1007/978-3-642-36309-2_42.

Koerbitz, W., Önder, I., & Hubmann-Haidvogel, A. (2013). *Identifying tourist dispersion in Austria by digital footprints* (pp. 495–506). Springer. https://doi.org/10.1007/978-3-642-36309-2_42.

Kolahkaj, M., Harounabadi, A., Nikravanshalmani, A., & Chinipardaz, R. (2020, 7). A hybrid context-aware approach for e-tourism package recommendation based on asymmetric similarity measurement and sequential pattern mining, *42*, 100978. Elsevier. https://doi.org/10.1016/J.ELERAP.2020.100978.

Koutras, A., Nikas, I., & Panagopoulos, A. (2019). Towards developing smart cities: Evidence from GIS analysis on tourists' behavior using social network data in the city of Athens.*Springer Proceedings in Business and Economics*, 407–418. https://doi.org/10.1007/978-3-030-03910-3_28.

Kurashima, T., Iwata, T., Irie, G., & Fujimura, K. (2013, 10). Travel route recommendation using geotagged photos. *Knowledge and Information Systems, 37*(1), 37–60. https://doi.org/10.1007/s10115-012-0580-z.

Li, X. (2013). Multi-day and multi-stay travel planning using geo-tagged photos. In *2nd ACM SIGSPATIAL International Workshop on Crowdsourced and Volunteered Geographic Information* (pp. 1–8). Association for Computing Machinery. https://doi.org/10.1145/2534732.2534733.

Liu, H., Wu, L., & Li, X. (2018, 3). Social media envy: How experience sharing on social networking sites drives millennials' aspirational tourism consumption, *58*(3), 355–369. https://doi.org/10.1177/0047287518761615.

Lu, X., Wang, C., Yang, J., Pang, Y., & Zhang, L. (2010). Photo2Trip: Generating travel routes from geo-tagged photos for trip planning. In *MM'10—Proceedings of the ACM Multimedia 2010 International Conference* (pp. 143–152). https://doi.org/10.1145/1873951.1873972.

Majid, A., Chen, L., Mirza, H., Hussain, I., & Chen, G. (2015, 1). A system for mining interesting tourist locations and travel sequences from public geo-tagged photos. *Data & Knowledge Engineering, 95*, 66–86. https://doi.org/10.1016/J.DATAK.2014.11.001.

Mamei, M., Rosi, A., & Zambonelli, F. (2010). Automatic analysis of geotagged photos for intelligent tourist services. In *6th International Conference on Intelligent Environments* (pp. 146–151). https://doi.org/10.1109/IE.2010.34.

Memon, I., Chen, L., Majid, A., Lv, M., Hussain, I., & Chen, G. (2015, 2). Travel recommendation using geo-tagged photos in social media for tourist. *Wireless Personal Communications, 80*(4), 1347–1362. https://doi.org/10.1007/s11277-014-2082-7.

Niblett, M., & Beuret, K. (2021). *Why travel?: Understanding our need to move and How it shapes our lives*. Bristol University Press. https://doi.org/10.51952/9781529216394.

Nikas, I., Koutras, A., Panagopoulos, A., & Vasileiadis, A. (2022). Exploring tourists' food and beverage spots in an urban destination using a spatial–temporal approach. *Springer Proceedings in Business and Economics*, 977–989. https://doi.org/10.1007/978-3-030-92491-1_58.

Pappas, N. (2021, 6). COVID19: Holiday intentions during a pandemic. *Tourism Management, 84*, 104287. https://doi.org/10.1016/J.TOURMAN.2021.104287.

Peng, X., & Huang, Z. (2017, 7). *A novel popular tourist attraction discovering approach based on geo-tagged social media big data* (Vol. 6, p. 216). Multidisciplinary Digital Publishing Institute. https://doi.org/10.3390/IJGI6070216.

Popescu, A., Grefenstette, G., France, E., & Moëllic, P.-A. (2009). Mining tourist information from user-supplied collections. In *18th ACM Conference on Information and Knowledge Management* (pp. 1713–1716). ACM Press. https://doi.org/10.1145/1645953.1646211.

Poulaki, I., & Nikas, I. (2021, 10). Measuring tourist behavioral intentions after the first outbreak of COVID-19 pandemic crisis. Prima facie evidence from the Greek market. *International Journal of Tourism Cities, 7*(3), 845–860. https://doi.org/10.1108/IJTC-09-2020-0218.

QGIS Development Team. (2009). QGIS Geographic Information System. *Open Source Geospatial Foundation.* Retrieved from http://qgis.osgeo.org.

R Core Team. (2022). R: A language and environment for statistical computing. *R Foundation for*. R Foundation for Statistical Computing. Retrieved from https://www.R-project.org/.

Reichenberger, I. (2017, 11). C2C value co-creation through social interactions in tourism. *International Journal of Tourism Research, 19*(6), 629–638. https://doi.org/10.1002/jtr.2135.

Sigala, M. (2018, 1). New technologies in tourism: From multi-disciplinary to anti-disciplinary advances and trajectories. *Tourism Management Perspectives, 25*, 151–155.

Song, W., & Lee, H. (2020, 1). An analysis of traveler need for and willingness to purchase airline dynamic packaging: A Korean case study. *Journal of Air Transport Management, 82*, 101735. https://doi.org/10.1016/J.JAIRTRAMAN.2019.101735.

Spyrou, E., Korakakis, M., Charalampidis, V., Psallas, A., & Mylonas, P. (2017, 3). A geo-clustering approach for the detection of areas-of-interest and their underlying semantics. *Algorithms, 10*(1), 35. https://doi.org/10.3390/A10010035

Straumann, R., Çöltekin, A., & Andrienko, G. (2014). Towards (re)constructing narratives from georeferenced photographs through visual analytics. *The Cartographic Journal, 51*(2), 152–165. https://doi.org/10.1179/1743277414Y.0000000079.

Sun, Y., & Fan, H. (2014). Event identification from georeferenced images. *Lecture Notes in Geoinformation and Cartography, 7*, 73–88. https://doi.org/10.1007/978-3-319-03611-3_5.

Wan, L., Hong, Y., Huang, Z., Peng, X., & Li, R. (2018, 11). A hybrid ensemble learning method for tourist route recommendations based on geo-tagged social networks, *32*, 2225–2246. Taylor & Francis. https://doi.org/10.1016/J.ELERAP.2020.100978.

Wood, S., Guerry, A., Silver, J., & Lacayo, M. (2013, 10). Using social media to quantify nature-based tourism and recreation. *Scientific Reports, 3*(1), 1–7. https://doi.org/10.1038/srep02976.

Xiang, Z., Magnini, V., & Fesenmaier, D. (2015, 1). Information technology and consumer behavior in travel and tourism: Insights from travel planning using the internet. *Journal of Retailing and Consumer Services, 22*, 244–249.

Zhang, W., Tan, G., Lei, M., Guo, X., & Sun, C. (2018, 1). Detecting tourist attractions using geo-tagged photo clustering (Vol. 3, pp. 3–16). Sage. https://doi.org/10.1177/2397200917752649.

Towards a Framework for Participative Innovation in Tourism

Susanne Marx and Lorena Arens

Abstract Open innovation is a common form of innovation activities in tourism. However, the design of the participation of multiple stakeholders in these activities is challenging. The present research develops a framework for describing participation in innovation activities in tourism to help address this challenge. The research process follows the method framework for Design Science Research with a design- and development-centered entry point. Firstly, a model is designed based on extant literature. Secondly, three case studies are used to amend the model and demonstrate the applicability for individual projects, inter-organizational projects, and the portfolio level from the perspective of destination management organizations (DMO). Interviews are used to gather the case study data. The interview data is analyzed by applying template analysis. The research advances the understanding of design opportunities for participative innovation activities in tourism from the DMO perspective. Eight stages of participation in innovation activities are identified. These are demonstrated in the case study DMOs for various types of projects. Though the qualitative nature of the research is a limitation, DMO managers and other tourism actors are provided with a framework for designing innovation activities and participation in these in their destinations. Future research should validate the framework quantitatively, especially towards incorporating the perspectives of other actors from the touristic ecosystem, e.g., political and public actors.

Keywords Open innovation · Tourism · Participation

JEL Classification L83

S. Marx (✉) · L. Arens
Stralsund University of Applied Sciences, Stralsund, Germany
e-mail: susanne.marx@hochschule-stralsund.de

© The Author(s), under exclusive license to Springer Nature Switzerland AG 2023

V. Katsoni (ed.), *Tourism, Travel, and Hospitality in a Smart and Sustainable World*, Springer Proceedings in Business and Economics,
https://doi.org/10.1007/978-3-031-26829-8_19

1 Introduction

The tourism sector is "a network industry par excellence" (Gustafsson et al., 2014, p. 1583). Network development and innovation are linked in tourism (Volgger, 2017), with various tourism service providers (TSPs) jointly delivering a networked product in a so-called tourism service network (Beritelli et al., 2016). Destination management organizations (DMOs) play a unique role in the development of the destination as network managers (Volgger & Pechlaner, 2014), facilitators of partnerships (World Tourism Organization (UNWTO), 2019) or even implementing "meta-management of the whole community" (Beritelli et al., 2016, p. 108). Crossing organizational boundaries in innovating is expected in the tourism sector (Marx & Klotz, 2021), making open innovation (OI) – a term coined by Chesbrough (2003) – a dominant form. OI in networks is identified for further research potential (West & Bogers, 2017). The participation of various actors impacts the results of innovation activities (Mahr et al., 2014), and participatory decision-making positively affects satisfaction (Black & Gregersen, 1997). However, the selection (Guertler & Sick, 2021) and motivation for involvement, e.g., of external participants (West & Bogers, 2014) remain challenging. With OI gaining importance and the DMO's responsibility for developing entire destinations, we ask: How could a framework for participative innovation in tourism be designed from a DMO perspective? Following the Design Science Research framework (Johannesson & Perjons, 2014), we firstly develop an initial model based on extant literature in Sect. 2. The case study methodology to amend and demonstrate the model is described in Sect. 3. The findings and discussion in Sect. 4 lead to developing a revised model for participation in innovation activities from a DMO perspective.

2 Literature Review

Innovation in tourism is discussed controversially and perceived as either high (Hjalager, 2010) or low intensity (Volgger, 2017). One reason is the different understanding of the term innovation, which can either be regarded as the output (a product or service) or the process of innovating (the innovation activity) (OECD/Eurostat, 2018). Moreover, innovation as an output can take different characteristics (Hauschildt et al., 2016), e.g., how new and new to whom. Researching the Community Innovation Survey (CIS)—a dataset often used to measure innovation in different sectors (Smith, 2005)—, Nordli (2017) concludes that questions on innovation seem to be challenging to interpret by tourism actors due to the specifics of this industry. An aspect explaining this difference is the networked nature of the sector (Beritelli et al., 2016; Gustafsson et al., 2014). The product in this industry is networked as perceived by the customer (Zhang et al., 2009), built by several elements delivered by mainly small and medium-sized enterprises (Zenker & Kock, 2020). The broad range of beneficiaries of innovation in tourism is reflected in the

World Tourism Organization's (2016) definition of innovation: "the introduction of a new or improved component which intends to bring tangible and intangible benefits to tourism stakeholders and the local community, improve the value of the tourism experience and the core competencies of the tourism sector and hence enhance tourism competitiveness and /or sustainability" (p. 14). The networked nature of the touristic offer suggests that innovation relies on the collaboration of various actors.

Network-centric innovation (Nambisan & Sawhney, 2011) or even broader OI (Chesbrough & Bogers, 2014) are observed trends in innovation activities in general. Chesbrough and Bogers (2014, p.17) define OI "as a distributed innovation process based on purposively managed knowledge flows across organizational boundaries". OI in inter-organizational settings is an area identified for further research (Sydow & Müller-Seitz, 2020; West & Bogers, 2017). Only 13.4% of tourism companies (NACE79–tour operators and travel agencies) develop service innovations only by themselves, according to the German CIS data 2016 (Center for European Economic Research (ZEW), 2018), whereas the average value describing all industries in the CIS data is much higher (61.6%). The reliance of tourism micro and small enterprises on collaboration and participation of multiple actors in innovation activities is confirmed by case studies (Marx & Klotz, 2021). Marx (2022) identifies three groups involved in innovation activities in tourism: entrepreneurs and their enterprises (TSPs), DMOs, and the ecosystem with a large variety of actors. These are involved innovations activities based on business and private relations (Marx & Klotz, 2021). To select and motivate various actors for participating in innovation activities is a challenge (Guertler & Sick, 2021; West & Bogers, 2014).

Since DMOs are network managers (Provan & Kenis, 2007; Volgger & Pechlaner, 2014) and network and innovation development are linked in the tourism sector (Volgger, 2017), the DMOs view on and engagement in innovation activities is relevant. The research question therefore is: How could a framework for participative innovation in tourism be designed from a DMO perspective?

DMOs design and implement various projects with multiple partners to bring their strategies into action (Marx, 2019). A project is a "temporary endeavor undertaken to create a unique product, service, or result" (Project Management Institute, 2021, p. 4). When involving several organizations striving "jointly to create a tangible product/service in a limited time" (Jones & Lichtenstein, 2008, p. 234), we speak of inter-organizational projects (IOPs). Projects or programs "managed as a group to achieve strategic objectives" are considered portfolios (Project Management Institute, 2021, p. 4). Marx (2022) suggests that the aforementioned three types of actors participate in individual innovation projects of TSPs and IOPs orchestrated by the DMO as a portfolio manager in a project network. In project networks, inter-organizational projects are frequently implemented based on inter-organizational relations (DeFillippi & Sydow, 2016; Sydow, 2022; Sydow & Windeler, 2020). Temporary endeavors are thus jointly implemented, building on non-temporary relations of the actors, with differing forms of engagement in each project.

Gausemeier et al. (2019) evaluate various process models for innovation activities and suggest three generic phases: ideation and selection, conception and development, and market entrance and dissemination. Involvement in innovation activities is referred to as participation, along with many terms synonymously, such as collaboration or co-creation (Ommen et al., 2016). The participation of various actors could occur at all these stages of innovation activities. In a business context, participation is related to making collective decisions (Hucker, 2008; Newig, 2011), not only focusing on the selection among alternatives but a varying degree of involvement at different stages of the decision process (Black & Gregersen, 1997): identifying the problem, ideating alternative solutions, choosing a solution, planning and implementing the solution, and evaluating the result. The differences between the three phases of innovation activities (Gausemeier et al., 2019) and the generic participative decision-making process (Black & Gregersen, 1997; Hucker, 2008) lay at the beginning and the end, with the decision process suggesting a problem identification initiating the process, which we translate into the identification of an innovation need or opportunity. From a participation perspective, the evaluation of the solution is suggested to be added to the innovation process resulting in six stages of participation in innovation activities:

- identification of innovation need/opportunity
- idea generation (development of alternative solutions)
- selection of solution
- development and implementation of solution
- marketing of solution/dissemination
- evaluation of solution.

Research on participative decision-making suggests that the level of involvement significantly positively affects satisfaction and performance (Black & Gregersen, 1997). These effects cumulate with involvement in all stages of the process (Black & Gregersen, 1997). The participation of tourism actors in innovation activities is associated with benefits stemming both from the innovation activity and output (Marx & Klotz, 2021).

Based on the view of the tourism innovation system as a project network, Fig. 1 summarizes that non-temporary actors (TSPs, DMOs, ecosystem) participate in temporary innovation activities (projects, IOPs, and portfolio) for a perceived benefit stemming from the innovation activity, or the innovation output.

3 Methodology

The research follows the method framework for Design Science Research (Johannesson & Perjons, 2014) with a design- and development-centered entry point (Peffers et al., 2008). It presents two phases: design and development, and demonstration. Firstly, the framework has been designed based on extant literature, as displayed in Fig. 1. Secondly, case studies from three DMOs are used to demonstrate the framework's applicability for individual TSP projects, inter-organizational projects (IOPs), and the portfolio level of the DMO. A (multiple) case study strategy is widespread in innovation research (Elsahn et al., 2020), suitable for exploratory studies with limited

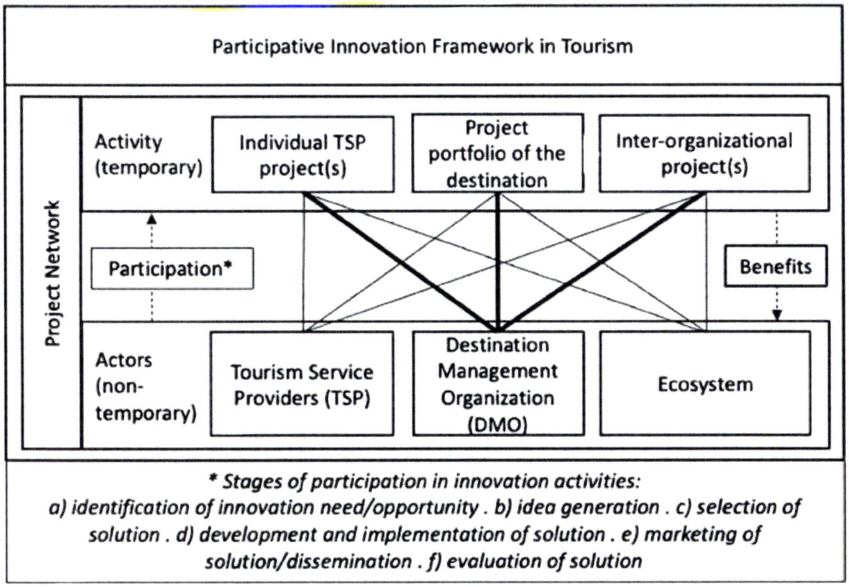

Fig. 1 Participative innovation framework in tourism–initial version

control of the research environment (Yin, 1994). The three DMOs were chosen based on the following criteria as exemplary cases (Elsahn et al., 2020):

- belonging to the same region to reduce regional differences (Germany)
- different scope of destination size (regional, city DMO, rural DMO)
- experienced in innovation activities (DMO1 received the German tourism prize, DMO2 best-practice example in collaborative financing, DMO3 best-practice health town)

Case 1 (DMO1) is a regional DMO of one of the most prominent destinations in Germany. The DMO is organized as an association, to a small extent financed by member organizations and, to a more significant extent, financed by public funds from the federal state. Case 2 (DMO2) is a local DMO of a larger city with outreach towards the neighboring communities, known for its specific collaboration with the touristic actors. The DMO consists of two parts: a municipal company part-financed by public funds from the municipality and income from touristic fees, and an adjunct public-owned limited company, financed by yearly fees from a network of touristic enterprises from the whole tourism value chain. Case 3 (DMO3) is a local DMO of a small coastal town, a municipal company, part-financed by public funds from the municipality and by income from touristic fees. The interviewees represent the top management level.

The interviews were semi-structured, recorded, and transcribed. Template analysis was applied to analyze the data (Brooks et al., 2015; King, 2004; Saunders et al., 2009). The two authors coded the interviews based on the code template. They discussed the resulting differences focusing on the reasons for the differences – a method to increase coding quality (Kuckartz & Rädiker, 2020). Coding by multiple researchers is recommended to increase data richness and reach an inter-coder agreement (Goffin et al., 2019). The interview questions were constructed around the participation of touristic actors in individual innovation activities of TSPs, IOPs, and a portfolio of projects from the perspective of the DMO.

To assess the participative innovation activities, we assess four dimensions: the actors (TSPs, DMO, ecosystem) focusing on the view of the DMO, the type of innovation activities (individual TSP projects, IOPs, portfolio), the benefits and the stages of participation. The template used to assess participation are the six stages suggested in Sect. 2.

4 Results

The interviewees mentioned diverse examples of innovation activities that can be allocated to individual and inter-organizational projects. Table 1 names the examples of innovation projects given by the three DMOs. The additional category of DMO-owned projects was added while coding the data. While most DMO-driven innovation projects involve various actors, we suggest distinguishing IOPs of several organizations formally engaged in the project to create a joint service (Jones & Lichtenstein, 2008) from projects owned and mainly designed and driven by the DMO. As DMOs are network organizations, these projects often provide a frame for other actors to develop their own innovative offers. For example, DMO2 creates and realizes innovative event concepts with artists or maritime service providers, while restaurants build on this frame to develop their products in this context. We describe this role of the DMO as enabling TSPs' innovation activities and add this form of participation to the model. The portfolio management is lifted on a layer above all projects in the model (Fig. 2).

Table 1 Project examples named by case DMOs

DMO	1	2	3
Individual TSP projects	• Exceptional concert at special location	• New temporary outdoor attraction	• Digital solution for beach chair rental
DMO-owned projects	• Exhibition at large fair	• Cultural events at various locations • Maritime large event • Educational project	• New event series • Infrastructure projects
IOPs	• Visitor cards across federal state • Port network	• Visitor card with several destinations	• Visitor card with several destinations • Infrastructure projects

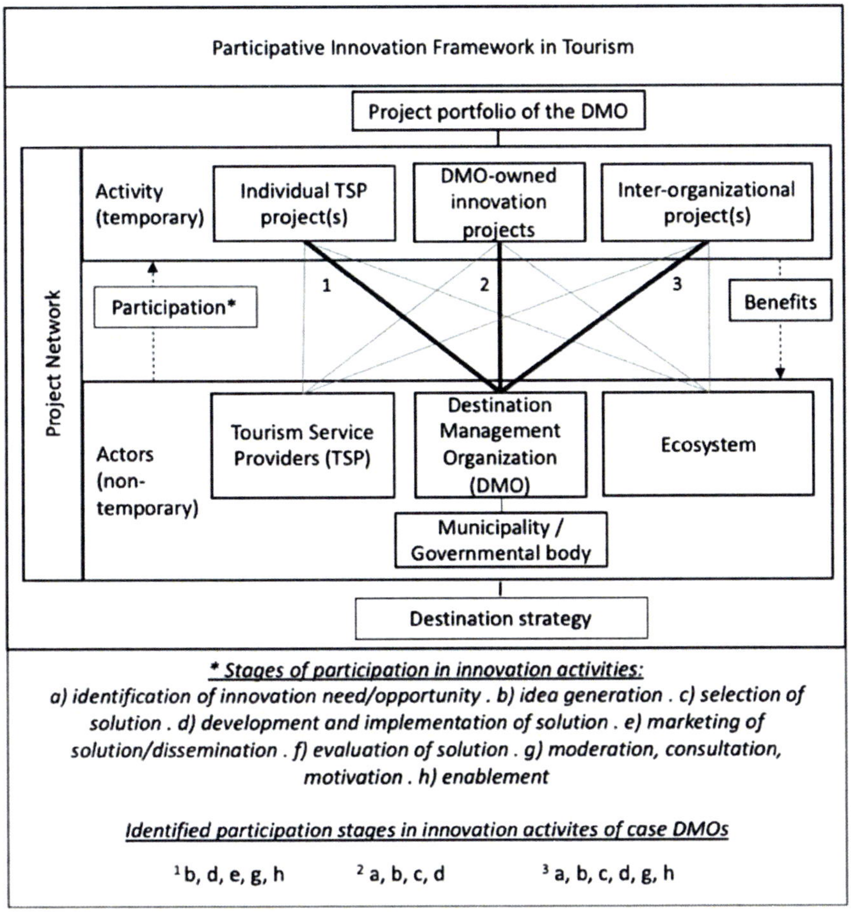

Fig. 2 Participative innovation framework in tourism – revised version

To answer the research question "How could a framework for participative innovation in tourism be designed from a DMO perspective?", we present and discuss the three temporary forms of innovation activities and the participation of DMOs in these separately. Table 2 demonstrates the stages of participation of the case DMOs in innovation activities. While the DMO representatives rarely mentioned the marketing of solutions, the evaluation was not mentioned at all.

TSPs commonly suggest individual projects to the DMO (e.g., DMO1 by an event company for an innovative high-level event, DMO2 by a hotel manager for an outdoor attraction). The collaboration with individual organizations is framed and limited by the duties of the DMOs as municipal companies or municipally financed organizations (all three case DMOs). The engagement of a DMO must not be for the benefit of a single, especially profit-oriented organization, but for the destination in

Table 2 Participation of DMOs in innovation activities

Activity	Individual TSP project	DMO-owned project	IOP
a. Identification of innovation need/opportunity		XX	XX
b. Idea generation	X	XXX	XX
c. Selection of solution		XX	X
d. Development and implementation of solution	XX	XXX	XX
e. Marketing of solution/dissemination	XX		
f. Evaluation of solution			
g. Moderation, consultation, motivation	XX		X
h. Enabling	XXX		XX
X = 1 DMO, XX = 2 DMOs, XXX = 3 DMOs			

total (DMO1: "*You cannot promote a single company in that way*", DMO2: "*We do not support any business model of a private entrepreneur*", DMO3: "*...act for everyone equally, and if we provide support to someone, we would have to be able to provide it in the same way for others, too*"). The participation of the case DMOs is related to idea generation (b), development and implementation (d), and marketing (e). Additionally, DMOs take an enabling role (h) where they support by organizing finances and administrative tasks or by delivering a conceptual framework for innovation activities (DMO1: "*We were basically an administrative unit in between*"). Moreover, DMOs moderate, consult, and motivate for the innovation activity (g) (DMO2: "*We are trying to moderate this to some extent with the administration, the citizens, and other tourism service providers, and to create a kind of advantage translation so that in the end it serves everyone and not just the individual who had the idea*", "*And that is how I understand our mission, to make things possible and to support and encourage entrepreneurs to implement even such crazy ideas*", DMO3: "*If someone comes to us with her or his ideas, then we are prepared to give some advice in some form or other*". The latter (g) is added to the model (Fig. 2).

For the DMO-owned projects, the DMOs identify the innovation need/opportunity (a) themselves (e.g., DMO3 a yearly revise of the event portfolio) or get ideas from the ecosystem (DMO1: Ideas from members' meeting and board of directors, DMO2 referring to tourists and citizens: "*there are many ideas, which are more or less spontaneously passed on to us*", DMO3: "*We get much input from outside. ... offers from people we know or do not know, that come up throughout the year, but also guests*"). No interviewee mentioned a specific process for a structured idea generation (b) for the DMO's projects. The idea generation is instead based on ad hoc project opportunities (DMO2: "*that is often at such short notice. That is when we see an opportunity because we might have gotten a call or read in the newspaper that funds are available for it*") or by direction from the municipal or governmental body (DMO1: "*there are many things where we are instructed by the state government*", DMO3 gets tasks for realizing infrastructure projects). Since the superior authority of the case DMOs seems to have significant importance on the innovation projects approached by the DMOs, we extract this actor from the ecosystem as a distinct type of actor in the model (Fig. 2). For the selection of solutions (c), DMOs check the compliance with DMO statutes and targets (DMO1: "*it has to have a touristic purpose and suit our statutes*"; in DMO3, the decision process depends on financial limits regulated in the statutes, DMO3: "*With such* ad hoc *activities, we check if that suits our destination, our target groups*", "*we decide this in our internal circle*" (their management team)). The DMOs are involved in developing and implementing innovation activities (d). The interviewees did not mention the marketing of solutions/dissemination (e) and the evaluation of solutions (f) regarding the DMO-owned projects. This does not reflect the marketing and development focus of DMOs (Presenza et al., 2005) and the requirement of reporting to their funding providers. We assume that this finding does not mean the absence of these stages, but that it demonstrates the lack of awareness of these stages of potential participation in the innovation process.

The DMOs are all engaged in inter-organizational innovation projects (Table 1). All three interviewees named the example of visitor cards for inter-organizational innovation activities. These digital cards are provided to tourists to access free or reduced-price public transport and special offers from TSPs. They also collect touristic taxes and data on tourists' behavior for the DMOs. The identification of the innovation need/opportunity (a) and idea generation (b) results from continuous relations in networks (DMO1: "*in conversations, in workshops... where things come up, where you then say, we will make a project out of it together*", DMO3 about a regional network with neighboring DMOs: "*to kick off joint projects*") or ideas from actors in the touristic ecosystem (for DMO1 by a port company). The importance of engagement in networks for innovation development confirms Volgger's (2017) view of network and innovation development as intertwined. The case DMOs participate in the selection of solutions (c) and their implementation (d) (e.g., DMO1: "*then you look at how you can get it financed... and then also engage in the implementation*", "*We also contribute technically*"). Also, in IOPs, DMOs engage in moderation, consultation, and motivation (g) (DMO2: "*So we try to make clear to the partners that it is only good to join such an idea, and we try to emphasize the advantages of the individual actor*"). Further, they participate in enabling (h) (DMO1: "*We facilitate the process*", DMO2: "*We have a pioneering role, we demonstrate how things can work*").

The statutes and the destination strategy guide the DMOs' portfolio management of innovation activities. DMOs create a framework for their work by creating several-year strategies for the destination's touristic development. This strategy is created in a highly participative manner, involving a broad range of actors (DMO2: "*Any concept that we make, which also has a certain half-life and needs the necessary acceptance, we always make with the participation of the administration, the politicians, the industry*", DMO3: "*That was done on quite a broad basis, we involved many actors in the destination*" (examples: hotels, restaurants, sports associations, cleaning companies)). While actors apart from the DMO participate in developing the tourism strategy, the participation of other actors in the DMO's portfolio selection decisions is regarded as limited. DMO2 states: "*We can still get better at that*" *and "There is not that kind of participatory process, which I would like to have.*" A reason mentioned is the perception of roles; DMO2 explained, the "*others are much focusing on themselves, they see the DMO responsible for innovation.*"

The touristic ecosystem involved in innovation activities apart from DMOs and TSPs is broad. Table 3 displays the actors mentioned by the case DMOs. This broad participation of actors confirms that OI with "purposively managed knowledge flows across organizational boundaries" (Chesbrough & Bogers, 2014, p. 17) is the dominant form also from a DMO perspective, in line with research on tour operators (Marx & Klotz, 2021). Innovation projects of the case DMOs often have an ad hoc character emerging from established relations with TSPs and the ecosystem. DMO3 reported selecting specific partners with whom there was a successful collaboration in the past. With the range of actors apparent, we can confirm the proposition of Marx (2022, p. 20) that "in a tourism innovation project network, both organizations and individuals participate in innovation activities in flexible compositions framed by business and private relationships."

The case DMOs associate benefits both with the participatory innovation activity and its output, in line with findings of Marx and Klotz (2021) and underlining the separation of innovation as an activity and as an output (OECD/Eurostat, 2018). The benefits from the innovation output depend on the nature of the project. The

Table 3 Ecosystem actors mentioned by the case DMOs in addition to TSPs

Actors	DMO1	DMO2	DMO3
Other DMOs	Regional Organizational Structures, members of the DMO	Neighboring DMOs	Regional and tourism associations,
Politicians	State Government, Minister-President	Politicians	Municipal committees (tourism committee, finance committee)
Public Administration/ Ministries	Ministry of Economics, Ministry of Transport, State Secretary	Administration, municipal partners	Municipality
Network organizations			Spa association, infrastructure association with neighboring municipalities
Tourists		Tourists	Tourists
Citizens		Citizens	
Other actors	Non-profit organizations, media agencies	Port Company	Sports club, cleaning company, architects, designers, artists, media agencies, photographer

case DMOs named especially the enhancement of the quality delivered to the guests. That increases the attractivity of the destination based on improved availability of data, marketing content and coverage, additional tailored touristic services, leading to income for the DMO (DMO2: "*We earn money directly through the tourism tax, which is the largest source of income [...]*"). Benefits associated with participatory innovation activities by the case DMOs are:

- Increase reputation (DMO3 "*one becomes more visible in the committees and in the ministries and with the federal state government*"),
- Increase creativity (DMO2 "*it [...] promotes diversity and individuality*"),
- Increase acceptance of solutions (DMO2: "*it enhances acceptance*"),
- Expand the network (DMO3: "*otherwise, we would probably never have come across this [person]*"),
- Leverage expertise (DMO2: "*He would not have been able to manage it, and I think he would not have done it without our support*").

Furthermore, regional structures and inter-municipal cooperation are strengthened due to participatory innovation activities. Smaller partners mainly profit from larger ones that give them guidance, support, and encouragement.

Benefits from participation are identified for developing the destination strategy by the case DMOs (e.g., DMO2 names increased acceptance, diversity of opinions, increased feeling of shared responsibility and summarizes: "*It would be crazy to come up with something by your own in privacy [...] Because if they are not involved, they bear no responsibility for it*"). However, these benefits seem not to be transferred to the same extent to participation in innovation projects or the portfolio level for implementing the destination's strategy.

The research finds proof that DMOs are engaged in (several) networks and relations from which single innovation projects emerge, confirming the proposition of Marx (2022) to regard the tourism innovation system as a project network. Tourism innovation activities are coordinated on three levels: individual projects by TSPs, DMO-owned projects, and IOPs. Participation in innovation activities in this project network can be designed in various stages, from identifying the innovation need to evaluating the solution or a moderating or enabling role along this process. Based on the findings from the literature on participative decision-making and innovation processes (Black & Gregersen, 1997; Gausemeier et al., 2019; Hucker, 2008) and the interview data, we suggest considering the participation of actors in tourism innovation activities in eight stages and three types of projects (TSP-owned, DMO-owned, IOPs) as displayed in the revised model in Fig. 2:

(a) identification of innovation need/opportunity
(b) idea generation
(c) selection of solution
(d) development and implementation of solution
(e) marketing of solution/dissemination
(f) evaluation of solution

(g) moderation, consultation, motivation
(h) enablement.

While various actors participate in the strategy development of the case DMOs, participation is not evident at the project portfolio level. Projects are used for operationalizing strategy (IPMA, 2016; Jamieson & Morris, 2008). Black and Gregersen (1997) suggest that participation in decision-making has positive effects and gets stronger, when possible at all stages. Therefore, we suggest considering participation opportunities at the portfolio level, in line with Marx (2019).

Overall, the engagement of DMOs in innovation activities is led by the municipal duties in the statutes of the DMO, superior authorities, and the destination's strategy. This strategy is developed in a highly participative manner, presented as the tourism concept for several years, and building the guideline for the DMO to move within. However, on a daily implementation basis, we found neither a formal innovation process nor systematic participation of various actors, but rather ad hoc emergence and evaluation of ideas. A distinct separation of identification of the innovation needs and the idea generation phase was not found in the interviews. Generic models of innovation processes start with idea generation (e.g., Gausemeier et al., 2019; Hansen & Birkinshaw, 2007). However, decision-making processes suggest differentiating identification of the need and the ideation (Black & Gregersen, 1997), which could help in the selection stage. Hinders for participation in the portfolio of innovation projects in a destination seem to be regulations or contradicting perspectives of actors. The case DMOs may not support single profit-oriented entities, their statutes determine specific public tasks, and they mention the conflict between the DMO's focus on the destination versus the TSPs' focus on their individual benefits. The latter is described as a tension typical in project networks (DeFillippi & Sydow, 2016), caused by the identity paradox that requires focusing on building trust and relationships instead of roles. Purposefully allowing for participation in the various innovation activities at project or portfolio level to implement the destination's strategy could be a way to foster the building of trust and relations.

5 Conclusion

The research advances the understanding of design opportunities for participative innovation activities in tourism and develops eight stages to describe diverse actors' engagement. The suggested model displays various opportunities for participation in innovation activities of TSPs, DMOs, and the touristic ecosystem on eight different stages and in three types of projects: TSP-owned, DMO-owned, and IOPs. The model is amended with the portfolio and strategy perspective. Practitioners can use the model to assess their innovation activities on available and potential participation opportunities. The research demonstrates such assessment from the perspective of three DMOs. Though the qualitative nature of the research and the focus on public-funded DMOs are limitations, DMO managers and other tourism actors are provided with a framework to compare with for designing innovation activities in their destinations. Awareness of the suggested eight stages of participation may help practitioners design the involvement of various actors more purposefully. Future research should validate the framework quantitatively, especially towards incorporating the perspectives of single TSPs and other actors from the touristic ecosystem, e.g., political and public actors. Another strand of research could evaluate which stages of participation in innovation activities affects satisfaction, building of trust, and relationships.

References

Beritelli, P., Buffa, F., & Martini, U. (2016). Logics and interlocking directorships in a multi-stakeholder system. *Journal of Destination Marketing and Management, 5*(2), 107–116. https://doi.org/10.1016/j.jdmm.2015.11.005

Black, J. S., & Gregersen, H. B. (1997). Participative decision-making: An integration of multiple dimensions. *Human Relations, 50*(7), 859–878.

Brooks, J., McCluskey, S., Turley, E., & King, N. (2015). The utility of template analysis in qualitative psychology research. *Qualitative Research in Psychology, 12*(2), 202–222. https://doi.org/10.1080/14780887.2014.955224

Centre for European Economic Research (ZEW). (2018). *Results of CIS 2016 for Germany.*

Chesbrough, H. (2003). *Open innovation: The new imperative for creating and profiting from technology*. Harvard Business School Publishing.

Chesbrough, H., & Bogers, M. (2014). Explicating open innovation: Clarifying an emerging paradigm for understanding innovation. In H. Chesbrough, W. Vanhaverbeke, & J. West (Eds.), *New frontiers in open innovation* (pp. 3–28). Oxford University Press.

DeFillippi, R., & Sydow, J. (2016). Project networks: Governance choices and paradoxical tensions. *Project Management Journal, 47*(5), 6–17.

Elsahn, Z., Callagher, L., Husted, K., Korber, S., & Siedlok, F. (2020). Are rigor and transparency enough? Review and future directions for case studies in technology and innovation management. *R and D Management, 50*(3), 309–328. https://doi.org/10.1111/radm.12412

Gausemeier, J., Dumitrescu, R., Echterfeld, J., Pfänder, T., Steffen, D., & Thielemann, F. (2019). *Innovationen für die Märkte von morgen [Innovations for markets of tomorrow]*. Carl Hanser Verlag.

Goffin, K., Åhlström, P., Bianchi, M., & Richtnér, A. (2019). Perspective: State-of-the-art: The quality of case study research in innovation management. *Journal of Product Innovation Management, 36*(5), 586–615. https://doi.org/10.1111/jpim.12492

Guertler, M. R., & Sick, N. (2021). Exploring the enabling effects of project management for SMEs in adopting open innovation—A framework for partner search and selection in open innovation projects. *International Journal of Project Management, 39*(2), 102–144. https://doi.org/10.1016/j.ijproman.2020.06.007

Gustafsson, E., Larson, M., & Svensson, B. (2014). Governance in multi-project networks: Lessons from a failed destination branding effort. *European Planning Studies, 22*(8), 1569–1586. https://doi.org/10.1080/09654313.2013.784588

Hansen, M. T., & Birkinshaw, J. (2007). The innovation value chain. *Harvard Business Review, 85*(6), 121–131.

Hauschildt, J., Salomo, S., Schultz, C., & Kock, A. (2016). *Innovationsmanagement [Innovation management]* (6th ed.). Franz Vahlen München.

Hjalager, A. M. (2010). A review of innovation research in tourism. *Tourism Management, 31*(1), 1–12. https://doi.org/10.1016/j.tourman.2009.08.012

Hucker, T. (2008). *Betriebliche Partizipation und gesellschaftlicher Wandel [Participation in the enterprise and societal change]*. Rainer Hampp Verlag.

IPMA. (2016). *Organisational competence baseline for developing competence in managing by projects*. International Project Management Association (IPMA®).

Jamieson, A.; Morris, P. (2008). Implementing strategy through programmes or projects. In R. Turner (Ed.), *Handbook of project management* (4th ed., pp. 27–45). Gower.

Johannesson, P., & Perjons, E. (2014). *An introduction to design science*. Springer.

Jones, C., & Lichtenstein, B. B. (2008). Temporary inter-organizational projects: How temporal and social embeddedness enhance coordination and manage Uncertainty. In S. Cropper, C. Huxham, M. Ebers, & P. Smith Ring (Eds.), *The Oxford handbook of inter-organizational relations* (pp. 231–255). Oxford University Press. https://doi.org/10.1093/oxfordhb/9780199282944.003.0009.

King, N. (2004). Using templates in the thematic analysis of text. In C. Cassell & G. Symon (Eds.), *Qualitative methods in organizational research* (pp. 256–270). Sage.

Kuckartz, U., & Rädiker, S. (2020). Fokussierte Interviewanalyse mit MAXQDA [Focussed interview analysis with MAXQDA]. *Springer Fachmedien Wiesbaden*. https://doi.org/10.1007/978-3-658-31468-2

Mahr, D., Lievens, A., & Blazevic, V. (2014). The value of customer cocreated knowledge during the innovation process. *Journal of Product Innovation Management, 31*(3), 599–615. https://doi.org/10.1111/jpim.12116

Marx, S. (2019). Organizational challenges of digitalization initiatives in tourism network management organizations. In W. Abramowicz & R. Corchuelo (Eds.), *Business Information Systems Workshops. BIS 2019. Lecture Notes in Business Information Processing* (Vol. 373, pp. 157–168). Springer. https://doi.org/10.1007/978-3-030-36691-9_14.

Marx, S. (2022). *Contributions to participative innovation in tourism network organizations*, Unp. diss., University of Rostock.

Marx, S., & Klotz, M. (2021). Entrepreneurship during crisis: Innovation practices of micro and small tour operators. *The International Journal of Entrepreneurship and Innovation, 2021*(November). https://doi.org/10.1177/14657503211061025.

Nambisan, S., & Sawhney, M. (2011). Orchestration processes in network-centric innovation: Evidence from the field. *Academy of Management Perspectives, 25*(3), 40–57.

Newig, J. (2011). Partizipation und neue Formen der Governance [Participation and new forms of governance]. In M. Groß (Ed.), *Handbuch Umweltsoziologie [Handbook of Environmental Sociology]*. Verlag für Sozialwissenschaften. https://doi.org/10.1007/978-3-531-93097-8.

Nordli, A. J. (2017). Measuring innovation in tourism with community innovation survey: A first step towards a more valid innovation instruments. *Scandinavian Journal of Hospitality and Tourism, 17*(4), 423–440. https://doi.org/10.1080/15022250.2016.1247382

OECD/Eurostat. (2018). *Oslo Manual 2018: Guidelines for collecting, reporting and using data on innovation* (4th ed.). OECD Publishing and Eurostat. https://doi.org/10.1787/9789264304604-en.

Ommen, N. O., Blut, M., Backhaus, C., & Woisetschläger, D. M. (2016). Toward a better understanding of stakeholder participation in the service innovation process: More than one path to success. *Journal of Business Research, 69*(7), 2409–2416. https://doi.org/10.1016/j.jbusres.2016.01.010

Peffers, K., Tuunanen, T., Rothenberger, M. A., & Chatterjee, A. S. (2008). A design science research methodology for information systems research. *Journal of Management Information Systems, 24*(3), 45–77.

Presenza, A., Sheehan, L., & Ritchie, J. B. (2005). Towards a model of the roles and activities of destination management organizations. *Journal of Hospitality, Tourism and Leisure Science, 3*(1), 1–16. https://doi.org/10.1017/CBO9781107415324.004

Project Management Institute. (2021). *A guide to the project management body of knowledge—PMBOK(R) Guide* (7th ed.). Project Management Institute.

Provan, K. G., & Kenis, P. (2007). Modes of network governance: Structure, management, and effectiveness. *Journal of Public Administration Research and Theory, 18*(2), 229–252. https://doi.org/10.1093/jopart/mum015

Saunders, M., Lewis, P., & Thornhill, A. (2009). *Research methods for business students* (9th ed.). Pearson Education.

Smith, K. (2005). Measuring innovation. In J. Fagerberg, D. C. Mowery, & R. R. Nelson (Eds.), *The Oxford handbook of innovation* (pp. 148–177). Oxford University Press.

Sydow, J. (2022). Studying the management of project networks: From structures to practices? *Project Management Journal, 53*(1), 3–7. https://doi.org/10.1177/87569728211061814

Sydow, J., & Müller-Seitz, G. (2020). Open innovation at the interorganizational network level—Stretching practices to face technological discontinuities in the semiconductor industry. *Technological Forecasting and Social Change, 155*(C), 119398. https://doi.org/10.1016/j.techfore.2018.07.036.

Sydow, J., & Windeler, A. (2020). Temporary organizing and permanent contexts. *Current Sociology, 68*(4), 480–498. https://doi.org/10.1177/0011392120907629

World Tourism Organization (UNWTO). (n.d.). *Conceptual framework*. Retrieved April 30, 2019, from http://www2.unwto.org/content/conceptual-framework-0.

Volgger, M. (2017). *Umsetzungskompetenz als Erfolgsfaktor in Tourismusdestinationen [Implementation competence as a success factor in tourism destinations]*. Springer Gabler.

Volgger, M., & Pechlaner, H. (2014). Requirements for destination management organizations in destination governance: Understanding DMO success. *Tourism Management, 41*(April 2014), 64–75. https://doi.org/10.1016/j.tourman.2013.09.001.

West, J., & Bogers, M. (2014). Leveraging external sources of innovation: A review of research on open innovation. *Journal of Product Innovation Management, 31*(4), 814–831. https://doi.org/10.1111/jpim.12125

West, J., & Bogers, M. (2017). Open innovation: current status and research opportunities. *Innovation: Management, Policy and Practice, 19*(1), 43–50. https://doi.org/10.1080/14479338.2016.1258995.

World Tourism Organization (UNWTO). (2016). *Report of the committee on tourism and competitiveness—Annex I.* (Issue CE/103/5 rev.2).

Yin, R. K. (1994). *Case study research—Design and methods* (2nd ed.). Sage.

Zenker, S., & Kock, F. (2020). The coronavirus pandemic—A critical discussion of a tourism research agenda. *Tourism Management, 81*(May), 104164. https://doi.org/10.1016/j.tourman.2020.104164

Zhang, X., Song, H., & Huang, G. Q. (2009). Tourism supply chain management: A new research agenda. *Tourism Management, 30*(3), 345–358. https://doi.org/10.1016/j.tourman.2008.12.010

A Conceptual Framework for Applying Social Signal Processing to Neuro-Tourism

Evgenia Gkintoni, Garefalea Kakoleres, Giorgos Telonis, Constantinos Halkiopoulos, and Basilis Boutsinas

Abstract Social Signal Processing is a new research field aiming at providing computers with the ability to understand human social signals, i.e., (dis)agreement, amusement, etc. Social signals are manifested through an array of nonverbal behavioral cues, such as gestures, postures. Neuro-tourism is an idea that combines the fields of Neuroscience and Tourism. Neuroscience, marketing, and computing methods like as Electroencephalography (EEG), Analysis of Physiological Responses, Virtual and Augmented Reality, and Eye-tracking, among others, enable us to monitor tourists' emotions, decision-making processes, and behavior. We claim that neuro-tourism in combination with Social Signal Processing Techniques, more efficiently explores the tourists' behaviors/preferences when they are browsing webpages or even when they are traveling. In this paper, we propose a Conceptual Framework for applying Social Signal Processing to Neuro-Tourism. The proposed Conceptual Framework takes into consideration various application areas within Tourism field and constitutes a roadmap of analyzing/exploiting social signals produced during social interactions or human–computer interactions.

Keywords Neuro-tourism · Social signal processing · Neuropsychology · Neuroscience · Tourism industry

JEL Classification D91 · Z32 · O32 · I12 · I31 · C80

1 Introduction

1.1 Neuroscience in Tourism Management–An Overview

Neuroscience is a discipline of the twentieth century that explores the nervous system. Cognitive neuroscience explores, among other issues, the brain bases of cognition, attention, memory, emotion, language, and movement. Neuroscience and cognitive

E. Gkintoni (✉) · G. Kakoleres · G. Telonis · C. Halkiopoulos · B. Boutsinas
Department of Business Administration, University of Patras, Patras, Greece
e-mail: evigintoni@upatras.gr

© The Author(s), under exclusive license to Springer Nature Switzerland AG 2023

V. Katsoni (ed.), *Tourism, Travel, and Hospitality in a Smart and Sustainable World*, Springer Proceedings in Business and Economics,
https://doi.org/10.1007/978-3-031-26829-8_20

science influenced a propensity to combine social sciences with neuroscience. The disciplines of neuroeconomics, neuromarketing, neurodecision making, neuroeducation, and neuromanagement are interconnected (Gkintoni et al., 2022; Halkiopoulos et al., 2022; Katsoni & Dologlou, 2017). Tourism management is a management discipline. Travel is a popular leisure activity that significantly contributes to economic expansion. Thus, it is essential to advance tourism management research (Panas et al., 2017; Panteli et al., 2021; Sarigiannidis et al., 2021). Neuroscience theories and practices may transform tourism. Using cognitive neuroscience technologies like EEG, EMG, PET, fMRI, and NIRS. Prof. Ma established Neuromanagement in 2006 to research brain activity when people face managerial and economic challenges. Neuromanagement contributes to management study by illuminating the process behind human social behavior. Physiological and brain data collected directly from the human body are more objective and reliable indicators of mood, attitudes, and cognitive performance. Feelings, attitudes, and cognitive processes are more objective than surveys and interviews (Halkiopoulos et al., 2021a, b, c). Traditional measurement techniques "may expose individuals to the danger of concealing their true thinking by manipulating their responses. Traditional self-report evaluations are insufficiently objective to be reliable. Self-reporting instruments, such as interviews, focus groups, and surveys might occasionally give false findings. Self-reported measures presume a rational, aware consumer. According to research, humans do not know why they have specific behaviors or thoughts. Their mental processes, such as observation, evaluation, and motivation, may never attain the level of conscious awareness. Utilizing multilayered evaluation to explain and anticipate behavior more precisely is the primary advantage of neuromanagement. This expands the use of neuroscience in management science concerns (Ma et al., 2014).

Neuroscience provides techniques for analyzing brain activity to understand neural system-behavior relationships (Lei et al., 2022; Zurawicki, 2010). Neuroscience offers marketers an opportunity to 'unlock the black box' of customers (Dooley, 2011), find their 'buying buttons', and learn how consumers' 'buyology' works (Lindstrom, 2008). Neuromarketing study shows a close link between the two. Neuro-tourism examines the brain mechanics behind visitor behavior, according to Ma et al. (2014). Neuro-tourism extends beyond tourism management using neuroscience using augmented reality and artificial intelligence. In other words, neurotourism blends tourism with neuroscience (Giudici et al., 2017). Multiple studies have confirmed that the tourism business uses neuroscientific methods (Neuro-sell, 2014). Sop (2021) noted that tourism researchers began using neuroscience methods in the 2010s. Most relevant tourist research has used EEG and eye-tracking but not fMRI (Sop, 2021). Only a few conceptual studies explore the necessity for a neuroscientific approach in tourist marketing and management.

1.2 *Behavioral Tracking Using Social Signal Processing–Basic Concepts*

Social Signal Processing is an emerging domain that aims at understanding and modeling social interactions though computerized analysis of nonverbal behavior (Pentland, 2004, 2007, 2010; Vinciarelli et al., 2009). Social signals include (dis)agreement, amusement, boredom, aggressiveness, empathy, hostility, and any other attitude toward others. Social signals are manifested through an array of nonverbal behavioral cues, such as gestures, interpersonal distance, body postures like look exchange, eyelashes, smiles, nods, crossed arms and so on.

Social Signal Processing encompasses several modern technologies using Artificial Intelligence techniques that can automatically detect and analyze humans' behavioral cues. Social Signal Processing is enabled using various technologies, e.g., by video capture, images, or capture of speech. Furthermore, specialized sensor devices can be used, such as the sociometric badge, for measuring movement, space, environment, and vocal behavior. Sensor devices can be used, e.g., sociometric signal. A further relatively no-interfering portable device is the sensor OpenBeacon provided by the joint vendor SocioPatterns (Cattuto et al., 2010). By using it, the movement the space and the surrounding can be imprinted providing a simple but effective method for their observation and analyses of natural social signals (Atzmueller, 2014).

Many decisions are largely based on or at least influenced by human-to-human interaction. Therefore, understanding the impact of social signals being sent during human interactions is very important. For instance, hand gestures and body postures have been shown to influence the perception of the employer, since it might signal confidence and dominance. In this paper, we claim that decisions are also influenced by human-to-computer interaction. Therefore, the analysis of social signals being sent during human–computer interactions, can be used to predict, or influence human's decisions/selections. The objective of such an analysis, based on SSP, is to be able to know what tourists feel during browsing to web pages and information systems promoting tourism information. This is very important since tourists usually interact with such web pages and information systems during all the five phases of tourism, i.e., Dreaming, Planning, Reservation, Experience, and Loyalty.

1.3 *The Key Idea*

Neuro-tourism has been recently introduced as the application of a combination of information technology and neuroscience techniques (Sop, 2021). There are different proposed interpretations of neuro-tourism concept. Firstly, neuro-tourism is proposed as a methodology that explores the neural mechanism underlying tourists' behaviors and emotions to measure tourist satisfaction and revisiting intention. To this end, neuroscience tools are used to measure happiness, satisfaction, reasons for revisiting, etc., based on neurophysiology data. Going beyond the latter interpretation, neuro-tourism is proposed to be a new form of tourism understood as implemented neural experience. Such interpretation is based on the claim that the convergence of neuroscience and e-tourism will potentially lead to the creation of the tourist experience through the implementation of sensations and emotions directly to the neural system (Panyik & Gonçalves, 2017).

Neuro-tourism has been applied to different problems (see Sop, 2021 for various neuroscientific experiments), i.e., to Tourism Management (Ma et al., 2014), to detect hotel preferences of the consumers (Tosun et al., 2016), to food retailing (Horská & Berčík, 2017), etc.

In this paper, we propose a Conceptual Framework for applying Social Signal Processing to neuro-tourism. The proposed Conceptual Framework takes into consideration various application areas within tourism field and constitutes a roadmap of analyzing/exploiting social signals produced during social interactions or human–computer interactions.

2 Literature Review

2.1 Neuro-Tourism—An Overview

Neuro-tourism combines neuroscience, psychology, sociology, and tourism to examine individual decisions. Neuro-tourism combines neuroscience with travel. Neuroscience research is being used in tourism alongside the social sciences (Gkintoni et al., 2021). Neuroscientific tools analyze happiness, satisfaction, and motives to collect neurophysiology data (Gkintoni & Dimakos, 2022). Using neuroscience approaches, cognitive and emotional processes may be captured during a tourism experience, allowing managers to monitor and plan more efficiently. Therefore, neurophysiology data should be included in management information systems. Psychological factors determine tourism. Psychological processes include people's reasons for traveling and choice of alternatives (Panteli et al., 2021; Panas et al., 2020). Psychology also relates to changes in people's thoughts and attitudes when traveling, their consequences on conduct, reactions to the frustrations and hardships of travel and holiday living, and the effects of tourism on the mental state and behavior of visitors and locals (Panas et al., 2022). Tourists buy products during the psychological shift between leaving and returning home.

2.2 Neuroscientific Methods for Tourism Behavior

Neuro-tourism (and neuromarketing) research instruments and methods are classified into two broad categories: tools that assess the body's responses to tourism and tools that measure the brain's responses to tourism. As estimation technologies, each methodology has advantages and disadvantages for capturing various signal types. The most important neuromarketing (and neuro-tourism) measurements based on bodily reactions are facial expressions. The human face stores a variety of emotional states that may be deciphered on two levels: apparent changes in expressions (such as a grin or a frown) and invisible micro-changes (e.g., small muscle contractions). These measurements have been found to be reliable markers of good or negative emotional responses. Eye-tracking technology. This gadget suggests the use of infrared beams to detect eye movement and pupil dilation while viewing an object or a person. Neuro-tourism offers several uses for eye-tracking. They are both a standalone instrument and a complement to other measures. The rate and direction of eye movement variations are useful indicators of attention, interest, and curiosity. They display the direction a person's eyes go when seeing a magazine advertising as well as the amount of time spent analyzing a certain feature (Alvino et al., 2020).

Electro-dermal activity. Electrodes placed to the fingertips are used to measure sweat and can offer a sensitive indicator of subconscious reactions. This instrument's signals grow in response to emotional stimulation of the neurological system. As individuals recall a specific memory, a significant increase in emotional arousal is observed. This instrument's limitation is its inability to distinguish between good and negative emotional consequences. Heart rate and respiration. This approach is based on detecting the heart's beating rate and breathing rate using chest-mounted sensors or a finger pulse meter. It has been shown that when attention rises, heart rate decreases. Rapid and deep respiration is associated with energy and enthusiasm, whereas shallow respiration might indicate concentration, tense anticipation, panic, or terror. Response time can be a helpful indicator of bodily arousal. It is vital to measure the pace of the brain's reaction to word correlations or visual judgments to comprehend how unconscious brain processes manifest in behavior. Reaction time metrics provide a straightforward and accessible method for evaluating the quality of ideas. Neurological methods based on brain reactions are more sophisticated and precise than bodily measurements, and these technologies need additional work and expense. These are the three most important neurological methods.

Functional magnetic resonance imaging (fMRI). This approach is used by researchers because it enables precise imaging of activity anywhere in the brain to assess blood flow. It monitors the rise in oxygen levels in the brain's blood flow and signals when the activity of particular brain regions increases. When portions of the brain become active, blood flows to those regions. Due to the presence of iron in blood, this flow may be tracked by a magnet that encircles the head of the individual being examined. They are essential for testing brain functioning models and theories. However, some neuromarketers believe that the fMRI's utility as a market research testing instrument is exaggerated due to its expense, complexity, and artificial testing setting. *Electroencephalographic recording (EEG).* This brain-scanning device utilizes sensors to detect the minute electrical impulses produced by brain activity. EEG measures the quality of a few electrical fields produced by brain activity at the scalp. It is the most popular approach since it requires minimum effort and flexible equipment. EEG is limited in that it cannot reliably measure major changes in the electrical activity of the brain.

Magnetoencephalography (MEG). This technique measures minute variations in magnetic fields generated by the brain. It is a multimillion-dollar piece of equipment that requires temperatures close to zero to function. It has been utilized in academic research, but not yet in a neuromarketing (or neuro-tourism) study.

Near-Infrared Spectroscopy (NIRS) and *Positron Emission Tomography (PET)* are the most widely used techniques in neuroscientific research nowadays.

Event-related potentials (ERPs). The EEG can also be used to obtain records pertaining to a particular occurrence following the introduction of a stimulus.

Transcranial magnetic stimulation (TMS). Modifying the activity of specific brain areas of interest by using the electric coil housed in a plastic case that is held on the subject's head are the most applicable neuroscience methods that record the brain activity millisecond-by-millisecond and produce much more accurate temporal data regarding when something occurs in the brain.

In addition to these methodologies (or tools), Bercea Olteanu (2015) presents *steady-state topography* (SST), a tool for viewing and quantifying the fast changes in human brain activity in response to a stimulus. This instrument measures the extra brain activity over the continuous exposure to the principal (stationary) oscillating visual stimulus; it provides a signal less corrupted by random interferences such as head movement and eye blinks (Zurawicki, 2010).

Several neuroscience techniques are applicable to several fields, including marketing, management, economics, and tourism. However, Sop (2021) revealed that EEG and eye-tracking were the most utilized research techniques in tourism. The review article by Scott et al. (2019) further demonstrates the predominance of the eye-tracking approach in tourist research. In addition, Li et al. (2021) suggests using EEG to explore how the cognitive process of emotion influences tourist behavior.

2.3 Social Signal Processing in Neuro-Tourism

Social Signal Processing is an emerging domain that aims at understanding and modeling social interactions through computerized analysis of nonverbal behavior. Individuals' decision-making is strongly affected by nonverbal behavior. Social signals include (dis)agreement, aggressiveness, empathy, hostility, and any other attitude toward others. Social signals are manifested through an array of nonverbal behavioral cues, such as gestures, interpersonal distance, body postures like look exchange, eyelashes, smiles, nods, crossed arms, and so on. Social Signal Processing encompasses several modern technologies using Artificial Intelligence techniques that can automatically detect and analyze humans' behavioral cues. Thus, Social Signal Processing aims at providing computers with the ability to sense and understand human social signals, using various technologies, e.g., by video capture, images, or capture of speech. Neuro-tourism has attracted considerable interest in recent years due to marketing, neuroscience, and neuropsychology. These techniques and instruments of these disciplines are utilized in the tourism industry. The purpose is to determine how tourists feel during their experience, various activities, and when interacting with specific local offers. Specifically, the neuro-tourism Information Framework encompasses all of neuroscience and facilitates integration of existing knowledge and databases of many types. These examples illustrate the opportunities and challenges of data mining across multiple tiers of neuroscience information and underscore the need for cultural and infrastructure changes if neuroinformatics is to fulfill its potential to advance our understanding of the brain. The neuroscientific results allow us to know the cognitive and emotional responses in real-time, giving us insight into what occurs in the mind and body of a person when they encounter tourism products and services.

In this paper we propose the application of Social Signal Processing, as an alternative method to achieve the main goal of neuro-tourism, i.e., to determine tourists' preferences and how tourists feel during their experience and when interacting with package holiday advertisements. To our knowledge, this is the first work investigating the application of Social Signal Processing to neuro-tourism.

3 The Proposed Conceptual Framework

In this paper, we propose a Conceptual Framework for applying Social Signal Processing to neuro-tourism (see Fig. 1). The key idea of the proposed framework is to investigate the social signals being sent by individuals after browsing webpages about a destination.

Social Signals Processing usually analyze social signals produced during social interactions. In the proposed Conceptual Framework, social signals produced during human–computer interactions are considered.

The proposed framework is based on extracting patterns of social signals related to specific responses of individuals. These patterns are frequent subsequences of social signals. The patterns can be extracted by applying Machine Learning algorithms, such as sequential pattern mining, string mining, association rule mining and process mining (see Kakoleres & Boutsinas, 2022 for a presentation of the application of such algorithms to sequences of social signals).

The input to these algorithms is social signal sequences, each of them related to a specific person. The extracted patterns are first checked to detect the corresponding responses they arouse, to various different content types (landscapes, hotels, offers, etc.). Then, the extracted patterns are stored in a Knowledge Base, along with the corresponding responses they arouse.

Then, the patterns of social signals stored in the Knowledge Base can be used to detect user preferences by comparing them to the sequence of social signals produced by an individual during browsing webpages about a destination. The responses assigned to the most similar pattern, determine user preferences straightforwardly. A lot of similarity measures proposed in the literature can be used to find the most similar pattern (Boutsinas & Papastergiou, 2008).

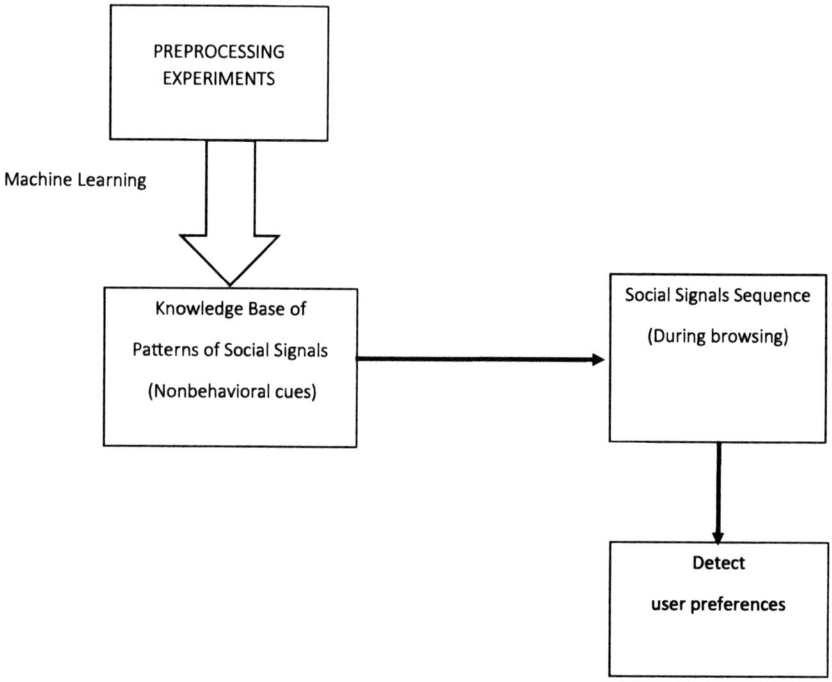

Fig. 1. The proposed conceptual framework

After automatically detecting user preferences, the content presented in the webpage can be adjusted accordingly, another related website could be recommended or any other marketing/communication technique could be adopted. In the following, we present an example use case scenario based on the proposed Conceptual Framework:

Step 1: Conduct a preprocessing experiment with several participants. The experiment is performed in a laboratory where a computer monitor, and a camera are located. The participants were previously asked to describe their preferences over ordinary versus beautiful landscapes. Several websites of local travel agencies and tourism bureaus, whose content concern package holiday advertisements, are chosen suitably, so that to include either ordinary or beautiful landscapes according to participant's preferences.

Step 1.1: The chosen websites are presented randomly to the participants through the computer monitor and they are asked to browse each website.

Step 1.2: The participants' gestures and postures when browsing each website are monitored by the camera and the data are recorded and processed automatically by SSP computer software. It is obvious that gestures, as facial expressions, are responsible for nonverbal communication. Consider how much information can be conveyed with a smile or a frown. While nonverbal communication and behavior can vary dramatically between cultures, facial expressions for many social signals, e.g., happiness, sadness, anger, and fear are similar throughout the world. shown that when it comes to sentimental situation, body is way more important that face as to communicating social signals.

Step 1.3: The automatic interpretation of participants' gestures and postures is realized reliably enough with the help of specific algorithms. State-of-the-art algorithms can interpret gestures and postures straight from a video projection. Open Pose (Dense Pose are two recent such systems. The interpreted gestures and postures are stored as a sequence of social signals, one sequence for each browsing period, i.e., website.

Step 1.4: Participants are asked to make a 'Ordinary' or 'Beautiful' choice, based on their preferences. Each sequence of social signals is stored along with the corresponding classification as 'Ordinary' or 'Beautiful'.

Step 2: A machine learning algorithm (e.g., PrefixSpan) is applied to extract frequent subsequences both for 'Ordinary' and for 'Beautiful' landscapes. The extracted frequent subsequences are stored as patterns in the Knowledge Base, along with the corresponding class, i.e., 'Ordinary' or 'Beautiful'.

Step 3: While a user is browsing a website, her gestures and postures are monitored by a camera are processed automatically by SSP computer software. The extracted sequence of social signals is continuously compared to the stored patterns. The most similar(s) pattern(s) is(are) detected.

Step 4: The most similar pattern determines if the website presents 'Ordinary' or 'Beautiful' landscapes. This information can be exploited accordingly by applying certain marketing/communication techniques, for instance, an automatic recommendation of similar websites including 'Beautiful' landscapes. Such automatic response can be based only on social signals produced by the user during browsing.

The proposed Conceptual Framework can also be used to determine tourists' preferences when they are traveling. The preprocessing experiments should be set up in a corresponding environment, in order to extract the frequent subsequences of social signals and then to store them in the Knowledge Base as patterns. Such an environment could be within or outside of the accommodation facility, for instance during staying at a hotel room. Then, the stored patterns can be used in analog to the above in order to detect user preferences.

4 Discussion and Conclusion

Tourist literature indicates that the number of research employing neuroscience approaches is rising, with most of these studies focusing on tourism marketing (Sop, 2021). In this regard, Laws and Scott (2015) stress that most studies on tourist behavior depend on self-reported questionnaires and that self-reported data have specific validity concerns. Koc and Boz (2014) further assert that participants' self-reported responses on a self-reporting questionnaire may not reflect the actual truth or cannot assess their 'true' consuming behavior. Under these conditions, Laws and Scott (2015) recommend collecting data with sensors that can measure the participant's heart rate, level of arousal, attentional concentration, etc. Li et al. (2021) proposes the EEG technique to investigate how the cognitive process of emotion also influences tourist behavior. Koc and Boz (2014), who created the phrase 'psychoneurobiochemistry of tourist marketing,' also suggest investigating neurotransmitters and hormones such as serotonin, dopamine, and melatonin to gain a deeper understanding of tourism consumers.

Given the ongoing disputes over conventional research methodologies, neuroscience approaches appear to be accepted in the tourism field. This is because integrating neuroscience techniques into tourism research facilitates collecting more precise data from respondents. Moreover, the data produced through these approaches circumvent the frequent problems associated with conventional research techniques, such as social desirability responding, common method bias. Consequently, appropriate investigations might yield surprising but more reliable results. In Li et al. (2021) study, for instance, the EEG equipment revealed that participants had the maximum degree of arousal when seeing advertising for non-conflict-ridden areas. Moreover, the eye-tracking equipment findings demonstrated that the fixation time was longer for conflict-ridden areas, indicating that the desire to discover new destinations and curiosity may motivate visitors to incur a modest risk. Aicher et al. (2016) eye-tracking experiment on TripAdvisor (Katsoni & Fyta, 2021), showed that rating symbols are more important than textual content and that excellent and negative reviews only partially impact an individual's judgment regarding a particular property.

In addition, the use of neuroscience techniques in tourism study enhances the multidisciplinary standing of tourism. As is well-known, tourism collaborates with several disciplines such as sociology, psychology, economics, history, anthropology, architecture. Goodrich (1995), and neuroscience may be regarded as the most recent collaborator of the tourist field. It is anticipated that tourism academics will collaborate with neuroscientists (Ma et al., 2014) to study the neuroscientific data of their research. In addition, neuroscientists will be required as reviewers in tourist publications for articles employing neuroscience methodologies.

The findings of relevant literature reviews (Sop, 2021) suggest that the use of neuroscience methodologies will grow among tourism researchers and that new research concepts or subjects, such as neuroethics, may emerge in the tourism field. Although neuroscience approaches have primarily been utilized in tourist marketing, they are also valuable for other study contexts, such as organizational behavior and management in the tourism sector.

The present work was a preliminary study and would be useful to consider for further analysis and synthesis of research studies in the field of neuro-tourism and capturing behavioral analysis through intelligent social pattern processing algorithms. Future research that can apply the present conceptual framework using social signal processing in combination with neuroscientific tools could provide an even more thorough description of the subject.

References

Aicher, J., Asiimwe, F., Batchuluun, B., Hauschild, M., Zöhrer, M., & Egger, R. (2016). Online hotel reviews: Rating symbols or text… text or rating symbols? That is the question!. In A. Inversini, R. Schegg (Eds.), *Information and communication technologies in tourism 2016*. Springer, Cham. https://doi.org/10.1007/978-3-319-28231-2_27

Alvino, L., Pavone, L., Abhishta, A., & Robben, H. (2020). Picking your brains: Where and how neuroscience tools can enhance marketing research. *Frontiers in Neuroscience, 14*. https://doi.org/10.3389/fnins.2020.577666

Atzmueller, M. (2014). Data mining on social interaction networks. *Journal of Data Mining & Digital Humanities*. https://doi.org/10.46298/jdmdh.7

Aviezer, H., Trope, Y., & Todorov, A. (2012). Body cues, not facial expressions, discriminate between intense positive and negative emotion. *Science, 338*(6111), 1225–1229.

Bercea Olteanu, M. D. (2015). Neuroethics and responsibility in conducting neuromarketing research. *Neuroethics, 8*(2), 191–202. https://doi.org/10.1007/s12152-014-9227-y

Boutsinas, B., & Papastergiou, T. (2008). On clustering tree structured data with categorical nature. *Pattern Recognition, 41*(12), 3613–3623.

Cao, Z., Simon, T., Wei, S. E., & Sheikh, Y. (2017). Realtime multi-person 2D pose estimation using part affinity fields. In *IEEE Conference on Computer Vision and Pattern Recognition (CVPR)*. https://doi.org/10.1109/cvpr.2017.143

Cattuto, C., Van den Broeck, W., Barrat, A., Colizza, V., Pinton, J.-F., & Vespignani, A. (2010). Dynamics of person-to-person interactions from distributed RFID sensor networks. *PLoS ONE, 5*(7), e11596. https://doi.org/10.1371/journal.pone.0011596

Dooley, J. (2011). *Software development and professional practice*. https://doi.org/10.1007/978-1-4302-3802-7

Gilmore, D. C., Stevens, C. K., Harrell-Cook, G., & Ferris, G. R. (1999). Impression management tactics. *The Employment Interview Handbook*, 321–336.https://doi.org/10.4135/9781452205519.n18

Gkintoni, E., Meintani, P.M., Dimakos, I. (2021). Neurocognitive and Emotional Parameters in Learning and Education Process. 14th Annual International Conference of Education, Research and Innovation, 8th–10th November, Seville, Spain. https://doi.org/10.21125/iceri.2021.0659

Gkintoni, E., Halkiopoulos, C., & Antonopoulou, H. (2022). Neuroleadership an asset in educational settings: An overview. *Emerging Science Journal 6*(4), 893–904. https://doi.org/10.28991/esj-2022-06-04-016

Gkintoni, E., & Dimakos, I. (2022). An overview of cognitive neuroscience in education. In *14th Annual International Conference on Education and New Learning Technologies*, 4th–6th July, Mallorca, Spain. https://doi.org/10.21125/edulearn.2022.1343

Goodrich, J. N. (1995). Book reviews: Tourism principles, practices, philosophies. *Journal of Travel Research, 34*(1), 88–89. In R. W. Mclntosh, C. R. Goeldner, & J. R. Brent Ritchie (Eds.), John Wiley and Sons, New York, 7th ed., 551 p. https://doi.org/10.1177/004728759503400111

Guidici, E., Dettori, A., & Caboni, F. (2017). Neurotourism: Futuristic perspective or today's reality? In *20th Excellence in Services International Proceedings 7–8 September 2017*. University of Verona. ISBN 978890432774.

Halkiopoulos C., Antonopoulou, H., Gkintoni E., & Giannoukou I. (2021a). An expert system for recommendation tourist destinations: An innovative approach of digital marketing and decision-making process. *International Journal of Innovative Science and Research Technology, 6*(4), 398–404. ISSN: 2456-2165.

Halkiopoulos, C., Katsouda, M., Dimou, E., & Panteli, A. (2021b). Cultural and tourism promotion through digital marketing approaches: A case study of social media campaigns in Greece. In V. Katsoni, C. van Zyl (Eds.), *Culture and tourism in a smart, globalized, and sustainable world*. Springer Proceedings in Business and Economics. Springer, Cham. https://doi.org/10.1007/978-3-030-72469-6_46

Halkiopoulos, C., Dimou, E., Kompothrekas, A., Telonis, G., & Boutsinas, B. (2021c). The E-Tour facilitator platform supporting an innovative health tourism marketing strategy. In V. Katsoni, C. van Zyl (Eds.), *Culture and tourism in a smart, globalized, and sustainable world*. Springer Proceedings in Business and Economics. Springer, Cham. DOI:https://doi.org/10.1007/978-3-030-72469-6_39

Halkiopoulos, C., Antonopoulou, H., Gkintoni, E., & Aroutzidis, A. (2022). Neuromarketing as an indicator of cognitive consumer behavior in decision-making process of tourism destination—an overview. In V. Katsoni, A.C. Şerban (Eds.), Transcending Borders in Tourism Through Innovation and Cultural Heritage. Springer Proceedings in Business and Economics. Springer, Cham. https://doi.org/10.1007/978-3-030-92491-1_41

Horská, E., & Berčík, J. (2017). *Neuromarketing in food retailing*. Wageningen Academic Publishers.

Kakoleres, G., & Boutsinas, B. (2022). A conceptual framework for applying social signal processing to decision-making in management. In *Proceedings of 7th Business & Entrepreneurial Economics Conference*, Plitvice Croatia (pp. 123–130)

Katsoni, V., & Dologlou, N. (2017). ICT applications in smart ecotourism environments. *SmartCities in the Mediterranean*, 225–244. https://doi.org/10.1007/978-3-319-54558-5_11

Katsoni, V., & Fyta, A. (2021). From Pausanias to Baedeker and Trip Advisor: Textual proto-tourism and the engendering of tourism distribution channels. *Turyzm/tourism, 31*(1), 11–19. https://doi.org/10.18778/0867-5856.31.1.11

Koc, E., & Boz, H. (2014). Psychoneurobiochemistry of tourism marketing. *Tourism Management, 44*, 140–148. https://doi.org/10.1016/j.tourman.2014.03.002

Laws, E., & Scott, N. (2015). Tourism research: Building from other disciplines. *Tourism Recreation Research, 40*(1), 48–58. https://doi.org/10.1080/02508281.2015.1005926

Lei, M., Chen, W., Wu, J., Zhang, Y., & Lv, Y. (2022). Neurophysiological measures in hospitality and tourism: Review, critique, and research agenda. *Journal of Hospitality & Tourism Research*, https://doi.org/10.1177/10963480221091117

Li, S., Lyu, T., Chen, M., & Zhang, P. (2021). The prospects of using EEG in tourism and hospitality research. *Journal of Hospitality & Tourism Research, 46*(1), 189–211. https://doi.org/10.1177/1096348021996439

Lindstrom, M. (2008). Playing the brand game. *Young Consumers, 9*(2). https://doi.org/10.1108/yc.2008.32109bab.001

Ma, Q. G., Hu, L. F., Pei, G. X., Ren, P. Y., & Ge, P. (2014). Applying neuroscience to tourism management: A primary exploration of neurotourism. *Applied Mechanics and Materials, 670–671*, 1637–1640. https://doi.org/10.4028/www.scientific.net/amm.670-671.1637

Neuro-Sell. (2014). How neuroscience can power your sales success. *Journal of Consumer Marketing, 31*(6/7), 555–556. https://doi.org/10.1108/jcm-05-2014-0960

Panas, G., Heliades, G., Halkiopoulos, C., Tsavalia, D., & Bougioura, A. (2017). Evaluation of Athens as a city break destination: Tourist perspective explored via data mining techniques. In V. Katsoni, A. Upadhya, & A. Stratigea (Eds.), *Tourism, culture and heritage in a smart*

economy. Springer Proceedings in Business and Economics. Springer, Cham. https://doi.org/10.1007/978-3-319-47732-9_6
Panteli, A., Kompothrekas, A., Halkiopoulos, C., & Boutsinas, B. (2021). An innovative recommender system for health tourism. In V. Katsoni, & C. van Zyl (Eds.), *Culture and tourism in a smart, globalized, and sustainable world*. Springer Proceedings in Business and Economics. Springer, Cham. https://doi.org/10.1007/978-3-030-72469-6_42
Panyik, E., & Gonçalves, J. (2017). Neurotourism: a paradigm shift in travel on the intersection of e-tourism and neuroscience. *Revista Turismo & Desenvolvimento*, *2*(27/28), 321–323. https://doi.org/10.34624/rtd.v2i27/28.20849
Panas, G., Vasiliadou, S., & Halkiopoulos, C. (2020). Data analysis evaluation of web technologies enhancing communication in tourism industry: Case study in Kefalonia Island. In: Katsoni V., & Spyriadis T. (Eds) *Cultural and Tourism Innovation in the Digital Era*. Springer Proceedings in Business and Economics. Springer, Cham. https://doi.org/10.1007/978-3-030-36342-0_14
Panas, G., Thrasidi, N., Halkiopoulos, C., & Gkintoni, E. (2022). Consumer behavior and cognitive factors in relation to gastronomic tourism and destination marketing in Greece. *Springer Proceedings in Business and Economics*, 655–677. https://doi.org/10.1007/978-3-030-92491-1_40
Pentland, A. (2007). Social signal processing [Exploratory DSP]. *IEEE Signal Processing Magazine, 24*(4), 108–111. https://doi.org/10.1109/msp.2007.4286569
Pentland, A. (2010). To signal is human. *American Scientist, 98*(3), 203. https://doi.org/10.1511/2010.84.203
Pentland, A. (2004). Social dynamics: Signals and behavior. In J. Triesch, & T. Jebara (Eds.), *Proceedings of the 2004 International Conference on Development and Learning*.
Sarigiannidis, C., Halkiopoulos, C., Giannopoulos, K., Giannopoulou, F., Politis, A., Boutsinas, B., & Kollias, K. (2021). Do hotels care? A proposed smart framework for the effectiveness of an environmental management accounting system based on business intelligence technologies. In V. Katsoni, & C. van Zyl (Eds.), *Culture and tourism in a smart, globalized, and sustainable world*. Springer Proceedings in Business and Economics. Springer, Cham. https://doi.org/10.1007/978-3-030-72469-6_41
Sop, S. A. (2021). The methodological contribution of neuroscience to tourism discipline. In *Proceedings of the Conference on Managing Tourism Across Continents (MTCON'20)* (pp. 351–360). https://doi.org/10.5038/9781955833011
Tosun, C., Ozdemir, S., & Cubuk, F. (2016). Usage of neuro-tourism methods in hotel preferences of the consumers. In *Proceedings of 2016 WEI International Academic Conference* (pp. 19–26).
Vinciarelli, A., Pantic, M., & Bourlard, H. (2009). Social signal processing: Survey of an emerging domain. *Image and Vision Computing, 27*(12), 1743–1759. https://doi.org/10.1016/j.imavis.2008.11.007
Zurawicki, L. (2010). *Neuromarketing*. https://doi.org/10.1007/978-3-540-77829-5

Applying Big Data Technologies in Tourism Industry: A Conceptual Analysis

Leonidas Theodorakopoulos, Constantinos Halkiopoulos, and Dimitris Papadopoulos

Abstract Tourism is the "heavy industry" of our country, as confirmed by the statistical data of many surveys. With the help of new technologies, the goal of attracting quality tourism and increasing per capita spending on products and services provided by the tourism industry is becoming more and more achievable. In the present work, the adoption of big data technologies in the field of tourism is examined through a review of the existing literature, with the main goal of gaining a deeper knowledge and understanding of the requirements of tourists in our country and improving the way of decision making. The paper also mentions the advantages and benefits of using big data technologies and popular methods used for sorting and contracting big data.

Keywords Big Data · Tourism industry · Demand forecasts · Factor model · LASSO model

JEL Classification Z32 · Z33 · O32 · O33 · L86

1 Introduction

The rate of information recovery in the era of new technologies is growing in all sectors and especially in the tourism industry and the use of big data can provide the possibility to produce new and more informed conclusions regarding the activity and behavior of tourists, which will certainly give a big boost to the tourism product and will benefit both tourists and the tourism industry (Weaver, 2021). Tourists using their mobile phones and other electronic devices leave behind digital traces on the Internet (Giotopoulos et al., 2019; Katsoni & Dologlou, 2017). Thus, a huge amount of data is available from each tourist concerning each stage of the trip (before, during, and after the trip). Social media such as Twitter, Facebook, Instagram provide most of the

---

L. Theodorakopoulos · C. Halkiopoulos (✉) · D. Papadopoulos
EDILAB, Department of Management Science and Technology, University of Patras, Patras, Greece
e-mail: halkion@upatras.gr

© The Author(s), under exclusive license to Springer Nature Switzerland AG 2023

V. Katsoni (ed.), *Tourism, Travel, and Hospitality in a Smart and Sustainable World*, Springer Proceedings in Business and Economics,
https://doi.org/10.1007/978-3-031-26829-8_21

data (Line et al., 2020). Due to a large amount of data available in the cloud, similar analyses are required to extract useful information (Halkiopoulos & Papadopoulos, 2022; Panas, et al., 2020). The most likely scenario for someone planning a trip is to seek help from the Internet either to buy tickets, find a place to stay, or look for attractions, places to eat, etc. (Ogbeide et al., 2020). Thus, those involved in the travel industry are increasingly turning to the exploitation of big data in order to improve their decision making processes as well as their overall performance (Ardito et al., 2019). For example, big data can be used to connect dispersed information from different sources and systems and add another dimension to the way of decision making, improving it (Li et al., 2018).

2 Literature Review

2.1 *Key Features of Big Data in Tourism*

Data volume is the main characteristic of big data and is the key answer to the inability of traditional systems to manage it. In tourism, the huge volume of data comes primarily from the different points of sale or from other traditional information channels, such as call centers, websites, shops, or even from customer relations (Halkiopoulos & Giotopoulos, 2022). The tourism big data content is generated on a daily or even hourly basis, and the aim is to extract useful information, understand it and transform it into upgraded/smart data, suitable to be used for policymaking and tourism planning (Volo, 2020; Yallop & Seraphin, 2020; Zhang, 2021).

Another key feature of big data, both in terms of cost and ease of use, is the diversity of data since the data can come from any accessible technology. By diversity, we describe the different forms of data, which cannot be stored in structured relational database systems (Mariani, 2019). That is, in systems based on a relational data model, in which data is grouped into tables with rows and columns, with each table/relationship relating to a "node name" (e.g., customer, product). The formats of big data can be documents, emails, text messages, images, graphics, videos, and any other type of data that can be generated by mobile phones, GPS signals, sensors, and machine registries (Gao, 2021). Such data has always existed and is classified as unstructured or partially structured. In fact, 80% of tourism-related information is in unstructured form, resulting in organizations relying on only 20% of structured data (Stylos et al., 2021). For example, asset management systems, web or blog content management systems, or customer relationship management systems can only use structured data, while data on customer preferences is in an unstructured or partially structured form, and requires innovative technologies to analyze it, with the ultimate goal of developing new or improved products and services (Centobelli & Ndou, 2019; Halkiopoulos et al., 2021).

The third key characteristic of big data is speed, which is referred to as the speed of response. Regarding tourism big data, there are three main components that define this speed (Stylos & Zwiegelaar, 2019). The first component is the continuous and complete collection, storage, and analysis of fast-moving big data streams, such as for example the metrics recorded by a sensor or the history of visits to web pages and the "clicks" from each visitor to a web page (Zhang & Dong, 2021). The second component is related to the timeliness and temporal validity characteristics of the information. More specifically, we need to capture, store, and use the big data within a certain time frame, depending on the type of information, since the value and relevance of some data are long-lasting, while the relevance of other data is lost in a very short time (Li, 2019). The third component is the speed with which big data must be stored and retrieved. In other words, the architecture for collecting, analyzing, and presenting the data must ensure that the overall processing is done in real-time (in this case in fractions of a second) and that this process is carried out consistently for thousands of new customers. In the tourism sector, for example, we are interested in being able to send the right offer to the right person at the right time, when that person arrives at a destination (Halkiopoulos et al., 2022; Lv et al., 2022). At the same time, it is important to be able to receive timely information when a dissatisfied customer of a hotel reports their dissatisfaction on social media instead of to the relevant hotel department (Xu et al., 2020). Also, taking, as an example travel businesses, airlines, timely management, and price changes can be done in a dynamic way using complex algorithms that take, directly, as input the real-time or near real-time customer behavior (Panas et al., 2022; Rahmadian et al., 2021).

Also, an additional characteristic is the validity of the data, which refers to the truth and accuracy of the data, taking into account the context and the speed at which things happen. The validity of big data lies in the impartiality, noise, and discontinuities in the data (Peterlin et al., 2021). That is, we are interested in whether the data collected and stored is useful and meaningful to the problem, and we are analyzing. Compared to volume and speed, validity and filiality in data analysis is the biggest challenge (Mariani and Baggio, 2021; Çeltek and Ilhan, 2020). When using big data to formulate strategies, it is necessary to perform proper data cleansing and ensure that there is no accumulation of "dirty data" in the processing systems (Hartmann and Lederer, 2021).

Finally, an important feature of big data is now considered to be the value of the data. The importance of big data in tourism can be described through its innovative applications in the tourism industry (Gössling, 2020). By using big data in tourism, we introduce the concepts of personalized strategy, personalized marketing, and targeted product design, which are extremely important opportunities in this sector. Using a series of interviews conducted within the travel industry, studies have shown how big data can be used to increase interaction for mutual benefit and reduce friction between disciplines, both within a company and within the entire bi, and personalization is a key pillar in the use of big data (Shams et al., 2022). To achieve personalization more effectively, large travel companies are required to work on information databases to collect the myriad of data generated by a consumer at different points. The second valuable application of big data in tourism is the customer-centric approach, in the sense that the customer must be at the heart of all efforts using data. However, if big data collection is seen as invasive or intrusive/intrusive, the consumer will not be happy, and loyalty will be lost. Nevertheless, all the evidence suggests that consumers are willing to accept the corresponding intrusion into their behaviors if the resulting product is more targeted and able to predict the full range of their needs (Dritsas et al., 2018).

2.2 *Advantages of Using Big Data in Tourism*

The gradual recognition of the benefits of the use of big data is expected both from the side of consumers and suppliers of tourism products. Personalized marketing and targeted product design are extremely important opportunities for both groups (Antonopoulou et al., 2022a). It is evident that big data can provide better, targeted, and profitable services and products to consumers (Giannoukou et al., 2022; Panteli et al., 2021; Katsoni et al., 2012). More specifically, big data analysts can capture information about consumers' interests from photos posted on Facebook or other social networks (e.g., a tourism agency can promote information about local cycling destinations or cycling clubs when they spot a photo of a bike).

Previous studies on tourism have mainly relied on surveys or expert opinions, which means that they have sampled the population as a whole and do not have actual data on all tourists (Solakis et al., 2022). In contrast, a study related to big data in tourism has tried to introduce data based on the actual actions of all users, rather than drawing information from survey samples. For example, we can consider the use of big data generated by tourists themselves (e.g., through mobile phones connected to the telecommunication network or bank cards connected to POS terminals), which increases knowledge regarding the market targeted by the tourism industry and acts as an adjunct in forecasting tourism demand.

The exploitation of the large volume of tourism data using innovative methods has the following advantages over traditional methodologies:

1. The use of big data is based on the actual actions of users, rather than on relevant surveys. In other words, the analysis of real actions replaces the analysis of answers to questions and statements of intent. Moreover, by collecting information from all possible sources, the number of samples from which knowledge is extracted is increased and more components are considered than in a conventional survey. The reliability of big data analysis enables us to consider all aspects of information in order to produce comprehensive results instead of unbiased conclusions resulting from the loss of information in the data samples of conventional surveys.
2. New information flows: The large amount of data related to tourism is a type of information that comes directly from tourists themselves. This data enriches the knowledge of tourism enterprises about the market share they are targeting and is particularly useful for analyzing and predicting consumer demand for different tourism products and services. Since tourism big data is mainly structured, it is possible to cross-reference them with other sources, such as social media and

open public data. The analysis of tourism big data can be compared with the internal data of each tourism company in order to determine whether the supply of tourism products/services in each area of a city is aligned with the demand for products/services by tourists.

3. Real-time data and direct prediction: one of the innovative uses of big data is related to the exploitation of data generated in real-time and describing the most recent activities before they are made available from official data sources: for example, Varian argued that real-time Google search is a good method to monitor consumer activity, as the correlation analysis performed by Google correlate has a 6-week lead from reported values. A notable example of using queries from the Google search bar for direct prediction is Google flu trends, which identifies potential flu outbreaks one to two weeks ahead of official health reports by tracking the appearance of flu-related search terms in the Google search engine. There are also many other studies that have used structured search engine data to extract predictions for the tourism sector. In addition to search engine queries, there are other types of real-time data streams that can be collected and analyzed: for example, such types of data streams can be data from credit card purchases, parcel/package delivery, and mobile phone usage, which are useful pieces of information. The ultimate goal of using real-time big data is to develop applications that are able to directly understand the pulse of the economy and respond to consumer needs with appropriate recommendations. It is important, however, that this takes place under controlled conditions and that it can be activated/deactivated at will at any time.

2.3 *Benefits of the Use of Big Data for Tourism Businesses*

Big data analytics is bringing changes to all industries, not just the usual retail industries, procurement and logistics departments, and the high-tech industry (Sarigiannidis et al., 2021). It is also changing the landscape in the world of hospitality and travel, as the respective businesses deal with a multitude of data from users, covering all possible sources of information (e.g., flight confirmations or a customer's room preferences) and creating every possible data correlation to discover otherwise unknown information. In addition, there are some significant changes in the use of big data as the cost of analytics platforms continues to decline and employees become increasingly familiar with how they can exploit it. Essentially, big data can be used to adapt marketing campaigns and find the inefficiencies in each business model. The analysis of big data can provide the necessary business insights and be a source of innovation for tourism organizations and the industry in general. The potential offered by the use of big data in tourism is enormous and tourism organizations should give it due importance (Antonopoulou et al., 2022b).

With the right approach, the tourism industry can learn a lot about consumer preferences and use this information to create the appropriate connections with each individual traveler. The ability to offer travelers the right service or product at the right time is critical. Without the right information and an appropriate targeting strategy, advertising will not lead to an outcome and therefore loses its value. The travel industry is changing at a rapid pace, which leads to the need for rapid data analysis and quick decision making. In the tourism sector, every desire of travelers must be met immediately in order to meet their relevant needs while they exist, which is what makes big data so important. With the intense growth in the volume and applications of big data, traditional tourism data and traditional analytical methods will have to interact with new data and new methodologies: for example, call centers will be linked to consumer reviews made online, loyalty programs will be linked to booking histories, and preferences for places to stay will be combined with social media conversations.

2.4 *Tourism Demand Forecasts Based on Big Data*

A widely held view is that big data can act as an adjunct to improve forecasting because its analysis can reveal hidden patterns, and hence that big data-based forecasts can improve decision making. Data mining techniques have been proposed as more suitable for big data-based forecasting. However, the choice of mining techniques as a forecasting tool is contradicted by the fact that data mining is concerned with static data rather than real-time or time series data, which in practice results in mining techniques being rarely used in tourism demand forecasting. (Theodorakopoulos, 2022) Traditional tourism demand forecasting techniques, when fed with big data, do not work efficiently due to the inherent characteristics of big data (volume, speed of access and processing, complexity, etc.) that make it difficult to process effectively. The unsuitability of traditional prediction techniques for processing big data applies even to data that is in a structured form. Applying traditional prediction methods to big data requires further simplification of already structured big data. One of the possible solutions is to shrink the big data and extract the important information in a suitable format, processable by traditional prediction models. Factor models are the most common and popular statistical analysis and data mining technique used for prediction, via big data. Neural networks and Bayesian models are two also popular options.

2.5 *Neural Networks*

Many researchers focus their interest on artificial neural network as a predictive tool. Probably the most well-known and widely used architecture of ANN is the Multi-Layer Perceptron (MLP) which is a feedforward artificial neural network. The MLP network consists of one input layer, the hidden layers, and the output layer (Fig. 1). Nodes in the hidden layers use an activation function (most commonly a sigmoid function) in order to provide the forecast through the output layer. Examples of neural

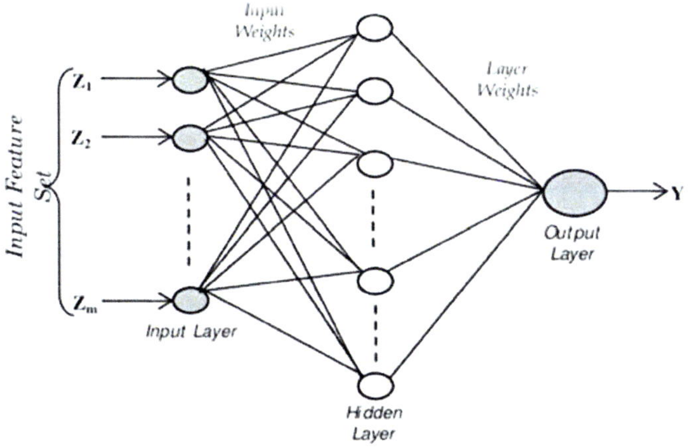

Fig. 1 Structure of multilayer perceptron (MLP) network

networks applications in tourism industry are included in the following studies (Bi et al., 2020; Çuhadar, 2020; Nguyen et al., 2021; Silva et al., 2019; Wang et al., 2020; Yao & Cao, 2020).

2.6 Bayesian Models

It is a classification technique based on Bayes' Rule which, in turn, is based on the Fundamental Theorem of Total Probability (Law of Total Probability) of Probability Theory (Assaf et al., 2019). The basis of this theorem is the total probability theorem of Bayes (total probability with an assumption of independence between predictors (Kulshrestha et al., 2020).

Simply put, a Naïve Bayes classifier assumes that the presence of a particular characteristic in a class is unrelated to the presence of any other characteristic (Assaf et al., 2019). For example, a fruit can be considered an apple if it is red, round, and about a centimeter in diameter. Even if these characteristics depend on each other or on the existence of other characteristics of an apple, the Naive Bayes classifier assumes that all of these characteristics independently contribute to the probability that this fruit is an apple (Song et al., 2021). Because of this assumption of independence of characteristics defining a class, which is often not true, the classifier is called Naïve. The Naive Bayes model is easy to construct and particularly useful for very large datasets (Ampountolas, 2019).

In particular, regarding the theoretical framework of the method, the classifier is based on Bayes' Theorem (or rule). This theorem provides a way to calculate the posterior probability $P(c|x)$ based on $P(c)$, $P(x)$, and $P(x|c)$ according to the equation

$$P(c|x) = \frac{\boldsymbol{P}(x|c)\boldsymbol{P}(c)}{\boldsymbol{P}(x)}$$

where:

- $P(c|x)$ is the a posteriori or posterior (that is, after the observation) knowledge about the probability of being in class c, which is the target to be recognized, given the predictor x, which is the characteristic(s) that we observe.
- $P(c)$ is the a priori or prior knowledge about the probability of class c.
- $P(x|c)$ is the probability of observing predictor x within class c.
- $P(x)$ is the prior probability of observing predictor x regardless of class.

Based on the above definition of conditional probability, we can now formulate Bayes' Theorem based on the above definition of conditional probability. Let be a sample space (sample space) partitioned into n categories or classes $E_1, E_2, \ldots, E_n$. Then, based on Bayes' Rule we can calculate the observation-based a posteriori probability based on the prior known, or a priori, probabilities as follows:

$$\Pr(E_i|B) = \frac{\Pr(E_i)\Pr(B|E_i)}{\sum_{k=1}^{n}\Pr(E_k)\Pr(B|E_k)}.$$

3 Methodology for Sorting and Condensing Big Data

Big data contains a wealth of information, which makes it difficult to manage, especially in terms of storage and analysis, as mentioned above. The biggest problem, therefore, for making tourism forecasts using structured big data is managing the huge amount of data. The two most popular methods used for sorting and condensing large amounts of structured data are the factor modeling and least absolute shrinkage and selection operator (LASSO) approaches.

3.1 Factor Model

The factor model is the most commonly used method for sorting and condensing structured big data. Some recent studies in the economics-related literature have focused on the use of factor models in terms of the predictions resulting from the use of large data sets (Antonopoulou et al., 2022c; Hsu et al., 2020; Li et al., 2017; Park et al., 2018; Qiu Zhang et al., 2017; Wen et al., 2021). In particular, they analyze the benefits regarding the extraction of forecasts after dimensionality reduction of independent component analysis (ICA) and sparse principal component analysis (SPCA), in combination with some other methods of estimation and data reduction. To evaluate the success of the use of big data, we could conduct a forecasting "contest" that involves the estimation of different types of baseline models, each constructed using a variety of specification approaches, estimation approaches, and econometric reference models (Li et al., 2021).

In a factor model, the correlations among n variables, x_1, …, x_n, for which T observations are available, are assumed to be entirely due to a few, $r < n$, latent unobservable variables, called factors (Lourenço et al., 2021). The link between the observable variables and the factors is assumed to be linear. Thus, each observation x_{it} can be decomposed as

$$x_{\text{it}} = \mu_i + \lambda_i' f_t + e_{\text{it}}$$

where μ_i is the mean of x_i, λ_i is an $r * 1$ vector, and e_{it} and f_t are two uncorrelated processes. Thus, for $I = 1, \ldots, n$ and $t = 1, \ldots, T$, x_{it} is decomposed into the sum of two mutually orthogonal unobserved components: the common component, $x_{\text{it}} = \lambda'_i f_i$, and the idiosyncratic component, $\xi_{\text{it}} = \mu_i + e_{\text{it}}$. While the factors drive the correlation between x_i and x_j, $j \neq i$, the idiosyncratic component arises from features that are specific to an individual x_i variable. Further assumptions placed on the two unobserved components result in factor models of different types (Camacho and Pacce, 2018).

3.2 *LASSO*

The LASSO method (Tibshirani, 1996) is an alternative method to the least squares method for fitting a linear model (Garrod & Almeida, 2021; Liu et al., 2021; Tian et al., 2021; Yang et al., 2022) Its use is mainly recommended in cases where the number of explanatory variables is greater than the number of available observations ($p > N$). The most important advantage of this method is its ability to produce sparse models, i.e., models that contain only a subset of the initially available factors (Pu et al., 2020). This leads to easier interpretation and reduced complexity in the final model since we end up with only those variables that play an important role in explaining the response variable (Chokethaworn et al., 2020). The LASSO estimator provides a solution to the following minimization problem:

$$\min_{\beta \in \mathbb{R}^p} \left\{ \frac{1}{2N} \sum_{i=1}^{N} \left(y_i - \beta_0 - \sum_{j=1}^{p} \beta_j x_{ij} \right)^2 \right\}, \text{ where } \sum_{j=1}^{p} |\beta_j| \leq s$$

where $s \geq 0$, is an upper bound on the sum of the absolute values of the model coefficients (Ippolito, 2022). The representation minimized by the LASSO estimator is identical to that of the least squares method, but now we impose an additional restriction on the coefficients. Thus, when s is large enough, we do not restrict the coefficients, thus resulting in the least squares estimator. On the contrary, when s decreases, we restrict the coefficients to a greater extent, with the result that several of them tend to 0 and some of them are exactly zero (Sun et al., 2022). The parameters are usually chosen through a process called Cross Validation (Hastie, 2001). By restricting the coefficients of the independent variables, we can achieve more prediction accuracy in our model. As mentioned, least squares estimators are unbiased but have a large variance. By placing a constraint on the parameters, we introduce a small amount of bias but achieve a significant reduction in their variance. Thus, the mean squared error of the estimators, which consists of the sum of the squared bias and the variance,

$$\mathrm{MSE} = \mathrm{bias}^2 + \mathrm{Variance}$$

usually decreases, leading to better predictions for the response variable. It is important before proceeding to calculate the estimator to standardize the values of the independent variables so that they have a mean of 0 and a variance equal to unity, as follows:

$$x_{ij} \leftarrow \frac{x_{ij} - \overline{x}_j}{\sqrt{\frac{1}{N}\sum_{i=1}^{N}\left(x_{ij} - \overline{x}_j\right)^2}}, \overline{x}_j = \frac{1}{N}\sum_{i=1}^{N} x_{ij}$$

for each $j = 1, \ldots, p$ and $i = 1, \ldots, N$. Thus, the resulting LASSO estimators will be independent of the units of measurement of each explanatory variable. For simplicity, we can also center the response variable y, so that $\overline{y} = 0$. Thus, we can omit the constant β_0 from the minimization of (1.2.1) because it follows that

$$\widehat{\beta}_0 = \overline{y} - \sum_{j=1}^{p} \widehat{\beta}_j \overline{x}_j$$

with $\overline{y}$ and $\overline{x}_j, j = 1, \ldots, p$, the mean values calculated on the original observations (unstandardized).

4 Steps Towards the Use of Big Data in Tourism Forecasting

Companies that capture big data and implement strategies based on it gain a competitive advantage, as the technology required to process big data is a barrier for many companies because of its complexity and cost. There are of course several steps in the process of capturing big data before making use of it (Li et al., 2021). More specifically the key steps are as follows:

- Mission/objective/purpose: The first step is to determine the purpose of using big data, which is to ensure business benefits emanating from its use. When collecting big data, it is important to be able to access it and know what is available so that we can subsequently determine what is valuable and important to the business (Theodorakopoulos et al., 2022). In other words, we need to know the possibilities that big data gives us, what exactly we are looking for and what its value of it is. It is therefore important that we set specific business objectives and not just get bogged down in the management of this very data (Assaf & Tsionas, 2019; Follett & Vander Naald, 2020; Theodorakopoulos, 2022).

- Visualizing big data: the second step is to make big data "open" and accessible to all users within a company/organization. This will enable those making statistical forecasts in tourism to determine, for example, the optimal quantities of a product and adjust their procurement and supply processes accordingly in order to maximize their efficiency. The purpose of data visualization is to find methods by which data could be efficiently collected from different sources (visual and non-visual) and presented in such a way that users can easily get information from it. In this way, forecasters will be able to use big data more effectively and directly (Jiao and Chen, 2019).
- Structuring the big data: the third step is to structure the data that is in unstructured form. More specifically, by structuring the data, we refer to organizing the big data according to the size and format specifications of traditional data, so that it becomes possible to arrange it appropriately in rows and columns in spreadsheet tables. Structured data is often called relational data because it is usually found in relational databases (Theodorakopoulos & Papadopoulos, 2022). On the other hand, unstructured data can be easily mapped into predefined fields. For example, the structured data of a call center includes numbers, dates, and groups of words and numbers called strings. This type of data represents about 20% of the total volume of big data. The challenge regarding unstructured data is the combination of its relative volume on top of all big data and the difficulty of analyzing it. That is, it is data that cannot be analyzed using standard procedures, while at the same time it constitutes the bulk of big data, since about 80% of it belongs to unstructured or partially structured data (Xie et al., 2020). To solve this problem, various software solutions have been designed to analyze unstructured data and extract important information (Liu et al., 2022). An example of a tool for managing large volumes of unstructured data is NoSQL (not only SQL), which is a database for storing unstructured and unformatted data. The technique used is that each row has a different unique set of columns to which it is mapped. For tourism forecasting, in this case, filtered structured big data is mainly used.

5 Conclusion

In conclusion, the use of big data technologies in the tourism industry is an emerging trend that creates expectations for a deeper understanding of tourism demand and adaptation of supply by tourism businesses in order to meet the demands of tourists and the profitable activity of tourism businesses. Also, big data can help the tourism industry to anticipate the “wants” and needs of tourists in our country and improve the decision making process and more. The combination of big data with cloud computing and neural network technologies can be the future for forecasting in the tourism industry. The immediate next step worth considering in future research is how the tourism industry will unite all available technologies, creating a directness with the tourist, increasing the added value in tourism through innovation and technology leading to the complete digital transformation of the tourism industry to Tourism 4.0 by developing a collaborative model, minimizing the negative impacts of tourism while improving the overall experience.

References

Ampountolas, A. (2019). Forecasting hotel demand uncertainty using time series Bayesian VAR models. *Tourism Economics, 25*(5), 734–756. https://doi.org/10.1177/1354816618801741

Antonopoulou, H., Mamalougou, V., & Theodorakopoulos, L. (2022a). The role of economic policy uncertainty in predicting stock return volatility in the banking industry: A big data analysis. *Emerging Science Journal, 6*(3), 569–577. https://doi.org/10.28991/esj-2022-06-03-011

Antonopoulou, H., Theodorakopoulos, L., Halkiopoulos, C., & Mamalougkou, V. (2022b). On the predictability of Greek systemic bank stocks using machine learning techniques. https://doi.org/10.20944/preprints202207.0462.v1

Antonopoulou, H., Mamalougou, V., & Theodorakopoulos, L. (2022c). The role of economic policy uncertainty in predicting stock return volatility in the banking industry: A big data analysis. *Emerging Science Journal, 6*(3), 569–577. https://doi.org/10.28991/esj-2022-06-03-011

Ardito, L., Cerchione, R., Del Vecchio, P., & Raguseo, E. (2019). Big data in smart tourism: Challenges, issues and opportunities. *Current Issues in Tourism, 22*(15), 1805–1809. https://doi.org/10.1080/13683500.2019.1612860

Assaf, A. G., & Tsionas, M. G. (2019). Forecasting occupancy rate with Bayesian compression methods. *Annals of Tourism Research, 75*, 439–449. https://doi.org/10.1016/j.annals.2018.12.009

Assaf, A. G., Li, G., Song, H., & Tsionas, M. G. (2019). Modeling and forecasting regional tourism demand using the Bayesian global vector autoregressive (BGVAR) model. *Journal of Travel Research, 58*(3), 383–397. https://doi.org/10.1177/0047287518759226

Bi, J. W., Liu, Y., & Li, H. (2020). Daily tourism volume forecasting for tourist attractions. *Annals of Tourism Research, 83*, 102923. https://doi.org/10.1016/j.annals.2020.102923

Camacho, M., & Pacce, M. J. (2018). Forecasting travellers in Spain with Google's search volume indices. *Tourism Economics, 24*(4), 434–448. https://doi.org/10.1177/1354816617737227

Çeltek, E., & Ilhan, I. (2020). Big data, artificial intelligence, and their implications in the tourism industry. In *Handbook of research on smart technology applications in the tourism industry* (pp. 115–130). IGI Global. https://doi.org/10.4018/978-1-7998-1989-9.ch006

Centobelli, P., & Ndou, V. (2019). Managing customer knowledge through the use of big data analytics in tourism research. *Current Issues in Tourism, 22*(15), 1862–1882. https://doi.org/10.1080/13683500.2018.1564739

Chokethaworn, K., Chaiboonsri, C., & Wannapan, S. (2020, November). A spatial analysis of international tourism demand model: The exploration of ASEAN countries. In *International Symposium on Integrated Uncertainty in Knowledge Modelling and Decision Making* (pp. 310–321). Springer. https://doi.org/10.1007/978-3-030-62509-2_26

Çuhadar, M. (2020). Modelling and forecasting inbound tourism demand to croatia using artificial neural networks: A comparative study. *Journal of Tourism and Services, 21*(11), 55–70. https://doi.org/10.29036/jots.v11i21.171

Dritsas, E., Livieris, I. E., Giotopoulos, K., & Theodorakopoulos, L. (2018, November). An apache spark implementation for graph-based hashtag sentiment classification on twitter. In *Proceedings of the 22nd Pan-Hellenic Conference on Informatics* (pp. 255–260). https://doi.org/10.1145/3291533.3291552

Follett, L., & Vander Naald, B. (2020). Explaining variability in tourist preferences: A Bayesian model well suited to small samples. *Tourism Management, 78*, 104067. https://doi.org/10.1016/j.tourman.2019.104067

Gao, H. (2021). Big data development of tourism resources based on 5G network and internet of things system. *Microprocessors and Microsystems, 80*, 103567. https://doi.org/10.1016/j.micpro.2020.103567

Garrod, B., & Almeida, A. (2021). On the strategic management of an events portfolio to extend tourists' length of stay: A LASSO approach. *Current Issues in Tourism*, 1–18. https://doi.org/10.1080/13683500.2021.2010673

Giotopoulos, K., Halkiopoulos, C., Papadopoulos, D., Antonopoulou, H., (2019). Towards bring your own device marketing policy. *International Journal of Technology Marketing, 13*(2). Inderscience Publishers: IJTMKT-10313. https://doi.org/10.1504/ijtmkt.2019.10023906

Giannoukou, I., Nikolopoulou, A., & Halkiopoulos, C. (2022). Tourists' and employees' psychology in the framework of strategic management in Greek tourism. *Springer Proceedings in Business and Economics*, 699–721. https://doi.org/10.1007/978-3-030-92491-1_42

Gössling, S. (2020). Technology, ICT and tourism: From big data to the big picture. *Journal of Sustainable Tourism, 29*(5), 849–858. https://doi.org/10.1080/09669582.2020.1865387

Halkiopoulos, C., Dimou, E., Kompothrekas, A., Telonis, G., Boutsinas, B. (2021). The E-tour facilitator platform supporting an innovative health tourism marketing strategy. In V. Katsoni, C. van Zyl (Eds.), *Culture and tourism in a smart, globalized, and sustainable world*. Springer Proceedings in Business and Economics. Springer. https://doi.org/10.1007/978-3-030-72469-6_39

Halkiopoulos, C., Antonopoulou, H., Gkintoni, E., & Aroutzidis, A. (2022). Neuromarketing as an indicator of cognitive consumer behavior in decision-making process of tourism destination—An overview. In V. Katsoni, A.C. Şerban (Eds.), *Transcending borders in tourism through innovation and cultural heritage*. Springer Proceedings in Business and Economics. Springer. https://doi.org/10.1007/978-3-030-92491-1_41

Halkiopoulos, C., & Giotopoulos, K. (2022). Tourism's use of web-based information systems and the influence of tourism trends. In V. Katsoni, A.C. Şerban (Eds.), *Transcending borders in tourism through innovation and cultural heritage*. Springer Proceedings in Business and Economics. Springer. https://doi.org/10.1007/978-3-030-92491-1_25

Halkiopoulos, C., Papadopoulos, D. (2022). Computational methods for evaluating web technologies and digital marketing techniques in the hospitality industry. In V. Katsoni, A.C. Şerban (Eds.), *Transcending borders in tourism through innovation and cultural heritage*. Springer Proceedings in Business and Economics. Springer. https://doi.org/10.1007/978-3-030-92491-1_24

Hartmann, K., & Lederer, M. (2021). The current state of Big Data research in tourism: Results of a systematic literature analysis. *Zeitschrift Für Tourismuswissenschaft, 13*(2), 209–226. https://doi.org/10.1515/tw-2021-0015

Hsu, C. Y., Chen, M. Y., Nyaupane, G. P., & Lin, S. H. (2020). Measuring sustainable tourism attitude scale (SUS-TAS) in an Eastern Island context. *Tourism Management Perspectives, 33*, 100617. https://doi.org/10.1016/j.tmp.2019.100617

Ippolito, P. P. (2022). Hyperparameter tuning. In *Applied data science in tourism* (pp. 231–251). Springer. https://doi.org/10.7717/peerj-cs.1005/table-1

Jiao, E. X., & Chen, J. L. (2019). Tourism forecasting: A review of methodological developments over the last decade. *Tourism Economics, 25*(3), 469–492. https://doi.org/10.1177/1354816618812588

Katsoni, V., & Dologlou, N. (2017). ICT applications in smart ecotourism environments. *Smart Cities in the Mediterranean*, 225–244. https://doi.org/10.1007/978-3-319-54558-5_11

Katsoni, V., Giaoutzi, M., & Nijkamp, P. (2012). Market segmentation in tourism: An operational assessment framework. *Quantitative Methods in Tourism Economics*, 329–352. https://doi.org/10.1007/978-3-7908-2879-5_18

Kulshrestha, A., Krishnaswamy, V., & Sharma, M. (2020). Bayesian BILSTM approach for tourism demand forecasting. *Annals of Tourism Research, 83*, 102925. https://doi.org/10.1177/0047287518759226

Li, D. (2019). Research on the application of big data technology in tourism industry. *International Journal of Frontiers in Engineering Technology, 1*(1). https://doi.org/10.4018/978-1-7998-1989-9.ch006

Li, J., Xu, L., Tang, L., Wang, S., & Li, L. (2018). Big data in tourism research: A literature review. *Tourism Management, 68*, 301–323. https://doi.org/10.1016/j.tourman.2018.03.009

Li, X., Law, R., Xie, G., & Wang, S. (2021). Review of tourism forecasting research with internet data. *Tourism Management, 83*, 104245. https://doi.org/10.1016/j.tourman.2020.104245

Li, X., Pan, B., Law, R., & Huang, X. (2017). Forecasting tourism demand with composite search index. *Tourism Management, 59*, 57–66. https://doi.org/10.1016/j.tourman.2016.07.005

Line, N. D., Dogru, T., El-Manstrly, D., Buoye, A., Malthouse, E., & Kandampully, J. (2020). Control, use and ownership of big data: A reciprocal view of customer big data value in the hospitality and tourism industry. *Tourism Management, 80*, 104106. https://doi.org/10.1016/j.tourman.2020.104106

Liu, A., Lin, V. S., Li, G., & Song, H. (2022). Ex ante tourism forecasting assessment. *Journal of Travel Research, 61*(1), 64–75. https://doi.org/10.1016/j.annals.2022.103445

Liu, H., Liu, Y., Li, G., & Wen, L. (2021). Tourism demand nowcasting using a LASSO-MIDAS model. *International Journal of Contemporary Hospitality Management*. https://doi.org/10.1108/ijchm-06-2020-0589

Lourenço, N., Gouveia, C. M., & Rua, A. (2021). Forecasting tourism with targeted predictors in a data-rich environment. *Economic Modelling, 96*, 445–454. https://doi.org/10.1016/j.econmod.2020.03.030

Lv, H., Shi, S., & Gursoy, D. (2022). A look back and a leap forward: A review and synthesis of big data and artificial intelligence literature in hospitality and tourism. *Journal of Hospitality Marketing & Management, 31*(2), 145–175. https://doi.org/10.1080/19368623.2021.1937434

Mariani, M. (2019). Big data and analytics in tourism and hospitality: A perspective article. *Tourism Review*. https://doi.org/10.1108/tr-06-2019-0259

Mariani, M., & Baggio, R. (2021). Big data and analytics in hospitality and tourism: A systematic literature review. *International Journal of Contemporary Hospitality Management*. https://doi.org/10.1108/ijchm-03-2021-0301

Nguyen, L. Q., Fernandes, P. O., & Teixeira, J. P. (2021). Analyzing and forecasting tourism demand in vietnam with artificial neural networks. *Forecasting, 4*(1), 36–50. https://doi.org/10.3390/forecast4010003

Ogbeide, G. C., Fu, Y. Y., & Cecil, A. K. (2020). Are hospitality/tourism curricula ready for big data? *Journal of Hospitality and Tourism Technology*. https://doi.org/10.1108/jhtt-09-2017-0081

Panas, G., Thrasidi, N., Halkiopoulos, C., & Gkintoni, E. (2022). Consumer behavior and cognitive factors in relation to gastronomic tourism and destination marketing in Greece. In V. Katsoni, A.C. Şerban (Eds.), *Transcending borders in tourism through innovation and cultural heritage*. Springer Proceedings in Business and Economics. Springer. https://doi.org/10.1007/978-3-030-92491-1_40

Panas, G., Vasiliadou, S., & Halkiopoulos, C. (2020). Data analysis evaluation of web technologies enhancing communication in tourism industry: Case study in Kefalonia Island. In V. Katsoni, T. Spyriadis (Eds.), *Cultural and tourism innovation in the digital era*. Springer Proceedings in Business and Economics. Springer. https://doi.org/10.1007/978-3-030-36342-0_14

Panteli, A., Kompothrekas, A., Halkiopoulos, C., &Boutsinas, B. (2021). An innovative recommender system for health tourism. In: V. Katsoni., & C. van Zyl (Eds.), *Culture and Tourism in a Smart, Globalized, and Sustainable World*. Springer Proceedings in Business and Economics. Springer, Cham. https://doi.org/10.1007/978-3-030-72469-6_42

Park, S., Hahn, S., Lee, T., & Jun, M. (2018). Two factor model of consumer satisfaction: International tourism research. *Tourism Management, 67*, 82–88. https://doi.org/10.1016/j.tourman.2018.01.007

Peterlin, J., Meško, M., Dimovski, V., & Roblek, V. (2021). Automated content analysis: The review of the big data systemic discourse in tourism and hospitality. *Systems Research and Behavioral Science, 38*(3), 377–385. https://doi.org/10.1002/sres.2790

Pu, Z., Du, H., Yu, S., & Feng, D. (2020, February). Improved tourism recommendation system. In *Proceedings of the 2020 12th International Conference on Machine Learning and Computing* (pp. 121–126). https://doi.org/10.1145/3383972.3384074

Qiu Zhang, H., Fan, D. X., Tse, T. S., & King, B. (2017). Creating a scale for assessing socially sustainable tourism. *Journal of Sustainable Tourism, 25*(1), 61–78. https://doi.org/10.1080/09669582.2016.1173044

Rahmadian, E., Feitosa, D., & Zwitter, A. (2021). A systematic literature review on the use of big data for sustainable tourism. *Current Issues in Tourism*, 1–20. https://doi.org/10.1080/13683500.2021.1974358

Sarigiannidis, C., Halkiopoulos, C., Giannopoulos, K., Giannopoulou, F., Politis, A., Boutsinas, B., & Kollias, K. (2021). Do hotels care? A proposed smart framework for the effectiveness of an environmental management accounting system based on business intelligence technologies. In V. Katsoni, C. van Zyl (Eds.), *Culture and tourism in a smart, globalized, and sustainable world*. Springer Proceedings in Business and Economics. Springer. https://doi.org/10.1007/978-3-030-72469-6_41

Shams, S. R., Vrontis, D., & Christofi, M. (2022). Stakeholder causal scope analysis–centered big data management for sustainable tourism. *Journal of Hospitality & Tourism Research, 46*(5), 972–978. https://doi.org/10.1177/1096348020978547

Silva, E. S., Hassani, H., Heravi, S., & Huang, X. (2019). Forecasting tourism demand with denoised neural networks. *Annals of Tourism Research, 74*, 134–154. https://doi.org/10.1016/j.annals.2018.11.006

Solakis, K., Katsoni, V., Mahmoud, A. B., & Grigoriou, N. (2022). Factors affecting value co-creation through artificial intelligence in tourism: A general literature review. *Journal of Tourism Futures*. https://doi.org/10.1108/jtf-06-2021-0157

Song, H., Liu, A., Li, G., & Liu, X. (2021). Bayesian bootstrap aggregation for tourism demand forecasting. *International Journal of Tourism Research, 23*(5), 914–927. https://doi.org/10.1002/jtr.2453

Stylos, N., & Zwiegelaar, J. (2019). Big data as a game changer: How does it shape business intelligence within a tourism and hospitality industry context?. In *Big data and innovation in tourism, travel, and hospitality* (pp. 163–181). Springer. https://doi.org/10.1007/978-981-13-6339-9_11

Stylos, N., Zwiegelaar, J., & Buhalis, D. (2021). Big data empowered agility for dynamic, volatile, and time-sensitive service industries: The case of tourism sector. *International Journal of Contemporary Hospitality Management*. https://doi.org/10.1108/ijchm-07-2020-0644

Sun, Q., Huang, X., & Liu, Z. (2022). Tourists' digital footprint: Prediction method of tourism consumption decision preference. *The Computer Journal*. https://doi.org/10.1093/comjnl/bxab210

Theodorakopoulos, L. (2022). *Big data analysis in humanities and economics with machine learning techniques and use of cloud computing technologies*. Doctoral dissertation, University of Patras, Greece. https://doi.org/10.12681/eadd/51196

Theodorakopoulos, L., Antonopoulou, H., Mamalougou, V., & Giotopoulos, K. C. (2022). The drivers of volume volatility: A big data analysis based on economic uncertainty measures for the Greek banking system. *Banks and Bank Systems, 17*(3), 49–57. https://doi.org/10.21511/bbs.17(3).2022.05

Theodorakopoulos, L., & Papadopoulos, D. (2022, July). The management of big data in online marketing: Challenges and opportunities. A theoretical review. In *10th International Conference on Contemporary Marketing Issues (ICCMI)*.

Tian, F., Yang, Y., Mao, Z., & Tang, W. (2021). Forecasting daily attraction demand using big data from search engines and social media. *International Journal of Contemporary Hospitality Management, 33*(6), 1950–1976. https://doi.org/10.1108/ijchm-06-2020-0631

Tibshirani, R. (1996). Regression shrinkage and selection via the lasso. *Journal of the Royal Statistical Society: Series B (methodological), 58*(1), 267–288. https://doi.org/10.1111/j.1467-9868.2011.00771.x

Volo, S. (2020). Tourism statistics, indicators and big data: A perspective article. *Tourism Review*. https://doi.org/10.1108/tr-06-2019-0262

Wang, L., Wu, B., Zhu, Q., & Zeng, Y. R. (2020). Forecasting monthly tourism demand using enhanced backpropagation neural network. *Neural Processing Letters, 52*(3), 2607–2636. https://doi.org/10.1007/s11063-020-10363-z

Weaver, A. (2021). Tourism, big data, and a crisis of analysis. *Annals of Tourism Research, 88*, 103158. https://doi.org/10.1016/j.annals.2021.103158

Wen, L., Liu, C., Song, H., & Liu, H. (2021). Forecasting tourism demand with an improved mixed data sampling model. *Journal of Travel Research, 60*(2), 336–353. https://doi.org/10.1177/0047287520906220

Xie, G., Qian, Y., & Wang, S. (2020). A decomposition-ensemble approach for tourism forecasting. *Annals of Tourism Research, 81*, 102891. https://doi.org/10.1016/j.annals.2020.102891

Xu, F., Nash, N., & Whitmarsh, L. (2020). Big data or small data? A methodological review of sustainable tourism. *Journal of Sustainable Tourism, 28*(2), 144–163. https://doi.org/10.1080/09669582.2019.1631318

Yallop, A., & Seraphin, H. (2020). Big data and analytics in tourism and hospitality: Opportunities and risks. *Journal of Tourism Futures*. https://doi.org/10.1108/jtf-10-2019-0108

Yang, Y., Fan, Y., Jiang, L., & Liu, X. (2022). Search query and tourism forecasting during the pandemic: When and where can digital footprints be helpful as predictors? *Annals of Tourism Research, 93*, 103365. https://doi.org/10.1016/j.annals.2022.103365

Yao, Y., & Cao, Y. (2020). A neural network enhanced hidden Markov model for tourism demand forecasting. *Applied Soft Computing, 94*, 106465. https://doi.org/10.1016/j.asoc.2020.106465

Zhang, J. (2021). Big data and tourism geographies–an emerging paradigm for future study?. In *Tourism spaces* (pp. 131–136). Routledge. https://doi.org/10.1080/14616688.2018.1519719

Zhang, J., & Dong, L. (2021). Image monitoring and management of hot tourism destination based on data mining technology in big data environment. *Microprocessors and Microsystems, 80*, 103515. https://doi.org/10.1016/j.micpro.2020.103515

Integration of Blockchain Technology in Tourism Industry: Opportunities and Challenges

Constantinos Halkiopoulos, Hera Antonopoulou, and Nikos Kostopoulos

Abstract Tourism is crucial to people's everyday life and the global economy. However, despite its significance, the conventional tourism business confronts significant development obstacles. Consequently, incorporating digital technology into the traditional tourism business in order to further cut costs and increase efficiency is regarded as an absolute necessity. As a nascent technology, blockchain has the potential to revolutionize the tourism sector since it provides a secure platform connecting the tourism business and its customers (employees and tourists). Through this endeavor, a comprehensive literature assessment on blockchain technology in smart tourism will be conducted. In addition, the Internet of Things (IoT) and large-scale analytics can be utilized to enhance the tourist experience. It is important to note that during the COVID-19 pandemic, the tourism industry's issues were more standardized. Therefore, the conventional tourism business should be swiftly modernized. From the perspective of smart tourism, blockchain can serve as a platform that reliably and efficiently connects travelers with tourism products.

Keywords Blockchain · Smart tourism · Tourism industry · Innovative platforms · Tourist experience · Data analytics

JEL Classification O32 · O33 · Z32 · L86

1 Introduction

New technologies have enabled significant advancements in information and communication in modern society. Internet and modern IT applications, which were merely concepts a few decades ago, have facilitated a substantial expansion in numerous industries. Modern information and communication technologies have become an

C. Halkiopoulos (✉) · H. Antonopoulou · N. Kostopoulos
EDILAB, Department of Management Science and Technology,
University of Patras, Patras, Greece
e-mail: halkion@upatras.gr

© The Author(s), under exclusive license to Springer Nature Switzerland AG 2023

V. Katsoni (ed.), *Tourism, Travel, and Hospitality in a Smart and Sustainable World*,
Springer Proceedings in Business and Economics,
https://doi.org/10.1007/978-3-031-26829-8_22

integral part of our daily lives. Moreover, this is because, as a result of the development of technology, we are able to better plan our obligations, perform tasks with greater ease, receive immediate, accurate, and timely information, and generally enhance the quality of our daily lives. As far as the tourism, travel, and hospitality industry is concerned, technological advancements and their applications have optimized many aspects of tourism to a great extent. Tourists are constantly seeking innovative modern IT applications (Halkiopoulos et al., 2021b) that will be beneficial and useful for their travel, tours, safety, and in general, an optimally good guide for each travel destination. It is understood that tourists travel and move in unfamiliar environments and seek information through modern IT platforms and applications to enhance their travel experiences (Dickinson et al., 2016; Giannoukou, et al., 2022).

Moreover, it often appears that technology can influence a person's expectations regarding their vacation, destination, and activities; thus, the search for any destination is directly dependent on an application (Halkiopoulos et al., 2020; Xiang & Fesenmaier, 2022). Likewise, the local tourism businesses of a location must prioritize the satisfaction of the visitors' needs based on their preferences (Halkiopoulos et al., 2021a). Therefore, each local tourism business must have the appropriate digital solutions in order to provide visitors with accurate information about the destination in question (Panteli et al., 2021). Tourists are no longer passive observers but actively contribute to the creation of value for themselves and the tourist destination (Binkhorst & Dekker, 2009). Improving the tourist experience is only possible if the visitor's interests and expectations are catered to (Benyon et al., 2013). Therefore, tourist destinations and local tourism businesses must be aware of the use and design of innovative digital technologies, as they contribute significantly to economic growth (Giotopoulos et al., 2019). Notables are the influence of digital technology and the significance of its incorporation as a means of informing the tourist public.

Due to the aforementioned factors, the traditional tourism industry is in dire need of new and innovative digital technologies to reduce costs and increase efficiency. Blockchain is an emerging technology that promises to revolutionize the tourism industry by providing a dependable solution for the valid, certified, and secure connection between the tourism industry, tourism product, and tourism audience (Önder & Gunter, 2020; Bamakan et al., 2021). However, the existing smart tourism solutions based on blockchain technologies, which we will describe below, are insufficient for resolving the fundamental challenges facing the tourism industry (Bulut, 2021). To implement the innovative solutions offered by blockchain technology in the tourism industry via smart technological tourism products, it is necessary to invest capital (Mohamed & Al-Jaroodi, 2019).

2 Literature Review

2.1 ICT and Tourism

The development and application of digital technologies are radically altering how people live, work, travel, and conduct business, thereby transforming and reshaping the tourism industry. The scope and adoption of digital technologies vary by nation, industry, organization, and geographical location. Digital marketing and e-commerce have received considerable focus to reach new markets, attract customers, and build a brand. However, while these technologies may increase market access and awareness, boost connectivity, and facilitate financial transactions, they are less effective at boosting productivity and innovation in a global marketplace that is becoming increasingly competitive. In the tourism industry, productivity-enhancing technologies such as cloud computing, data analytics, and revenue management software have been generally poorly adopted. In contrast, innovative technologies such as augmented reality and geotagging create, adapt, and deliver new visitor products, services, and experiences in an increasingly innovative manner (Halkiopoulos & Papadopoulos, 2022).

Tourism businesses operate on a global market while simultaneously offering locally as part of a unique tourist destination offer (Panas et al., 2017). The geographical distribution of businesses is only constrained by the allure and accessibility of locations. Digital businesses, such as online travel agencies and accommodation platforms, have revolutionized tourism by connecting tourism products and services to customers in real-time and non-real-time from anywhere in the world (Katsoni et al., 2012).

It is noteworthy that there has been a recent surge in the implementation of new commercial applications and/or software around the world in order to expand the tourism industry. Presently, tourists use the Internet on their mobile devices, which affords them extraordinary opportunities. Numerous tourist destinations offer technological systems, such as mobile applications and sensors, to assist tourists in interacting with their environment for information search and navigation, cultural attraction interpretation. At the same time, tourism businesses can target customers worldwide due to the unique opportunities afforded by new technologies and innovative digital marketing techniques, like neuromarketing, for redesigning tourism products to meet the needs and desires of travelers (Halkiopoulos et al., 2022).

Due to the increased use of new technologies, the increased demand for flexible, personalized options and high-quality information has personalized leisure and tourism behavior. Through new technologies and social and economic evaluations (e.g., social media platforms like Facebook, Twitter, and blogs), it is possible to exchange information and evaluations about the destination, the quality of services in hotels and restaurants, and the purchasing of local goods. This flexibility in traveler options and service offerings has dramatically increased travelers' demands for high-quality, cost-effective services and products (Tussyadiah et al., 2017). In this manner, most tourism organizations strive to keep up with the latest technological trends (Sarigiannidis et al., 2021).

Moreover, in a world of digitization and limitless possibilities, it is no longer a thing of the past to find new ways to optimize travel and promote appealing services (Solakis et al., 2022). Innovative strategies can now result in highly sophisticated applications with extraordinary capabilities for travelers, stimulating games, and virtual reality applications (Katsoni & Dologlou, 2017). Clearly, technological trends are constantly evolving, and as a result, competition in the tourism and travel application development industry is exceptionally fierce. According to Think with Google Travel Trends, the amount of time spent watching mobile TV travel diaries and blogs has increased significantly over the past two years. Numerous mobile users downloaded and utilized a travel-related smartphone application. Before embarking on a vacation, travelers want to know everything about their destination, from the entertainment options to the room specifications. Likewise, 55% of travelers claim they must consult an excessive number of sources of travel information before making a decision (Halkiopoulos & Giotopoulos, 2022). Fortunately, recent technological advancements can help achieve this objective, and the use of modern technology in the creation of travel applications is gaining popularity (Katsoni & Fyta, 2021). Without a doubt, mobile application development is the industry's lifeblood. Although travel agents, suppliers, and some customers prefer booking on a desktop computer, the majority of users prefer to conduct the majority of their transactions on their mobile devices.

Because of this, the majority of people search for and book their travel tickets, food, and lodging online (Panas et al., 2022). As a result of the widespread use of the Internet, the tourism and hospitality industry has undergone significant transformations. To achieve transparency and security in transactions, many travel companies, such as Expedia Group, BCD Travel, Uber, Ola, and Airbnb, have replaced their traditional business models with Consumer-to-Consumer (C2C) models in the digital age. Additionally, the tourism industry has a significant demand for innovative platforms that integrate technology, money, and knowledge. TUI and a number of other businesses have already begun using blockchain technology to implement functions such as ticket booking and payment processing. Several companies, including Expedia, CheapAir, Webjet, and One-Shot Hotels, accept Bitcoin for travel-related transactions, such as booking and reserving tickets.

3 Blockchain Technology in Tourism Industry

The globe is undergoing a rapid transformation, to which all sectors of the economy are adapting. In addition, tourism is not an exception to this norm. The travel industry's digital footprint is constantly expanding. It utilizes modern technologies, such as blockchain technology (Stamatiou et al., 2022; Johng et al., 2018; Stallone et al., 2021). Blockchain is regarded as one of the twenty-first century's most significant discoveries (Nakamoto, 2017). Beyond the financial industry (Antonopoulou et al., 2022; Theodorakopoulos et al., 2022; Gousteris et al., 2022), blockchain technology has the potential to be utilized in a vast array of industries (Hussien et al., 2021; Kar & Navin, 2021; Kostopoulos et al., 2022) which has attracted the interest of researchers. Blockchain technology was developed to serve as the central protocol for preserving Bitcoin (Bentov & Kumaresan, 2014; Kishigami et al., 2015). This technology's concept is focused on the organization and storage of data according to a specified logic (Fernandez-Vazquez & de la Fuente, 2022). The data and information are encrypted as opposed to being recorded and stored in a central server database (Morkunas et al., 2019). In user-shared distributed databases, a duplicate is created on each network-connected node. Tourism is one of the industries where blockchain technology can be utilized (Melkić & Čavlek, 2020; Puthal et al., 2018). The potential of blockchain technology is currently being researched and utilized by tourism professionals worldwide (Treiblmaier, 2022). Examples include the use of blockchain technology in payments in the form of cryptocurrencies, minimizing transaction costs, the application of cryptocurrencies for company loyalty programs, reward programs for the tourist public, facilitating the storage and verification of customer data, as well as in reducing intermediaries, and, therefore, transaction costs, which can also have a positive effect on the profit of the tourist public as well as companies, which can increase revenue (Valeri & Baggio, 2020). Below is a list of services where blockchain technology can be utilized with great success and maximize the outcome for both tourists and the tourism business (Ozdemir et al., 2020).

3.1 Booking Administration and Blockchain

Numerous Internet booking agencies dominate the travel sector, acting as mediators between customers and services such as hotels, boat and helicopter charter companies, and leisure services, among others. A minimum 15% commission is charged by agencies for their services, increasing the price of the product for both the customer and the service provider (Pilkington, 2017). In a system of decentralized booking markets, these intermediaries are no longer required, and tourism service providers communicate directly with the traveling public. Additionally, transactions, pricing, and potential fees are transparent, resulting in a better experience for tourism consumers (Ming & Wei, 2021).

3.2 Credentials and Blockchain

Identifying the tourist population is vital for all facets of the tourism sector, particularly for assuring the safety of transportation and border control certification. Through a zero-knowledge protocol (zero-knowledge proof) that enables the documentation of identity documents without the need to reveal sensitive information, blockchain technology enables the public to be tracked at every stage of a trip, saving time and money. As a result, blockchain has penetrated the tourism sector in a significant way, with tourism heavyweights like as Marriott, TUI, AXA, and others adopting the technology to protect and optimize the tourism product supplied in various service areas (Li et al., 2021).

3.3 Customer Loyalty Programs and Blockchain

In general, businesses in the travel industry offer loyalty incentives. Standard perks include of airline miles, free stays, etc. However, users frequently feel the rewards system to be limiting. Digital tokens present a chance to improve the security, exchangeability, and transparency of these fees. By incorporating tokens into reward programs and so transforming them into a decentralized value network, the consumer can now receive the reward of their choosing. Now, products and services offered by various companies are transparent and comparable, prizes can be readily transferred between reward programs, and reward points can be transformed into digital assets (Banerji et al., 2021).

3.4 Tracking of Luggage and Blockchain

Lost luggage is a common occurrence in the travel industry and a significant issue for both passengers and businesses. The problem stems from the fact that baggage in transit from point A to point B is handled by multiple parties, including airlines, security, ground employees, and others. Blockchain technology can facilitate a decentralized and transparent network for the tracking of baggage. Thus, the interested parties are able to discover the luggage without having to utilize complicated databases. Moreover, employing novel contract technology, the performance of tasks and compensations can be determined and automated (Mul et al., 2021).

3.5 Insurance for Travel and Blockchain

The nature of travel insurance necessitates the use of high-quality data, and smart contracts' ability to automate choices based on this data make blockchain technology suitable for administering travel insurance. Using blockchain in cases of claims against corporations for missing luggage or flight delays, for instance, a smart contract utilizing experimental data can track the claim's path at any time (Kar & Navin, 2021).

3.6 Internet of Things, Big Data, and Blockchain

Moreover, the Internet of Things (IoT) and large amounts of analytical data can be leveraged to enhance tourists' sightseeing experiences (Qian et al., 2018). However, they are unable to convince tourists to become repeat consumers or to try new products. From the perspective of smart tourism, blockchain will likely serve as a platform for establishing a reliable connection between travelers and attractions. The blockchain appears to be a promising supplement for visitor incentives (Luo & Zhou, 2021).

3.7 Blockchain and Transactions

Recent advancements in online payment alternatives appear to be another aspect that is beginning to influence the expansion of travel applications. The influence of blockchain technologies and cryptocurrencies on economies has been enormous. The tourism industry appears to be an early user of this new wave of e-commerce, which is gradually surpassing its commerce on mobile apps (Garg et al., 2021). Understandably, tourists desire a secure currency that is widely accepted across borders. Through its simplicity, the concept of blockchain technology adheres to the fluidity of movement. Such as removing intermediaries and increasing direct fund transfer (Robinson et al., 2022). Perhaps therefore big travel businesses like Expedia are integrating cryptocurrency payments into their trip offerings (Seebacher & Schüritz, 2017).

4 Blockchain Applications in Tourism Industry

4.1 LockTrip App

LockTrip (LockTrip–where travel meets the Blockchain!) is a complex project with several application layers. At the initial level, travel is purchased and hosted via the LockTrip.com platform. Due to the sophisticated capabilities of blockchain technology, it supposedly enables users to book hotels and vacation rentals for less than leading competitors such as Booking.com, Expedia, and Airbnb. It features a fully decentralized booking engine–the LOC Ledger (based on the Ethereum VM)–and a marketplace that gradually integrates with the decentralized LOC Ledger at a second level. In addition, property owners and hoteliers can integrate their short-term rentals and rooms into the decentralized LOC ecosystem in a single step via the platform. Additionally, the platform features an integrated exchange mechanism that changes any foreign currency into LOC at the moment of booking. LockTrip is also free to use for both hosts and guests, regardless of the number of properties or revenue made.

4.2 SmartTrip App

The SmartTrip platform is a decentralized ecosystem that connects travelers and service providers, providing all the functionality necessary for safe, comfortable, authentic travel. Most individuals find it difficult to plan a vacation due to the various factors that complicate the process, such as foreign currencies, language hurdles, and locating a service that suits their needs. The platform is intended to be the ultimate solution to the problems that plague travelers and travel service providers: it is an ecosystem that enables all users to interact directly, organize trips quickly and efficiently, save money, and engage in a dynamic community. The Smart Trip platform will not only provide a vast array of services, ranging from boat rentals to safaris, but will also serve as a valuable knowledge repository. It is an entirely new travel concept that minimizes research time and maximizes the authenticity of the experience (Ozdemir et al., 2020).

4.3 GOureka App

GOeureka (GOT), a travel technology firm utilizing blockchain technology to develop a hotel booking platform, announced the introduction of the alpha version of its platform, which enables consumers to book hotel rooms with no commission and no hidden fees, in addition to receiving loyalty rewards. GOeureka is among the first in the industry to test bookings utilizing blockchain technology via smart contracts on Ethereum, which is made possible in part by the massive existing hotel infrastructure. The Board of Directors of GOeureka is comprised of C-Level executives from Accor Hotels and Agoda, among others in the worldwide hotel business. The rapid expansion of the number of bookable hotels through GOeureka (400,000) gives it a considerable advantage over competitors within and outside the blockchain industry.

Several booking sites dominate the hotel reservation market, charging commissions of 10–30% on top of the room pricing. This fee is assessed in the booking site's backend, with little cost transparency for the consumer. Consequently, the retail price for a single room varies significantly based on the transactions between the hotel, the bed bank wholesaler, and the internet booking site.

GOeureka streamlines and reduces the cost of hotel bookings, providing the consumer with a fair and consistent price and facilitating transactions at the hotel. By eliminating intermediaries and allowing the hotel to deal directly with the consumer, incidental fees are eliminated. The consumer reaps the benefits and receives considerably better prices as a result. Similarly, hotels can shorten the supply chain to increase efficiency and minimize costs.

4.4 Etherisc App

Etherisc (https://etherisc.com) is a platform for developing decentralized insurance applications. Etherisc is developing blockchain-based decentralized apps for several sectors of the insurance industry, including airline delay insurance. In addition, the company focuses on reducing inefficiencies, such as excessive processing fees and lengthy claims processing times, through the use of blockchain technology. Etherisc has already produced six innovative decentralized insurance applications; some of them also collaborate in the tourism industry.

4.5 Beenest App

Beenest is a decentralized network allowing crypto enthusiasts to share their homes. Similarly, to Airbnb, users of Beenest can reserve homes using the company's Bee Token. App for Blockchain Insurance: Together with WeTrust, the company will build blockchain-based insurance for Beesnest homeowners.

5 Smart Contracts in Tourism Industry

5.1 Definition

The use of blockchain technology to generate fully automated contracts, i.e., agreements that are implemented without human intervention, is one of its most promising applications (Wang et al., 2021). In the technology context, such agreements are frequently referred to as smart contracts (Karimov, 2019). Specifically, smart contracts are conventional contracts implemented in computer code and designed to operate in a blockchain context (Wu et al., 2021). Blockchain will automatically validate, execute, and enforce the contract terms between the parties. These contracts are referred to as "smart" because they can partially or completely execute themselves. Consequently, a smart contract is a piece of code that is placed in a blockchain, activated by transactions, and then reads and writes data based on this blockchain. The six defining criteria of a smart contract (Ølnes et al., 2017; Savelyev, 2017) are as follows:

1. fully electronic nature,
2. software implementation,
3. heightened assurance,
4. conditional nature,
5. self-performance, and
6. self-sufficiency.

In conclusion, a smart contract that executes on the blockchain operates autonomously. If the provisions of the contract are met, payments or other exchanges are made in accordance with the terms of the contract. Similarly, payments may be withheld if the contract terms are not satisfied if they are subscribed to the smart contract. On a decentralized network of computers on the blockchain, smart contracts are performed as programmed, eliminating the dangers associated with illegal alterations. The contract operates automatically, exchanging value and payments between parties without recourse to attorneys or courts. Nodes and their actions in the blockchain, for instance, are timestamped and cannot be altered. This gives a great foundation for contracts, as all contract modifications are timestamped (Aggarwal & Kumar, 2021).

Contracts can be preserved (and new versions can be generated) while retaining older versions (as well as accurate timestamps on all edits and revisions). Because the ledger cannot be manipulated, it provides a more accurate representation of the processes that occurred and encourages all parties involved to be more truthful about the transactions that occurred. All the aforementioned demonstrates that the smart contract goes much beyond traditional contracting paradigms (Rashideh, 2020).

5.2 *Using Smart Contracts*

Many businesses have begun to develop applications utilizing smart contracts based on blockchain technology (Feng et al., 2019; Zhao et al., 2019). As an example, Ascribe enables numerous artists to assert ownership of their work and distribute limited edition prints. Ascribe leverages blockchain to track the provenance of all original compositions and transactions within them. It also features a marketplace where artists can promote, and individuals can buy and sell art online. Then, we discuss two potential use cases, Airbnb and Uber, as well as how smart contracts will impact enterprises (Chowdhury, 2019).

5.3 *Airbnb is Now bAirbnb*

Airbnb is now the largest provider of rooms in the world, based on market value and rooms held. However, lodging suppliers receive only a portion of the value they generate. Airbnb maintains all renter and customer data, which creates privacy concerns. International payments are processed through intermediary authorities, which collect commissions on each transaction and major swaps from the top. The blockchain concept, however, opens the door to new business opportunities, such as the member-only ownership cooperative. Specifically, bAirbnb is a distributed application consisting of smart contracts that hold home listings on a blockchain. The bAirbnb app provides a clean interface where owners may upload property information and images. The platform keeps track of the reputation scores of suppliers and tenants in order to improve the judgments of both parties. The user experience of bAirbnb is identical to that of Airbnb, with the exception that peer-to-peer communication on the network is conducted via encrypted and secure messages that are not maintained in a central database, as is the case with Airbnb. Only you and the room's owner are able to read these messages.

Using Internet of Things (IoT)-connected smart locks, property is accessed. When you arrive, your phone can sign a message containing evidence of payment using your public key, and the smart lock will unlock for you. Therefore, the owners are not required to provide you with a key or visit the property. In addition, both you and the owner now save money on Airbnb fees, there are no foreign exchange fees for international contracts, and settlements are secured immediately. Lastly, rental history data is secure and confidential only between individuals who have agreed to it. This is the true value-sharing economy. Customers and service providers both benefits.

5.4 *To BUber from Uber*

Airbnb, Uber, Lyft, and TaskRabbit are examples of companies that participate in the sharing economy. In the case of Uber, cars and drivers are brought together on a common platform, and their services are then resold to individuals in need of a trip. They provide exceptional service; however, it has certain restrictions. Drivers are notoriously erratic or violent. All are susceptible to price increases. In addition, truckers lack collective negotiating strength. As a result, Uber takes 20% of the value of what is created, which is disproportionately high. During this process, they are also collecting information about where individuals are and where they are going, which might be exploited for commercial gain in the future. In the event that Uber is constructed using blockchain technology. Let's refer to a blockchain-based application as blockchain Uber or BUber. So that anyone may obtain a copy of the BUber app, similar to Uber, and log in similarly to how they do with the Uber app. BUber understands the type of car you desire, your destination, and your desired speed, and instantly matches you with a car that satisfies these characteristics. BUber also features a built-in payment system, similar to Ethereum and Ether or bitcoin on the Bitcoin network, thus there is no need for a centralized intermediary to process payments. In addition, there is no need for a central middleman for further capabilities, as the distributed application can perform any action, including paying for gasoline, discussing liability, purchasing insurance, etc.

6 Discussion and Conclusion

According to the preceding article, incorporating blockchain technology into the traditional tourism business in order to further cut costs and increase efficiency is a must. As a nascent technology, blockchain has the potential to revolutionize the tourism sector since it provides a secure platform connecting the tourism business and its customers (employees and tourists). In this paper, a literature review on the usage of blockchain technology in smart tourism was conducted. In addition, the Internet of Things (IoT) and detailed analytics can help enhance the tourist experience. It is important to note that during the COVID-19 pandemic, the tourism industry's issues were more standardized. Therefore, the conventional tourism business should be swiftly modernized.

As stated previously, millions of dollars are continually invested in tourism and travel to develop new technology and applications that enhance the visitor experience. Furthermore, this is the case since international tourism generates large annual earnings. According to surveys and statistical data, visitors are always searching for new apps for their travels, they download apps prior to and during their excursions, and their whole tourist experience is highly dependent on such applications. Referring to the Greek region, the offered statistics indicate that tourism in Greece, and cruises in particular, contribute significantly to the Greek economy. Therefore, any investment in this sector could only be advantageous.

From the perspective of smart tourism, blockchain can serve as a platform that reliably and efficiently connects travelers with tourism products. According to the findings, blockchain can be utilized for more than just financial services and can bring solutions to a wide range of problems, including hurdles and challenges linked to trust and security. Blockchain technology is still in its infancy, and its standards and versions are continually growing. However, assessments indicate that blockchain technology research and tourism applications are expanding. Current research trends in the tourism business indicate that blockchain technology is mostly employed for safe data exchange, access control, and payment security. Therefore, blockchain's promise remains mostly untapped. Smart contracts, which allow the automation of activities on a blockchain network, could be utilized more often. Most of the studies could also include a prototype implementation or implementation-specific discussion.

In conclusion, it is important to note the potential obstacles that may arise; the deployment of blockchain technology is still in its infancy. Consensus exists regarding its potential, but it is impossible to predict where it will go and how it will affect businesses such as tourism. There are two primary obstacles in this field. The first is identification of the individual. This is due to the fact that technology, by its very nature, may pose a considerable barrier, especially when personal information is involved. Therefore, it is impossible to regulate personal information on the Internet by legislation such as the GDPR in the EU or the CCPA in the USA. Standardization may also prove to be a formidable barrier. Although blockchain is the underlying technology, each firm, organization, or group can develop its own network and include its own algorithms (like with cryptocurrency), making intersystem communication challenging. The rapid acceptance of this technology in numerous industries and the altering behavior of customers, which increases their acquaintance with these concepts, leads us to the conclusion that it will become a more common notion in our lives as it adapts to the changing demands and problems of users.

References

Antonopoulou, H., Mamalougou, V., & Theodorakopoulos, L. (2022). The role of economic policy uncertainty in predicting stock return volatility in the banking industry: A big data analysis. *Emerging Science Journal, 6*(3), 569–577. https://doi.org/10.28991/esj-2022-06-03-011

Aggarwal, S., & Kumar, N. (2021). Blockchain 2.0: Smart contracts. In *The Blockchain technology for secure and smart applications across industry verticals* (pp. 301–322). https://doi.org/10.1016/bs.adcom.2020.08.015

Bamakan, S. M. H., Babaei Bondarti, A., Babaei Bondarti, P., & Qu, Q. (2021). Blockchain technology forecasting by patent analytics and text mining. *Blockchain: Research and Applications, 2*(2), 100019. https://doi.org/10.1016/j.bcra.2021.100019

Banerji, D., Rashideh, W., Arora, B., & Pratihari, A. R. (2021). Application potential of Blockchain technologies in the travel and tourism industry. In T. Choudhury, A. Khanna, T.T. Toe, M. Khurana, & N. Gia Nhu (Eds.), *Blockchain applications in IoT ecosystem.* EAI/Springer Innovations in Communication and Computing. Springer. https://doi.org/10.1007/978-3-030-65691-1_19

Bentov, I., & Kumaresan, R. (2014). How to use bitcoin to design fair protocols. Lecture Notes in Computer Science (pp. 421–439). https://doi.org/10.1007/978-3-662-44381-1_24

Benyon, D., Quigley, A., O'Keefe, B., & Riva, G. (2013). Presence and digital tourism. *AI & Society, 29*(4), 521–529. https://doi.org/10.1007/s00146-013-0493-8

Binkhorst, E., & Den Dekker, T. (2009). Agenda for co-creation tourism experience research. *Journal of Hospitality Marketing & Management, 18*(2–3), 311–327. https://doi.org/10.1080/19368620802594193

Bulut, E. (2021). Blockchain-based entrepreneurial finance: Success determinants of tourism initial coin offerings. *Current Issues in Tourism, 25*(11), 1767–1781. https://doi.org/10.1080/13683500.2021.1980505

Chowdhury, N. (2019). Ethereum and smart contracts. In *Inside blockchain, bitcoin, and cryptocurrencies* (pp. 93–108). https://doi.org/10.1201/9780429325533-5

Dickinson, J. E., Hibbert, J. F., & Filimonau, V. (2016). Mobile technology and the tourist experience: (Dis)connection at the campsite. *Tourism Management, 57*, 193–201. https://doi.org/10.1016/j.tourman.2016.06.005

Feng, Q., He, D., Zeadally, S., Khan, M. K., & Kumar, N. (2019). A survey on privacy protection in blockchain system. *Journal of Network and Computer Applications, 126*, 45–58. https://doi.org/10.1016/j.jnca.2018.10.020

Fernandez-Vazquez, S., & de la Fuente, D. (2022). The forgotten cryptocurrencies: Beyond bitcoin. In *Implementing and leveraging blockchain programming* (pp. 193–203). https://doi.org/10.1007/978-981-16-3412-3_13

Garg, P., Gupta, B., Chauhan, A. K., Sivarajah, U., Gupta, S., & Modgil, S. (2021). Measuring the perceived benefits of implementing blockchain technology in the banking sector. *Technological Forecasting and Social Change, 163*, 120407. https://doi.org/10.1016/j.techfore.2020.120407

Giannoukou, I., Nikolopoulou, A., & Halkiopoulos, C. (2022). Tourists' and employees' psychology in the framework of strategic management in Greek tourism. In V. Katsoni, A.C. Şerban (Eds.), *Transcending borders in tourism through innovation and cultural heritage.* Springer Proceedings in Business and Economics. Springer. https://doi.org/10.1007/978-3-030-92491-1_42

Giotopoulos, K., Halkiopoulos, C., Papadopoulos, D., & Antonopoulou, H. (2019). Towards bring your own device marketing policy. *International Journal of Technology Marketing, 13*(2). Inderscience Publishers: IJTMKT-10313. https://doi.org/10.1504/ijtmkt.2019.10023906

Gousteris, S., Stamatiou, Y. C., Halkiopoulos, C., Antonopoulou, H., & Kostopoulos, N. (2023). Secure distributed cloud storage based on the blockchain technology and smart contracts. *Emerging Science Journal, 7*(2), 469–479. https://doi.org/10.28991/esj-2023-07-02-012

Halkiopoulos, C., Antonopoulou, H., & Gkintoni, E. (2021a). Consumer behavioral data analysis for decision making process: Cognitive and emotional factors. *International Journal of Innovative Science and Research Technology.* ISSN: 2456-2165.

Halkiopoulos, C., Dimou, E., Kompothrekas, A., Telonis, G., & Boutsinas, B. (2021b). The E-tour facilitator platform supporting an innovative health tourism marketing strategy. In V. Katsoni, C. van Zyl (Eds.), *Culture and tourism in a smart, globalized, and sustainable world*. Springer Proceedings in Business and Economics. Springer. https://doi.org/10.1007/978-3-030-72469-6_39

Halkiopoulos, C., Antonopoulou, H., Gkintoni, E., & Aroutzidis, A. (2022). Neuromarketing as an indicator of cognitive consumer behavior in decision-making process of tourism destination—An overview. In V. Katsoni, A.C. Şerban (Eds.), *Transcending borders in tourism through innovation and cultural heritage*. Springer Proceedings in Business and Economics. Springer. https://doi.org/10.1007/978-3-030-92491-1_41

Halkiopoulos, C., Giotopoulos, K., Papadopoulos, D., Gkintoni, E., & Antonopoulou, H. (2020). Online reservation systems in e-business: Analyzing decision making in e-tourism. *Journal of Tourism, Heritage & Services Marketing (JTHMS), 6*(1), 9–16. ISSN: 2529-1947. https://doi.org/10.5281/zenodo.3603312

Halkiopoulos, C., & Giotopoulos, K. (2022). Tourism's use of web-based information systems and the influence of tourism trends. Springer proceedings in business and economics, 407–426. https://doi.org/10.1007/978-3-030-92491-1_25

Halkiopoulos, C., & Papadopoulos, D. (2022). Computational methods for evaluating web technologies and digital marketing techniques in the hospitality industry. Springer proceedings in business and economics, 387–406. https://doi.org/10.1007/978-3-030-92491-1_24

Hussien, H. M., Yasin, S. M., Udzir, N. I., Ninggal, M. I. H., & Salman, S. (2021). Blockchain technology in the healthcare industry: Trends and opportunities. *Journal of Industrial Information Integration, 22*, 100217. https://doi.org/10.1016/j.jii.2021.100217

Johng, H., Kim, D., Hill, T., & Chung, L. (2018). Using blockchain to enhance the trustworthiness of business processes: A goal-oriented approach. In *2018 IEEE International Conference on Services Computing (SCC)*. https://doi.org/10.1109/scc.2018.00041

Kar, A. K., & Navin, L. (2021). Diffusion of blockchain in insurance industry: An analysis through the review of academic and trade literature. *Telematics and Informatics, 58*, 101532. https://doi.org/10.1016/j.tele.2020.101532

Karimov, E. (2019). How does it work? Components of blockchain. In *Blockchain in swift*. https://doi.org/10.1007/978-1-4842-5382-3_2

Katsoni, V., & Dologlou, N. (2017). ICT applications in smart ecotourism environments. In *Smart cities in the mediterranean* (pp. 225–244). https://doi.org/10.1007/978-3-319-54558-5_11

Katsoni, V., & Fyta, A. (2021). From pausanias to baedeker and trip advisor: Textual proto-tourism and the engendering of tourism distribution channels. *Turyzm/tourism, 31*(1), 11–19. https://doi.org/10.18778/0867-5856.31.1.11

Katsoni, V., Giaoutzi, M., & Nijkamp, P. (2012). Market segmentation in tourism: An operational assessment framework. In *Quantitative methods in tourism economics* (pp. 329–352). https://doi.org/10.1007/978-3-7908-2879-5_18

Kishigami, J., Fujimura, S., Watanabe, H., Nakadaira, A., & Akutsu, A. (2015). The blockchain-based digital content distribution system. In *2015 IEEE Fifth International Conference on Big Data and Cloud Computing*. https://doi.org/10.1109/bdcloud.2015.60

Kostopoulos, N., Antonopoulou, H., & Halkiopoulos, C. (2022). Blockchain technology as an asset for innovative educational applications. A systematic review. In *EDULEARN22 Proceedings* (pp. 5783–5789). https://doi.org/10.21125/edulearn.2022.1358

Li, J., He, Q., Liang, R., & Jiang, B. (2021). Smart tourism identity authentication service based on blockchain and decentralized identifier. In H.N. Dai, X. Liu, D.X. Luo, J. Xiao, & X. Chen (Eds.), *Blockchain and trustworthy systems. BlockSys 2021*. Communications in Computer and Information Science (Vol. 1490). Springer. https://doi.org/10.1007/978-981-16-7993-3_42

Luo, L., & Zhou, J. (2021). BlockTour: A blockchain-based smart tourism platform. *Computer Communications, 175*, 186–192. https://doi.org/10.1016/j.comcom.2021.05.011

Melkić, S., & Čavlek, N. (2020). The impact of blockchain technology on tourism intermediation. *Tourism, 68*(2), 130–143. https://doi.org/10.37741/t.68.2.2

Ming, Q., & Wei, J. (2021). On the high-quality development of "blockchain+" enabled smart tourism, *Smart Tourism, 2*(2). https://doi.org/10.54517/st.v2i2.1731

Mohamed, N., & Al-Jaroodi, J. (2019). Applying blockchain in industry 4.0 applications. In *2019 IEEE 9th Annual Computing and Communication Workshop and Conference (CCWC).* https://doi.org/10.1109/ccwc.2019.8666558

Morkunas, V. J., Paschen, J., & Boon, E. (2019). How blockchain technologies impact your business model. *Business Horizons, 62*(3), 295–306. https://doi.org/10.1016/j.bushor.2019.01.009

Mul, S., Philip, A., Correia, M., & Gadhikar, L. (2021). Baggage tracking using RFID and blockchain technology. In *4th Biennial International Conference on Nascent Technologies in Engineering (ICNTE), 2021* (pp. 1–5). https://doi.org/10.1109/ICNTE51185.2021.9487770

Nakamoto, N. (2017). Centralised bitcoin: A secure and high performance electronic cash system. *SSRN Electronic Journal.* https://doi.org/10.2139/ssrn.3065723

Ølnes, S., Ubacht, J., & Janssen, M. (2017). Blockchain in government: Benefits and implications of distributed ledger technology for information sharing. *Government Information Quarterly, 34*(3), 355–364. https://doi.org/10.1016/j.giq.2017.09.007

Önder, I., & Gunter, U. (2020). Blockchain: Is it the future for the tourism and hospitality industry? *Tourism Economics, 28*(2), 291–299. https://doi.org/10.1177/1354816620961707

Ozdemir, A. I., Ar, I. M., & Erol, I. (2020). Assessment of blockchain applications in travel and tourism industry. *Quality & Quantity, 54*(5–6), 1549–1563. https://doi.org/10.1007/s11135-019-00901-w

Panas, G., Heliades, G., Halkiopoulos, C., Tsavalia, D., & Bougioura, A. (2017). Evaluation of Athens as a city break destination: Tourist perspective explored via data mining techniques. In V. Katsoni, A. Upadhya, & A. Stratigea (Eds.), *Tourism, culture and heritage in a smart economy.* Springer Proceedings in Business and Economics. Springer. https://doi.org/10.1007/978-3-319-47732-9_6

Panas, G., Thrasidi, N., Halkiopoulos, C., & Gkintoni, E. (2022). Consumer behavior and cognitive factors in relation to gastronomic tourism and destination marketing in Greece. In V. Katsoni, A.C. Şerban (Eds.), *Transcending borders in tourism through innovation and cultural heritage.* Springer Proceedings in Business and Economics. Springer. https://doi.org/10.1007/978-3-030-92491-1_40

Panteli, A., Kompothrekas, A., Halkiopoulos, C., & Boutsinas, B. (2021). An innovative recommender system for health tourism. In V. Katsoni, C. van Zyl (Eds.), *Culture and tourism in a smart, globalized, and sustainable world.* Springer Proceedings in Business and Economics. Springer. DOI:https://doi.org/10.1007/978-3-030-72469-6_42

Pilkington, M. (2017). Can blockchain technology help promote new tourism destinations? The example of medical tourism in Moldova. *SSRN Electronic Journal.* https://doi.org/10.2139/ssrn.2984479

Puthal, D., Malik, N., Mohanty, S. P., Kougianos, E., & Das, G. (2018). Everything you wanted to know about the blockchain: Its promise, components, processes, and problems. *IEEE Consumer Electronics Magazine, 7*(4), 6–14. https://doi.org/10.1109/mce.2018.2816299

Qian, Y., Jiang, Y., Chen, J., Zhang, Y., Song, J., Zhou, M., & Pustišek, M. (2018). Towards decentralized IoT security enhancement: A blockchain approach. *Computers & Electrical Engineering, 72*, 266–273. https://doi.org/10.1016/j.compeleceng.2018.08.021

Rashideh, W. (2020). Blockchain technology framework: Current and future perspectives for the tourism industry. *Tourism Management, 80*, 104125. https://doi.org/10.1016/j.tourman.2020.104125

Robinson, P., Ramesh, R., & Johnson, S. (2022). Atomic crosschain transactions for ethereum private sidechains. *Blockchain: Research and Applications, 3*(1), 100030. https://doi.org/10.1016/j.bcra.2021.100030

Sarigiannidis, C., Halkiopoulos, C., Giannopoulos, K., Giannopoulou, F., Politis, A., Boutsinas, B., & Kollias, K. (2021). Do hotels care? A proposed smart framework for the effectiveness of an environmental management accounting system based on business intelligence technologies. In V. Katsoni, C. van Zyl (Eds.), *Culture and tourism in a smart, globalized, and sustainable world*. Springer Proceedings in Business and Economics. Springer. https://doi.org/10.1007/978-3-030-72469-6_41

Savelyev, A. (2017). Contract law 2.0: 'Smart' contracts as the beginning of the end of classic contract law. *Information & Communications Technology Law, 26*(2), 116–134. https://doi.org/10.1080/13600834.2017.1301036

Seebacher, S., & Schüritz, R. (2017). Blockchain technology as an enabler of service systems: A structured literature review. In *Exploring services science* (pp. 12–23). https://doi.org/10.1007/978-3-319-56925-3_2

Solakis, K., Katsoni, V., Mahmoud, A. B., & Grigoriou, N. (2022). Factors affecting value co-creation through artificial intelligence in tourism: A general literature review. *Journal of Tourism Futures*. https://doi.org/10.1108/jtf-06-2021-0157

Stallone, V., Wetzels, M., & Klaas, M. (2021). Applications of blockchain technology in marketing—A systematic review of marketing technology companies. *Blockchain: Research and Applications, 2*(3), 100023. https://doi.org/10.1016/j.bcra.2021.100023

Stamatiou, Y. C., Halkiopoulos, C., Giannoulis, A., & Antonopoulou, H. (2022). Utilizing a restrictedaAccess e-learning platform for reform, equity, and self-development in correctional facilities. *Emerging Science Journal, 6*, 241–252. https://doi.org/10.28991/esj-2022-sied-017

Theodorakopoulos, L., Antonopoulou, H., Mamalougou, V., & C. Giotopoulos, K. (2022). The drivers of volume volatility: A big data analysis based on economic uncertainty measures for the Greek banking system. *Banks and Bank Systems, 17*(3), 49–57. https://doi.org/10.21511/bbs.17(3).2022.05

Treiblmaier, H. (2022). Blockchain and tourism. In *Handbook of e-tourism* (pp. 475–495). https://doi.org/10.1007/978-3-030-48652-5_28

Tussyadiah, I. P., Jung, T. H., & tom Dieck, M. C. (2017). Embodiment of wearable augmented reality technology in tourism experiences. *Journal of Travel Research, 57*(5), 597–611. https://doi.org/10.1177/0047287517709090

Valeri, M., & Baggio, R. (2020). A critical reflection on the adoption of blockchain in tourism. *Information Technology & Tourism, 23*(2), 121–132. https://doi.org/10.1007/s40558-020-00183-1

Wang, X., Ni, W., Zha, X., Yu, G., Liu, R. P., Georgalas, N., & Reeves, A. (2021). Capacity analysis of public blockchain. *Computer Communications, 177*, 112–124. https://doi.org/10.1016/j.comcom.2021.06.019

Wu, J., Liu, J., Zhao, Y., & Zheng, Z. (2021). Analysis of cryptocurrency transactions from a network perspective: An overview. *Journal of Network and Computer Applications, 190*, 103139. https://doi.org/10.1016/j.jnca.2021.103139

Xiang, Z., & Fesenmaier, D. R. (2022). Travel information search. *Encyclopedia of Tourism Management and Marketing*. https://doi.org/10.4337/9781800377486.travel.information.search

Zhao, G., Liu, S., Lopez, C., Lu, H., Elgueta, S., Chen, H., & Boshkoska, B. M. (2019). Blockchain technology in agri-food value chain management: A synthesis of applications, challenges and future research directions. *Computers in Industry, 109*, 83–99. https://doi.org/10.1016/j.compind.2019.04.002

Application of Importance–Performance Analysis in Determining Critical Smart Hotel Technology Amenities: An Examination of Customer Intentions

Georgiadis Mauroudis, Maria Tsourela, and Dimitris Paschaloudis

Abstract As technology changes, smart hotels must develop new amenities, capabilities, and service innovations that are important to customers. A hotel's technology amenities can be defined as extra supplies or facilities provided to guests within or outside their guestrooms for no extra charge. The availability of technology amenities is on the rise and evolving rapidly, and many types of them used in smart hotels are identified in the literature. This study explores the smart hotel technology amenities identified in the literature, in an international tourist hotel. The study collected customers responses and applied an importance-performance analysis (IPA), using the self-explained matrix, to indicated which the order they should be addressed. Based on the research results, the author discusses some useful implications and contributes to raising awareness of the importance of technological amenities and services for the future visiting intentions of hotel consumers.

Keywords Smart hotel · Technology amenities · Customer intentions · Service innovation

JEL Classification Z32 · Z33

1 Introduction

The past decade has seen a spectacular development of the Internet worldwide. "Smart people" built the "smart artifacts" to live a "smarter life". "Things" have changed, thus changing our interaction with them. Societies are characterized by an extreme pace of growth of technology. Individuals tend to replace personal computers and laptops with smartphones and tablets, as the latter ones can do almost the same daily activities, while being much smaller and easy to carry at all times. This technological evolution has led to the development of smart applications.

G. Mauroudis · M. Tsourela (✉) · D. Paschaloudis
International Hellenic Univeristy, Serres, Greece
e-mail: mt@ihu.gr

© The Author(s), under exclusive license to Springer Nature Switzerland AG 2023

V. Katsoni (ed.), *Tourism, Travel, and Hospitality in a Smart and Sustainable World*, Springer Proceedings in Business and Economics,
https://doi.org/10.1007/978-3-031-26829-8_23

In hospitality and tourism industry, adopting technological innovations can prove to be of a great importance for achieving competitive advantage. Technological innovations adoption can create smarter destinations by facilitating improvement and restructuring processes for tackling challenges such as seasonality and overcrowding (Maravilla & Gantalao, 2019). "The smart hotel concept emerged around 2008 and has attracted attention in recent years" (Jin, 2018). "Smart hotels are intelligent hotels with a range of new information and communication technologies (ICT) (Wu & Cheng, 2018), using AI and service concept to provide customers with a new experience of intelligence" (Xu, 2018). "Through global computer networks, smart hotels directly manage and integrate their technical systems and technological operations" (Basyuk et al., 2014). "These integrate contemporary IT, such as the IoT, cloud computing, mobile Internet, smart devices, and big data (Xu, 2018), to provide customers with improved service experiences and far higher levels of personalization" (Lai & Hung, 2018).

Due to constant technology developments, smart hotels are prompt to create new amenities and service innovations that create added value to consumers. A hotel's technology amenities can be defined as extra supplies or facilities provided to guests within or outside their guestrooms for no extra charge (Yang et al., 2021). Current and potential technology amenities are of high priority in the tourism and hospitality industry, thus various types of them currently used in smart hotels are identified in the literature. This study explores the smart hotel technology amenities identified in the literature, in an international tourist hotel. The study collected customers responses and applied an importance-performance analysis (IPA), using the self-explained matrix, to indicated which the order they should be addressed. Based on the research results, the author discusses some useful implications and contributes to raising awareness of the importance of technological amenities and services for the future visiting intentions of hotel consumers.

2 Literature Review

With the advancements in IoT and the Internet of everything (IoE), hotel operations and services are heading toward smart environments (Buhalis & Sinarta, 2019; Leung, 2020; Webster & Ivanov, 2020). Leung (2020) indicated that different stakeholders have a different definition of smart hotels determined by the technology task or role it holds in the organization.

Smart hotels are developing rapidly around the world (Jaremen et al., 2016) and they are critical to the hospitality industry (Lai & Hung, 2018). The gradual emergence of smart hotels using AI has changed customers' purchasing intentions and businesses' operational processes (Leung, 2020). Sigala (2018) states that many technological developments, such as artificial intelligence, industrial web, Internet of things, smart devices, robots, drones, sensors, virtual, and augmented reality, make the change and innovation inevitable in the provision of services in the tourism industry.

"Smart tourism, however, also attracts attention as a technological innovation where people exchange information using Internet technologies, exchange of information becomes faster and applications such as virtual reality and augmented reality are used in tourism" (Hunter et al., 2015). "Smart tourism is a new concept that enables the use of information communication technology in tourism sector by destinations, businesses and tourists, allowing large amounts of data to be transformed into value-creating structures" (Gretzel et al., 2015).

Recently, some vendors make use of sensor networks (hereafter sensors) to collect data, and beacons to enhance communication with customers (Buhalis & Leung, 2018). Leonidis et al. (2013) proposed an intelligent hotel room platform for improving the ambient environment during the guest stay and provide required travel information. Marriott International presented the IoT guestroom that offers guests the opportunity to handle in-room amenities through beacons and sensors. These capabilities can elaborate in-room information through sensors, such as where the individual is located inside the room and accustom room related data such as temperature and light level to enhance accommodation conditions. Accordingly, additional data interconnections to external information, such as weather in the wider area and on the spot, as well traffic reports can enhance the overall perceived positive experience. Automated hotel operations with collaborative international and external marketing strategy can co-create a holistic and coherent experience to hotel guest (Buhalis & Leung, 2018). Jen Hotel, which is part of the Shangri-la hotel chain, employed autonomous robot to delivery hotel orders, as well as services directly to guests during their visits. Similarly, Henn-na Hotel in Japan presented the first robotic hotel which has integrated automatic automated customer services. Robot concierge at the Hilton hotel connected to a cognitive system with machine learning abilities that can on one hand provide hotel and travel related recommendations to hotel guests, and on the other hand learn and improve its database via interaction with guests. Robots for cleaning and room service are currently on trial in using around the world.

3 Technology Amenities in Smart Hotels

Neuhofer et al. (2015) stated that the use of smart tourism technologies in hospitality businesses is an effective tool in increasing the guest experiences and providing personalized service and draw attention to the use of smart technologies in the synchronization of the services provided in the sections such as collection, storage, reception, guest relations, rooms section of the customers. Conyette (2015) states that while developing technologies from the twentieth century to the twenty-first century significantly affect the activities of the travel agencies, the effectiveness of these enterprises which have an active role in the past decreased and the efficiency of mobile and wearable smart technologies increased.

Even though mobile services and applications are highly used in hospitality and tourism industry over the years, it must be noted that during the last decade there has been a rise in recorded studies over this area. Atay and Yalçinkaya (2019) "in their study on the evaluation of smart hotel applications in Istanbul, concluded that the concept of a smart hotel was applied in two themes as mobile applications and personalized services in 4- and 5-star hotels within the scope of the research". "The user profile notes and personalized room service applications are included in the research under the theme of personalized services with online check-in and online check-out applications under the theme of mobile applications" (Atay & Yalçinkaya, 2019).

"A hotel's TA can be defined as extra supplies or facilities provided to guests within or outside their guestrooms for no extra charge" (Vallen & Vallen, 2009). "The availability of TA is on the rise and evolving rapidly (Bilgihan et al., 2016), and many types of TA used in smart hotels are identified in the literature". "These include free Wi-Fi (Eriksson & Fagerstrøm, 2018; Sheck, 2015), face or voice recognition (Liu, 2018; Liu et al., 2020; Morosan, 2020; Revfine, 2020), service robots (De Kervenoael et al., 2020; Liu et al., 2020; Zhong et al., 2020), smart TV (Bilgihan et al., 2016; Melián-González & Bulchand-Gidumal, 2016), smart room keys (Ozturk et al., 2016), and VR headsets (Roberts, 2015). Touchscreen panels allow hotel guests to control their lighting, temperature (Shoenfeld, 2019), music (Dutilly et al., 2015), and curtains (Lai & Hung, 2018)". "Mobile devices can provide proper check-in and check-out processes to enhance customers' _hotel stay experience (Leung, 2020), customers may also use mobile apps for convenient self-service" (Keith et al., 2015; Ozturk et al., 2016). Smart "things" are of great interest in hospitality and tourism industry, and hotels that employ smart gadgets are now able to provide personalized experience to the customers (Jangid, 2019), such as smart mirror, rollable TV, smart glass in the shower (EHL Insights, 2019; Prabhu, 2019). In hospitality, it is important to keep abreast of the latest technology available to maximize energy saving by using occupancy sensors, infrared sensors, and motion sensors (DePinto, 2017). Table 1, as adopted by Yang et al. (2021), summarizes smart hotel technology amenities identified in the literature.

4 Importance-Performance Analysis (IPA) Concept

Importance-performance analysis (IPA) is a universal known procedure for measuring service quality. The concept of IPA was introduced in 1977 by Martilla and James. "Essentially, the idea of IPA comes from the theory of customer satisfaction as a function of the expectation on important attribute and the judgment of attribute performance" (Martilla & James, 1977). "The underlying assumption in IPA is the relationship between importance attribute and attribute performance toward customer satisfaction is linear and symmetric" (Martilla & James, 1977). Thus, "IPA can reveal the lacuna between customer expectation on the importance and assessment of the performance of services. The objective is to identify which attributes or its combination gives more impact toward customer satisfaction and leads to the repetitive customer purchase behavior" (Matzler et al., 2004).

Table 1 Smart hotel technology amenities (Yang et al., 2021)

Technology amenities	Description	Authors
Wi-Fi	High-speed, stable and complimentary Wi-Fi	Bilgihan et al. (2016), Eriksson and Fagerstrøm (2018)
Touchscreen panel	Tablet-controlled room system, including lights, music, temperature, curtains, etc.	Dutilly et al. (2015), Keith et al. (2015) Lai and Hung (2018)
Mobile devices	Guests operate room controls, including check-in/check-out devices, smart keys, temperature, lighting, window blinds, thermostat, through mobile phone apps	Lai and Hung (2018), Ozturk et al. (2016), Shoenfeld (2019)
Smart gadgets	Smart mirror, rollable TV, smart glass in shower, etc.	EHL Insights (2019), Prabhu (2019)
Voice recognition	Voice commands used to make AI assistant execute tasks, such as voice-based control of in-room devices and travel assistance (e.g., Tmall Genie, Amazon Alexa, Google Home, Apple Siri)	Liu (2018) Revfine (2020)
Facial recognition	Facial images used for authentication/verification to gain access to guest areas and services (e.g., guestroom, pool, gym, business center, food and beverages, entertainment products, etc.)	Liu et al. (2020), Morosan (2020), Revfine (2020)
Robots	Delivery robots, cleaning robots, etc.	De Kervenoael et al. (2020), Liu et al. (2020), Zhong et al. (2020)
Sensors	Motion sensor, occupancy sensor, infrared sensor, etc.	DePinto (2017)
SMART TV	Digital TV providing hotel services and information	Bilgihan et al. (2016), Meli´an-Gonz´alez and Bulchand-Gidumal (2016)
VR headsets	Allows guest to visit the tourist spots all around the world and play VR games	Roberts (2015)
Smart room key	Mobile key	Ozturk et al. (2016)

"The IPA consists of a pair of coordinate axis where the 'importance' (y-axis) and the 'performance' (x-axis) of the different elements involved in the service are compared" (Martilla & James, 1977). Each of the quadrants combines the importance and the performance assigned by the customers/user given element of the service and possesses a different value in terms of management and the respective mean of self-stated raw importance and attribute performance data is the original point of this IPA matrix. Each quadrant suggests a different marketing strategy.

The acceptance of IPA as the measurement of service quality is corroborated by extensive application of IPA, largely in the tourism sector, followed by banking sector, education sector, electric and electronic sector, automotive sector, and many other. It offers information indication possible consumer satisfaction/dissatisfaction over the features of the consumed service.

The four quadrants in importance-performance analysis are characterized as (Martilla & James, 1977):—"A. Concentrate here–high importance, low performance: requires immediate attention for improvement and are major weaknesses;—B. Keep up with the good work–high importance, high performance: indicate opportunities for achieving or maintaining competitive advantage and are major strengths;—C. Low priority–low importance, low performance: are minor weaknesses and do not require additional effort;—D. Possible overkill–low importance, high performance: indicate that business resources committed to these attributes would be overkill and should be deployed elsewhere" (Martilla & James, 1977).

5 Methodology

The study collected customers responses and applied an importance-performance analysis (IPA), using the self-explained matrix, to indicated which the order they should be addressed. The survey was conducted during June 2022 in Greece. A total of 383 valid questionnaires were received, using a non-probability convenience sample. There was no reward connected to completing the questionnaire. Potential guests were asked to report importance levels connected to the 11 items related to smart hotel technology amenities. The response set for each item was a 5-point Likert-type scale, with possible response options ranging from 1 (not at all important) to 5 (extremely important). Respondents were also asked to report performance levels connected to the same set of 16 items. The response set for each item was a 5-point Likert-type scale, with possible response options ranging from 1 (not at all performed) to 5 (extremely performed).

6 Results

In order to approach the importance that guests assign to different IoT appliances of the hotel, as well as know the value of performance, means, and standard deviations for the aggregated data were calculated. The overall importance means ranged from 2.64 to 4.94 and the overall performance means ranged from 3.18 to 4.74 (Table 2). The overall grand means for importance and satisfaction were 3.78 and 3.79,

Table 2 Importance—performance ratings

	Corresponding number	Importance		Performance		GAP	Quadrant
		Mean	SD	Mean	SD		
Wi-Fi	1	4.83	0.56	4.74	0.66	0.09	2
Touchscreen panel	2	3.65	0.74	3.83	0.89	−0.18	3
Mobile devices	3	4.94	0.55	3.37	1.23	1.57	1
Smart gadgets	4	4.09	1.17	3.74	1.05	0.35	1
Voice recognition	5	3.87	0.93	4.04	0.93	−0.17	2
Facial recognition	6	2.64	0.98	3.18	0.69	−0.54	3
Robots	7	3.66	0.46	3.54	0.47	0.12	3
Sensors	8	2.97	1.25	3.26	0.54	−0.29	3
SMART TV	9	3.25	0.74	3.75	0.85	−0.5	3
VR headsets	10	3.18	0.84	3.74	1.29	−0.56	3
Smart room key	11	4.53	1.34	4.36	0.84	0.17	2

respectively. The position of the grid is determined by these two overall mean values for attribute importance and attribute importance (Fig. 1).

According to Table 2, the smart hotel technology amenities that individuals mentioned as the most important were: Wi-Fi, Mobile devices, Smart gadgets, Smart room key, and Voice recognition. On the other hand, the ones mentioned as less important were touchscreen panel, facial recognition, robots, sensors, SMART TV, and VR headsets.

In evaluating the performance IoT appliances, the Wi-Fi was the highest (4.74), followed by the Smart room key (4.36) and Voice recognition (4.04). The rest were evaluated as low performance IoT appliances, with facial recognition (3.18) being the lowest-rated performance attribute.

The values obtained for the standard deviation (SD) are considered low, indicating that the variations of the responses are low. This measure reflects the variation of dispersion of a set of data around the mean, i.e., the greater or less variability of results (Silva & Fernandes, 2011).

Importance-performance matrix is represented in Fig. 1. The results are spread over 3 quadrants. In Quadrant A there were 2 amenities. Quadrant A shows hotels' weaknesses. Guests valued these amenities but showed low satisfaction with the hotels' performance. Hotels should focus on them immediately, in order to improve guest's satisfaction. In Quadrant B were 3 amenities. These are of the highest importance for guests and of high satisfaction, judging from the good performance evaluations. Hotels should keep up the good work here and continue to meet the needs of

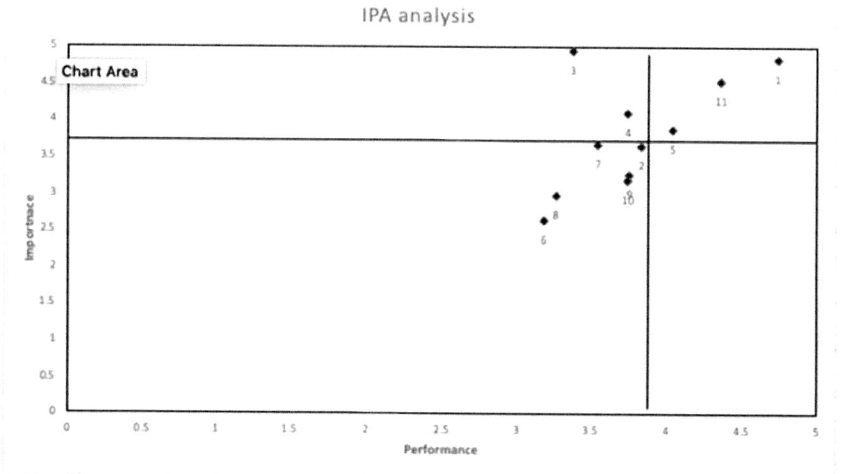

Fig. 1 Importance-performance analysis (IPA) of smart hotel amenities

their guests. Amenities included in Quadrant C, are of low priority and need no additional action. Guests neither evaluated them as important nor rated them as of high performance. Six amenities were included here. In Quadrant D, no IoT appliances appeared.

7 Conclusion

Importance-performance analysis (IPA) is a very straightforward and powerful tool that can help hotel managers reveal the attributes that need improvement in order to enhance overall guest satisfaction. Generally, it offers multiple advantages in smart hotel amenities assessment. The distance of importance and satisfaction from each factor is also shown in Fig. 2, which reveals the difference. This is helpful for guiding hotel managers prioritize every single amenity.

Hotels must offer high guest satisfaction in their effort to create and establish a competitive advantage. To do so, hotel managers must be able to lighten up vital amenities performance that lead to value creation and ultimately to guest satisfaction. Hotel guests are satisfied when the hotels' performance matches their expectations. However, many studies have evaluated either guests' expectations or actual amenities performance only, rather than both. There is a gap n literature over guests' expectations and hotels performance, which impedes professionals to create and follow a solid marketing strategy.

Current research's results though, are not to be overstated. It would be useful to examine many possible decisive factors that influence hotel guests' loyalty and investigate ways that this technique can provide the means to hotels to distribute their resources precisely.

Acknowledgements This research was implemented for the completion of the project of Mr Georgiadi Mauroudi in the Master in Public Business Administration (MPA) of the Department of Business Administration of the International Hellenic University.

References

Atay, L., & Yalçinkaya, P. (2019). İstanbul' Daki Akilli Otel Uygulamalarinin Değerlendirilmesi. *Manas Journal of Social Studies, 8*(1), 667–678.

Basyuk, D., Myts, I., & Kohan, O. (2014). The introduction of the intellectual property concept in the hotel industry of Ukraine. Retrieved from http://dspace.nuft.edu.ua/jspui/bitstream/123456789/13896/3/The%20introduction%20of%20the%20intellectual%20property%20concept%20in%20the%20hotel%20industry%20of%20Ukraine.pdf

Bilgihan, A., Smith, S., Ricci, P., & Bujisic, M. (2016). Hotel guest preferences of in-room technology amenities. *Journal of Hospitality and Tourism Technology.*

Buhalis, D., & Leung, R. (2018). Smart hospitality—Interconnectivity and interoperability towards an ecosystem. *International Journal of Hospitality Management, 71*, 41–50.

Buhalis, D., & Sinarta, Y. (2019). Real-time co-creation and nowness service: Lessons from tourism and hospitality. *Journal of Travel & Tourism Marketing, 36*(5), 563–582.

Conyette, M. (2015). 21 century travel using websites, mobile and wearable technology devices. In *Athens: ATINER'S Conference Paper Series.*

de Kervenoael, R., Hasan, R., Schwob, A., & Goh, E. (2020). Leveraging human-robot interaction in hospitality services: Incorporating the role of perceived value, empathy, and information sharing into visitors' intentions to use social robots. *Tourism Management, 78*, 104042.

DePinto, J. (2017). Occupancy sensing improves hospitality. Retrieved from https://www.linkedin.com/pulse/occupancy-sensing-improves-hospitality-jesse-depinto

Dutilly, D. T., Browne, J., Ingeneri, N. P., & Monroe, S. R. (2015). System and method for using a touchscreen as an interface for music-based gameplay. Retrieved from https://patentimages.storage.googleapis.com/55/f8/42/5a1278a98dd7d3/US8961309.pdf

EHL Insights. (2019). Three top tech gadgets for hotels at CES 2019. Retrieved from https://hospitalityinsights.ehl.edu/top-tech-gadgets-for-hotels-ces-2019

Eriksson, N., & Fagerstrøm, A. (2018). The relative impact of Wi-Fi service on young consumers' hotel booking online. *Journal of Hospitality & Tourism Research, 42*(7), 1152–1169.

Gretzel, U., Werthner, H., Koo, C., & Lamsfus, C. (2015). Conceptual foundations for understanding smart tourism ecosystems. *Computers in Human Behavior, 50*, 558–563.

Hunter, W. C., Chung, N., Gretzel, U., & Koo, C. (2015). Constructivist research in smart tourism. *Asia Pacific Journal of Information Systems, 25*(1), 103–118.

Jangid, D. (2019). 5 must-have technologies you should be using for your hotel. eHotelier. Retrieved from https://insights.ehotelier.com/insights/2019/02/08/5-must-have-technologies-you-should-be-using-for-your-hotel/

Jaremen, D., Jędrasiak, M., & Rapacz, A. (2016). The concept of smart hotels as an innovation on the hospitality industry market-case study of Puro Hotel in Wrocław. *Zeszyty Naukowe Uniwersytetu Szczecińskiego Ekonomiczne Problemy Turystyki, 36*(4).

Jin, K. L. (2018). Analysis and study on the influence factors of smart hotel-a case study of payment mode. *Unpublished Postgraduate Dissertation, Zhejiang Gongshang University, China.*

Keith, M. J., Babb, J. S., Lowry, P. B., Furner, C. P., & Abdullat, A. (2015). The role of mobile-computing self-efficacy in consumer information disclosure. *Information Systems Journal, 25*(6), 637–667.

Lai, W. C., & Hung, W. H. (2018). A framework of cloud and AI based intelligent hotel.

Leonidis, A., Korozi, M., Margetis, G., Grammenos, D., & Stephanidis, C. (2013, December). An intelligent hotel room. In *International Joint Conference on Ambient Intelligence* (pp. 241–246). Springer.

Leung, R. (2020). Hospitality technology progress towards intelligent buildings: A perspective article. *Tourism Review, 76*(1), 69–73.

Liu, C. (2018). China's Alibaba to open its first automated hotel. Retrieved from https://asia.nikkei.com/Business/Companies/China-s-Alibaba-to-open-its-first-automated-hotel.

Liu, C., Hung, K., Wang, D., & Wang, S. (2020). Determinants of self-service technology adoption and implementation in hotels: The case of China. *Journal of Hospitality Marketing & Management, 29*(6), 636–661.

Maravilla, V. S., Jr., & Gantalao, C. S. (2019). E-tourism adoption of the travel agencies in Cebu City, Philippines. *Journal of Economics and Business, 2*(3), 555–568.

Martilla, J. A., & James, J. C. (1977). Importance-performance analysis. *Journal of Marketing, 41*(1), 77–79.

Matzler, K., Bailom, F., Hinterhuber, H. H., Renzl, B., & Pichler, J. (2004). The asymmetric relationship between attribute-level performance and overall customer satisfaction: A reconsideration of the importance–performance analysis. *Industrial Marketing Management, 33*(4), 271–277.

Melián-González, S., & Bulchand-Gidumal, J. (2016). A model that connects information technology and hotel performance. *Tourism Management, 53*, 30–37.

Morosan, C. (2020). Hotel facial recognition systems: Insight into guests' system perceptions, congruity with selfimage, and anticipated emotions. *Journal of Electronic Commerce Research, 21*(1), 21–38.

Neuhofer, B., Buhalis, D., & Ladkin, A. (2015). Smart technologies for personalized experiences: A case study in the hospitality domain. *Electronic Markets, 25*(3), 243–254.

Ozturk, A. B., Bilgihan, A., Nusair, K., & Okumus, F. (2016). What keeps the mobile hotel booking users loyal? Investigating the roles of self-efficacy, compatibility, perceived ease of use, and perceived convenience. *International Journal of Information Management, 36*(6), 1350–1359.

Prabhu, A. (2019). Conditions and influencers of hotel room purchases in the digital era. Retrieved from http://ijrar.com/upload_issue/ijrar_issue_20543893.pdf

Revfine. (2020). How can voice control benefit the hospitality industry? Retrieved https://www.revfine.com/voice-control-hospitality-industry/

Roberts, D. (2015). Your next hotel could offer a virtual reality experience. Retrieved from http://fortune.com/2015/09/09/marriott-virtual-reality/

Sheck, S. (2015). Why offering free WiFi to your customers is wise. Retrieved from https://smallbiztrends.com/2015/11/small-businesses-need-offer-free-wifi-customers.html

Shoenfeld, N. A. (2019). Pass-through convenience cabinet for hotel or similar public accommodation. Retrieved from https://patentimages.storage.googleapis.com/cb/41/72/65a2625d39f4e4/US20190082829A1.pdf

Sigala, M. (2018). New technologies in tourism: From multi-disciplinary to anti-disciplinary advances and trajectories. *Tourism Management Perspectives, 25*, 151–155.

Silva, F. H., & Fernandes, P. O. (2011). Importance-performance analysis as a tool in evaluating higher education service quality: The empirical results of ESTiG (IPB). In *In the 17th International Business Information Management Association Conference* (pp. 306–315).

Vallen, G. K., & Vallen, J. J. (2009). *Check-in check-out: Managing hotel operations* (8th ed., pp. 146–153).

Webster, C., & Ivanov, S. (2020). *Robots in travel, tourism and hospitality: Key findings from a global study*. Zangador.

Wu, H. C., & Cheng, C. C. (2018). Relationships between technology attachment, experiential relationship quality, experiential risk and experiential sharing intentions in a smart hotel. *Journal of Hospitality and Tourism Management, 37*, 42–58.

Xu, X. (2018, August). Research on the construction and development of smart hotel from the perspective of serving customers. In *2018 2nd International Conference on Education Science and Economic Management (ICESEM 2018)* (pp. 975–978). Atlantis Press.

Yang, H., Song, H., Cheung, C., & Guan, J. (2021). How to enhance hotel guests' acceptance and experience of smart hotel technology: An examination of visiting intentions. *International Journal of Hospitality Management, 97*, 103000.

Zhong, L., Sun, S., Law, R., & Zhang, X. (2020). Impact of robot hotel service on consumers' purchase intention: A control experiment. *Asia Pacific Journal of Tourism Research, 25*(7), 780–798.

Implementation of Digital Marketing Techniques in Smart Tourism

Constantinos Halkiopoulos, Hera Antonopoulou, and Konstantinos Giotopoulos

Abstract The evolution of technology has changed how tourism businesses promote their products and services and the nature of marketing from traditional to digital. The purpose of applying digital marketing techniques is to increase the number of potential tourists for choose tourism products/services. In addition, the inclusion of digital techniques in tourism marketing contributes so that tourism businesses can advertise their tourism goods with different media and tools, allowing choosing different promotion methods depending on each business's requirements. As a result, tourism businesses can use various media to promote their products, establish their brand, attract new consumers, and increase their profits. With digital marketing, new opportunities are presented to tourism business units, and the modern tourism industry is developing. In the context of this research, the importance of digital marketing techniques in the field of tourism is highlighted. The data collected was primarily using quantitative methodology. The sample consisted of 1200 young adults (18–32 years old) who were asked to answer questions about using digital marketing applications in the tourism sector during the Covid-19 pandemic. Through the research, specific digital marketing techniques such as websites, promotional videos, and informative blogs appear to be reliable tools for the promotion of the tourist product and related services as well as for the preference of new tourist applications/platforms in terms of savings of time. Therefore, the results, among others, highlight the importance of modern digital marketing techniques for their systematic application in the tourism industry.

Keywords Digital marketing · Smart tourism · Tourism industry · Innovative platforms · Tourist experience · Data analytics

JEL Classifications O32 · O33 · Z32 · L86

C. Halkiopoulos (✉) · H. Antonopoulou · K. Giotopoulos
EDILAB, Department of Management Science and Technology, University of Patras, Patras, Greece
e-mail: halkion@upatras.gr

© The Author(s), under exclusive license to Springer Nature Switzerland AG 2023
V. Katsoni (ed.), *Tourism, Travel, and Hospitality in a Smart and Sustainable World*, Springer Proceedings in Business and Economics,
https://doi.org/10.1007/978-3-031-26829-8_24

1 Introduction

The growing number of technological innovations indicated the fourth Industrial Revolution. In the past ten years, the use of high technology, including digitization, 3D printer technology, artificial intelligence applications, nanotechnology, and intelligent systems, has gained substantial traction in all industries, particularly tourism. Since 1800, the industrial revolution in Europe has through at least three evolutionary phases (1.0, 2.0, 3.0) because of evolving technology and advances and has recently evolved into version 4.0 (Industry 4.0). Tourism, one of the most afflicted businesses at now due to the covivirus-19 epidemic, offers a vast area of application for Industry 4.0. Therefore, it is now referred to as "smart tourism". Tourism businesses increasingly use digital platforms and advanced digital communication methods to advertise, promote tourism products, and reach potential tourists/customers (Cheung & Leung, 2022). Potential tourists can swiftly examine 3D graphics and photographs of the chosen tourist accommodation, the location and design of the room with the associated amenities, the appearance of the facilities, and the wellness units, such as the spa, that they will utilize (Sobrino, 2022). The tourism industry of any nation can gain a competitive advantage by adopting and implementing a smart tourism system and by utilizing cutting-edge digital marketing tools and digital technologies in general (Cunha et al., 2022).

2 Literature Review

2.1 e-Tourism

The technological advances and the 4.0 industrial revolution have facilitated the increasing connectivity of business units to current online information systems management systems. The application of contemporary technology to tourism provides amenities for consumers, businesses, and suppliers. The following widespread information systems and technical applications greatly contribute to the tourism industry:

- Reservation information systems that give businesses with up-to-date information regarding hotels, modes of transportation, itineraries, tickets, etc.
- Airlines and shipping firms utilize electronic tickets for the simplest purchase and reservation of tickets at the lowest feasible price.
- Geographic information systems, which collect data about a location, identify geographic areas for tourism development in order to evaluate the potential of a tourist destination.
- Information systems for international distribution, specialized databases for the electronic market of tourist products and services, and other intelligent functions.

The correct application of web technologies can substantially contribute to the growth and spread of digital tourism (Panas et al., 2020). Tourists can obtain information about tourism sites through web research. Consumers of tourism items are able to rate the goods they purchased and the location they visited, as well as read reviews, allowing them to make the best choice possible (Nadda et al., 2020; Panas et al., 2022).

The electronic display of tourism services leads to the development and maintenance of better relationships between tourist enterprises and consumers, as these services are provided more efficiently and at lower costs. Tourism marketing can function in a more targeted manner, classify the demands of tourists, and meet them to the greatest extent feasible, while simultaneously enhancing the image of the business unit (Alves et al., 2020). When a tourism business is featured on the Internet, it has its own website, which it may design as it sees fit, and it can attract more clients (Katsoni & Dologlou, 2017). Consumer preference enables a tourism enterprise to become competitive, realize its potential, thrive, establish itself in the tourism industry, and even expand abroad (Majumder, 2019).

Multiple advantages can result from digital promotion in the tourism industry for the traveling public, as summarized below:

- Providing access 24/7.
- Accessibility from home and anywhere else.
- Quickness and ease of usage.
- The opportunity to compare various service-providing regions.
- The ability to compare pricing for tourist places and products.
- The lowering of corporate operational expenses and the decrease of tourist expenses.
- The direct communication between suppliers of tourist items and tourists, without the involvement of travel agents.
- The convenience and speed of credit card payment methods.
- The opportunity to promote a tourism enterprise on a worldwide scale.
- The enhancement of services as a result of intense competition.
- The enhancement of products and services based on comments and observations made on company websites.
- The application of user-friendly research methods to examine customer behavior and determine their needs.
- The increase in tourist-related product sales and the expansion of clients.

The advancement of technology has influenced every area of our life, including travel. It influences people's travel, their desired destinations, and their interactions with travel agents. Younger generations are more exposed to technology, travel more frequently, and largely shape the future of travel and tourism industry advancements (Çınar & Weiss, 2019). Consumers are now searching for new methods to enhance their travel experience, as the demand for activity-packed vacations has increased in recent years (Korstanje, 2015). In addition, most tourists have their full travel itinerary planned, including hotel and flight reservations, car rentals, navigation using digital maps, etc. It is noteworthy that a new sort of traveler has emerged in recent years: the "business traveler", who travels for business purposes but also engages in the same activities as pleasure visitors (Shin & Xiang, 2022).

The travel industry continues to make significant efforts to meet the needs of modern travelers, who expect innovative tourist services (Panteli et al., 2021), to find them rather than seek them out (Halkiopoulos et al., 2021b). In the near future, tourists are anticipated to spend more money on attractions than on shopping and nightlife (Mancinelli, 2022). Tourism providers must optimize the equilibrium between their offers and bookings in order to assimilate the major benefit for both parts, general public and tourism industry sector.

2.2 *Digital Promotion Tools in Tourism Industry*

Traditional advertising has given way to digital advertising as a result of the advancement of technology, which has altered how tourist businesses market their products and services. Thus, businesses reach their intended demographic, and a marketing campaign can be deemed more effective. Advertising's objective is to increase the number of visitors to the company's website and convert them into potential customers. Since digital advertising is the most cost-effective kind of promotion, while applying neuromarketing techniques (Halkiopoulos et al., 2022), it is easier to convince a tourist to make a reservation online. Also strengthened are the product's brand and the company's trade name. In addition, it contributes to the fact that tourism firms can sell their tourism products using a variety of strategies and tools, allowing businesses to select from a variety of promotion methods based on their individual needs (George, 2021). Various media can be utilized by tourism firms to market their products, establish their brand, attract new customers, and boost their revenues.

At this point, it is worthwhile to summarize the digital marketing techniques that tourist businesses already use or may use in the future. Creating a website that displays the attributes of tourism products and promotes them directly to consumers or that aims to inform the tourist public, followed by online reservations and online payments, is a method of promotion. The tourism company registers a domain name, i.e., a unique web address, in order to promote its business. The creation of such a website does not impose significant fees on the firm.

Creating and leveraging a list of email addresses and sending potential consumers offers and vacation packages is a great digital marketing strategy for promoting tourist firms. Adoption of the mailing list not only fosters trusting relationships with existing customers, but also attempts to establish positive interactions with new consumers. Pay-per-click advertising is another form of online promotion when a traveler searches for information about a destination on the Internet, search engines direct him to pages that display advertising messages connected to the terms the user has entered. Numerous travel companies utilize it to sell their vacation packages. When a tourist clicks on an ad, he is redirected to the company's website, where he is required to pay the agreed-upon price (Maurer, 2022).

Affiliate marketing based on human resources through the Internet, which aims to promote tourism products and attract visitors to the company's website, is an essential digital marketing promotion strategy. This means that the visibility of the products can be improved, sales can be boosted, online bookings can grow, and the company's image can be enhanced in general. In addition, banner advertising is commonly used by tourism businesses to promote their services on websites offering related services (Minghetti, 2022).

Blog/Vlog Marketing is also a widely used digital marketing strategy, since every tourist business unit that has developed a website to promote its goods also attaches a blog where it publishes articles/videos/stories about the tourism industry's changes and advances. The Blog/Vlog publications inform the tourist audience, who can remark positively or adversely. It is a digital marketing tool that enables businesses to identify the demands of tourists, comprehend how they think, learn what they think about a product and their preferences, and so discover ways to enhance it (Garanti, 2022).

Viral marketing is an innovative digital marketing approach used for specialized and focused promotion that many tourism businesses employ to promote their own products and services. It is implemented similarly to how a virus spreads. Consequently, a targeted message travel rapidly. It is effective advertising since tourism-related themes are very engaging and attract a huge audience of people seeking relevant information.

Tourism businesses must also employ a specialized digital marketing technique, namely the optimization of website promotion through search engine marketing (SEM) to increase their visibility on engine results pages (Halkiopoulos & Giotopoulos, 2022). In addition, the strategy described above involves the optimization of technical features of a website using SEO (Halkiopoulos & Papadopoulos, 2022). Lastly, Customer Relationship Management Systems (e-CRM), with the aid of advanced technology, mobile devices, and call centers, provide a common platform for the tourism company to communicate with potential customers and optimize digital marketing processes, user service, and tourism industry sales (Yuce, 2021).

2.3 *Social Media Marketing in Smart Tourism*

Tourism business units were among the first to recognize the need of digital marketing tools for product planning, promotion, and sales. For tourism product and service promotion, social networks are an indispensable digital marketing tool. The global effect of social media, the expanding usage of the Internet, and the need for people to connect, seek knowledge, and be informed are the primary reasons why businesses should utilize social media (Buhalis, 2022).

Therefore, tourism businesses can use social media to make strategic decisions and advertise their products using implementation strategies and practical applications. Websites, forums, and blogs/vlogs enable users to connect and interact with others, as well as exchange opinions, critique, and evaluate services, and suggest ways to improve them. By publicizing prior vacationers' ideas and opinions, they can provide knowledge that is accessible to users worldwide without the need for travel companies. Tourism items can include a tourist package, which includes a plane ticket, a hotel reservation, excursions, and anything else associated with the upcoming location (Tung et al., 2019).

All firms in the tourism industry are increasingly engaged on social media to improve their brand image (Polat, 2022). In addition, businesses frequently use social media for market research in order to develop their products and respond to client needs as effectively as possible. It is important to note that many tourism-related businesses, such as hotels and restaurants, take advantage of the services provided by social media and rely on the physical location of the user. These services make it possible to display information on hotels, restaurants, monuments, public transportation, and all other nearby, mapped, and within the user's reach locations (Halkiopoulos et al., 2021a).

A traveler who utilizes social media and must make decisions about a tourist location will be influenced by the identification of a need, the identification of products and services that will satisfy this need, the evaluation of alternatives, and ultimately the selection of the most effective option. Internet information influences tourists to acquire more knowledge and make better decisions. Social media platforms such as Facebook and Instagram assist the dissemination of information regarding tourism products, services, and destinations (Qumsieh-Mussalam & Tajeddini, 2019). Due to the intangible character of tourism products, they cannot be evaluated prior to consumption. Therefore, the opinions of others have a significant impact on the purchasing decisions of tourists visiting a tourist location (Chhabra, 2021). Moreover, product promotion boosts a company's efficiency and sales. In terms of consumer behavior, knowledge, and decision-making processes, the use of social media has been demonstrated to be an effective digital marketing promotion technique (Mathew & Soliman, 2021).

Undoubtedly, social media is an integral component of the majority of people's daily lives. Numerous travelers post images and videos of their adventures online, which can encourage new visitors. Today, travelers select destinations, hotels, restaurants, and attractions based on the reviews and experiences of other travelers (Kontis et al., 2019). On social media, where people engage freely and share their thoughts and experiences, and information seeking is believed to be easier.

2.4 Digital Marketing in Tourist Agencies

In modern times, sales of trip packages through advanced and intelligent Internet information systems are increasing, while traditional sales are declining. The use of current digital marketing technologies may constitute a challenge to travel agencies, as they may result in a decline in customers, but they may also generate new opportunities. Choosing a tourist product entail searching for it, preselecting it, confirming its availability, and finalizing the package. Through the Internet, travel businesses are able to provide their services to customers around the globe. In addition to online reservations and sales, the Internet can be used to investigate the design of tourist packages or the optimization of existing products (De Lemos et al., 2020). In recent years, travel businesses such as Travelocity and Expedia (Figueiredo et al., 2022) have developed their tourism activities exclusively through the digital environment (Dewi, 2020).

Social media are now advertising platforms that allow for targeted advertising. For instance, advertising messages on the Facebook network are unquestionably effective at reaching the intended demographic. Additionally, depending on the demands of each business, Instagram advertisements that are brief but compelling might boost sales and build brand recognition among consumers. Since social media allows us to choose to whom and where ads are displayed, digital marketing for travel firms can be beneficial if they target a specific audience (Katsoni et al., 2012). In the first phase of an internet advertising campaign, the target audience and objectives should be determined (Zhang, 2021). Next are the plans for the online promotion of tourist products and the utilization of the websites of tourist agencies, as well as the ongoing improvements/optimizations for the duration of the advertising campaign (Halkiopoulos et al., 2021c).

With digital marketing, travel operators may connect with potential tourists to tourist sites more readily (Saura et al., 2020). A journey may be inspired by photographs of pals. A travel agency should aim to invite friends on social networks, proceed to add links pointing to the website and emails, offer incentives and contests, and post high-quality material that is always focused on a certain tourist location. With digital marketing, operators of travel agencies can supply even elderly users with effective platforms (Sharma et al., 2020). Travel businesses should use social media to advertise travel destinations, employ hashtags, and encourage their employees to share images from the trips they plan and take (Khairani, 2021).

2.5 Tourist Industry Digital Strategies Applied by the Hellenic State

In recent years, Greek tourism product has been increasingly advertised through digital tools/platforms and with the help of social media (Katsoni & Stratigea, 2016). Regions and municipalities recognize the need to showcase different landscapes through digital platforms and have proceeded to create smart solutions to get in touch with their potential tourists.

The South Aegean Region is the first organization of the Greek State that had participated in a global competition with advertising material it had created for its tourism product. A lot of information is provided in the advertising campaign for these Greek islands, such as, for example, the services that one can take advantage of various experiences that the visitor can obtain, but also places where they can be guided in order to get to know the history of each place. Social media promotion is done with audio-visual material, photos, and videos, which aim to highlight each place in combination with its natural beauty, cultural culture, customs, and traditions (Perinotto & Soares, 2020).

The advertising campaign for the South Aegean Region took part in an open public competition and had as its theme "Promotion of natural heritage, culture, and tourism of the South Aegean Region". This specific advertising campaign presented and promoted the South Aegean tourism product by highlighting the tourist heritage, aiming to strengthen the position and competitiveness of these islands. The results of this advertising activity have been recognized in recent years since the number of island visitors has increased. This advertising campaign aims for tourists to get to know the islands of the South Aegean and their culture, and it seems that this goal has been achieved since the number of visitors to the islands of the South Aegean has increased despite the conditions created by the Covid-19 pandemic the last two years (Ketter & Avraham, 2021).

In addition, the South Aegean Region is particularly active in creating and promoting advertising messages for the islands that make it up and the possibilities they offer to their visitors. In 2016, like no other campaign, the Aegean islands aimed at research and data analysis, tourism product marketing strategy, competitiveness planning, and communication plan for each island.

For this year, the third summer after the study of the effects of the Covid-19 pandemic, the Greek islands seem to be a popular destination for Greek and foreign tourists, with the number of bookings until the end of the summer at very high levels. A new advertising campaign will be launched for the eastern Aegean islands, Samos, Chios, Lesbos, Kos, and Leros. The name of the campaign is Greek4ForYou, to support these islands' communication and promotion. The creation of advertising spots to highlight the five islands, the beaches, the activities, the food, the customs, and the customs are necessary for foreign visitors to get to know Greek hospitality. Digital promotion, too, is a tool chosen to promote these islands more quickly. Social media is leading in this, and videos and photos show the beautiful Greek landscapes. The successful tourist campaign "Greek4ForYou", which concerns the five islands of the Eastern Aegean (Chios, Lesvos, Leros, Samos, and Kos), will continue in 2023 with a unique tourist promotion program abroad as well as the Greek market with funding from the Greek Ministry of Immigration and Asylum.

3 Research Design–Methods

3.1 Purpose of the Research

The research conducted for the purpose of this study is primary and quantitative. The sample consisted of 1200 young persons between the ages of 18 and 32, who were asked to respond to questions regarding the use and implementation of digital

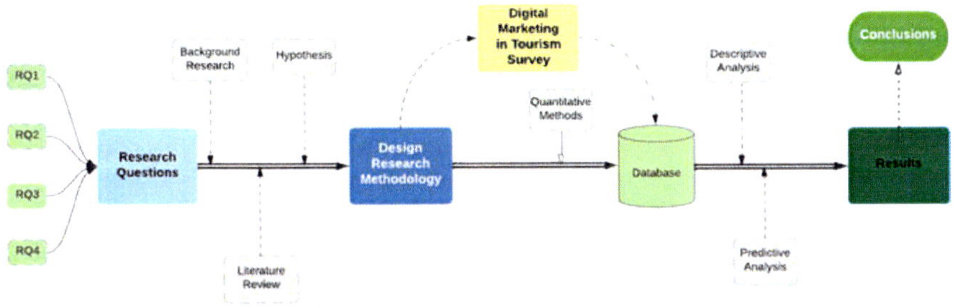

Fig. 1 Flowchart of research methodology

marketing applications/techniques in the tourist industry using a Google Forms-created electronic questionnaire. The questionnaire contains three sections. The initial section consists of questions regarding Internet usage and digital marketing. The second section comprises of digital marketing and tourism-related questions. In the third and last section of the questionnaire, four questions pertain to the respondents' demographic information. In the first phase, research data were searched, and then they were coded, connected, and categorized. This was accomplished using the statistical software SPSS Statistics. The questions were administered to the respondents via the Internet platform. The next section provides the research questions of the study and the flowchart of research methodology (Fig. 1).

Research Questions:

- [RQ1] Does Digital Marketing enhance the image of a business tourism unit?
- [RQ2] Does the use of Digital Marketing in the tourism industry contribute to the promotion of tourist destinations?
- [RQ3] Which Digital Marketing promotional tools for advertising tourism products and services are preferred?
- [RQ4] Do consumers of tourism products prefer innovative tourism applications and platforms to find IT, save time and optimize destination search?

3.2 *Descriptive Analysis*

The data analysis step involved pre-processing, coding, and entering data acquired in the field into a knowledge database. Each category of variables is analyzed using a different type of statistical analysis, both descriptive and inductive. Additionally, each group of variables was analyzed using a different statistical approach (descriptive/inductive). To investigate the possibility of a relationship between categorical variables, both descriptive (frequency tables–percentages, bar graphs, and pie charts) and inductive methods (parametric Pearson Chi-Square and non-parametric Likelihood ratio Fisher's exact test and Linear-by-Linear Association) were used. A correlation study was conducted to determine if there is statistical evidence of a relationship between two variables. If $p < 0.05$, there is a linear relationship between the variables under consideration; however, if $p > 0.05$, the null hypothesis is accepted, and the presence of a linear relationship between the tested variables is therefore rejected.

3.3 Sample Characteristics

Most of the sample, 52.9%, are women, while men make up 47.1%. Regarding the age of the people who make up the research sample, the most significant percentage, 32.5% (N = 390), is made up of people aged 21–23, followed by a percentage of 20.83% (N = 250) and corresponds to people aged 27–29, while 20.00% (N = 240) are aged 24–26, 18.50% (N = 222) aged 18–20, and finally 8.17% (N = 98) aged 18–20 years. In the question related to the educational level of the respondents, more than half of the sample, 61% (N = 732), are graduates of higher education, followed by 21% (N = 252) of the respondents who are graduates of secondary education. Fewer are the respondents who hold a master's degree, a percentage corresponding to 18% (N = 216) of the sample. Regarding the professional status of the sample, 51% (N = 612) of the sample answered that they work full-time, followed by 38% (N = 456) of the sample who declared part-time work. An 11% (N = 132) of the sample declared casual employment or unemployed (Fig. 2).

Regarding the claim if smart tourism attracts more tourists, 56% (N = 672) of the sample of respondents agree with the above proposition. To the question about whether the use of digital marketing in the tourism industry contributes to the promotion of tourist destinations [RQ2], 79% (N = 948) of the sample answered positively as well as to the question about whether respondents receive updates online about tourism products and services, respectively, 56% (N = 672) of the sample wishes to receive updates on tourist goods electronically. Completing the first set of key questions, most of the respondents, more specifically 77% (N = 924), agree that digital marketing generally enhances the profile of a tourism business [RQ1].

Fig. 2 Demographic characteristics (age range)

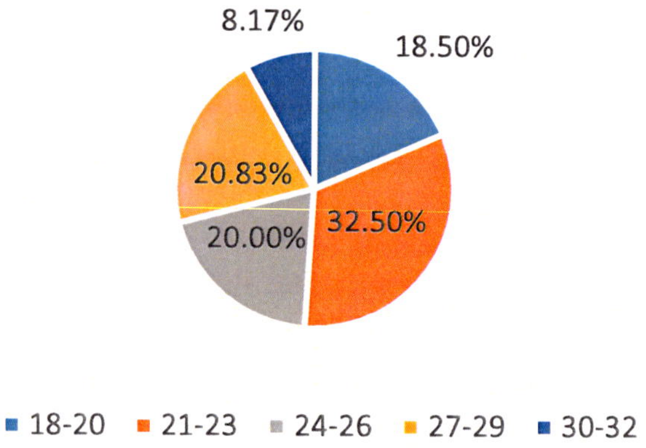

4 Results

The results based on the correlation analysis are presented in the following tables. Specifically, Table 1 presents the results for the two variables, the influence of advertisements [Q10: How much do you think you are influenced by advertisements displayed on social media?] displayed on social media and compelling information about online shopping [Q11: Do you think that through digital marketing, you are well informed about your online purchases before you make them?] before respondents make them. According to the linear correlation coefficient ($r = 0.372$, $p = 0.000$), we consider a weak positive linear correlation between them.

According to the results in Table 2, there seems to be statistical significance between questions [Q14] and [Q15]. The Internet and social media are a source of information for tourist destinations [Q14], as well as for purchasing tourist products/services [Q15] from the study sample ($p = 0.014$) [RQ4].

Suppose we hypothesize (Table 3) that the various variables evaluating the degree of preference for promotional tools of tourism products/services [Q17a, Q17b, Q17c, Q17d] are related to respondents' travel information via the Internet and social media [Q14]. In that case, the resulting results are the following [RQ3]. This hypothesis is acceptable for most variables [Q17b, Q17c, Q17d] since the value of p for all three variables evaluating the degree of preference for promotional tools of tourism products/services, which are websites [Q17b], promotional videos [Q17c], and blogs [Q17d] are less than the $p < 0.05$ value indicating that there is statistical significance

Table 1 Correlation matrix [variables Q10, Q11]

		Q10	Q11
Q10	Pearson correlation	1	0.372**
	Sig. (2-tailed)		0.000
	N	1200	1200
Q11	Pearson correlation	0.372**	1
	Sig. (2-tailed)	0.000	
	N	1200	1200

** Correlation is significant at the 0.01 level (2-tailed)

Table 2 Correlation matrix [variables Q14, Q15]

		Q14	Q15
Q14	Pearson correlation	1	−0.154*
	Sig. (2-tailed)		0.014
	N	1200	1200
Q15	Pearson correlation	−0.154*	1
	Sig. (2-tailed)	0.014	
	N	1200	1200

* Correlation is significant at the 0.05 level (2-tailed)

Table 3 Correlation matrix [variables Q14, Q17a, Q17b, Q17c, Q17d]

		Q14	Q17a	Q17b	Q17c	Q17d
Q14	Pearson correlation	1	0.088	0.289**	0.124*	0.167**
	Sig. (2-tailed)		0.164	0.000	0.048	0.008
	N	1200	1200	1200	1200	1200
Q17a	Pearson correlation	0.088	1	0.321**	0.257**	0.227**
	Sig. (2-tailed)	0.164		0.000	0.000	0.000
	N	1200	1200	1200	1200	1200
Q17b	Pearson correlation	0.289**	0.321**	1	0.477**	0.468**
	Sig. (2-tailed)	0.000	0.000		0.000	0.000
	N	1200	1200	1200	1200	1200
Q17c	Pearson correlation	0.124*	0.257**	0.477**	1	0.501**
	Sig. (2-tailed)	0.048	0.000	0.000		0.000
	N	1200	1200	1200	1200	1200
Q17d	Pearson correlation	0.167**	0.227**	0.468**	0.501**	1
	Sig. (2-tailed)	0.008	0.000	0.000	0.000	
	N	1200	1200	1200	1200	1200

* Correlation is significant at the 0.05 level (2-tailed)
** Correlation is significant at the 0.01 level (2-tailed)

as well as a weak positive correlation between the variables, due to the respective values of the coefficient $r = 0.289$, $r = 0.124$, and $r = 0.167$. Regarding e-mails [Q17a] as a promotional tool for tourism products/services, there is no statistical significance in the above correlation.

Table 4 shows the hypothesis that the variables agree with the proposition that the use of electronic media in tourism [Q19] is related to the trust in tourism product/service advertisements [Q20a, Q20b, Q20c] accepted by respondents with electronic way. We accept this hypothesis for two of the three variables since the p-value is $p = 0.049$ and $p = 0.031$ for the agreement expressed by the respondents that the use of electronic media in the tourism sector offers 24-h access [Q20b] and provision of organized information [Q20c], respectively. The Pearson coefficient is $r = 0.123$ and $r = 0.135$, respectively, so there is a weak positive correlation between the variables. Regarding the easy and quick update [Q20a], there is no statistical significance regarding this correlation ($p > 0.005$).

Table 5 presents the correlation results of the use of digital marketing [Q21] in the tourism industry that contributes to the promotion of tourism destinations and attracts more tourists from smart tourism [Q22] [RQ2]. In addition, there is a correlation between the variables of the use of digital marketing in the tourism industry that contributes to the promotion of tourist destinations and the fact that tourism business advertisements through social networks also contribute to the effective promotion of tourist destinations [Q23] [RQ2].

Table 4 Correlation matrix [variables Q19, Q20a, Q20b, Q20c]

		Q19	Q20a	Q20b	Q20c
Q19	Pearson correlation	1	0.046	0.123*	0.135*
	Sig. (2-tailed)		0.468	0.049	0.031
	N	1200	1200	1200	1200
Q20a	Pearson correlation	0.046	1	0.755**	0.670**
	Sig. (2-tailed)	0.468		0. 000	0.000
	N	1200	1200	1200	1200
Q20b	Pearson correlation	0.123*	0.755**	1	0.679**
	Sig. (2-tailed)	0.049	0.000		0.000
	N	1200	1200	1200	1200
Q20c	Pearson correlation	0.135*	0.670**	0.679**	1
	Sig. (2-tailed)	0.031	0.000	0.000	
	N	1200	1200	1200	1200

* Correlation is significant at the 0.05 level (2-tailed)
** Correlation is significant at the 0.01 level (2-tailed)

Table 5 Correlation matrix [variables Q21, Q22, Q23]

		Q21	Q22	Q23
Q21	Pearson correlation	1	−0.346**	−0.215**
	Sig. (2-tailed)		0.000	0.001
	N	1200	1200	1200
Q22	Pearson correlation	−0.346**	1	0.252**
	Sig. (2-tailed)	0.000		0.000
	N	1200	1200	1200
Q23	Pearson correlation	−0.215**	0.252**	1
	Sig. (2-tailed)	0.001	0.000	
	N	1200	1200	1200

** Correlation is significant at the 0.01 level (2-tailed)

The research carried out in the context of this research work contributed to the following results. The vast majority of the sample, i.e., 97% (N = 1164) of it, use the Internet daily, and regarding the length of stay on it, the majority of the sample, 71% (N = 852), remains on the Internet for over 5 h. The most popular search engine used by the majority of the sample (89%, N = 1068) is Google, and the device that the respondents (83%, N = 996) mainly use to connect to the Internet lately is their smartphone. Regarding the use of social networks, more than half of the sample (69%, N = 828) use Instagram, and a percentage of 47% (N = 564), use Facebook. Regarding the frequency with which they use the Internet to make online purchases, 60% (N = 720) of the sample answered often. About 87% (N = 1044) of the sample agree, either to a lesser or greater extent, with the proposition that businesses use the Internet and social media for advertising and that digital marketing enhances the image of a business tourism unit [RQ1]. As a promotional tool for tourism products and services, respondents prefer websites, followed by e-mails, promotional videos, and blogs [RQ3]. 80% of the sample (N = 960) agree with the proposition that the use of electronic media in the tourism sector offers easy and fast information [RQ4]. The majority of the sample, 79% (N = 948), responded positively to the question regarding the use of digital marketing in the tourism industry contributing to the promotion of tourist destinations [RQ2]. More than half of the sample, 58% (N = 696) of them, agreed with the proposition that consumers of tourism products prefer new tourism applications and platforms to save time and money [RQ4]. The variables evaluating the degree of preference for promotional tools of tourism products and services related to the respondents' information about travel through the Internet and social media are websites, promotional videos, and blogs [RQ3]. Finally, there is a correlation in the use of digital marketing in the tourism industry that contributes to the promotion of tourist destinations and attracting more tourists, but also to the fact that tourism business advertisements through social networks enhance the image of a tourism business unit [RQ1].

5 Discussion and Conclusion

The main topic examined in this research paper is that digital marketing presents opportunities and challenges to organizations. This study listed and explained how modern businesses use all the tools. Digital media provides markets with custom frameworks and interaction models that coordinate operations from consumers' locations, including their activities at the time. Therefore, digital marketing enhances a large part of the business actions in order to optimize the goals of the respective business at all levels. An additional contribution of digital marketing is the promotion of social interconnectivity. Therefore, as brand engagements extend to digital platforms, a forum for focused consumer interaction emerges where they can socialize with the brand.

This research attempted to highlight the importance of applying digital marketing techniques in the tourism industry through traditional data collection methods such as questionnaires and self-report scales. Since the demand in the years, we are going through is imperative to apply innovative techniques through the field of artificial intelligence and intelligent algorithms, but also the field of neuromarketing with the application of neuroimaging tools, for the analysis of behavioral patterns and choices in the field of tourism, we consider that these issues and the method of data collection are part of the limitations of the present study.

Concluding and from the point of view of future implications, we could emphasize that on the one hand, the tourism businesses and, on the other hand, the government bodies responsible for the activity in the tourism sector should try to improve the services provided and to upgrade the tourism product. In this context, businesses' lifelong training and information regarding innovative tools (Dritsas et al., 2018) and practices for the advantages of digital marketing in tourism for the optimal promotion of tourism products and services is an inescapable need and a required field to target their business actions (Solakis et al., 2022). Additionally, beyond the dynamic promotion of continuing education in the business field of tourism regarding digital marketing and the acquisition of digital skills, marketers must also familiarize themselves with soft skills (Gkintoni et al., 2022), explicitly acquiring leadership skills (Antonopoulou et al., 2020, 2021). Also, each business unit should determine someone responsible (social media manager) in order to deal with social networks and its website. The social media manager should also contact consumers to answer and resolve their queries. The tourist consumer should be constantly informed about the positive elements that digital marketing offers to the tourism industry, which will also contribute to their familiarity with modern online digital tools and digital marketing techniques to have meaning and substance to the tourism industry of smart tourism.

References

Alves, G. M., Sousa, B. M., & Machado, A. (2020). The role of digital marketing and online relationship quality in social tourism. In *Digital marketing strategies for tourism, hospitality, and airline industries* (pp. 49–70). https://doi.org/10.4018/978-1-5225-9783-4.ch003

Antonopoulou, H., Halkiopoulos, C., Barlou, O., & Beligiannis, G. N. (2021). Transformational leadership and digital skills in higher education institutes: During the COVID-19 pandemic. *Emerging Science Journal, 5*(1), 1–15. https://doi.org/10.28991/esj-2021-01252

Antonopoulou, H., Halkiopoulos, C., Barlou, O., & Beligiannis, G. (2020). Leadership types and digital leadership in higher education: Behavioural data analysis from University of Patras in Greece. *International Journal of Learning, Teaching and Educational Research, 19*(4), 110–129. https://doi.org/10.26803/ijlter.19.4.8

Buhalis, D. (2022). Smart tourism. In *Encyclopedia of tourism management and marketing*. https://doi.org/10.4337/9781800377486.smart.tourism

Cheung, M. L., & Leung, W. K. S. (2022). Digital Community. *Encyclopedia of Tourism Management and Marketing*. https://doi.org/10.4337/9781800377486.digital.community

Chhabra, D. (2021). Smart and sustainable marketing frameworks in heritage tourism. In *Resilience, authenticity, and digital heritage tourism* (pp. 74–92). https://doi.org/10.4324/9781003098836-5

Çınar, K., & Weiss, D. (2019). Digital marketing and virtual tourism to enhance destination accessibility. In *Sustainable destination branding and marketing: Strategies for tourism development* (pp. 198–204). https://doi.org/10.1079/9781786394286.0198

Cunha, C. R., Lopes, L., & Mendonça, V. (2022). Immersive digital marketing for smart cities focusing tourism. In *Smart innovation, systems and technologies* (pp. 605–614). https://doi.org/10.1007/978-981-16-9268-0_51

De Lemos, F. F., Vieira, D., & Pinho, T. (2020). Digital marketing strategies for accessible tourism. In *Digital marketing strategies for tourism, hospitality, and airline industries* (pp. 71–92). https://doi.org/10.4018/978-1-5225-9783-4.ch004

Dewi, N. P. R. C. (2020). Digital marketing strategy on travel tourism businesses in marketing 4.0 era. *International Research Journal of Management, IT and Social Sciences*. https://doi.org/10.21744/irjmis.v7n3.898

Dritsas, E., Livieris, I. E., Giotopoulos, K., & Theodorakopoulos, L. (2018). An apache spark implementation for graph-based hashtag sentiment classification on twitter. In *Proceedings of the 22nd Pan-Hellenic Conference on Informatics* (pp. 255–260). https://doi.org/10.1145/3291533.3291552

Figueiredo, G. V. M., & Neto, J. P. de B. (2022). Digital marketing as a driver of tourism. In *Advances in hospitality, tourism, and the services industry* (pp. 214–230). https://doi.org/10.4018/978-1-7998-8306-7.ch011

Garanti, Z. (2022). Digital marketing. In *Encyclopedia of tourism management and marketing*. https://doi.org/10.4337/9781800377486.digital.marketing

George, R. (2021). Digital marketing in tourism and hospitality. In *Marketing tourism and hospitality* (pp. 369–415). https://doi.org/10.1007/978-3-030-64111-5_12

Gkintoni, E., Halkiopoulos, C., & Antonopoulou, H. (2022). Neuroleadership an asset in educational settings: An overview. *Emerging Science Journal, 6*(4), 893–904. https://doi.org/10.28991/esj-2022-06-04-016

Halkiopoulos, C., Antonopoulou, H., Gkintoni E., & Giannoukou, I. (2021a). An expert system for recommendation tourist destinations: An innovative approach of digital marketing and decision-making process. *International Journal of Innovative Science and Research Technology, 6*(4), 398–404. ISSN: 2456-2165.

Halkiopoulos, C., Dimou, E., Kompothrekas, A., Telonis, G., & Boutsinas, B. (2021b). The e-tour facilitator platform supporting an innovative health tourism marketing strategy. In V. Katsoni, C. van Zyl (Eds.), *Culture and tourism in a smart, globalized, and sustainable world*. Springer Proceedings in Business and Economics. Springer. https://doi.org/10.1007/978-3-030-72469-6_39

Halkiopoulos, C., Katsouda, M., Dimou, E., & Panteli, A. (2021c). Cultural and tourism promotion through digital marketing approaches. A case study of social media campaigns in Greece. In V. Katsoni, C. van Zyl (Eds.), *Culture and tourism in a smart, globalized, and sustainable world*. Springer Proceedings in Business and Economics. Springer. https://doi.org/10.1007/978-3-030-72469-6_46

Halkiopoulos, C., Antonopoulou, H., Gkintoni, E., & Aroutzidis, A. (2022). Neuromarketing as an indicator of cognitive consumer behavior in decision-making process of tourism destination—An overview. In V. Katsoni, A.C. Şerban (Eds.), *Transcending borders in tourism through innovation and cultural heritage*. Springer Proceedings in Business and Economics. Springer. https://doi.org/10.1007/978-3-030-92491-1_41

Halkiopoulos, C., & Giotopoulos, K. (2022). Tourism's use of web-based information systems and the influence of tourism trends. In V. Katsoni, A.C. Şerban (Eds.), *Transcending borders in tourism through innovation and cultural heritage*. Springer Proceedings in Business and Economics. Springer. https://doi.org/10.1007/978-3-030-92491-1_25

Halkiopoulos, C., & Papadopoulos, D. (2022). Computational methods for evaluating web technologies and digital marketing techniques in the hospitality industry. In V. Katsoni, A.C. Şerban (Eds.), *Transcending borders in tourism through innovation and cultural heritage*. Springer Proceedings in Business and Economics. Springer. https://doi.org/10.1007/978-3-030-92491-1_24

Katsoni, V., & Dologlou, N. (2017). ICT Applications in smart ecotourism environments. In *Smart cities in the mediterranean* (pp. 225–244). https://doi.org/10.1007/978-3-319-54558-5_11

Katsoni, V., & Stratigea, A. (Eds.). (2016). *Tourism and culture in the age of innovation*. Springer Proceedings in Business and Economics. https://doi.org/10.1007/978-3-319-27528-4

Katsoni, V., Giaoutzi, M., & Nijkamp, P. (2012). Market segmentation in tourism: An operational assessment framework. In *Quantitative methods in tourism economics* (pp. 329–352). https://doi.org/10.1007/978-3-7908-2879-5_18

Ketter, E., & Avraham, E. (2021). #StayHome today so we can #TravelTomorrow: Tourism destinations' digital marketing strategies during the Covid-19 pandemic. *Journal of Travel & Tourism Marketing, 38*(8), 819–832. https://doi.org/10.1080/10548408.2021.1921670

Khairani, A. (2021). Empirical investigation on how digital content marketing tourism affects customer engagement. *Journal of Tourism, Leisure and Hospitality*. https://doi.org/10.48119/toleho.995179

Kontis, A.-P., Skoultsos, S., & Papayiannis, D. (2019). *Tourism marketing channels in digital evolution era: Online travel agencies in Greek tourism industry* (pp. 1303–1310). Springer Proceedings in Business and Economics. https://doi.org/10.1007/978-3-030-12453-3_149

Korstanje, M. E. (2015). Heritage in the digital era: Cinematic tourism and the activist cause. *Annals of Tourism Research, 55*, 186–187. https://doi.org/10.1016/j.annals.2015.08.007

Majumder, T. (2019). Tourism in the digital ERA. *Emperor Journal of Applied Scientific Research, 01*(06), 01–07. https://doi.org/10.35337/eijfmr.2019.1601

Mancinelli, F. (2022). Digital nomads. In *Encyclopedia of tourism management and marketing*. https://doi.org/10.4337/9781800377486.digital.nomads

Mathew, V., & Soliman, M. (2021). Digital content marketing and tourism consumer behavior-model. In *PsycTESTS Dataset*. https://doi.org/10.1037/t81717-000

Maurer, C. (2022). Digital marketing in tourism. In *Handbook of e-tourism* (pp. 1311–1333). https://doi.org/10.1007/978-3-030-48652-5_82

Minghetti, V. (2022). Digital divide in tourism. In *Encyclopedia of tourism management and marketing*. https://doi.org/10.4337/9781800377486.digital.divide.in

Nadda, V., Chaudhary, H. S., & Arnott, I. (2020). Cloud computing in tourism. In *Digital marketing strategies for tourism, hospitality, and airline industries* (pp. 141–155). https://doi.org/10.4018/978-1-5225-9783-4.ch007

Panas, G., Thrasidi, N., Halkiopoulos, C., & Gkintoni, E. (2022). Consumer behavior and cognitive factors in relation to gastronomic tourism and destination marketing in Greece. In V. Katsoni, A.C. Şerban (Eds.), *Transcending borders in tourism through innovation and cultural heritage*. Springer Proceedings in Business and Economics. Springer. https://doi.org/10.1007/978-3-030-92491-1_40

Panas, G., Vasiliadou, S., & Halkiopoulos, C. (2020). Data analysis evaluation of web technologies enhancing communication in tourism industry: Case study in Kefalonia Island. In V. Katsoni, T. Spyriadis (Eds.), *Cultural and tourism innovation in the digital era*. Springer Proceedings in Business and Economics. Springer. https://doi.org/10.1007/978-3-030-36342-0_14

Panteli, A., Kompothrekas, A., Halkiopoulos, C., & Boutsinas, B. (2021). An innovative recommender system for health tourism. In V. Katsoni, C. van Zyl (Eds.), *Culture and tourism in a smart, globalized, and sustainable world*. Springer Proceedings in Business and Economics. Springer. https://doi.org/10.1007/978-3-030-72469-6_42

Perinotto, A. R. C., & Soares, J. R. R. (2020). Photographic image, credibility, and consumption of tourism in the digital era. In *Digital marketing strategies for tourism, hospitality, and airline industries* (pp. 93–110). https://doi.org/10.4018/978-1-5225-9783-4.ch005

Polat, E. (2022). Digital marketing and digital marketing applications in the asian tourism industry. In *Handbook of technology application in tourism in Asia* (pp. 115–130). https://doi.org/10.1007/978-981-16-2210-6_5

Qumsieh-Mussalam, G., & Tajeddini, K. (2019). Innovation in tourism destination marketing. In *Tourism, hospitality and digital transformation* (pp. 165–174). https://doi.org/10.4324/9780429054396-10

Saura, J. R., Reyes-Menendez, A., & Palos-Sanchez, P. R. (2020). The digital tourism business. In *Digital marketing strategies for tourism, hospitality, and airline industries* (pp. 1–22). https://doi.org/10.4018/978-1-5225-9783-4.ch001

Sharma, A., Sharma, S., & Chaudhary, M. (2020). Are small travel agencies ready for digital marketing? Views of travel agency managers. *Tourism Management, 79*, 104078. https://doi.org/10.1016/j.tourman.2020.104078

Shin, S., & Xiang, Z. (2022). Digital tourism text. *encyclopedia of tourism management and marketing*. https://doi.org/10.4337/9781800377486.digital.tourism.text

Sobrino, R. G. (2022). Digital communications in tourism marketing strategies. In *International case studies in tourism marketing* (pp. 213–217). https://doi.org/10.4324/9781003182856-26

Solakis, K., Katsoni, V., Mahmoud, A. B., & Grigoriou, N. (2022). Factors affecting value co-creation through artificial intelligence in tourism: A general literature review. *Journal of Tourism Futures*. https://doi.org/10.1108/jtf-06-2021-0157

Tung, V. W. S., Cheong, T. M. F., & To, S. J. (2019). Tourism management in the era of smart mobility: A perspective article. *Tourism Review, 75*(1), 283–285. https://doi.org/10.1108/tr-05-2019-0180

Yuce, A. (2021). the impact of digital transformation and virtual reality in tourism marketing. In *Handbook of research on IoT, digital transformation, and the future of global marketing* (pp. 1–14). https://doi.org/10.4018/978-1-7998-7192-7.ch001

Zhang, T. (2021). *Creative digital marketing campaigns for hospitality and tourism organizations*. https://doi.org/10.4135/9781529798937

The Trade Fair Industry in Transition: Digital, Physical and Hybrid Trade Fairs. The Case of Thessaloniki

Eleana Kostopoulou, Spyros Avdimiotis, and Dimitris Kourkouridis

Abstract The subject of this paper is the transition of exhibition activity from physical to digital and hybrid exhibitions, with particular emphasis on how the COVID-19 pandemic will affect the future of trade fairs. According to the literature, the COVID-19 global pandemic health crisis has shown that technology has contributed to the flexibility of trade fairs, as digital trade fairs have been an effective means of adapting to the adverse conditions imposed by the pandemic. Nevertheless, it seems that exhibitors continue to show a preference for physical trade fairs, while recognizing the benefits of digital ones. In the context of the present paper, the case study of Thessaloniki was examined, conducting primary research with the method of the questionnaire to exhibitors of TIF—HELEXPO SA in order to investigate the views of the participants toward the physical, digital and hybrid trade fairs, but also the investigation of their views on the future of exhibition activity in the post-COVID era. The results show that exhibitors recognize the benefits of digital trade fairs and digital tools that can be used, but do not want to replace physical trade fairs.

Keywords Trade fairs · Exhibitions · Digital trade fairs · Hybrid trade fairs

JEL Classifications The article belongs to the following JEL category: O. Economic development · Innovation · Technological change and growth · O3 innovation · Research and development · Technological change · Intellectual property rights · O33 technological change: Choices and consequences · Diffusion processes

---

E. Kostopoulou · S. Avdimiotis
Department of Organization Management, Marketing and Tourism, International Hellenic University, 57400 Thessaloniki, Greece

D. Kourkouridis (✉)
School of Spatial Planning and Development, Aristotle University of Thessaloniki, 54636 Thessaloniki, Greece
e-mail: kourkouridis@yahoo.gr

© The Author(s), under exclusive license to Springer Nature Switzerland AG 2023

V. Katsoni (ed.), *Tourism, Travel, and Hospitality in a Smart and Sustainable World*, Springer Proceedings in Business and Economics,
https://doi.org/10.1007/978-3-031-26829-8_25

1 Introduction

Although the impact of digitization has changed the format of exhibitions in recent years, physical trade fairs continue their main purpose, which is to bring buyers and sellers together in the same place (Kourkouridis & Frangopoulos, 2022). The crisis of the COVID-19 pandemic has led in several cases to a mandatory transition to digital trade fairs, proving to be an effective means of adapting to the adverse conditions imposed by the pandemic. However, available research data shows that exhibitors still show a preference for physical trade fairs, even though they recognize many of the advantages of digital ones (Tradeshow Logic, 2020; Nella, 2021; UFI, 2022).

The subject of this paper is the transition of exhibition activity from physical to digital and hybrid exhibitions, with a particular focus on how the COVID-19 pandemic will affect the future of trade fair operations. The purpose of the research carried out was to investigate the views and attitudes of trade fair participants, i.e., trade exhibitors, toward physical, digital and hybrid trade fairs, but also to explore their views on the future of exhibition activity in the post-COVID era. The research focuses on the case of Thessaloniki, where there is a long tradition of exhibition activity (Kourkouridis et al., 2019; 2023). In order to investigate this specific issue, a primary quantitative survey was carried out among exhibitors who participated in trade fairs in Thessaloniki.

2 Literature Review

2.1 Digital and Hybrid Trade Fairs

The Internet has evolved tremendously in recent years and has become an integral part of everyday business life. Many companies now see the Internet as a means to expand their reach and be able to sell globally without having to physically expand into new markets. This realization has caused significant changes in communication and coordination channels as well as in the organization of work. The new Internet-based economy, with its capacity for cross-border exchanges, has given rise to a number of cyber-societies (e.g., virtual communities and virtual organizations) while in recent years an exciting new communication channel has emerged in the form of 'virtual exhibitions' (Lee-Kelley et al., 2004). 'Virtual'—or 'digital' or 'online'—trade fairs can be defined as those (Gani et al., 2021: 288): *'held in cyberspace, where all types of organisations (from small to large) use computer-mediated information technology (IT) with web-based capability can participate'*. Accordingly, Gottlieb and Bianchi (2017: 17) describe them as: *'a type of virtual event, where exhibitors and visitors connect with one another* via *the virtual environment (Internet), regardless of geographic location, to interact and exchange information'*.

A typical digital trade fair often includes a virtual exhibit hall, where users enter with specific permissions. Exhibitors can build virtual booths to display information about their goods or services, just as they would at a physical trade fair. They may also include other elements, such as online conferences, webinars or other educational presentations. Companies can also use tracking mechanisms that allow them to determine the flow of traffic in the digital exhibition (Gottlieb & Bianchi, 2017). Attendees can visit online virtual halls and booths to get information about a company's profile, products and services, while interactions between exhibitors and visitors take place entirely in a multimedia environment consisting of written, visual and audio evidence (Geigenmüller, 2010).

Digital trade fairs offer innovative services to both exhibitors and visitors (Geigenmüller, 2010). Among their advantages are the fact that they allow interaction between exhibitors and visitors through a virtual environment, overcoming geographical barriers related to the exchange of physical or virtual goods. That is, in this way, the limitations of space, time and location are avoided (Gani et al., 2021). Research by Gottlieb and Bianchi (2017) found that the main motivations for participating in digital trade fairs are to increase sales revenue, reduce costs, access new or different markets and build brand credibility and organizational legitimacy among visitors attending such events. The findings suggest that the future of digital trade fairs will likely depend on the technological development of the systems that support them.

However, there are also some significant challenges. For example, their effectiveness is still questioned, while at the same time there is an increased risk of data leakage (Wei, 2022). Also, according to Schiefelbein (2018), virtual meetings are not taken with the same level of seriousness as in-person meetings because the participants are behind screens and technology. However, she believes that with the right tools and tips, digital meetings can be just as, if not more, productive than face-to-face meetings.

While the question of whether digital trade fairs will replace face-to-face trade fairs is still debatable, in some cases the two formats may be combined to achieve greater market coverage and better performance (Gani et al., 2021), forming the so-called 'hybrid events'. These events include '*any meeting or event with at least one group of face-to-face participants that digitally connects with participants in another or multiple locations*' (Nilsson, 2020: 12). These events require organizers to expand their creativity, plan their strategy and integrate technology with traditional live events to create new types of experiences (Nilsson, 2020).

Sarmento and Simões (2019) focused their study on the interaction between the combination of digital and physical trade fairs. They cited the importance of physical interaction, direct participation and sensory pleasure from physical trade fairs, while emphasizing value creation from digital trade fairs. The findings of the study highlighted the general role that trade fairs play in facilitating companies' interactions with existing and potential customers. In the case of the hybrid trade fairs, it is developed on physical and virtual platforms, where companies promote business relationships and create learning experiences and customer loyalty. Participants seek solutions to problems and often innovation is a consequence of engagement and learning processes. However, while the physical trade fair is instrumental in human face-to-face interaction, i.e., creating informal networks of contacts and face-to-face interactions, digital trade fairs are highlighted as a catalyst for enhancing interactivity and connectivity before and after the physical trade fair.

2.2 *Recent Developments in the Era of the COVID-19 Pandemic*

At the beginning of 2020, it was taken for granted that trade fair plans would be executed on the same terms as in previous years. However, the outbreak of the COVID-19 pandemic, as early as February 2020, led to a number of trade fair cancelations (Nella, 2021). This had a significant impact on the exhibition industry. For example, in Germany, a country with a highly developed exhibition activity, trade fairs were forced to stop for almost 19 months (between March 2020 and September 2021). In 2020, this caused a 70% loss of sales for trade fair organizers alone. The total economic loss amounts to 42 billion euros to date. For 2021, again, a dramatic loss is looming: by the end of September more than two-thirds of the 380 planned trade fairs had to be canceled. A year without trade fairs means financial losses, too, for more than 40% of exhibitors (Hochheim, 2021).

Globally, the Global Association of Exhibition Industry (UFI, 2021a) reports that global exhibition industry revenue for 2020 was down 68%, compared to 2019. More specifically, 2020 revenue represented only 23% of these in 2019 in Central and South America, 24% in the Middle East and Africa, 27% in the Asia–Pacific region and 32% in Europe and 36% in North America. Exhibitions have a direct impact on many sectors in the regions where they take place—not just the exhibition industry (venues, organizers and service providers), but all related sectors such as accommodation, restaurants and transport. Taking all these sectors into account, it is estimated that at least €200 billion of total exhibition-related output was not produced in 2020, including €80 billion in North America, €65 billion in Europe and €46 billion in the Asia–Pacific region. This equates to 2.4 million full-time jobs affected worldwide.

In the USA, the Center for Exhibition Industry Research reports that the growing number of trade fairs canceled due to the COVID-19 crisis will result in a loss to the economy of $14 billion to $22 billion. As of March 15, 2020, at least 50 B2B events have announced their cancelations, amounting to a total loss of $318 million in organizer revenue, with a loss to the economy of $1.8 billion also taking into account the direct expenditures of attendees and exhibitors (Palmer, 2020).

However, in addition to the aforementioned cancelations and large revenue losses, we find that since the start of the pandemic an increasing number of traditional trade fairs that used to bring participating exhibitors from the same or related industries from around the world to a physical location, took their experience online (Wei, 2022).

It should be mentioned that during the pandemic period an additional advantage was added to digital exhibitions, that of social distancing. Digital trade fairs can avoid cross-border movement of people and reduce the possibility of contamination, and cross-border spread of mutant strains caused by the movement of people. Under the unstable situation of the pandemic, digital trade fairs can meet the needs of traders from areas with severe virus spread to participate in exhibitions and can greatly reduce the risk of infection, ensuring the safety of exhibitors' lives (Wei, 2022).

Nella's (2021) research focuses on the 'forced transformation' of international tourism exhibitions to digital or hybrid ones during the pandemic, presenting the results of a relevant quantitative study among tourism exhibitors from Greece. The results showed that 32% were satisfied with the event, while 27% expressed limited satisfaction with their virtual event experience. The main reasons for dissatisfaction included the limited scope of a virtual event compared to a physical one, limited networking opportunities due to a lack of representatives from specific target markets and technical difficulties encountered both on their side and on the organizers' platform. Compared to physical exhibitions, 61% of participants said that digital exhibition was less effective, while only 8% said it was equally effective. In terms of the extent to which their involvement actually contributed to enhancing their networking and extroversion, 27% said the contribution was significant, while 38% said it was limited or non-existent. Nella (2021) concludes that it appears that organizers' matchmaking systems play an even more critical role in digital and hybrid events. In addition, user-friendly platforms and punctuality in appointment times are additional prerequisites for successful virtual events.

2.3 *Estimates for the Evolution of the Exhibition Industry in the Post-COVID Era*

As described above, the COVID-19 pandemic has dramatically affected the exhibition industry. The question that reasonably arises, however, concerns the evolution of the exhibition industry in the post-COVID era. According to the latest Global Exhibition Barometer by the Global Association of Exhibition Industry (UFI, 2022), the recovery of the exhibition industry is accelerating, with global exhibition revenues expected to reach 71% of 2019 levels in 2022. As in the previous edition of the Global Exhibition Barometer, released in July 2021 (UFI, 2021b), the 'impact of the COVID-19 pandemic on business' remains the most pressing business issue, although this has decreased to 19% of respondents, up from 29% six months ago.

A second question arising from the developments of the exhibition industry due to the pandemic concerns the form that exhibitions will take in the future. Physical trade fairs will continue to be essential for exhibitors in the post-COVID era. However, trade fairs will be increasingly complemented by virtual event formats: this is the conclusion of a current analysis of the German trade fair industry before, during and after the COVID-19 pandemic by the German Institute for Economic Research. The research was based on a poll of more than 2,000 exhibitors, including nearly 1,600 from the manufacturing sector and more than 530 from the wholesale sector. The questions were about exhibitors' general plans to participate in the future trade fairs. Research shows impressively that exhibitors continue to rely on physical trade fairs. Exhibitors with digital trade fair experience also wish to take part in physical trade fairs again (Hochheim, 2021).

The Global Association of Exhibition Industry (UFI, 2022) also tried to examine the specific question about the future of exhibitions in the post-COVID era. The survey found that 80% of respondents agree that COVID-19 confirms the value of face-to-face events and expect the sector to recover quickly (with 38% saying 'yes, definitely' and 42% to indicate 'likely'). However, 13% of respondents are not sure. Also, 44% believe there will be fewer international physical exhibitions and, overall, fewer participants (with 6% saying 'yes, definitely' and 38% saying 'likely'), while 30% are not sure. Also, 73% believe there is a push toward hybrid events and more digital elements at events (with 24% saying 'yes, definitely' and 49% saying 'likely'), while 15% are not sure. Finally, just 10% agree that virtual events will replace physical events (with 2% saying 'yes, definitely' and 9% saying 'most likely'), while 18% are unsure and 54% indicates 'definitely not'.

3 Methodology

3.1 Study Area

The study area of the research is the city of Thessaloniki, which has a long tradition of events in the form of trade fairs, during the Byzantine and Ottoman Empires. However, the most important stage in the exhibition activity of the city—in its modern form—was the conception of the idea for what is known today as the Thessaloniki International Exhibition (TIF) during the interwar period (Kourkouridis et al., 2016). On April 30, 1925, permission was granted by the then Minister of National Economy for the establishment of the exhibition (Kourkouridis et al., 2018). The current facilities of TIF—HELEXPO SA extend over 180,000 m^2 of which 62,000 m^2 are covered exhibition, conference and other spaces (TIF—HELEXPO, 2014). The modern period of operation of TIF—HELEXPO is characterized by the rapid development of trade fairs (Tsiftelidou et al., 2017).

The exhibition activity in Thessaloniki received a big blow from the economic crisis, but it was overcome very quickly, which proves that it is a resilient economic activity. In the current period, the exhibition activity in Thessaloniki, as well as all over the world as described above, is going through a period of even greater crisis due to the COVID-19 pandemic (Kourkouridis, 2021). The question that is reasonably asked here is how the exhibition activity in Thessaloniki will evolve in the post-COVID era.

3.2 Research Objectives-Research Questions

The main objective of the primary research is to investigate the attitudes and opinions of the exhibitors participating in the Thessaloniki trade fairs toward physical, digital and hybrid trade fairs, but also to investigate their views on the future of exhibition activity in the post-COVID era. In more detail, the research questions we asked are the following:

Research Question 1. To what extent are exhibitors willing to participate in hybrid trade fairs in the future?

Research Question 2. To what extent has the COVID-19 pandemic affected exhibitors' intention to participate in digital trade fairs?

Research question 3. What do exhibitors think will be the future of trade fairs after the COVID-19 pandemic?

3.3 Data Collection Method

For the needs of the research, a questionnaire was created, which consists of a total of 18 questions and is divided into three sections. The first section includes 6 questions related to the profile of the respondents, as well as the profile of their companies. The second section included 7 questions about participants' views on digital, hybrid and physical trade fairs. Finally, the third part of the questionnaire includes 5 questions about respondents' views on the future of trade fairs after the COVID-19 pandemic.

The questionnaires were self-completed by the participants, while the questions are all closed type and multiple choice. Also, 9 of the questions were based on the Likert scale.

The target group of the research conducted was TIF—HELEXPO SA exhibitors who participated in trade fairs in Thessaloniki.

3.4 Research Process

The survey was conducted in March 2022, specifically from March 21 to 30, 2022. A total of 1,500 invitations were sent to TIF—HELEXPO trade exhibitors to participate in the survey, of which 188 finally participated. Invitations were sent by email and included a brief description of the purpose of the survey as well as the link to the questionnaire in the Google Forms.

After the responses were collected, they were appropriately coded and entered into the IBM SPSS Statistics software for analysis.

4 Results

4.1 Research Participants

Table 1 presents the basic information about the survey participants, where we see that 61.2% of them are men, while 38.8% are women. Additionally, 34% are between 41 and 50 years old, 25% between 31 and 40 and 22.3% between 51 and 60. Also, 37.8% are business owners and 18.1% are employees. 56.4% of the participants stated that the company has been operating for more than 20 years and 61.2% that it employs more than 8 employees. Finally, 34.6% of businesses deal with products/goods, 17% provide services and 20.2% a combination of the two.

4.2 Advantages and Disadvantages of Digital Trade Fairs

Figure 1 presents the participants' views on the advantages of digital trade fairs. In general, it seems that the majority of respondents agree with the opinion that money saving (49.5% strongly and 24.5% very strongly), time saving (46.8% strongly and 17.6% very strongly), health safety (36.2% strongly and 50.5% very strongly) and the easy exchange of information and data (31.9% strongly and 11.2% very strongly) are advantages of digital trade fairs compared to physical ones. Opinions about the possibility of expanding the market reach through digital trade fairs compared to physical ones appear to be more widely distributed in the responses, compared to the previous advantages examined, where the opinions were clearer.

Figure 2 presents the participants' views on the disadvantages of digital trade fairs. The vast majority of survey participants agree with the opinion that the lack of face-to-face contact (61.7% very strongly and 31.4% strongly), the lack of development of personal relationships (56.4% very strongly and 35.1% strongly), the lack of contact with the product (50% very strongly and 31.9% strongly) and the potential technology problems (36.7% strongly and 20.7% very strongly) are disadvantages of digital trade fairs.

4.3 Hybrid Trade Fairs

The respondents' willingness to participate in hybrid trade fairs, as shown in Fig. 3, ranges at moderate levels. Specifically, 31.9% stated a moderate willingness to participate in hybrid trade fairs, 30.3% a strongly, 21.3% a little, 9.0% a very strongly and 7.4% stated not at all.

From Fig. 4, it can be seen that 49.5% of participants moderately agree with the view that hybrid trade fairs have a negative impact on exhibitors. Thus, a neutral attitude toward hybrid trade fairs is evident. This neutral attitude is also evident in

Table 1 Basic information about the survey participants

Participants					
Gender					
Men	Women				
61.2%	38.8%				
Age					
18–30	31–40	41–50	51–60	60+	
11.7%	25%	34%	22.3%	6.9%	
Position					
Owner	Manager	Business executive	Employee	Other	
37.8%	17.6%	25%	18.1%	1.6%	
Years					
Less than 5	5–10	11–20	More than 20		
3.7%	14.4%	25.5%	56.4%		
Number of employees					
Less than 3	3–5	6–8	More than 8		
10.6%	15.4%	12.8%	61.2%		
Activity					
Products/goods	Services	Combination of products/services	Organization	Public body	Other
34.6%	17.0%	20.2%	4.8%	6.4%	17.0%

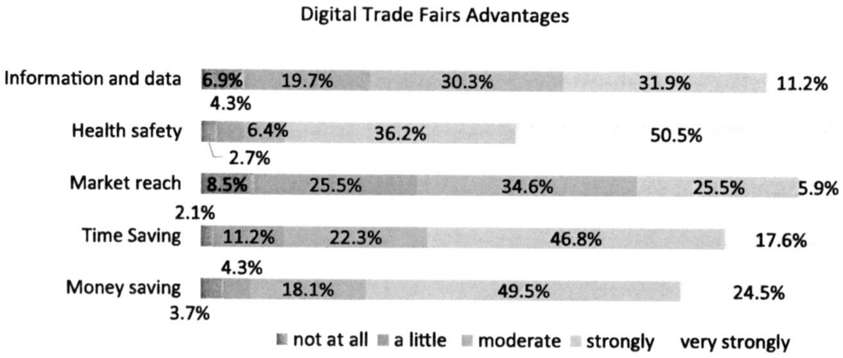

Fig. 1 Digital trade fair advantages

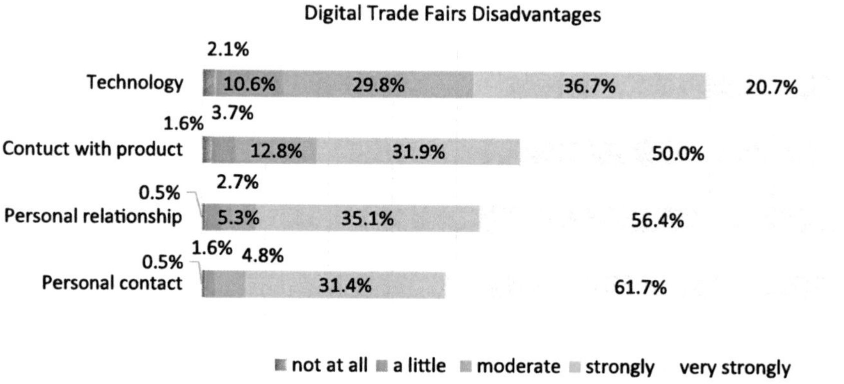

Fig. 2 Digital trade fairs disadvantages

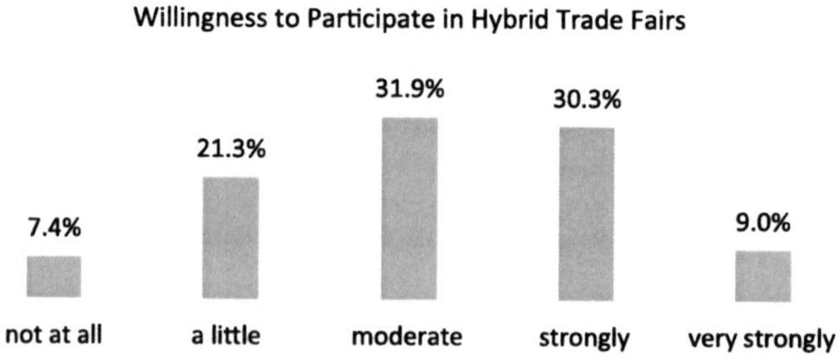

Fig. 3 Willingness to participate in hybrid trade fairs

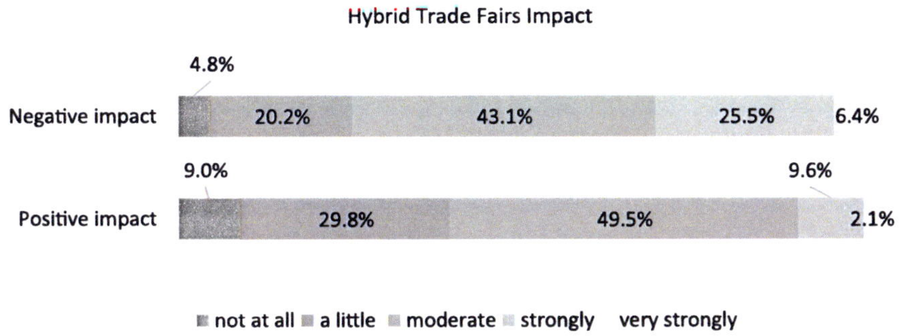

Fig. 4 Positive and negative impact of hybrid trade fairs on exhibitors

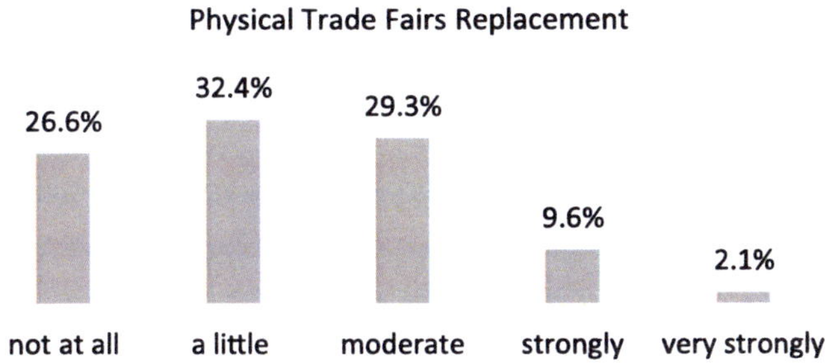

Fig. 5 Degree of agreement that hybrid trade fairs will replace physical trade fairs

the case where 43.1% of respondents moderately agree with the opinion that hybrid trade fairs have a positive impact on exhibitors.

In Fig. 5, it can be seen that 32.4% of the survey participants agree a little with the opinion that hybrid trade fairs will completely replace physical ones, while 29.3% agree moderately. A fairly significant percentage (26.60%), however, does not agree with this opinion at all.

4.4 *Effect of COVID-19 and Future of Trade Fairs*

From Fig. 6, it appears that the COVID-19 pandemic has significantly affected the intention of exhibitors to participate in physical trade fairs (37.8% strongly and 27.7% very strongly). On the other hand, the effect of the pandemic on the intentions to participate in digital trade fairs seems to be divided among the different answers.

Figure 7 shows that the majority of respondents (64.4%) believe that after the pandemic the exhibition activity will return to physical trade fairs. However, a fairly significant percentage (32.4%) consider the future of exhibition activity to be hybrid, while only 3.2% consider it to be digital. Also, 71.8% said that after the pandemic

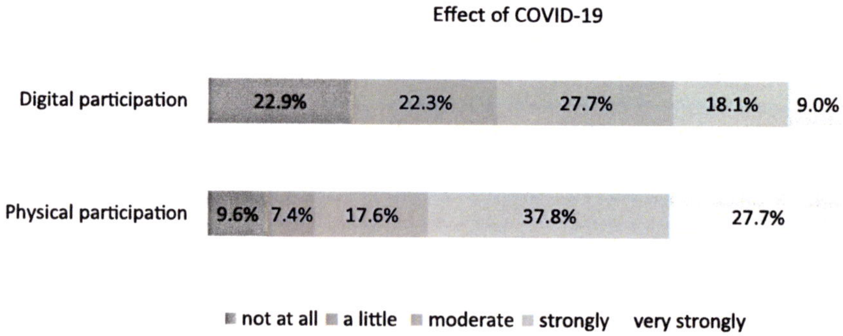

Fig. 6 Effect of COVID-19 on the intention to participate in physical and digital trade fairs

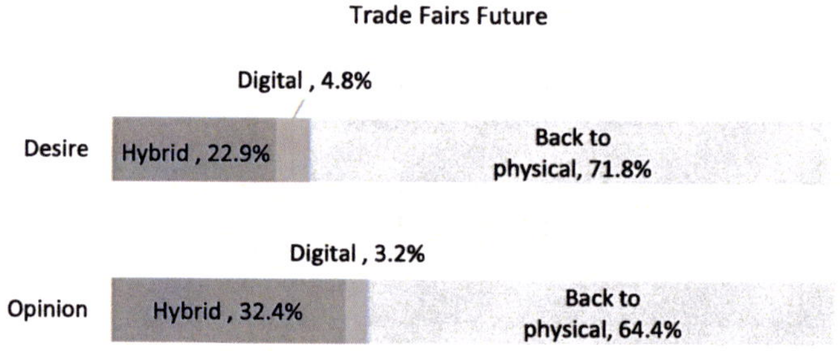

Fig. 7 Participants' views and desire for the future of trade fairs

they want exhibition activity to return to physical trade fairs. The desire for hybrid trade fairs was expressed by a significant percentage of 22.9%, while only 4.8% expressed the desire for digital trade fairs.

4.5 Cross-Tabulation Analysis

In this section, the results obtained from the cross-tabulation analysis are presented. Specifically, tables were created for the 'gender' and 'age' variables with the variable regarding the exhibitor's intention to participate in hybrid trade fairs. This analysis was done in order to investigate whether the gender and age of the exhibitors affect their intention to participate in hybrid trade fairs. The results are presented in Tables 2 and 3.

In Table 2, we notice that for men the answer 'strongly' has the largest percentage. Specifically, 29.6% of men stated that they would strongly like to participate in hybrid trade fairs in the future, while 59.6% of those who answered 'strongly' were men.

Table 2 Gender * participation in hybrid trade fairs cross-tabulation

			Participation in hybrid					Total
			Not at all	A little	Moderate	Strongly	Very strongly	
Gender	Man	Count	10	29	33	34	9	115
		% within gender	8.7%	25.2%	28.7%	**29.6%**	7.8%	100.0%
		% within participation in hybrid	71.4%	72.5%	55.0%	**59.6%**	52.9%	61.2%
		% of total	5.3%	15.4%	17.6%	18.1%	4.8%	61.2%
	Woman	Count	4	11	27	23	8	73
		% within gender	5.5%	15.1%	**37.0%**	31.5%	11.0%	100.0%
		% within participation in hybrid	28.6%	27.5%	**45.0%**	40.4%	47.1%	38.8%
		% of total	2.1%	5.9%	14.4%	12.2%	4.3%	38.8%

Table 3 Age * participation in hybrid trade fairs cross-tabulation

			Participation in hybrid					Total
			Not at all	A little	Moderate	Strongly	Very strongly	
Age	18–30	Count	0	3	7	10	2	22
		% within age	0.0%	13.6%	31.8%	**45.5%**	9.1%	100.0%
		% within participation in hybrid	0.0%	7.5%	11.7%	**17.5%**	11.8%	11.7%
		% of total	0.0%	1.6%	3.7%	5.3%	1.1%	11.7%
	31–40	Count	4	8	19	13	3	47
		% within age	8.5%	17.0%	**40.4%**	27.7%	6.4%	100.0%
		% within participation in hybrid	28.6%	20.0%	**31.7%**	22.8%	17.6%	25.0%
		% of total	2.1%	4.3%	10.1%	6.9%	1.6%	25.0%
	41–50	Count	6	19	15	19	5	64
		% within age	9.4%	**29.7%**	23.4%	**29.7%**	7.8%	100.0%
		% within participation in hybrid	42.9%	**47.5%**	25.0%	**33.3%**	29.4%	34.0%
		% of total	3.2%	10.1%	8.0%	10.1%	2.7%	34.0%
	51–60	Count	4	6	14	12	6	42
		% within age	9.5%	14.3%	33.3%	**28.6%**	14.3%	100.0%
		% within participation in hybrid	28.6%	15.0%	23.3%	**21.1%**	35.3%	22.3%
		% of total	2.1%	3.2%	7.4%	6.4%	3.2%	22.3%
	60+	Count	0	4	5	3	1	13
		% within age	0.0%	30.8%	**38.5%**	23.1%	7.7%	100.0%
		% within participation in hybrid	0.0%	10.0%	**8.3%**	5.3%	5.9%	6.9%
		% of total	0.0%	2.1%	2.7%	1.6%	0.5%	6.9%

Accordingly, 31.5% of women stated that they would strongly like to participate in hybrid trade fairs in the future, while 40.4% of those who answered 'strongly' were women. For women, the answer 'moderate' gathered the largest percentage. Specifically, 37.0% of women stated that they moderately wish to participate in hybrid trade fairs in the future, while 45.0% of those who answered 'moderately' were men. Accordingly, 28.7% of men stated that they moderately wish to participate in hybrid trade fairs in the future, while 55.0% of those who answered 'moderately' were men.

From the above, we find that the differences between genders are small, with a slight preference of men to participate in hybrid trade fairs, however, we consider them to be negligible.

From Table 3, we observe a slight difference regarding the 18–30 age group in their willingness to participate in hybrid trade fairs in the future. Specifically, 45.5% of exhibitors aged 18–30 years stated that they would strongly like to participate in hybrid trade fairs in the future, while 17.5% of those who answered ‘strongly’ belong to this age group. So, we can say that there is a small correlation of willingness to participate in hybrid trade fair and age, where exhibitors in the 18–30 age group show a slightly greater willingness.

5 Conclusion

The subject of this paper is the transition of exhibition activity from physical to digital and hybrid exhibitions, with a particular focus on how the current COVID-19 pandemic will affect the future of trade fairs. The purpose of the paper was to investigate the opinions of trade fair participants, specifically in the case of the Thessaloniki trade fairs, against physical, digital and hybrid trade fairs, but also to investigate their views on the future of exhibition activity post-COVID time. For this reason, a quantitative survey was conducted on a representative sample of exhibitors participating in trade fairs in Thessaloniki.

The study of the literature shows that the evolution of technology brought about changes in the form of exhibitions, with digital and hybrid trade fairs being new forms of exhibition activity. The global health crisis of the COVID-19 pandemic has shown that technology has contributed to the flexibility of trade fairs, as digital events have been an effective means of adapting to the adverse conditions imposed by the pandemic. Nevertheless, it seems that exhibitors continue to show a preference for trade fairs with a physical presence, of course recognizing the advantages of digital ones (Tradeshow Logic, 2020; Nella, 2021; Hochheim, 2021; UFI, 2022). Exhibitors' expectations for the future of trade fairs appear from the available research data to converge toward a return to their traditional form with a physical presence, but at the same time supported by virtual events and digital elements. However, the replacement of physical trade fairs by digital ones does not seem to exist as a possibility (UFI, 2022).

From the research conducted, we identified similar findings to the above research results. Regarding the first Research Question of the survey, which concerns the willingness of exhibitors to participate in hybrid trade fairs in the future, we found that it ranges at moderate levels. Although in a different question they recognized some advantages of digital compared to physical trade fairs: money saving, time saving and health safety. At the same time, however, they recognize perhaps to a greater degree, the disadvantages: lack of personal contact, lack of personal relationship, lack of product contact and technology problems.

In relation to the second Research Question of the survey on the effect of the COVID-19 pandemic on the intention of exhibitors to participate in digital trade fairs, no clear conclusion was found, as the responses were divided among the answers.

Regarding the third Research Question that concerns the exhibitors' views on the future of trade fairs after the pandemic, it was found that the majority of exhibitors believe that after the pandemic exhibition activity will return to physical trade fairs. The willingness of the exhibitors to return to physical trade fairs was also established in the question regarding their desire for the future of exhibitions, where the majority stated that after the pandemic, they wish the exhibition activity to return to physical trade fairs.

However, a fairly significant percentage believes that the future of exhibition activity is hybrid trade fairs, while a minimal percentage believes that it is digital trade fairs. These findings are similar to those identified in the literature (Tradeshow Logic, 2020; Nella, 2021; Hochheim, 2021; UFI, 2022). Exhibitors are increasingly recognizing that hybrid trade fairs can combine the advantages of physical and digital trade fairs.

Finally, from the cross-tabulation analysis, no significant effect of the gender variable was found on the exhibitors' intention to participate in hybrid trade fairs in the future. A small effect on exhibitors' intention appears to exist from the age variable, where exhibitors belonging to the younger age group (18–30 years) of the sample show a slightly more positive attitude toward participating in hybrid trade fairs than older exhibitors.

In summary, we can say that exhibitors recognize the advantages of digital trade fairs, however, they do not wish to replace trade fairs with a physical presence. These results were expected, both because of the corresponding findings from similar studies and because of the undisputed superiority of the benefits of physical trade fairs. However, the future of exhibitions is likely to feature more digital elements. The COVID-19 pandemic has perhaps marked the beginning of a new generation of trade fairs that will combine the advantages of physical and digital trade fairs.

References

Gani, M. O., Takahashi, Y., Faroque, A. R., Mortazavi, S., & Alam, M. Z. (2021). Virtual trade show: Past assessment, present status, and future prospects. *Journal for International Business and Entrepreneurship Development, 13*(3–4), 286–310.

Geigenmüller, A. (2010). The role of virtual trade fairs in relationship value creation. *Journal of Business & Industrial Marketing, 25*(4), 284–292.

Gottlieb, U., & Bianchi, C. (2017). Virtual trade shows: Exhibitors' perspectives on virtual marketing capability requirements. *Electronic Commerce Research and Applications, 21*, 17–26.

Hochheim, H. (2021). Ifo survey: German exhibitors continue to rely on real trade fairs. Retrieved from: https://www.auma.de/en/media/reports/Ifo-survey-German-exhibitors-continue-to-rely-on-real-trade-fairs. Accessed the 4th of July 2022, at 10:32.

Kourkouridis D., Fragopoulos I., Karanikolas N., & Seretis S. (2016) Exhibition, city and crisis: A negotiated relationship. A first approach. In *Proceedings of the 14th National Conference of the Hellenic Regional Science Society (ERSA-GR): Productive Reconstruction and Regional Development*. Regional Development Institute - Panteion University, Athens, June 24–25.

Kourkouridis, D., Dalkrani, V., Pozrikidis, K., & Frangopoulos, I. (2018). Hosted buyers program (HBP)—Tourism development and the city TIF-HELEXPO HBP for the period 2014–2016. In *Innovative approaches to tourism and leisure* (pp. 537–551). Springer.

Kourkouridis, D., Dalkrani, V., Pozrikidis, K., & Frangopoulos, Y. (2019). Trade fairs, tourism and city: Thessaloniki international fair and the concept of honoured countries. *Tourismos, 14*(2), 30–56.

Kourkouridis, D. (2021). *Sociospatial dimensions of exhibitions, city and interactive dynamics*. Doctoral Dissertation, Department of Spatial Planning and Development Engineering, Aristotle University of Thessaloniki (No. GRI-2021-30680).

Kourkouridis, D., & Frangopoulos, I. (2022). *Trade fair*. Edward Elgar Publishing.

Kourkouridis, D., Frangopoulos, Y., & Kapitsinis, N. (2023). Socio-economic effects of trade fairs on host cities from a citizens' perspective: The case of Thessaloniki Greece. *International Journal of Event and Festival Management, 14*(1), 113–133. https://doi.org/10.1108/IJEFM-10-2022-0078.

Lee-Kelley, L., Gilbert, D., & Al-Shehabi, N. F. (2004). Virtual exhibitions: An exploratory study of Middle East exhibitors' dispositions. *International Marketing Review, 21*(6), 634–644.

Nella, A. (2021). Tourism exhibitions in the pandemic era: State of play and initial valuation of their transformation to virtual and hybrid events. In *9th International Conference on Contemporary Marketing Issues, Virtual Conference*, 3–5 September, (pp. 304–307).

Nilsson, L. (2020). *Hybrid events: Breaking the borders. Transferring your hybrid event into an engaging and inclusive experience for different audiences and stakeholders*. Master Thesis, Master of Business Administration, Degree Program in International Business Development. Lathi, Finland: LAB University of Applied Sciences.

Palmer, A. (2020). *Coronavirus-related exhibition cancellations will cost up to $22 Billion in economic impact*. Retrieved from: https://www.northstarmeetingsgroup.com/News/Industry/CEIR-Economic-Trade-Show-Impact-Coronavirus-COVID-19. Accessed the 4 th of July 2022, at 10:40.

Sarmento, M., & Simões, C. (2019). Trade fairs as engagement platforms: The interplay between physical and virtual touch points. *European Journal of Marketing, 53*(9), 1782–1807.

Schiefelbein, J. (2018). *Leveraging virtual team meetings: four tips for success*. MISTI Training Institute.

TIF–HELEXPO. (2014). Corporate profile February 2014. TIF–HELEXPO S.A.

Tsiftelidou, S., Kourkouridis, D., & Xanthopoulou-Tsitsoni, V. (2017). Assessment of impact-contribution of cultural festival in the tourism development of Thessaloniki. In *Tourism, culture and heritage in a smart economy* (pp. 411–424). Springer.

Tradeshow Logic. (2020). *Redefining value for today's exhibitors & sponsors* (Vol. 2). Virtual Event Satisfaction, September 2020. Tradeshow Logic, Marietta, GA.

UFI. (2021a). UFI releases figures stating the impact of COVID-19 on the global exhibition economy in 2020. Media Release. The Global Association of Exhibition Industry, Steinbeis Transfer Centre, Ravensburg, Germany.

UFI. (2021b). The global exhibition Barometer July 2021. Ravensburg, Germany: The Global Association of Exhibition Industry, Steinbeis Transfer Centre.

UFI. (2022). The global exhibition Barometer February 2022. The Global Association of Exhibition Industry, Steinbeis Transfer Centre, Ravensburg, Germany.

Wei, Z. (2022). Opportunities and challenges of virtual trade shows amid the Covid-19 pandemic—take China as an example. *Sino-US English Teaching, 19*(2), 60–64.

Role of Instagram Influencers on Destination Image: A Netnographic Study

Hulisi Binbasioglu, Yasemin Keskin Yilmaz, and Mehmet Baris Yilmaz

Abstract The main purpose of this study is to reveal how the posts of popular travel influencers who produce special content for Cappadocia, an attractive destination in Turkey, determine the image and perception of Instagram users about the region. For this purpose, user comments (emotions, thoughts, expectations, etc.) on the posts of influencers about Cappadocia on Instagram were examined by the netnography method. The analysis was applied to the comments made by the social media users on posts of three travel influencers who visited Cappadocia between 1 January and 15 August 2022 and shared with related tags. Considering the findings obtained from the social media user' comments, the posts of the travel influencers about Cappadocia activated the emotions of the users and contributed positively to the perception of the region.

Keywords Social media · Instagram influencers · Destination image · Netnography · Cappadocia

JEL Classification L83 · M31

1 Introduction

Social media has made it possible for regular people to create their brands by engaging with and sharing stimulating content. They are often referred to as social media influencers. According to Hearn and Schoenhoff (2016), "*social media influencers work to generate a form of "celebrity" capital by cultivating as much attention as possible and crafting an authentic "personal brand"* via *social networks, which can subsequently be used by companies and advertisers for consumer outreach*" (p. 194).

H. Binbasioglu (✉) · Y. K. Yilmaz
Malatya Turgut Ozal University, Malatya, Turkey
e-mail: hulusi.binbasioglu@ozal.edu.tr

M. B. Yilmaz
Inonu University, Malatya, Turkey

© The Author(s), under exclusive license to Springer Nature Switzerland AG 2023

V. Katsoni (ed.), *Tourism, Travel, and Hospitality in a Smart and Sustainable World*, Springer Proceedings in Business and Economics,
https://doi.org/10.1007/978-3-031-26829-8_26

The term "social media influencers" refers to people who have a large following of online users who can sway consumers' opinions and decision-making processes in their favour as trustworthy tastemakers. (De Veirman et al., 2017). In marketing activities on online platforms, influencers take the place of opinion leaders (Walter & Brüggemann, 2020) and are new mediators of word-of-mouth communication (Litvin et al., 2008).

The social media influencer strategy was first used in the fashion and beauty industries, but it has since spread to practically every economic activity, with travel and tourism as the industries where influencers have gained particularly significant prominence (Femenia-Serra & Gretzel, 2020). Much research shows that social media influencers have a positive and significant effect on destination image and purchase intention (Caraka et al., 2022; Gholamhosseinzadeh et al., 2021; Magno & Cassia, 2018). One research indicates that 49% of consumers rely on influencer recommendations while 40% of them have bought items after seeing them on Twitter, YouTube, or Instagram (Digital Marketing Institute, 2021). Moreover, Instagram is one of the most popular social platforms that enables influencers to share their attractive picture and videos. So, with 65% of consumers say they followed their favourite influencer on the image and video-driven social media channel Instagram (Rakuten Advertising, 2020).

It is very crucial for social media influencers to choose visually impressive places in terms of interaction. The post's popularity on social media such as Instagram may be explained by the way that existing aesthetic standards have ingrained the theme's narrative basis (Smith, 2021). Many social media users and influencers are drawn to Turkey's Cappadocia region because of its visual splendour, which includes hot-air balloons and fairy chimneys. Visitors who took a hot-air balloon trip or choose to watch the balloons put a high value on Cappadocia's distinctive natural features in their reviews (Özkan-Erciyas & Yıldız, 2018). Göreme National Park and the Rock Sites of Cappadocia were recorded in 1985 as a UNESCO World Heritage Site, spanning the area between the cities of Nevsehir, Urgup, and Avanos, the sites of Karain, Karlik, Yesiloz, Soganli, and the underground cities of Kaymakli and Derinkuyu. They are situated on the central Anatolian plateau within a volcanic landscape sculpted by erosion to form a succession of mountain ridges, valleys, and pinnacles (WHC, 2022). The historical region of Cappadocia in Central Anatolia welcomed approximately 1.9 million tourists until the first seven months of 2022, increasing by 121 per cent compared with the same period of last year (Hurriyetdailynews, 2022).

Researchers' attention has recently been drawn to social media influencers and the role that they play in the online world (Dwivedi et al., 2018). However, while there is increasing research on tourism and social media influencers (Femenia-Serra & Gretzel, 2020; Gretzel, 2018b), there is a lack of research on the netnographic approach to comments on influencers' posts in the tourism industry. This study aims to evaluate the role of Instagram influencers on destination image at the popular tourist destination of Cappadocia in Turkey. The online communications between three foreign travel Instagram influencers who have visited and posted about Cappadocia as a destination and the social media users who commented on their posts were studied. To better understand the impact on the formation of destination images, this study examines how the promoted destination image is perceived by social media users.

2 Literature Review

Tiago et al. (2019) findings indicated that influencer-generated content primarily combines user and professional-generated content, but because it is offered through a storytelling technique, it is more akin to user-generated content. To promote destinations to the target group, social media influencers are being employed more frequently by tourism businesses and destinations to create more effective and trendy content (Motahar et al., 2021). Besides, influencers independently share travels with their followers and shape their opinions about destination images on Instagram, YouTube, etc. It is essential for influencers to maintain a strategic relationship with their followers so that they can keep attracting tourism businesses and destinations (van Driel & Dumitrica, 2021).

Traditional word-of-mouth influence is constrained by its reliance on face-to-face interaction, but in the Internet age, social media influencers are unconstrained in how they disseminate information (George et al., 2021). Social media influencers have become effective actors that can reshape information flows (Femenia-Serra et al., 2022). As an information source, social media travel influencers share information with their followers on unexplored locales, cultures, and travel requirements (Asan, 2022). Influencers play a crucial role in the brand strategy of destinations since they may improve the perception of such places (Femenia-Serra & Gretzel, 2020).

Influencers who act like professional or amateur advertisers, brand ambassadors, and friends boost the purchase probability to create and market their social media pages (Jiménez-Castillo & Sánchez-Fernández, 2019). It is common practice to link the popularity and success of social media influencers to the followers' perception of how trustworthy the influencers are (Munar, 2011). Engagement is the key tool to this popularity and success. Engagement metrics such as the number of likes, comments, and shares of posts on social media sites like Facebook and Instagram, reflect the number of interactions compared to an influencer's followers (Myers et al., 2022). On the other hand, Jang et al. (2021) state that the idea of engagement can be put into practice in a variety of different ways when considering Instagram as a whole. For example, the level of engagement could be evaluated based on the number of times an influencer responds to the comments made by their followers and other social media users.

Lin and Huang (2006) found by analysing messages posted on the website that the desire to travel to Greece increased after reading blog comments about the country. The results of Gholamhosseinzadeh et al. (2021) netnographic study into the attributes of a destination's image as determined by social media communication showed that the influencers' practices were capturing followers' interest, encouraging followers' self-discovery, involvement, and information exchange. The followers were astounded and thrilled by this finding, and some even said they intended to go to the Japanese archipelago. Furthermore, negative comments from displeased and/or incensed social media users may also hurt the influencers' strategies and lessen the other users' motivation to travel, since negative material has a greater effect than good content (Pop et al., 2022).

3 Methodology

3.1 *Objective and Importance of the Study*

The increasing influence of the media on consumption culture with digitalization has transformed individuals who accept and adopt new consumption values into active, productive individuals who take an active role in the marketing process through social media. Especially in the tourism industry, the ability of individuals, who are also tourists, to reflect their feelings, thoughts, experiences, perceptions, and expectations in the new media environment has allowed businesses and brands to capture consumer insights. In this context, studies on consumers' sharing their thoughts and feelings with their expressions gain value. This study, which is structured on destination image, is important in terms of determining the perspectives, perceptions, feelings, and thoughts of users towards a destination regarding the content produced by popular travel influencers on Instagram. On the other hand, it is obvious that presenting the image of Cappadocia, which is an important destination in the Turkish tourism market, through the eyes of travel influencers, who are opinion leaders, from their perspectives, will contribute to the preference of many tourists to the region.

The main subject of the study is the role of the posts made by travel influencers famous on Instagram about Cappadocia in tourists' perception of Cappadocia's image. The main purpose of the study is to reveal how popular travel influencers who produce and publish content specific to Cappadocia, an important cultural destination of Turkey within the scope of tourism marketing, determine the image and perspective of Instagram users who are also tourists.

Within the scope of the determined purpose, answers to the following questions are sought:

1. Do famous travel influencers on Instagram contribute to Instagram users' image of Cappadocia?
2. How do users react (positively/negatively) to the Cappadocia content posts of famous travel influencers on Instagram?

3.2 *Methodology of the Study*

In this study, user comments (feelings, thoughts, expectations, etc.) on the posts of travel influencers famous on Instagram about Cappadocia, one of Turkey's popular tourism destinations, were analysed by the netnography method, one of the qualitative research methods. Consumer behaviour has a complex and variable structure. For this reason, marketing communication strategies should not be determined by interpreting the ordinary responses of consumers, but rather an approach to understanding and interpreting the behaviour of consumers in their social and cultural reality should be adopted. Kozinets (2006) explains netnography as a method that allows the researcher to wear an "invisible jacket" and offers the opportunity to examine users' interactions without disturbing them. Due to these characteristics, netnography was preferred in the study. The preference for this method also shows the methodological limitation of the research.

Varnalı (2012) states that netnography, which enables the generation of contextual and cultural insights, has the ability to provide brands with a wide range of data about consumer groups when carefully planned and comprehensively implemented. The netnography method, which allows exploratory research to be conducted, has been introduced to the literature as a result of the applicability of ethnographic methods in the virtual environment. Ethnography, an anthropological research method, "*identifies, understands and describes the stories, symbols, spiritual values, social orders, historical fears, excitements, goals, daily life flows, in short, the subtexts of culture that guide the behavior of the social group under study.*" (Varnalı, 2012, p. 28). With digital transformation, the virtual and Internet-based version of the ethnographic method for the same purpose is netnography. The method developed by Kozinets to understand the world of consumers in the virtual environment is used in e-tourism marketing as digital cultural insight techniques in basic issues such as e-WOM, online reviews, collection, analysis, and interpretation of data on online communities (Kozinets, 2020). There are five stages of the netnographic research process developed by Kozinets (2006, p. 281):

- Entrée: the identification of the research questions and specific virtual community
- Data collection and analysis
- Ensuring trustworthy interpretations
- Conducting ethical research
- Providing opportunities for the members to give feedback.

In the study, all travel influencers who shared posts with the tags *#cappadocia, #gocappadocia, and #visitcappadocia* on Instagram between 1 January 2022 and 15 August 2022 were followed; three travel influencers were determined by considering criteria such as having the highest number of followers, allowing commenting users to interact, the number and quality of comments made. Cappadocia posts of three foreign travel influencers and user comments under them were recorded between 15 and 17 August 2022. The fact that three influencers included in the study were limited to the most commented Cappadocia posts.

While collecting and storing data, the posts shared by the influencers that received the most comments during the relevant period were analysed. As a result of the analysis, it was determined that the high number of user comments in some posts were activities such as campaigns and advertisements made by the influencer to its followers. For this reason, while the posts with the most comments were shared for a purpose such as an advertisement or promotion were defined, they were excluded from the analysis, and the post with the next most comments was included in the analysis.

Cappadocia photo or video posts presented by the influencers from their perspectives have been analysed in terms of how they determine the feelings, thoughts, expectations, and perceptions of those who comment as Instagram users, whether they are followers or not, on the image of Cappadocia. In this sense, the study is important in terms of reflecting the opinions and expectations of potential tourists on the perception of Cappadocia in depth. The framework of this research involves questions like "What do potential tourists who have visited Cappadocia before or have not yet visited feel and think about all kinds of posts about the destination by travel influencers who are famous for sharing photos and videos of the places, they have travelled on Instagram?".

Instagram was determined as the most appropriate platform for the study. In addition to the positive impact of its algorithm based on image and video sharing, Instagram is the most popular medium in the pre-travel decision-making process, especially for young social media users (Varkaris & Neuhofer, 2017). In this context, the medium of the research was limited to Instagram due to variables such as being suitable for visual expression elements in terms of destination image as a new media tool, having an increasing number of users (approximately 1.5 billion), being the fourth fastest growing platform globally (Wearesocial, 2022).

On the other hand, UGC, which is called user-generated content in social media channels, can affect the image of touristic destinations and the travel intention and decision-making process of tourists (Leung et al., 2013). Instagram, which rises together with UGC in tourism marketing, is the most popular visual-centred platform for potential tourists in searching and evaluating destinations (Varkaris & Neuhofer, 2017), as well as affecting tourism behaviour. In this context, the UGC photo and video posts of Instagram's famous travel influencers provide academics and practitioners with insights into how tourists associate, interpret, or perceive the destination of Cappadocia. Therefore, an exploratory netnography was deemed appropriate to investigate how travel influencers construct or determine destination images through photo/video-based UGC.

It is important to comply with ethical standards in netnography research. Considering the ethical dimension, the posts and comments used in this study were all taken from digital influencers' public Instagram accounts, where all their followers are aware that their messages are available to anyone. In addition, the researchers used non-intrusive (passive) observation techniques to collect data without any social contact with Instagram users who are online community members. In the netnography method, researchers need to be familiar with the mechanisms of online conversations, to be able to navigate a big mass of unstructured information, interpret those assigning meanings to them, and detect the cultural connections that link content together (Addeo et al., 2019).

3.3 Sample

The universe of this study is the comments on the posts about Cappadocia made by popular travel influencers on Instagram on their profile pages. Criterion sampling, one of the purposeful sampling types, was preferred as the sample, and accordingly, the comments on the photo or video posts of the top three travel influencers who visited Cappadocia between 1 January 2022 and 15 August 2022 with any of the tags *#cappadocia, #gocappadocia, #visitcappadocia* were determined as the sample. The post with the most comments among these posts was preferred. The basic understanding of criterion sampling is to study situations that meet predetermined criterion or criteria. The criterion mentioned here can be determined by the researchers (Yıldırım & Şimşek, 2013). The criteria in the study are that the influencers should post with the tags determined during the visit to Cappadocia in the selected date range and the number of followers should be at least ten thousand.

Three profile accounts that share posts by the established criteria are @khemric (403.000 followers), @johnzimmer_ (88.600 followers), and @fienandolivier (37.600 followers). In the profile of each influencer, 209 comments related to three posts about Cappadocia were selected for the research and related to the research questions. Comments containing exceptional circumstances (such as advertisements, promotional messages, or messages of compliment to the influencer) and comments independent of the research, such as the responses of the travel influencer to users' comments, were excluded. At this point, user comments under each of the three posts were repeatedly read and categorized.

4 Results

Within the scope of the research, it was determined that the influencer Kristi Hemric with the username @khemric shared a video, John Zimmer with the username @johnzimmer_ shared a video, and Fien and Oliver couple with the username @fiendandolivier shared four photos in one post. Information on the post content of the influencers is given in Table 1. As a result of the research, data were obtained from the comments on the Cappadocia posts of three Instagram travel influencers. Six different main themes (admiration, like, positive emoji, surprise, curiosity, and missing) were identified in the data obtained from the comments. These categories are discussed in detail below under headings.

4.1 Cappadocia-Related Post Comment Analysis of @khemric

A visual analysis map was created according to the codes prepared for user comments on the posts subjected to analysis. Before, the image of the @khemric post (Image 1)

Table 1 Information on Cappadocia-themed posts of influencers

Username	Number of followers[a]	Type of post	Date of post	Number of comments on post[b]
@khemric	403.000	Video (reels)	7 August 2022	173
@johnzimmer_	88.600	Video (reels)	8 June 2022	82
@fienandolivier	37.600	Photo	31 July 2022	70

[a] Number of followers recorded on 15 August 2022
[b] Number of all comments recorded on 15 August 2022

and information about the image were presented. The visual analysis map *of* the Cappadocia post of the first-ranked influencer with the username @khemric is given in Fig. 1.

Image 1 Cappadocia image of @khemric (Access date: 15.08.2022)

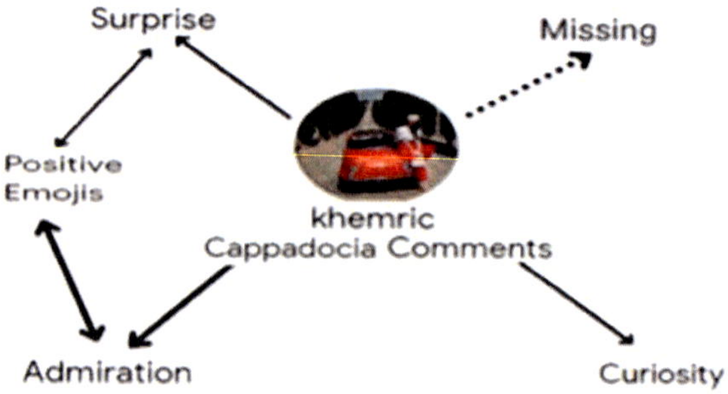

Fig. 1 Visual analysis map of posted comments of @khemric. *Note* Arrows are determined from thick to thin according to the relationship level

The most commented Cappadocia post of @khemric, the first travel influencer determined within the scope of the research, was a reel video shared on 7 August 2022 and received 5524 likes. In the video, @khemric walks backward towards hot-air balloons in a general Cappadocia landscape. The reels video is accompanied by a suitable song (sweet dreams) for the destination. The description section of the post is "*Everyone's Looking for Something! I've travelled the world and the 7 Seas, but no place has exceeded my expectations like Cappadocia supporters!*".

When the comments received by the users on the related reels video were examined, it was determined that the most common code as comment message content was *admiration*. The *admiration* code was also interrelated with *positive emojis*, and this admiration was interpreted with the expressions "*Like a dream, Epic, Like a movie trailer, Wonderful, Magic, Absolutely breathtaking,* etc.". The second most common code was *surprise* towards the destination. It was determined that users posted expressions such as "*OMG That is surreal, Incredible, Unreal, Amazing, Cappadocia is such a fairy-tale, This is stunning,* etc." with *positive emojis*. Another code appears as a *curiosity*. It was inferred from the comments that some of the users were curious about the natural and cultural geography of Cappadocia. This curiosity was expressed with comments "*Tell us more about Cappadocia, This is definitely on my list, Added to my list,* etc.". Finally, it was seen that there was a *missing* code in the analysis. Although low in number, the comments of users who had visited Cappadocia before were found to contain *missing*. The expressions "*Making me miss Cappadocia, Memories, Now I wanna be there*" support the *missing* code.

*4.2 Cappadocia-Related Post Comment Analysis of @johnzimmer_*

The next Instagram profile belongs to a couple with the username @johnzimmer_. On 8 June 2022, the Cappadocia post with the most comments (2387 likes) was a video of reels accompanied by music (Image 2).

In the reels post, which is a compilation of photos and videos they took at different points in Cappadocia, there are various poses of the influencer couple such as between hot-air balloons, at Uçhisar Castle, in the sunset view of Ürgüp, by the poolside of a cave hotel, on local rugs. In the text of the post, they post "*That's your sign to visit this magical place. Location: Cappadocia, Turkey*". The visual analysis map of the user comments for this post is given in Fig. 2.

When the analysed post comments of the influencer couple named @johnzimmer_ were examined, it was seen that there were codes of *like, admiration, curiosity, and positive emojis*. The most common code was *like*. And the *like and positive emojis* were mutually related. User comments regarding the *like* code were interpreted as "*Best place in the world, Love this, Beauty, So gorgeous Lovely,* etc.". *Admiration* was defined as the second code. Again, it was observed that the *admiration* code and *positive emojis* were mutually related and used as expressions of admiration.

Image 2 Cappadocia image of @johnzimmer_ (Access date: 15.08.2022)

Fig. 2 Visual analysis map of posted comments of @johnzimmer_. *Note* Arrows are determined from thick to thin according to the relationship level

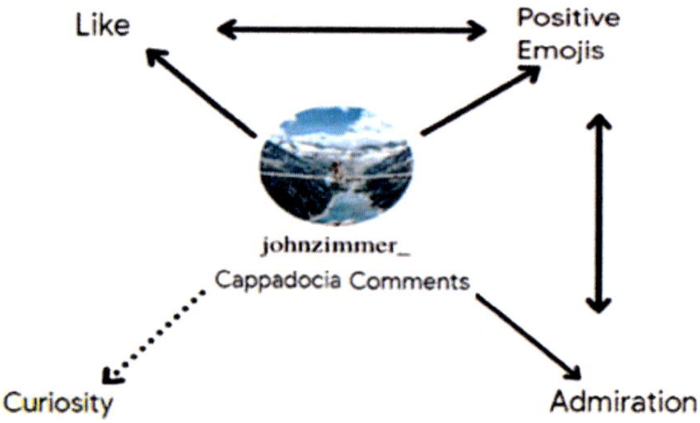

The comments of "*Dream place, Magical, Stunning, Absolutely incredible, Really magical place, So amazing,* etc." could be expressed as examples of *admiration* code. Finally, the code of *curiosity* was also identified in the comments, albeit to a lesser extent. The statements "*We are visiting soon can't wait, Maybe we will be there this year, Oh I definitely would love to visit Cappadocia,* etc." reflect the users' sense of *curiosity*. It was observed that the *positive emojis* code was widely used in most of the comments and supported the other codes.

Image 3 Cappadocia image of @fienandolivier (Access date: 15.08.2022)

4.3 Cappadocia-Related Post Comment Analysis of @fienandolivier

The most commented Cappadocia post on the profile account of the influencer couple (Fien and Olivier in Image 3), who define themselves as a travel couple on their Instagram profile, is presented in Fig. 3.

The post consisted of four photographs of the couple in the landscape of the fairy chimneys and next to a classic car. The post was dated 31 July 2022, received 532 likes, and no background music was used in the post. In the description text of the post, "*Create your own reality. Don't wait for things to happen, make them happen. Work towards your dreams and turn them into reality, stop waiting, the time is now.*" The visual analysis map prepared for the comments of the Cappadocia post subject to the analysis of the influencer couple is given in Fig. 3.

In the comments of @fienandolivier's post subject to analysis, the *admiration* coded as the comment message content is the most common code and *positive emojis* were seen to be mutually related to *admiration.* The expressions "*Wow stunning, Obsesssssed, Looks so dreamy, This place looks so cool, Wauw, Magical Cappadocia,* etc." were used with *positive emojis* and reinforced the feeling of admiration. The *like* code was also included in the comments on the post. Again, the expressions in the *like* code related to *positive emojis* were determined as "*Love this photo, Love this, Beautiful, This is so cute, Love the colours in this,* etc.".

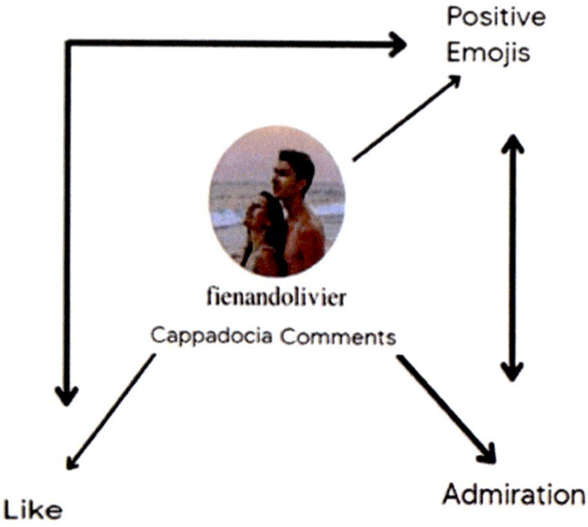

Fig. 3 Visual analysis map of posted comments of @fienadnolivier. *Note* Arrows are determined from thick to thin according to the relationship level

5 Conclusion

Destination image, which is affected by all types of information acquired by tourists (Kuhzady & Ghasemi, 2019), has gained a different dimension and importance with the creation of new communication technologies in the tourism industry. With the opportunity of individuals to create and publish their content due to the nature of social media, travelling tourists have also restructured the concept of destination image by producing content about the destination during or after the travel. Especially on Instagram, which is defined as "visual social media" (Gretzel, 2018a), travel influencers are likely to be effective in the destination image formation, travel decisions, or behaviour of potential tourists. Tourism businesses and DMOs can build ongoing relationships with influencers to create an impressive brand image and promotion activities. The starting point of this study is to investigate the role of travel influencers in the destination image perceptions of Instagram users who are also tourists.

There are different studies on destination image in tourism literature. Research on the image and branding of tourism destinations through DMOs has attracted attention (Kuhzady & Ghasemi, 2019; Lojo et al., 2020; Molinillo et al., 2017; Nixon, 2022; Stepchenkova & Zhan, 2013). However, the fact that only one study has been conducted on how travellers, who are defined as travel influencers on Instagram and have high followers with their travel posts, determine the perspective and perception of Instagram users on the image of a destination can be considered as the reason for preferring the research topic. In the study conducted by Gholamhosseinzadeh et al. (2021), comments on the content published by travel bloggers on different social media platforms were analysed by the netnography method, and content analysis was also applied to the posts. According to the findings, the comments of two French influencers' posts regarding the Okinawa destination in Japan, the followers were excited and surprised, while the perception of the destination was by the image presented by the influencers.

In the study conducted by Tavakoli and Wijesinghe (2019), the netnography was still not used sufficiently in tourism research, and in the studies that were used, it was addressed from a single perspective (either textual or visual), and Instagram was the least preferred among online platforms for data collection. From this point of view, the fact that the data obtained from Instagram were analysed with netnographic analysis by the research questions makes the study authentic.

The research findings reveal those travel influencers, who are digital influencers on destination image, positively affect the perspectives of the users on the destination. It has been observed that the visual posts of travel influencers with UGC regarding Cappadocia, one of the important tourist destinations that include Turkey's cultural and natural attractions, stimulate the emotions of the users, and determine the perception of the region.

Considering the netnographic findings obtained from user comments. It was stated that six basic codes play a role in image perception. The codes determined as *admiration, like, positive emojis, surprise, curiosity, and missing*, respectively, show that the perception of Cappadocia is positively shaped. In particular, it was found that positive emojis intensively support the codes of *admiration, like, and curiosity*. Cappadocia posts through the eyes of travel influencers created the most admiration among users. Then, the surprise of not being aware of the destination before and the surprise of the natural beauties of the destination strengthened the perception of the image of Cappadocia. In addition, the *curiosity* to visit the region and see it on-site was also reflected in the comments. It is thought that the *curiosity* code will positively affect the users' travel decision-making process. The desire of the users who have visited the destination before to visit again, even if it is low, is determined by the *missing* code.

Regarding the research questions of the study, it is more likely to say that travel influencers on Instagram determine the image of Cappadocia of the users. On the other hand, travel influencers play a positive role in the image perception of users regarding the Cappadocia destination. One of the important findings of the current research is that social media influencers in tourism marketing are particularly effective in creating an image of a destination. This study is based on data obtained from Instagram. In future studies, research on online destination image creation can be carried out in different media and different online communities. In addition, following the changing and transforming dynamics of the influencers who meet the concept of "influencer, guide", especially the travellers in tourism, by the academics and practitioners working in this field will contribute to the studies to determine the travel decision process and behaviour of the tourist. The current research provides an addition to existing destination image literature and updates it concerning the process' external stimuli. This also satisfies the demand made by numerous researchers to examine the role of Instagram influencers in the tourism industry. Destination marketing managers focus on this research to develop marketing strategies or begin utilizing influencer marketing.

As with any study, this research is not without its limitations. The current study focused only on three Instagram posts of travel influencers and their followers about Cappadocia in a limited period. Studies on destination image in tourism research can be performed on different samples. The more widespread use of netnography in tourism research, which is useful in analysing the perspectives, perceptions, expectations, and value judgments of social media users in online environments, will also bring insights to practitioners about tourist behaviours. In future studies, it would be advisable to uncover the differences or similarities in destination image by conducting comparative studies to examine it from an intercultural perspective.

References

Addeo, F., Paoli, A. D., Esposito, M., & Bolcato, M. Y. (2019). Doing social research on online communities: The benefits of netnography. *Athens Journal of Social Sciences, 7*(1), 9–38. https://doi.org/10.30958/ajss.7-1-1

Asan, K. (2022). Measuring the impacts of travel influencers on bicycle travellers. *Current Issues in Tourism, 25*(6), 978–994. https://doi.org/10.1080/13683500.2021.1914004

Caraka, R. E., Noh, M., Lee, Y., Toharudin, T., Tyasti, A. E., et al. (2022). The impact of social media influencers Raffi Ahmad and Nagita Slavina on tourism visit intentions across millennials and zoomers using a hierarchical likelihood structural equation model. *Sustainability, 14*, 524. https://doi.org/10.3390/su14010524

De Veirman, M., Cauberghe, V., & Hudders, L. (2017). Marketing through Instagram influencers: The impact of number of followers and product divergence on brand attitude. *International Journal of Advertising, 36*(5), 798–828. https://doi.org/10.1080/02650487.2017.1348035

Digital Marketing Institute. (2021). 20 surprising influencer marketing statistics. Retrieved from https://digitalmarketinginstitute.com/blog/20-influencer-marketing-statistics-that-will-surprise-you. Accessed 20 August 2022.

Dwivedi, Y. K., Kelly, G., Janssen, M., Rana, N. P., Slade, E. L., & Clement, M. (2018). Social media: The good, the bad, and the ugly. *Information Systems Frontiers, 20*(3), 419–423. https://doi.org/10.1007/s10796-018-9848-5

Femenia-Serra, F., & Gretzel, U. (2020). Influencer marketing for tourism destinations: Lessons from a mature destination. In J. Neidhardt & W. Wörndl (Eds.), *Information and communication technologies in tourism 2020* (pp. 65–78). Springer Nature. https://doi.org/10.1007/978-3-030-36737-4_6

Femenia-Serra, F., Gretzel, U., & Alzua-Sorzabala, A. (2022). Instagram travel influencers in #quarantine: Communicative practices and roles during COVID-19. *Tourism Management, 89*, 104454. https://doi.org/10.1016/j.tourman.2021.104454

George, R., Stainton, H., & Adu-Ampong, E. (2021). Word-of-mouth redefined: A profile of influencers in the travel and tourism industry. *Journal of Smart Tourism, 1*(3), 31–44. https://doi.org/10.52255/smarttourism.2021.1.3.6

Gholamhosseinzadeh, M. S., Chapuis, J.-M., & Lehu, J.-M. (2021). Tourism netnography: How travel bloggers influence destination image. *Tourism Recreation Research*. https://doi.org/10.1080/02508281.2021.1911274

Gretzel, U. (2018a). Influencer marketing in travel and tourism. In M. Sigala & U. Gretzel (Eds.), *Advances in social media for travel, tourism and hospitality: New perspectives, practice and cases* (pp. 147–156). Routledge.

Gretzel, U. (2018b). Tourism and social media. In C. Cooper, W. Gartner, N. Scott, & S. Volo (Eds.), *The sage handbook of tourism management* (pp. 415–432). Sage.

Hearn, A., & Schoenhoff, S. (2016). From celebrity to influencer—Tracing the diffusion of celebrity value across the data stream. In P. D. Marshall & S. Redmond (Eds.) (2015), *A companion to celebrity* (pp. 194–212). John Wiley & Sons/Blackwell.

Hurriyetdailynews. (2022). Some 1.9 mln tourists visit Cappadocia. Retrieved from https://www.hurriyetdailynews.com/some-1-90-mln-tourists-visit-cappadocia-175954. Accessed 17 August 2022.

Jang, W., Kim, J., Kim, S., & Chun, J. W. (2021). The role of engagement in travel influencer marketing: The perspectives of dual process theory and the source credibility model. *Current Issues in Tourism, 24*(17), 2416–2420. https://doi.org/10.1080/13683500.2020.1845126

Jiménez-Castillo, D., & Sánchez-Fernández, R. (2019). The role of digital influencers in brand recommendation: Examining their impact on engagement, expected value, and purchase intention. *International Journal of Information Management, 49*, 366–376. https://doi.org/10.1016/j.ijinfomgt.2019.07.009

Kozinets, R. V. (2020). E-tourism research, cultural understanding and netnography. In Z. Xiang, et al. (Eds.), *Handbook of e-tourism* (pp. 1–16). Springer Nature. https://doi.org/10.1007/978-3-030-05324-6_43-1

Kozinets, R. V. (2006). Click to connect: Netnography and tribal advertising. *Journal of Advertising Research, 46*(3), 279–288. https://doi.org/10.2501/S0021849906060338

Kuhzady, S., & Ghasemi, V. (2019). Pictorial analysis of the projected destination image: Portugal on Instagram. *Tourism Analysis, 24*(1), 43–54. https://doi.org/10.3727/108354219X15458295631954

Leung, D., Law, R., Van Hoof, H., & Buhalis, D. (2013). Social media in tourism and hospitality: A literature review. *Journal of Travel & Tourism Marketing, 30*(1–2), 3–22. https://doi.org/10.1080/10548408.2013.750919

Lin, Y. S., & Huang, J. Y. (2006). Internet blogs as a tourism marketing medium: A case study. *Journal of Business Research, 59*(10), 1201–1205. https://doi.org/10.1016/j.jbusres.2005.11.005

Litvin, S. W., Goldsmith, R. E., & Pan, B. (2008). Electronic word-of-mouth in hospitality and tourism management. *Tourism Management, 29*(3), 458–468. https://doi.org/10.1016/j.tourman.2007.05.011

Lojo, A., Li, M., & Xu, H. (2020). Online tourism destination image: Components, information sources, and incongruence. *Journal of Travel & Tourism Marketing, 37*(4), 495–509. https://doi.org/10.1080/10548408.2020.1785370

Magno, F., & Cassia, F. (2018). The impact of social media influencers in tourism. *Anatolia, 29*(2), 288–290. https://doi.org/10.1080/13032917.2018.1476981

Molinillo, S., Liébana-Cabanillas, F., & Anaya-Sánchez, R. (2017). Destination image on the DMO's platforms: Official website and social media. *Tourism & Management Studies, 13*(3), 5–14. https://doi.org/10.18089/tms.2017.13301

Motahar, P. S., Rokhshad, R., & Mura, P. (2021). Social media influencers' visual framing of Iran on YouTube. *Tourism Recreation Research.* https://doi.org/10.1080/02508281.2021.2014252

Munar, A. M. (2011). Tourist-created content: Rethinking destination branding. *International Journal of Culture, Tourism and Hospitality Research, 5*(3), 291–305. https://doi.org/10.1108/17506181111156989

Myers, S., Sen, S., Syrdal, H., & Woodroof, P. (2022). The impact of Persuasion knowledge cues on social media engagement: A look at pet influencer marketing. *Journal of Marketing Theory and Practice.* https://doi.org/10.1080/10696679.2022.2093224

Nixon, L. (2022). Do DMOs promote the right aspects of the destination? A study of Instagram photography with a visual classifier. In J. L. Stienmetz, B. Ferrer-Rosell & D. Massimo (Eds.), *Information and communication technologies in tourism 2022* (pp. 174–186). Springer. https://doi.org/10.1007/978-3-030-94751-4_16

Özkan-Erciyas, F., & Yıldız, M. S. (2018). Analysis of the customer experiences with netnographic investigation: A Cappadocia region five-star hotels case. *Journal of Travel and Hospitality Management, 15*(3), 590–602. https://doi.org/10.24010/soid.490705

Pop, R.-A., Săplăcan, Z., Dabija, D.-C., & Alt, M.-A. (2022). The impact of social media influencers on travel decisions: The role of trust in consumer decision journey. *Current Issues in Tourism, 25*(5), 823–843. https://doi.org/10.1080/13683500.2021.1895729

Rakuten Advertising. (2020). 2019 Influencer marketing global survey—Consumers. Retrieved from https://www.iab.com/wp-content/uploads/2019/03/Rakuten-2019-Influencer-Marketing-Report-Rakuten-Marketing.pdf. Accessed 20 August 2022.

Smith, S. P. (2021). Landscapes for "likes": Capitalizing on travel with Instagram. *Social Semiotics, 31*(4), 604–624. https://doi.org/10.1080/10350330.2019.1664579

Stepchenkova, S., & Zhan, F. (2013). Visual destination images of Peru: Comparative content analysis of DMO and user-generated photography. *Tourism Management, 36*, 590–601. https://doi.org/10.1016/j.tourman.2012.08.006

Tavakoli, R., & Wijesinghe, S. N. (2019). The evolution of the web and netnography in tourism: A systematic review. *Tourism Management Perspectives, 29*, 48–55. https://doi.org/10.1016/j.tmp.2018.10.008

Tiago, F., Moreira, F., & Borges-Tiago, T. (2019). YouTube videos: A destination marketing outlook. In A. Kavoura, et al. (Eds.), *Strategic innovative marketing and tourism* (pp. 877–884). Springer. https://doi.org/10.1007/978-3-030-12453-3_101

van Driel, L., & Dumitrica, D. (2021). Selling brands while staying "Authentic": The professionalization of Instagram influencers. *Convergence, 27*(1), 66–84. https://doi.org/10.1177/1354856520902136

Varkaris, E., & Neuhofer, B. (2017). The influence of social media on the consumers' hotel decision journey. *Journal of Hospitality and Tourism Technology, 8*(1), 101–118. https://doi.org/10.1108/JHTT-09-2016-0058

Varnalı, K. (2012). Dijital kabilelerin izinde—Sosyal medyada netnografik araştırmalar [Search of digital tribes—Social netnographic studies]. MediaCat Kitapları.

Walter, S., & Brüggemann, M. (2020). Opportunity makes opinion leaders: Analyzing the role of first-hand information in opinion leadership in social media networks. *Information, Communication & Society, 23*(2), 267–287. https://doi.org/10.1080/1369118X.2018.1500622

Wearesocial. (2022). Digital 2022: Another year of bumper growth. Retrieved from https://wearesocial.com/uk/blog/2022/01/digital-2022-another-year-of-bumper-growth-2/ Accessed 15 August 2022.

WHC. (2022). World Heritage Convention—Göreme National Park and the Rock Sites of Cappadocia. Retrieved from https://whc.unesco.org/en/list/357/. Accessed 17 August 2022.

Yıldırım, A., & Şimşek, H. (2013). Sosyal bilimlerde araştırma yöntemleri [Research methods in social sciences]. Seçkin Yayınları.

Gamification and Storytelling Enhancing Successful Wine and Food Tourism Products

Dália Liberato, Pedro Liberato, and Marta Nunes

Abstract This research focuses on the proposed tourism offer gamification and storytelling in the Douro region. Currently, the use of games in the tourism sector is beginning to be a strategy applied in several destinations. However, the application of gamification to the tourism sector has not always been successful. There are shortcomings in travelers' engagement due to the underestimation of tourism facts/events or resources of the regions. The main objective is to combine gambling with storytelling as part of wine and food tourism, to ensure greater tourist involvement. Gamification associated with storytelling can disseminate and enhance the tourism offer in a region, creating value in tourism experience. Considering the research methodology, it was considered relevant to implement a qualitative approach, using a focus group interview script, with residents, as well as the use of semi-structured interviews applied to Farms and the Douro Museum. The main results highlight the development of the region's tourism resources, which are crucial in a process of international promotion, increasing tourism demand by proposing a game as a driver of tourism and gambling associated with storytelling, and a key element in involving the player, telling the story, and spreading the word about the destination. Studies and projects dealing with gamification and storytelling in tourism are unknown, particularly in low-density inland destinations, being crucial to propose projects on this topic, suitable to similar destination typologies.

D. Liberato (✉) · P. Liberato · M. Nunes
School of Hospitality and Tourism, Polytechnic Institute of Porto, Vila do Conde, Portugal
e-mail: dalialib@esht.ipp.pt

P. Liberato
e-mail: pedrolib@esht.ipp.pt

M. Nunes
e-mail: martaisabel.nunes@gmail.com

D. Liberato · P. Liberato
CiTUR Researcher (Centre for Tourism Research, Development, and Innovation), Vila do Conde, Portugal

© The Author(s), under exclusive license to Springer Nature Switzerland AG 2023
V. Katsoni (ed.), *Tourism, Travel, and Hospitality in a Smart and Sustainable World*, Springer Proceedings in Business and Economics,
https://doi.org/10.1007/978-3-031-26829-8_27

Keywords Gamification · Storytelling · Cultural heritage · Wine and food tourism · Tourism experience · Douro region

JEL Classification Z32

1 Introduction

In recent years, there has been a demand for differentiating, unique, and authentic experiences where the sharing of knowledge of the destination presents a wide range of information scattered on the Internet, either through blogs, websites, social networks, or applications of the destination itself. However, there is a lack of content that is part of the identity that tells the story that talks about the characters that were part of the essence of the land and its people. The streets, the places, the people, and the characters tell stories within the city's own history. In this perspective, this project aims to explore a holistic aspect of tourism in a destination, specifically, the city of 'Peso da Régua', based on gamification associated with storytelling.

The concept of gamification has been applied for several years, however, Burke (2014) states that there is no exact definition of the term, although there are common characteristics, namely motivating people to change behaviors, develop skills, achieve goals, and drive innovation. The goals of storytelling are identical to those of gamification, but the way they are achieved are different. While gamification uses the reasoning process of game creation and game mechanics to engage users and solve problems (Zichermann & Cunningham, 2011), storytelling seduces through story and has profound effects on the way people interact with each other (Mccabe & Foster, 2006).

Xu et al. (2017) point out that tourism gamification can contribute to more rewarding interactions and higher satisfaction levels, as well as increase brand recognition and destination loyalty. Regarding the tourism sector, Xu et al. (2017) identify two types of games: social games based on social media, mainly used for brand activation, to attract potential customers and to build the image of a destination or a company; and location-based mobile games, used to provide experiences to locals and tourists at the destination in a more entertaining and informative way. However, these games are often not successful. Furthermore, according to Xu et al. (2017), gamification can benefit the tourism industry in the following areas: (i) increase brand recognition; (ii) enhance tourist experiences; (iii) improve player engagement; (iv) improve customer loyalty; (v) provide entertainment to the tourist experience; and (vi) can be used as an internal company function in that it can motivate employees.

Studies on gamification associated with storytelling in tourism are still scarce, as mentioned in the literature review, the field of study is more directed toward education, behavioral organization, and marketing. In this sense, some of the attributes of the study were obtained using secondary data, which allowed to know the 'state of the art' storytelling articulated with a literature review on gamification.

2 Literature Review

Zichermann and Cunningham (2011) define gamification as "the reasoning process of creating games and game mechanics to engage users and solve problems." Burke (2014) notes that while there is no exact definition of the term, there are common characteristics and presents gamification as "the use of game mechanics and experience design to digitally engage and motivate people to achieve their goals." The goal of gamification is to motivate people to change behaviors, develop skills, drive innovation, as well as support players to achieve goals. In recent years, it has been proposed as a solution to engage people in individual and socially sustainable behaviors, such as exercise, sustainable consumption, and education (Koivisto & Hamari, 2014). Already in recent years, gamification has broadened its scope and become an important applied research topic in other sectors such as health and well-being, sustainability, labor productivity, tourism, and cultural heritage (Araújo et al., 2019). However, although the academic literature offers little understanding of gamification, there are three aspects that allow a development in this regard namely (i) new knowledge about the design and management of gaming experiences, (ii) intertwined with the emergence of social media and the development of cell phones, laptops, tablets, and (iii) the increased interest in providing engaging experiences (Robson et al., 2015). Technology allows users to create information from various sources on their mobile devices, customize their profile through apps and social networks, and interact in a dynamic way with their context (Buhalis & Foerste, 2015). For Robson et al. (2015), there are four types of people involved in gamification experiences. These authors were based on Pine and Gilmore's (1998) two dimensions, active or passive participation, in which player participation describes to what extent the individual is actively or only passively involved, and player connection which describes the type of environmental relationship that connects the individual to the experience (absorption/immersion); furthermore, in absorption, the experience occupies the person's mind and immersion, one becomes part of the experience itself, physically or virtually (Fig. 1).

There are several elements that make up a game, as well as several ways to classify them. Schell (2012) created a model called the Elementary Tetrad, which divides the main elements into four distinct categories and which are part of the game experience. The categories are (i) aesthetics, which is related to the individual's sensory experience, what he hears, sees, and feels, it is the category that has a more direct relationship with the player experience; (ii) mechanics, which comprises the rules and procedures, i.e., the system that makes the game work; (iii) story, which is based on the narrative, the sequence of events that occurs in the game; and (iv) and technology, which encompasses both the materials and the media used. Also, for Hunicke et al. (2004) game elements can be grouped into three categories: mechanics, dynamics, and aesthetics, and these categories represent in a game the rules, the system and fun, respectively. Mechanics describe the specific components of a game and remain constant throughout the experience, not changing from one player to another.

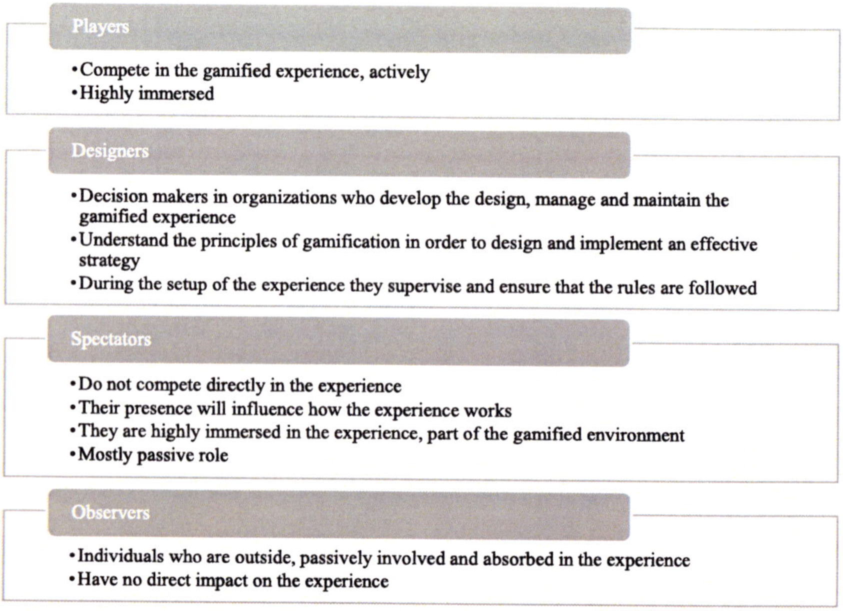

Fig. 1 Type of users involved in gamification experiences. *Source* Robson et al. (2015), Adapted

Regarding the tourism sector, the concept of gaming is no stranger, and according to the World Travel Market Report (2011), gamification is a trend for the coming years in tourism, attracting consumers of all ages. The Amadeus report (2015) entitled Future Traveler Tribes 2030, Understanding Tomorrow Traveler states that in the coming years travel trends will be determined mainly by the intensive use of technology, social criteria, cultural criteria, as well as the ease of travel management and travel (mobility). Game mechanisms can create positive experiences in tourism: fun, thrill, excitement, pleasure, sense of accomplishment, however, although gamification can be applied in any industry, the hospitality sector is not making the most of gamification as other industries are doing (Negruşa et al., 2015). In turn, the FCT (2019) mentions that technological growth in tourism, with the integration of several research areas, aligned with the technological development axes, namely at the level of gamification, can attract an increasing demand in Portugal, considering the three phases of the capture: capture to visit, support to their experience during their visit(s) and the post-visit. It is important to have information systems able to store, manage, and present information that allows the analysis of tourism activity, since information is the main element between the various actors in the tourism value chain, such as tourists, tour operators, and decision-makers, generating information flows between the various actors. Optimizing the use of information and communication technologies (ICT) allows organizations to improve the flow of tourism information, improve response times to requests from outside, and enhance their development in an increasingly competitive society (Ramos, 2008). Consumers are becoming increasingly powerful and can determine elements of their tourism products. Simultaneously, they are more sophisticated and experienced and, therefore, more difficult to please. It is critical to create differentiating strategies at the strategic and tactical plan level to ensure that ICT-driven opportunities and challenges are leveraged by tourism organizations to improve their innovation and competitiveness (Buhalis & Law, 2008).

For Cheong et al. (2014), gamification does not mean simply using game elements, just as these are not solely responsible for the player experience, but rather the interaction of design, techniques, and game conception, and how this interaction of all element's merges with the goals of the gamified activity, referring that tourists have at their disposal several applications for smartphones, however, often their content is not complete or updated, not observing feedback from the user. Generally, the best information about a city is shared by locals or recent visitors.

At all stages of the trip, information must be timely and relevant, especially during the trip. When consumers are out of their comfort zone, they may not speak the language and are unfamiliar with the local geography. From the perspective of enhancing the tourist experience, information should be highly personalized and related to the individual context (Buhalis & Law, 2008; Liberato et al., 2018).

Xu et al. (2016) report that games, in recent years, are used by some tourism sectors as a marketing tool to promote tourism destinations, offering an opportunity to create informative, and entertaining environments for successful brand awareness, interaction, and communication. Games offer benefits in several areas, such as entertainment, education (interpretation), and co-creation of tourism experiences, allowing to increase tourists' interest in the destination, provide experiences and knowledge that are not otherwise available and provide a personal experience during the visit. Simultaneously, it can involve other players and residents, allowing to further enhance this experience and influence people to interact with destinations on a large scale. Liberato et al. (2018) corroborate the same idea and state that information and communication technologies (ICT) can bring numerous benefits to the tourism industry, namely in the promotion of tourist destinations, where tourist activities can be monitored, advertised, and marketed, supporting tourists before and during their stay in the respective destinations. Furthermore, travelers share their experiences online with family, friends, and the public through email messages, using social media such as Facebook and Twitter, and posting on travel blogs (Chandralal et al., 2015).

Motivation plays an important role in gaming. In this sense, Xu et al. (2013) identified several motivations to play a tourism game and created a pyramid called "Motivation to Play Tourism Games". The main motivations at the base of the pyramid are curiosity and exploration. Tourists very often want to know and get practical information about the destination and set out to discover it. These two motivations can coexist because they are multidimensional. At the third level, they identify the need for socialization which reflects the desire to establish social relationships with both tourists and residents. The intrinsic motivations (fun and fantasy; challenge and conquest) seem to support the flow of the game, allowing playing in a virtual and real environment (fantasy and fun). The authors stress that DMOs should consider the player's motivation when using gamification. Information is key, but also the opportunity to socialize with other tourists and residents is important. The flow of the game should not be ignored, in that the need to provide pure fun to the travel experience, stimulates that experience, and allows tourists to engage with the destination in an interactive way. In turn, at the top of the pyramid comes the need for challenge and achievement. A tourism game can include a challenge, such as visiting several places in a limited time, identifying monuments, among others. As a reward for scoring achievements, games can accumulate points or badges, or other rewards. The difficulty level can vary, and the challenge should lead to a reward for the player's performance.

According to Weber (2014), playing tourist destination games with the help of a smartphone will bé one of the major developments in the tourism and gaming industry in the coming years. This type of game is based on the location of the tourist and is played specially to encourage more engagement and enhance local experiences in a more fun and informative way. One of the most innovative and interesting ways to transform the phases of a trip into unforgettable life experiences is through gamification, and some tourism managers use gamification in their marketing strategy with the aim of creating a culture of well-being and fun for their customers and employees, allowing the creation of engagement (Weber, 2014). Also, Weber (2014) presents ten pioneering practices in the application of gaming and gamification concepts, recommended for pre-, during and post-travel, to create innovative products and services for the travel and tourism industry. They are location-based games: location-based games, via mobile devices, that allow tourists to take interactive tours of the place visited, bringing the story to life and create playful interactions between the visitor and the tourist destination; gamification tour guides: through applications such as Foursquare, which is used in both urban and rural environments. Visitors explore the city through a scavenger hunt, allowing them to collect points that can be redeemed for vouchers or discounts, or compete with other players; gaming in theme parks: theme parks are great for implementing gamification because they provide fun and playful experiences; players need to collect as many coins as possible while running away from the witch, but at the same time, the children teach the witch to save money so that she no longer needs to steal it; gaming in cultural heritage: gamification produces new opportunities for visitors to draw on historical facts and information in an interactive and immersive way, creating a greater connection with history and fostering a different engagement than more classical systems. There is scientific evidence to show that playing evokes a sense of being present in earlier times and contacting the people who lived in those times; transmedia storytelling: for DMO's, it is important to engage visitors before, during and after the trip. As the game is multifaceted, it can be combined with other techniques, such as transmedia storytelling that uses social media platforms to create a narrative; gamified restaurant experience: the fast-food chain MacDonald's created the game Pick'n Play, whose objective is to interact with digital billboards to get free hamburgers; gamification in hospitality: for Weber (2014), gamification can be used in hospitality to motivate employees and engage customers; gamified flying experience: in the airline industry, there are already several examples of gamification, such as brand loyalty systems through benefits for those who make more miles with a particular airline; virtual cultural heritage: through virtual worlds, the visitor can learn about a culture without the negative impact of mass tourism and is also very useful for people with accessibility issues; virtual travel: the integration of games into tourism environments is increasingly used for those who want to 'see' the world without leaving home or if they are undecided where to go and like to experience/experience destinations in advance.

Storytelling is part of human nature, is possibly the oldest form of verbal communication, and has profound effects on how people interact with each other, Mccabe and Foster (2006). This method or procedure is referred to, in different areas of application, as storytelling (Lukosch et al., 2011). Creating unique stories through storytelling to create extraordinary experiences is increasingly important for the tourism industry (Mei et al., 2020). There are still few studies that show how storytelling contributes to the development of a tourism destination, as well as the effects of joint construction of a story in terms of the ability of a destination's stakeholders to cooperate (Gravili et al., 2017). However, those in existence argue that it is a differentiating element in tourism products, especially for tourists who are increasingly looking for distinctive products, contributing to their tourism experiences (Gonçalves et al., 2018), as it brings memories to life and immortalizes times and places, awakening the five senses (look, hearing, touch, taste, and smell). It can be a way to process emotions (Line, 2019), it presents a catalyzing effect, not only because it is a unique product (Gonçalves et al., 2018), but also because it bridges the gap between residents and tourists, bringing them closer in a peculiar way that the traditional side of tourism would never do. Mora and Livat (2013) corroborate with the idea of tourists' strong connection to places, which automatically contributes to conferring greater value to the product, because storytelling adds value to destinations, promoting historical and cultural heritages (Bassano et al., 2019). In addition to presenting a differentiating and catalyzing effect, storytelling can confer additional authenticity to a place (Hannam & Ryan, 2019).

Also, wine producers tell many stories about wine, their properties and even their ancestors. These stories are part of the activities and sensations that translate into the experiences of wine tourists who also become storytellers by telling their travel stories, witnessing their experiences (Mora & Livat, 2013). In a society where all destinations claim to be unique, but end up being mere copies of others, stories become one of the few forms of exclusive 'sale' that places have at their disposal. In this way, the narratives have the power to add a special and differentiating touch to tourist destinations (Bryon, 2012).

Stories have the unique power to persuade and motivate because they appeal to our emotions and create empathy. For Hsu (2008), the best stories told from generation to generation captivate audiences and emotions can be inextricably linked to their own characters. Story content has a direct and positive effect on empathy and persuasion, proving that a story can engage and inspire participants (Chi-Hua & Chi-Chun, 2020). Concomitantly, memories are also reconstructed through intangible storytelling practices, and storytelling can confer additional authenticity to a material place (Hannam & Ryan, 2019). Destination storytelling is seen as a strategy to enhance the reputation of regions, how municipalities and cultural organizations understand storytelling can have a significant impact on the success of a region. Place storytelling allows different bodies to tell the stories about their places, allowing to highlight storytelling management as a strategic communication, involving stakeholders in the process of building place identity (Bassano et al., 2019).

The storytelling associated with a region is more than a simple 'tell a story' or a mere chronology of facts. Bassano et al. (2019), also encompass communication, as well as (…) "all the processes of seducing a tourist, the cultural and ritual exchanges between visitor and host, the tourist narratives on which the identities of people, places and brands are built" (Wichels, 2013). For example, libraries, archives, and museums are increasingly using digital technologies to tell stories and create mobile applications (apps) for cultural heritage content (Basaraba et al., 2019). Building and sharing stories consist of proper narrative management, which can be understood in three ways (Gravili et al., 2017).

Interaction management defined as the ability to develop horizontal and collaborative processes between all subjects potentially involved in the narrative project; content management understood as the ability to make the best use of storytelling by the narrator, with the aim of increasing the emotional connection with the destination itself, making their stories compelling and compelling; crisis management, with a view to responding individually, quickly and effectively to all complaints and/or comments that could damage the brand value of the destination.

Gastronomy occupies the third motivation for which tourists decide to visit a tourist destination, behind cultural interest and nature, and is an essential component of history, tradition, and identity, as well as offering enormous potential to stimulate local, regional, and national economies and promote sustainability and inclusion (UNWTO, 2016). According to the UNWTO (2016), gastronomic tourism emerges as an indispensable resource that adds value and provides solutions to the need for destinations to differentiate themselves and offer exclusive products. In this perspective, it lists five reasons that justify this constantly growing phenomenon, namely the need for a destination to differentiate itself and create a unique commercial proposal, in the search for the authentic. At the same time, destinations are using gastronomy to attract tourists, allowing them to direct the tourist flow to less visited destinations, as well as to design an effective communication strategy using a narrative that easily appeals to the emotional side of the visitor and offers deeper and more meaningful experiences, capable of leaving a more lasting impression. Wine and gastronomy can be expressive of a regional culture. This relationship is extremely significant for tourism because of the possibilities of establishing an association of wine and the wine landscape with a strong regional identity (Hall & Sharples, 2003).

Minasse (2020) summarized the main attractions identified, which are presented below (Table 1).

Lopes et al. (2018) summarized the activities framed in wine tourism (Table 2).

In turn, wine tourism in the Douro region presents a different set of modalities from those existing in the national territory, because it is more diversified both in forms and in the type of infrastructure. The production base continues to be the wine production farms, units of great family tradition or acquired by economic groups. The luxury

Table 1 Main gastronomic attractions

Attractive	Description
Courses and learning activities	Courses and educational activities of different lengths that teach about food or beverage production and preparation
Tasting and matching	Tastings of different products (olive oil, chocolate, honey, cheese, wine, beer, cachaça, and typical dishes) that can be offered in different spaces and establishments
Food establishments	Bars, restaurants, nightclubs, bakeries, and other out-of-home eating establishments that, by some characteristic (location, building, association with a movie or book, distinctive cuisine) can attract visitors
Product commercialization establishments	Places dedicated to selling food products (ingredients or ready-to-eat products), such as municipal markets, wine cellars, and chocolate stores
Gastronomic events	Feasts, festivals, and other gastronomic events. Mostly, there are entertainment programs
Free fairs, gastronomic fairs, and street food	Fairs held in public spaces to sell gastronomic products (food and drink). Street food and food trucks
Production sites	Food production spaces such as cheese dairies, wineries, chocolate factories, and stills with a visit
Food museums	Institutions dedicated to safeguarding, conserving, and exhibiting collections related to different food practices
Typical food and beverages	Traditional dishes, delicacies, and drinks of a locality, offered at events and establishments
Routes, itineraries, and tourist circuits	Previously defined routes that have a gastronomic theme or have a high incidence of gastronomic elements

Source Minasse (2020)

lodgings and tourist complexes create innovative concepts and take advantage of the unique scenic surroundings in which they are located, promoting the 'look' of those who stay overnight, betting on quality and high prices, as well as the extension of the stay, offering activities which invite involvement with the landscape (Vaz, 2008). For Martins et al. (2012), the authenticity and uniqueness of products are concerns of most destinations. It is natural that these places seek to offer unique or rare to be found experiences, to maintain their competitiveness and tourism differential. It stands out the involvement of the bodies responsible for tourism in the development of the gastronomic product, facilitating the interaction between producers, suppliers, consumers, and the local community.

Table 2 Activities framed in wine tourism

Economic	Buying wine and local products
Cultural	Visits to museums, interpretive centers of wine and vine culture, wine tasting, and tasting of local products at the wineries
Tourism	Visits to wineries, cellars, vineyards, monuments, natural heritage, participation in wine fairs and festivals
Social	Visitors meet the local experiences, traditions, and the inhabitants of that region
Sports	Horseback riding, cycling, balloon rides, and extreme sports
Health and wellness	Obtaining antioxidant properties, their consumption, and their derivatives for therapeutic purposes in wellness/spa center component
Pedagogical	Visitors can learn something about wine and its culture, within a logic of obtaining educational experiences

Source Lopes et al. (2018)

3 Methodology

In the Douro region (NUT III), the culture of vines and wine is the fundamental economic base of the region's economy, presenting a fundamental brand and image on a national and international scale. The region is also marked by the extensive area of agricultural and horticultural production, such as apple, grape, cherry, potato, chestnut, almond, and olive. The valorization of the territory, allowing the enjoyment of the historical and cultural heritage, and preservation of its authenticity; the urban regeneration; the economic enhancement of the natural and rural heritage is part of one of the five strategic axes proposed by the Tourism Strategy 2027, by the national DMO.

The Douro Demarcated Region (RDD),[1] the oldest regulated wine region in the world, runs along the Douro River and its tributaries over an area of about 250,000 hectares between Barqueiros and Barca d'Alva. It represents the first institutional model of organization of a wine region, created to regulate the production of Porto wine. Currently, the RDD circumscribes the Controlled Denomination of Origin of Porto and Douro wines. The cultural landscape of the Alto Douro Vinhateiro (ADV)[2] combines the nature of the Douro River valley, with steep slopes, poor and rugged soils, with human activity, allowing the creation of a unique ecosystem, where the characteristics of the terrain are taken advantage of in an exemplary way, with the shaping of the landscape in terraces, preserving it from erosion and allowing the cultivation of vines. The vineyards throughout the Douro Valley date back to Antiquity, the region has been producing wine for over 2000 years, however, it is from the sixteenth century that quality viticulture assumes importance, especially with the signing of the Methuen Treaty in 1703, through which Porto Wine began to benefit

[1] Source http://www.museudodouro.pt/regiao-demarcada-do-douro.

[2] Source https://www.unescoPortougal.mne.pt/pt/temas/proteger-o-nosso-patrimonio-e-promover-a-criatividade/patrimonio-mundial-em-Portougal/alto-douro-vinhateiro.

Table 3 Residents

N. No. interview	Age	Gender	Profession
R1	37	Female	Coordinator of the UA Local Learning Center
R2	33	Female	Guide in the Douro Museum
R3	60	Male	EDP Employee and Writer
R4	69	Male	Journalist
R5	42	Female	Responsible for the Educational Service of the Douro Museum
R6	44	Female	Responsible for the Educational Service of the Douro Museum

Source Authors

from preferential customs rates on exports to England and from the second half of the seventeenth century, began to be produced and exported in quantity, mainly by the hand of English producers and traders (Pereira et al., 2001). In 2008, the Douro Valley was considered, by the OMT, a destination of excellence, being part of the 77 'Wonders of Nature of the World' (CIM Douro, 2015). The methodology adopted for the development of this work was qualitative, with the design of a focus group to residents reiterating some key ideas that interconnect the two previous processes, identifying the places and characters that could serve as a basis for the design of the game. According to Coutinho (2018), qualitative research uses a multiplicity of methods in a naturalistic and interpretive way, therefore, it allows a better understanding about the opinions of the interviewees and the data is richer. Qualitative methodology generally privileges the analysis of microprocesses, through the study of social, individual, and group actions (Martins, 2004). The focus group script was built according to seven themes, established in a funnel perspective, from general to particular: I-Information about the interview, II-Socio-demographic data of the interview sample, III-Tourism in 'Peso da Régua', IV-Perception about the Region, V-Perception about the wine, VI-Perception about the game, and VII-Conclusion. The focus group was coded using the letter 'R', followed by an increasing number predefined by the moderator as shown in the following table (Table 3).

After the application of the focus group, other research methods were used, namely semi-structured interviews. The use of these qualitative methods aims to provide clues for the development of the gamified game.

The focus group interview script was prepared with the purpose of associating the themes and the respective general and specific objectives, creating interconnections between the themes, and starting from the general to the specific.

Table 4 Focus group themes and objectives

Themes	Specific goals
Perceptions about the region	Ascertain residents' perception of the attractions and differentiating factors of 'Peso da Régua' in the regional context Identify the material and immaterial heritage, relevant in 'Peso da Régua' Identify the dominant tourism products in the region Identify the activities/events that are characteristic of the region and that ensure greater participation
Perception about wine	To understand the potential and capacity of differentiation of the Douro as a wine tourism destination To investigate the involvement of stakeholders in initiatives that enhance wine tourism and gastronomic tourism at the destination
Perception about the game	Understand the importance of play as an enabler of the region's heritage Specify the significant tourist attractions of 'Peso da Régua' to incorporate into the game Identify famous figures associated with 'Peso da Régua' Record the narratives of the characters and their main characteristics Listening to the stories/legends, words, traditions, and icons of 'Peso da Régua'
Finishing	Understand if there were any important aspects not covered To thank them for their collaboration, availability, and contribution to the research

4 Results

The main thematic findings of the focus group are presented in Tables 4, 5 and 6.

5 Conclusion

The Douro is very rich in intangible heritage. The interviewees highlight some of the legends associated with the region, the very nature of the people of Peso da Régua, their connection to the river, the people from Ovar who come to sell salt. The Galicians who came from Galicia to build the schist walls. The story of 'Ferreirinha' and the Baron of Forrester. The history of Porto wine associated with pork meat. Stories associated with the various occupations, including the Muslim occupation, where many legends of this Arab imaginary, the hermitages, the chapels, the rocks, appear. There are popular words and sayings that characterize the region, a lexicon that has to do with the vineyard, with the landscape, and is based on Galician references.

The floods and their marks are also referenced and can also be part of the narrative. As Barreto (2018) refers, worse than navigation accidents were the floods. In Régua, in Vila Nova de Gaia, and on the 'Ribeira' area (Riverside) in Porto, on the walls of

Table 5 Focus group perceptions about wine

Subcategories	Examples/evidence
What distinguishes the Douro from other wine tourism destinations	"The Douro, the oldest wine region in the world, world heritage and Humanized Landscape" [R1, R4, R5, R6] "Pioneer demarcation, it is our great bastion, a region that is still very diversified where, next to a farm with all the latest technology, with the whole mechanized process, we have a micro producer who still uses the techniques his grandfather taught him, and the visitor appreciates this diversity. The Douro can never be a mass destination, it is a region that favors a certain calmness, a slowness in the visit, and a quietness that never exists elsewhere" [R2] "The exclusive character of our main product, which is wine, i.e. wine as an agricultural product made with the soul, with the heart, there is a great interconnection between the man who produces the wine and the wine that results from what that man has made. In the man/agricultural product relationship, the wine takes on a unique character of passion, of intensity, and I think that those who visit us feel this, feel that they are in a unique place, which produces by itself a unique product, which is the Porto wine. The landscape is unique. What distinguishes us as a landscape is the fact that it results from an action that man made" [R3]
Stakeholder involvement (public/private initiatives to enhance the development of wine and food tourism in the region)	"The gastronomic weeks. The professional school of Ródão and the hotel management school of Lamego have been developing some projects, including the most recent was the creation of a specific cookie to accompany Porto wine, to give it a new look, a new presentation" [R2] "The two schools, which have been able to create and preserve gastronomy, as for the dissemination of what is happening, I think that the Douro suffers from a problem of scale, at the level of gastronomy. The Douro still must move from a parochial vision to a more comprehensive vision of the region" [R3] "The sweetness of the chestnut. There was in Loureiro the chestnut festival. I've seen it translated in Vidago Palace. Adapt the old recipes)" [R4] "The food weeks and fairs" [R5, R6]

Source Authors

Table 6 Focus group perceptions of the game

Subcategories	Examples/evidence
The game as an enhancer of the region's heritage	"Bringing together innovation, new technologies, knowledge, sharing, history, and people. We intend to reach people and involve gastronomy, handicrafts, because we even try to include all the people so that the visitors have to do and say 'I'm going here…' with the strategy of the playful and pedagogical related to education, this is to take me to know something is like the result of entering a store and then receiving a candy from Régua, if you pass the plateau and answer the questions because it has a narrative receive a bottle of Porto wine as a prize, why not, a voucher with a bowl of tripe on Wednesday, so that these traditions that are very important are not lost." [R1] "Yes, combining playfulness, that sense of discovery, motivates the visitor in a different way, in what way? The very circulation of people and the fact that the game provides a tool for knowledge, for interpretation, so it has all the value." [R2]. "The game involves people, so it brings people. Yes." [R3]. "All initiatives are positive and welcome; this is one that excels in originality." [R4] "To look for associations in the landscape. Make the association between elements." [R5] "In the educational service, it is important that people know the whole approach of the territory, to include observation would be very interesting" [R6]
Attractions to incorporate into the game	"D. Antónia is a product of excellence. (…) the landscape, the Museum, the religious part (…) One would pray and ask whoever took the wine in the Rabelo boat to come back. It took days and days, and it was a struggle of great strength. S. Leonardo de Galafura, with the church and we could see the patchwork. And the inhabitants, the reguenses are the main ones, without them there would be no history, there would be nothing." [R1] "The pier, or a more open point, that allows one to read the landscape. The Douro vineyard landscape is not only the vineyard, the monoculture of the vineyard, but it is the mosaic, the patchwork that has a little patch of vines and then a border of olive trees and to the side a little Mediterranean forest, a little chapel up there, a road that provides a view of this mosaic (…) the Museum, (…) the bridges, not only the pedestrian one, but the funny story that the one on the road to the Douro is the only one that allows us to see the landscape. the Museum, (…) the bridges, not only the pedestrian one, but the funny story that has the fact that today's road bridge was built for the train that never crossed it to the other side and that became a road bridge in the meantime (…) the CP station, with the history of the train, with the candy boxes at the exit. The Douro house and the connection with the Porto Wine Institute." [R2] "The Cruzeiro Church has a facade by Nicolau Nasoni and has the pattern of the Nossa Senhora da Agonia (Our Lady of the Agony), which has that name for a very interesting reason: the Rabelo boats that came from Porto and made the bend in the willows were facing the Cruzeiro Church. The trip of the Rabelo boat to Porto was itself an epic, because when they left, they didn't know if they would return alive, the trip was very dangerous, there was no guarantee that they would return, and they came face to face with the church of the Cruzeiro. The human figure of D. Antonia, the figure of that woman, can be a tourist product of excellence" [R3] "(…) the game is Important for its own diffusion, as a means of promoting the Douro, its usefulness as a means of creating an appetite for the Douro, a way of generating a desire to consume the Douro, or to consume Régua in this case. (…) to identify 5 aspects I would list from the start, the landscape, the river and its slopes, navigability, fishing, the very nature of the river, its history, the rabelo boat, there is a whole universe of things that can be connected and that the game should value. It would also be ungrateful to talk about Régua without mentioning the inhabitants (…) the knowledge of the ox carts, the stained-glass window of the Douro house. D. Antónia, (…) the inhabitants have three main vectors, the residents, and the emigrants from Galicia and Ovar, and there are few places in our country that have such a defined mark, we are a mixture of three cultures, the people from Ovar and the Galicians who came from the Ribeiro area, my great-grandfather was Galician and most of the influential families here in Régua are from Ovar." [R4] "The landscape, the sweets, the river issue, the season." [R6]

(continued)

Table 6 (continued)

Subcategories	Examples/evidence
Artists associated with 'Peso da Régua'	"Obviously João de Araújo Correia, Agustina, has the particularity that we urgently need to take advantage of, that she is buried in the cemetery of Godim, 'Peso da Régua', this can also be turned into a tourist product, we have several writers, we have Alves Redol, Custódio José Vieira, Manoel de Oliveira, with films about the Douro." [R3, R5, R6]
Historical characters to incorporate into the narrative	"João de Araújo Correia spoke very well. (…) 'Ferreirinha' and her contribution. Her context is one that favored her, she founded the soup kitchen, that's what gave rise to the patronage of Godim, which still exists today, the hospitals she created." [R4] "A contemporary of D. Antónia, the Baron of Forrester, one always talks about maps, about studies of viticulture, one, a pioneer of photography in the Douro, also had a house. It is inevitable to talk about D. Antonia and not mention the Baron of Forrester. The characters of the land, the 'arrais', the 'feitor' of the farm, the 'regedor', who was almost the village sheriff. And they were figures that existed in every village." [R2] "He proposed the creation of the general company of the Vineyards of the upper Douro and put a hero who is General Silveira, great military chief of the French invasions and was one of those responsible for Napoleon not having won the invasions. He played a major role in resisting the French troops." [R3] "The Régua sweets, which are part of our history and tell our story so well." [R1] "Not to forget the objects, the landmark, the train, the Rabelo boat, the baskets, the objects connect to the place, more than the characters." [R6]
Cultural weight and stories to tell	"1st Personality João de Araújo Correia, has many short stories about the characters of the Douro, his Portraits of the social life of the Douro, of the small experiences, he was a doctor, but he wasn't always at the doctor's office, he was a João Semana, he traveled a lot around here, And he wrote a lot about the people of the Douro, my favorite is "Contos Bárbaros", but there is "Sem Método", he has a series of short stories about the life in Régua, he founded a newspaper, he spoke very well, he had tenderness. The stories are all associated with 'Ferreirinha'" [R4] Baron Forrester, a very complex personality. There is a duality of the man of science and the man of art, he was an artist, he painted, he photographed, and he never stopped being in the know about wine issues, the defense of the farmers and the region, the purity of the wine. He is credited with having found a cure for oidium. The rest are characters like, the arrais, the sailor from the Douro, they would go round the boat between the river and the sea, the figure of the feitor as the authority. [R2] "Obligatorily Don Antonia, vivid narrative that enchants, and combining that with the fact that she's a woman, it's very interesting." [R3]
Legends, stories, traditions, icons	"The nature of the people of the Régua, their connection to the river, made by the people from Ovar who come to sell salt, who come here to fish and then become traders, is a story that marks the Régua, the Galicians, who came mainly from the south of Galicia and who came to build the schist walls of the Douro, they came to revolutionize and stayed here, now there are families here, José Vaz Osório, the Corrajes, the Monteiros came, here if you want to highlight some history, the very nature of the people who live here. (…) As far as gastronomy is concerned, there was an idea that Porto wine had meat in it, in a certain period people from the south said they wouldn't drink Porto wine because it had meat in it, and this was connected because in the rich houses people would take a leg of pork/ham and put it in Porto wine and boiled it, it was like a ham. That Porto wine that was left over was drunk by the farm workers." [R4] "The Muslim occupation." [R2]

(continued)

Table 6 (continued)

Subcategories	Examples/evidence
Proper vocabulary, popular sayings	"Until the middle of the twentieth century, "Is there salt in Régua?", …There is salt in Régua". The origin of the word is different …in the beginning it has to do with the presence of people from Ovar, salt in the seventeenth, eighteenth, nineteenth century, was a highly valued and rare commodity, when a boat came from the coast that arrived at the Régua, the "almocreves", all the merchants, would rush to come quickly to have the guarantee to negotiate, the first was much more coveted, we don't know very well, why… and, therefore, there was a hurry to get to the Régua. [R4] "There is a lexicon that has everything to do with the vineyard. More specific regional expressions and I highlight them as well, going for the reference to the Galicians that still persists in the older generations." [R2]
Old traditions that remain/or can be revived	"The Wine Fair is more than 100 years old and is very interesting. The festivals and pilgrimages are all to be continued, even if some modernisms are used, some traditions related to wine, which should be maintained, the samarras fair, the blanket fairs, the cheese fair, the Terrincho cheese, and the goldsmiths, who came from Gondomar. The gold tradition has to do with the people of Minho, why they used to use pounds of gold, the pounds of gold had to do with the products that were sold to England and were paid in pounds, which had no commercial value, but had value for their specific weight of gold. Régua has always had cyclical periods linked to fat cows and lean cows" [R1, R3, R5, R6]

Source Authors

riverside houses or on bridge pillars, macabre memories are still visible: the most important flood marks.

So did the 'Rabelo', who ended their working life in the 1950s, replaced by the railway and the road. In the late 1980s, tourists started lurking the Douro and Gaia merchants had their 'Rabelos' built or repaired (Barreto, 2018). The interviewees emphasize the importance of this icon. Although there is a 'Rabelo' boat in front of the Museum, there could be one on the pier that would allow the visit and explain its operation. As Barreto (2018) refers, the trip on the 'Rabelo' boats, the rituals had some joy. One could hear the wind in the sails, the wood creaking, the water splashing (...), the almost absolute silence, the shipwright's voice giving orders, one talked to the shores, gave errands, received requests and orders.

In the study carried out, it was found that the interviewees recognize the figures of D. Antónia Adelaide Ferreira, Joseph James Forrester and João de Araújo Correia. Two nineteenth century contemporaries who were friends, loved and fought for the Douro, each in his own way. 'Ferreirinha', as everyone affectionately calls her, had a huge weight in the region, left a huge legacy, helped the poor, fought for the region's wine. Baron Forrester, in love with the Douro and the river that would kill him, a very complex personality. João de Araújo Correia, a storyteller who portrays the people of the Douro, considered a João Semana, or Holy Man, erudite and who knows the people of the Douro like no one else. Finally, the 'Rebuçadeira', who personifies these people. A fighter and a reference figure for those who visit the city.

Destination managers should understand that this phenomenon is fundamental (Gonçalves et al., 2018) and that strategies should be developed and bet on differentiating elements such as gamification and storytelling to attract more tourists, enhance their tangible and intangible heritage and engage residents.

If on the one hand the Douro is rich in expressions, legends and stories that can be valuable tourist resources, with a unique and captivating story, told by the very people who helped build it through the ages. On the other hand, the tourist is looking for unique and memorable experiences, wants to know the history, get in touch with the locals, taste the local cuisine and wines. In short, the tourist wants to engage with cultural heritage in a holistic sense. The tourist is not a passive consumer, he is the configurator of his own trip, seeking tourist experiences that meet his needs and motivations (Turismo de Portugal, 2016).

Thus, it is important to mention that the research contemplates two of the three levels of the new marketing strategy recommended for Destination Portugal (Turismo de Portugal, 2016) namely (i) traffic generation for a digital platform that integrates contents related to the national offer. Communication content should be based on the one hand, on distinctive aspects of the Portuguese tourism offer, in this case the material and immaterial culture of the Douro. On the other hand, they should explore narratives about distinctive, authentic Portuguese products with international relevance and prestige such as wine and olive oil, for example, (ii) presentation of tourist experiences according to the motivation and interests of demand, facilitating the formatting of experiences according to a territorial perspective. The dynamics of content production at this level should be ensured by the Regional Destinations, which should develop content with the specific goal of converting visitors into leads.

However, the third level can also be present, as it is the (iii) materialization of the sale through a transactional platform that aggregates the product offer of the companies on a regional basis. In this sense, the game can stimulate the purchase of typical regional products.

This gamification proposal associated with storytelling can enhance the relationship with people and privilege the quality of the experience through a product based on narratives and content, given the differentiating elements of the destination. History and wine have bequeathed to 'Peso da Régua' its natural condition of Capital of the Douro Demarcated Region. In this sense, and beyond the vineyards, its landscapes, the river, the Douro is also its people, its history. What the river started, man completed: both working together, made the region (Barreto, 2018). The world of the vineyard is a world of stories. Just as we taste the wine, the gastronomy, the landscape, we taste stories and each story, each one with its plot has its characters, with which we identify, surprise, or learn. 'Peso da Régua' and the Douro Region are a narrative that can be told through the game.

Acknowledgements The authors acknowledge the financial support of CiTUR, R&D unit funded by the FCT—Portuguese Foundation for the Development of Science and Technology, Ministry of Science, Technology and Higher Education, under the scope of the project UID/BP/04470/2020.

References

Araújo, N., Gomes, R. A., & Cardoso, L. (2019). La Gamificación aplicada al Sector Turístico: Análisis sistemático sobre la base de datos Scopus. *International Journal of Marketing, Communication and New Media, 7*(12), 5–23.

Barreto, A. (2018). Douro, Rio., Gente e Vinho. (2nd Ed.) Relógio D'Água Editores.

Basaraba, N., Conlan, O., Edmond, J., & Arnds, P. (2019). Digital narrative conventions in heritage trail mobile apps. *New Review of Hypermedia and Multimedia, 25*, 1–30.

Bassano, C., Barile, S., Piciocchi, P., Spohrer, J., Iandolo, F., & Fisk, R. (2019). Storytelling about places: Tourism marketing in the digital age. *Cities, 87*, 10–20.

Bryon, J. (2012). Tour guides as storytellers—From selling to sharing. *Scandinavian Journal of Hospitality and Tourism, 12*(1), 27–43.

Buhalis, D., & Foerste, M. (2015). SoCoMo marketing for travel and tourism: Empowering co-creation of value. *Journal of Destination Marketing & Management, 4*(3), 151–161.

Buhalis, D., & Law, R. (2008). Progress in information technology and tourism management: 20 years on and 10 years after the Internet—The state of eTourism research. *Tourism Management, 29*, 609–623.

Burke, B. (2014). Gamify: How gamification motivates people to do extraordinary things. ISBN 978-1-937134-86-0 (ebook).

Chandralal, L., Rindfleish, J., & Fredy, V. (2015). An application of travel blog narratives to explore memorable tourism experiences. *Asia Pacific Journal of Tourism Research, 20*(6), 680–693. https://doi.org/10.1080/10941665.2014.925944

Cheong, C., Filippou, J., & Cheong, F. (2014). Towards the gamification of learning: Investigating student perceptions of game elements. *Journal of Information Systems Education, 25*(3), 233–245.

Chi-Hua, L., & Chi-Chun, L. (2020). The effects of empathy and persuasion of storytelling via tourism micro-movies on travel willingness. *Asia Pacific Journal of Tourism Research, 25*(4), 382–392. https://doi.org/10.1080/10941665.2020.1712443

CIM DOURO. (2015). Estratégia Integrada de Desenvolvimento Territorial da Região do Douro (2014–2020). Disponível em https://www.norte2020.pt/sites/default/files/public/uploads/programa/EIDT-99-2014-01-020_Douro.pdf. Consultado a 05 de maio de 2020.

Coutinho, C. P. (2018). *Metodologia de Investigação em Ciências Sociais e Humanas—Teoria e Prática* (2ª). Almedina.

FCT. (2019). Agenda Temática de Investigação e Inovação Turismo, Lazer e Hospitalidade, FCT. https://doi.org/10.34621/fct.edicoes.agendastematicas-2

Future Foundation/Amadeus Traveller Trend Observatory (ATTO). (2015). Future traveler tribes 2030. Understanding tomorrow traveller. Disponível em https://amadeus.com/documents/en/retail-travel-agencies/research-report/amadeus-future-traveller-tribes-2030-report.pdf

Gonçalves, J. C, Seabra, C., & Silva, C. (2018). Histórias de cultura. O poder do Storytelling em destinos de Turismo Cultural. *Cadernos de Geografia, 37*, 113–120.

Gravili, S., Rosato, P., & Iazzi, A. (2017). Managing DMOs through storytelling: A model proposal for network and value co-creation in tourism. *International Business Research, 10*, 8–21.

Hall, C., & Sharples, L. (2003). The consumption of experiences or the experience of consumption? An introduction to the tourism of taste in food tourism around the world (pp. 1–24). Elsevier Butterworth-Heinemann.

Hannam, K., & Ryan, E. (2019). Time, authenticity and photographic storytelling in The Museum of Innocence. *Journal of Heritage Tourism, 14*(5–6), 436–447. https://doi.org/10.1080/1743873X.2019.1622707

Hsu, J. (2008). *The secrets of storytelling—Our love for telling tales reveals the working of mind* (pp. 46/61). Scientific American Mind.

Hunicke, R., Leblanc, M., & Zubek, R. (2004). MDA: A formal approach to game design and game research. *AAAI Workshop—Technical Report* (No. 1).

Koivisto, J., & Hamari, J. (2014). Demographic differences in perceived benefits from gamification. *Computers in Human Behavior, 35*, 179–188. https://doi.org/10.1016/j.chb.2014.03.007

Liberato, P., Alén-González, E., & Liberato, D. (2018). Digital technology in a smart tourist destination: The case of Porto. *Journal of Urban Technology, 25*(1), 75–97. https://doi.org/10.1080/10630732.2017.1413228

Line, M. (2019). Storytelling: A way for winter adventure guides to manage emotional labour. *Scandinavian Journal of Hospitality and Tourism, 19*(1), 66–81. https://doi.org/10.1080/15022250.2017.1411827

Lopes, A., Silva, C., Seabra, C., & Abrantes, J. L. (2018). Wine routes: Development of wine regions and local communities. *International Journal of Multidisciplinarity in Business and Science, 4*(5), 39–44 https://hrcak.srce.hr/205405

Lukosch, S., Klebl, M., & Buttler, T. (2011). Utilizing verbally told stories for informal knowledge management. *Group Decision and Negotiation, 20*, 615–642. https://doi.org/10.1007/s10726-011-9237-7

Martins, H. (2004). Metodologia Qualitativa de Pesquisa. *Educação e Pesquisa, 30*(2), 289–300. https://doi.org/10.1590/S1517-97022004000200007

Martins, U., Batista, M., & Carlos, C. (2012). A gastronomia como produto turístico-cultural: Uma análise das boas práticas de destinos internacionais. *Revista Turismo & Desenvolvimento, 17/18*, 52–64.

Mccabe, S., & Foster, C. (2006). The role and function of narrative in tourist interaction. *Journal of Tourism and Cultural Change, 4*, 194–215. https://doi.org/10.2167/jtcc071.0

Mei, X. Y., Hågensen, A.-M.S., & Kristiansen, H. S. (2020). Storytelling through experience scape: Creating unique stories and extraordinary experiences in farm tourism. *Tourism and Hospitality Research, 20*(1), 93–104. https://doi.org/10.1177/1467358418813410

Minasse, M. H. S. G. G. (2020). Food tourism as research object: Analysis of publications in Brazilian journals (2005–2017). *Revista Brasileira de Pesquisa em Turismo, 14*(1), 92–111. Epub March 23, 2020. https://doi.org/10.7784/rbtur.v14i1.1669

Mora, P., & Livat, F. (2013). Does storytelling add value to fine Bordeaux wines? *Wine Economics and Policy, 2*(1), 3–10. ISSN 2212-9774. https://doi.org/10.1016/j.wep.2013.01.001

Negruşa, A., Toader, V., Sofică, A., Tutunea, M., & Rus, R. (2015). Exploring gamification techniques and applications for sustainable tourism. *Sustainability, 7*, 11160–11189.

Pereira, P., Dantandreu, R., & Nascimento, J. C. (2001). *Portugal, Património Mundial—Objetos singulars*. Objetos Universais.

Pine, B. J., II., & Gilmore, J. H. (1998). Welcome to the experience economy. *Harvard Business Review, 76*(4), 96–104.

Ramos, C. (2008). A integração dos sistemas de informação e do turismo, o caso IMPACTUR.

Robson, K., Plangger, K., Kietzmann, J., McCarthy, I., & Pitt, L. (2015). Is it all a game? Understanding the principles of gamification. *Business Horizons*. https://doi.org/10.1016/j.bushor.2015.03.006

Schell, J. (2012). The art of game design: A book of lenses. CRC Press. http://www.sg4adults.eu/files/art-game-design.pdf

Turismo de Portugal. (2016). Turismo 2020: Plano de Ação para o desenvolvimento do Turismo em Portugal. https://travelbi.turismodeportugal.pt/ptpt/Paginas/turismo-2020-plano-de-acao-para-o-desenvolvimento-do-turismo-em-portugal.aspx

UNWTO. (2016). Red de Gastronomía de la OMT, Plan de Acción, 2016/2017, UNWTO, Madrid. Disponível em http://cf.cdn.unwto.org/sites/all/files/docpdf/gastronomyactionplanprint2esweb.pdf

Vaz, A. I. I. G. S. L. (2008). O enoturismo em Portugal: Da "Cultura" do vinho ao vinho como cultura. PhD thesis, Universidade de Lisboa, Portugal.

Weber, J. (2014). Game and gamification in tourism. *Digital Tourism Think Thank*. file:///C:/Users/formador/Downloads/Gamification-in-Tourism-Best-Practice%20.pdf

Wichels, S. (2013). A experiência sensorial na comunicação turística, Imprensa da Universidade de Coimbra. ISSN 0870-4112. https://doi.org/10.14195/0870-4112_11_20

World Travel Market (2011). Industry report. https://www.comunicatur.info/wpcontent/uploads/2011/11/onsite_wtm_industry_report_2011.pdf

Xu, F., Tian, F., Buhalis, D., & Weber, J. (2013). Marketing tourism via electronic games: Understanding the motivation of tourist players. In *2013 5th International Conference on Games and Virtual Worlds for Serious Applications (VS-GAMES)* (pp.1–8), Poole. https://doi.org/10.1109/VS-GAMES.2013.6624235

Xu, F., Tian, F., Buhalis, D., Weber, J., & Zhang, H. (2016). Tourists as mobile gamers: Gamification for tourism marketing. *Journal of Travel and Tourism Marketing, 33*(8), 1124–1142 https://doi.org/10.1080/10548408.2015.1093999

Xu, F., Buhalis, D., & Weber, J. (2017). Serious games and the gamification of tourism. *Tourism Management, 60*, 244–256. https://doi.org/10.1016/j.tourman.2016.11.020

Zichermann, G., & Cunningham, C. (2011). *Gamification by design: Implementing game mechanics in web and mobile apps*. O'Reilly Media.

Tourism Destinations, Hospitality and Diversification

A Review of Destination Personality Literature: Focus on Articles Published in the Last Decade (2013–2022)

Athina Nella

Abstract The purpose of this paper is to provide a review of the most recent literature in the field of destination personality. As the importance of symbolic attributes in tourist decision making is increasing destination personality (DP) or destination brand personality (DBP), becomes a key construct in destination branding. DP has been linked with destination image, self-congruence and behavioral intentions, with a positive relationship being supported in most cases. Due to increased research interest in the field, an updated literature review of destination personality literature was considered as valuable. The review was performed with the PRISMA protocol and focused on articles published in high-impact journals during the last decade; these were analyzed in terms of methodologies and measurement scales used, destinations and major findings. Results showed that quantitative methodologies prevailed. Relevant research is not restricted to specific geographical regions but comes from various destinations all over the world. Aaker (Journal of Marketing Research 34:347–356, 1997) 5-dimensional brand personality scale (BPS) consists a reference point in the process of developing DP measurement scales. There is no consensus on the use of a single DP measurement scale, and most researchers acknowledge the need of DP scales to be destination-specific, culture-specific or even tourism-product specific. The most commonly used dimensions were sincerity, excitement, sophistication, conviviality, competence and vibrancy; the number of measurement items ranged from 7 to 26 while the average sample in quantitative studies was 480 respondents. More research in the field with the use of mixed methods is expected to portray better the evolving nature of destination personalities.

Keywords Destination personality · Brand personality · Destination branding

JEL Classification L83

A. Nella (✉)
Hellenic Open University, Patras, Greece
e-mail: nella.athina@ac.eap.gr

© The Author(s), under exclusive license to Springer Nature Switzerland AG 2023

V. Katsoni (ed.), *Tourism, Travel, and Hospitality in a Smart and Sustainable World*, Springer Proceedings in Business and Economics,
https://doi.org/10.1007/978-3-031-26829-8_28

1 Introduction: Brand Personality and Destination Brand Personality

Although brand personality (BP) has been acknowledged as an important branding and differentiation tool, until recently there were limited studies referring to destination BP (Bekk et al., 2016; Kumar & Nayak, 2014). Though, during the last decade, significant contributions have been made. This contribution aims to review destination personality literature published in the last decade (2013–2022).

Patterson (1999) has identified more than 15 different definitions of brand personality (BP); he defined brand personality as consumers' emotional response to a brand, through which brand attributes are personified and used to differentiate between alternative offerings. Although brand personality has been widely studied for general consumer goods, it seems not to have been adequately studied for tourism and travel products (Kim & Lehto, 2013).

Jennifer Aaker's (1997) study formed the basis for measuring brand personality in various product categories and settings. Based on the theory of the Big Five Model, Aaker provided a rigorous measurement instrument consisting of five dimensions (sincerity, excitement, sophistication, ruggedness and competence) and 42 personality traits (brand personality scale-BPS). Though, Hankinson (2001) considers branding applications in the context of destinations more challenging, due to the complex nature of the tourism product. Extending Aaker's (1997) definition of brand personality to the tourism context, Ekinci and Hosany (2006) defined destination personality (DP) as the set of human characteristics associated with a destination.

2 Literature Review

One of the earliest studies focusing on destination personality comes from Ekinci and Hosany (2006), who identified three dimensions of DP, namely sincerity, excitement and conviviality. The authors also concluded that DP has a positive impact on perceived destination image and intention to recommend the destination. Murphy et al. (2007) suggested that tourists can articulate different destination brand personalities for each region. According to Usakli and Baloglu (2011), self-congruity is a partial mediator on the relationship between destination personality and tourist's behavioral intentions.

Recently, destination personality has been studied in relation to various tourism marketing concepts, such as destination branding (Apostolopoulou & Papadimitriou, 2015; Hultman et al., 2017), destination image (Chen & Phou, 2013; Souiden et al., 2017; Xie & Lee, 2013), destination brand equity (Salehzadeh et al., 2016), positioning (Hanna & Rowley, 2019), involvement (Liu et al., 2021), destination attachment (Huang et al., 2017) and behavioral intentions (Baloglu et al., 2014; Khazaei

Pool et al., 2018). Another interesting aspect of the relevant literature refers to the relationship of destination personality with self-congruity (Usakli & Baloglu, 2011), self-identity or tourist-destination identification (Huaman-Ramirez et al., 2021; Hultman et al., 2015; Zhang et al., 2022).

Leung and Law (2010) suggested that prior DP studies can be grouped into 7 dimensions: disposal, biological, intrapsychic, cognitive, social, adjustment in personality research and plus brand personality. Moreover, Usakli and Baloglu (2011) provided a comparative analysis of DP studies published in the period 2006–2009. Later on, Kumar and Nayak (2014) summarized DP studies of the period 2006–2014. A comparison of 15 DP studies is also portrayed in the study of Dickinger and Lalicic (2016) also provided a presentation of 15 DP studies. The most common classification criteria in the above mentioned literature reviews included methodology, geographical context of the study, sample, year and journal of publication. An updated DP literature review would add more to existing knowledge, given the increasing research interest in the field.

3 Methodology

In order to fulfill the aim of this contribution, the basic principles and steps of the PRISMA protocol (Page et al., 2022) were followed. The PRISMA protocol has been recently applied in other tourism studies (e.g., Papavasileiou et al., 2022; Ram, 2021). The initial step was to define clearly the research objective of this literature review, i.e., to provide an updated mapping of scientific papers focusing strictly on destination (brand) personality.

After defining the primary purpose of this literature review, definition of eligibility (inclusion and exclusion) criteria followed. Inclusion criteria included (a) strict primary focus on the concept of destination personality and not on similar, broader or linked concepts and (b) research articles published in high-impact scientific journals in the fields of tourism or marketing during the period 2013–2022. Exclusion criteria included (a) publications in books, reports, conference proceedings and other sources of scientific knowledge and (b) scientific articles not written in English. Information sources, search strategy, selection process as well as synthesis methods were also determined, according to the PRISMA protocol.

For the purpose of this chapter and in order to identify articles relevant to destination brand personality, Google Scholar was used as the primary data base. The search terms used were “destination” AND “personality”, so as to avoid results with studies with broader or alternative focus. Google Scholar was the primary information source for the identification of DP literature. Produced search results exceeded 130 for the period 2006–2022. The lists of Q1 or Q2 journals in subject areas “tourism, leisure and hospitality management” and “marketing” of Scimago Journal & Country Rank were chosen as a filter for high-impact journals.

Thus, 43 contributions published in Q1 or Q2 tourism and marketing journals were included to form the basis of this analysis; all of the selected articles were available in English and published in the period 2013–2022, thus confirming that destination personality is a field mainly formed in the last 10 years.

4 Results

Based on predefined synthesis methods, these 43 scientific papers were classified and analyzed in terms of key findings, destination(s) under focus, journal and year of publication, methodology and sample (where applicable). Our analysis started with a classification per journal and year of publication. In order to provide an overview of research interest in the field, the total number of published articles in Q1 and Q2 Scimago-listed journals was examined initially per year (Fig. 1). The average number of published studies per year (4.3) during the last decade is substantially increased in comparison with the previous period of 2006–2012, where the average number of published articles per year was 1.4. These 43 published articles were published in 24 tourism and marketing journals of the Q1 and Q2 Scimago list (Table 1).

The next step of our analysis included a classification of articles per geographical focus and research method (Table 2). Starting with the geographical distribution of DP papers, research seems not to be entirely focused in a limited number of regions but comes from various destinations all over the world. The most popular destinations under examination in DP studies were China, UK, Australia, South Korea and Italy. Comparative studies were also identified in the DP literature (e.g., De Moya & Jain, 2013; Hanna & Rowley, 2019; Vinyals-Mirabent et al., 2019).

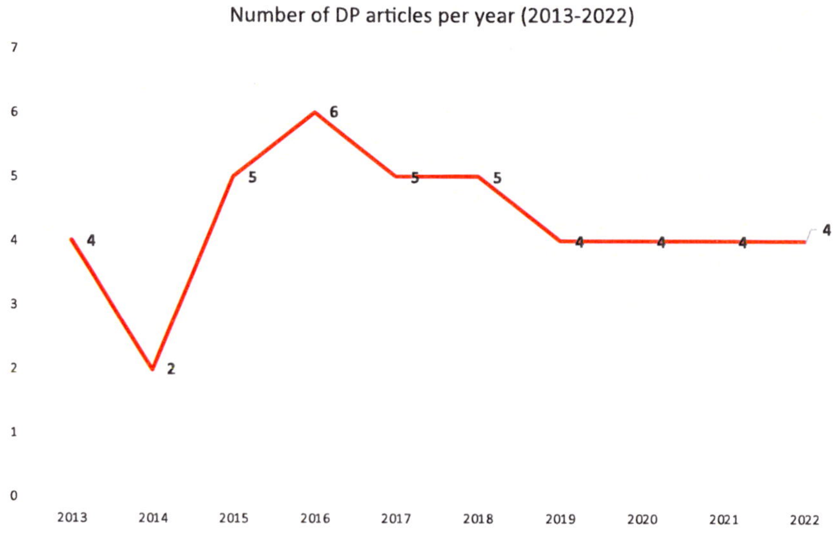

Fig. 1 Number of DP articles per year of publication (2013–2022)

Table 1 Number of DP articles per journal

Q1/Q2 Scimago-listed journal	Number of articles	Year(s) of publication
Tourism Management	**6**	**2013, 2016, 2017, 2019, 2019, 2021**
Journal of Travel and Tourism Marketing	4	2013, 2014, 2017, 2018
Journal of Destination Marketing and Management	5	2015, 2017, 2018, 2021, 2022
Journal of Travel Research	3	2013, 2015, 2016
International Journal of Tourism Cities	2	2020, 2022
Information Technology and Tourism	2	2016, 2019
Tourism Management Perspectives	2	2014, 2016
Journal of Strategic Marketing	2	2020, 2021
Journal of Hospitality and Tourism Management	2	2017, 2020
Tourism Analysis	1	2018
International Journal of Tourism Research	1	2016
Tourism Review	1	2016
Annals of Tourism Research	**1**	**2022**
Journal of Vacation Marketing	1	2021
Cities	1	2022
Journal of Marketing Management	1	2019
Journal of Hospitality and Tourism Research	1	2018
Psychology and Marketing	1	2017
Current Issues in Tourism	1	2015
Asia Pacific Journal of Tourism Research	1	2015
Public Relations Review	1	2013
Journal of Business Research	1	2015
Tourism and Hospitality Research	1	2018
Journal of International Consumer Marketing	1	2020

The majority of studies (over 70%) used quantitative methods, while approximately 10% used qualitative techniques, 7% used mixed methods and an even lower percentage was based on literature reviews. In studies employing quantitative techniques, some of the most popular methods included exploratory and confirmatory factor analyses, regression models and structural equation modeling. In studies

Table 2 DP studies of the period 2013–2022, destinations examined, methods and sample used

	Authors	Destinations	Methods used	Sample
1	Chen and Phou (2013)	Cambodia-Multinational sample	EFA, CFA, structural model	428
2	Xie and Lee (2013)	Beijing, China	EFA, CFA, structural model	500
3	Kim and Lehto (2013)	South Korea, US sample	CATPAC for website analysis, EFA, CFA	480
4	De Moya and Jain (2013)	Mexico, Brazil—Facebook content analysis	Computer-aided content analysis	–
5	Kumar and Nayak (2014)	Literature review	Literature review	–
6	Baloglu et al. (2014)	Jamaica	Subgroup analysis, regression analysis	312
7	Guiry and Vequist (2015)	South Korea, US sample for medical tourism	Factor analysis, t-tests, logistic regression analysis	1588
8	Pereira et al. (2015)	Portugal, golf destination	Qualitative techniques	46
9	Hultman et al. (2015)	Most recently visited destination-Taiwan sample	EFA, CFA, structural model	490
10	Papadimitriou et al. (2015)	Greece-Greek sample	EFA, CFA, structural model	361
11	Apostolopoulou and Papadimitriou (2015)	Greece-Greek sample	Principal component factor analyses, ANOVA, regression analyses	568
12	Gomez Aguilar et al. (2016)	UK visitors to 2 specific Spanish destinations	CFA, reliability analysis, model fit	329
13	Dickinger and Lalicic (2016)	Vienna—international visitors	Survey findings comparing to social media reviews, text-mining approach	599
14	Bekk et al. (2016)	Spain, Fuerteventura	Principal component factor analysis, CFA, multiple regression analysis, model fit	308
15	Kumar (2016)	India	EFA, CFA, SEM	356
16	Salehzadeh et al. (2016)	Pool, Iran	Reliability analysis, CFAs, SEM	367
17	Matzler et al. (2016)	Slovakia	PLS-SEM	2.116
18	Huang et al. (2017)	China—Yangshuo	EFA, CFA, SEM	337

(continued)

Table 2 (continued)

	Authors	Destinations	Methods used	Sample
19	Souiden et al. (2017)	Canadians concerning Dubai	EFA, CFA, SEM	173
20	Kim and Stepchenkova (2017)	Jeju Island, South Korea	Content analysis—blogs	–
21	Pan et al. (2017)	Mainland Chinese travelers	Content analysis—reviews, EFA, CFA, multiple regression	394
22	Hultman et al. (2017)	Switzerland, UK airport visitors	CFA, ERLS	261
23	Kumar and Nayak (2018)	India, foreigners	Principal component analysis	329
24	Chi et al. (2018)	Sardinia, non-resident tourists (both international and domestic)	CFA, SEM	1266
25	Khazaei Pool et al. (2018)	Iran, foreign tourists who chose a historical-Islamic destination	CFA, SEM	397
26	Jeon et al. (2018)	US, Italy and Peru websites, US citizens	CFA, SEM	433
27	Xu and Tavitiyaman (2018)	Japan (Hong Kong tourists)	Quantitative methods	–
28	Vinyals-Mirabent et al. (2019)	Content analysis of websites, 12 European urban destinations	Content analysis—websites	12 destinations
29	Sertkan et al. (2019)	350 destinations	Regression analysis, seven-factor model—recommender systems	350 destinations
30	Zhang et al. (2019)	Urban destinations	Data analysis tool, machine learning, algorithm	672
31	Hanna and Rowley (2019)	10 European capital cities—website content	Content analysis—websites	–
32	Hsu and Scott (2020)	Macao	PLS-SEM	450
33	Šagovnović and Kovačić (2020)	Novi-Sad, Serbia	Factor analysis, ANOVA	203
34	Sharifsamet et al. (2020)	London, UK	CFA, bootstrapping analysis	515
35	Yang et al. (2020)	Glasgow, UK (Chinese sample)	PLS-SEM	226

(continued)

Table 2 (continued)

	Authors	Destinations	Methods used	Sample
36	González-Rodríguez et al. (2021)	China, Spain	PLS-SEM	1.132
37	Huaman-Ramirez et al. (2021)	Aix-en-Provence, France	PLS-SEM	212
38	Pan et al. (2021)	Chinese and U.S. sample	CFA, SEM	1.092
39	Liu et al. (2021)	Tibetan pilgrimage destination	EFA, CFA, SEM	348
40	Yang et al. (2022)	Auckland and Glasgow	PLS-SEM, PLS–Henseler's multigroup analysis	544
41	Wang et al. (2022)	China	Neuroscience experiment	32
42	Pasquinelli et al. (2022)	4 Italian cultural destinations	Content analysis	–
43	Auemsuvarn et al. (2022)	Thailand	EFA, CFA	411

employing quantitative techniques or mixed methods, the average sample was 480 subjects.

The issue of measuring destination personality and identifying the most suitable dimensions and traits was the next focus of the analysis. One of the most critical issue in DP studies was how to measure this complex construct; various scales and BPS dimensions were used for this purpose. In most of the initial studies of the field Aaker's (1997), 5 dimensions and 42 traits of BPS formed the basis of measurement scales for DP. Though, a significant number of other researchers tried to test and use the BPS in respect with tourism destinations. Some of the authors tried to use the BPS in its original form as a starting-point, while others upon judgment eliminated items/traits of the BPS not relating to the specific destination. This explains the great variation in terms of DP measurement items used in various DP studies (9–42 items of the BPS). Other researchers used mixed methods and a multi-stage identification of traits most relevant to the specific destination or tourism product (e.g., open questions, interviews with experts, content analysis, use of previous DP scales already validated for the specific culture or destination).

The majority of studies identified DP dimensions ranged from 3 to 5, but there were also measures ranging from 2 to 7 dimensions. Some of the most commonly identified measurement dimensions identified across geographical contexts were the following: sincerity, competence, excitement, sophistication and ruggedness. Though a great variety of other dimensions were identified for various destinations (e.g., masculinity, sacredness, conformity, liveliness, family orientation, activeness, dependability, philoxenia, tranquility, viciousness, etc.), as shown in Table 3.

Relevant literature shows that dimensions of destination personality may vary according to the destination under examination. Some authors stress the importance

Table 3 Identified DP measurement dimensions and traits/items

Authors	Destinations	DP dimensions finally proposed and items included in models
Chen and Phou (2013)	Cambodia-Multinational sample	Sincerity, excitement, sophistication, ruggedness, contemporary (5 dimensions, 18 items)
Xie and Lee (2013)	Beijing, China	Sincerity, competence, excitement, sophistication, ruggedness (5 dimensions, 13 items)
Kim and Lehto (2013)	South Korea, US sample	Excitement, competence, sincerity, sophistication, ruggedness, uniqueness and family orientation (7 dimensions, 39 items)
Guiry and Vequist (2015)	South Korea, US sample for medical tourism	Sincerity, competence, ruggedness (3 dimensions, 11 items)
Hultman et al. (2015)	Most recently visited destination-Taiwan sample	Excitement, sophistication, activeness, dependability, philoxenia, ruggedness (6 dimensions, 26 items)
Papadimitriou et al. (2015)	Greece-Greek sample	Sincerity, excitement (2 dimensions, 8 items)
Apostolopoulou and Papadimitriou (2015)	Greece-Greek sample	Sincerity, excitement (2 dimensions, 8 items)
Gomez Aguilar et al. (2016)	UK visitors to two specific Spanish destinations	Sincerity, competence, ruggedness, sophistication, excitement (5 dimensions, 26 items)
Dickinger and Lalicic (2016)	Vienna—international visitors	Sincerity, competence, ruggedness, sophistication, excitement (5 dimensions)
Bekk et al. (2016)	Spain, Fuerteventura	Sincerity, excitement and sophistication (3 dimensions, 15 items
Kumar (2016)	India	Well mannered, vibrancy, creativity, conformity, viciousness (5 dimensions, 23 items)
Matzler et al. (2016)	Slovakia	Responsibility, activity, aggressiveness, simplicity, emotionality (5 dimensions, 12 items)

(continued)

Table 3 (continued)

Authors	Destinations	DP dimensions finally proposed and items included in models
Huang et al. (2017)	China—Yangshuo	Excitement, competence, charming, outdoorsy (4 dimensions, 22 items)
Souiden et al. (2017)	Canadians concerning Dubai	Sentimental, competence (2 dimensions, 7 items)
Pan et al. (2017)	mainland Chinese travelers	Competence, sacredness, vibrancy, femininity, excitement (5 dimensions, 18 items)
Hultman et al. (2017)	Switzerland, UK airport visitors	Excitement, sophistication, activeness, dependability, philoxenia, ruggedness (6 dimensions, 24 items)
Kumar and Nayak (2018)	India, foreigners	Courteousness, vibrancy, conformity, liveliness, viciousness, tranquility (6 dimensions, 23 items)
Chi et al. (2018)	Sardinia, non-resident tourists (both international and domestic)	Conviviality, sophistication, vibrancy (3 dimensions, 9 items)
Zhang et al. (2019)	Urban destinations	Temperament, competence, mood, attitude (4 components, 32 items)
Šagovnović and Kovačić (2020)	National and international tourists in Novi-Sad, Serbia	Sincerity, competence, excitement, sophistication, ruggedness (5 dimensions, 24 items)
Sharifsamet et al. (2020)	London, UK	Competence and contemporary-feminine (2 dimensions, 8 items)
Pan et al. (2021)	Chinese and U.S. sample	25-item destination gender scale (DGS) is assembled with 4 sub-dimensions of masculinity (i.e., dominance, vigor, courage and competence) and 4 sub-dimensions of femininity (i.e., grace, softness, gorgeousness and kindheartedness)
Huaman-Ramirez et al. (2021)	Aix-en-Provence (southern France), French sample	Aaker (1997)'s 5 dimensions confirmed, 15-item-scale

of developing culture-specific, validated destination personality scales (Auemsuvarn & Ngamcharoenmongkol, 2022; Kumar, 2016; Pan et al., 2017), while others emphasize the importance of developing customized scales for each destination (e.g., Huang et al., 2017). As research in the field evolves, scholars seem to be less hesitant to develop, use and propose more customized scales and this can be seen as a sign of progress in the field.

5 Conclusions

Destination personality is a complex construct drawing increasing research interest during the last decade. The use of PRISMA protocol led us to the analysis of 43 scientific papers. The identified articles were comparatively presented in terms of main findings, measurement scales and dimensions of DP, methods and samples used, destinations, journal and year of publication. Aaker's (1997) study provided the initial basis for developing DP measurement scales, but recently scholars are less reluctant to develop culture-specific, destination-specific or even tourism-product-specific measurement scales. Despite variations in dimensions and measurement items/traits used, the DP literature is becoming richer and enhancing its contribution in destination branding. The links of DP with other marketing constructs such as destination image, self-congruity and behavioral intentions were also clearly shown. In most cases, quantitative research methods and tools were employed. This review confirms that DMO's and NTO's should invest on creating distinctive destination personalities, based on their overall tourism product and strategies. Market research provides appropriate qualitative and quantitative tools for the identification of perceived destination personality.

References

Aaker, J. L. (1997). Dimensions of brand personality. *Journal of Marketing Research, 34*(3), 347–356. https://doi.org/10.1177/002224379703400304

Apostolopoulou, A., & Papadimitriou, D. (2015). The role of destination personality in predicting tourist behaviour: Implications for branding mid-sized urban destinations. *Current Issues in Tourism, 18*(12), 1132–1151. https://doi.org/10.1080/13683500.2013.878319

Auemsuvarn, P., & Ngamcharoenmongkol, P. (2022). Destination personality: A dimensions analysis and a new scale development in Thailand. *International Journal of Tourism Cities*. https://doi.org/10.1108/IJTC-07-2021-0144

Baloglu, S., Henthorne, T. L., & Sahin, S. (2014). Destination image and brand personality of Jamaica: A model of tourist behavior. *Journal of Travel & Tourism Marketing, 31*(8), 1057–1070. https://doi.org/10.1080/10548408.2014.892468

Bekk, M., Spörrle, M., & Kruse, J. (2016). The benefits of similarity between tourist and destination personality. *Journal of Travel Research, 55*(8), 1008–1021. https://doi.org/10.1177/0047287515606813

Chen, C. F., & Phou, S. (2013). A closer look at destination: Image, personality, relationship and loyalty. *Tourism Management, 36*, 269–278. https://doi.org/10.1016/j.tourman.2012.11.015

Chi, C. G. Q., Pan, L., & Del Chiappa, G. (2018). Examining destination personality: Its antecedents and outcomes. *Journal of Destination Marketing & Management, 9*, 149–159. https://doi.org/10.1016/j.jdmm.2018.01.001

De Moya, M., & Jain, R. (2013). When tourists are your "friends": Exploring the brand personality of Mexico and Brazil on Facebook. *Public Relations Review, 39*(1), 23–29. https://doi.org/10.1016/j.pubrev.2012.09.004

Dickinger, A., & Lalicic, L. (2016). An analysis of destination brand personality and emotions: A comparison study. *Information Technology & Tourism, 15*(4), 317–340. https://doi.org/10.1007/s40558-015-0044-x

Ekinci, Y., & Hosany, S. (2006). Destination personality: An application of brand personality to tourism destinations. *Journal of Travel Research, 45*(2), 127–139. https://doi.org/10.1177/0047287506291603

Gomez Aguilar, A., Yaguee Guillen, M. J., & Villasenor Roman, N. (2016). Destination brand personality: An application to Spanish tourism. *International Journal of Tourism Research, 18*(3), 210–219. https://doi.org/10.1002/jtr.1997

Gonzalez-Rodriguez, M. R., Díaz-Fernandez, M. C., Bilgihan, A., Shi, F., & Okumus, F. (2021). UGC involvement, motivation and personality: Comparison between China and Spain. *Journal of Destination Marketing & Management, 19*, 100543

Guiry, M., & Vequist, D. G., IV. (2015). South Korea's medical tourism destination brand personality and the influence of personal values. *Asia Pacific Journal of Tourism Research, 20*(5), 563–584. https://doi.org/10.1080/10941665.2014.904804

Hankinson, G. (2001). Location branding: A study of the branding practices of 12 English cities. *Journal of Brand Management, 9*(2), 127–142. https://doi.org/10.1057/palgrave.bm.2540060

Hanna, S., & Rowley, J. (2019). The projected destination brand personalities of European capital cities and their positioning. *Journal of Marketing Management, 35*(11–12), 1135–1158. https://doi.org/10.1080/0267257X.2019.1647274

Hsu, F. C., & Scott, N. (2020). Food experience, place attachment, destination image and the role of food-related personality traits. *Journal of Hospitality and Tourism Management, 44*, 79–87. https://doi.org/10.1016/j.jhtm.2020.05.010

Huaman-Ramirez, R., Merunka, D., & Maaninou, N. (2021). Destination personality effects on tourists' attitude: The role of self-congruity and ambiguity tolerance. *Journal of Strategic Marketing*, 1–25. https://doi.org/10.1080/0965254X.2021.1874488

Huang, Z., Zhang, C., & Hu, J. (2017). Destination brand personality and destination brand attachment—The involvement of self-congruence. *Journal of Travel & Tourism Marketing, 34*(9), 1198–1210. https://doi.org/10.1080/10548408.2017.1330171

Hultman, M., Skarmeas, D., Oghazi, P., & Beheshti, H. M. (2015). Achieving tourist loyalty through destination personality, satisfaction, and identification. *Journal of Business Research, 68*(11), 2227–2231. https://doi.org/10.1016/j.jbusres.2015.06.002

Hultman, M., Strandberg, C., Oghazi, P., & Mostaghel, R. (2017). The role of destination personality fit in destination branding: Antecedents and outcomes. *Psychology & Marketing, 34*(12), 1073–1083. https://doi.org/10.1002/mar.21047

Jeon, H., Ok, C., & Choi, J. (2018). Destination marketing organization website visitors' flow experience: An application of Plog's model of personality. *Journal of Travel & Tourism Marketing, 35*(4), 397–409

Khazaei Pool, J., Khodadadi, M., & Asadi, A. (2018). The impact of congruence between self-concept and destination personality on behavioural intentions for visiting an Islamic-historical destination. *Tourism and Hospitality Research, 18*(3), 378–387. https://doi.org/10.1177/1467358416663820

Kim, H., & Stepchenkova, S. (2017). Understanding destination personality through visitors' experience: A cross-cultural perspective. *Journal of Destination Marketing & Management, 6*(4), 416–425. https://doi.org/10.1016/j.jdmm.2016.06.010

Kim, S., & Lehto, X. Y. (2013). Projected and perceived destination brand personalities: The case of South Korea. *Journal of Travel Research, 52*(1), 117–130. https://doi.org/10.1177/0047287512457259

Kumar, V. (2016). Examining the role of destination personality and self-congruity in predicting tourist behavior. *Tourism Management Perspectives, 20*, 217–227. https://doi.org/10.1016/j.tmp.2016.09.006

Kumar, V., & Nayak, J. K. (2014). The measurement & conceptualization of destination personality. *Tourism Management Perspectives, 12*, 88–93. https://doi.org/10.1016/j.tmp.2014.09.002

Kumar, V., & Nayak, J. K. (2018). Destination personality: Scale development and validation. *Journal of Hospitality & Tourism Research, 42*(1), 3–25. https://doi.org/10.1177/1096348014561027

Leung, R., & Law, R. (2010). A review of personality research in the tourism and hospitality context. *Journal of Travel & Tourism Marketing, 27*(5), 439–459. https://doi.org/10.1080/10548408.2010.499058

Liu, L., Zhang, X., & Cheng, L. (2021). Antecedents and impacts of the pilgrimage destination personality in the Tibetan region: Case from Larong Wuming Buddhist Academy. *Journal of Vacation Marketing*. https://doi.org/10.1177/13567667211070013

Matzler, K., Strobl, A., Stokburger-Sauer, N., Bobovnicky, A., & Bauer, F. (2016). Brand personality and culture: The role of cultural differences on the impact of brand personality perceptions on tourists' visit intentions. *Tourism Management, 52*, 507–520. https://doi.org/10.1016/j.tourman.2015.07.017

Murphy, L., Benckendorff, P., & Moscardo, G. (2007). Linking travel motivation, tourist self-image and destination brand personality. *Journal of Travel & Tourism Marketing, 22*(2), 45–59. https://doi.org/10.1300/J073v22n02_04

Page, M. J., Moher, D., & McKenzie, J. E. (2022). Introduction to PRISMA 2020 and implications for research synthesis methodologists. *Research Synthesis Methods, 13*(2), 156–163. https://doi.org/10.1002/jrsm.1535

Palmatier, R. W., Houston, M. B., & Hulland, J. (2018). Review articles: Purpose, process, and structure. *Journal of the Academy of Marketing Science, 46*(1), 1–5. https://doi.org/10.1007/s11747-017-0563-4

Pan, L., Lu, L., & Zhang, T. (2021). Destination gender: Scale development and cross-cultural validation. *Tourism Management, 83*, 104225. https://doi.org/10.1016/j.tourman.2020.104225

Pan, L., Zhang, M., Gursoy, D., & Lu, L. (2017). Development and validation of a destination personality scale for mainland Chinese travelers. *Tourism Management, 59*, 338–348. https://doi.org/10.1016/j.tourman.2016.08.005

Papadimitriou, D., Apostolopoulou, A., & Kaplanidou, K. (2015). Destination personality, affective image, and behavioral intentions in domestic urban tourism. *Journal of Travel Research, 54*(3), 302–315. https://doi.org/10.1177/0047287513516389

Papavasileiou, E., Parry, E., & Stergiou, D. (2022). Generations in the tourism workforce: A systematic literature review and meta-analytic insights. In *Academy of Management Proceedings* (Vol. 2022, No. 1, p. 16034). Academy of Management. https://doi.org/10.5465/AMBPP.2022.16034abstract

Pasquinelli, C., Trunfio, M., Bellini, N., & Rossi, S. (2022). Reimagining urban destinations: Adaptive and transformative city brand attributes and values in the pandemic crisis. *Cities, 124*, 103621. https://doi.org/10.1016/j.cities.2022.103621

Patterson, M. (1999). Re-appraising the concept of brand image. *Journal of Brand Management, 6*(6), 409–426. https://doi.org/10.1057/bm.1999.32

Pereira, R. L., Correia, A. H., & Schutz, R. L. (2015). Towards a taxonomy of a golf-destination brand personality: Insights from the Algarve golf industry. *Journal of Destination Marketing & Management, 4*(1), 57–67. https://doi.org/10.1016/j.jdmm.2014.12.003

Ram, Y. (2021). Metoo and tourism: A systematic review. *Current Issues in Tourism, 24*(3), 321–339. https://doi.org/10.1080/13683500.2019.1664423

Rojas-Méndez, J. I., Papadopoulos, N., & Alwan, M. (2015). Testing self-congruity theory in the context of nation brand personality. *Journal of Product & Brand Management, 24*(1), 18–27. https://doi.org/10.1108/JPBM-05-2014-0618

Šagovnović, I., & Kovačić, S. (2020). Influence of tourists' sociodemographic characteristics on their perception of destination personality and emotional experience of a city break destination. *International Journal of Tourism Cities, 7*(1), 200–223. https://doi.org/10.1108/IJTC-05-2020-0105

Salehzadeh, R., Pool, J. K., & Soleimani, S. (2016). Brand personality, brand equity and revisit intention: An empirical study of a tourist destination in Iran. *Tourism Review, 71*(3), 205–218. https://doi.org/10.1108/TR-02-2016-0005

Sertkan, M., Neidhardt, J., & Werthner, H. (2019). What is the "Personality" of a tourism destination? *Information Technology & Tourism, 21*(1), 105–133.

Sharifsamet, S., Jin, H. S., & Martin, B. (2020). Marketing destinations: The impact of destination personality on consumer attitude. *Journal of Strategic Marketing, 28*(1), 60–69. https://doi.org/10.1080/0965254X.2018.1485726

Souiden, N., Ladhari, R., & Chiadmi, N. E. (2017). Destination personality and destination image. *Journal of Hospitality and Tourism Management, 32*, 54–70. https://doi.org/10.1016/j.jhtm.2017.04.003

Usakli, A., & Baloglu, S. (2011). Brand personality of tourist destinations: An application of self-congruity theory. *Tourism Management, 32*(1), 114–127. https://doi.org/10.1016/j.tourman.2010.06.006

Vinyals-Mirabent, S., Kavaratzis, M., & Fernández-Cavia, J. (2019). The role of functional associations in building destination brand personality: When official websites do the talking. *Tourism Management, 75*, 148–155. https://doi.org/10.1016/j.tourman.2019.04.022

Wang, J., Li, M., Li, C., Li, D., & Lin, G. (2022). Revisit the formation of destination brand personality. *Annals of Tourism Research, 95*, 103408. https://doi.org/10.1016/j.annals.2022.103408

Xie, K. L., & Lee, J. S. (2013). Toward the perspective of cognitive destination image and destination personality: The case of Beijing. *Journal of Travel & Tourism Marketing, 30*(6), 538–556. https://doi.org/10.1080/10548408.2013.810993

Xu, J. B., & Tavitiyaman, P. (2018). How young tourists are motivated: The role of destination personality. *Tourism Analysis, 23*(1), 165–170. https://doi.org/10.3727/108354217X15143857878723

Yang, S., Isa, S. M., & Ramayah, T. (2022). Does uncertainty avoidance moderate the effect of self-congruity on revisit intention? A two-city (Auckland and Glasgow) investigation. *Journal of Destination Marketing & Management, 24*, 100703. https://doi.org/10.1016/j.jdmm.2022.100703

Yang, S., Isa, S. M., Ramayah, T., Blanes, R., & Kiumarsi, S. (2020). The effects of destination brand personality on Chinese tourists' revisit intention to Glasgow: An examination across gender. *Journal of International Consumer Marketing, 32*(5), 435–452. https://doi.org/10.1080/08961530.2020.1717400

Zhang, C., Huang, Z., Cao, F., & Chen, X. (2019). Recognise me from outside to inside: Learning the influence chain of urban destination personalities. *Tourism Management, 70*, 390–403. https://doi.org/10.1016/j.tourman.2018.09.005

Zhang, S., Kim, K., Yim, B. H., Hyun, B., & Chai, W. (2022). Destination personality and behavioral intention in Hainan's Golf tourism during COVID-19 pandemic: Mediating role of destination image and self-congruity. *Sustainability, 14*(11), 6528. https://doi.org/10.3390/su14116528

Agritourism: A Lever for the Development of the Greek Countryside. The Case of Nymphaeum of Florina/Greece

Konstantina Boulouta and Georgios Karagiannis

Abstract This paper attempts to map agritourism and outline its process that has been taken up today in Greece. It sets questions regarding agritourism development issues and their relationship with the area under investigation, which is Nymfaio in Florina. Initially, it refers to characteristics and activities that have taken place, in areas where agritourism is developing, as a tangible reality during the last seventy (70) years, for developed agritourism countries of Europe and a new experience of almost fifty (50) years, for the Greek tourist economy and society. It records the "new tourist reality" in Greece, detecting the peculiarities of Nymfaio, an agritourism destination, an area of exceptional importance with a preserved traditional settlement and also a preserved historical place, at an altitude of 1,346 m in Vitsi area. On the one hand, it is the area for entertainment and resting of tired people of the cities, who prefer to visit it, and on the other hand, it gives local people the possibility to "stimulate their rural income." The two aforementioned different starting points are the one, which can also specify the future of agritourism in the "Nymfaio of Florina" and obviously omens positively the agritourism's progress of development.

Keywords Agritourism development · Ecotourism · Rural tourism · Nymfaio destination · New tourist reality

JEL Classification Z32

1 Introduction

Agritourism in Greece is considered to have gone through a course for some decades—if we want to connect the beginnings of its development, with the first

---

K. Boulouta (✉)
Faculty of Food Sciences, University of West Attica, Campus Egaleo Parc, Athens, Greece
e-mail: boulouta@uniwa.gr

G. Karagiannis
Department of Statistics and Insurance Science, University of Piraeus, Piraeus, Greece

© The Author(s), under exclusive license to Springer Nature Switzerland AG 2023
V. Katsoni (ed.), *Tourism, Travel, and Hospitality in a Smart and Sustainable World*, Springer Proceedings in Business and Economics,
https://doi.org/10.1007/978-3-031-26829-8_29

agricultural collective efforts, in rural space. A course that raises many and reasonable questions such as what is the position that tourism holds internationally and what about its position/contribution to the Greek economy? What are some of the main characteristics of "agritourism"? Are there any other sectors which serve agritourism? What is the difference between "agritourism" and other alternative forms of tourism, and finally what is the dynamic of agritourism development in Nymfaio, Florina?

The answer to all these questions presupposes the mapping of tourist and agritourism area, providing the possibility to understand the essence and the content of "agritourism." Although it is not easy to codify many answers to the questions about "agritourism," we start by exploring the first concepts of tourism, since tourism is the industry in which "agritourism" was and will be based for the next decades. In fact, it seems that this concern is shared almost entirely by the country's tourism experts, and even by those who in the recent past expressed the fear that "Greeks may be transformed into waiters of Europeans."

The new Tourism Reality which has been created due to rapid global political developments, changes and incredible technological advances and their global spread and expansion during the last century, is the framework in which tourism moves globally. Alternative forms, such as agritourism, seem to be the future for tourism's survival. Strategies and tactics must be explored, policies must be adopted, so that agritourism in Greece will multiply the benefits from the tertiary and the primary sector of the economy, when it uses them. Our focus on "agritourism," as an alternative "form of tourism," is due to the fact that agritourism gives life again to the countryside and the rural area (Georgopoulos, 2004). Finally, we submit our scientific assessments and proposals within the area Nymfaio in Florina/Greece. We record the agritourism experiences by personal interviewing locals-residents of the area, while data and information related to Nymfaio have been selected by agritourism/ecotourism brochures, multifold brochures and historical books.

2 Tourism the Most Successful Industry. Forecasts by the World Tourism Organization (WTO)

Tourism, one of the world's largest industries, has a global character and a relatively short lifespan, since essentially as a local, national and international phenomenon, it made its appearance, for Greece, the first serious steps, from the 1950s. From then until today, we come across studies, researches, monographs and a rich articleography with countless published articles, and many book publications, which aim to map the field of tourism and outline the path it has been taken, in Greece and internationally (Polo, 2022).

Etymologically, the term comes from the English word "tour" which means trip or round (a correspondence also exists in the French language). The subdivisions of the term "tourism" are many and depend on the criterion used (e.g., the form of travel, the means of transport, the purpose of the tour on an individual or collective level, etc.) (Lytras et al., 1997).

However, apart from all these, tourism has something deeper and rougher, where it was able to open windows of hope, for a man's way out of the thick spiritual darkness of the era into the socioeconomic light. If this represents those who are responsible for the Greek tourism, it also creates a heavy debt of our own. And if, of course, no one knows what weight our moral standing has—beyond rhetoric and inflated words—toward tourism, what is certain is that it projects ideals, provides the most powerful expressive opportunities and possibilities, and promotion of our country, with such a high mandate. This should be the realization of our responsibility (Oikonomou & Karagiannis, 2022).

Tourism has emerged as a major economic activity worldwide. In recent decades, world tourism has grown faster. The World Tourism Organization (WTO) found itself in the unpleasant position of declaring a decrease, in the number of tourists for 2020, compared to the forecasts, due to coronavirus appearance. With an announcement made on 6.3.2020, it reviews its forecasts for the global number of tourists in 2020, estimating that it will decrease by 1–3% compared to 2019, due to COVID-19 crisis, against an initial forecast for an increase of 3–4% (as the Table 1 shows). This reduction had an effect of 30–50 billion dollars (reduction) compared to 2019. Table 1 shows development of world tourism under normal conditions, before the appearance of the corona virus.

Table 1 Tourism toward 2030

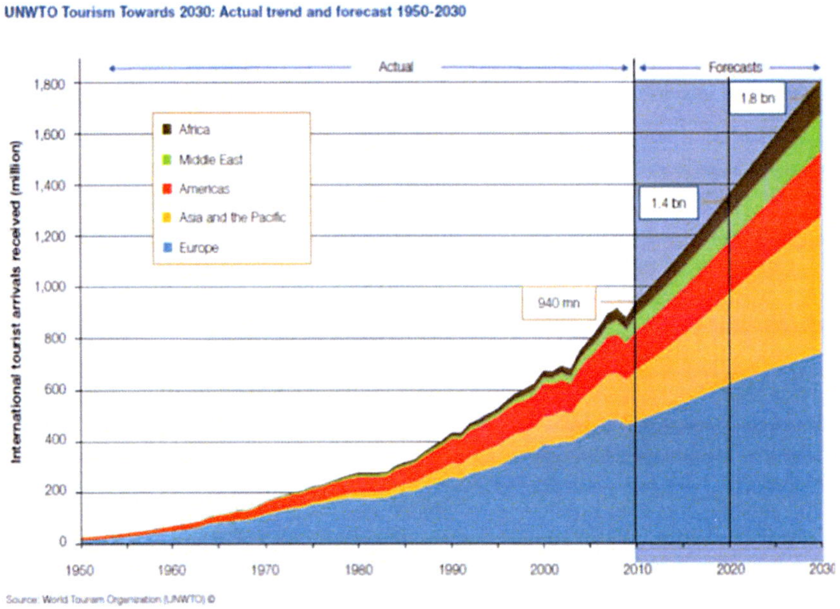

At the same time, one can conclude that the dynamism and the established upward trend, even in tourism, depend on unexpected factors considered as global events, such as the emergence of the coronavirus. That is why the WTO found itself in the unpleasant position of reversing all tourism indicators for 2020, in relation to the forecasts (Delitheou & Karagianni, 2020).

2.1 Conceptual Approach, Defining Tourism

Hunziker and Krapf gave an extensive conceptual approach to tourism, which includes travel, accommodation, permanent residents, the sum of relationships, etc. (Vanhove, 2019). On the other hand, there is the conceptual approach from another point of view, recording tourism to include all trips lasting more than 24 h, for reasons of entertainment, work, studies, and health. The typical concerns on this subject are

- the movement of people with temporary abandonment of their place of permanent residence and which is done for reasons of professional or psychosomatic satisfaction
- the individual or group movement of people to other places, for a short or longer period, for leisure purposes, without professional obligations and without the prospect of permanent settlement
- the movement of people to other places, causing in those places, a multitude of relationships and events
- the movement of people to other places, even for a long period of time, which should not exceed one year
- the movement of people, in addition to leisure or professional activity, which also includes activities related to various purchases of goods and cultural interactions
- the movement of people, can also have the character of an overnight stay beyond the permanent residence and for a place at a distance of more than 100 km, within the national borders of the same country (Karagiannis & Exarchos, 2006).

Considering the above reflections, we detect more than one bibliography where they mention definitions such as Today tourism is defined as an independent phenomenon, synonymous with consumption and free time. It presented a speed of development, a spatial spread and a social acceptance at a global level, which tends to be considered as a banal and integral part of modern everyday life (Varvaresos, 2000). On the contrary, Adrian Bull claims that it is neither a phenomenon, nor a simple set of disciplines. It is a human activity, including human behavior, use of resources and contacts and relationships with other people, economies and environments (Bull, 2002).

In general, the definition appears understandably as a complex, socioeconomic phenomenon, with various dimensions, which—beyond the economic—manifest as a set of effects, which are presented before, during and after a short- or long-term leisure trip, and it exerts psychological, social, ecological and political influences, both on the traveler himself and on the inhabitants of the area or the places he visits.

Therefore, for most of the scientists, tourism is a "global institution" that is governed by millions of interactions, has a history and is defined as a "thematic unit" of a society of people, who feel members of this institution as they are tourists–travelers (Karagiannis, 1997).

3 Alternative Forms of Tourism. Greek Agritourism Development

According to Professor Paul Bernecker from the Vienna University of Economics, the object of tourism is defined, in the sense of tourist exploitation, of the nature and culture of a landscape, with the aim of creating ideal conditions, successfully meeting the requirements of the authorities of tourism. Every place or every region, based on their natural and cultural aspects, can be the object of tourism exploitation. Factors that attract tourists are multiple, but appear quite uniform.

As the half of the planet's population lives in cities, the human need becomes more intense to spend time in nature. It is noted that 20% of international travel is motivated by the natural environment of the destination. Additionally, it has become a modern trend among all ages of visitors, the need for travel experiences "live like a local," to have experiential authentic experience, something that both agro- and ecotourism include in their main characteristics (Dologlou, 2017). The great importance of nature and the natural landscape in tourism must once again be an object of thought. The fact that nature and the natural landscape have the ability to provide the visitor with the conditions for rest and relaxation proves the enormous importance of nature in tourism (Bernecker, 1985). Then, the natural environment also includes the necessary conditions, which must be observed even as prohibitory provisions—to prevent intense urbanization, with the consequence of destroying the natural environment and reconstructing short-term effects within the natural conditions (flora-fauna—aesthetics of the landscape) for the—allegedly—pleasant stay of the tourist. However, in agritourism staying in the rural nature is subject to conditions, having as basic the demonstration by tourists of unlimited respect for the natural environment and therefore the sovereignty of the protection of natural environment (Kovos, 2021).

The characteristics of alternative forms of tourism set conditions and terms for tourist infrastructure and superstructure such as

- to have accommodation of all kinds and categories
- to have a developed transportation network and a network of modern means of transport
- to have sufficient means of entertainment, amusement, sports, feeding and buy food items ("shopping")
- to provide services by appropriately trained and educated staff in the "destination locations"

- to have a sense of security in relation to people and in relation to natural phenomena, etc. (Kokkosis et al., 2011). Greece is suitable for developing alternative forms of tourism, as each region has a rich natural wealth, architecture, historicity, classical and modern, unique and unique local products.

3.1 Ecotourism–Agritourism, Principles, Enterprises

Ecotourism highlights the exploitation and utilization of sensitive ecosystems, which are made up of soil-geomorphological formations (volcanic landscapes, canyons, etc.), wetlands with their rare fauna and flora, (Prespes which are not far away from Nymphaeum, a water paradise, with cloudy landscapes, sun on the snowy mountain tops, beech forests, reedbeds, with 260 species of birds, with 23 species of fish and 1,300 species of plants). The National Parks with rare fauna and flora (Valia Kalda with the brown bear, "National Park of Pindos" which is considered one of the greatest and most pristine Parks in Greece, with the aim to protect the rich flora and fauna). Also, the old centured unique trees with traditional settlements with rare folk-art architecture (Zagoria, villages of Pelion, Nymphaeum, etc.) and, in general, areas of particular natural beauty. Ecotourism ultimately emerged as a reaction against the negative consequences of massive tourism, which caused irreparable damage to the natural environment.

The development of ecotourism aims to • highlight and exploit rare habitats, • protect and utilize of habitats on a scientific basis, • maintain the ecological balance, • exploration of fauna and flora, • watching wild birds, • wandering in nature, • exploring virgin forests, • painless and simple routes in the habitats. Ecotourism provides to local residents an alternative form of economic activity, in return for the restrictions imposed by the protection regime on the areas. Some identify ecotourism with agritourism—not wrongly—since the dominant concern of agritourism is the promotion and exploitation of ecotourism (Fennell, 2001).

Agritourism emerged as a concept, shortly after the end of World War II, when some developed countries (USA and some Western European countries) began to direct some of their domestic tourists to countryside, to enjoy tourism products and services provided by the farmers, referring to their agricultural incomes of their rural areas, with their "agritourism units" or "agritourism enterprises." This concept initially started in Western Europe and later noted in Western Europe later moved to Canada, Australia, Israel and elsewhere. In Greece, ecotourism makes its appearance in the 1980s, marked by the country's entry into the European Economy Community (EEC). Finally, the term "agritourism" seems to take later another conceptual content such as (a) the reception and hospitality of tourists on a farm, where the guests participate in farmers everyday life and in agricultural activities and (b) the creation of tourist accommodation, either in the form of small touristic units (quest houses) or in the form of rented rooms in the rural settlements where bed and breakfast is offered (Papastavrou & Karameris, 1996).

Agritourism units and enterprises had and still have a family or cooperative character with the aim, among others to • increase farmers' income, • the development of the local economy and society, • promotion and preservation of the architectural and cultural heritage of the countryside, • promotion and preservation of the general local natural and man-made environment. The European experience and reality have so far proven that agritourism operates in a framework that is defined by the protection of the natural environment, respect for folk architecture and folk tradition, the revival of ritualistic old customs, unfailing human contact and others, from which benefit people of European megacities. It is affected by the socioeconomic characteristics of each destination (Broccardo et al., 2017). This form of vacation is particularly widespread in countries where the concept of farm is intertwined with agricultural holdings (Austria–Germany) and guests participate in the daily routine of farmers and agricultural activities (Lagos, 2005). The Greek experience certainly moves into the same context.

Agritourism differs from the general tourism, because it is an activity that takes place and is carried out exclusively in rural areas, is provided mainly by farmers to cities inhabitants and has the meaning of local tourism that is connected with the countryside aiming to increase the low agricultural income. It is connected, in harmony with the main occupation of rural people, it functions as a complement to their main occupation and moves outside and beyond the logic of competition in terms of the utilization and exploitation of natural resources and human resources. In fact, agritourism is considered to be a mild form, agricultural exploitation, that develops in rural communities, and gives special importance to the development of mountainous and disadvantaged areas.

3.1.1 Agritourism's Contribution to the Local Community

For developed tourism countries in Europe, agritourism is considered as the most important strategic choice for the future of the rural world, because it is the one that—by carrying out various activities—contributes to

- the retention and permanent settlement of the rural population in the countryside
- strengthening the social connection and social relations of the local population
- to create jobs for young farmers
- in the long-term economic and social development of mountainous and disadvantaged areas
- to preserve, safeguard, protect and highlight the traditional rural landscape (e.g., national forests, olive groves, habitats, etc.)
- to preserve, safeguard, protect and highlight the architectural and residential element (e.g., traditional settlements, monasteries, castles, towers, historical buildings, etc.)
- to the overall promotion of areas and "elements" of special natural beauty
- to the quality improvement, promotion and dissemination of the products of the agricultural household industry/crafts and handicrafts

- to the tourism promotion of unknown and new areas of the countryside
- to improve the standard of living of rural people and especially the rural world.

Being so popular on the international stage, agritourism mobilizes local agencies, who have found that, with the infrastructure projects in the provision of tourist services in the countryside, they support the rural population, stimulate the common interest of the local population and visitors and revitalize the countryside (Exarchos & Karagiannis, 2004).

4 Agritourism Development in "Nymphaeum"

The comparative advantages of Nymphaeum, compared to others for the development of agritourism, are many. We only mention the most important ones: geographical location, topography and the geographical relief, climatic conditions, huge and significant number of flora and fauna species, rich tradition of distinct local customs and traditions, rich variety of traditional agricultural products, multitude of handicrafts and wood carvings of the traditional settlement. All this gives unlimited possibilities for agritourism development in Nymphaeum and the wider area. Through agritourism today, a significant number of guesthouses have been created, as well as a series of other basic infrastructures and activities, such as, e.g., environmental education centers, trail networks, outdoor activities, which—indeed—shape a new landscape for rural tourism, which aims that the visitor gets to know, among other things, the agricultural activities, the local products, the typical traditional cuisine, the daily life of the inhabitants, the cultural elements of the area, the authentic characteristics of the man-made and natural environment. The development of agritourism requires a reorientation of tourism and tourist economy. State and citizens must have as a primary objective the ecological dimension in all their movements and interventions (Karagiannis, 2007).

Nymfaio is characterized as the "dream" village of Florina, as it is included among the ten most beautiful villages in Europe. It has been characterized as a "preservable traditional settlement" and a tour of the place, shows its unchanged image, but also its local identity. Nymphaeum is perched on Mount Vici, at an altitude of 1340 m. It is actually located at the southern end of the prefecture of Florina, at a distance of less than one hour from the city of Florina (iefimerida.gr & News Room 7/12/2021). The settlement of Nymphaeum, as a general presentation according to Mr. Lypitka Nik, is an impressive natural landscape, elegant, aristocratic, as a settlement. It is considered one of the best mountain destinations, since it has magnificent mansions with handmade furniture, fine wines and local flavors. Walking on its cobblestones, one ends up through paths, in the beech forests, and meets bears that live in the "Arktouros" Bear Protection Center, which is unique in Greece. Nymphaeum, a preserved traditional settlement and also a preserved historical place, was settled around 1385 by Vlachs, i.e., Latin-speaking native Greek Macedonians, who for

1,400 years, guarded the neighboring Egnatia Odos and then fled to the inaccessible mountains, after having fierce battles with the Ottoman invaders. According to Mr. Lypitkas Nik interview on 5/6/2022. Few residents of Nymphaeu assured us of the revision of the forecasts, regarding the tourist traffic in their settlement, in the last two years.

With the strategic goal of our paper, to outline all the qualifications for the agritourism development of Nymphaeum, we also talked with Boutaris Georgios for a number of years (Interview 7/6/2022). He himself agreed with the data recorded by us, drawn from secondary sources of information. With reference to the population data: In 2001, 415 inhabitants were enumerated against 240 in 1991 and 132 inhabitants in 2011. Regarding the settlement's infrastructure, there are adequate water and sewage networks, telecommunications networks, fixed wireless telecommunications and two mobile telephone antennas. It is a fact that approximately 10,000 sq. m. the roads are cobbled and on the ring road. There is a three-floor Museum of Silver and Goldsmithing that has been incorporated into Arcturos, the European brown bear protection center, also a Museum of History and Folklore, as an authentic copy of a traditional mansion with precious heirlooms. There is a community hostel of social functions and creation of new volunteers for the protection of Wild Nature and a medical clinic. There is the European brown bear conservation center under "Arcturos" where 15 brown bears live. Also there is a mountain center HAN Thessaloniki, which is a traditional settlement, in the forest, where 250 young people are trained every summer.

The modern Conference and Education Center of the Aristotle University of Thessaloniki and the Community is available for scientific and business conferences or other events. It has an amphitheater named Rigas Feraios, a round table room, classrooms and offices. There is the valuable Library. There is also an open-air car parking at the entrance of the settlement, a second car parking at the eastern exit of the village and a third one. There are basketball, volleyball and tennis courts. Also marked walking and riding trails in the forest. The Holy Church of Agios Nikolaos, a monumental three-bed wooden-roofed Basilica, was restored from its pedestals and brightly decorated. A very powerful and flexible machine clears all the village roads from heavy snow. Five books were published about Nymphaion. Three Nymphaeum exhibitions were held in the municipalities of Thessaloniki and Larisa, and 40,000 information leaflets were distributed. The stone-paved road Nymphaeum–Arktouros was built. Three Aqueducts in Nymphaeum were auctioned and put into operation. The Nymphaeum Visitable Wildlife and Recreation Park in 1,200 fenced acres of community forest with deer, roe deer and game, which will be gradually released, has been auctioned (Exarchos & Karagiannis, 2004).

4.1 The General-Specific Tourist Infrastructure of Nymphaeum

Nymphaeum receives 50,000 visitors and 50,000 schoolchildren from all over Greece every year. The European Union, in a pan-European competition, honored Nymphaeus with the European Distinction of Village Renaissance in 2000. In the general infrastructure, for the tourist development of an area, as for Nymphaeum, the following are included, among others: the road network, the electricity supply networks, water supply, telecommunications, the sewage network, the collection of waste and even the possibilities of hospitalization in "Bodosakio" of Ptolemais, a relatively nearby area. The special infrastructure includes the main and auxiliary accommodation (questhouse), catering businesses. So today, in the settlement of Nymphaeum, there are ten to eleven traditional guesthouses, and rentals, a capacity of around 300 beds in total! 80% of hostels operate with three stars and 20% with fewer stars. There are also five restaurants, four standard coffee bars and four traditional shops. Tourism operates throughout the year, so there is no seasonality effect, and mainly the best time periods for guesthouse owners are the three-day holidays, August, Christmas, Easter. The foreign customers come from Israel, Germany and Holland. Today's overnight prices range from 70 to 130 Euros. The information was obtained following an interview with Iatridis Giannis (Interview 7/6/2022). The verticalization of local production provides the possibility of selling local products of excellent quality and taste such as: the red pepper, the famous strawberry, dairy products such as "Florinella" cheese, jams, spoon sweets, local liqueur, local Trahanas and the noodles and wood carvings.

4.1.1 Highlights

At this critical time, perhaps more critical than many can imagine for Tourism, the bundle of agritourism offer should be improved as a modern production mechanism of a new component extremely interesting, since, of course, it will include factors such as The Natural, Historical Cultural–Ethnological, Morals and Customs, Cultural events, etc. Then follows the pricing policy of the offered tourist services, but mainly the profile of the agritourism area (idyllic, quiet, pleasant, cosmopolitan, social and attractive). Still the way of reception in the settlement and in the units providing hospitality, with the basic condition of the training of the staff, for the consolidation of the tourism policy. In addition, drastic measures must be taken to protect the natural and cultural environment, which will contribute decisively to its preservation, but at the same time to its utilization thanks to human existence, regardless of whether he has the status of a tourist or a resident of a tourist host country. However, an excessive and uncontrolled touristic development can cause pollution of the environment, disruption of the ecological balance, etc., and ultimately lead to a degradation of the quality of life rather than an upgrade of it in the area where it is attempted (Igoumenakis, 1997).

4.2 *Short "SWOT" Analysis*

A corrective development perspective is revealed through a "SWOT" analysis. In this way, the special characteristics of the area under investigation will be highlighted. It is assumed that after tabulating the data, a new action plan can be implemented (Delitheou et al., 2022).

"SWOT" analysis Nymphaeum

Strengths	Weaknesses
Recognition that Nymphaeum is one of the ten best villages in Europe	Nature: The difficult accessibility under bad weather conditions
Nature: The picturesqueness of the settlement	Services: Water supply problematic in the summer months
The settlement is the safest tourist shelter in the area	The non-administrative functional autonomy of the settlement
The fine local products and the local Culture	The difficulty of repairing damages when they occur since the Municipality is 15 km from Nymphaeum
Opportunities	**Threats**
Increase in the share of the agritourist market and to the wider area	Installation of wind turbines is prohibited
New prospects for the development of forms of tourism: walking–cycling	Low municipal financing of the municipality in short-term projects
Development of Conference—Seminar Tourism at the Nikeio School organized by the Nymphaeum Hoteliers Association	Reduced demand for jobs due to limited capacity in beds
Spa in the baths area if it joins a European development program	Increase in energy and gasoline prices decrease tourist demand and unsustainable businesses

5 Conclusion

Greece is a recognized destination worldwide and therefore with a diversification and enrichment of the offerings options as agritourism to the visitor which can be part of a targeted policies on what in addition we will recommend to visitors. The conditions for agritourism development, with a focus on Nymphaeum, are indicative "examples" of tourist places, from a "quick approach" to the whole issue. There are many agritourism sites in Greece, but in Nymphaeum there can be real agritourism development, after it is investigated in a systematic, thorough and convincing manner, and leaves no room for failure. If they really want to make rational use of qualitative agritourism, as a distinguishable alternative form of tourism, it will be a huge comparative advantage of the area over others, aptly called "abandoned countryside" with traditional societies and communities. In such a "plan" it must be evolved dynamically, making use of all the comparative advantages that Nymphaeum has (nature, traditions, culture, arts, archaeological, historical and architectural monuments), offering the rural population the potential to create new forms of supplementary occupations in the tourism sector and placing it on the Greek "alternative tourism" map.

Policymakers have to develop effective ecotourism programs, as a strategic tool, to incorporate visibility and successful promotion of any agritourism destination with respect to cultural and national identity. Municipality, region and local tourism agencies should play an important role where they have to take the marketing of this particular identity of each locality. Creating a website with tasteful content, including general information on the location and advertising of local production and processing plants, can make an effective contribution in this direction. Ecotourism businesses need to be profitable, to be sustainable and make good use of all the funds available in specific E.U. programs. Undoubtedly, we have to follow the new trends in the fastest growing international tourist market.

References

Bernecker, P. (1985). *Manual-introduce to tourism*, Notes for courses. Prof. Economical University of Wienn.

Broccardo, L., Culasso, F., & Truant, E. (2017). Unlocking value creation using an agritourism business model. *Sustainability* (Switzerland), *9*(9).

Bull, A. (2002). Touristic economy (p. 15). Publications Klidarithmos.

Dalgish, C. (2018). Is Ecotourism a 'magic Bullet' for sustainable development?—An anthropological essay. medium.com. Retrieved from https://medium.com/environmental-intelligence/is-ecotourism-a-magic-bullet-for-sustainable-development-an-anthropological-essay-c60c1062aaff

Delitheou, B., Karagianni, M., & Michalaki, E. (2022). Yachting, Review of Decentralization local Government and Regional Development (T. No 107), Athens, pp. 69–82.

Dologlou, N. (2017). Agrotourismos Ecotourismos, Concepts and Reality, Jean Monnet European Centre of Excellence, Athens, 2017.

Exarchos, G., & Karagiannis, St. (2004). *Agritourism* (pp. 68–70). Crete's Technological Research Center.

Fennell, D. A., & Smale, B. J.(1992). Ecotourism and natural resource protection. *Tourism Recreation Research*, *17*(1), 21–32.

Fennell, D. A. (2015). *Ecotourism* (4th ed.). Routledge. ISBN 978-0-203-38211-0.

Fennell, D. A. (2001). *Ecotourism* (pp. 30–36). Publications Ellin.

Georgopoulos A. (2004). *Preface to the Book which theme is Agritourism* (p. 2). Publications Technological Research Centre of Crete.

iefimerida.gr & News Room. (2021).

Igoumenakis, N. (1997). *Touristic economy* (pp. 167–177). Publication Interbooks.

Karagiannis, St., & Exarchos, G. (2006). *Tourism* (p. 42) Publications Creta's Center of Technological Research.

Karagiannis, S. (1997). *Hotels organization and operation, (A Volume)* (pp. 59–60). Publications Ellin.

Karagiannis, St. (2007). *Kerkini: Example of ecological tourism policy implementation (monograph)* (pp. 28–32). Publications DR.A.T.TE.

Kokkosis, H., Chartas, P., & Gryba, E. (2011). *Special and alternative forms of tourism* (pp. 247–249). Publications Kritiki.

Kovos, D. (2021). *Environment: Emergence of environmental philosophy and protection* (p. 29). Floriniotikes Publications.
Lagos, D. (2005). *Touristic economies* (p. 77). Publication Kritiki.
Lytras, P., Lainos, I., & Koumelis, Th. (1997). *Interpretive dictionary of tourist terms* (pp. 132–133). Publications Interbooks.
Note of the Local Union of Municipalities and Communities of N. Florina. (2005). *Man—History—Nature, Nymphaeum* (pp. 35–37).
Oikonomou, A., & Karagiannis, S. (2022). *Circular tourism and sustainability* (p. 87). Under publication.
Papastavrou, A., & Karameris, A. (1996). *Mountain tourism issues* (p. 22). Forestry Policy Laboratory, A.U.TH.
Polo, A. (2022). Preface in the Book under publication with theme Circular tourism, the writers: Agisilaos Oikonomou and Stefanos Karagiannis (p. 2).
Delitheou, B., & Karagianni, M. (2020). *Local humanities* Local development-sustainability (p. 105). Thesaloniki, ISBN:978-960-454-232-1. Publications Erodios.
Vanhove, N. (2019). *Economics of tourism destinations* (p. 30). Publications Gutenberg.
Varvaresos, S. (2000). *Tourism-economic approaches* (p. 13). Publications Propompos.

Interviews

Iatridis Giannis. Hotelier in Nympaeum, Interview 7/6/2022.
Lypitkas Nik. Educational, Deputy Mayor "AETOU" Municipality President of the Area Council and Secretary of the Union of Municipalities and Communities of Florina, Interview 5/6/2022.
Mpoutaris Georgios. President of Nympaeum Municipality and official of Municipality of Aminteo, Interview 7/6/2022.
Natsis Nik. Director of Secondary Education of Florina, Interview 9/6/2022.

Consumers' Perceptions Regarding Film Tourism at the Level of the Tourist Destination Transylvania

Diana Foris, Florina-Nicoleta Dragomir, and Tiberiu Foris

Abstract The film-generated tourism is a relatively recent phenomenon that has grown mainly due to the influence of mass media and social media, because film and television are a strong force in attracting tourists, directly or incorporated into the marketing strategy of the tourist destination. Film and television are tools that can be used to shape the image of a tourist destination and to encourage tourism. The aim of the paper is to identify the opinions of young consumers regarding the travel and the tourism generated by the film at the level of the tourist destination Transylvania. In this respect, a quantitative marketing research was conducted. The results of the study highlighted the positive attitude of young consumers toward film tourism and the potential benefits of this form of tourism regarding the development of tourist destinations. The results of the study are released for the managers of tourist destinations in order to develop attractive and successful destinations. The paper also offers suggestions for future research directions.

Keywords Film tourism · Tourist destinations · Quantitative research · Transylvania

JEL Classification Z11 · Z31 · Z32

1 Introduction

If until some time ago, film tourism was considered only a form of niche tourism (Faber et al., 2011), in the last decade, film tourism is seen as a form of cultural tourism and it is considered one of the dynamic directions of tourism development (Kim & Reijnders, 2017). The motivation for travel is one of the most important

D. Foris (✉) · F.-N. Dragomir · T. Foris (✉)
Transilvania University of Brasov, Brasov, Romania
e-mail: diana.foris@unitbv.ro

T. Foris
e-mail: tiberiu.foris@unitbv.com

© The Author(s), under exclusive license to Springer Nature Switzerland AG 2023
V. Katsoni (ed.), *Tourism, Travel, and Hospitality in a Smart and Sustainable World*, Springer Proceedings in Business and Economics,
https://doi.org/10.1007/978-3-031-26829-8_30

components of the definition of tourism, thus when film and tourism are combined, a by-product known as film tourism is obtained (Kumawat et al., 2022). Roesch considers that "film tourism is a specific pattern of tourism that drives visitors to see screened places during or after the production of a feature film or a television production" (Roesch, 2009). The research topic regarding film tourism shows more and more interest (Beeton, 2001). Popular culture tourism is a form of tourism that results from popular cultural phenomena such as movies, books, and music. Popular culture can have an important contribution in creating the image of the destination, thus contributing to the growth of the brand image of tourist destinations and media industries. Movies and TV series have become powerful tourist motivators, so that just the simple fact that a destination appears in a movie can increase the interest of many potential tourists (Domínguez-Azcue et al., 2021). The purpose of this paper is to identify the opinions of young consumers regarding travel and film tourism at the level of the tourist destination Transylvania.

2 Literature Review

Popular culture tourism destinations arise from the demand of popular culture consumers who experience the need to connect to stories related to their interest in certain movies, books, and music (Light, 2007). In the current context, when tourism has become accessible to an increasing number of people (cost, efficient means of transport, and so on), which allows fans to visit destinations presented in books, films, and music, but also in the media (mass media, social media), the impact of this form of tourism can be found both at the level of the development of a destination, but also on the increase of its competitiveness. Because destination images influence tourist behavior, a destination must be differentiated and positioned positively in tourists' minds (Pike & Ryan, 2004). Movies can have an impact on destination image if the location plays a role in a movie (Hudson & Richie, 2006). The film can promote an area still largely unknown to national and international tourists (Bencivenga et al., 2015).

Analyzing the literature, it can be seen that in the tourism industry film is used as a way of attracting tourists to certain destinations and also the growing interest in creating local tourist attractions based on film (Kolasinska, 2020). Film tourism, as a form of cultural tourism, is becoming a phenomenon of global importance (Cardia et al., 2019), as an alternative to the traditional form of visiting tourist attractions, with thematic visits being organized, even routes based on films or characters from movies. Hudson and Ritchie believe that almost all destination marketing organizations (DMOs) could be more proactive with film tourism (Hudson & Richie, 2006). The destination management has an essential role in its development because destinations must provide value for tourists (Foris et al., 2020). Challenges for managing and marketing popular culture tourism could include capacity issues in accommodation spaces but also the lack of supply of tourist experiences and attractions associated with this type of tourism (Connell, 2012). In this sense, popular culture tourism destinations must have a more strategic planning approach to tourism development, be proactive, respond to community needs, incorporate implementation into a planning process (Hall & Page, 2007). In this sense, the management and the marketing related to film tourism, respectively, the way in which the managers of the destination marketing organizations manage the development of the design in order to build a destination brand, requires more attention from specialists. The destination marketing organizations can engage in a variety of marketing activities both before and after a film's release: during and after a film's release, media attention can be drawn to the film's location, during a film's screening, the promotion of the destination can be done during the screening in order to draw attention about the destination, but most of the marketing activities related to film tourism can be done after the release of a film. At this stage, the challenge for DMOs is to convert audience interest in a film into a commitment for a future visit and capitalize on the additional visitors brought in by the film (Hudson & Richie, 2006). In addition, it is important to include guided tours and film walks in the activity offer of the destination. Tourist attractions, museums, accommodation, and food establishments used in films are often promoted to the public in order to generate tourism activities. It is important to have a website dedicated to the film that links the film to locations and location tours (Croy & Walker, 2003). Also, the promotion of this information through mass media and social media must be included in the strategy for promoting the tourist destination. However, film tourism is a complex and dynamic concept, and success also depends on a number of factors outside the control of a destination (Hudson & Richie, 2006). Managers of many destinations must understand that having a film or television series in their region will automatically increase tourism and that this will be to the advantage of the tourism destination (Beeton, 2006). In an increasingly competitive tourism market, placing the destination in movies and TV shows is a strategic marketing approach that increases awareness, improves the destination's image, and drives the number of tourists. Film tourism offers tourism destinations the opportunity to generate significantly higher income, tourist visits, and economic development (Hudson & Richie, 2006).

3 Methodology

Considering the research problem, namely to identify the opinions of the young consumers—students regarding the travel and tourism generated by the film at the level of the tourist destination Transylvania, the research objectives were set as follows:

(O1) Identification of the potential of the film tourism at the level of the Transylvania destination;

(O2) Identifying the opinions of the young consumers regarding film tourism at the level of the tourist destination Transylvania and their interest in the tourism generated by the film.

In order to fulfill the first objective O1, films were identified whose action takes place in Transylvania, but also films for which Transylvania was chosen as a filming destination.

In order to implement the second objective O2, a quantitative marketing research was carried out based on the questionnaire method. The questionnaire was structured in four sections and includes a number of 11 questions. The first section refers to the behavior of tourists regarding the activities they carry out in a tourist destination and how often they travel. The second section refers to the habits of young people regarding watching movies. The third section is about film-generated tourism—to find out whether young consumers are aware of the existence of film-generated tourism and how interested they are in practicing this type of tourism. The last section of the questionnaire seeks to find out the opinion of young consumers regarding the potential of film tourism at the level of the Transilvania destination.

The data collection was carried out online through a questionnaire using Google Forms, between 03.21 and 05.20.2022. The link that hosts the questionnaire was distributed to students from the Faculty of Food and Tourism (the only faculty with this profile in Romania), within Transilvania University of Brașov, who are considered to be young lovers of tourism, but also future specialists in this field. 104 responses were recorded and processed.

4 Results and Discussion

(O1) Identification of the potential of the tourism film at the level of the Transilvania destination.

Transylvania is one of the most famous regions in Romania, especially because of the stories that link it to the famous legend of Count Dracula. However, this region in central Romania is much more, encompassing stunning landscapes, castles, fortresses, fortified churches, and beautiful towns.

Through the lens of film tourism, Transylvania, Romania is very well known for the myth of Dracula. "Count Dracula is a fictional character created by Bram Stoker. It was inspired by the real historical figure Vlad the Impaler. The British writer Bram Stoker could easily consult at the Royal Library in London some of those Saxon engravings from the fifteenth century, which were also found in the collections of the British Museum, in which Vlad the Impaler is described as a monster, a vampire who drinks human blood and a great lover of cruelty. He probably also had access to the book History of Moldavia and Wallachia by Johann Christian Engel, which describes Vlad the Impaler as a bloody tyrant, which probably gave him the idea to take the prince of Wallachia as the model for his fictional character, "Dracula". Although the novel departs from reality, describing Vlad Țepeș as a mythological character—a vampire, he is the cause of the notoriety that the Romanian ruler enjoys even today. The first screen adaptation of the novel was made in Germany in 1922, the film was called "Nosferatu, Eine Symphonie des Grauens" (Nosferatu—Symphony of Horror), starring the actor Max Schreck.

The last screen adaptation—and the most faithful and the best known—was filmed in 1992. The film—"Bram Stoker's Dracula" is directed by Francis Ford Coppola and has a cast of established Hollywood actors (some of them unknown at the time), starting with Gary Oldman—Count Dracula, with Anthony Hopkins—Dr. Van Helsing, Winona Ryder, Keanu Reeves, and Cary Elwes, who later also stars in "Shadow of the Vampire" (Shadow of the Vampire) with John Malkovich and William Dafoe, a film about the filming of "Nosferatu".

The mythical character Dracula is primarily known for impaling beggars, thieves, etc. Hence, the name was "Vlad Țepeș" (Dracula, 2022).

Dracula has enjoyed enormous popularity since its publication and spawned an extraordinary vampire subculture in the second half of the twentieth century (Melton, 1999). There have been over 200 films made that feature Count Dracula (and several hundred others that have vampires as a subject). Over 1,000 novels have been written about Dracula or vampires, along with a multitude of cartoons, comics, and television programs (Light, 2007). At the center of this, vampire subculture is the place of the myth, namely Transylvania, a destination that has become almost synonymous with vampires (Gelder, 1994). As one writer noted, "What calls another land above such mystic visions of shrouded, misty woods; of driverless wagons plying treacherous, unexplored roads to hidden castles; of black-cloaked figures prowling through moonlit cemeteries in the cold of night?" (Brokaw, 1976). Thus, over a century after it was written, the Dracula myth continues to structure the ways in which Transylvania is understood in the popular Western imagination. By extension, Romania has become "Dracula's country" (Light, 2007).

Nowadays, the Hotel Transylvania film series represent famous films about Romania. Both are American 3D computer-animated fantasy comedy films produced by Sony Pictures. Hotel Transylvania tells the story of Count Dracula, the owner of a hotel called Hotel Transilvania. This is the only monster hotel in the world.

Among other films directed in Transylvania with international success, the following can be listed:

- "The Nun: The Mystery of the Convent" (2018): British director Corin Hardy chose Romania, more specifically Transylvania, as the filming destination—the script presents the story of a nun who committed suicide in a monastery and the mystery that a team from the Vatican is sent to decipher. "The Nun" was filmed at Corvin Castle, Bethlen Castle in Criş, in the tunnels of Mogoşoaia, in Sighisoara, at Copşa Mare, at the Parliament Palace.
- "A Christmas Prince", "The Christmas Prince: A Royal Wedding", "The Princess Switch" (2017): The Netflix miniseries was filmed in several historical monuments in Romania: Peleş Castle, Bragadiru Palace, Cotroceni National Museum, University of Medicine and Pharmacy Carol Davila (UMF), and Aurel Vlaicu Airport. The story is a classic one, although the land of Aldovia is imaginary: a prince and a common girl fall in love and build a fairy tale life together.
- The Brothers Bloom (2008): Rachel Weisz, Adrien Brody, and Mark Ruffalo are the protagonists of a sensational comedy, "Brothers Bloom", which was also filmed in Romania (in Bucharest, Comarnic, Sinaia, and Constanța).

- Ghost Rider: Spirit of Vengeance (2011): Nicholas Cage filmed this action movie at the Corvin Castle, also known as the Huniazis Castle, in Hunedoara.

(O2) Identifying the opinions of the young consumers regarding film tourism at the level of the tourist destination Transylvania and their interest in the tourism generated by film.

Regarding the number of the trips made in the last 12 months, 39.4% of the respondents made between 2–3 trips, 30.8% made a single trip, and 29.8% made more than 4 trips. Therefore, most of the respondents made more than 2 trips in a year.

Respondents travel for relaxation and vacation (81.7%), visiting family/friends (49%) but also for cultural reasons (music, art, film, folklore, and so on) (30.8%), as well as health (16.3%), business (8.7%), and pilgrimage (4.8%).

During the trips, the main activities carried out by the respondents were: visiting tourist attractions (72.1%), nature activities (hiking, horse riding, hiking, and so on) (58.7%), relaxation (SPA, beach) 49%, fun (club, pub, cafe) 38.5%, shopping 37.5%, and cultural activities (23.1%). It is found that young consumers are much more interested in visiting tourist attractions than participating in cultural activities, which implies on the part of the managers of the destinations a more careful approach in creating an offer of more attractive cultural tourism experiences for young people, using also new technologies.

Regarding the opinion of the respondents about the importance of the development of popular culture tourism at the level of a tourist destination, 88.5% of them consider it very important.

Regarding the movie viewing habits, 54.8% of the respondents watch on average about 1–3 movies per week, 21.2% watch between 3 and 5 movies per week, and 11.5% watch more than 5 movies per week. Only 12.5% of them do not regularly watch movies. So, the films are of interest to the young generation.

Regarding the preferred movie genres, the most popular (58.7%) are comedies, followed by romance/drama (56.7%), and action/thriller (52.9%). 40.4% of respondents approve of fantasy/sci-fi films, 29.8% of biographical/historical films, and 25% of mystery/horror films. The most unpopular category is that of psychological films (1%) (Fig. 1).

It is noticed that romance/drama movies together with fantasy/sci-fi movies are among the most preferred movie genres, therefore they inspire the most film-generated tourism. The results confirm those of Roesch (Roesch, 2009).

Even though 61.5% of the respondents had not heard of film tourism, when asked to indicate places/tourist attractions that are related to international film productions they visited, most respondents 67.3% indicated Bran Castle (referring to the 2012 film Hotel Transylvania, where the main character is connected to the legend of Count Dracula), 65.4% Peleș Castle (referring to the 2017 film A Christmas Prince), 28.8% Corvin/Hunyadi Castle (2011 film Ghost Rider: Spirit of Vengeance), and 24% indicated Corvin/Hunyadi Castle, Sighisoara, Bethlen Castle in Criş (2018 film The Nun). None of the respondents indicated an additional location to those presented

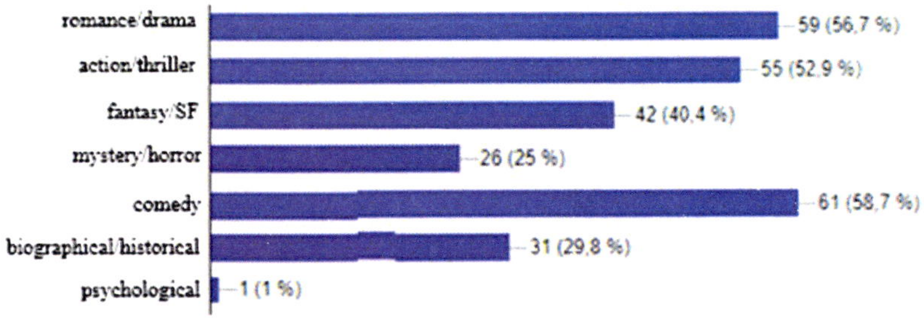

Fig. 1 Favorite movie genres

as answer options, which shows that the Transylvanian destination marketing organizations could be more proactive regarding film tourism. The results confirm those of Hudson and Ritchie (Hudson & Richie, 2006).

The authors wanted to know what would be the reasons that would lead the respondents to visit Transylvania for the purpose of film tourism. In this sense, the activities included in the field of film tourism, classified by Beeton, were used: visiting a place where a film was shot, visiting places featured in films, participating in a thematic tour related to films, participating in a film festival, and attending film premieres (Beeton, 2005) (Fig. 2).

The respondents showed the greatest interest in visiting places featured in movies, places where a movie was shot, but also in attending movie premieres. It should be noted that most respondents are interested in participating in a thematic tour related to the film (36.5%), in visiting some places where films were shot (34.6%) but also in participating in a film festival.

At the suggestion of the authors to indicate a Romanian film which would make them visit certain places in Transylvania that are related to that film, most of the respondents did not do this. Only 28 respondents provided an answer to this question, and only 16 respondents indicated Romanian films, as follows: "Drumul oaselor" (3 answers), "Dacians" (3 answers), "Miami Bici" (2 answers), "Murmurs" (2 answers), "Kiss me!" (2 answers), "BD la munte si la mare" (1 answer), Baltagul (1 answer),

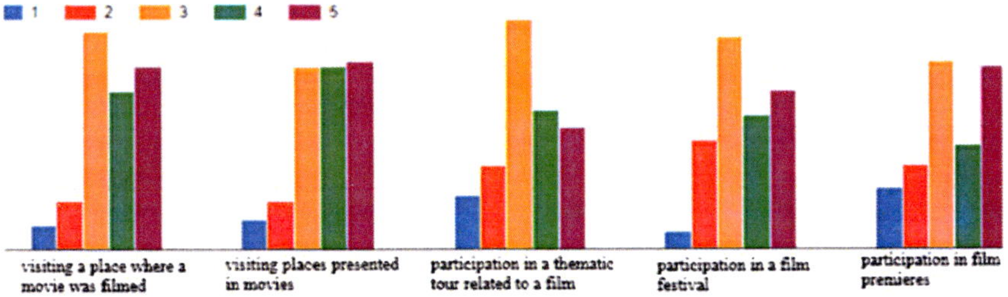

Fig. 2 Motivational factors regarding the practice of activities specific to film tourism. *Notes* 1—total disinterest, 5—total interest

"Oglinda. The beginning of the truth" (1 answer), and "Selfie 69" (1 answer). The other 12 respondents said that they did not have any Romanian film in mind that they could point to as an answer. It can be seen that young people are little interested and little familiar with Romanian films. Therefore, Romanian films are not sufficiently attractive for young tourists to constitute a motivation to travel.

Regarding the opinion of the respondents regarding the promotion of Transylvania domestically and internationally as a tourist destination generated by the film, most of them (75%) totally agree. This fact may suggest the confidence that students have in the natural, cultural, and historical potential that Transylvania could capitalize on as a filming location in international film production.

5 Conclusion

Film tourism shows more and more interest among specialists, being a form of tourism that can make an important contribution to increasing the brand image of tourist destinations, but also to generating added value for tour operators.

The results of the study highlighted the positive attitude of young consumers toward film tourism and the potential to consider this form of tourism regarding the development of tourist destinations.

Regarding the reasons that would lead young tourists to visit Transylvania for the purpose of film tourism, the greatest interest was to visit places shown in films, places where a film was shot, but also to participate in film premieres. It should also be noted that most young people are interested in participating in a thematic tour related to film, in visiting some places where films were shot, but also in participating in a film festival.

The films with international production are the ones that could present an attraction for the practice of film tourism, Romanian films not being attractive enough for young tourists to determine a motivation to travel.

The results of the research reveal one and another important aspect to highlight. The concept of the tourism film is not sufficiently well known among consumers, thus requiring increased attention from the managers of tourist destinations in its development and promotion. Moreover, destination marketing organizations in Transylvania could be more proactive in terms of film tourism, results that confirm those of Hudson and Ritchie (Hudson & Richie, 2006).

Transylvania can be promoted domestically and internationally as a tourist destination generated by the film. Since today tourists are increasingly informed through the Internet, the existence of dedicated blogs could be a valuable source of obtaining the necessary information to increase awareness of the objectives that can be visited in different tourist areas or with film tourism potential.

The results are useful for managers of tourist destinations in order to develop attractive and successful destinations, but, at the same time, also for tour operators that develop tourism products based on capitalizing on film tourism. Regarding the theoretical implication, this study helps to strengthen existing studies on the tourism generated by the film at the level of the tourist destination and provides the basis for future research. The authors aim to continue research on this topic, including qualitative studies, and to expand research to other destinations.

References

Beeton, S. (2001). Smiling for the camera: The influence of film audiences on a budget tourism destination. *Tourism Culture and Communication, 3*(1), 15–25.
Beeton, S. (2005). Understanding film induced tourism. *Tourism Analysis, 11*, 181–188.
Beeton, S. (2006). *Film-induced tourism* (p. 64). Channel View Publications.
Bencivenga, A., Chiarullo, L., & Colangelo, D. (2015). Film tourism in Basilicata. *Almatourism—Journal of Tourism, Culture and Territorial Development, 6*(4), 241–260.
Brokaw, K. (1976). *A night in Transylvania: The Dracula scrapbook* (p. 12). Grosset and Dunlap.
Cardia, G., Jones, A., & Gavelli, D. (2019). Innovating and diversifying cultural tourism in Europe through smart movie tourism in UNESCO sites and destinations. The case study of a famous project. In W. V. Katsoni & M. Segarra-Oña (Eds.) *Smart tourism as a driver for culture and sustainability*. Springer Proceedings in Business and Economics (pp. 3–14). Springer.
Connell, J. (2012). Film tourism—Evolution, progress and prospect. *Tourism management*, 1007–1029.
Contele Dracula. (2022). Retrieved from https://ro.wikipedia.org/wiki/Contele_Dracula. Accessed 18 May 2022.
Croy, W. G., & Walker, R. D. (2003). Rural tourism and film: Issues for strategic regional development. In D. Hall, L. Roberts & M. Mitchell (Eds.) *New directions in rural tourism* (pp. 115–133). Ashgate Publishing Ltd.
Domínguez-Azcue, J. P., Tapia, G., Mercadé Melé, P., & Almeida García, F. (2021). New perspectives of film tourism: A comparative study (2014–2020). *Journal of Tourism Analysis, 28*(2), 93–120.
Faber, F., & Tomala, K. (2011). Film Tourism. In A. Papathanassis (Ed.), *The long tail of tourism: Holiday niches and their impact on mainstream tourism* (pp. 149–158). Gabler Verlag.
Foris, D., Florescu, A., Foris, T., & Barabas, S. (2020). Improving the management of tourist destinations: A new approach to strategic management at the DMO level by integrating lean techniques. *Sustainability, 12*(23), 10201.
Gelder, K. (1994). *Reading the vampire*. Routledge.
Hall, C. M., & Page, S. J. (2007). The geography tourism and recreation: Environment, place and space. *Tourism Management, 28*(3), 935–936.
Hudson, S., & Richie, B. (2006). Promoting destinations via film tourism: An empirical identification of supporting marketing initiatives. *Journal of Travel Research, 44*(4), 387–396.
Kim, S., &Reijnders, S. (2017). Asia on my mind: Understanding film tourism in Asia. In S. Kim & S. Reijnders (Eds.) *Film tourism in Asia. Perspectives on Asian tourism* (pp. 1–18). Springer.
Kolasinska, A. (2020). Opportunities for the development of film tourism in Kraków. *Studies of the Industrial Geography Commission of the Polish Geographical Society, 34*(3), 53–66.
Kumawat, H., Shekhar, C., & Kothari, A. (2022). Impact of film tourism on Udaipur. *Journal of Emerging Technologies and Innovative Research, 9*(3), 423–432.
Light, D. (2007). Dracula tourism in Romania: Cultural identity and the state. *Annals of Tourism Research, 34*(3), 746–765.
Melton, J. (1999). *The vampire handbook* (2nd ed.). Visible Ink Press.
Pike, S., & Ryan, C. (2004). Destination positioning analysis through a comparison of cognitive, affective and conative perceptions. *Journal of Travel Research, 42*(4), 333–342.
Roesch, S. (2009). *The experiences of film location tourists* (p. 6). Channel View Publications.

Projecting an Image: TV Series Amor Amor and Lua de Mel and the Impacts of Film-Induced Tourism in Penafiel

Ana Ferreira, Maria Carlos Lopes, and Elisa Alen

Abstract The image of a touristic destination is a complex construction based on different sources of information. Among them we can highlight the films and television series as they are an excellent tool for promoting tourism for the territories where they are filmed. The appearance of a particular area in a film or television series can have a huge effect on the number of visitors of an already existing place and create a new kind of tourism to the area and generate a boost for the local economy. The main objective of this article is to analyze the impact of the soap opera "Amor, Amor" and "Lua de Mel" that was broadcast at prime time in a Portuguese television channel, in the image of the city of Penafiel as a touristic destination. We intend to assess the short-term impacts of the great media exposure felt by the general population and, in particular, by the various stakeholders, on the influx of domestic tourists to this city; and the way this influx of tourists is being fostered in order to generate medium- and long-term revenue turning the small town of Penafiel into a tourist hotspot. We have done some literature review as well as an exploratory and quantitative investigation. The data collection was carried out with a questionnaire designed to be answered online by the inhabitants of Penafiel. The results of the analysis highlight the importance of films and series as important sources of information giving visibility to destinations that are still little known or not at all. Write the abstract of your paper here. It should not be more than 250 words.

A. Ferreira (✉)
School of Hospitality and Tourism, Polytechnic Institute of Porto, Vila do Conde, Portugal
e-mail: anaferreira@esht.ipp.pt

A. Ferreira · M. C. Lopes
CiTUR Researcher (Center for Tourism Research, Development, and Innovation), Portugal, Portugal

M. C. Lopes
School of Hospitality and Tourism, Polytechnic Institute of Portalegre, Portalegre, Portugal

E. Alen
Faculty of Business Sciences and Tourism, University of Vigo, Ourense, Spain

© The Author(s), under exclusive license to Springer Nature Switzerland AG 2023
V. Katsoni (ed.), *Tourism, Travel, and Hospitality in a Smart and Sustainable World*, Springer Proceedings in Business and Economics,
https://doi.org/10.1007/978-3-031-26829-8_31

Keywords Film-induced tourism · TV series · Destination image · Impacts · Penafiel

JEL Classification Z32

1 Introduction

The image is fundamental in the creation of the imaginary of a territory, and it has an increasingly decisive influence on the tourists' choice of a destination. The films and television series broadcast by the media show the most striking elements of the localities and, in this way, help individuals to create an image of the destination, becoming influencers of the chosen destinations (Tudor, 2015). Thus, TV and cinema became part of the main sources of tourist publicity.

This work aims to understand the impact of the television series Amor Amor and Lua de Mel on tourism in the city of Penafiel and the opportunities and challenges presented to the organizations of the destination. This research covers both the benefits and negative effects that can arise from the identification of a television series with a touristic destination.

Based on a review of the literature on film-induced tourism, in particular the evidence of the importance of films and series as important sources of information for the creation and modification of destination images, we intend to identify, from the perspective of residents, the positive effects and negative aspects of this type of tourism for the city of Penafiel.

2 Literature Review

2.1 Storytelling, Set-Jetting, and Film-Induced Tourism

Today, tourism is unanimously recognized as one of the world's main economic activities. To prove this, the United Nations World Tourism Organization (UNWTO) predicts that once the effects of the pandemic have been overcome, tourism activity will continue to grow and by 2030 it will be twice as high as it was in 2012, when, for the first time, the world registered one billion tourists.

The development of this activity has led it to integrate an increasingly broad set of elements that act as supporting elements in the strategy of valuing and promoting the offer and competitiveness of tourist destinations. At the same time, those elements may expose it to several human and natural conditioning factors.

2.1.1 Destination Storytelling

Nowadays, tourists want to connect with their destination. They want to live memorable experiences. To effectively promote engaging travel experiences, it is necessary to mix experiences, memories, and people. We are aware of the importance of destination storytelling as part of marketing local experiences, satisfying the needs of travelers who seek unique and memorable experiences.

In the travel and tourism context, it's not hard to imagine the positive impact of good storytelling: After all, memorable travel experiences are all about stories—stories of the local communities and their cultural traditions, stories told by different sources or in different ways.

Cinema and TV as multimedia storytelling convey a narrative journey with images and allow unique travel experiences. Sharing destination stories without limiting travelers' imagination helps them create their own travel experiences at the destination.

As Bassano states "storytelling about places is recognized as a tool to enhance the reputation of regions as they compete for tourism and economic development" (Bassano et al., 2019, p. 10).

In fact, the same author points out that "Storytelling can be managed in a local service system to enhance brand competitiveness" and "The way local governments and cultural organizations understand and encourage storytelling about places ("place storytelling") can have a significant impact on the success of a region competing for tourism and development spending." (Bassano et al., 2019, p. 10), which allows us to affirm the growing importance of film-induced tourism as storytelling tourism.

2.1.2 Set-Jetting

According to Joliveau, "set-jetting can be considered as a new and rapidly growing form of tourism. Set-jetting involves visiting real-world film locations from Hollywood blockbusters by tourists and fans. The narrative of the book—and of the film—takes the audience through a series of cultural sites throughout the world (e.g., Paris, Rome, Jerusalem) which, in turn, stimulated the curiosity of the public" (Joliveau, 2009, p. 37).

Thus, set-jetting turns movies or books locations into tourist attractions because visitors who practice set-jetting accumulate extensive information about films or books related to the places, developing an image of the territory they are attracted to.

In fact, as Joliveau says "through set-jetting, fantasy produces a strong effect on reality. Once located on a map, the fictional place becomes attractive and a potential source of real profits. For a long time, cities have been developing strategies in order to attract film crews, which will later attract film tourists" (Beeton, 2005; Joliveau, 2009, p. 37).

2.1.3 Film-Induced Tourism

Arguably, film became the dominant art form of the twentieth century. Although films are not produced so that people will visit destinations as tourists, they enhance the awareness, appeal, and profitability of locations through the power of imagery and fantasy of story that they portray. Authors such as Hudson and Ritchie, Beeton, Grihault, Riley et al., Tooke & Baker and Butler refer to this phenomenon of tourists visiting particular destinations because of their association with particular films.

The most cited definition of film-induced tourism is the one given by Hudson and Ritchie. According to those authors, "tourist visits to a destination or attraction as a result of the destination being featured on television, video, DVD, or the cinema screen." (Hudson & Ritchie, 2006, p. 256).

In fact, most authors include film-induced tourism in cultural tourism and agree that the term "tourism" includes both film and television-induced tourism. Kim and Reijnders (2018) refer film-induced tourism as a social and cultural phenomenon that results from individuals visiting certain locations due to their relationship with a film or a television series (Kim & Reijnders, 2018).

2.2 *Film-Induced Tourism: The Impact of TV Series in Destination Promotion*

As Domínguez-Azcue et al. say "films and TV series continually show the places, streets, and cities where they are located" (Domínguez-Azcue et al., 2021, p. 8). Although not generally produced with the intention of attracting tourists to a destination, films and TV series tend to influence viewers indirectly as a background message, helping to imagine a destination and reflecting on a very important improvement in the image. They have the capacity to help us imagine a destination, and they help the individual design a destination image according to the information perceived from these media.

In addition, many researchers claim that certain films/series raise awareness about the locations where they are shot, providing bonds with the destination. On the other hand, much of the film literature suggests that people can experience a place indirectly identifying with the characters, which helps reduce previous insecurity and increase the knowledge of potential tourists and their confidence in the destination. If potential tourists feel familiar with the destination, they are likely to have a more accurate knowledge of the destination offers, and this gives them confidence, which can help them choose their final trip. Based on the image of a destination created by television, more and more tourists decide to visit the place in order to see the hotspots that appear in the movie or series, such as parks, cafes, restaurants, among others. This phenomenon often extends to a wider area, eventually leading to visiting the entire surrounding region. Therefore, film-induced tourism can be a very effective marketing tool for destinations that cannot afford strong travel advertising campaigns.

2.3 The Economic and Socio-Cultural Impacts of Tourism

Tourism has an impact on people's lives whether they are tourists or residents. In this work we will analyze specifically the influence of tourist activity on residents' lives.

Residents can enjoy and take advantage of the development of tourism. On the other hand, their daily life may be affected by tourist activity which may place some constraints causing socio-cultural and environmental changes on the host community life. (Thetsane, 2019).

About the economic impacts, García et al. (2015) argues that these are the most positive for residents. This statement corroborates the opinion of other authors who consider that tourism promotes job creation (Del Chiappa et al., 2018), allows to generate more revenue and taxes and increases the profits of the owners of catering, accommodation, and commerce (Pramanik & Ingkadijaya, 2018); enables investment in accessibility, different infrastructure and public services (Alén et al., 2018); creates more opportunities to do business within the local community related to tourism activity, even indirectly (García et al., 2015; Postma & Schmuecker, 2017), and promotes an improvement in the well-being and quality of life of residents (Mathew & Sreejesh, 2017; Lopes, 2020).

However, negative economic factors are also pointed out by residents, especially instability in revenues due to the seasonality of tourist activity which leads to more negative issues—workers in the tourism sector become unemployed in the low season period; the increase of prices due to speculation in tourism activity, namely in real estate cost (high rents and the difficulty to purchase own housing by locals) (Li et al., 2017) better (Li et al., 2017; Pramanik & Ingkadijaya, 2018).

It should be noted, however, that the economic impact of tourism activity on residents of a community has a more positive impact on those working directly on tourism, rather than on those who have no connection to the sector (Woo et al., 2018).

In social terms, a positive relationship may be established between residents and tourists. Tourist activity may bring some benefits for the community, such as more establishments, infrastructures, and social facilities; more and better social events where residents can enjoy their social life; more services offered to communities, among others which promote a better quality of life for community residents (Zach & Hill, 2017). On the other hand, the residents' quality of life can be deeply affected. Residents can feel contempt and insecure due to the increase of vandalism and crime, abusive consumption of alcohol and/or drugs, the increase in prostitution (Joo et al., 2019), among others, caused by the excessive concentration of tourists (Almeida & Araújo, 2017; García et al., 2015; Woo et al., 2018).

These impacts are associated with cultural issues. In this context, tourism offers a wide range of cultural activities; it triggers in the community and in local authorities a sense of pride and of belonging to the community; it raises awareness of the importance of restoration and preservation of historical monuments, old buildings and diverse places (Kozak et al., 2013; Pratama & Mandaasari, 2020), as well as of the need to revitalize and continue ancient traditions, customs and folklore of the community, promoting contact with new cultures (Liberato et al., 2018).

Regarding the negative impacts of the cultural dimension, Öztürk et al. (2015) highlight the permeability to foreignisms to which the mother tongue is subject due to contact with the mother tongue of tourists; loss of authenticity; mercantilization of culture; and loss of identity, these especially highlighted in the studies carried out about film-induced tourism by Beeton (2021) and Thelen et al. (2020).

3 Methodology

This study is based on a theoretical approach based on the literature review with the objective of systematizing the relations between television, destination image, and tourist destination, in order to evaluate the impact of television series on the tourism of the destination that serves as a backdrop, from the perspective of inhabitants.

In this investigation, we implemented a sample questionnaire to 224 residents of Penafiel. The questionnaire, implemented as a Google forms, is divided into two parts, in a total of nine closed questions: in the first part the identification of the resident, as age, gender, academic qualifications and professional activity; in the second part we try to associate these television series with tourism in Penafiel, in particular, questioning the socio-cultural and economic impacts from the perspective of local residents. The analysis of the results, made with the application of variance analysis (ANOVA), allowed us to determine the importance of the television series *Amor Amor* and *Lua de Mel* to publicize and promote tourism in the city of Penafiel.

4 Results, Capitalize the First Letter of Every Word in the Title

4.1 Individual Profile of Residents

Analyzing Table 1, at the global level, our sample consists mostly of female individuals (54.0%). Regarding age, we would like to point out that it was a condition for inclusion in the study that all respondents should be 15 years old and above. According to the table under analysis, we also verified that age was grouped into age categories, with the objective of obtaining homogeneous amplitude classes. The last class was aggregated because the remaining ones contained few observations. After this codification, we observed that it is the class of the elderly that represents the smallest number of respondents, that is, those whose age, in this classification, is above 65 years-old represent approximately 3.1% of the respondents. The most representative class is the one of adults (35–44 years) with 33.5% of respondents. Locals with secondary school and a degree predominate, 30.8 e 30.4%, respectively, 11.2% have completed "Master's" or "PhD," and 26.3% of inhabitants have no more than basic education. 20.51% have a liberal profession, 19.6% work in trade and

Table 1 Individual profile of inhabitants in Penafiel

Inhabitants		Frequency	%
1. Gender	Female	121	54.0
	Male	103	46.0
2. Age	15–24	22	9.8
	25–34	49	21.9
	35–44	75	33.5
	45–54	17	7.6
	55–64	54	24.1
	65+	7	3.1
3. Qualifications	Junior school	59	26.3
	Secondary school	69	30.8
	Degree	68	30.4
	Master/PhD	25	11.2
	Other	3	1.3
4. Professional activities	Catering	19	8.5
	Hospitality	6	2.7
	Tourist activities	8	3.6
	Transp. Communications	6	2.7
	Trade	44	19.6
	Education	36	16.1
	Liberal Profession	45	20.1
	Other	58	25.9
	DA	2	9
Total		224	100.0

Source Own elaboration

16.1% have a profession in the education sector. Only 8.5% of respondents work in catering 3.6% in tourism activities and 2.7% in hospitality.

4.2 Residents' Opinion About the Importance of TV Series to the Destination Penafiel

Analyzing now the results obtained for Penafiel as a touristic destination, it is possible to show in Table 2 that although most residents (54.9%) do not watch TV series, and 92.0% consider that the images transmitted by the series are important for the promotion of the city as a tourist destination, and 84.8% report that the TV series *Amor Amor* and *Lua de Mel* have contributed to the increase of tourists/visitors in the city.

Table 2 Residents' opinion

	TV Series viewer (%)	The images of the series publicize the city (%)	TV series increase tourism in the city (%)
Yes	45.1	92.0	84.8
No	54.9	7.1	6.3
DA		0.9	8.9
Total (224)	100.0	100.0	100.0

Source Own elaboration

4.3 *Film-Induced Tourism: Economic, Social, and Cultural Impacts in Penafiel*

Convinced that TV series encourage tourism in Penafiel we present an analysis of the impacts of this type of tourism on the residents' life in the city. We have used the analysis of variance (ANOVA) based on which we could cross the information and understand which variables most impact the results and understand which characteristics of the sociodemographic profile of the sample interfere in their personal evaluation of the impacts of this tourism.

In this work, we will only refer to the statistically more significant connections, between the profile of residents in Penafiel and the relation to the social, cultural, and economic impacts.

The most positive aspect pointed out in the social issues was "the national and international recognition of the city," reported by 73% of respondents; and culturally, the increase of "the collective self-esteem of the community" is reported by 87% of respondents. The most important economic impact was the "opportunity for economic growth of the city" for 41% of inhabitants.

Analyzing, in particular, the social impacts, we realize that the increase of "national and international projection of the city" and "the spirit of community", are the most valued for 52% of respondents. The age groups that contributed the most to the response rate mentioned above were 25–35 years old, and 55–64 years old, 57% of whom were women and 43% were men. The individuals that have valued these social impacts the most have secondary school and degree (47 and 37%, respectively); and their professions are mostly related to commerce, education, and others. 65% of these respondents are not TV series viewers.

Evaluating residents' opinions on the cultural impacts of the film-induced tourism in Penafiel, in the 175 valid surveys, "collective self-esteem of the community" was the most highlighted still being pointed out other cultural aspects as "national and local pride," "regional values and traditions," and "the collective self-esteem of the community". 51% of the respondents who considered this set of cultural aspects are between 55 and 64 years old (male and female). 73% of these individuals have only junior and secondary school and half of them work in education or are liberal professionals. 64% of these respondents are *Amor Amor* and *Lua de Mel* series viewers.

Finally, when we analyze the economic impacts of film-induced tourism, "investment opportunities for the local and regional economy" and "opportunity for economic dynamics of the city" were the most highlighted aspects. However, in "increased revenues in the various local economic activities" and "opportunity for economic boosting of the city" also be referred. The age group that most influenced these results was 35–44 years old (71%), and 55–64 years old (59%). There were no significant differences in gender-related responses. The inhabitants who highlight "increase in revenues in the various local economic activities" mostly have a degree (57%) and those who highlight the "opportunity for economic boosting of the city," curiously are the ones who have only junior school or those who have, master/PhD, both with the same percentage (32%). Also, most of these respondents have liberal professions (64%), and 63% of them aren't television series viewers.

After analyzing the sociodemographic profile of the population surveyed, there opinion about the TV series' importance to the tourism in Penafiel and its positive impacts in the city and the region, we will then present the reasons pointed out to do not encourage this type of tourism. It should be noted that of the total of the 224 respondents, 155 (69.2%) reported not recognize any negative impact of tourism in the city and is own life, so from now on, we will analyze the negative reasons that appoint the 69 residents surveyed.

According to the data collected we can ensure that tourism is seen by most residents as the cause of the increase in the cost of living, especially those in the age groups of 25–34 and 35–44, both in the case of men and women with levels of education between basic, secondary, or even bachelor's degree, This finding is in line with the work of Mandić et al. (2017) who concluded that film tourism has increased the price of products in general and the furniture market in particular. In addition, we can affirm that most residents don't consider tourists intruders in their life or in their community or that tourism in Penafiel left out the local community, except for 18.8% of individuals between 54 and 65 years of age and 15.9% of women surveyed.

5 Conclusion

In the analysis presented we would like to point out that among all positive impacts listed by the residents in Penafiel, the most highlighted are from the cultural dimension. The percentage of respondents who have reported at least one item of this dimension was 78%. The social dimension was marked by 73%, and the economic dimension was the one in which fewer positive impacts of film-induced tourism were reported—only 50% of respondents consider that there is at least one positive economic aspect in tourist activity in Penafiel, influenced by the television series.

This conclusion is in line with that mentioned by Beeton (2016), Kim et al. (2017) and Croy et al. (2018) who found that films or other television programs do not always have positive economic effects for areas of activity other than the film market.

In this work, we found that the social and cultural dimensions are those to which residents recognize the most positive impacts of film-induced tourism, which is in line with the work of Eslami et al. (2018) who claim that social and cultural aspects have a positive influence in the satisfaction of people's lives in a community, thereby improving their quality of life, and ultimately having a very positive effect on the way the population in question views tourism activity and tourists themselves.

Filiz and Yılmaz (2017) consider that the demographic characteristics of residents have a significant influence on their support for the development of tourism in the community. In our study, we found that residents aged between 55 and 64 years were the one who highlighted the most positive aspects of the tourist activity in the city of Penafiel, in all dimensions.

Also, it was possible to verify for the cultural dimension that the "opportunity for economic diversification of the city" was selected, mostly, by individuals who had a degree, which corroborates the one mentioned by Pillai et al. (2019) that proved that high degree of qualification and younger individuals support film-induced tourism, because it represents new job opportunities for the community.

Regarding the influence that the profession of residents has in the support they give to the development of tourism activity, it is not possible to measure the effective relationship from the results of our study. Although Woo et al. (2018) and Horecký and Blažek (2019) claim that residents who have tourism-related professions have a more positive attitude toward the development of tourism, we can only say that residents of Penafiel who positively assess the effects of tourism are those who play liberal or education role.

Despite the negative factors signalized, in general, the opinions of residents regarding the tourist activity promoted by television series are positive, because they consider that tourism enriches and improves the life of the local community (García et al., 2015).

References

Alén, E. (2018). Reflections on "assessing tourist behavioral intentions through perceived service quality and customer satisfaction". *Journal of Global Scholars Marking Scice 28*(4), 358–362.

Almeida, P., & Araújo, S. (2017). *Introdução à Gestão de Animação Turística* (2nd ed.). Lidel.

Baloglu, S., & McCleary, K. W. (1999). A model of destination image formation. *Annals of Tourism Research, 26*(4), 868–897.

Bassano, C., Barile, S., Piciocchi, P., Spohrer, J. C., Iandolo, F., & Fisk, R. (2019). Storytelling about places: Tourism marketing in the digital age. *Cities, 87*, 10–20.

Beeton, S. (2005). Film-induced tourism. In *Film-Induced Tourism*. Channel view publications.

Beeton, S. (2021). The film-induced tourism experience. In *Routledge handbook of the tourist experience* (pp. 315–327). Routledge.

Boğan, E., & Sariişik, M. (2016). Yerel halkin turizm faaliyetine yönelik görüş ve algilamalarinin belirlenmesi üzerine alanya'da bir araştirma. *Kastamonu Üniversitesi İktisadi ve İdari Bilimler Fakültesi Dergisi, 12*(2), 325–342.

Butler, R. W. (1990). The influence of the media in shaping international tourist patterns. *Tourism Recreation Research, 15*(2), 46–53.

Butler, R. (2004). Geographical research on tourism, recreation and leisure: origins, eras and directions. *Tourism Geographies, 6*(2), 143–162.

Croy, W. G., Kersten, M., Mélinon, A., & Bowen, D. (2018). Film tourism stakeholders and impacts. In *The Routledge handbook of popular culture and tourism* (pp. 391–403). Routledge.
Del Chiappa, G., Atzeni, M., & Ghasemi, V. (2018). Community-based collaborative tourism planning in islands: A cluster analysis in the context of Costa Smeralda. *Journal of Destination Marketing & Management, 8*, 41–48.
Domínguez-Azcue, J., Almeida-García, F., Pérez-Tapia, G., & Cestino-González, E. (2021). Films and destinations—Towards a film destination: A review. *Information, 12*(1), 39. MDPI AG. https://doi.org/10.3390/info12010039
Eslami, S., Khalifah, Z., Mardani, A., & Streimikiene, D. (2018). Impact of non-economic factors on residents' support for sustainable tourism development in Langkawi island, Malaysia. *Economics & Sociology, 11*(4), 181–197.
Filiz, A., & Yılmaz, Ö. D. (2017). Turizm gelişim yaklaşımları kapsamında yerel halkın turizme yönelik algı ve tutumları: Mazı Köyü-Türkbükü karşılaştırmalı analizi. *Hitit Üniversitesi Sosyal Bilimler Enstitüsü Dergisi, 10*(2), 1799–1824.
García, F. A., Vázquez, A. B., & Macías, R. C. (2015). Resident's attitudes towards the impacts of tourism. *Tourism Management Perspectives, 13*, 33–40.
Gartner, W. C. (1989). Tourism image: Attribute measurement of state tourism products using multidimensional scaling techniques. *Journal of Travel Research, 28*(2), 16–20.
Hançer, Ş, & Manci, A. R. (2017). Yerel halkın turizme bakış açısının belirlenmesi üzerine bir araştırma: Diyarbakır. *Journal of Tourism and Gastronomy Studies, 70*, 91.
Horecký, J., & Blažek, M. (2019). Dependent work and internship. *Central European Journal of Labour Law and Personnel Management, 2*(2), 7–20.
Hudson, S., & Ritchie, J. R. B. (2006). Promoting destinations via film tourism: An empirical identification of supporting marketing initiatives. *Journal of Travel Research*, 44.
Joliveau, T. (2009). Connecting real and imaginary places through geospatial technologies: Examples from set-jetting and art-oriented tourism. *The Cartographic Journal, 46*(1), 36–45. https://doi.org/10.1179/000870409X415570
Joo, D., Cho, H., & Woosnam, K. M. (2019). Exploring tourists' perceptions of tourism impacts. *Tourism Management Perspectives, 31*, 231–235.
Kilinç, T., & Calişkan, U. Kırsal Alanlarda Yerel Halkın Film Turizmi Algıları ve Destek Tutumları Arasındaki İlişki: Muğla Yeşilyurt Örneği. *İzmir İktisat Dergisi, 36*(2), 329–353.
Kim, S., & Reijnders, S. (Eds.). (2018). *Film tourism in Asia*. Springer. https://doi.org/10.1007/978-981-10-5909-4
Kim, S., & Park, E. (2021). An integrated model of social impacts and resident's perceptions: From a film tourism destination. *Journal of Hospitality & Tourism Research*, 10963480211011641.
Kim, S., Kim, S., & Han, H. (2019). Effects of TV drama celebrities on national image and behavioral intention. *Asia Pacific Journal of Tourism Research, 24*(3), 233–249.
Kim, S., Kim, S., & Oh, M. (2017). Film tourism town and its local community. *International Journal of Hospitality & Tourism Administration, 18*(3), 334–360.
Kozak, N., Kozak, M. A., & Kozak, M. (2013). *Genel Turizm, İlkeler-Kavramlar* (p. 13). Detay Yayıncılık.
Li, S., Li, H., Song, H., Lundberg, C., & Shen, S. (2017). The economic impact of on-screen tourism: The case of The Lord of the Rings and the Hobbit. *Tourism Management, 60*, 177–187.
Liberato, D., Alén, E., Liberato, P., & Dominguez, T. (2018). Governance and cooperation in Euroregions: Border tourism between Spain and Portugal. *European Planning Studies, 26*(7), 1347–1365. https://doi.org/10.1080/09654313.2018.1464129
Lopes, M. C. (2020). Relação entre Turismo e Qualidade de Vida no Turismo Sénior. Avaliação na Eurorregião. (Tese de Doutoramento). Universidade de Vigo. Vigo
Lu, T., Zhang, F., & Wu, F. (2018). Place attachment in gated neighbourhoods in China: Evidence from Wenzhou. *Geoforum, 92*, 144–151.
Mandić, A., Petrić, L., & Pivčević, S. (2017). Film as a creative industry constituent and its impacts on tourism development: Evidence from Croatia. *Tourism in Southern and Eastern Europe, 4*(1), 335–348.

Mathew, P. V., & Sreejesh, S. (2017). Impact of responsible tourism on destination sustainability and quality of life of community in tourism destinations. *Journal of Hospitality and Tourism Management, 31*, 83–89.
Özyurt, P. M. (2018). Turizm gelişiminin yerel halkın bireysel ve toplumsal yaşam kalitesi üzerine etkilerinin incelenmesi.
Ozturk, A. B., Ozer, O., & Çaliskan, U. (2015). The relationship between local residents' perceptions of tourism and their happiness: A case of Kusadasi, Turkey. *Tourism Review, 70*(3), 232–242.
Pillai, S. K. B., Kamat, K., Scaglione, M., D'Mello, C., & Weiermair, K. (2019). Film tourism and its impact on residents' quality of life: A multi logit analysis. In *Best practices in hospitality and tourism marketing and management* (pp. 181–199). Springer.
Postma, A., & Schmuecker, D. (2017). Understanding and overcoming negative impacts of tourism in city destinations: Conceptual model and strategic framework. *Journal of Tourism Futures.*
Pramanik, P. D., & Ingkadijaya, R. (2018, April). The impact of tourism on village society and its environmental. In *IOP Conference Series: Earth and Environmental Science* (Vol. 145, No. 1, p. 012060). IOP Publishing.
Pratama, I. G. S., & Mandaasari, I. C. S. (2020). The impact of tourism development on the economic, cultural and environmental aspects of local communities. *International Research Journal of Management, IT and Social Sciences, 7*(1), 31–36.
Ramkissoon, H. (2020). Perceived social impacts of tourism and quality-of-life: A new conceptual model. *Journal of Sustainable Tourism*, 1–17.
Roesch, S. (2009). *The experiences of film location tourists* (Vol. 42). Channel View Publications.
Thelen, T., Kim, S., & Scherer, E. (2020). Film tourism impacts: A multi-stakeholder longitudinal approach. *Tourism Recreation Research, 45*(3), 291–306.
Thetsane, R. M. (2019). Local community participation in tourism development: The case of Katse Villages in Lesotho. *Athens Journal of Tourism, 6*(2), 123–140.
Tudor, G. C. (2015). Film tourism—A successful journey for New Zealand. The way towards being the world's top film tourism destination. *Cactus Tourism Journal, 12*(2), 45–53.
Woo, E., Uysal, M., & Sirgy, M. J. (2018). Tourism impact and stakeholders' quality of life. *Journal of Hospitality and Tourism Research, 42*, 260–286.
Yoon, Y., Kim, S. S., & Kim, S. S. (2015). Successful and unsuccessful film tourism destinations: From the perspective of Korean local residents' perceptions of film tourism impacts. *Tourism Analysis, 20*(3), 297–311.
Zach, F. J., & Hill, T. L. (2017). Network, knowledge, and relationship impacts on innovation in tourism destinations. *Tourism Management, 62*, 196–207.
Żemła, M., & Zawadzki, P. (2014). Film-induced tourism: Basic relations between films and tourism from tourism destinations perspective. *Ekonomiczne Problemy Turystyki [Economic Problems of Tourism], 4*(28), 203–221.

Tourism Destinations *"Where You Can Engage in Battles with the Romans"*: The Role of Reenactment in Shaping a Destination Image

Simona Mălăescu

Abstract Tourists prefer destinations well known for reenactment representation for various reasons. Our preliminary qualitative–quantitative study on how reenactors perceive their role in the public agora and the visitors' experience at destination revealed that reenactors actually play an active role in modeling the experience tourists have at the destination and in some cases, over decades, the tourism destination in itself. The results revealed both a direct and an indirect role of reenactors in developing a heritage site as a tourism destination. The direct impact is resulting from the participation in the decision-making boards of local administrative structures, museum's managerial teams, attracting funding and developing tourism strategies for the site. It also resides in founding reenactment groups that becomes inspirational and reenactment festivals that attract exponentially more visitors to the site. Their indirect role comes from the design of the reenactment activities, the tourist animation, the level of involvement of tourists during the activities and the emotions tourists have at the destination. The study also revealed the reenactors' perspective of the importance of the context in which they act. Different countries and regions have different policies regarding the approval of festivals near a site and funding the reenactment. Differences could be explained by the differences in the life cycle stages of the tourism destinations or their long-term strategy. The practical implications of the study reside in helping the local authorities to design new levers of developing a heritage tourism destination and to understand the long-term implications of making from reenactment performances funding a priority.

Keywords Cultural heritage · Heritage tourism · *Tourism area life cycle model* · Reenactment festivals · Heritage sites · Tourism destination development

JEL Classification D91 Role and effects of psychological, emotional, social, and cognitive factors on decision making

S. Mălăescu (✉)
Babeş-Bolyai University, Cluj-Napoca, Romania
e-mail: simona.malaescu@ubbcluj.ro

© The Author(s), under exclusive license to Springer Nature Switzerland AG 2023
V. Katsoni (ed.), *Tourism, Travel, and Hospitality in a Smart and Sustainable World*, Springer Proceedings in Business and Economics,
https://doi.org/10.1007/978-3-031-26829-8_32

1 Introduction

Although the occurrence of live heritage interpretation in form of reenactment at heritage sites dates several decades back, only recent literature brings evidence of the role reenactment performances play in the experience tourists have in heritage tourism destinations and the development of tourism (Carvalho et al., 2020; Malaescu, 2021, 2022; Waldemar et al., 2022). Previous literature had examined extensively either the reenactment phenomenon per se in terms of reenactors' motivation, authenticity, historical accuracy or educational impact (Agnew, 2004; Gapps, 2003; Hall, 1994; McCalman, 2004; McCalman & Pickering, 2010; Popa, 2016; Zotica & Mălăescu, 2015), or the national/regional discourse and the role of visits at heritage sites in terms of self-identity genesis or national affects exultation (Malaescu, 2021; Park, 2010; Popa, 2016; Tian et al., 2020; Zhou et al., 2020). At the beginning of the last decade, studies emphasized the link between this events and the positive dynamic of tourism (Rojas, 2010; Rojas-Rabaneda, 2012). The studies focusing on heritage tourism link the events held in heritage sites in general with economic gains and increased flux in visitors (Carvalho et al., 2020; Santa Cruz et al., 2020). Recent literature brought evidence about the role of visitors' motivation for reenactment performances at heritage sites in the growing dynamics of tourists and tourists' place attachment genesis (Malaescu, 2021) and later, in tourists' destination loyalty formation (Malaescu, 2022). Waldemar et al. (2022) analyzed the importance of reenacting past events and ancient industrial traditions of iron smelting held in Nowa Slupia on attracting tourists and the development of tourism services. The role of reenactors in creating a certain type of experience—more or less authentic or historically accurate with respect to historical events or with more open-ended outcome (Gapps, 2003)—was widely researched (Agnew, 2004; Gapps, 2003; Hall, 1994; McCalman & Pickering, 2010). However, no study (as far as we know) spotted the multiple hypostases of reenactors' role in developing a tourism destination or nuancing a certain destination image.

2 Literature Review

Previous studies pointed out the fact that historical reenactment events are organized with the aim of attracting "the maximum number of visitors, that is, to attract tourism" (Rojas-Rabaneda, 2012, p. 75). Before, Rojas (2010) argue that these expectations are usually met, these events having the capacity to attract audiences and consequently, the fairs hosting these performances grow in number of visitors from one edition to another (Rojas-Rabaneda, 2012). However, if we set the bar higher, and the aim becomes the genesis of the visitors' destination loyalty, Santa Cruz et al. (2020) argued that in order to guarantee the formation of destination loyalty the perception of the event and the evaluation of the elements of the destination are also relevant, apart from these simple motivations. The assessment process of the elements of the destinations' is taking place in the context where heritage interpretation has to solve the crisis of the previous decade where using designed heritage interpretation based on immobile devices and non-interactive content was summed up by Howard (2003, p. 260) as "dead heritage interpretation". The *dead heritage interpretation* label appeared as opposed to live heritage interpretation (using human resources such as tour-guides, cultural animators, reenactment performances, storytelling or community theater) (Howard, 2003). Only a part of the local authorities and site managers, however, saw the reenactment representations as a mean to attract more tourists, consequently budgeted, and funded this type of events. These heterogeneous perspectives regarding the role of reenactment remain in play nowadays.

Several studies emphasized the economic contribution of events held in a heritage destination: from generating an increased flux of visitors to a longer average length of stay manifested by tourists and finally, the genesis of their loyalty to the destination (Carvalho et al., 2020).

More than a decade ago, Agnew (2004) resumed the joy and enthusiasm for reenactment spanning from amateurs to scholars as

> As anyone who has swabbed decks and gone aloft knows, reenactment is fun. It indulges the twin passions of work and play, which are generally divorced from each other. It licenses dressing up, pretending and improvising, casting oneself as a protagonist of one's own research, and getting others to play along. (...) History enthusiast gather weekly to enact past events, television history programs are aired to good ratings, living museums hire costumed performers, civic governments sponsor local performances on historical themes, tourists "follow the steps" of earlier travellers, and academics venture into public history. Reenactment thus spans diverse history-themed genres—from theatrical and "living history" performances to museum exhibits, television, film, travelogue and historiography. (...) Perhaps symptomatic of a broader public interest in history reenactment has gained urgency in the West during the past decade, drawing on the combined resources of reenactors, government agencies, academic advisers, event planners, corporate sponsors, merchants, and spectators. (Agnew, 2004, pp. 327–328)

The motivation of individuals traveling from their home to different places involving themselves in reenactment activities encompasses a large range, from merely fun and leisure pursuit to unique and profound ways of satisfying personal of professional aims. In the later position are the scholars, with historians considering reenactment a method to test some assumptions used in historiography (Agnew, 2004). Memorialists also used reenactment as an investigation method, preferring the term "foot-stepping" (Richard Holmes quoted in McCalman & Pickering, 2010, p. 14).

In respect to reenactors, Howard (2003) distinguished two categories of reenactors: the scholarly group who seek greater authenticity and "the so called *the Booze and Bash Brigade*, who unashamedly enjoy weekends dressed up, drinking rather heavily and having a fight. Perhaps this group has realized that the search for authenticity is completely vain unless done with seventeenth-century minds and with death on the battlefield a real possibility!" (Howard, 2003, p. 258) he said. Studies show that reenactors rarely, however, spoke about how they perceive their role (if existent) in the public agora.

In the tourists' case, recent literature examining the experience that tourists have at heritage sites stress the importance of the level of engagement tourists have with the site (Vieira et al., 2020). The emotions experienced by tourists at heritage sites are of great importance in relation with the genesis of destination loyalty (Rahman et al., 2021; Sánchez-Sánchez et al., 2020). Considering this findings, the site managers or the event planners responsible for the design of activities and the level of involvement they design and provide for tourists in the framework of the agenda of the events could make a significant difference. Evidence of the use of reenactment representations of festivals as part of the strategies to attract tourists as a site bring Waldemar et al. (2022) regarding the reenactment of antique blacksmithing, gilding, bronzing and other crafts held in the proximity of the Cultural-Archaeological Centre in Nowa Slupia. Studies show that visitors are nowadays more engaged that ever at the end of two decades of experiential tourism (Haddouche & Salomone, 2018; Mueke & Schänzel, 2019; Tripadvisor, 2019).

In respect of tourists' involvement, festival managers and the reenactment performances planners can tune the fine-tuning of the design of the activities on a continuum. At one end of this continuum, one could find the maximum involvement of the tourists in the activities reenacted. At the other end of the spectrum, they can choose simply to reenact a certain event or competition in front of a passive audience, which remains passive during the entire day of scheduled activities.

3 Methodology

The present study reveals the preliminary results of a more complex research on the interactive rapport between reenactment, the heritage site as a tourism resource and visitors. We used a mix of qualitative–quantitative methodology based on both intercept and online survey with 123 reenactors and 294 visitors and participants to reenactment events in 20 different heritage tourism destinations in Romania. In addition to the survey data some conclusions are drown based on participant observation at reenactment festivals or smaller reenacting events in Romania such as *Apulum* (Alba Iulia), *Dac Fest Sub Semnul lupului* (English: *Dac Fest Under the wolf sign,* Măgura Uroiului), *Sighişoara Medieval Festival,* Romania National Day on 1 December (Alba Iulia), *History Days* and *Cluj Napoca Days* (Cluj Napoca).

We used both close-ended items questionnaires and semi-structured interviews on reenactors and questionnaires and participant observation on tourists measuring multiple individual variables like destination affective loyalty, place attachment and other destination/event-related variables. The semi-structured interview guide, the main themes investigated were motivation for reenactment, perceived role of reenactment in public agora, Romanian reenactment characteristics and the most memorable experience involving audience.

4 Results and Discussions

The role the reenactors play in the development of a tourism destination needs to be understood, in the natural context of different policies (at national, regional or local level) regarding funding and management, the required conditions for heritage preservation vs tourism exploitation, especially in the case of heritage sites situated in small scale settlements. In these situations, the local authorities' perspective becomes relevant. Reenactment festivals take place in different contexts, urban or rural, in settlements with different economic and social development strategies, with different perspectives of the local authorities on the role that heritage sites should play in the development of the areas as attracting visitors and building a tourism destination for example. These differences in perspectives regarding how the sites should be managed and financed are reflected sometimes in the perspective regarding the reenactment activities at a site in general, and in particular, local authorities' policies regarding two elements pertaining to the development of reenactment-based tourism activities. The first considers the pertinence of holding reenactment festivals near the conservation area of a site. The second pertains to authorities' views on budgeting financial support. On these issues both, our field data and literature revealed different policies, and the topic seems to be a relevant issue for reenactors. The reenactors from Great Britain, Italy and Bulgaria all brought up in their interviews the issue of different policies: "*We fought with the authorities for years, for the permit* [n.a. for the festival *Natale di Roma*] *(...) In Italy the authorities are not very keen to give permit, if we ask for Ostia Antiqa (...) they say "No". They prefer to hire the area for music and concerts because of the money* [the gesture for money of rubbing fingers] *(...) in Britain they are softer on some things, for example they give you the permit but not any resources"* (*Romanitas*, Italy). In the British context: "[Informant 1]: *We don't really have anything like this* [authors' note: referring to an international festival like Dac Fest, Romania] *in Britain* [Informant 2] *It is a budgeting thing, Here* [n.a. in Romania] *the authorities wants to show the heritage (...)* [Informant 1] *In Britain it use to be quite a few multi-group festivals, the funding dried up, the local council as the main organizer reduced their budget year on year, the first thing to go is the silly guys who dressed like Romans for the weekend basically. (...) being able to participate here* [a.n: Dac Fest] *and see it [//Informant 2: "Alive and kicking"] as wonderful as it is ... I genuinely hope that this doesn't change."* (*Legio Secunda Augusta*, Great Britain). *"Reenactment here is more popular. Local authorities prioritize the festivals, rather than in Bulgaria. So, reenactment here is more important as a whole.* (*Legio I Italica*, Bulgaria). One vice-president of a Romanian reenactment association explained for their local context: "*In my opinion the most prominent characteristic of the reenactment in Romania is the organization and functioning of reenactment festival based on contracts, for a certain amount of money. Abroad this is rare. They don't really spend* [n.a. referring to authorities spending money on reenactment performances]. *Rather the reenactors are spending money; in our case is the other way around. (...) reenactment started in Romania with people that did not had the financial capacity to make their own costumes or equipment. A good*

Roman equipment could cost around 2000 euro and you cannot produce it by yourself. This was the logic. Abroad, the reenactors are the ones who manage to make it by themselves. In general, terms they have their own gear and then the participation to a festival is not financially compensated. The expenses payed abroad are the one generated by travel costs. So the financial terms are different in Romania when it comes to reenactment." In France, some reenactors reported that in certain archaeological sites, the festival still take place in the site, but once every two year. The year of the pause is dedicated to the advancement of the archaeological excavations. Reenactors consider that reenactment performances held in or near a historical site gain in authenticity and profoundness. *"From here, from Măgura Uroiului, one can see Orăştiei Mountains, where the Dacians citadels were located. A Roman road passed by this place, here was Petris settlement, considerably enlarged by Romans, so we are in a historical site" (...) The festival at Măgura Uroiului is unique, considering the scale, the location in a special archaeological site"* (referent for the General Direction of Heritage Monuments' Management and Tourism Promotion (GDHMMTP) and member of *Terra Dacica Aeterna* (TDA) Reenactment Association quoted by Scarlat, 2022).

In our opinion, the causality mentioned by the member of *Apulum Guard* should be framed in the general working hypothesis that explains why some local authorities are more inclined to allow reenactment activities in the best location and are budgeting reenactment and reenactment festivals:

(a) Differences in the long-term development strategies for the area where the heritage site is located and the role tourism should play in its development. In the situations that tourism is perceived like a sustainable lever for the development of the area authorities are probably more incline to allow reenactment festivals in the best historically accurate locations and budget funding for reenactment. The deciding boards positioning a heritage site as part of the tourism resources of an attractive tourism destination are the ones more generous with the funding of reenactment (festivals, the logistics of the associations and reenactors' participation).

(b) Differences in the stages of the life cycle as a tourism destination registered by different destinations in which these reenactment performances take place. Considering Butlers' (1980) *Tourism Area Life Cycle Model,* some reenactment festivals take place in locations that could be appreciated to be in their involvement or even exploration stage and that could explain why authorities decide to be supportive with activities like reenactment. Considering the Romanian contexts (varying from region to region, or municipality to municipality), indeed authorities in areas that could be evaluated as being in their involvement, or even exploration stage as a tourism destination (like the area Simeria–MăguraUroiului–Orăstie–Orăştie Mountains, Drobeta-Turnu Severin, Zalău-Porolisum, Zimnicea, etc.) are reported to be supportive and budgeting on the long term reenactment performances and festivals as a part of their tourism development strategy. Other heritage destinations hosting reenactment performances are already in their development stage (like Alba Iulia, Sighişoara and ClujNapoca), but reenactment was budgeted in some cases, for nearly two decades in their development strategies. At the other end of a continuum, when reenactors request that a highly appreciated international reenactment festival (of 2000–3000 reenactors) like *Natale di Roma* to take place in the historical center of Rome (Circus Maximus, via Colosseum) no wonder the attitude of the local authorities is different. Rome is one of the most notorious tourism destinations in Europe from centuries, probably past its stagnation stage in terms of its life cycle, and prioritizing the limitation of a peak tourism flux in the historical center is not surprising. Nevertheless, this hypothesis does not cover all situations so future literature should explore other correlations.

4.1 The Direct Impact of Reenactors on the Development of a Heritage Site as a Tourism Destination

In some occasions, the reenactors contribute to the development of a heritage site as a tourism destination in the most direct manner, as members of the local administrative structure or the management of museums. They also contribute by initiatives such as founding a reenactment festival at an archaeological site, a place of an important battle or, as founder of a reenactment festival that lately attracts tourists, etc. One member of the Svishtov Council of tourism (Bulgaria) spoke about his personal relationship with the Roman heritage in Svishtov, the creation of the first reenactment group and the first Roman Period festival in Bulgaria and the traction of tourists due to the presence of this activities: *"20 years ago (:) in the town that I live it was a Roman legionary fortress Legio Prima Italica. Actually, it was excavated for half a century by different teams. When I was a child, I helped every summer the persons of the excavation team. Then when I was an associate professor etcetera etcetera in university, I was elected as a municipal counsellor. So, from that position I convinced the council and the mayor to develop it because it was just an excavation. (...) And now I succeeded to take a European project, a big European project and to redesign the whole area. Now, the place is as one Italian guy said: a jewel from the Danube ... like rebuilt the legionary fortress, so, after that I realized that I need some reenactment group for activities there but not only there ...* [I: In Bulgaria] *Yes, In Bulgaria and many similar places where there is a general need for reenactment, just for visibility, to have a chance to communicate with the locals, for tourism purposes. Step by step we established one group, and then the group became bigger, and then now most of the members are not from my town but from bigger cities around, so something natural. And also, I would like to emphasize that another groups emerged from our main group. [...] People split from us; create other groups and most of our competitors started from my association. It is a natural process and it occurs not only in my association. (...) We are the first reenactment group in Bulgaria, now the biggest one ... and also we established the first historical festival (a.n: Eagle on the Danube) like this one* (a.n: *Dac Fest*, Măgura Uroiului) *in my home town, Svishtov, in 2008*

(...) one of the oldest Roman period Festival in the Balkans (...) In Bulgaria now the best and the biggest" (53 years old, Council of Tourism Svishtov).

Another reenactor spoke about the first initiative of creating Dac Fest reenactment festival: "*Everything started here, at Măgura Uroiului, from the initiative of a professor of Art School from Cluj Napoca Paul Cheptea, which founded a reenactment association* [n.a. long-time president of TDA]" (reenactor TDA and referent for GDHMMTP quoted by Scarlat, 2022).

One reenactor from *Asociația culturală pentru Istorie Vie* explained how the reenactment started in Alba Iulia as a lever to attract tourists: "*I am working for the Alba Iulia Museum and I was the initiator of the reenactment program. We had also projects, around 7 projects that we applied at the Culture Ministry for funding. At the museum we have created a reenactment department that I am in charge with (...) we combine museum activity with this type of activity. Therefore, it is part of the job. We noticed that it gains traction. If the situation was different, we would have probably stopped after one year. However, especially because it did gain interest we gathered so much volunteers in all those years now we have around 100 covering all [historical] periods. As long as we have the public, the interest from the Mallory and County's Council ... we try to honour the invites.*" (44 years old, *Asociația culturală pentru Istorie Vie/Apulum guard*).

4.2 The Level of Involvement of the Audience in the Reenacted Activities

We are tempted to believe that if the historical battles and events are unique, the reenactment of the same event should be identical in different reenactment festivals. Our results show that although the historical facts are common and accurately depicted in reenactment performances, the atmosphere created, and the interactive framework design for the activity is different depending on several sources of differentiations: (1) the perspective and decisions of local political and museums' authorities; (2) the performers' own perspective on their educational mission; (3) their reenactment experience and (4) the perspective on their profession and their mission of museums' employers. The topics more exposed to variations in the level of tourists' involvement in reenacted activities and activities' design are the reenactment of the life style characteristic for different historical periods, battle gear presentation, the writing, dressing, beauty cosmetics preparation, hobbies, art crafts, etc. One reenactor from TDA said "*First of all, I tend to consider that for the audience the reenactment performances should represent* [deeper] *knowledge: Okay, you can come today at a historic festival and maybe you did not know that an armour in real life could weight 3–4–6 kg. Maybe you did not know that a sword, in real life or a spike is more difficult to use if you don't have the necessary experience. We are happy when somebody comes, sees, tries it and goes with more information than he previously had*". (45 years old, TDA). Another spoke about its philosophy of multisensory teaching and processing the information through experimentation: "*you are educating the people in an audio-visual and tactile manner because you can give them parts of the equipment while you explain how they were made, for example, swords, blades, different types of tools used in blacksmithing, carpentering.*" (17, student, TDA). Some festivals are more engaging when it comes for audience than others. At *Dac Fest*—the festival considered the more experimental-based, visitors are invited to write on wax tables and papyrus, make a Roman mosaic, to engage in Roman children' plays, make their own antique jewelry, sewing, feel the armors of a centurion or try Roman patricians' make-up. They also can witness the reenactment of battles and rituals (varying from edition to edition), army training, gladiator's fights, blacksmiths' work, ceramics, marble and wood carving, the preparations of recipes from *De re coquinaria* recipes book of Marcus Gavius Apicius.

The quantitative data revealed that reenactors consider that their performances have a role firstly in enhancing the level of public awareness about the importance of certain historical events (Table 1). The average value (on a 7 points Likert scale) of the appreciation that their role is strictly educational (as opposed to contribute in the sapping of individual's identities or enhancing patriotic feelings, etc.) is lower, but the dispersion of the values is bigger. The average value for the visitor's sample, however, is slightly higher for the perceived (and limited to) educational role. Reenactors also considered the prioritization of reenactment performance over professional and personal life as common, the average appreciation of reenactment as becoming a lifestyle was over the medium value (Table 1). The opinions regarding the role of reenactment in the public agora and the role in their personal life were not dependent on age, level of education or number of years of reenactment. The correlational analysis revealed small indexes of correlation and indicated an indirect relationship of the age and number of years of reenactment activities with the level of prioritization of reenactment activities over personal life. In the open-ended items tourists explained how the see the impact of reenactment activities on the destination: "*Alba it is the only place you can see a cannon firing*" (student, 22); (about Suceava) "*Due to the citadel's restauration and the apparition of this festival the city became more populated in August, it become one of the strengths of the destination.*" (student, 22); (about Sighișoara) "*The medieval festival became sinonimous with the city*" (student, 20).

Table 1 Descriptive statistics of the perceived role of re-enactment by sample (reenactors and visitors)

Sample	Reenactors				Visitors			
Variable	Mean	Std. Dev.	Min	Max	Mean	Std. Dev.	Min	Max
Educational role	3.71	2.44	1	7	4.83	1.66	1	7
Awareness role	5.44	1.81	1	7	4.65	1.71	1	7
Prioritization	4.81	2.00	1	7	–	–	–	–
Lifestyle	4.50	2.21	1	7	–	–	–	–

Considering the forth lever of differentiation (the perspective of museums' employees on their mission and profession) one reenactor from Alba Iulia (a heritage site with multi-layered historical periods heritage from Roman period to Modern history) explained: "*We see reenactment as part of our work. Some of us are working for the Alba Iulia Museum [...], and we reenact also the WWI, now we cover the Coronation in 1922 also. We have also something on the medieval period, there are historical characters that we have to embody (...) and we have been doing that for roughly 12 years and apart from that, we have our job description and professional attributions to fulfil. This is extra, like a hobby that we do apart from our daily duties.*" (*Apulum Guard*).

4.3 Places Where You Can Engage in Battles with Romans

More than a decade ago McCalman (2004) and Cook (2004) interrogated on the possibly of reenactment to play a more dynamic role and to represent a more accessible way in conveying history knowledge. This is the case of Alba Iulia where during *Apulum* the reenactors narrating had the brilliant marketing idea to transcend the limit between reenactment and tourism animation and make a tradition from inviting children in the audience to reinforce the Dacian defeated army in the aftermath of a massive battle reenactment moment between the Romans and Dacians. We interviewed the reenactor presenting the public demonstrations during the festival. On one hand, he revealed his role in both the local museum management team and the strategic reenactment department he runs, and on the other hand, the fact that building on the success of the idea of allowing the children in the audience to "get involved in the battle with the Romans" became a strategic idea. He quoted some testimonials of parents traveling every year since the first festival, from all of the country, hundreds of kilometers (for example, Craiova), for a weekend break in Alba Iulia in order for their children to assist the events, but more important, participate in "the battle against the Romans." We also observed in the 2018s edition that the presenter, in his public discourse, enhanced this reputation of the destination always stressing the uniqueness of the opportunity in this particular destination *as the only place where you can "engage in battle with the Romans"* (Fig. 1).

The separation between reenactment and the following tourism animation was evident for adults, but not so evident for children. The children take it very seriously. They wait and prepare for this moment the whole year being more prepared for those unique moments each edition. The actual reenactment moment of the battle from *Tapae*, the second war (105–106) that Dacians fought against Romans, ends sadly with Dacians being decimated. Observing the popularity of the initiative in 2018, indeed, the moment the audience has been invited to come and fight on the Dacians' side, the younger wing of the audience, being on their marks, crawled from their seats on the battlefield armed with more or less period-appropriate weapons. They did not just fought with the Romans (and not just once), they were cast, by their own will and consideration in two groups (labeled *"punished"* and *"unpunished"*), were

Fig. 1 At *Apulum* Festival (2018). In the aftermath of the second *Tapae* battle the audience was invited to fight the Romans on Dacian's behalf. In the center *"the punished"* team of children

engaged in serious dialogs (displaying their previous knowledge from the festival and historical facts learned in school) and others not so serious for the amusement of the audience: Presenter: *"What do we say when we want to resurrect the Romans?* Children (chorus): *Resuscita.* Presenter: *What do we say if we want to engage in battle with the Romans?* Children (chorus): *Pugnate.* Presenter: *We have a special mission for you, <the punished>. What is our special mission?* One child in the microphone: *To defeat the Romans.* Presenter: *No, to run five laps of the stadium.* [Children confused and lacking in enthusiasm are taking the position to run.] The presenter: *Relax, I am joking* [children relax] … *three laps.* [Some children protest frustrated and confused, audience laughter]."

For an advised observant eye, the multitude of weapons children had, some of them with their porphyry clothing and almost Roman-like sandals in leather threads (Fig. 1, left side) could not help himself to wonder: How did they happen to have/wear these at the right time and from where did they purchase them? The fact that children come battle and historical period appropriate dressed only confirm the tradition, their anticipation of the event and intentionally dress attendance to the festival. Following a tour of the festival the same observer could notice the multitude of mobile shops selling swords and shields, Dacian hats and symbols, wolf battle banners etc. So firstly comes the consistent tradition of the activities designed crystallizing the target group, secondly this attracts shops, vendors, and then more and more visitors and vendors expanding the festival and increasing the range of tourists targeted groups. We had the chance to observe numerous other Dacians–Romans battles reenacted, some presented even by the same presenter, but in none of these occasions, the audience has been invited to fight the Romans. *Apulum* deliberately remains "the only place you can engage in battle with the Romans". Alba Iulia—it is a city where the Vauban

Citadel was selected as the competitive advantage for the tourism destination along with the promise of the brand: the fact that here was decided the unification of all Romanian provinces as a single modern state. These already come with their tourists' targeted groups. For an urban medium size tourism destination extending the targeted subgroups firstly by adding a reenactment festival, and later by increasing its appeal to the families with children is not a small thing to achieve.

Our findings match the previous research conclusions that the emotions and the level of involvement of the tourists in the destination have a direct impact on the destination loyalty genesis (Rahman et al., 2021; Sánchez-Sánchez et al., 2020). Moreover, the level of involvement and the activities design are decisions that reenactors take sometimes individually, sometimes at the level of the entire group depending on their group negotiated perceived vision and mission.

References

Agnew, V. (2004). Introduction: What is reenactment? *Criticism, 469*(3). ProQuest Research Library. Retrieved from www.academia.edu/1009391/What_is_Reenactment?. Accessed 15 November 2018.

Butler, R. W. (1980). The concept of a tourism area cycle of evolution: Implications of resources management. *The Canadian Geographer, 24*(1), 6–12.

Carvalho, A., Fernandes, J., & Moreira, V. (2020). The importance of cultural events for the promotion of the territory: The case study of the medieval fair in Torre de Moncorvo. *Smart Innovation, Systems and Technologies, 2*, 26–41. https://doi.org/10.1007/978-981-33-4260-6_3

Cook, A. (2004). The use and abuse of historical reenactment: Thoughts on recent trends in public history, *Criticism, 469*(3) [Online]. Retrieved from https://www.jstor.org/stable/23127329?seq=7#metadata_info_tab_contents. Accessed the 1st December 2018 at 12.30.

Gapps, S. (2003). Authenticity matters: Historical re-enactment and Australian attitudes to the past. *Australian Cultural History*, 23. Retrieved from https://trove.nla.gov.au/work/39329703?q&versionId=52170563. Accessed 15 November 2018.

Haddouche, H., & Salomone, C. (2018). Generation Z and the tourist experience: Tourist stories and use of social networks. *Journal of Tourism Futures, 4*(1), 69–79.

Hall, D. (1994). Civil war reenactors and the postmodern sense of history. *Journal of American Culture, 17*(3), 17–23.

Howard, P. (2003). *Heritage: Management, interpretation, identity*. Continuum.

Malaescu, S. (2021). Place attachment genesis: The case of heritage sites and the role of reenactment performances. In V. Katsoni & C. van Zyl (Eds.), *Culture and tourism in a smart, globalized, and sustainable world*, Springer Proceedings in Business and Economics (pp. 435–449). https://doi.org/10.1007/978-3-030-72469-6_29

Malaescu, S. (2022). Visitors at heritage sites: From the motivation to visit to the genesis of destination affective loyalty. In V. Katsoni & A. Şerban (Eds.), *Transcending borders in tourism through innovation and cultural heritage*, Springer Proceedings in Business and Economics (pp. 571–586), *8th International Conference of the International Association of Cultural and Digital Tourism*, IACuDiT 2021 Hydra Island September 2021 through 3 September 2021. https://doi.org/10.1007/978-3-030-92491-1_34

McCalman, I. (2004), The little ship of horrors: reenacting extreme history. *Criticism, 46*(3). Retrieved from https://www.jstor.org/stable/23127328. Accessed 1 December 2018.

McCalman, I., & Pickering, P. (2010). *Historical reenactment: From realism to the affective turn*. Palgrave Macmilian.

Mueke, V. R., & Schänzel, A. H. (2019). A tourism inflex: Generation Z travel experiences. *Journal of Tourism Futures, 2*, 127–141.
Park, H. J. (2010). Heritage tourism emotional journeys into nationhood. *Annals of Tourism Research, 37*(1), 116–135. https://doi.org/10.1016/j.annals.2009.08.001
Popa, C. N. (2016). The significant past and insignificant archaeologists. Who informs the public about their 'national' past? The case of Romania. *Archaeological Dialogues, 23*(1), 28–39. https://doi.org/10.1017/S1380203816000064
Rahman, M. S., Abdel Fattah, F. A. M., Hussain, B., & Hossain, M. A. (2021). An integrative model of consumer-based heritage destination brand equity. *Tourism Review, 76*(2), 358–37325. https://doi.org/10.1108/TR-12-2019-0505
Rojas, A. (2010). The uses of cultural heritage in Catalonia: The history re-enactment as way of preservation, exhibition and local development. In *Proceedings of the 2nd International Conference on Heritage and Sustainable Development, 22–26 June*. Green Lines Institute for Sustainable Development, Évora, Portugal.
Rojas-Rabaneda, A. (2012). Cultural events, cultural heritage, tourism and new media. *International Journal of Web Based Communities, 8*(1), 73. https://doi.org/10.1504/ijwbc.2012.044683
Sánchez-Sánchez, M. D., De-Pablos-Heredero, C., & Montes-Botella, J. (2020). A behaviour model for cultural tourism: Loyalty to destination. *Economic Research-EkonomskaIstraživanja*. https://doi.org/10.1080/1331677X.2020.1838313
Santa Cruz, G., Lopez-Guzman, T., Pemberthy Gallo, L.S., & Rodriguez-Gutierrez, P. (2020). Tourist loyalty and intangible cultural heritage: The case of Popayán, Colombia. *Journal of Cultural Heritage Management and Sustainable Development, 10*(2), 172–18819. https://doi.org/10.20944/preprints201703.0063.v1
Scarlat, D. (2022) Excursie la daci și romani, pe Măgura Uroiului, înapoi cu două milenii. *Jurnalul.ro*. Retrieved from https://jurnalul.ro/timp-liber/calatorii/excursie-la-daci-si-romani-pe-magura-uroiului-inapoi-cu-doua-milenii-905217.html. Accessed 5 August.
Tian, D., Wang, Q., Law, R., & Zhang, M. (2020). Influence of cultural identity on tourists' authenticity perception, tourist satisfaction, and traveler loyalty. *Sustainability, 12*(16). https://doi.org/10.3390/su12166344
Tripadvisor. (2019). Experiential travel trends: New skills, health & wellness. *Family Activities Come Into Focus*. Retrieved from https://www.tripadvisor.com/blog/experiential-travel-trends-health-wellness-family/. Accessed 21 January 2021.
Vieira, E., Borges, A. P., Rodrigues, P., & Lopes, J. (2020). The role of intangible factors in the intention of repeating a tourist destination. *International Journal of Tourism Policy, 10*(4), 327–3502020. https://doi.org/10.1504/IJTP.2020.112639
Waldemar, C., Jolliffe, J., & Guz, A. (2022). Heritage event as tourist attraction: The case of Dymarki Swietokrzyskie, Poland. *GeoJournal, 87*(4), 2975–2992. https://doi.org/10.1007/s10708-021-10407-4
Zhou, M., Yan, L., Wang, F., & Lin, M. (2020). Self-congruity theory in red tourism: A study of Shaoshan City, China. *Journal of China Tourism Research*, 1–18. https://doi.org/10.1080/19388160.2020.1793859
Zotica, M., & Mălăescu, S. (2015). The reenactment as tourism exploitation through heritage interpretation of heritage sites in Transylvania. *Studia Universitatis Babeș-Bolyai Seria Geographia, 60*(1), 157–175.

Accessible Tourism Businesses in the Cross-Border Region of Greece—Republic of Northern Macedonia: The Case of Greece

Dimitris Kourkouridis, Vicky Dalkrani, and Asimenia Salepaki

Abstract The purpose of the present study is to evaluate the performance of the existing accessible tourist services and facilities of the tourism businesses in five Regional Units of Northern Greece. Quantitative research was conducted in 134 selected tourism businesses. The research was conducted as part of the project "ALTER TRIP: Alternative Touristic Experience" of the program INTERREG IPA CBC Greece—Republic of Northern Macedonia 2014–2020. The results showed that generally the situation in the study area is moderate, regarding the accessibility of the tourism businesses. The goal is always for accessibility to be universal and ubiquitous, so we conclude that there is room for improvement in the accessibility of tourism businesses in the study area. The most encouraging finding of the survey is the fact that those involved in tourism in the study area are willing to be educated and trained about accessibility, and also, they are willing to invest money in their businesses' accessibility. The article concludes with some suggestions for improving the accessibility of tourism businesses in the study area, based on the barriers that emerged from the research.

Keywords Accessibility · Accessible tourism · Tourism business

JEL Classification The article belongs to the following JEL Category: Z. Other Special Topics, Z3. Tourism Economics, Z32. Tourism and Development

D. Kourkouridis (✉)
Business and Exhibition Research and Development Institute, 54636 Thessaloniki, Greece
e-mail: kourkouridis@yahoo.gr

V. Dalkrani
Business and Exhibition Research and Development Institute, R&D Manager TIF—HELEXPO SA, 54636 Thessaloniki, Greece

A. Salepaki
School of Spatial Planning and Development, Aristotle University of Thessaloniki, Greece, 54636 Thessaloniki, Greece

© The Author(s), under exclusive license to Springer Nature Switzerland AG 2023
V. Katsoni (ed.), *Tourism, Travel, and Hospitality in a Smart and Sustainable World*, Springer Proceedings in Business and Economics,
https://doi.org/10.1007/978-3-031-26829-8_33

1 Introduction

Disability is a phenomenon very prevalent in society and, in one way or another, familiar to all people, an element of their direct and indirect social environment (Harisis, 2020). It is a fact that globally over a billion individuals are disabled in some way. This equates to around 15% of the global population, with up to 190 million (3.8%) people aged 15 and older experiencing substantial difficulties in functioning (WHO, 2021). In the European Union, people with disabilities account for around 15% of the population. This indicates that around 80 million individuals with disabilities reside in all EU nations, deprived of possibilities to engage fully in social and economic life due to physical and mental obstacles. Moreover, because of socio-demographic changes associated to aging society, adapting built environment—including hotels—to the requirements of older and disabled people is becoming increasingly important (Szewczyk, 2015).

In all societies, there are still barriers that prevent people with disabilities from exercising their rights and freedoms equally. These obstacles make it difficult for them to participate fully in the activities of their societies. The design of tourist facilities is considered the main obstacle preventing disabled tourists from vacationing and traveling. The limitations and difficulties faced by these people are not only part of the operational limitations, but also of the inability of the society—in this case also of the tourism businesses—to take into account their needs and peculiarities.

The present study concerns the performance of the existing accessible tourist services and facilities, specifically of the tourism businesses of the Greek part of the cross-border area of Greece—Republic of Northern Macedonia. As part of the evaluation, a quantitative survey was conducted using a properly designed questionnaire, in 134 selected tourism businesses in the study area. The research was conducted as part of the project "ALTER TRIP: Alternative Touristic Experience" of the program INTERREG IPA CBC Greece—Republic of Northern Macedonia 2014–2020. ALTER TRIP's major purpose is to assist in transforming the intervention area into a tourism attraction that is accessible to everybody, without exceptions. Furthermore, the program aims to promote the cross-border region's sustainable tourist growth.

2 Literature Review

2.1 *People with Disabilities*

The term "disability" is defined by multiple interest groups and professions in different ways, making it almost impossible to achieve a single theoretical definition or a single model that explains disability as a phenomenon. According to the World Health Organization, human functioning issues are divided into three categories that are all interconnected: (WHO, 2011: 5):

Impairments are problems in body function or alterations in body structure—for example, paralysis or blindness.

Activity limitations are difficulties in executing activities—for example, walking or eating.

Participation restrictions are problems with involvement in any area of life—for example, facing discrimination in employment or transportation.

The term "disability" refers to problems in one or more of the above three areas of functioning (WHO, 2011). Most disability statistics now recognize that the multidimensional concept consists of (a) mobility, (b) hearing, (c) vision, (d) cognitive/learning, (e) mental health, and (f) sensitivities and long-term health conditions (Darcy & Dickson, 2009). However, the way disability is usually presented and understood, as a concept and as a situation, through the current perception, that is, almost exclusively as a fact of personal tragedy or as a personal misfortune, obscures its essence and makes it difficult to deal with (Harisis, 2020).

Disability is a social phenomenon, governed by specific relationships and connections within society and as such must be understood and addressed. One way of underestimating and misunderstanding disability as a social condition and phenomenon is its frequent identification with "impairment," with the result that disability is understood only in the context of medical, or narrowly biological. But disability and impairment, although closely linked, are not the same thing. Impairment is a certain loss or dysfunction of a mental, physiological, or anatomical structure, which may be the result of heredity, illness, or trauma accident. Disability, on the other hand, is the limitation or lack (which of course results from impairment) of the ability to perform an activity or series of activities that are necessary for a person to be fully integrated into society. Impairment, disease, genetic problem, or physical trauma is a biological fact, but the restriction of activity is a social condition, which depends on the stage of development, the structure and the relationships that govern each society (Anastasiou & Kauffman, 2013; Freund, 2001; Harisis, 2020; Samaha, 2007; Terzi, 2004).

2.2 *Accessible Tourism*

The term "accessibility" refers to the characteristics of the environment, which allow all citizens without discrimination of gender, age, and other characteristics (physique, strength, perception, etc.) to be able to independently and safely approach and use the infrastructure, services (conventional and electronic) and goods offered (NCDP, 2008). Access discussion focuses on enabling social participation by providing access requirements (e.g., ramps, tactile ground surface indicators, wayfinding signage, etc.) (Darcy & Dickson, 2009).

Accessibility is very important because people with disabilities in an environment with access to building infrastructure, transport, and information and communication technologies, can develop, move, study, and work on an equal footing (NCDP, 2008). Recognizing that an individual with a multidimensional access problem will require multiple levels of access provision is one of the complexities associated with such a construct. A person with cerebral palsy, for example, may have a mobility dimension that necessitates the use of a wheelchair or crutches, as well as a communication dimension that necessitates the use of a communication board due to associated speech impairment. However, the complexity concern is much more complicated: This person's access needs are considerably distinct from those of a person with arthritis, who may necessitate a continuous pathway with handrails to assist in weight bearing, seats to provide resting areas, universal handles on doorways, and taps designed to assist reduced dexterity (Darcy & Dickson, 2009).

Accessible tourism is a set of products, methods, and services that are designed, promoted, distributed, and provided to meet specific accessibility needs and requirements of visitors regardless of age or ability (ELEVATOR, 2018), while the dynamics of this market is very promising (Alén et al., 2012; Domínguez et al., 2013; Dwyer & Darcy, 2010; Luiza, 2010). It is true that accessible tourism has grown from the idea of providing accommodations or adaptations so that persons with disabilities can participate in tourism, to a concept of quality tourism for all, with the knowledge that accessibility is an integral aspect of that quality (Darcy et al., 2020; WHO, 2016).

The concept of accessible tourism evolved along with the approach to disability. It no longer related simply to disabled visitors, but rather to the adaptation or accommodation of the tourism sector to make it accessible to all population groups. The current trend is to combine the concept of accessible tourism with the overall concept of Universal Design by incorporating other more global features such as (a) an increase in the number of people involved; (b) a link between disability and elderly people; (c) accessibility as a helpful component for as many people as possible; and (d) Universal Design or Design for All as a means of attaining accessibility (WHO, 2016).

Academic and politics interest in advancing the concept of accessible tourism has led to more comprehensive definitions of the concept that have evolved to be wider and more inclusive in recent years. These new definitions of accessible tourism include the following (WHO, 2016: 18):

> *Accessible tourism is the ongoing endeavor to ensure tourist destinations, products and services are accessible to all people, regardless of their physical limitations, disabilities or age. It encompasses publicly and privately owned tourist locations. The improvements not only benefit those with permanent physical disabilities, but also parents with small children, elderly travelers, people with temporary injuries such as a broken leg, as well as their travel companions.*

It should also be noted that there are several misconceptions around accessible tourism and accessibility in general. The most common myth is that accessible tourism only concerns people with disabilities. Despite the fact that people with disabilities have the highest need for accessibility, accessible tourism caters to a wide range of customers with varying access requirements (not always visible), which can be caused by impairment, illness, injury, age, stature, foreign language proficiency, or culture. Accessible locations shall be prioritized by anybody who has difficulty completely accessing, utilizing, or enjoying tourist services and facilities in a comfortable, safe, and autonomous manner (ELEVATOR, 2018; WHO, 2016).

It is a fact that in developed countries there has been rapid progress in the field of tourism for people with disabilities, at least in recent decades, but literature shows that there are still many difficulties and barriers for the provision of accessible tourist services. McKercher and Darcy (2018) propose a comprehensive framework for the various constraints that affect people with disabilities' capacity to travel. Barriers are categorized into four categories, moving from broad to specific. Each tier indicates a higher level of specificity, starting with general restrictions and progressing to the examination of highly personal impairment effects. More specifically, the tiers that McKercher and Darcy (2018) propose are the following: (a) common to all people—structural, interpersonal, intrapersonal, interest, (b) common to all people with disabilities—ignorance, attitude information, industry, the individual, and (c) unique to specific disabilities.

Barriers encountered in the lower layers must generally be overcome before upper tier barriers may be addressed. However, because tourism involvement is multifaceted and continuous, concerns at one tier may have an impact on obstacles at the next (McKercher & Darcy, 2018).

This approach, although complex, takes into account the holistic approach to accessibility. It is a fact that the literature focuses heavily on the accessibility of people with mobility problems, neglecting other segments of the population. However, this theoretical approach takes into account exactly what "Designing for All" involves.

2.3 *Greek Legislation—Designing Instructions "Designing for Everyone"*

The basic Law for the accessibility of the built environment in Greece is the Law 4067/2012 "New Building Regulation." This law states that the horizontal and vertical, autonomous and safe access by people with disabilities and reduced mobility and their service in all outdoor and indoor areas of buildings, must be ensured in accordance with the Designing Instructions of the Ministry of Environment, Energy and Climate Change "Designing for Everyone." The goal of the Designing Instructions "Designing for Everyone" is to remove architectural barriers and anything that prevents the autonomous movement and living of people with disabilities—and people with reduced mobility in general—inside buildings or outdoors. Tourism businesses are included in Chapter: "Buildings used for the public" of the Guidelines.

The instructions state that the entrance of each building is the most decisive element of its accessibility, at the same time determining the degree of autonomy of the movement of its users. By accessibility, we mean the possibility provided to all users of the space—even to people with disabilities—for autonomous and safe entry and exit from it, as well as use of all services offered to it. In order to ensure the accessibility of the main entrance of buildings by people with disabilities, but also all people with reduced mobility in general, the following should apply: (a) road–sidewalk connection; (b) sidewalk–building connection; (c) accessible entrance door; (d) door equipment; and (e) signage.

The installation of ramps or walkable inclined plane is recommended, that is any level that allows people and/or wheelchairs to access a building by connecting two different level surfaces. If this is not possible, then one of the following mechanical lifting systems must be selected: (a) elevator; (b) vertical movement lift; or (c) ladder lift.

Accessibility routes are also an important element of the accessibility of the buildings. Accessible route means the shortest, continuous, safe and unobstructed route, of suitable width and coverage material followed by a person with a disability and/or reduced mobility to move from one point to another. It includes level paths, ramps of appropriate slope and/or lifting mechanisms, as well as the appropriate signage in forms accessible to people with disabilities (sound, light, and embossed form).

Another necessary condition for a public building or public space to be accessible is the existence of a sanitary facility, properly designed and with the necessary equipment to serve every category of users. The term "sanitary facilities" means an area that includes sinks, urinals, and areas with W.C. basin. Each building used by the public must have a certain number of sanitary facilities, depending on their frequency of use and surface area. In all these buildings, there will be at least one toilet per floor specially designed for wheelchair users. In the case of existing buildings where it is not possible to construct separate toilets to serve wheelchair users, then it is essential that the designated toilets be designed to serve all categories of users—including wheelchair users—and provide more appropriate equipment.

3 Methodology

3.1 *Study Area*

The study area of the research includes five Regional Units of Northern Greece: (a) Thessaloniki, (b) Kilkis, (c) Serres, (d) Pella, and (e) Florina, in the cross-border area of Greece—Republic of Northern Macedonia. The population of the regional units is as follows: Thessaloniki 1,074,827; Kilkis 80,419; Serres 176,430; Pella 139,371; and Florina 51,841. Table 1 presents some useful information about the tourism businesses in the study area, while Table 2 presents some additional information about hotels.

We observe that the Regional Unit of Thessaloniki, as the largest, has the most tourism businesses. It should also be noted that the Regional Unit of Thessaloniki has a metropolitan character, as Thessaloniki is located in it (second largest city in the country) (Frangopoulos et al., 2009), while the other Regional Units have a regional character. Also, the number of tourism businesses reflects the tourism development of Thessaloniki, in contrast to the other regional cities, which are not developed to this extent.

Table 1 Tourism businesses in the study area

Region unit	Type of business	Number of businesses	Turnover		Change 2021/2020
			2020	2021	
Thessaloniki	Accommodation	575	119,257,578	313,535,224	162.9
Kilkis	Accommodation	14	2,498,168	1,945,346	−22.1
Serres	Accommodation	47	6,072,205	4,147,811	−31.7
Pella	Accommodation	149	8,093,087	8,026,094	−0.8
Florina	Accommodation	42	1,876,262	1,440,927	−23.2
Thessaloniki	Food Businesses	5,258	375,799,108	418,543,381	11.4
Kilkis	Food Businesses	582	12,233,395	11,779,502	−3.7
Serres	Food Businesses	1,334	27,631,551	30,166,255	9.2
Pella	Food Businesses	1,073	22,523,096	25,199,675	11.9
Florina	Food Businesses	489	9,026,355	10,293,018	14.0

Source Hellenic Statistical Authority, 2022

Table 2 Hotels in the study area

Region unit		5*	4*	3*	2*	1*	Total
Thessaloniki	Units	15	31	42	24	31	143
	Rooms	2,113	2,099	2,314	812	740	8,078
	Beds	3,924	3,937	4,469	1,540	1,570	15,440
Kilkis	Units	–	–	9	4	4	17
	Rooms	–	–	252	133	55	440
	Beds	–	–	515	388	127	1,030
Serres	Units	–	6	20	6	1	33
	Rooms	–	299	548	179	4	1,030
	Beds	–	706	1,142	392	9	2,249
Pella	Units	–	7	32	14	9	62
	Rooms	–	207	671	254	166	1,298
	Beds	–	430	1,422	564	337	2,753
Florina	Units	1	3	17	10	2	33
	Rooms	38	59	343	136	43	619
	Beds	82	132	715	281	14	1,334

Source Hellenic Chamber of Hotels, 2020

3.2 *Research Sampling*

The target population of the present research is the tourism businesses in the study area. Specifically, the businesses to which the research focused belong to the following categories: (a) accommodation (hotel, rooms to let—apartments, boarding house, guest house, etc.); (b) food and beverage (restaurant, cafe, bar, etc.); (c) sports activities; and combination of the above-mentioned categories.

Initially, information and data were collected for the study population, and a database consisting of 524 tourism businesses was created. A message was sent to these companies via e-mail, describing the research and its purpose, as well as the importance of their responses. The message also included the link for the survey questionnaire, which was created in Google Form. Finally, 134 of these tourism businesses were selected, which were the sample of the survey.

3.3 Data Collection Method

For the needs of the present quantitative research, a questionnaire was created, which consists of three parts. In Part A, there are questions about basic information of the tourism businesses participating in the research, Part B consists of nine questions about accessible and disabled service of the tourism businesses, and Part C consists of five questions about staff training in serving people with disabilities.

4 Results

4.1 Basic Information

Table 3 shows the results regarding the basic information of the tourism businesses participated in the survey. The results showed that 55.22% of the businesses are based in the Regional Unit of Thessaloniki, since it is bigger than the others and also includes the Metropolitan Region of Thessaloniki, which is the second bigger city in Greece. 44.78% of the businesses employ less than five employees, so they are small businesses, while 29.85% are large, as they employed more than 15 employees. Most of the tourism businesses (52.24%) have been operating more than 15 years, so they are quite experienced in the tourism sector. The results also showed that most of the businesses (61.19%) serve customers both from Greece and abroad. Finally, the sample of the survey consisted mostly of accommodation businesses (52.24%) and food and beverage businesses (32.84%), while there was a small sample of businesses engaged in sports activities (2.99%).

Table 4 presents some additional information for the accommodation, food and beverage, and sport activities businesses. It seems that most of the accommodation businesses belong to the category of 3 stars (51.43%). Moreover, 31.43% of them have more than 40 rooms, while 25.71% of them have less than 10. An important finding is the fact that 42.86% of the accommodation businesses do not have any room for people with disabilities.

Table 3 Basic information of the tourism businesses

Regional Units				
Thessaloniki	Serres	Pella	Kilkis	Florina
55.22%	8.96%	8.96%	8.96%	17.9%

Number of employees			
Less than 5	From 5 to 10	From 11 to 15	More than 15
44.78%	14.93%	10.45%	29.85%

Years of operation			
Less than 5 years	From 5 to 10 years	From 11 to 15 years	More than 15 years
17.91%	17.91%	11.94%	52.24%

Costumers' origin		
Greece	Abroad	Both
34.33%	4.48%	61.19%

Type of business				
Accommodation	Food and beverage	Sports activities	Combination	Other
52.24%	32.84%	2.99%	7.46%	4.48%

Table 4 Basic information of the accommodation, food and beverage, and sport activities businesses

Accommodation					
Category					
1 star	2 stars	3 stars	4 stars	5 stars	None
2.86%	11.43%	51.43%	14.29%	17.14%	2.86%
Capacity					
Less than 10 rooms	From 10 to 20 rooms	From 21 to 30 rooms	From 31 to 40 rooms	More than 40 rooms	
25.71%	14.29%	14.29%	14.29%	31.43%	
Number of Rooms for People with Disabilities					
None	1 room	2 rooms	3 rooms	4 rooms	More than 4 rooms
42.86%	11.43%	14.29%	5.71%	11.43%	14.29%

Food and beverage		
Type		
Restaurant/tavern	Café	Other
63.64%	27.27%	9.09%

(continued)

Table 4 (continued)

Food and beverage

Considering accessibility when placing table seats

Yes	No	I don't know
68.18%	13.64%	18.18%

Sports activities

Organization of activities for people with disabilities

Yes	No	I don't know
50.00%	50.00%	0.00%

Difficulties in organization of activities for people with disabilities

No difficulties	Lack of information/training of the staff	Lack of proper infrastructure	Other difficulties
0.00%	100.00%	0.00%	0.00%

Reason for not organized activities for people with disabilities

Did not occur	No information/training of the staff	No proper infrastructure	Other reasons
100.00%	0.00%	0.00%	0.00%

As far as food and beverage businesses are concerned, most of them (63.64%) are restaurants or taverns, while 27.27% of them are cafés. Another important thing that emerged from the survey is that 68.18% of the food and beverage businesses claimed that they considered accessibility when placing table seats. This means that they provided a lot of free space for the movement of people with wheelchairs or mobility problems, inside their businesses.

All the businesses engaged in sports activities that they do organize activities for people with disabilities, claimed that they faced difficulties due to lack of information/training of the company's staff. On the other hand, the businesses engaged in sports activities that they do not organize activities for people with disabilities, claimed that it just did not occur to organize such activities.

4.2 Accessibility—*Serving People* with Disabilities

The second part of the survey regards the accessibility of the tourism businesses and the serving of people with disabilities. 77.14% of the participants claimed that they rarely serve people with disabilities, while 17.17% of them claimed that they often serve people with disabilities. Moreover, 5.71% of them claimed that they never serve people with disabilities. Thus, the most of businesses serve people with disabilities, even on a rare basis.

Figure 1 summarizes the results obtained for the accessibility of tourism businesses, in terms of the built environment. A very important thing that emerged from

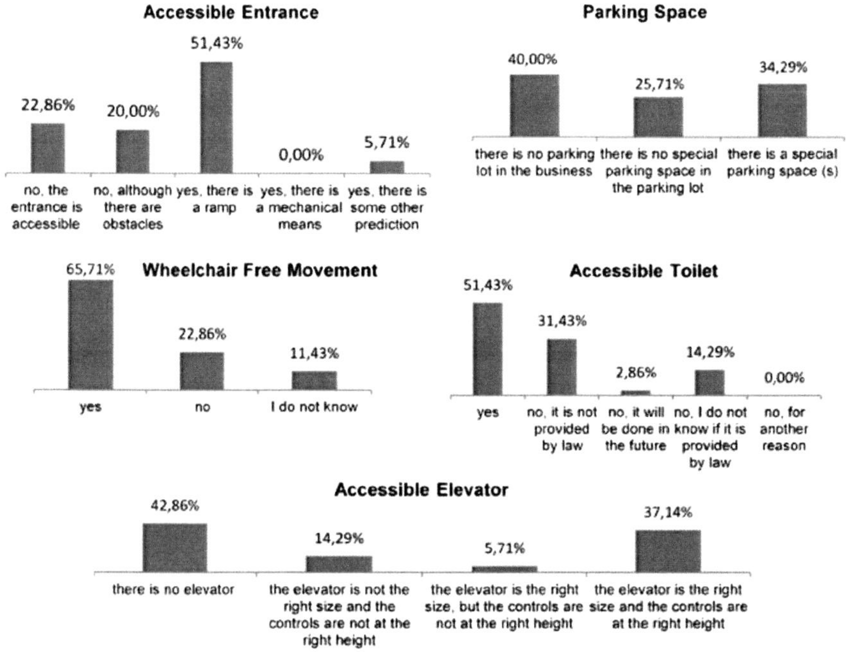

Fig. 1 Accessibility of tourism businesses built environment

the survey is that 22.86% of the tourism businesses do not have any special provision in their entrances, although there are altitude differences or other obstacles. This means that these businesses are not accessible at all for wheelchair users or people with reduced mobility. On the other hand, though, we found that 51.43% of the businesses have a ramp, and 5.71% of them have some other prediction for their entrances' accessibility.

Moreover, 65.71% of the participants claimed that a wheelchair can move freely inside their businesses, while 22.86% claimed that it cannot and 11.43% of the participants do not know if a wheelchair can move freely in their business.

About parking spots, it was found that 25.71% of the businesses do not have a special parking space for people with disabilities in their parking lots, while 34.29% do have a special parking space (or parking spaces). Moreover, 40.00% of the participants do not have a parking lot in their businesses.

Also, 42.86% of the businesses do not have an elevator. But, 26.09% of the businesses that do have an elevator, it is not accessible: it is not the right size, or the controls are not at the right height, or both. On the other hand, 65.22% of the businesses that do have an elevator, it as accessible for people with disabilities (it is the right size and the controls are at the right height).

Finally, 51.43% of the businesses have a toilet for people with disabilities and 2.86% claimed that it will be done in the future, because it is provided by legislation.

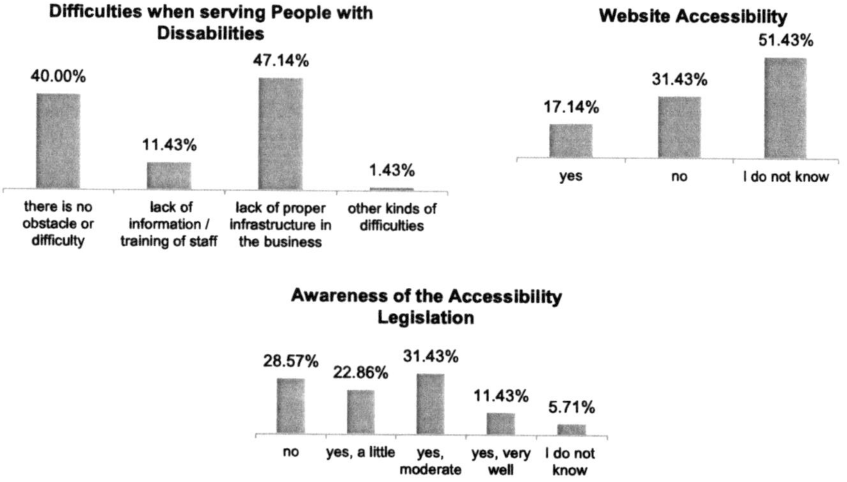

Fig. 2 Difficulties and obstacles when serving people with disabilities

31.43% do not have a toilet for people with disabilities because it is not provided by legislation, but another 14.29% do not have a toilet for people with disabilities and do not know if it is provided by law.

Figure 2 presents the difficulties and obstacles when serving people with disabilities. The main difficulties when serving people with disabilities are mostly—for the 47.14% of the participants—due to lack of proper infrastructure in their businesses. 11.43% of the participants claimed that there are difficulties due to lack of information/training of their staff, and 1.43% that there are other kind of difficulties. It should be noted though, that 40.00% of the businesses claimed that they do not face any obstacle or difficulty when serving people with disabilities.

The majority of the participants (65.71%) claimed that they do know the accessibility legislation even a little. But, a fairly significant percentage (28.57%) claimed that no one in their businesses knows the legislation, while 5.71% do not know if anyone in their businesses knows the legislation.

Finally, website accessibility is not something that the participants are familiar with. Specifically, 51.43% of them do not even know if their website is accessible, while the 31.43% claimed that their website is not accessible. Only 17.14% of the businesses have accessible websites.

Figure 3 summarizes the results obtained regarding the training of the staff of tourism businesses in matters of serving people with disabilities. The results showed that the staff knowledge about serving people with disabilities is moderate for the 40.00% of them, while 25.71% of them claimed that their knowledge is very good, and 2.86% that it is excellent. 17.14% believe that their knowledge is a little good, and 14.29% that it is not good at all.

Their experience in serving people with disabilities is also moderate for the 40.00% of the participants, while 28.57% claimed that it is very good, and 2.86%

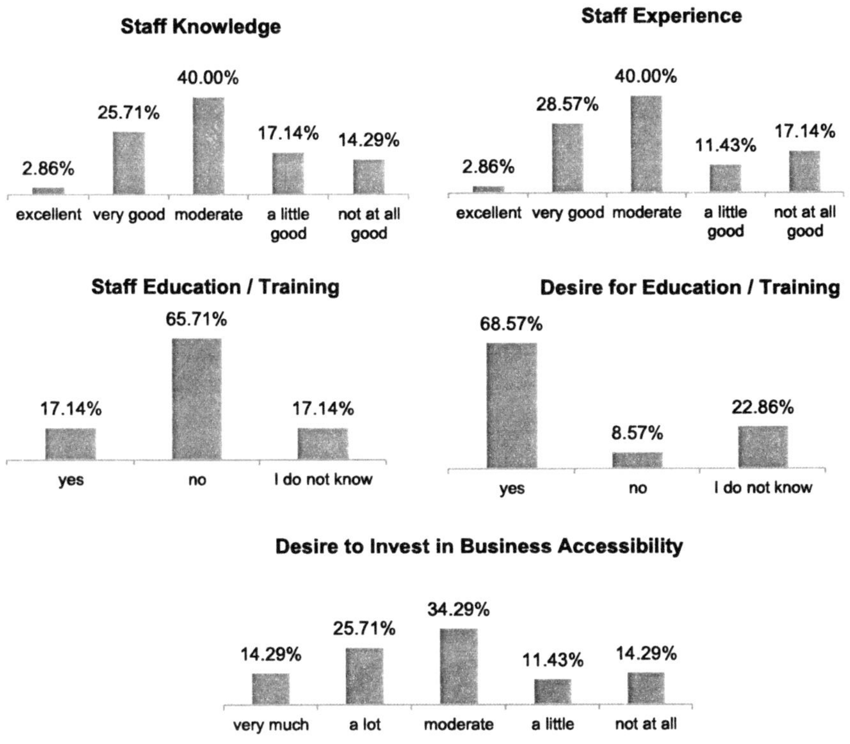

Fig. 3 Staff training in serving people with disabilities

that it is excellent. But, 11.43% of the participants believe that their experience is a little good, and 17.14% of them that it is not good at all.

It is also found that 65.71% of the businesses have never attended a seminar or other educational/informational activity on accessibility for people with disabilities, while just only the 17.14% have attended. 17.14% of the participants claimed that they do not know if anyone on their businesses have ever attended a seminar or educational/informational activity on accessibility.

A very important finding is that most of the participants (68.57%) want to be educated and trained about accessibility. Just 8.57% claimed that they do not want this kind of education/training, while 22.86% claimed that they do not know if they want it.

Finally, the participants are willing to invest money in their business' accessibility. Specifically, 25.71% of them claimed that the want a lot to invest in accessibility, 14.29% that they want it very much, and 34.29% that they moderate want it. On the other hand, 11.43% of the participants claimed that they want a little to invest in accessibility, and 14.29% of them that they do not want it at all.

Table 5 Pearson chi-square results for accessibility factor and staff factor

Variable	Pearson chi-square
Accessibility factor	
Regional unit	0.000578
Number of employees	0.000060
Years of operation	1.585E-8
Type of business	0.000905
Staff factor	
Regional unit	0.000184
Number of employees	0.081
Years of operation	0.000179
Type of business	0.000023

For the purposes of the analysis, two new variables were created ("Accessibility Factor" and "Staff Factor"). Accessibility Factor emerged as the average of the variables related to the accessibility of tourism businesses: (a) special provision at the entrance, (b) wheelchair free movement, (c) parking for people with disabilities, (d) accessible elevator, (e) toilet for people with disabilities, and (f) website accessibility, and Staff Factor as the average of the variables related to the staff of tourism businesses: (a) legislation awareness, (b) staff knowledge, and (c) staff experience. Table 5 presents the results obtained from the chi-square test of independence statistical hypothesis for these factors in relation to four variables: (a) regional unit, (b) number of employees, (c) years of operation, and (d) type of business. The results showed that Accessibility Factor is related to all the variables, while Staff Factor is related to all variables except from number of employees.

5 Conclusion

5.1 *Research Conclusions*

The purpose of the present research was to evaluate the performance of the existing accessible tourism services and facilities, specifically of the tourism businesses of the study area (regional units of (a) Thessaloniki, (b) Kilkis, (c) Serres, (d) Pella, and (e) Florina). As part of the evaluation, a quantitative survey was conducted using a properly designed questionnaire, in 134 selected tourism businesses in the study area.

The results showed that generally the situation in the study area is moderate, regarding the accessibility of the tourism businesses. The first important thing that emerged is that a significant number of the accommodation businesses (almost 43%), do not have any room for people with disabilities. Therefore, in these facilities people with disabilities are virtually excluded. Another drawback is the fact that almost

23% do not have any special provision in their entrances, although they are not accessible. This is also an exclusion factor from the facilities. Moreover, there is no special parking spot for people with disabilities in almost 26% percent of the businesses, while their elevators—when there is one—is not accessible in almost 26% of the businesses. Also, the website accessibility was found to be an issue that the participants are not familiar with. We also found that almost 51% of the businesses have a toilet for people with disabilities and another 3% will have in the future due to legislation provision. But this is also an exclusion factor from the facilities that do not have a toilet for people with disabilities.

The above-mentioned obstacles are also reflected in the responses of the participants who stated in a percentage of almost 47% that they face difficulties when serving people with disabilities due to lack of proper infrastructure in their businesses.

But we should also note some positive facts that emerged from the research. The first thing regards 66% of all participants claimed that a wheelchair can move freely inside their businesses. Moreover, the majority (almost 66%) stated that they know the accessibility legislation—even a little. While there are some positive results, it should be noted that the goal is always for accessibility to be universal and ubiquitous. Therefore, we conclude that there is room for improvement in the accessibility of tourism businesses in the study area.

Regarding the staff knowledge and experience in serving people with disabilities, it was found to be moderate for the 40% of the participants. Perhaps the most encouraging finding of the survey is the fact that those involved with tourism in the study area are willing to be educated and trained about accessibility (almost 66% of them), and they are also willing to invest money in their businesses' accessibility (almost 60% want it either a lot or very much).

5.2 Suggestions for Improving the Accessibility of Tourism Businesses

Participation in tourism is an inalienable right of all people, and no one should be excluded from it. In an inclusive society, barriers and obstacles must be removed, thus allowing everyone to experience unhindered travel experiences. Tourism businesses are perhaps the most important segment of the tourism market, so their accessibility is a key factor in the decision of people with disabilities to participate in tourism.

Given the ever-increasing size of the market for people with disabilities and the elderly, targeting specific segments of the population should be a priority for tourism businesses. It should also be understood that the accessibility of tourism businesses is not limited to specific population groups, but potentially concerns all tourists. Thus, accessibility is an important quality factor for tourism businesses, offering a significant competitive advantage.

The analysis carried out in this survey shows that achieving accessibility is not a time consuming or costly process. What is most important is the proper information of the entrepreneurs about the specific needs that they have to serve. The most important points that tourism entrepreneurs should focus on are the following:

Ensure that their businesses have accessible entrance. In case there are altitude differences at the entrance of the business, there are simple and cheap solutions, such as the installation of a ramp.

Ensure that there is sufficient free space inside the business for wheelchair free movement. Modifications in this case may simply involve a change in the layout of the equipment.

For accommodation businesses, they need to provide at least one accessible room. The investment required to convert an accessible room is not large. It is enough to provide free space for wheelchair movement, to place suitable handles in the required places and to ensure that the equipment of the room (bed, tables, etc.) are at the appropriate height.

For food and beverage businesses, they are required to provide free space for the free movement of wheelchairs, to have equipment at the appropriate height (tables, etc.) and to have a toilet for the disabled.

Tourism businesses that have a parking space should provide parking spot for people with disabilities, which will be as close as possible to the entrance of the business, and there will be an unobstructed route to get there.

Tourism businesses that have an elevator should ensure that it is accessible. If there is no possibility of replacement, then there are cheaper solutions, such as wheelchair lifts.

Tourism businesses, like any other public service, must have toilets for the disabled, which will meet the requirements and specifications of accessibility.

The staff of tourism companies should be properly trained to serve people with disabilities. There are many training programs available on the internet and in many cases, they are low cost or no cost at all.

By ensuring the above, tourism businesses gain a significant competitive advantage, while contributing to the social integration of people with disabilities.

References

Alén, E., Domínguez, T., & Losada, N. (2012). New opportunities for the tourism market: Senior tourism and accessible tourism. *Visions for global tourism industry: Creating and sustaining competitive strategies*, 139–166.

Anastasiou, D., & Kauffman, J. M. (2013). The social model of disability: Dichotomy between impairment and disability. *The Journal of Medicine and Philosophy: A Forum for Bioethics and Philosophy of Medicine, 38*(4), 441–459.

Darcy, S., & Dickson, T. J. (2009). A whole-of-life approach to tourism: The case for accessible tourism experiences. *Journal of Hospitality and Tourism Management, 16*(1), 32–44.

Darcy, S., McKercher, B., & Schweinsberg, S. (2020). From tourism and disability to accessible tourism: A perspective article. *Tourism Review, 75*(1), 140–144.

Domínguez, T., Fraiz, J. A., & Alén, E. (2013). Economic profitability of accessible tourism for the tourism sector in Spain. *Tourism Economics, 19*(6), 1385–1399.

Dwyer, L., & Darcy, S. (2010). Economic contribution of tourists with disabilities: An Australian approach and methodology. In D. Buhalis & S. Darcy (Eds.), *Accessible tourism: Concepts and issues*. Channel View Publication.

ELEVATOR. (2018). ELEVATOR Report 3.3 Accessible Tourism Guide. Policy Makers. ELEVATOR Consortium.

Frangopoulos, I., Dalakis, N., & Kourkouridis, D. (2009). Urban structure and mobility in a modern city: The example of the submerged tunnel of Thessaloniki through the empirical investigation of the citizens opinion. *International Journal of Sustainable Development and Planning, 4*(4), 333–344. https://doi.org/10.2495/SDP-V4-N4-333-344

Freund, P. (2001). Bodies, disability and spaces: The social model and disabling spatial organisations. *Disability & Society, 16*(5), 689–706.

Harisis, A. (2020). *Motor disabilities and social policy in Greece: A first approach*. National Institute of Labor and Human Resources. (in Greek).

Hellenic Chamber of Hotels. (2020). Hotel sector in Greece 2020—By region—prefecture—Island. Retrieved from https://www.grhotels.gr/ksenodocheiako-dynamiko-elladas-2020-ana-perifereia-nomo-nisi/. Accessed 9 June 2022.

Hellenic statistical authority. (2022). https://www.statistics.gr/en/home/. Accessed 9 June 2022.

Law 4067/2012. New building regulation. (OGG 79/02.04.2012).

Luiza, S. M. (2010). Accessible tourism—The ignored opportunity. *Annals of Faculty of Economics, 1*(2), 1154–1157.

McKercher, B., & Darcy, S. (2018). Re-conceptualizing barriers to travel by people with disabilities. *Tourism Management Perspectives, 26*, 59–66.

Ministry of Environment. Energy and climate change. *Designing instructions 'Designing for Everyone'*. Athens.

NCDP. (2008). Designing a disability policy. Athens: National Confederation of Disabled People—Greece (in Greek).

Samaha, A. M. (2007). What good is the social model of disability? *The University of Chicago Law Review, 74*(4), 1251–1308.

Szewczyk, I. (2015). Accessible Szczyrk hotels' features for disabled tourists. *Tourism in Southern and Eastern Europe, 3*, 369–382.

Terzi, L. (2004). The social model of disability: A philosophical critique. *Journal of Applied Philosophy, 21*(2), 141–157.

WHO. (2011). World report on disability 2011. World Health Organization.

WHO. (2016). Manual on accessible tourism for all: Principles, tools and best practices—Module I: Accessible tourism—Definition and context. World Health Organization.

WHO. (2021). Disability and health. World Health Organization. Retrieved from https://www.who.int/news-room/fact-sheets/detail/disability-and-health. Accessed 9 June 2022.

Predicting the Impact of Cruise Ships on Island Destinations

Oscar Trull Domínguez, Cesar Gómez-Palacios, and Angel Peiró-Signes

Abstract The increase in tourism and population in highly touristic places produces an enormous influence on the resources of the destinations, both economically and environmentally. In general, the indicators that are used to monitor the number of tourists in a destination measure the flow of visitors but do not pay attention to the specific daily situation. The arrival of large cruise ships notably increases the pressure on island destinations. The human pressure indicator (HPI) makes it possible to know the instantaneous situation of the population in highly touristic places, and especially on islands, including both tourists and residents. In this work, we have implemented a new HPI model that includes the influence of cruise ships arriving at islands and we apply its study to the Hawaiian Islands, where we evaluate the real pressure exerted on them, so that we can appreciate the influence of cruise ships on them. The index can be used to better manage destination capacity and resources. The article describes the calculation methodology and the conclusions reached with the application.

Keywords Human pressure indicator · Tourism · Cruise ships · Destination management

JEL Classification L83 · F6 · Q56

1 Introduction

Tourist destinations have shown constant growth in the number of visitors for several decades. The means of transport have evolved and improved, and, more specifically, recreational transport is able to offer increasingly accessible travel options for these destinations. The Hawaiian Islands are an example of this tremendous growth. The islands are located in the Polynesian area of the Pacific Ocean, and the number of arrivals of transoceanic flights continues to increase, in addition to domestic flights. The number of visitors to the islands has increased exponentially, rising from 100,000

O. T. Domínguez · C. Gómez-Palacios · A. Peiró-Signes (✉)
Universitat Politècnica de València, Valencia, Spain
e-mail: anpeisig@omp.upv.es

© The Author(s), under exclusive license to Springer Nature Switzerland AG 2023
V. Katsoni (ed.), *Tourism, Travel, and Hospitality in a Smart and Sustainable World*, Springer Proceedings in Business and Economics,
https://doi.org/10.1007/978-3-031-26829-8_34

visitors in 1960 to more than 3,000,000 in 2013. It is estimated that by 2030 this figure will increase to 7,000,000 (Dator, 2019).

The increase in the influx of visitors translates into greater use of transport, energy, and use of the islands' own resources. To measure the impact produced by tourism on the Islands, movement flow indicators are usually used, such as the arrival of travelers, days of stay, and other economic indicators, which are analyzed separately from the resident population indicators. These indicators can be obtained from the Hawaii Tourism Dashboard (Hawaii Tourism Dashboard, 2022).

The use of specific indicators, such as the daily or monthly human pressure indicator (DHPI, MHPI), offers the advantage of being able to unify both ways of measure in a single criterion and measure the real population in tourist destinations.

To use the HPI on the Hawaiian Islands, there is a need to adapt it to the particular circumstances of the Islands, including the arrival of large transoceanic cruise ships. Cruise ships represent a substantial increase in travelers on a one-off basis that is not considered in the HPI. Although cruise travelers stay for a short period of time, the number of resources allocated to those visitors can be very high. In this way, we can also observe and predict the impact produced by cruise ships on the islands in terms of human pressure.

Additionally, the global confinement due to the pandemic caused by COVID-19 has produced an anomalous situation in the time series related to tourism, since it was completely paralyzed. For nearly two years, cruise ships were prohibited from docking in Hawaiian ports. This represents an opportunity to assess the performance of the HPI as a global indicator of human influence on the Islands.

The objective of this article is to apply a new HPI, adapted to the Hawaiian Islands, and to predict the effect of cruise ships on the islands.

The article is divided into the following sections: After the introduction, Sect. 2 analyzes the literature related to the work; Sect. 3 develops the methodology and applies it to the case study: the Hawaiian Islands. In Sect. 4, the results are shown and analyzed, and in Sect. 5, the conclusions are presented.

2 Literature Review

The growth of world tourism is closely associated with economic growth in terms of not only the number of travelers who travel but also the capital that is available for their enjoyment, resulting in economic growth in tourist destinations (Risso, 2018). In 2002, it was reported that there were about 700 million international travelers worldwide, while about 10 million people worked in the industry (Mason, 2015).

Tourism in Hawaii is no exception to this global growth. The number of visitors has grown gradually since the post-war years, suffering a brief hiatus due to the 2007 financial crisis (Nelson et al., 2011). Since the crisis, growth has been disproportionate, with the number of visitors almost doubling in just 10 years.

One of the problems caused by this increase in tourist pressure can be seen in the capacity of the inhabitants of the Islands to purchase new homes (Moore, 2019).

A study by Jordan et al. (2019) determines that the impact produced by tourism in Hawaii translates into a two-way evolution, generating emotions and stress. Saito (2013) analyzes the expenses of the visitors to the island of Hawaii (Big Island) and finds that 72.9% of such expenditure is on accommodation, restaurants, tours, and rental cars. But the most important statistic is that 44.7% of the water consumed on the island is consumed by visitors.

Deery et al. (2012) conclude that the impact produced by the growth of tourism is not only related to visitors. Residents' sense of distance from tourist activities also increases as well as the perceptions of the dependence of tourism, among other personal perceptions. To measure the impact of humans on these destinations, it is necessary to use a single indicator that counts both the movements of visitors and the activity of residents. Sometimes, the way of presenting the particular culture of a destination can even be offensive (Gertner, 2019).

In terms of energy resources, this growth is a serious problem. Hawaii is highly dependent on oil. About 90% of the energy used and produced in Hawaii comes from oil, as other alternatives have not been deployed (Dator, 2019). Moreover, this oil must be imported from overseas. This problem was already reported in the 1970s, when there was a warning about the increase in the price of crude oil, in addition to the need to introduce new energy production plants (Sathaye et al., 1979). Despite the fact that Hawaii has the ability to produce electricity from renewable energy in large quantities, it always chose to generate in a non-renewable way (Kaya & Yalcintas, 2010; Yalcintas & Kaya, 2009). In 2014, more than 86% of the electrical energy produced came from non-renewable sources, but in 2018, this quantity had decreased to 73% (2019 Hawaii Energy Facts & Figures, 2019). About 60% of the energy produced was destined for tourism in 1997 (Tabatchnaia-Tamirisa et al., 1997).

Cruise tourism is a growing sector. The large shipping companies are increasingly introducing new ships with greater capacity and with greater tourist attractions on board (Ramanathan & Ramanathan, 2016).

Cruise ship tourism in Hawaii has a long tradition, with an occupancy rate that sometimes exceeds the capacity planned by the ships (Lee & Ramdeen, 2013). The State of Hawaii has its own particularities, since the Passenger Vessel Services Act (PVSA) of 1886 is applicable to cruises in Hawaii and requires foreign cruises to spend a mandatory amount of time in foreign ports. Thus, companies such as Norwegian Cruise Line America (NCLA) operate their ships under the US flag (Mak et al., 2010). However, the business associated with cruise tourism that the islands obtain is a bit uncertain. Pratt and Blake (2009) perform a comparison of the total income produced by cruise tourism. Hawaii DBEBT[1] declares an income of $US 381.07 million, including direct and indirect income, while Pratt and Blake state that this income is only US $85.9 million. It is therefore not easy to determine the impact of cruise tourism on the islands. Some kind of indicator is needed that can help determine the indicated values.

[1] DBEDT stands for Department of Business, Economic Development & Tourism in Hawaii and provides all economic information regarding Hawaii through the economic data warehouse.

The human pressure indicator (HPI) relates visitors' movements with resident population, measuring the instantaneous demographic load in a region. It was developed to measure this load in the Balearic Islands (Riera Font & Mateu Sbert, 2007b). Two fundamental aspects allow the use of this indicator in a reliable way: The entries to and exits from the islands are easily countable. That is why this indicator finds applicability as a demographic measure in the Hawaiian Islands.

3 Methodology

The study carried out consists first of determining a new model for calculating human pressure that includes the influx of passengers by cruise ships, and then this model is applied to the Hawaiian Islands to check its impact on the islands.

3.1 Formulation of the Human Pressure Indicator

The HPI counts the existing population instantly in a region, including the resident population and the circumstantially existing population in the place. The most developed way to measure pressure is through the use of a daily indicator (DHPI), although it is also possible to use the monthly human pressure indicator (MHPI). Equation (1) describes the calculation for DHPI.

$$DHPI_{y,d,i} = RESP_{y,d,i} + SEASP_{y,d,i} \tag{1}$$

where $DHPI_{y,d,i}$ is the HPI, expressed in its daily form, for the year y and day d related to the island i. $RESP_{y,d,i}$ is defined as the estimated resident population at that instant related to the same parameters as before. $SEASP_{y,d,i}$ refers to the seasonal contribution of visitors to the island.

The term associated with the resident population is calculated using the population census. Due to the fact that the indicator used here has a resolution of days and the population census is published annually, it is necessary to perform an estimation of the population in the year of study. Equation (2) shows how to obtain this term.

$$RESP_{y,d,i} = \left\{ \begin{array}{ll} pop_{y,i} & \text{when } d = \text{July, 1st} \\ pop_{y,i} + \frac{pop_{y+1,i} - pop_{y,i}}{n_{days_y}} & \text{otherwise} \end{array} \right\} \tag{2}$$

In Hawaii, the population census is conducted on July 1, which is used as the annual reference. Thus, $pop_{y,i}$ stands for the estimate of the population in year y of the island i. In the intermediate time, it is calculated as the proportional increase in that year between the census of one year and the next, with n_{days_y} being the length of the year measured in days.

The seasonal population term measures the short-term variability in the population due to the arrival and departure of non-residents—generally tourists. It can be seen in Eq. (3).

$$SEASP_{y,d,i} = NFL_{y,d,i} + \vartheta_{y,i} + SEASP^*_{y,d,i} \tag{3}$$

The term $\vartheta_{y,i}$ is used to adapt the traffic data within the year and avoid the possible jump that would occur between July 1 of one year and the previous June 30, due to the adjustment produced by the census, as described in Eq. (4). $SEASP^*_{y,d,i}$ is an auxiliary estimate of the seasonal population described in Eq. (6).

$$\vartheta_{y,i} = \left(\frac{TRF_{y,d,i}}{TRF_{y,i}}\right)\left(SEASP^*_{y,06/30,i} - SEASP^*_{y-1,06/30,i} - NFL_{y,d,i}\right) \tag{4}$$

$NFL_{y,d,i}$ stands for the net passenger flow (difference between travelers arriving on the islands and those leaving), and $TRF_{y,d,i}$ stands for passenger traffic (the total sum of the number of passengers arriving and departing). Cruise tourism is of fundamental importance in the islands, although passengers do not spend the night, since visitors disembark in the morning and embark on the cruise again in the afternoon. However, during this period, many travelers enter the islands, exerting human pressure on them. To account for this situation, the number of travelers arriving on the cruise has been added to $\vartheta_{y,i}$, resulting as in Eq. (5).

$$\vartheta_{y,i} = V_{CRUISE_{y,d,i}} + \left(\frac{TRF_{y,d,i}}{TRF_{y,i}}\right)\left(SEASP^*_{y,06/30,i} - SEASP^*_{y-1,06/30,i} - NFL_{y,d,i}\right) \tag{5}$$

$NFL_{y,i}$ and $TRF_{y,i}$ stand for yearly aggregated net passenger flow and passenger traffic, respectively. $V_{CRUISE_{y,d,i}}$ indicates the visitors arriving by cruise in a day d on island i in year y. The calculation of the auxiliary seasonal population $SEASP^*_{y,d,i}$ described in Eq. (6) serves to weight the effect produced by the population flow at the time of the census, so that there may be people who are included in the census but who are not really on the islands at the time of the census, such as people who may be living on the islands, either for work or long-term stays, even though they are not officially included in the census. The calculation method requires establishing a reference before the census (established on the last Wednesday of the month before the census) and adjusting it with the net flow of passengers and the estimated traffic between that date and the last Wednesday of the month after the census. Thus, the estimation is performed according to Eq. (6).

$$SEASP^*_{y,d,i} = 0, \text{ if d is the last Wednesday of June, year y.}$$

Otherwise,

$$SEASP^*_{y,d,i} = SEASP^*_{y,d-1,i} + NFL_{y,d,i} - TRF_{y,d,i}\left(\frac{NFL^*_{y,i}}{TRF^*_{y,i}}\right), \text{ with}$$

$$NFL^*_{y,i} = \sum_{\text{last Wed June of year y}}^{\text{last Wed Aug of year y}} NFL_{y,d,i},$$

$$TRF^*_{y,i} = \sum_{\text{last Wed June of year y}}^{\text{last Wed Aug of year y}} TRF_{y,d,i}. \quad (6)$$

As the census in Hawaii is conducted on July 1, the last Wednesday in June is taken as a reference and is adjusted with the weighting between the net flow of passengers and the passenger traffic between that date and the last Wednesday of August following the census ($NFL^*_{y,i}$ and $TRF^*_{y,i}$).

The daily measurement of human pressure provides punctual values. Sometimes it is interesting to be able to observe from a more long-term point of view. In that case, the monthly indicator (MHPI) is used. On this occasion, values accumulated throughout the month are used for all the variables.

3.2 *Case Study: Hawaii Islands*

The State of Hawaii is an archipelago of volcanic islands located in Polynesia within the Pacific Ocean. It is made up of more than 100 islands and islets and is divided into five counties: Hawaii (Big Island), Honolulu, Maui, Kauai, and Kalawao. The arrangement of the counties and their main islands is shown in Fig. 1. Due to the remote location of these islands, access to them is primarily by plane, although access by cruise ships is increasingly important. There are three international airports (Honolulu HNL, Hilo ITO, and Kailua-Kona KOA) and another series of smaller airports for domestic flights. Many ports have access to ships, but six ports have been especially designed for cruise docking.

As previously mentioned, this census is updated on July 1 of each year. Figure 2 shows the population growth in the different counties of Hawaii as well as the overall state. Continuous growth can be observed since 1960, with the population having almost tripled by 2020. Honolulu County's population represents approximately 80% of the state's population. However, the non-resident population that goes to the islands usually makes domestic movements between them, so the measurement of human pressure must be carried out separately for each island.

Data on arrivals at airports are shown in Fig. 3. These data are represented monthly for a better understanding and reflect arrivals in each of the counties, regardless of the airport at which the passengers arrive. Stability in arrivals has been observed

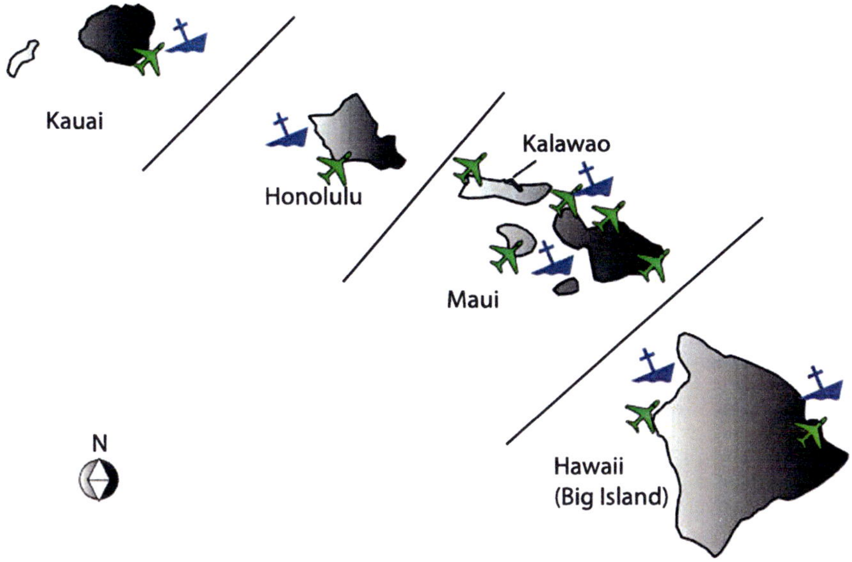

Fig. 1 State of Hawaii. Counties and airports located in Hawaii. *Source* Own design

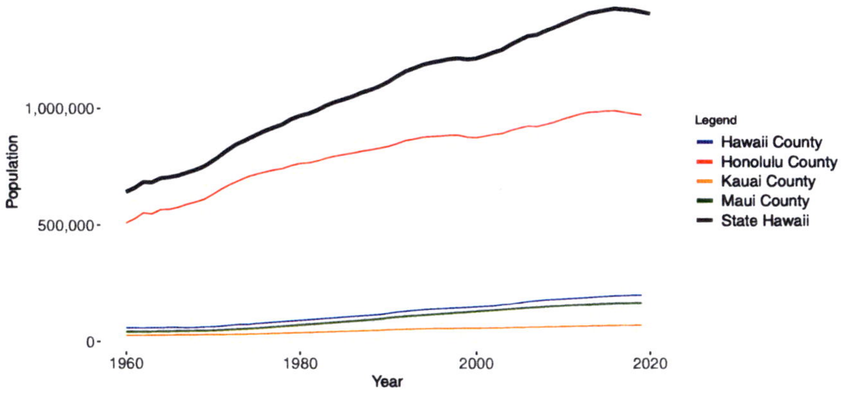

Fig. 2 Population grew in all Counties in Hawaii from 1960 to 2020. *Data Source* Hawaii DBEDT

since 1990, although it is appreciated perfectly as the arrivals occur seasonally. In recent years, however, the number of travelers arriving on the islands has grown significantly, mainly due to improvements in airplanes, which are bigger and more efficient, transporting more people (Dator, 2019). Finally, in 2020, the COVID-19 pandemic occurred, and flights to the islands were almost completely stopped.

Hawaii also has a large number of ports where large cruise ships dock continually. A daily estimate of the number of visitors arriving on the islands by cruise ship is

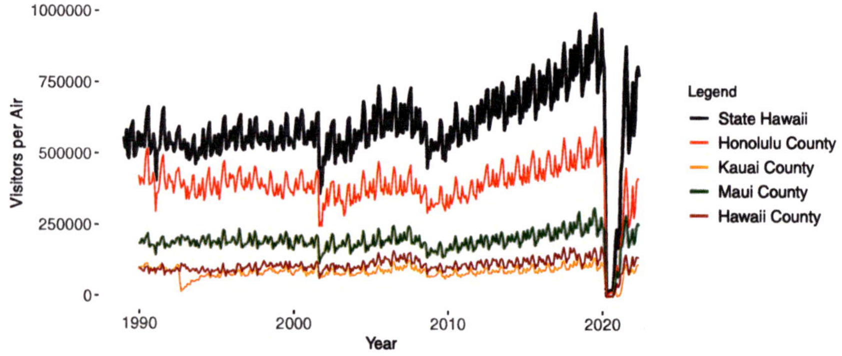

Fig. 3 Monthly visitors arrived in Hawaii islands per air. *Data source* Hawaii DBEDT

shown in Fig. 4. There is an increase in the number of arrivals in terms of both the number of travelers and frequency of arrivals, only broken by the pandemic.

The vast majority of cruises come from outside the Hawaiian Islands and circulate between them without varying the number of travelers. Only the passengers have been considered, and not the crew of the ships. In addition, the Pride of America (NCLA) is an inland cruise, moving between the islands, with travelers arriving on the islands. In the past, the NCLA shipping line had more cruises that were finally used in other destinations.

In the period from March 2020 to January 2022 inclusive, the government of the Hawaiian Islands prohibited the docking of cruise ships on the islands. As of January

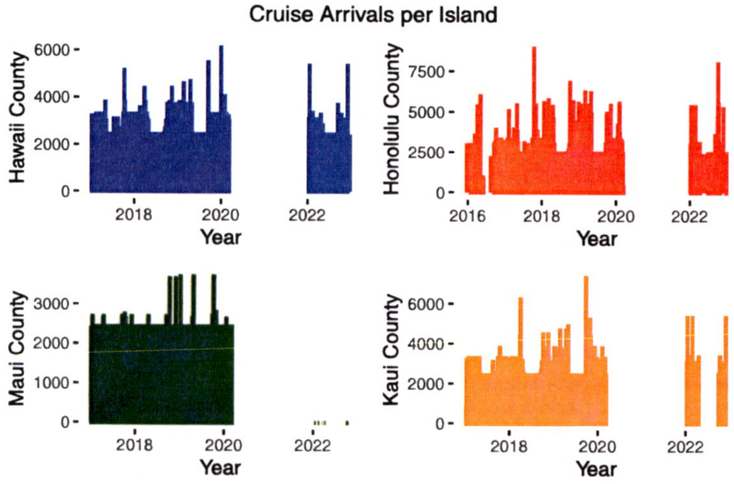

Fig. 4 Daily visitors per cruise to Hawaii counties. *Data source* Hawaii DBEDT

2022, shipping companies increased the number of ships available, and the Pride of America resumed its inter-island route.

4 Results

The impact of these cruises on the HPI is notable, especially in the counties with smaller populations. Figure 5 shows the results obtained for the MHPI, so that it is possible to observe the effect produced by these cruises. We have decided to use this indicator to show the graphical comparisons because it is easier to observe than the DHPI.

We show the original model compared to the improved model with the inclusion of cruise visitors. The objective of this comparison is to be able to observe how the effect of cruises is remarkable, especially in those islands with smaller populations and access by air. To statistically determine the effect of introducing the cruises, the distance between the original DHPI and the original DHPI that includes the cruises has been calculated. For this analysis, a specific R software package has been used (Mori et al., 2012). The results are summarized in Table 1.

It can be shown that the p-values of all t-tests are around 0, meaning that the time series are different, and that distances can reach up to 10%. This analysis confirms that the inclusion of cruise travelers represents a significant difference with respect to the original model.

It can be verified from the graphs in Fig. 5 that the new indicator that includes visitors arriving by cruise ship is always higher, and that, on occasions, this difference can become very notable, especially on islands whose population is very small, such as Kauai County. In more populated islands, such as Honolulu, the influence is

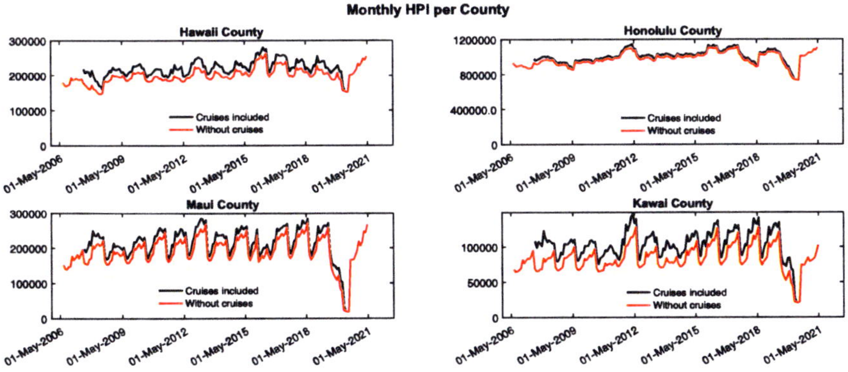

Fig. 5 Monthly Human Pressure Indicator for each County in Hawaii. Red line stands for the original MIPH from (Riera Font & Mateu Sbert, 2007b), and black line stand for the improved model. *Source* Own design

Table 1 Time series differences for the DHPI in the counties when using cruises information

County	t-test p-value	LP distance	Normalized compression-based distance (%)
Hawaii	~0	32,515	9.7
Honolulu	~0	33,041	9.6
Maui	~0	30,024	8.9
Kauai	~0	35,031	10.7

smaller since it receives a large number of visitors by airplane, which dilutes its effect (although it is significant). Hawaii and Maui Counties also show a high impact from cruise ship arrivals, although to a lesser extent than Kauai county.

In Kauai County, the arrival of cruise ships causes an increase of up to 10% in the population present daily on the island, with the consequent need for infrastructure and resources.

An analysis of the graph allows us to observe the behavior in all the islands in a similar way, where the presence of the arrivals of travelers is clearly visible on the most important holidays (Christmas, summer, etc.). Unlike the dashboard, with a single indicator, we can appreciate the evolution of the human presence on the islands.

Finally, we analyze the evolution of DHPI in the period of lockdown. The period of confinement in Hawaii began on March 18 with the limitation of flights to so-called essential ones only (Blane & Ushio, 2020). On March 19, the docking of cruise ships in the archipelago was also prohibited (Carrega, 2020). Subsequently, it can be seen that the indicator clearly reflects a decrease in human pressure in this period. In the less populated islands, this period is practically flat, although the island of Honolulu still shows activity. In this period, the number of travelers arriving via cruises falls to zero, while that due to flights reaches 98%. The indicator, however, counts the resident population as well as the seasonal variation and shows variations of 10% for Hawaii and Honolulu counties and about 25% for Maui and Kauai. Recovery occurs differently in Honolulu from the rest of the islands. Honolulu is the main receiver of visitors, and the arrival of travelers on essential flights occurs on this island.

After the end of the confinement period, a progressive recovery is observed after a jump in activity from the starting point of the confinement. But the two lines overlap because cruises were not allowed till the beginning of 2022.

Finally, the impact of the presence of cruise ships can be related to measures of resources available on the islands. Specifically, an analysis of the energy used in the islands and its relationship with the proposed indicators was carried out, using the two versions (original and with cruise ship accounting).

The time series of electricity sold in all sectors of each island and the series of electricity sent to the utilities of each County (both provided by the Hawaii DBEDT) were used. To relate them to the indicators, the distance between the two time series was measured, using Dynamic Time Warping (DTW), a commonly used method for comparing time series (Berndt & Clifford, 1996). It is a technique used to evaluate

Table 2 Distance between electricity series and HPI indicators for each County

County	Electricity total KWH sold all sectors		Electricity sent to system by electric utility companies	
	No cruises	With cruises	No cruises	With cruises
Hawaii	1.4943	1.4940	1.5994	1.5991
Honolulu	9.5530	9.5520	10.0110	10.0110
Maui	1.5771	1.5768	1.6687	1.6684
Kauai	0.6086	0.6083	0.6364	0.6361

Distances DTW $\times$ (10^{10})

sequences that do not present temporal synchronization. By comparing the series with both indicators, we can check whether the distance decreases when the new indicator is available.

Table 2 shows the results. The distances on the time series are shown to be in the order of 10^{10} (for example, for the Electricity Total KWh Sold All Sectors series and the MHPI without intersections, the distance is $1.4943 \cdot 10^{10}$). The order of the value is not relevant, but it is important to observe the reduction of distances with the use of the new indicator.

This reduction in distance is due to the fact that the indicator is related to the measured variable, which is what we wanted to check.

5 Conclusion

In this article we use an evolved version of DHPI and MHPI (Riera Font & Mateu Sbert, 2007a), which includes the effect of large cruise ships. Cruises produce massive arrivals at ports on these islands that are not correctly reflected in the indicator. HPI is adapted to the particular circumstances of the Hawaiian Islands and is applied to the period between 2017 and 2020. The application of this indicator shows how the arrival of cruise ships is significant. In the islands with smaller populations and less movement, the impact produced by visitors from cruise ships can reach 10% of the population.

Its application to the lockdown period shows how the influence of human pressure does not occur in the same way on the islands, as on the most populated islands this decrease in pressure is smaller.

With the analysis of the DTW distance between the HPI and energy production variables such as electricity, it has been verified that this new indicator is related to said variables. In the future, this would make it possible to establish a direct relationship between the indicator and the energy measures to determine the impact produced by the presence of cruise ships on the resources of the islands.

With this result, we can determine by means of a single indicator the influence of humans on tourist destinations, which can be related to other economic, social, and environmental indicators to justify possible actions by authorities. This indicator, however, is limited to calculations in tourist destinations where movements can be precisely determined, such as islands.

References

2019 Hawaii Energy Facts & Figures (Issue July). (2019). Retrieved from https://energy.hawaii.gov/wp-content/uploads/2019/07/2019-FF_Final.pdf

Berndt, D. J., & Clifford, J. (1996). Finding patterns in time series: A dynamic programming approach. In *Advances in knowledge discovery and data mining* (pp. 229–248).

Blane, S. K., & Ushio, E. (2020). Nighttime curfew to go into effect starting Friday, March 20 in support of emergency efforts; Mayor's tells visitors Kaua'i is on vacation. *The County of Kaua'i*. Retrieved from https://web.archive.org/web/20210420004650/https://www.kauai.gov/Portals/0/Mayor/PIO/NR031820KEMA.pdf?ver=2020-03-18-140021-373

Carrega, C. (2020, March 19). Despite no reported coronavirus, Hawaii won't let cruise ship passengers disembark—ABC News. *ABC News*. Retrieved from https://abcnews.go.com/US/reported-coronavirus-hawaii-cruise-ship-passengers-disembark/story?id=69683821

Dator, J. (2019). Tourism in Hawaii 1776–2076. In *Jim Dator: A noticer in time* (pp. 23–35). Springer.

Deery, M., Jago, L., & Fredline, L. (2012). Rethinking social impacts of tourism research: A new research agenda. *Tourism Management, 33*(1), 64–73. https://doi.org/10.1016/j.tourman.2011.01.026

Gertner, R. K. (2019). The impact of cultural appropriation on destination image, tourism, and hospitality. *Thunderbird International Business Review, 61*(6), 873–877. https://doi.org/10.1002/tie.22068

Hawaii Tourism Dashboard. (2022). Department of Business, Economic Development & Tourism, State of Hawaii. Retrieved from http://dbedt.hawaii.gov/visitor/tourism-dashboard/

Jordan, E. J., Spencer, D. M., & Prayag, G. (2019). Tourism impacts, emotions and stress. *Annals of Tourism Research, 75*(February 2018), 213–226. https://doi.org/10.1016/j.annals.2019.01.011

Kaya, A., & Yalcintas, M. (2010). Energy consumption trends in Hawaii. *Energy, 35*(3), 1363–1367. https://doi.org/10.1016/j.energy.2009.11.019

Lee, S., & Ramdeen, C. (2013). Cruise ship itineraries and occupancy rates. *Tourism Management, 34*, 236–237. https://doi.org/10.1016/j.tourman.2012.03.009

Mak, J., Sheehey, C., & Toriki, S. (2010). The passenger vessel services act and America's cruise tourism industry. *Research in Transportation Economics, 26*(1), 18–26. https://doi.org/10.1016/j.retrec.2009.10.004

Mason, P. (2015). Tourism impacts, planning and management: Third edition. In *Tourism impacts, planning and management: Third edition* (3rd ed.). Routledge. https://doi.org/10.4324/9781315781068

Moore, C. D. (2019). Hawaii: Priced out of paradise. *California Journal of Politics and Policy, 11*(1). https://doi.org/10.5070/P2cjpp11142409

Mori, U., Mendiburu, A., & Lozano, J. (2016). Distance measures for time series in R: The TSdist package. *The R Journal, 8*(2), 451. https://doi.org/10.32614/RJ-2016-058

Nelson, L. A., Dickey, D. A., & Smith, J. M. (2011). Estimating time series and cross section tourism demand models: Mainland United States to Hawaii data. *Tourism Management, 32*(1), 28–38. https://doi.org/10.1016/j.tourman.2009.10.005

Pratt, S., & Blake, A. (2009). The economic impact of Hawaii's cruise industry. *Tourism Analysis, 14*(3), 337–351. https://doi.org/10.3727/108354209789704977

Ramanathan, R., & Ramanathan, U. (2016). A new rational IPA and application to cruise tourism. *Annals of Tourism Research, 61*, 264–267. https://doi.org/10.1016/j.annals.2016.10.004

Riera Font, A., & Mateu Sbert, J. (2007a). Aproximación al volumen de turismo residencial en la Comunidad Autónoma de las Illes Balears a partir del cómputo de la carga demográfica real. *Estudios Turísticos, 174*, 59–71.
Riera Font, A., & Mateu Sbert, J. (2007b). Aproximación al volumen de turismo residencial en la Comunidad Autónoma de las Illes Balears a partir del cómputo de la carga demográfica real [Approximation to the volume of residential tourism in the Autonomous Community of the Balearic Islands from the. *Estudios Turísticos, 174*, 59–71.
Risso, W. A. (2018). Tourism and economic growth: A worldwide study. *Tourism Analysis, 23*(1), 123–135. https://doi.org/10.3727/108354218X15143857349828
Saito, O. (2013). Resource use and waste generation by the tourism industry on the big island of Hawaii. *Journal of Industrial Ecology, 17*(4), 578–589. https://doi.org/10.1111/jiec.12007
Sathaye, J., Ruderman, H., Chan, P., & Estrada, H. (1979). Economic impacts of two energy projections.
Tabatchnaia-Tamirisa, N., Loke, M. K., Leung, P., & Tucker, K. A. (1997). Energy and tourism in Hawaii. *Annals of Tourism Research, 24*(2), 390–401. https://doi.org/10.1016/s0160-7383(97)80008-4
Yalcintas, M., & Kaya, A. (2009). Conservation versus renewable energy: Cases studies from Hawaii. *Energy Policy, 37*(8), 3268–3273. https://doi.org/10.1016/j.enpol.2009.04.029

What Motivates Qualified Scuba Divers to Visit the Maltese Islands?

Simon Caruana and Francesca Zammit

Abstract Malta attracts a significant number of scuba divers every year with the aim of visiting the Maltese waters. In 2019 they amounted to almost 7% of all the visitors to the Maltese Islands. Malta has been identified as one of the best diving destinations within the Mediterranean. Yet this influx and other stakeholders pose a challenge to its sustainability which suggests the need for a management policy. One way to ensure this is to examine the factors that motivate divers to come and dive in the Maltese waters. To gain a better understanding, a quantitative approach was selected. Data was collected from the main stakeholders in the scuba diving industry. This was compared with data obtained from existing literature that looked at the various factors that may motivate divers to visit a particular place, including the Maltese Islands. The outcome suggests that water quality (in terms of pollution and consequently fish life) is a determining factor. Other significant factors were the sea temperature which one could associate with a higher level of comfort when diving in warmer waters. The variety of the diving sites was also indicated as significant. Interestingly, many of the respondents commented on the need to have clear regulation with respect to the protection of marine life and indicated that enforcement is a key factor in preserving the marine environment. This suggests that divers place much more importance towards sustainability today hence becoming a significant motivating factor in selecting the region to visit.

Keywords Scuba diving · Destination choice · Motivating factors

JEL Classification Z310 · Z320 · Z330 · Z390

S. Caruana (✉)
Institute of Tourism Studies, Luqa, Malta
e-mail: simon.caruana@its.edu.mt

F. Zammit
University of Malta, Msida, Malta

© The Author(s), under exclusive license to Springer Nature Switzerland AG 2023

V. Katsoni (ed.), *Tourism, Travel, and Hospitality in a Smart and Sustainable World*, Springer Proceedings in Business and Economics,
https://doi.org/10.1007/978-3-031-26829-8_35

1 Introduction

The scope of this study is to analyse the pull motivational factors of qualified scuba divers that choose the Maltese islands as their destination. Through this, the different types of dives that qualified scuba divers choose to practise in Malta will be identified.

It is hoped that the results will provide a clearer view of what is Malta's forte when it comes to scuba diving and hence what the Maltese Islands can compete with, via-à-vis other leading scuba diving countries. Moreover, the outcomes of this research will also provide a better understanding of where measures and restrictions need to be enforced to preserve the underwater marine environment. This will enhance the scuba diving industry while keeping both the diving sector and marine sustainability balanced.

2 Literature Review

2.1 Motivation to Travel

According to Pearce (2011), there are several reasons why tourists choose to travel. These motivational factors are the so-called push factors. There are two main push factors common among tourists. These two need to evade the everyday practice and pressure, and the wish to strive for leisure occasions and personal rewards (Mannell & Iso-Ahola, 1987).

Pearce (2011) argues that the tourist travel is about exploring something new, learn about new cultures and interaction with the local people. Reference is made to the theoretical framework of Mannell and Iso-Ahola (1987), where they discuss how the motivational factors for travel are to leave the everyday burden behind and gain personal and relational rewards. This may include relaxation, discovering, new challenges, interacting with different people and other factors.

2.2 Motivation to Scuba Dive

Wong et al. (2000) explain different motivational factors that encourage one to dive, starting with the leisure factor. This motivational factor could be both intrinsic and extrinsic. Intrinsic motivation is when the diver wants to feel enjoyment and give himself satisfaction, for instance the feeling of being adventurous. Extrinsic motivation gives the diver the want to accomplish something externally, such as socializing with new divers (Wong et al., 2000). Research by Meyer et al. (2003), in Florida, suggested that intrinsic motivation is more seen in women who dive, while men are pushed more by extrinsic motivation. For instance, when comparing the males' answers with the females', men stated that they dived mainly for the risk, to see shipwrecks and to collect curious objects. Female divers stated that they did it to prove themselves and get more self-confident, to challenge themselves and to discover more about the underwater while being creative (Meyer et al., 2003).

Motivation also seems to differ on the experience and diving qualifications held by the diver. Bentz et al. (2016) conducted a study in the Azores. Their results suggest that specialist divers seek to dive in locations that offer a completely unusual and unique experience. At times specialist divers practise diving to learn and spend time with their buddies or family, but this is more common among the generalist divers (Bentz et al., 2016).

A study in Australia focused on what motivated people to learn and practise scuba diving. The results suggested that the risky venture and the interest to understand the aquatic environment were the main reasons (Tschapka, 2006). These divers could be classified as "hard core divers" as they dive for the challenge. Some divers, however, only dive when they are on holiday and these are called the "tourist divers". Furthermore, then there are those who are starting from scratch and are eager to learn how to practise scuba diving, which are clustered as "potential divers". The motivational factors for these types of divers vary as they all have different experiences (Wong et al., 2000).

2.3 *Niche Tourism*

Robinson and Novelli (2005) argue that whereas before tourism was looked at as an amalgam of one solid sector with its confined markets and ordinary tourists, now it has changed. Mass tourism always had in it a mixture of extremely diverse understandings, preferences and behaviours. The difference is that now, this diversity is being catered for through the concept of niche tourism. Niche tourism is divided in macro niches such as cultural, urban, environmental, rural and others, and segmented again into micro-niches for instance heritage, religious (cultural) and business, exhibition (urban) (Robinson & Novelli, 2005) (Fig. 1).

Till this day there are no formal definitions of what can be stated as "niche tourism". Dinis and Krokaver (2016) agree with Robinson and Novelli (2005) that the term niche tourism could be described as a product that the market offers in a well-timed reaction to mass tourism. The niche tourism product has two sides, the consumption side and the production side. The consumption side is what the tourist is demanding, and since niche tourism is a small specific product, tourists seek to obtain value, significance and prestige through their own personal experience. Dinis and Krokaver (2016) continue to argue that the producers are those who create a niche tourism product and not the touristic demand. The production side benefits from niche tourism to win a new cost-effective market in the right time and use it as competition to other tourism products (Dinis & Krakover, 2016).

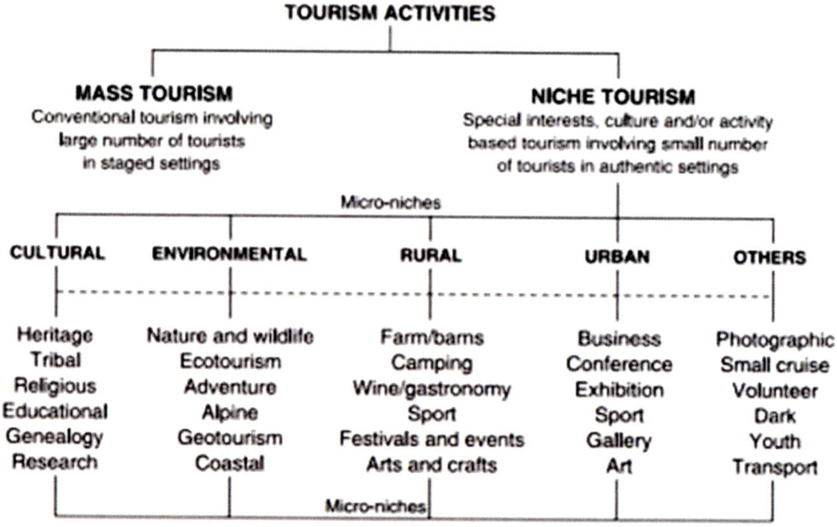

Fig. 1 Macro- and micro-niches of tourism (Robinson & Novelli, 2005)

2.4 Niche Tourism in Malta

The Tourism and Economic Development Directorate (2017), within the Ministry of Gozo, indicates that niche tourism markets are key to vary the product and abate the impacts of seasonality. They also ease the strain on tourism flashpoints by spreading tourists to less frequently visited spots. Invigorating demand through the low season can enhance tourism sustainability on the Maltese Islands. Examples of niche tourism markets that the Maltese Islands could offer, include religious tourism, gastronomy, film location tourism, adventure and nature tourism and more niches that cater for special interest markets (Tourism & Economic Development Directorate, 2017).

Benur and Bramwell (2015) refer to Malta as a good example to when choosing a destination to merge mass tourism and niche tourism markets. Albeit Malta is the main island and the largest in size; Gozo has a lot of touristic niche markets to offer. Yet, the number of tourists visiting Gozo is by far less than those visiting Malta.

One of Gozo's main allure to tourists is the diving sites and their quality. Nevertheless, it also attracts tourists by its walking possibilities that let the tourists experience Gozo's habitual way of life, its natural landscape and heritage sites. When comparing the two main islands, Gozo, due to its nature, has moved towards developing niche tourism markets rather than mass tourism like Malta. Nevertheless, one of Malta's main touristic visits is for its heritage and culture, which can be considered as a niche market. However, looking at the number of visitors visiting the capital city Valletta, given that it is a frequent stop for cruise liners, it might be more considered as mass tourism (Benur & Bramwell, 2015).

The tourism industry is a fundamental sector of the Maltese economy, contributing to over 25% to the nation's gross domestic product (WTTC, 2018). However, Bull and Weed (1999) argued that Malta had come to the situation where the tourism infrastructure is shattering the actual features which pulled the tourists towards the islands primarily. To tackle this problem, Metaxas (2009) suggests that both the development of small lucrative niche markets and the administration of tourism behaviour should be thought out more profoundly for Malta to be able to share its treasures with higher quality.

2.5 *Scuba Diving Industry as a Niche Market*

Scuba diving has developed significantly as a niche tourism product attracting numerous visitors who share the common interest in underwater life. Underwater, the diver can ascend and descend as s/he wishes without gravity defining his movement whilst surrounded by the colourful scenes and marine life. Thanks to technology, scuba diving became easier to practise than before and the marine environment developed into a relaxing destination for scuba divers (Musa & Dimmock, 2013) (Fig. 2).

Dimmock and Musa (2015) consider the Scuba Diving Tourism System (SDTS), to be formed from four main elements (Fig. 3). The core element of the system is the marine environment, and the three other elements being the scuba diving industry itself, the scuba divers and the host community which all depend on the marine environment.

Fig. 2 Scuba diving system framework (Dimmock & Musa, 2015)

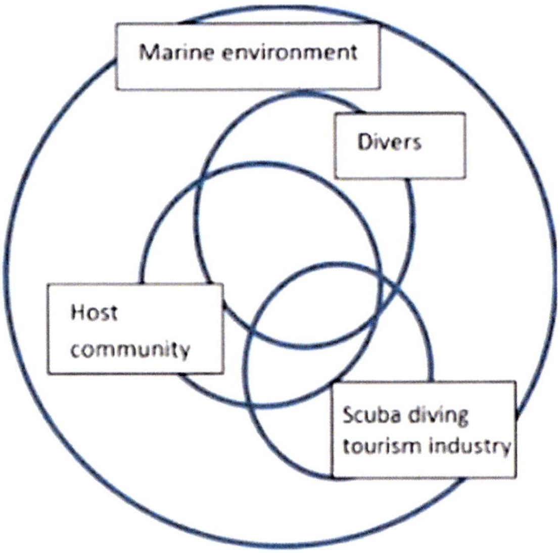

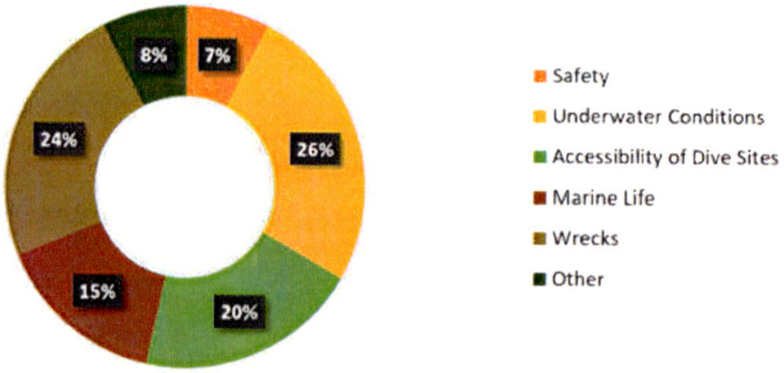

Fig. 3 Motivation to dive

The Professional Association of Diving Instructors (PADI), in the PADI 2019 Worldwide Corporate Statistics, stated that from 1967 until 2018, it had issued more than 27 million diver qualifications around the globe. No matter how many times they practise this sport, out of their driving force and anticipations of marine environments, scuba divers generate a demand for scuba diving tourism. As the divers' diving recurrence increases, their motivation changes as they generally seek more challenging diving, such as wreck dives and cave dives (Dimmock & Musa, 2015).

Scuba diving could be seen from a sustainable ecotourism point of view. The scuba diving industry could benefit the locals for instance by creating employment for marine guides, seafarers and dive masters. Moreover, divers who are ready to pay to dive in marine protected areas could financially support the management and protection roles for these sites. Involving the community in scuba diving tourism also helps with the protection and preservation of the marine environment as they learn to understand its importance. Unfortunately, the local communities do not always benefit financially; thus more opportunities with greater outcomes need to be developed (Dimmock & Musa, 2015).

The scuba diving tourism system is entirely dependent on the marine environment and its use and management. The quality of the marine flora and fauna, the weather elements and the water environment, all affect the experience of the diver which give a challenge to the operator to satisfy his clients. Therefore, it is essential that human impact in these areas is and remains very low. Local authorities finance monitoring and research studies to apply rules that protect the marine environment, such as precluding overfishing. Divers are also responsible for their impact; hence, they avoid touching marine plants and other species with their fins or body (Dimmock & Musa, 2015).

Scuba diving has helped to develop marine protected areas and spread awareness to safeguard the marine environment. Research suggests that in the Caribbean, much of the support towards managing the marine environment comes from tourism. Another example is the Great Barrier Reef Marine Park, which was formerly founded due to coastal spots being visited by a lot of divers and becoming hotspots for aquatic tourism. With time, Queensland coastline started gaining popularity as well with scuba divers in Australia. Until the year 2003, tourists visiting Australia for scuba diving purposes were already amounting to five percent of the country's inbound leisure industry market (Dimmock & Cummins, 2013).

Along the years, scuba diving related tourism expanded into other areas, apart from the Caribbean and Australia. The Red Sea in the Middle East, the Maldives and other island groups in the Indian Ocean, Malaysia, Indonesia, Thailand and neighbouring regions and others. Specific sites found across the globe such as shipwrecks, gained fame around the mid-twentieth century, greatly with leisure divers. Moreover, several countries began to use demobilized ships to create artificial reefs fur scuba diving purposes and originate a diving lure. Eventually, other exploration possibilities were also ascertained such as caves and sinkholes (Dimmock & Cummins, 2013).

2.6 Scuba Diving in Malta

PADI describes Malta as an island with a great diversity of underwater plants and wildlife thanks to the warm Mediterranean Sea that surrounds the Maltese archipelago. Diving is far more common during the hot months of summer, although diving in Malta is appropriate throughout the whole year as water temperatures in January still reach 14 °C. As Visit Malta website (2020) states, the islands' various reefs, caves and wrecks make diving alluring to first-timers, novice, advanced, as well as technical divers. Divers can easily engage in a recreational dive while appreciating the marine flora and fauna as the Maltese Islands provide excellent underwater visibility, which makes it exceptional also for photography (PADI Travel, n.d.).

Wrecks around the Maltese islands make scuba diving even more interesting. For instance, near Manoel Island one can find the X131 wreck, also known as Carolita. This used to serve as a barge in Malta but was damaged in 1942 and sank. Other popular wrecks are the tugboat Rozi found in the west coast of Malta, the Cominoland and the Karwela vessels that were sunk at Xatt l-Ahmar in Gozo purposely for diving and many more. Such sites have made divers appreciate more the marine environment and spread awareness for further protection and conservation (Rich, 2007).

PADI Travel promotes some of the Maltese Islands' best dive sites. These include the Island of Comino which is described as a small island that its water hosts a variety of underwater fauna. Its depth reaching a maximum of 15 m makes it one of the best dive sites even for beginners. There is also the wreck of Um el Faroud, which was sunk as an attraction in 1998 and the HMS Maori which sunk in 1942 at the Grand Harbour of Valletta. The wrecks lie at 33 and 16 m, respectively. As for the island of Gozo, PADI Travel mentions Reqqa Point, which its aquatic inhabitants never fail to amaze the divers with some predator and prey actions, and the Double Arch Reef which features two rising arches from 40 to 18 m.

In terms of environmental protection, there are 18 sites in Malta that fall within the Natura 2000 designation. The sites are either Sites of Community Importance (SCIs), which protect marine environments and species or Special Protected Areas

Table 1 Estimated number of diving tourisms (MTA, 2019)

	2010	2015	2016	2017	2018	2019
Total inbound tourists	1,338,840	1,783,364	1,965,928	2,273,837	2,598,690	2,753,240
Number of tourists motivated to visit Malta for scuba diving			101,700	117,300	127,350	131,382
Number of tourists engaging in scuba diving during trip	64,300	113,322	142,800	155,300	169,850	177,645

(SPAs), which protect aquatic species. Some of these sites in Malta became Marine Protected Areas (MPAs) to protect, for instance, the Maltese top shell which is a sea snail, the loggerhead turtle, the bottlenose dolphin, three specific seabirds that propagate in Malta, cave and reef environments and more (ERA, Marine Protected Areas, n.d.).

The Malta Tourism Authority (MTA) publishes an annual report with a profile of diving travellers that visit Malta. Data for 2019, (Table 1), from a total of over 2.7 million visitors, over 130,000 (4.8%) visited the islands with the main motivation being to dive. If one adds those who decide to try diving, the number increases to 170,000 (6.5%) (MTA Profile of Diving Travellers, 2019).

2.7 Conclusion

In The Profile of Diving Travellers 2019, (MTA) document, there is little specific data indicating what attracted divers to the Maltese Islands. Was it the underwater marine life, wrecks, caves, reefs or other factors? The next step attempts to provide answers, that is, what motivates divers to visit the Maltese Islands.

3 Methodology

3.1 Introduction

For this study, a quantitative approach has been adopted. Williams (2007) states that quantitative research method is explicit in its assessing and investigation since its developed on existing hypotheses. The research itself is not dependent of the investigator, and hence, it generates a connotation through objectivity discovered in the accumulated information. The outcomes from a quantitative research can be prognostic, explicatory and validating. Therefore, this type of research method benefits this study of scuba diving and the motivational factors of the scuba divers as they can be explored in an analytical way while helping the research process to obtain confirmative and realistic results. Moreover, this study is an exploratory one. This means it is an important instrument to help finding out what is going on and request new perceptions to uncover more information (Saunders et al., 2003). In this study, participants' answers were viewed and compared together to provide an unbiased measure of actuality as said by Williams (2007).

Bahari (2010) suggests that a deductive approach is an approach whereby a connection between theory and research in which the research is carried out with reference to hypotheses and suggestions gathered from the theory. A deductive approach is useful for this research study as it helps to move from a general idea which in this case it is scuba diving, to something more specific which are the motivational factors that pull the divers to Maltese dive sites.

3.2 Primary Data

For this study, questionnaires were chosen as a tool to collect primary data. This made it possible to reach a larger number of participants and evaluate their answers through a bigger sample which makes the results more valid. Hox and Boeije (2005) explains how the use of structured questionnaires requires gathering a large quantity of variables from a big and respective sample of participants. Since this study is to identify the motivational factors that pull scuba divers to visit the Maltese Islands, foreigners who do not reside in Malta were chosen for the sample of the research.

3.3 Collection of Data

The questionnaires included a set of questions related to scuba diving sites in Malta and the experience and opinions of the qualified scuba divers that visited the Maltese islands. Both open-ended and close-ended questions were used. Most of the questions were close ended to help keep the answers related to the subject. Reja et al. (2003) explain how such questions provide very limited responses, yet extremely consistent and although information obtained from them is relatively constricting, they are easy to evaluate and deduce. However, although kept to a minimum, open-ended questions were used at the end to allow the participants to articulate their point of view without their answer being interpreted in a biased way by the researcher (Reja et al., 2003).

Prior to distributing the questionnaire, a pilot questionnaire was performed to determine the clarity of the questions and obtain the final set of questions before commencing the primary research. Yet, all questions were clear enough and no changes were made. Subsequently, the questions were sent to the participants via email through the help of some scuba diving schools in Malta and Gozo and various participants were reached through social media. Identities of the participants were kept anonymous. All participants were given a participant consent form explaining the process and details. Furthermore, for data protection purposes, all research data was destroyed once it had been used for the study.

3.4 Data Analysis

For the purpose of this research, cross-tabulation technique was used to analyse the data collected, using the Jeffreys' Amazing Statistics Program (JASP) software. This method collects data to deliver back descriptive statistics, contingency tables, correlation tests and others. It helps to find patterns, tendencies and frequencies within raw data (Love et al., 2019). Once all questionnaires were sent back, the researcher needed to put the data in the JASP programme. The programme would then extract and analyse all the data. This process then set up assigned variable names, types, and titles that were needed. For instance, variables with age, sex, satisfaction levels and other trends (Wagenmakers et al., 2018). The findings from the primary data of this research were extracted through this software. This provided a simpler method to analyse the collected data.

3.5 Limitations

Quantitative research has its own limitations as the questionnaire is mainly made up of close-ended questions. This means that the participants need to choose their answer from the choice given by the researcher (Lundy, 1996). To overcome this, the use of a few open-ended questions was used to reduce the possibility of the outcome being biased. Moreover, due to the COVID-19 pandemic, not many tourists were travelling to the Islands, thus face-to-face contact was very limited with the divers.

4 Analysis and Results

4.1 Demographics

The study was based on 103 questionnaire responses from non-Maltese-resident scuba divers holding different scuba diving qualifications. Overall, there were 33 females and 70 males. The demographic distribution was summarized in the Table 2.

With regards to nationality, the UK has the highest share with 58%. Germany has the second largest share with 16%, with Italy at 6%, Spain and France account for 2% each. All the others make up the remaining 16%.

Table 2 Demographic distribution

Age	Gender	Frequency	Percent	Valid percent	Cumulative percent
18–30 years	Female	6	40.00	40.00	40.00
	Male	9	60.00	60.00	100.00
	Total	15	100.00		
31–50 years	Female	17	35.42	35.42	35.42
	Male	31	64.58	64.58	100.00
	Total	48	100.00		
51–65 years	Female	9	24.32	24.32	24.32
	Male	28	75.68	75.68	100.00
	Total	37	100.00		
Over 65 years	Female	1	33.33	33.33	33.33
	Male	2	66.67	66.67	100.00
	Total	3	100.00		

Table 3 Participants diving qualifications

Level	Amount per level	Total
Open water	7	103
Advanced open water	26	
Rescue diver	20	
Dive master	10	
Instructor	40	

In terms of diving qualifications, as shown in Table 3, the majority were diving instructors, followed by advanced open water and then rescue diver qualification. It is important to point out that the "Open Water" qualification is the "entry-level diving qualification" and poses limits in terms of what and where to dive. For example, the maximum depth allowed is 18 m, and they are not allowed to enter overhead environments (e.g. caves and wrecks).

4.2 *Motivation to Dive*

When looking at what motivates divers to engage in scuba diving in Malta, underwater conditions such as visibility and water temperature were the most common by 26%. This proves right what Meisel-Lusby and Cottrell (2008) said that one of the main attractions to dive is that of the water quality that one is diving in.

The second most popular motivational factor that prompted divers to engage in scuba diving in Malta was the wrecks by 24%. This response reflects the views of both Visit Malta (2020), and Rich (2007) that wrecks found in the Maltese islands make scuba diving even more interesting and attract several divers of all levels. Accessibility of dive sites was another popular motivational factor shared among the divers (20%).

Marine life (15%) seemed to be less of a motivational factor in Malta when comparing it to other countries. In Australia, this is a main motivator as said by Tschapka (2006). Safety had the lowest percentage of 7%.

The participants were given the choice to name other motivational factors that "pulled" them to dive in Malta and among these 8%, the most common answer was the quality provided by the diving schools. This may not appear to be a high value; however, it echoes Dimmock and Musa's (2015) argument. When operators concentrate on efficiency and concern to well-being and security throughout scuba diving excursions, it betters the experience of the divers.

Participants were asked what type of dive they enjoyed the most. As shown in Fig. 4, wreck dives and fun dives (dives based on reefs and marine life) were those enjoyed the most. Both dives shared almost the same amount, 38 for wrecks and 37 for fun dives, wreck dives were preferred by males. On the other hand, females preferred fun dives the most compared to all the other dives. This is what Meyer et al. (2003) said, where males dive for the risk, to see wrecks and collect objects, whereas females dive to challenge themselves and explore the underwater world.

This could also be seen by the fact that no females that completed this questionnaire answered that they enjoyed technical dives and while only one enjoyed a deep dive the most. This seems to corroborate the notion that males are pushed through extrinsic motivation, while females are pushed through intrinsic motivation.

Dimmock and Musa (2015) stated that most of the time divers seek more challenging dives such as wrecks and caves, when their diving is frequent. This is verified in Fig. 5. Diving instructors who tend to the ones with most dives were the ones that practised diving the most, with wrecks and deep dives being the most common. On the other hand, those holding an open water qualification practised mostly fun dives.

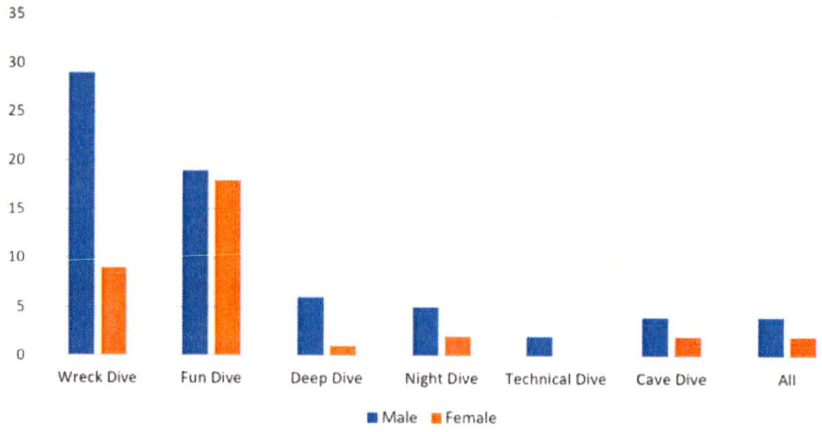

Fig. 4 Frequency of most enjoyed dives

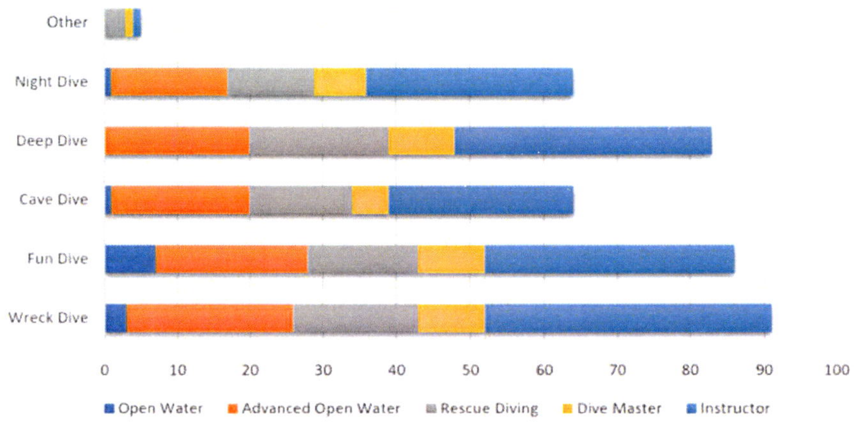

Fig. 5 Dive type preferences

This proves that the more dives the diver does, the more s/he starts to look for dives that offer a challenge and something unique.

Additionally, the participants were asked if the main motivator to visit Malta was diving. 76 of the respondents responded with a "yes", while 27 responded "no". Those who answered with a "no" were asked if diving was still part of the plan when visiting and from the 27 that previously answered "no", 23 responded that diving was part of the plan, while four answered that they did not plan to dive when visiting.

This suggests that although many of the qualified divers that come to Malta are mainly motivated to visit because of diving, most of those who do not find diving as the main motivator, still include diving as part of their plan.

4.3 *What Sets Diving in Malta Apart*

Dive site variety and ease of access were the most popular answer (Fig. 6). Having a wide variety of diving sites which please all type of divers from novice to very experience in such a small area adds to the divers' experience.

With most of the sites having an easy access, this continues to add to the divers' satisfaction. This also prevents them from having to walk long distances while carrying heavy equipment on their backs and reduces extra costs that divers would have to pay if they need to use a boat to get to the dive site.

Diving conditions were the second most common answer among the divers with 22%. Diving conditions included factors such as visibility, water quality, weather and climate. Having good visibility allows the divers to enjoy and appreciate better the beauty that the underwater world has to offer even when diving deep. This makes diving in Malta great for photography as mentioned in the literature review by PADI

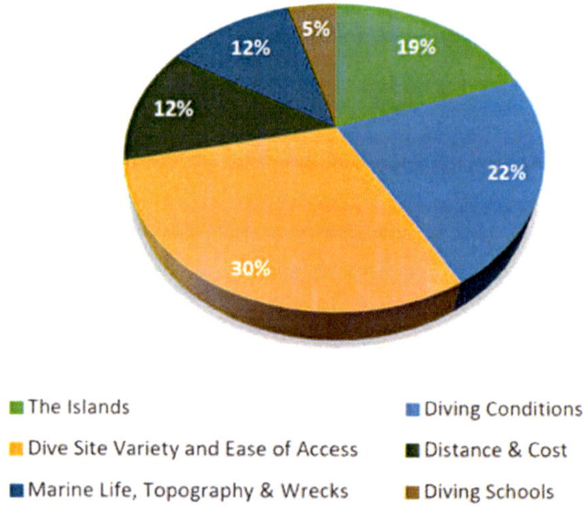

Fig. 6 What sets diving in Malta apart

Travel. The latter also states that the climate in Malta allows diving to be possible all year round as the water temperature during the cold months still reaches 14°.

The third factor that followed was the islands themselves with 19%. The Maltese culture, people and atmosphere set diving in Malta apart according to some of the participants who completed this questionnaire. The friendliness and kindness of the locals, the landscapes and the atmosphere that the Islands provide are very relaxing for the divers. Furthermore, since the Islands are small, divers said that it is very convenient as there are restaurants and other retail outlets all close by. Moreover, this makes it easy to combine a family holiday and a diving holiday together.

This seems to follow the proposed tourism system framework by Dimmock and Musa (2015) right. The diving schools, the hotels, restaurants, bars and the local people all work together to provide the divers with the best experience not just underwater but also during their stay on the Islands.

To further support this literature, participants in this study were asked if they visited Malta more than once and practised scuba diving and what has motivated them to come back. As presented in Table 4, from 103 participants, 12 answered that they only visited Malta and dived once, while 91 respondents said they have visited more than once. Being asked if they would visit Malta again to dive, 87 answered yes, 3 answered no and 13 maybe.

This was followed by a query as to would be the main reason for visiting again. Forty-one percent answered that they visited Malta again because of the people and friends they made in their previous visit, the culture and lifestyle of the locals and the atmosphere of the islands such as the landscapes.

This also reflects what was previously said in the literature review by Bentz et al. (2016) that at times divers practise diving to frequent their buddies or friends. Another 40% responded that they revisited the Islands for the diving conditions and

Table 4 Frequency of visits

Visiting Malta to dive	Frequency	Counts	Total
How many times have you visited Malta and dived?	Once	12	103
	More than once	91	
Do you plan on visiting Malta again to dive?	Yes	87	103
	No	3	
	Maybe	13	

dive site varieties. Distance and cost, and underwater life and scenery shared the least percentage with 10 and 9%, respectively.

4.4 *Ways to Improve Diving in Malta*

Finally, respondents were asked regarding more can be done. More than half of the participants were neutral if Malta can offer more, while seven disagreed. On the other hand, 37 respondents agreed that the Maltese Islands can offer more to diving (Table 5). The majority of those who answered in the affirmative stated that some dive sites could be better developed by introducing facilities such as showers and toilets. Others argued that there should be some parking spaces specifically for divers and improve the quality of the roads. Another popular comment was to improve access to sites in winter as in this season many ladders are removed, which access difficult if not impossible. Others suggested to add more shallow wrecks and introduce bio-rock reefs to increase marine life. Less common suggestions were to publish more historic information mainly about wrecks.

When asked if there should be any changes or restrictions to improve diving in Malta, marine life seemed to be the most common issue that divers were worried about. Suggestions were made to create marine reserves to sustain marine life. Moreover, many participants stated that regulations are already there, but they are not enforced. Hence, enforcement on illegal fishing especially in marine protected areas (MPAs) was suggested.

Table 5 Improving diving

Changes and improvements	Level	Frequency	Total
Do you agree that Malta can offer more when it comes to diving?	Agree	37	103
	Disagree	7	
	Neutral	59	
Do you think there should be any changes, restrictions or improvements in the way Malta offers diving experiences	Yes	30	103
	No	64	
	Maybe	9	

The issue of marine life was followed by that of dive site management. If not well managed diving will go from a niche tourism market to a mass tourism risking the ruining of both the underwater environment as well as the tourism product itself. Several sites are experiencing overcrowding from the divers. Therefore, participants suggested to limit the number of divers per group.

Furthermore, participants were asked if they would recommend diving in Malta to others. 96 out of 103 answered yes, two answered no and five maybe. This question was followed by an open-ended question asking to explain why they would or would not recommend Malta to others.

From those who answered yes, 41% said they would recommend diving in Malta for its diving conditions and diving variety. This included reasons such as visibility, underwater quality, easy access to dive sites and having a variety of dive sites for divers with different levels. Another 38% said they would recommend Malta for the friendliness and kindness that the locals share, the beauty of the islands themselves, historical and other attractions, for being a good place where to combine a diving holiday together with a family holiday, the quality of the staff at the diving centres and for the awards Malta won as a diving destination. 17% would recommend Malta as cost of life is fair and it is cheap to get to, while 4% would recommend it for the wrecks and marine life.

Those who answered maybe explained that Malta used to be a beautiful island, however, it is now being overbuilt and losing its charm and identity, becoming too commercialised and overcrowded. Others mentioned that this only depends on what the diver is looking for as there are better dive sites in the world that one can explore. Two participants said that they would not recommend Malta as it is now overdeveloped, and marine life got very poor. The other participant explained that the number of divers is increasing too much, and diving school facilities offer a poor-quality service (Fig. 7).

These results show that although Malta has great diving conditions and offers a wide variety of different dive sites which please several divers, actions should be taken to sustain marine life and manage the diving market not to become a mass tourism product.

4.5 Summary

The findings provide a clear picture of what motivated qualified scuba divers to explore the Maltese waters. When looking at what sets apart diving in Malta from other countries, results showed that the main factors are the variety of types of dives in such a small area, the ease of accessibility to the dive site, together with the diving conditions such as visibility, water quality and climate.

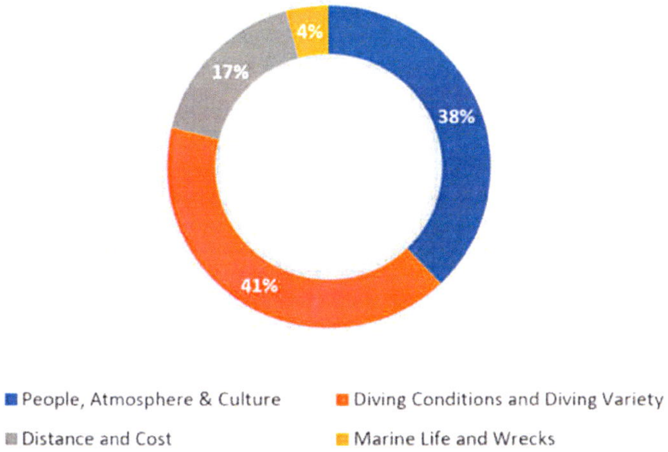

Fig. 7 Why Malta?

5 Conclusions and Recommendations

5.1 Introduction—Motivation to Dive

The main research question of this study was to identify the motivational factors that pull qualified scuba divers to visit and engage in scuba diving in Malta. The findings showed that the most common motivational factor among the participants was the visibility of the water and its temperature. This was followed by the wrecks and ease of access at the dive sites. In the literature it was mentioned by Meisel-Lusby and Cottrell (2008) that marine life is one of the main attractions for divers, however, the findings for Malta show otherwise. Therefore, it is recommended that more protection of fish and enforcement in sensitive areas is carried out, to increase and improve marine life.

Moreover, it was observed that scuba divers in Malta enjoy mostly wreck dives and fun dives. Wreck dives were preferred by men and fun dives by women. Wrecks were enjoyed for their history and their ease of access, while fun dives were enjoyed for the underwater topography and reefs rather than the marine life.

5.2 What Sets Diving in Malta Apart

The variety of diving sites and the ease of access to get to the diving spots were the most common answers among the participants. It is imperative that dive site accessibility should be maintained and improved.

The prevalent diving conditions such as water quality, visibility, weather and climate are also a significant factor. This has to be seen in the light that it makes diving in winter in Malta possible. Interestingly, 19% of the participants think that the Islands themselves set apart diving in Malta. They make reference to the locals, their friendliness, atmosphere, landscapes and the Maltese culture. Having amenities close by (e.g. groceries and restaurants) nearby makes it easy to combine a family holiday with a diving holiday.

Additionally, it was also concluded that the divers that returned to Malta, apart from coming back because of the diving conditions, were principally back because of the people and friends they made before, the culture, atmosphere and landscapes. As stated by Dimmock and Musa (2015), in the literature, diving tourism does not depend only on the marine environment, and hence, it is recommended that Maltese culture and especially landscapes are protected as these are must considered as part of the divers' experience.

5.3 *Ways to Improve Diving in Malta*

Once again marine life was raised as main issue. There are many ways of improving this. Malta should have marine reserves, enforce the restrictions the existing regulations and introduce bio-rock reefs to sustain marine life. All these would certainly contribute to improve the existing underwater environment.

5.4 *Future Research*

The findings analysis and comments from the participants suggest that more research needs to be carried out. This includes in-depth qualitative research at certain dive sites, particularly those frequently visited, to obtain information on how diving tourism could be made more sustainable while still attracting divers. Strong comments left by participants suggest that future research should investigate why Malta is lacking marine life. Moreover, as this study was mostly dominated by male participants, future research could be gathering female and other (e.g. families, senior citizens, etc.) focus groups and analyse their intrinsic and extrinsic motivation to dive. These would all help towards improving the diving tourism sector while keeping it sustainable.

References

Bahari, S. F. (2010). Qualitative versus quantitative research strategies: Contrasting epistemological and ontological assumptions. *Sains Humanika, 52*(1).

Bentz, J., Lopes, F., Calado, H., & Dearden, P. (2016). Understanding diver motivation and specialization for improved scuba management. *Tourism in Marine Environments, 12*(1), 35–49.

Benur, A., & Bramwell, B. (2015). Tourism product development and product diversification in destinations. *Tourism Management*, 213–224.

Bull, C., & Weed, M. (1999). Niche markets and small island tourism: The development of sports tourism in Malta. *Managing Leisure, 4*, 142–155.

Dimmock, K., & Cummins, T. (2013). History of scuba diving tourism. *Scuba Diving Tourism, 40*, 14.

Dimmock, K., & Musa, G. (2015). Scuba diving tourism system: A framework for collaborative management and sustainability. *Marine Policy, 54*, 52–58.

Dinis, A., & Krakover, S. (2016). Niche tourism in small peripheral towns: The case of Jewish heritage in Belmonte, Portugal. *Tourism Planning & Development, 13*(3), 310–332.

ERA. (n.d). Marine protected areas, environment and resources authority. Retrieved from https://era.org.mt/topic/marine-protected-areas-2/. Accessed 7 January 2022.

Hox, J. J., & Boeije, H. R. (2005). Data collection, primary versus secondary.

Love, J., Selker, R., Marsman, M., Jamil, T., Dropmann, D., Verhagen, J., Ly, A., Gronau, Q. F., Smira, M., Epskamp, S., & Matzke, D. (2019). JASP: Graphical statistical software for common statistical designs. *Journal of Statistical Software, 88*(2), 1–17.

Lundy, P. (1996). Limitations of quantitative research in the study of structural adjustment. *Social Science & Medicine, 42*(3), 313–324.

Mannell, R. C., & Iso-Ahola, S. E. (1987). Psychological nature of leisure and tourism experience. *Annals of Tourism Research, 14*, 314–331.

Meisel-Lusby, C., & Cottrell, S. (2008). Understanding motivations and expectations of scuba divers. *Tourism in Marine Environments, 5*(1), 1–14.

Metaxas, T. (2009). Place marketing, strategic planning and competitiveness: The case of Malta. *European Planning Studies, 17*(9), 1357–1378.

Meyer, L. A., Thapa, B., & Pennington-Gray, L. (2003). An exploration of motivations among scuba divers in North Central Florida. In Schuster, Rudy, comp. (Ed.), *Proceedings of the 2002 Northeastern Recreation Research Symposium. Gen. Tech. Rep. NE-302* (Vol. 302, pp. 292–295). US Department of Agriculture, Forest Service, Northeastern Research Station, Newtown Square, PA.

MTA. (2019). Achievements. Malta Tourism Authority. Retrieved from www.mta.com.mt/en/achievements. Accessed 4 March 2021.

Musa, G., & Dimmock, K. (Eds.). (2013). *Scuba diving tourism*. Routledge.

PADI. (2019). Worldwide Corporate Statistics. Data 2013–2018. PADI. Retrieved from: www.padi.com/sites/default/files/documents/2019-02/2019%20PADI%20Worldwide%20Statistics.pdf. Accessed 20 December 2021.

PADI Travel. (n.d). Diving in Malta. PADI. Retrieved from http://travel.padi.com/d/malta/. Accessed 4 January 2021.

Pearce, P. L. (2011). Travel motivation, benefits and constraints to destinations. *Destination Marketing and Management: Theories and Applications*, 39–52.

Reja, U., Manfreda, K. L., Hlebec, V., & Vehovar, V. (2003). Open-ended versus close-ended questions in web questionnaires. *Developments in Applied Statistics, 19*(1), 159–177.

Research Unit. (n.d). The profile of diving travellers 2019. Malta Tourism Authority. Retrieved from www.mta.com.mt/en/file.aspx?f=34514. Accessed 5 January 2022

Rich, S. (2007). Scuba diving as Mediterranean Culture. Preservation and Presentation of Gozo's Maritime Heritage.

Robinson, M., & Novelli, M. (2005). Niche tourism: An introduction. In *Niche tourism: Contemporary issues, trends and cases* (pp. 1–11).

Saunders, M., Lewis, P., & Thornhill, A. (2003). Research methods for business students. Prentice Hall: Financial Times.

Tourism and Economic Development Directorate. (2017). Supporting marketing initiatives for specific niche markets in Gozo, Ministry for Gozo. Retrieved from mgoz.gov.mt/en/Documents/Departments/Tourism%20and%20Economic%20Developmnt/Marketing%20Opportunities%202017%20-%20Notes%20to%20applicant.pdf. Accessed 13 December 2021.

Tschapka, M. K. (2006). *Involvement, motivations and setting preference of participants in the adventure tourism activity of scuba diving*. University of Canberra.

Visit Malta. (2020). Dive sites. Malta Tourism Authority. Retrieved from www.visitmalta.com/en/dive-sites. Accessed 30 December 2021.

Wagenmakers, E. J., Love, J., Marsman, M., Jamil, T., Ly, A., Verhagen, J., Selker, R., Gronau, Q. F., Dropmann, D., Boutin, B., & Meerhoff, F. (2018). Bayesian inference for psychology. Part II: Example applications with JASP. *Psychonomic Bulletin & Review, 25*(1), 58–76.

Williams, C. (2007). Research methods. *Journal of Business & Economics Research (JBER), 5*(3).

Wong, K. M., Thirumoorthy, T., & Musa, G. (2000). Scuba diving motivation. *Studies*, 60.

World Tourism and Travel Council (WTTC). (2018). *Travel and tourism—Economic impact 2018, Malta*. World Travel and Tourism Council website. Retrieved from https://www.wttc.org/-/media/files/reports/economic%20impact%20research/countries%202018/malta2018.pdf. Accessed 3 May 2019.

The Authentic Marathon Swim: Understanding Swimmers' Experiences and Behavior in a Small-Scale Sport Tourism Event

Stella Leivadi and Eugenia Tzoumaka

Abstract The Authentic Marathon Swim is a small sport tourism event which consists of an open-water swimming event, and it is organized by the Municipality of Istiea-Edipsos at Pefki in the island of Evoia, Greece. The event includes six races (14.5-km, 10-km, 5-km, 3-km, 1.5-km, and a 500-m kids' race). It has historical significance because it replicates a 2.500-year-old event that of Skyllias and his daughter Hydna who are considered according to Herodotus the first marathon swimmers. This event has a short but dynamic history of 3 years, all of them under the pandemic circumstances. Nevertheless, the last version of the event (2022) attracted 170 participants (55 children) from 14 different countries. The purpose of this paper to describe the participants' experience and behavior. This was investigated through a questionnaire which included questions related to the quality of their experience with the event; the location and its amenities; spending (accommodation, food, retail) while at the event; and future intentions regarding the event and the location as well as other administrative questions (e.g., location of residence, travel companions, location and duration of stay, etc.). Overall, the findings of this study revealed that this is a successful event which contributes not only to the image of the location but it also has a significant economic impact. In addition, the results also confirm previous research, which claims that positive word-of-mouth and the behavioral intentions of active tourists are affected positively by the destination image (i.e.,Kaplanidou and Vogt, Journal of Sport & Tourism 12:183–206, 2007; Kaplanidou and Gibson, Journal of Sport & Tourism 15:163–179, 2010;Kaplanidou et al., Journal of Sport Management 26:237–248, 2012;; Zhang et al., Tourism Management 40:213–223, 2014). In conclusion, findings have practical applicability to policy makers (local government and officials) as well as to event marketers.

Keywords Rural sport events · Small-sized sport events (SSSEs) · Active sport tourists · Sport tourism · Recurring events · Swimming

JEL Classification Z32 Tourism and Development

S. Leivadi (✉) · E. Tzoumaka
Deree, The American College of Greece, Athens, Greece
e-mail: sleivadi@acg.edu

© The Author(s), under exclusive license to Springer Nature Switzerland AG 2023

V. Katsoni (ed.), *Tourism, Travel, and Hospitality in a Smart and Sustainable World*, Springer Proceedings in Business and Economics,
https://doi.org/10.1007/978-3-031-26829-8_36

1 Introduction

In 2020, sport tourism is appraised at $323,420 million in the overall global tourism market which accounts for 12.12% and this percentage is doubled if the developed countries are taken into consideration (Kadam & Deshmuk, 2021). The European sport tourism market is the largest globally (36.8% market share), with the North American and the Asian-Pacific regions following in the second and third place, respectively (Vynz, 2020). In addition, the growth of sport tourism is projected by 2030 to be at $1,803,704 million. The most important factors which contribute to this growth are among others the increased popularity of sports, the growing sports participation, the global increase in the number of available athletic events, the developing sport infrastructure, the rise in the available income, and the growth in the government's initiatives aiming to promote sports tourism (Kadam & Deshmuk, 2021). In the context of the current study, in Greece, sport tourism is expected to grow annually at a rate of 41–45% with an average of over five sports events taking place each weekend of the year (Greek Travel Pages [GTP], 2019). In terms of the most popular activity in the Greek sport tourism industry, it is considered to be running (Greek Travel Pages [GTP], 2019). The benefits of sport events to the hosting community are along the lines of economic, social, and environmental impacts, with the academic literature focusing more on the economic impact (e.g., Chersulich et al., 2020).

Moreover, most of the current literature is focused on mega and major sport events and their impact on the hosting community (e.g., Agha & Taks, 2015; Gibson et al., 2018; Kaplanidou, et al., 2012; Taks et al., 2015) while smaller sport events such as the context of the current study are considered to exhibit significant potential, for example, for the direct economic impact and overall sustainable tourism development (Gibson et al., 2012). Finally, Costa and Chalip (2005) claim that sport events organized in rural areas such as the current study's location are underrepresented in the academic literature and therefore that there is a need for more relevant studies. However, there is some literature (e.g.,Hallmann & Breuer, 2011; Kaplanidou, 2020) suggesting that sport events boost the destination image of the host community, but at the same time more research is needed to establish the direct and indirect benefits to those communities. The authors of the present study are committed to this purpose, having also investigated the various impacts of another sport event organized in a rural area, that of Zagori Mountain Running event organized at Zagorochoria taking place in the poorest region of Europe, the Epirus prefecture (Tzoumaka & Leivadi, 2021; Tzoumaka et al., 2022). This study claims that recurring small-scale sport events organized in small rural destinations could pursue sustainable tourism development, through the economic impact which is clearly demonstrated. However, before this is established, an understanding of the participants' profile as well as their needs and expectations is needed.

Historical Background of the Event

Ancient historical accounts reveal that Skyllias and his daughter Hydna were the very first "marathon swimmers". Their story claims that in 480 B.C. Persians arrested them with the purpose to use them for an extensive exploration of underwater treasures from ancient Greek shipwrecks based on their ability as proficient divers.

However, as soon as the divers realized that Persians would use them against their compatriots in the upcoming naval battle in Artemision, they decided to slice the ropes from the anchors of the Persian ships and then swam toward Artemision to inform the Greek naval forces. These achievements have been recorded by the father of history Herodotus, who praises them for their swimming ability and their contribution to the triumph of the Greek naval forces. Therefore, this is considered as the first ever ultra-long-distance swim (approximately 80 cables or 9 nautical miles) which was recorded 2500 years ago.

It is among the aims of the Authentic Marathon Swim sport event to promote the history of the land as well as the historical meaning of the unique accomplishment of the first marathon swimmers in an effort to attract the much-needed exposure of the area of Artemision especially after the devastating wild fires of 2021.

Purpose of the Study

The purpose of the present study is to explore and describe the participants' profile and behavioral characteristics as well as their experience with the sport event and the local community. This was investigated through a questionnaire which included questions related to the quality of their experience with the event; the location and its amenities; spending (accommodation, food, retail) while at the event; and future intentions regarding the event and the location as well as other administrative questions (e.g., location of residence, travel companions, location and duration of stay etc.).

2 Literature Review

2.1 *Sport Tourism*

The Gibson (1998) definition of sport tourism is the most cited in the literature. That definition claims that travel which is based on individuals' leisure time and which takes them away from their community is temporary in nature and the purpose is to take part in physical activities or attend events which are associated with these type of activities. More recent attempts (e.g., Weed, 2006, 2009) recognized the more complex nature of the construct and at the same time its capability of having several possible meanings.

Historically, the first exhibition of sport as a touristic venture goes back to Ancient Greece, as Greeks stopped all war activities to travel to either participate or attend as spectators the Ancient Greek Games with the most popular destination being the ancient city of Olympia with the opportunity of the Olympic Games (Kurtzman, 1993).

Finally, the work of Van Rheenen et al. (2017) is considered by the literature as the most complete academic work to establish the term. Several (517) studies on sport tourism were used in a meta-analysis which resulted in five dimensions: those that are relevant to the concept "dimensions" refer to sport as ***motivation*** for travel which involves travel (***space***), there should be at least an overnight stay (***time***), the tourist is involved in a unique ***experience*** involving emotions, and it also results in the development of an economic sector (***economic motivation***).

Types of Sport Tourists

The most dominant distinction in the literature of sport tourists is that of passive and active tourists (Gibson et al., 2018; Standeven & De Knop, 1999). This distinction is mostly relevant to the economic impact of a sport event. Gibson et al. (2018) defined the active sport tourist as the one who while on vacation physically participates in an organized sporting activity as an athlete or as a participant in a tourism activity which involves physical exertion (e.g., hiking, canoeing etc.). On the other hand, the passive tourist is the one who while on holiday is a spectator of a sport event and/or visits a sport facility, a museum, or both. As it is indicated by Kadam and Deshmuk (2021), it seems that in terms of the market share, passive tourists account for more of it (59%). However, active sport tourists are also considered to have great impact on the hosting community mostly in economic terms (Achilleos et al., 2021; Drakakis et al., 2021).

Types of Sport Tourism Events

The typology of sport tourism events is primarily divided into two types: major/mega sport events and small-scale sport events (SSSEs). The first type includes events such as the Olympics, the Football World Cup, European Championships of any sport, Grand Slam tennis tournaments, and the like, which are one-time events for the hosting destination, and in most cases each version is hosted in different locations (Gratton et al., 2000). These events are mostly evaluated by their economic success and the degree to which they raise awareness, short or long term, of the hosting destination (Comerio & Strozzi, 2019). On the other hand, small-scale sport events are organized at a regional level, mostly in rural places, and take place every year in the same destination (Csobán & Serra, 2011). Their overall impact and reach are generally small, "confined" to the local community but very substantial for them (Gammon, 2011; Wafi et al., 2020). Their importance also lies to the fact that they are viewed as a viable alternative for sustainable tourism development (Gibson et al., 2012; Higham & Hinch, 2018; Malchrowicz-Mośko & Poczta, 2018).

As Wong and Tang (2016) have claimed, SSSEs much like major/mega events can not only attract many tourists but media attention too. Furthermore, there exist just a few studies that focus on the perceived value and its relation to tourist satisfaction and loyalty especially in the context of sport tourism (Aicher et al., 2015; Kim et al., 2013). However, there is a need to first understand the destination, the sport event, and its participants. The current study aims at exactly for a new and very small still sport event such as the Authentic Marathon Swim.

3 Methodology

The Sport Event

In honor of the first long-distance swimmers to ever be recorded, the Authentic Marathon Swim is taking place in the straits of Artemision (the location of the historical naval battle between the Greek and Persian forces) every July since 2020. This open-water swimming event provides amateur and elite swimmers with the opportunity to replicate the historical achievement of Skyllias and Hydna by swimming over the shipwrecks of the historical Artemision naval battle.

This sport event consists of the context of the current exploratory study, and it is a recurring two-day swimming event, taking place in early July at Pefki in the island of Evia with approximately 500 residents (ESLSTAT Census Bureau, 2011). This event has a short but dynamic history of three years, all of them under the pandemic circumstances. Nevertheless, the last version of the event (2022) attracted 170 participants (55 children) from 14 different countries. Pefki received in the last 2 days of the event 700 and 1000 visitors, respectively, who attended the cultural festivities surrounding the sport event. Furthermore, 110 individuals were hosted by the organizer for an average of 2 nights' stay. Finally, the event was covered by a team of 15 journalists.

Study Participants

The sample consisted of active sport tourists, who participated in one of the six races (14.5, 10, 5, 3, 1.5 km) with some of them participating in more than one swim race. The total number of responses was 62 from two versions of the event (years 2021–2022), all of them Greek nationals who were the vast majority of the event participants.

Their profile as indicated by the analyses was as follows:

- In terms of their ***Gender***, 48.3% were males and 51.7% females
- The average ***Age*** was 46.2 years (minimum 19—maximum 69)
- Regarding their ***Family Status***, 16.7% were single, 60.0% married or cohabitating, 20.0% divorced, and 3.3% widowed
- The average ***Number of Children*** was 1.25 (minimum 0—maximum 4)
- All of them (100%) were ***employed***
- The ***Average Household Income*** was 1,537.81€ per month
- Finally, in terms of their ***Educational Level***, 18.3% had a lyceum diploma (primary and secondary education), 8.3% an undergraduate education/degree, 20.0% had an undergraduate degree completed, and 3.3% had postgraduate studies.

The Instrument and Data Collection Process

To explore the basic demographic, behavioral characteristics and event experience, as driven by the participants, we adapted the methodology of Gratton et al. (2006). A self-administered questionnaire was employed which was pilot tested, and after all the necessary corrections (typing errors, use of more understandable expressions, esthetics and question progressing, etc.), the final version was uploaded on Google Forms®.

The potential study participants were informed regarding this research effort at the time of their registration for the races, also taking permission to use their e-mail for this purpose. The week after the completion of the event they received an e-mail with the link to the Google form. The timing of the sendoff of the questionnaire was considered appropriate as there were a number of questions in relation to their experience which could not be answered a priori.

The e-mail also communicated a lottery prize (one 50€ gift card from the largest chain of technology and entertainment department stores) used as incentive. The participants would get in a lottery draw for the prize, if they completed the survey and provided a valid e-mail address. Most of the participants provided their e-mail but there were a few who did not.

The questionnaire included questions requiring information on:

- Demographic and behavioral characteristics;
- Event and location evaluation;
- The number of visitors in the host area and the proportion of those making use of commercial accommodation;
- The number of nights those using commercial accommodation staying in the host area and the average accommodation cost per night;
- The average daily expenses spent on food of those participating in the event;
- The total commercial expenses spent in the host area;
- Other information (e.g., traveling alone or with others, awareness of event and local amenities etc.);
- Contact Details.

Data Analysis

Descriptive statistics (frequencies, mean, standard deviation, minimum and maximum value) as well as graphical representation of certain variables were used in order to have an initial account of the profile and experience of the event participants. Further methodological suggestions are provided in the conclusion section.

4 Results

The demographic characteristics of the event participants were presented in the previous section, and the analysis revealed an evenly balanced sample gender wise:

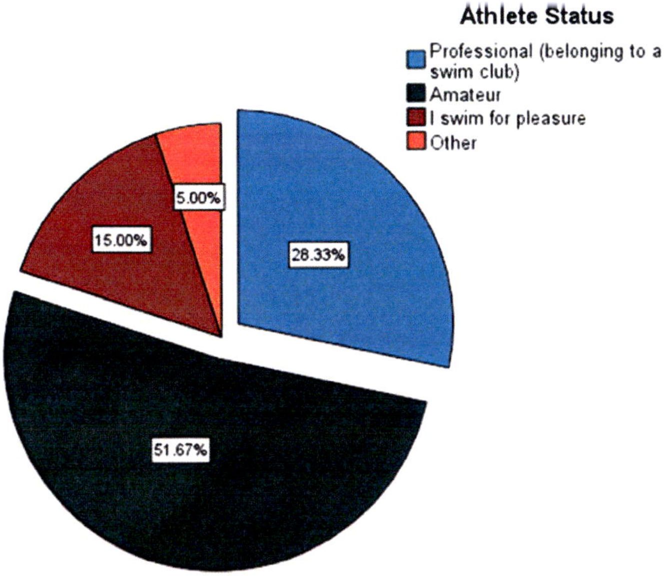

Fig. 1 Athlete status ("How do you perceive yourself as a swimmer?")

highly educated, in good professional and financial position, and in a typical for the Greek society familial position which means most of them being married in a small family which includes in most cases a single child. Finally, in terms of age, the participants reported a wide range (19–69) with the average being in the middle age category (46.2 years of age).

Participants' Athletic Profile

The following two figures conclude the information provided by the participants in relation to their athletic profile. In the first one (Fig. 1), it is obvious that most of them (53.67%) were amateur swimmers, followed by the swimmers who belong to a swim club (28.33%). Some of the participants reported that they swim for pleasure, and a small percentage chose the "Other" option which reveals that they considered themselves as something else than the options provided to them in the questionnaire.

Complimentary to the above information was also the reported data regarding their years of experience in the sport of swimming which shows that most of them (36.7%) had between 6 and 20 years of experience, while there were a lot of "novice" (1–5 years) swimmers and the rest of them had more than 20 years of experience (mean of swimming experience 17.7 years, SD = 15.2). Additionally, the swimmers reported that most of them (61.7%) have participated in the swim event more than once (Table 1).

Furthermore, Fig. 2 reveals that the most popular event to participate was the 3-km race (31.67%) followed by the 10-km race (28.33%) with a very close third the race of 1.5 km (26.67%). It is obvious that the most demanding race of the 14 km was the

Table 1 Average number of years of involvement in swimming

Years of experience	Percentage (%)
1–5	30
6–10	25
11–20	11.7
21–30	11.6
31–40	16.7
41+	10

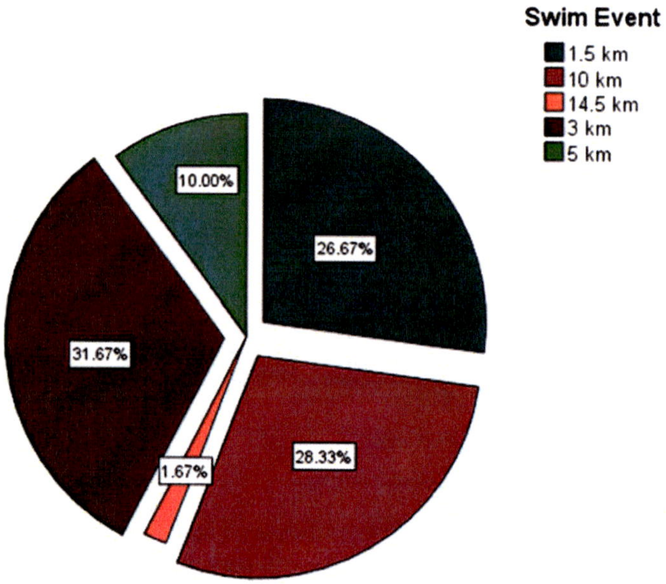

Fig. 2 Participation per event race

least "popular event" with 1.67% of participation. It is very interesting to note that in the latest version of the event (2022), there was a German famous marathon swimmer who participated in both the 10- and 14.5-km races and he succeeded the 1st and 2nd places, respectively. This is an accomplishment which is very rarely achieved, especially with the two races being separated by a day, and one worth mentioning through some data collection of qualitative nature.

Awareness of Location and Event Participation

The event's location is obvious that it was known to most of the event participants as it is indicated in Table 2 below. In addition, the sport event attracted quite a few first-time visitors (56%), while the rest of the participants have visited the area more than once.

Table 2 Participants' awareness of location and sport event

Awareness	Percentage (%)
Pefki as tourism destination	73.3
Local history	82
First-time visitors	56.7
Visited at least once before	13.3
Visited at least once before (up to 8 times)	30

The results also revealed that participants found the needed information regarding the event through the internet and the social media (65.85%), while the rest of them were informed through local authorities and from other athletes, coaches, or friends (Fig. 3).

Event Experience Evaluation

Participants were asked to evaluate their experiences in their current visit in relation to the event using a 7-point Likert-type scale. The results revealed (Table 3) moderate to high satisfaction in 8 different areas with the highest being the "Personal Safety" (mean = 5.7) and very close to the first result second being "Event Prestige" and "Event Atmosphere" both at a mean of 5.6. It seems that the least favorable area was the "Registration Fee" (mean = 4.3) which shows that participants find the fee required for their participation as somewhat expensive. In addition, when participants were asked to freely express any problems/issues they might have experienced while at the event, 66.7% of them reported that they did not experience any problem during

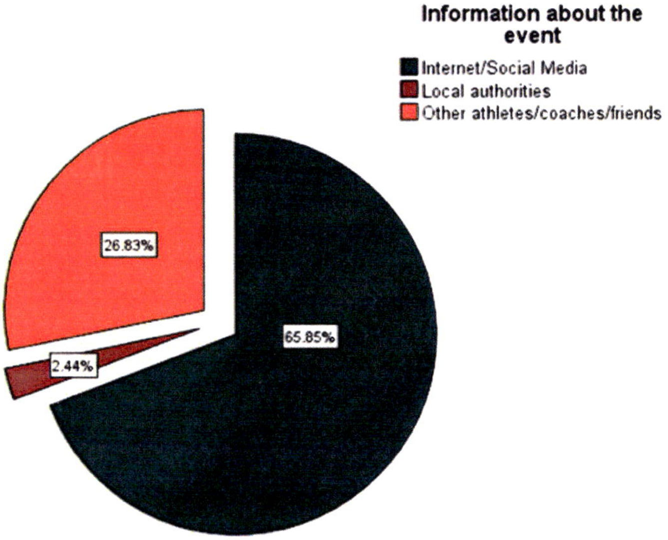

Fig. 3 Source of information regarding the event

Table 3 Participants' evaluation of experience with the sport event

Awareness	Mean
Event prestige	5.6
Competition level	5.4
Registration fee	4.3
Value for money	4.9
Ease of travel	5.2
Event atmosphere	5.6
Personal safety	5.7
Event destination	5.6

the event and the rest of them offered suggestions which had to do mostly with details in the organization of the event (Table 3).

Location Experience Evaluation

Furthermore, participants were asked to evaluate their experience in relation to their visit in the area using a 7-point Likert type scale. The results revealed (Table 4) moderate to high satisfaction in 11 different areas with the highest being the "Clean and Green Environment" (mean = 5.6) and very close to the first result the second being "Local Hospitality" with a mean of 5.3. It seems that the least favorable area was the "Retail Market" (mean = 4.4) which was verified from some informal discussions the researchers had with the participants who reported that there weren't too many options if they wanted to buy souvenirs or a local product. In addition, results revealed that participants were moderately satisfied with "Tourist Support & Information" (mean = 4.5).

Table 4 Participants' overall evaluation of experience in relation to the organization of the sport event

Awareness	Mean
Local infrastructure	4.8
Accommodation quality	4.7
Food quality	5.1
Entertainment infrastructure	4.5
Leisure infrastructure	4.6
Retail market	4.4
Tourist support and information	4.5
Local hospitality	5.3
Clean and green environment	5.6
Sights (in general)	4.7
Cultural and historical sights	4.8

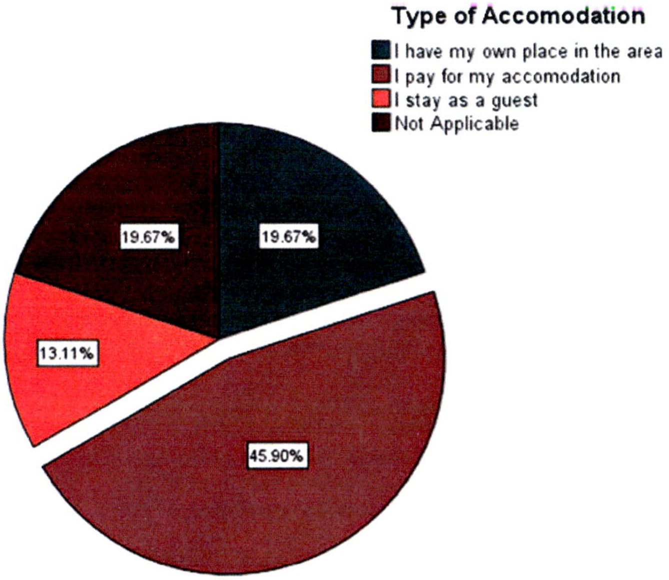

Fig. 4 Type of accommodation

Accommodation Experience

It was important for the purpose of the study and especially for the impact that the event had in the local community to investigate the type of accommodation the participants used in order to take part in the event. The results revealed (Fig. 4) that most of them paid for their accommodation (45.9%). There was also a 39.34% of the participants who answered either that they own a place in the area (19.67%) or that they were residents of Pefki or a close by area (e.g., Istiaia, etc.) which did not require booking an accommodation (19.67%). The rest of them were hosted by local residents. Furthermore, when participants were asked how they found their accommodation, most of them reported that this happened through an online travel agency (e.g., Booking, etc.), independent internet search, personal suggestions, or the Airbnb platform.

Figure 5 reveals that most participants stayed at least one night (78.3%), with the biggest percentage (33.3%) staying for 2 nights (mean = 1.72). An interesting finding here was that of the participants staying for 7 or more than 7 nights in the area (13.4%) which indicates that some of them combined their participation in the event with a vacation.

Spending

This study was also interested to explore the economic impact of the event in the local community. Participants revealed that they spent on average approximately 77.85€ per day on accommodation and food and 52€ overall in retail (Table 5). This

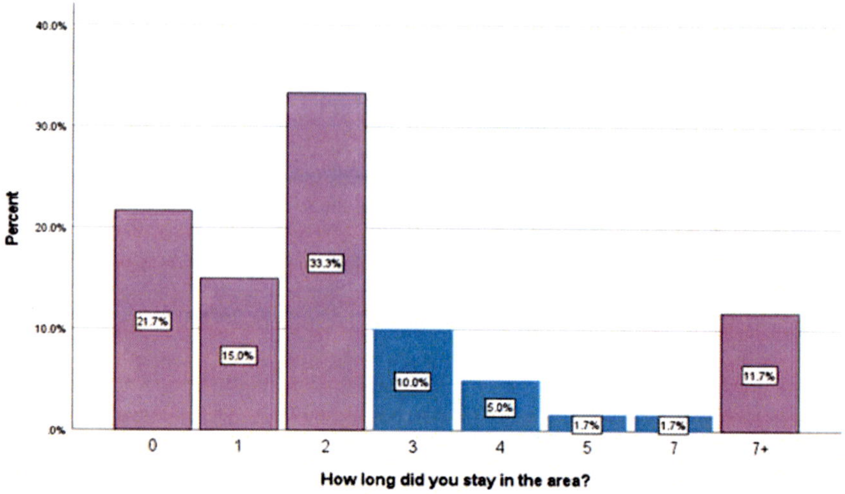

Fig. 5 Number of night stays in the area

Table 5 Spending

Type of spending	Mean
Accommodation/day	52.85€
Food/day	25€
Overall retail	52€

accounts for approximately 185€ spent per person. If we take into consideration that this year's event participants were 80, this accounts for a total of approximately 10,000–14,000€ spent by the participants on the local economy over a period of 2–3 days. This amount is considered as a conservative estimate which represents only the direct economic benefit due to the sport event. Local authorities organized cultural events for both days of competition and their estimate was that the number of visitors was 700–1,000.

5 Conclusion

The main purpose of this study was to explore and describe the participants' profile and behavioral characteristics as well as their experience with the sport event and the local community. As it was indicated by the results, the average participant for the Authentic Marathon Swim was of any gender, well educated, employed, family man/woman with at least one child, amateur but experienced swimmer who is swimming short distances (1.5, 3 & 5 km) and is returning to this event for at least a second time. In terms of their evaluation of the event and the local community, it seems that they are satisfied with both but they are expecting improvement in all areas. More specifically, areas to be improved in relation to the event itself are registration fee, value for money and ease of travel. In terms of the services offered by the local community, participants suggested that areas of improvement include the retail market, tourist support and information, and leisure infrastructure. Finally, regarding accommodation, most participants stayed for two nights with a few of them making a short vacation (7 or 7 + days stay) out of their participation in the event with almost half of them paying for their accommodation.

Understanding the profile of a sport tourism event participants as well as investigating their actual experience with the event and the local community is a crucial first step in this kind of research for theoretical and practical purposes. It is an important step before anyone can proceed to the next step which could be to explore structural relationships such as emotional experiences and emotional value, functional value and destination loyalty, tourist satisfaction etc. There is prior research examining the relationship between emotional value and its consequences which means that the more amplified is satisfaction as well as destination loyalty the more tourist behaviors can be modified (e.g., Jeong, Kim & Kim, 2020).

The above theoretical considerations are of valuable practical importance especially to destination marketers who are interested to improve the competitiveness of a sporting destination in an effective and efficient manner. This could mean that destination marketers' efforts could be focused on the emotional experiences of the participants in order to meet their needs and desires. It should be mentioned here that in an open-ended question in relation to this, some participants suggested that there should be an opening "get to know you" night event. Furthermore, this could even influence the training of the event staff and volunteers in order to provide the expectations and the positive emotional experience that participants are seeking. In the end, this will contribute to the development of a sporting event destination through the strengthening of tourist satisfaction, the perceived value, and destination loyalty.

The authors are aware of some limitations in relation to this study which include its exploratory nature, short number of responses, and the size of the event with a short but dynamic history organized and managed under the difficult circumstances of the pandemic. In future, passive tourists should also be included especially for the estimation of the economic impact and primary data should be complemented for robustness and generalizability of the results with secondary data which also offer great insights into the event and the participant experience.

Acknowledgements This research was funded by the Centre of Excellence in Food, Tourism & Leisure of Deree—The American College of Greece.

We express our gratitude to Active Media Group for the financial and operational support.

We thank the two graduates of Deree—The American College of Greece, Antonis Apostolopoulos and Dimitris Eveidis, who zealously participated in the instrument preparation and data collection phases.

References

Achilleos, A., Makrominas, M., Markides, C., Alexandrou, R., Konstantinidis, A., Papacosta, E., Constantinides, P., Zikouli, E., & Tselepos, L. (2021). Promoting active sports tourism through technology and evaluating its economic impact: Experiences from Cyprus. *Journal of Sport & Tourism, 25*, 297–315. https://doi.org/10.1080/14775085.2021.1965009

Agha, N., & Taks, M. (2015). A theoretical comparison of the economic impact of large and small events. *International Journal of Sport Finance, 10*, 199–216.

Aicher, T. J., Karadakis, K., & Eddosary, M. M. (2015). Comparison of sport tourists' and locals' motivation to participate in a running event. *International Journal of Event and Festival Management, 6*(3), 215–234.

Chersulich, T., Marko Perić, A., & Wise, N. (2020). Assessing and considering the wider impacts of sport-tourism events: A research agenda review of sustainability and strategic planning elements. *Sustainability 12*(11), 4473. https://doi.org/10.3390/su12114473

Comerio, N., & Strozzi, F. (2019). Tourism and its economic impact: A literature review using bibliometric tools. *Tourism Economics, 25*, 109–131. https://doi.org/10.1177/1354816618793762

Costa, C. A., & Chalip, L. (2005). Adventure sport tourism in rural revitalisation–An ethnographic evaluation. *European Sport Management Quarterly, 5*, 257–279. https://doi.org/10.1080/16184740500190595

Csobán, K. V., & Serra, G. (2011). The role of small-scale sports events in developing sustainable sport tourism–A case study of fencing. *Applied Studies in Agribusiness and Commerce, 8*, 17–22. https://doi.org/10.19041/Abstract/2014/4/3

Drakakis, P., Papadaskalopoulos, A., & Lagos, D. (2021). Multipliers and impacts of active sport tourism in the Greek region of Messinia. *Tourism Economics, 27*, 527–547. https://doi.org/10.1177/1354816620902328

Gammon, S. (2011). Sports events: Typologies, people and place. In S. J. Page & J. Connell (Eds.), *The Routledge Handbook of Events* (2nd ed., pp. 104–118). Routledge.

Gibson, H. J. (1998). Sport tourism: A critical analysis of research. *Sport Management Review, 1*, 45–76. https://doi.org/10.1016/S1441-3523(98)70099-3

Gibson, H. J., Kaplanidou, K., & Kang, S. J. (2012). Small-scale event sport tourism: A case study in sustainable tourism. *Sport Management Review, 15*, 160–170. https://doi.org/10.1016/j.smr.2011.08.013

Gibson, H. J., Lamont, M., Kennelly, M., & Buning, R. J. (2018). Introduction to the special issue active sport tourism. *Journal of Sport & Tourism, 22*, 83–91. https://doi.org/10.1080/14775085.2018.1466350

Gratton, C., Dobson, N., & Shibli, S. (2000). The economic importance of major sports events: A case-study of six events. *Managing Leisure, 5*(1), 17–28.

Gratton, C., Shibli, S., & Coleman, R. (2006). The economic impact of major sports events: A review of ten events in the UK. *The Sociological Review, 54*(2_suppl), 41–58.

Greek Travel Pages [GTP]. (2019). *The Impact of Sports Tourism & Future Trends—Opinion.* Retrieved June 16, 2922, from https://news.gtp.gr/2019/08/20/the-impact-of-sports-tourism-future-trends-opinion/.

Hallmann, K., & Breuer, C. (2011). Images of rural destinations hosting small-scale sport events. *International Journal of Event and Festival Management, 2*, 218–244. https://doi.org/10.1108/17582951111170290

Higham, J., & Hinch, T. (2018). Sport tourism development. Channel view publications.

Jeong, Y., & Kim, S. (2019). A study of event quality, destination image, perceived value, tourist satisfaction, and destination loyalty among sport tourists. *Asia Pacific Journal of Marketing and Logistics, 32*(4), 940–960.

Jeong, Y., Kim, E., & Kim, S. K. (2020). Understanding active sport tourist behaviors in small-scale sports events: Stimulus-organism-response approach. *Sustainability, 12*(19), 8192.

Kadam, A., & Deshmukh, R. (2021). Sports tourism market by product (Football/Soccer, Cricket, Motorsport, Tennis, and Others), type (Domestic and International), and category (Active and

Passive): global opportunity analysis and industry forecast 2021–2030. *Allied Market Research.* Retrieved August 24, 2021, from https://www.alliedmarketresearch.com/sports-tourism-market-A13076.

Kaplanidou, K. (2020). Sport events and community development: Resident considerations and community goals. *International Journal of Sports Marketing and Sponsorship, 22*, 53–66. https://doi.org/10.1108/IJSMS-05-2020-0082

Kaplanidou, K., & Gibson, H. J. (2010). Predicting behavioral intentions of active event sport tourists: The case of a small-scale recurring sports event. *Journal of Sport & Tourism, 15*, 163–179. https://doi.org/10.1123/jsm.24.5.544

Kaplanidou, K., & Vogt, C. (2007). The interrelationship between sport event and destination image and sport tourists' behaviours. *Journal of Sport & Tourism, 12*, 183–206. https://doi.org/10.1080/14775080701736932

Kaplanidou, K., Jordan, J. S., Funk, D., & Ridinger, L. L. (2012). Recurring sport events and destination image perceptions: Impact on active sport tourist behavioral intentions and place attachment. *Journal of Sport Management, 26*, 237–248. https://doi.org/10.1123/jsm.26.3.237

Kim, S. H., Holland, S., & Han, H. S. (2013). A structural model for examining how destination image, perceived value, and service quality affect destination loyalty: A case study of Orlando. *International Journal of Tourism Research, 15*(4), 313–328.

Kurtzman, J. (1993). Inaugural address—sports tourism international council. *Journal of Sport & Tourism, 1*, 5–17. https://doi.org/10.1080/10295398608718526

Malchrowicz-Mośko, E., & Poczta, J. (2018). A small-scale event and a big impact—Is this relationship possible in the world of sport? The meaning of heritage sporting events for sustainable development of tourism—Experiences from Poland. *Sustainability, 10*(11), 4289.

Standeven, J., & de Knop, P. (1999). *Sport tourism*. Human Kinetics.

Taks, M., Chalip, L., & Green, B. C. (2015). Impacts and strategic outcomes from non-mega sport events for local communities. *European Sport Management Quarterly, 15*, 1–6. https://doi.org/10.1080/16184742.2014.995116

Tzoumaka, E., & Leivadi, S. (2021). *Investigating the impact of the sport event destination image on consumer outcomes of active sport tourists.* Paper presented at the 9th International (Virtual) Conference on Contemporary Marketing Issues (ICCMI).

Tzoumaka, E., Leivadi, S., & Kaplanidou, K. (2022). Recurring rural destination sport events: A study on participants' direct spending. *International Journal of Financial Studies*. In press.

Van Rheenen, D., Cernaianu, S., & Sobry, C. (2017). Defining sport tourism: A content analysis of an evolving epistemology. *Journal of Sport & Tourism, 21*, 75–93. https://doi.org/10.1080/14775085.2016.1229212

Vynz, R. (2020). *Global sports tourism market—analysis and forecast (2018–2024)*. Retrieved August 21, 2021, from https://www.vynzresearch.com/consumer-goods/sports-tourism-market.

Wafi, A. A., Chiu, L. K., & Ibrahim, J. A. (2020). Organizing small-scale sports events: issues and challenges. In A. Fotiadis & C. Vassiliadis (Eds.), *Principles and practices of small-scale sport event management* (pp.127–146). IGI Global. https://doi.org/10.4018/978-1-7998-4757-1.ch007

Weed, M. (2006). Sports tourism research 2000–2004: A systematic review of knowledge and a meta-evaluation of methods. *Journal of Sport & Tourism, 11*(1), 5–30.

Weed, M. (2009). Progress in sports tourism research? A meta-review and exploration of Futures. *Tourism Management, 30*(5), 615–628.

Wong, I. A., & Tang, S. L. W. (2016). Linking travel motivation and loyalty in sporting events: The mediating roles of event involvement and experience, and the moderating role of spectator type. *Journal of Travel & Tourism Marketing, 33*(1), 63–84.

Zhang, H., Fu, X., Cai, L. A., & Lu, L. (2014). Destination image and tourist loyalty: A meta-analysis. *Tourism Management, 40*, 213–223. https://doi.org/10.1016/j.tourman.2013.06.006

Systematic Revision of the Literature on Satisfaction in Marinas and Nautical Ports

Fernando Toro Sánchez, Pedro Palos Sánchez, and Nélson Manuel da Silva de Matos

Abstract In recent years, the development of nautical tourism has been experiencing growth in the area of the Mediterranean countries and the European Atlantic coast, being a tourism sector with a high potential for both spending and developing innovative products involving technological development. Nautical tourism focuses its operations on the network of leisure ports and marinas scattered along the coastline and inland in rivers and lakes, with a heterogeneous variety depending on the type, size, public or private management, the services provided to users, and their involvement with the tourism development of the area where it is located. Likewise, there is no uniform definition of the characteristics and functions that a marina should have, despite the existence of international institutions and associations that regulate this activity, presenting the sector with a lack of unified criteria and procedures. This work presents a Systematic Literature Review (SLR) with a selection of 60 studies and references from 2005 to August 2022 using the bases of World of Science and Scopus. Exploring the types of questionnaires and methodologies used in the sector of marinas with the aim of finding homogeneous directions of study facilitates analysis based on the use of advanced statistical methodologies and helps to understand this sector for future studies in this area.

Keywords Sailing tourism · Marina user satisfaction · Nautical tourism marketing · Tourism sustainability · Systematic literature review (SLR)

JEL Classification L83: Sports · Gambling · Restaurants · Recreation · Tourism

F. T. Sánchez (✉)
Department of Market Research and Marketing, University of Sevilla), Seville, Spain
e-mail: fennantoro@gmail.com

P. P. Sánchez
Department of Financial Economics and Operations Management, University of Sevilla), Seville, Spain

N. . M. da Silva de Matos
ESGHT, (Universidade Do Algarve), Faro, Portugal

© The Author(s), under exclusive license to Springer Nature Switzerland AG 2023

V. Katsoni (ed.), *Tourism, Travel, and Hospitality in a Smart and Sustainable World*, Springer Proceedings in Business and Economics,
https://doi.org/10.1007/978-3-031-26829-8_37

1 Introduction

The nautical tourism sector is studied from three different perspectives according to its industry: the boarding of sailboats and motorboats owned, the rental of boats (charter), and the cruise industry (Luković, 2012). All of them have experienced high growth over the last thirty years (Çakıroğlu, 2019; Vlasic et al., 2019), with the service activity of marinas closely associated with boat rentals and yacht owners (Luković, 2012), in such a way that the traditional yacht clubs (Chen et al., 2016) focus their activity toward yacht owners while the business of the new marinas is more oriented toward yacht charter tourism (Nowacki, 2015).

Europe has a coastline of 68,000 km (European Environment Agency, 2020[1]), with a total of 4.500 marinas, 1.75 million berths, and 6.3 million yachts (Vlasic et al., 2019) with around 222.000 berths in France, 130.000 in Spain, as well as a similar number in Italy, 14.000 in Croatia, and 9.000 in Greece (Chen et al., 2016) as the most prominent countries within the European Union.

Basically, the European marina market is concentrated in five different areas, which different port configuration characteristics given the differences in coastline and climate (Kizielewicz & Luković, 2013):

- The Mediterranean Sea (the European coastline)
- European Atlantic coast (below the Arctic Circle)
- The Baltic Sea and countries around the Arctic Circle
- Continental European part (inland waters/freshwaters)
- The Black Sea.

The term "Nautical" comes from the Greek ναυτικός (naftikós) which means boat or sailor. In turn, the adoption of the word marina to define this type of facility is a term of Italian origin and means *a small port that receives recreational boats* (Klarić et al., 2015). An accurate definition of a marina related with the basic services required by this type of user is that offered by the "National Association of Motor and Boat Manufacturers"[2] in 1928 as a place where boaters can properly moor their boats, launch them, service them, stock up on of fuel and other navigational needs, taking a hot shower, eating on land, and being close to commercial, communications, and traffic infrastructure (Luković & Bilić, 2007, 116).

Nautical tourism is not limited to the possession of a boat and the use of marine resources, but can be framed with all those other forms of tourism that have an economic, social, and environmental impact on local development (Bizzarri & Foresta, 2011), positively affecting the economies of nearby towns with the generation of direct and indirect income (Kopke et al., 2008). For this reason, the indicators limited to the number of berths of the infrastructure of a port are insufficient to know the real and potential value of the marina. There are two determining factors related to the maritime area where it is located, which are the depth of the water in the

[1] https://www.eea.europa.eu/themes/water.
[2] https://www.nmma.org/about-us.

moorings and the effect of the tides: the minimum depth must be 2,44 m, which will affect the number of boats and type of these that can be moored in each marina.

Therefore, a full service marina requires (Smythe, 2010):

- appropriate depth of water for the draft of sailboats,
- sufficient land for dry boat storage,
- repair services,
- parking access,
- complementary services.

There is no uniform definition of the characteristics and functions that a marina should have, despite the existence of international institutions and associations that regulate this activity, presenting the sector with a lack of unified criteria and procedures. In the same way, the works developed on this subject are scarce and present diverse approaches, most of them observed as case studies aimed at various topics such as user experience (Benevolo & Spinelli, 2021; Paker & Vural, 2016; Silveira & Santos, 2014), the sustainability of facilities and their integration into the environment (Andres et al., 2017; Yang et al., 2014), the application of new technologies (Maglić et al., 2021; Rafał et al., 2022), tourism capacity (Benevolo & Spinelli, 2021; Payeras et al., 2011; Vlasic et al., 2019), or the situation within the legal framework where they fit (Alcover et al., 2011; Škorić et al., 2012).

The main objective of this work is to find works related to the general satisfaction of the services of the users of marinas that have been published in recent years with the idea of establishing solid elements that favor future quantitative and qualitative studies, supported by scientific methods that support advanced statistical techniques and predictive methods. In this direction, the following specific objectives are described as follows:

SO1: Find works on satisfaction in marinas based on quantitative studies and surveys

SO2: highlight the impact of the publications made on this topic.

SO3: Find consistent study factors on which to base further studies.

2 Literature Review

2.1 *Satisfaction in Marinas*

Users of Nautical Ports and marinas demand a series of services that are essential for the practice of their activity in this type of facility. They are usually described as a series of attributes that rather do not allow user dissatisfaction (Shen et al., 2021), which are grouped into the services intrinsic services of the marina, destination experiences on land, charter products, products of support to marina, and basic destination attributes (Mikulić et al., 2015), being essential the degree of perceived satisfaction even considering the bad weather conditions that are not usually an obstacle to the activity enjoyment (Arabadzhyan et al., 2021; Jovanovic et al., 2013), basically with attributes such as nature, the feeling of freedom, and flexibility that navigation brings (Butowski, 2018).

Scientifics works on satisfaction in marinas such as those by Gracan and Uran (2002); Mikulic et al. (2015) and Lam-González (2020) have focused on the degree of functionality using Likert scales without understanding how different marina attributes influence boating experiences (Shen et al., 2021) and that the combination of these in a destination within the user experience drives the level of tourist satisfaction (Albayrak et al., 2013;Toro-Sánchez, 2022). According to Pizam et al. (2016), Likert scales tend to offer positive-biased responses and assume the linear relationship between satisfaction and dissatisfaction when the use of advanced statistical methodologies such as structural equations, observing that this relationship is not linear (Jin et al., 2015).

There are study models on tourist satisfaction that seek to solve the combined analysis of the attributes of the experience, distinguishing these into three types:

(a) *Basic attributes* that determine the minimum requirements of the experience so that if they are not met, dissatisfaction appears gradually, in the same way that if these attributes appear in the appropriate utility, the degree of satisfaction remains stable (Jin et al., 2015) bearing in mind that dissatisfaction generally causes negative word of mouth (Alegre & Garau, 2011). Specifically, in the case of marinas, it would be the sanitary services, the qualities of the employees, the water and electricity connections on the pontoons, as well as the cleanliness of the waters on which they are located and the space conditions applicable to these.
(b) *One-dimensional attributes* that improve the experience to the extent that the user can enjoy it (Palumbo, 2015) such as the activities offered by the port in the form of restaurants, supermarkets, tourist information offices, and recreational activities.
(c) *Attractive attributes* which are not expected by users, so that they do not show dissatisfaction if they do not appear within the experience (Alegre & Garau, 2011; Jin et al., 2015). These attributes are more linked to external infrastructures in relation to the destination such as means of transport, connectivity, cultural offer as well as the existence of social networks linked to the experience that facilitate and expand communication as well as the use of new technologies and digital usability in the user experience.

In the same context, the services provided by the marinas can be classified as *primary*—related to the mooring itself, *secondary*—electricity and water connections, administrative procedures, gasoline supplies, parts, and management of sailors, and additionally *entertainment and tourist activities*—supermarket (Klarić et al., 2015). In spite of the degree of quality with which they are appreciated in this series of services, they influence the perception of satisfaction, especially in the education of the port staff and their friendliness as the main factor (Margariti et al., 2017) together with the dimensions that describe the quality of the berths and guarantee of the equipment (Gracan et al., 2016).

It is important not to forget crucial aspects such as those related to the prices of both general and adjacent services and the legislation related to navigation that concerns each harbor (Škorić et al., 2012), which can clearly present themselves as barriers to demand due to the heterogeneity shown in both aspects, even presenting as a differential factor in both flexible pricing policies and those adapted to the required services (Vlasic et al., 2019), as well as more lax legislation and permissive legislation toward the enjoyment of nautical experiences (Alcover et al., 2011). The price can distinguish between *ordinary* and *privileged* services, the most important factor for owners being an applicable and understandable fee base based on a calendar scheme (Çakirglu, 2019).

An accentuated dissatisfaction or even frustration is normally more expressible than satisfaction (Mikulić et al., 2015); for this reason it is important to determine the critical factors in the experiences of marinas, beyond the models that simply focus on levels of satisfaction, and satisfaction and exceeding expectations, given the high component of basic service that characterizes them. Therefore SERVQUAL (Parasuraman et al., 1985), ECOSERVE (Khan, 2003), and RENTQUAL (Ekiz et al., 2010) models measure the different dimensions of the quality of a service treating them as a global satisfaction of the service instead of the individual satisfaction of each one of the attributes of the global service, so the expectation and improvement of each attribute in particular cannot be explained with the use of these models (Shen et al., 2021).

Kuzma (2003) offers a more primal view of services in marinas, although it is shown at a practical level dividing services specifically in terms of destinations of use—boats, people, and transport—in such a way that it can be oriented toward the target audiences in question and matter of its use, depending on the degree of development and its integration in the environment and the real needs of the users. On the other hand, the degree of professional preparation (Lam-González, 2019) is really destined toward the provision of the specific service, managing resources efficiently and with user orientation (Fig. 1).

2.2 Sustainable Development

In the tourist context, the Marinas offer to the local entities a great opportunity for development that can benefit from them and a growth model (Luković, 2012), with a diversity of types of marinas depending on how they are integrated or not in the environment in which they are framed. Basically, two types can be schematized by classification (Kizielewicz & Luković, 2013):

- *Non-Integrated Marina Models*: Marinas on the Island of Greece
- *Integrated Marina Models*:
  - within the city (Frapa, Horta (Azores), Split (Croatia), Marinas in the city of Athens, Italian Marinas)

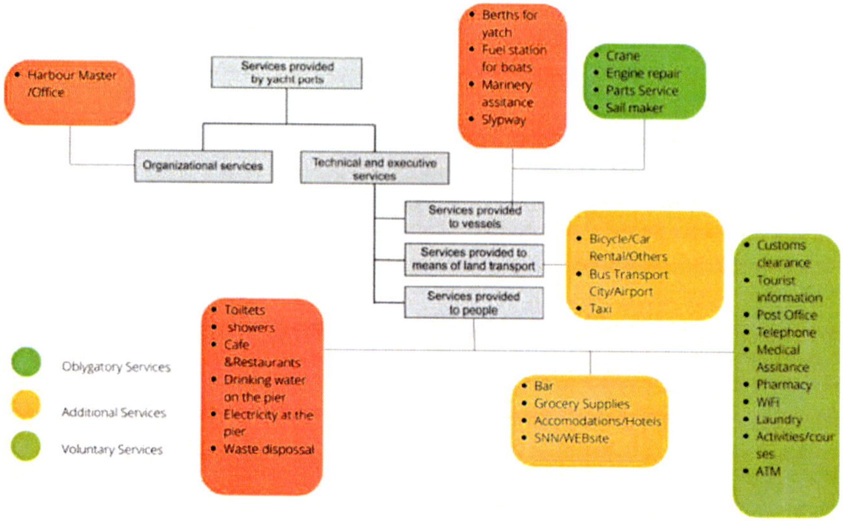

Fig. 1 Service in Marina model. *Source* Kuzma (2003), Lam-Gonzalez (2019) and authors' own elaboration

- within a tourist environment (Benalmádena, Puerto Sherry, Costa Azul in France, Palma de Mallorca).

Functionally, the most used development model is through the transfer of public bodies—who generally hold ownership of the coastal land—transferred to other public and private government bodies that jointly or may not manage the activity of the port as a natural asset. (Clark & Munro, 1994). Thus, service management toward a tourism market orientation with a focus on boaters and the destination's own tourists, affected or not to navigation, is presented as crucial for the development of the marina as a tourist product (Luković & Bilić, 2007) and continuous development in improving the quality of life and employment opportunities of the integrated social environment in the geographical area where it is located (Klarić et al., 2015). Manifestly, this development model is expressed in the EU's 2020 objectives in The European Strategy for Coastal and Maritime Tourism[3] where it concludes that:

> Coastal and maritime tourism needs an ambitious policy framework. The Commission, Member States, regional and local authorities, industry and other stakeholders must take targeted action in coherence with EU policies that have an impact on this sector.

The assessment of the environment by opinion groups and managers dismisses the importance of the attractiveness of the landscape and neighboring tourism resources and does not give it much importance (Yang et al. 2014).

---

[3] https://www.eea.europa.eu/policy-documents/a-european-strategy-for-more.

2.3 Touristic Development

Despite the perspective of the marina as a tourist destination (Luković, 2012), it is possible to approach the nautical tourism offer from a purely product-service approach (Benevolo & Spinelli, 2021) with a multi-offer vision (Silveira & Santos, 2014) and poly-functional activities (Luković & Bilić, 2007). Specifically, nautical tourism in marinas, both as regular boaters-owners and as charter yachts (Besteiro Rodríguez, 2004), offers a great opportunity given its level of spending compared to other types of tourism (Payeras et al., 2011) and very similar to another source of tourism with high potential such as golf (Alcover et al., 2011) with high levels of daily spending—94% more than traditional tourism—and longer stay in destination—11,8 days compared to 8,9 days of traditional holiday tourism in Spain (Payeras et al., 2011), which allows it to be classified as a premium product.

Frequently, the marina and sports port sector is limited to a single value proposition (Heron & Juju, 2012) ignoring the possibility of approaching new markets, while the destination concept does not respond to the real needs of nautical tourism (Lam-González, 2019), so a vision of segmented marketing—lacking in this sector (Paker & Vural, 2016)—is presented as an essential tool within tourism development.

The yachting tourism experience is significantly influenced by five attributes, including basic marina services, destination experiences on land, rental products, marina support products, and basic destination attributes (Mikulić et al., 2015). Curiously, this type of user does not value the tourist offer around the port in a consistent way (Yang, L et al., 2014), with which the need to establish a segmentation of the different users when tourism development actions are established in the marinas and that they are not limited, especially taking into account the public ownership of the land where they are located.

Benevollo and Spinelli (2021) argued that the segmentation in the different users of the marinas can be based psychologically on the benefit it brings to the user in their motivations or based on the attributes of the service itself, so it is important again to detect which of the elements of the marina offer are important for each type of user and how they are identified by them as essential, appreciated, or indifferent. In the same way, changes in user behavior are the forces that should direct the tourist offer (Poon, 1993) and in this case nautical (Silveira & Santos, 2014) given its ascetic and fun potential. For tourism development, it is necessary to generate links between the marina and its surroundings with strategies that connect the outstanding attractions of nearby cities (Nebot et al., 2017) as well as their natural environment together (Martín & Yepes, 2019).

Another solution that appears open to lower-income users is the yacht rental market (Chen et al., 2016), which allows the growth of nautical tourism in marinas, together, on the other hand, with the mega-car sector, yachts (Pallis & Lekakou, 2004), where the marinas require both technical and legal specifications to be able to provide such services and the cooperation of the different stakeholders for the development of this business (Corres, 2007).

3 Methodology

The research method used is the Systematic Literature Review (SLR) (Ramey & Rao, 2011; Schaller et al., 2019) and in particular following the guidelines of similar studies related to recent dates in the tourism field (Abarca et al., 2020; Perdomo-Verdecia et al., 2022; Rojas-Sánchez et al., 2022). In recent years, more and more articles have been found on the user satisfaction in marinas, although in a number certainly scarce and with diverse methodologies and generally focused only on expert panels (Martín & Yepes, 2019, 2022; Payeras et al., 2011) and few of them taking into consideration the opinions of marina users (Jovanovic et al., 2013; Mikulić et al., 2015; Shen et al., 2021). The heterogeneity of the bibliography and the thematic focus of the articles especially in the case study lead us to justify the use of SLR based on the following points (Abarca et al., 2020; Ramey & Rao, 2011) (Fig. 2).

This work raises the following questions to give rise to research (RQ):

RQ1: What are the main areas and topics of current research in Marina user satisfaction?
RQ2. What are the main constructs that have been investigated in current research on Marina user satisfaction?
RQ3. What are the gaps in existing research and possible areas for future research?

Then, we use the secondary sources of data through the search in the different platforms, Web of Science (WoS) and Scopus, where we use the search terms related to the objective of the work that deals with the satisfaction of users in marinas and marinas, finding few or no references using such terms. Using the terms of searches terms more amplios give the scarcity of publications found generating a pilot search as detailed:

- WoS: marina satisfaction + nautical tourism marinas + yacht tourism: 27 documents from 2011 to August 2022
- Scopus; (TITLE-ABS-KEY (marinas AND tourism) AND TITLE-ABS-KEY (nautical AND tourism) AND TITLE-ABS-KEY (Yacht AND tourism) 35 documents from 2005 to August 2022

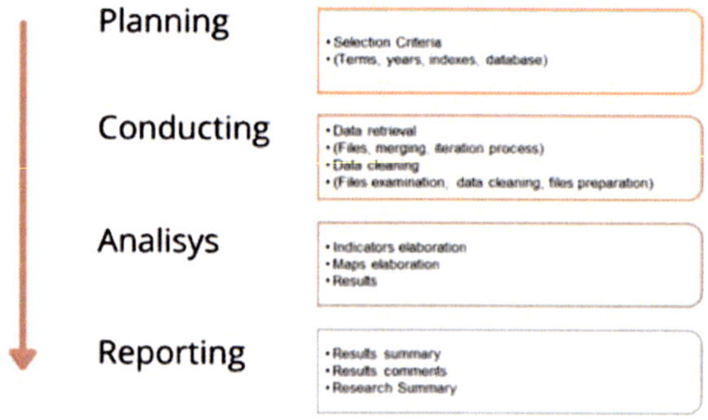

Fig. 2 Research methodology SLR. *Source* (Abarca et al., 2020; Ramey & Rao, 2011; Schaller et al., 2019)

In addition to the references taken from the first analysis on the publications found, we consider including a number of 8 more references from the Google Scholar reference base. The Google Scholar search also ensured that no important work had gone undetected in the WoS and Scopus searches, summarized in 60 publications. The results obtained were analyzed using the same terms as in the other searches.

In a first view, we observe the existence of a large number of articles directly related to a specific topic that is the Sustainability of Marinas. The most profound analysis of these publications does not lead to the conclusion that the vast majority focus on technical studies far from the purpose enunciated on the satisfaction of users, although obviously, sustainability and social commitments are the aspects to be considered within our objective although far from the service perception of the users.

Specifically, it is a total of 33 articles. Although we discard our work objectives, 8 of them are mentioned as relevant due to the relationship they may have in different studies more focused on the sustainability of marinas and as a complement to our theme addressed (Table 1).

Table 1 Relevant articles on Marina sustainability

Data Base	Authors	Article title	Citations
Wos	Favro and Grzetic (2008)	Nautical tourism–the advantages and effects of development	5
Wos	Bizzarri and La Foresta (2011)	Yachting and pleasure crafts in relation to local development and expansion: Marina di Stabia case study	7
Wos	Andres et al. (2017)	Marine pollution in the nautical seaports in Croatia by the effluent of tourists	57
Wos	Kovacic et al. (2016)	Construction of nautical tourism ports as an incentive to local development	12
Scopus	Rojo (2009)	Economic development versus environmental sustainability: The case of tourist marinas in andalusia	7
Scopus	Carević et al. (2014)	Basic plan dimensions of marinas in Croatia	4
Wos	Martín and Yepes (2019)	The concept of landscape within marinas: Basis for consideration in the management	15
Wos	Dinwoodie; Tuck	Sustainable development of maritime operations in port	62

4 Results

The publications obtained from the databases will be later analyzed according to the aspects they deal with. First of all, we observe the scarcity of articles found in a broad reference deal with user satisfaction in marinas. The following graph shows the number of articles found in the last twelve years (Fig. 3).

Likewise, of the articles found, the important weight of the issuing of Croatian origin stands out—46% of the number of publications—in relation to the rest of European countries, of which Spain, Italy, and Poland also stand out; the latter very much focused on articles related to inland navigation (Fig. 4).

Specifically, the table shown below details the articles obtained through the metho dology described and on the basis of data used with the indicated search terms, where a list of 39 publications is offered. This relationship affects the description of some

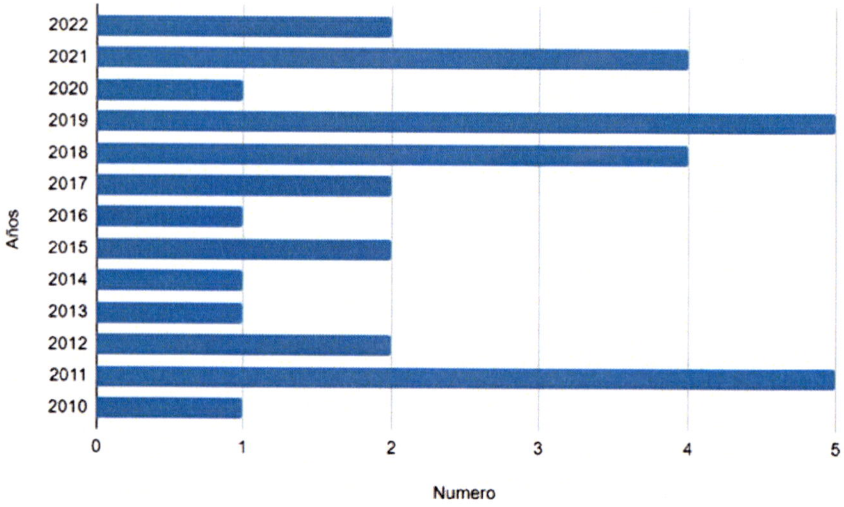

Fig. 3 Number in publications in Marina satisfaction from 2010

Fig. 4 Percentage of publications in Marina satisfaction in the from 2010 by type

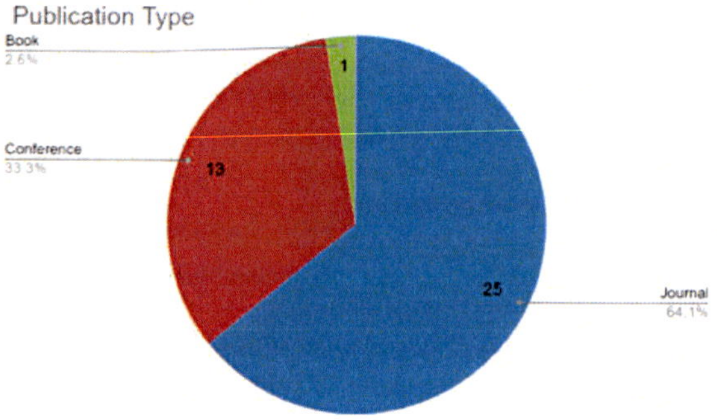

of the specific objectives initially set in this study, such as how the publication deals with a specific type of study, quantitative or qualitative, while detailing the type of methodology used in each of the studies. On the other hand, the relationship between the existence of a questionnaire within the study also stands out, given the importance revealed for the study by deciphering the indicators and constructs for future studies (Table 2).

The table below shows, in order of importance, the relationships of impact journals and decreasing order that contain articles and the number of them, on the subject of user satisfaction in marinas (Table 3).

5 Discussion and Conclusions

In this section, we propose to answer the different questions raised in our analysis.

Q1 What are the main areas and topics of current research in Marine user satisfaction?

Table 1 shows publications highlighted in this research along with the development of sustainability itself; within the specific analysis filter applied to the satisfaction of marina users, we find six groups of study areas, specifically: confidence in the service, quality of the service, supply and demand, sustainability but understood as a component of satisfaction and in relation to social responsibility, tourism and technology, and social networks. They are detailed in the following table and related to the articles that deal with the different areas (Table 4).

Trust in the service is shown above all in studies developed in panels of experts and descriptive analyzes (Santos & Perna, 2018; Mundula et al., 2020) and above all the aforementioned work by Parker and Vural (2016) that proposes the segmentation of users that are addressed with quantitative analysis with main components by Benevolo and Spinelli, (2021) and (Shen et al., 2021) in the Kano satisfaction model, both implementing user questionnaires in their analysis. At this point, it is interesting to cite the work of Jovanovic et al. (2013) who, with quantitative methods, address dissatisfaction closely related to the area of service quality, also using questionnaires in their analysis.

The quality of the services, different from the trust in which more personal factors are appreciated (Margariti et al., 2017; Mikulić et al., 2015), is addressed in the selection in comparative studies either between different case studies or in reference to the applicable regulations. As a work where quantitative methods are used, mention the one by Mikulić et al. (2015).

Sustainability is especially addressed in the case studies and in its impact on the local population (Ugolini & Ivaldi, 2017; Mundula et al., 2020) and in the degree of user satisfaction, although not showing it as representative (Shen et al., 2021).

Several articles on supply and demand are also detailed, addressing different sectors, highlighting Lam-González, (2019) and Haas and Heiner (2011), specifically where the niche of large yachts is contemplated.

Table 2 List of articles that address the theme of Satisfaction in Marinas

	Data base	Authors	Article title	Test	Tech. type	Methodology	Article title–SciELO
1	Wos	Santos and Perna (2018)	Yachts Passing by the West Coast of Portugal—What to Do to Make the Marina and the Destination of Figueira da Foz a Nautical Tourism Reference?	N	DELPHI	SWOT	5
2	Wos	Mundula et al. (2020)	Smart Marinas. The Case of Metropolitan City of Cagliari	N	Descriptive	Cluster	0
3	Wos	Lazarus et al. (2021)	Yachts and marinas as hotspots of coastal risk	N	Descriptive		2
4	Wos	Haas and Heiner (2011)	Marina Management for Super Yachts and Cruise vessels	N	Descriptive		4
5	Wos	Dowling (2013)	Nautical Tourism in the Pacific	N	Descriptive	Generalistic	0
6	Wos	Rafal et al. (2022)	Conceptual framework of a Global Yacht Positioning System in Poland	N	Descriptive	Case	0
7	Wos	Benevolo and Spinelli (2021)	Benefit segmentation of pleasure boaters in Mediterranean marinas: A proposal	Y	Quantitative	CPA	5
8	Wos	Parker &Vural	Customer segmentation for marinas: Evaluating marinas as destinations	Y	Qualitative	Cluster	87

(continued)

Table 2 (continued)

	Data base	Authors	Article title	Test	Tech. type	Methodology	Article title–SciELO
9	Wos	Kovacic et al. (2018)	Nautical Tourism in Croatia and in Portugal in the Late 2010's: Issues and Perspectives	N	Descriptive		2
10	Wos	Vukovicand Mišura (2019)	Croatian Maritime Code Reform - Charter and Contract of Nautical Berth	N	Descriptive	Legal	0
11	Wos	Ugolini and Ivaldi (2017)	Tourist Ports and Yachting: The Case of Sardinia	Y	Descriptive	Case	1
12	Wos	Martin-Vazquez and Yepes (2022)	Impact Analysis of Marinas on Nautical Tourism in Andalusia	Y	DELPHI	SWOT	1
13	Wos	Shen et al. (2021)	**Perceived importance of and satisfaction with marina attributes in sailing tourism experiences**	Y	Quantitative	Kano model	4
14	Wos	Almendarez-Hernandez et al., 2020	Economic valuation of sportfishing in the surroundings of Cerralvo Island, Baja California Sur, Mexico using the travel cost method	Y	Qualitative	Travel cost method	1

(continued)

Table 2 (continued)

	Data base	Authors	Article title	Test	Tech. type	Methodology	Article title–SciELO
15	Google Scholar	Payeras et al. (2011)	THE YACHTING CHARTER TOURISM SWOT: A BASIC ANALYSIS TO DESIGN MARKETING STRATEGIES	Y	Descriptive	SWOT	15
16	Google Scholar	Alcover et al., 2011	The economic impact of yacht charter tourism on the Balearic economy	Y	Descriptive		69
17	Google Scholar	ÇAKIROĞLU (2019)	A PROPOSED PROJECTION OF THE MARKETING TOOLS FOR THE TRNC MARINAS IN THE INTERNATIONAL MARKETS	N	Descriptive	Generalistic	
18	Google Scholar	Dikeç, and Töz (2017)	The impact of perceived service quality on satisfaction: An application on Marina customers in Turkey	Y	Quantitative	LINKERT 5	3
19	Wos	Mikulić et al. (2015)	**Critical Factors of the Maritime Yachting Tourism Experience: An Impact-Asymmetry Analysis of Principal Components**	Y	Quantitative	CPA	41

(continued)

Table 2 (continued)

	Data base	Authors	Article title	Test	Tech. type	Methodology	Article title–SciELO
20	Google Scholar	Jovanovic et al. (2013)	WHAT DEMOTIVATES THE TOURIST? CONSTRAINING FACTORS OF NAUTICAL TOURISM	Y	Quantitative	component exploratory factor analysis (EFA),	65
21	Scopus	Poljičak et al. (2022)	Nautical Tourism-Case Study in the Republic of Croatia	N	Qualitative	Regression	
22	Scopus	Bulum et al. (2021)	Competition Issues in the Croatian Seaport Sector Regarding the Provision of Nautical Tourism Services	N	Qualitative	Description	
23	Scopus	Maglić et al., 2021	Application of smart technologies in croatian marinas	N	Qualitative	Generalistic	3
24	Scopus	Lukovic et al. (2021)	Marina business in SME approaches [Poslovanje marina u pristupima malog i srednjeg poduzetnitva]	N	Qualitative	Generalistic	
25	Scopus	Popielarczyk (2019)	Using a small UAV for nautical ports inventory	N	Qualitative	Case	
26	Scopus	Benevolo and Spinelli (2019)	The use of websites by Mediterranean tourist ports	Y	Qualitative	2QCV3Q	7
27	Scopus	Padovan and Voglar (2019)	Marina operator liability insurance in croatian and Slovenian law and practice	N	Qualitative	Classification	

(continued)

Table 2 (continued)

	Data base	Authors	Article title	Test	Tech. type	Methodology	Article title–SciELO
28	Scopus	Petrinović and Mandić (2019)	Can marinas qualify as places of refuge?	N	Qualitative	Description	
29	Scopus	Padovan et al. (2018)	Security and enforcement of marina operator's claims: Croatian, Italian and Spanish law perspectives	N	Qualitative	Description	1
30	Scopus	Ivanić et al. (2018)	Nautical tourism: Generator of Croatian economy development	N	Qualitative	Classification	6
31	Scopus	Monteiro et al. (2017)	A microcluster approach applied to the case of the nautical tourism sector of the Algarve region (Portugal)	Y	Qualitative	Cluster	1
32	Scopus	Lam González andDe León-Ledesma (2015)	European nautical tourists: Exploring destination image perceptions	Y	Quantitative	LINKERT 5	13
33	Scopus	Stipanović et al. (2012)	The development concept in function of the competitiveness of marina frapa rogoznica [Koncepcija razvoja u funkciji konkurentnosti marine Frapa Rogoznica]	Y	Quantitative	Case	2

(continued)

Table 2 (continued)

	Data base	Authors	Article title	Test	Tech. type	Methodology	Article title–SciELO
34	Scopus	Franjić et al.	System concept of experience in nautical tourism	N	Qualitative		
35	Scopus	Kasum et al. (2011)	Nautical tourism	N	Quantitative	Correlations	2
36	Scopus	Mateos et al. (2010)	The marinas as infrastructures of it supports of the nautical recreational activities in Andalusia [Los puertos deportivos como infraestructuras de soporte de las actividades náuticas de recreo en Andalucía]	N	Descriptive	Generalistic	9
37	Scopus	Horak et al. (2006)	Competitiveness of Croatian nautical tourism	N	Descriptive		12
38	Scopus	Miočićey et al. (2006)	Internet business of the Croatian marinas	Y	Descriptive		
39	Scopus	Stipanović et al.	Development strategies functioning as ACI marina Umag competitiveness	Y	Descriptive	Case	2

Table 3 List of impact journals of Satisfaction in Marinas. Source: Observatorio de Calidad y Rendimiento de la Universidad de Salamanca

Group authors	Number	ISSN	JCR	WoS	CES indicators	SJR	Scopus	Cat scopus
Journal of Travel & Tourism Marketing	2	10,548,408	1,741	Q2	Social sciences-general	3,49	Q1	Business, Management-Accounting
International Journal of Tourism Research	1	10,992,340	1,095	Q3	Social sciences-general	1,064	Q1	Business, Management-Accounting
Tourism Economics	2	13,548,166	0,392	Q4	Social sciences-general	0,674	Q1	Business, Management-Accounting
Journal of Coastal Research	1	0,749,008	0,852	Q4	Environment-ecology	0,672	Q2	-Earth-Planetary Sciences-Environmental Sciences
Journal of Marine Science and Engineering	1	10,232,796	0,298	Q4	Engineering	0.235	Q3	-Earth-Planetary Sciences-Environmental Sciences

Table 4 Main subjects related in each publication

	Confidence of service	Service quality	Offer and demand	Sustainability	Tourism	TIC & SSN
1	x	x	x	x	x	
2	x	x	x	x	x	x
3	x	x	x			
4		x	x		x	
5		x	x		x	
6		x	x	x	x	x
7	x	x	x		x	
8	x	x	x	x	x	
9			x	x	x	
10		x				
11			x	x		
12		x		x	x	
13	x	x	x		x	
14		x	x		x	
15					x	
16		x	x	x	x	
17			x		x	
18	x	x	x	x		
19	x	x	x	x	x	
20	x				x	
21			x		x	
22		x	x		x	
23		x		x		x
24		x	x	x		
25		x		x	x	
26		x			x	x
27	x					
28		x		x		
29	x	x				
30		x			x	
31			x		x	
32						
33			x		x	
34	x		x		x	
35		x			x	

(continued)

Table 4 (continued)

	Confidence of service	Service quality	Offer and demand	Sustainability	Tourism	TIC & SSN
36		x	x	x	x	
37						
38					x	x
39		x			x	

Technology and the use of social networks is a little-cited area of work, although many articles point out the need to promote work in this line (Mundula et al., 2020; Paker & Vural, 2016). It is worth highlighting the interesting work on positioning by Rafał et al. (2022) as a proposal that connects the activity of several freshwater ports and the use of social networks by Benevolo and Spinelli (2019).

Q2. What are the main constructs that have been investigated in current research on Marina user satisfaction?

As previously mentioned, the satisfaction of the user of the marinas is observed by variables above all related to the quality of the service and the confidence generated, with indicators closely related to the specific services offered by the marina, as detailed above in Fig. 1, and above all those related to essential services such as the quality of the moorings, sanitary services, water quality, attention and professionalism of the staff, etc. It seems correct that any study model should approach the different indicators progressively in terms of satisfaction or dissatisfaction with the services to the extent that the user identifies them as essential or not within their recreational navigation activity.

The different constructs have been contemplated in the studies using different methodologies that are detailed in Table 2 and that we summarize in this Fig. 5

Q3. What are the gaps in existing research and possible areas for future research?

The following figures show the types of analysis carried out (Fig. 6) and which of them contemplate questionnaires (Fig. 7).

In general, it is observed that few studies are carried out with quantitative analyzes and the vast majority of these are based on panels of experts. In general, a deficit is observed in the proposal of questionnaires toward the users of the marinas and even propose new data sources with more segmented users of origin while the satisfaction indicators are extended toward the proposals of tourist services and related to the social interaction (social networks) and the use of technologies in the sector.

The selection of the literature shows, as a whole, a scarce development of the theme, especially focused on the application of advanced study methodologies that require more exhaustive previous developments, such as an account where the ideal indicators are determined that explore the satisfaction of the users of the marinas.

Contemplating the users of the marinas as a diverse and segmented compendium (Paker & Vural, 2016) in the face of both economic and environmental sustainability of the tourist offer (Bizzarri & Foresta, 2011) of the marinas and marinas, the studies focused on these. These objectives require a user orientation beyond the opinions

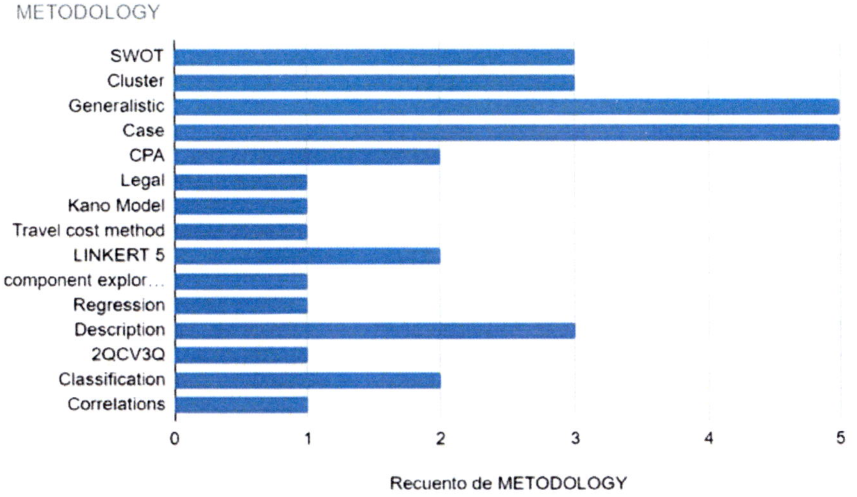

Fig. 5 Frequency of methodologies used in the different studies. *Source* authors

Fig. 6 Analysis types

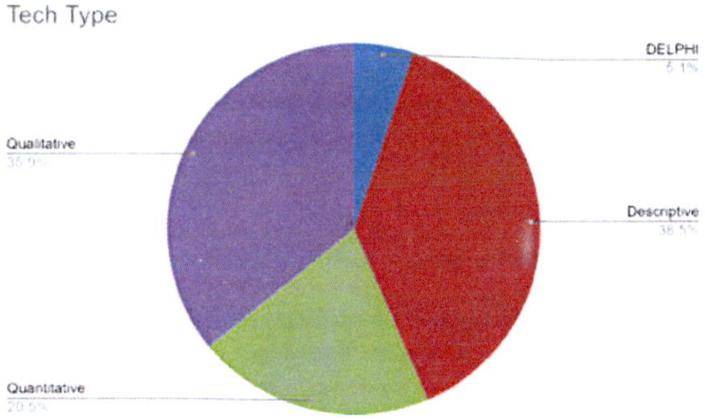

Fig. 7 Frequency of questionnaire

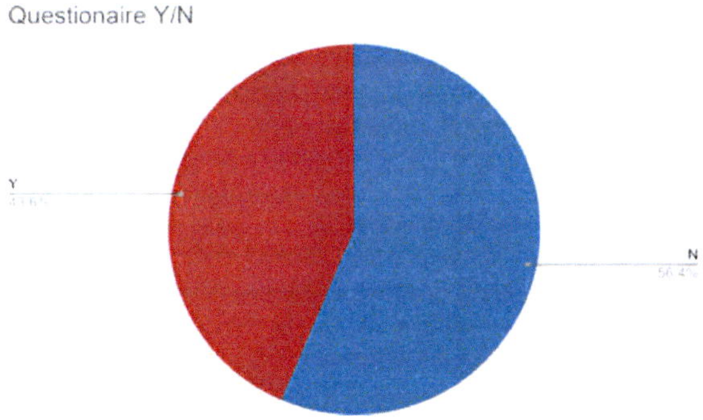

of experts and stakeholders based on Delphi-type methodologies and descriptive analyzes of the sector, which are the basis of the bibliography found in this study. Also, a clear distinction and cl of the services at the time of being assessed within a questionnaire expressed a clearer vision of the critical points in the final assessment of user satisfaction with a clear vision toward the use of statistical methodologies advanced as structural equations and partial least squares with extensive development already in social sciences.

In the same way, the systematic review developed in this study fails to observe a specific work model on which to determine studies on satisfaction, although there are references that can be very encouraging for future methodological development, such as works already developed in the hotel sector (Alcover et al., 2011) or the similarities found with another niche in the tourism market such as golf tourism (Payeras et al., 2011), both endowed with extensive literature.

References

Abarca, V. M. G., Palos-Sanchez, P. R., & Rus-Arias, E. (2020). Working in virtual teams: A systematic literature review and a bibliometric analysis. *IEEE Access, 8*, 168923–168940.

Albayrak, T., Aksoy, Ş., & Caber, M. (2013). The effect of environmental concern and scepticism on green purchase behaviour. *Marketing Intelligence & Planning*.

Alcover, A., Alemany, M., Jacob, M., Payeras, M., García, A., & Martínez-Ribes, L. (2011). The economic impact of yacht charter tourism on the Balearic economy. *Tourism Economics, 17*(3), 625–638.

Alegre, J., & Garau, J. (2011). The factor structure of tourist satisfaction at sun and sand destinations. *Journal of Travel Research, 50*(1), 78–86.

Andres, M. A., Madariaga, E., Delgado, O., & Martínez, J. E. (2017). Marine pollution in the nautical seaports in Croatia by the effluent of tourists. In European Transport-Trasporti Europei (p. 64).

Arabadzhyan, A., Figini, P., García, C., González, M. M., Lam-González, Y. E., & León, C. J. (2021). Climate change, coastal tourism, and impact chains–a literature review. *Current Issues in Tourism, 24*(16), 2233–2268.

Benevolo, C., & Spinelli, R. (2019). The use of websites by Mediterranean tourist ports. *Journal of Hospitality and Tourism Technology, 10*(2), 190–204.

Benevolo, C., & Spinelli, R. (2021). Benefit segmentation of pleasure boaters in Mediterranean marinas: A proposal. *International Journal of Tourism Research, 23*(1), 134–145.

Besteiro Rodríguez, B. (2004). El desarrollo del turismo náutico en Galicia. Cuadernos de turismo.

Bizzarri, C., & Foresta, D. (2011). Yachting and pleasure crafts in relation to local development and expansion: Marina di Stabia case study. *WIT Transactions on Ecology and the Environment, 149*, 53–61.

Bulum, B., Pijaca, M., & Primorac, Z. (2021). Competition in the croatian seaport sector regarding the provision of nautical tourism services. *YARS, 24*, 117.

Butowski, L. (2018). An integrated AHP and PROMETHEE approach to the evaluation of the attractiveness of European maritime areas for sailing tourism. *Moravian Geographical Reports, 26*(2), 135–148.

Çakiroğlu, H. (2019). A proposed projection of the marketing tools for the trnc marinas in the international markets. *Journal of International Social Research, 12*(63).

Chen, J., Balomenou, C., Nijkamp, P., Poulaki, P., & Lagos, D. (2016). The sustainability of yachting tourism: A case study on Greece.

Clark, C. W., & Munro, G. R. (1994). Renewable resources as natural capital. In *Investing in natural capital: The ecological economics approach to sustainability* (pp. 343–361). Island Press.
Corres, A. J. (2007). Greek Maritime Policy and the discreet role of shipowners' associations. *Research in Transportation Economics, 21*, 221–255.
Dainys, J., Gorfine, H., Mateos-González, F., Skov, C., Urbanavičius, R., & Audzijonyte, A. (2022). Angling counts: Harnessing the power of technological advances for recreational fishing surveys. *Fisheries Research, 254*, 106410.
Dikeç, G., & Töz, A. C. (2017). The impact of perceived service quality on satisfaction: an application on marina customers in Turkey.
Dowling, R. (2013). Nautical tourism in the Pacific. In *Nautical tourism* (pp. 189–198). Wallingford UK: CABI.
Ekiz, E., Hussain, K., & Ivanov, S. (2010). Investigating marketing opportunities of a politically challenged island destination: The case of North Cyprus. *Marketing Island Destinations, 65.*
Favro, S., Kovačić, M., & Gržetić, Z. (2008). Nautical tourism the basis of the systematic development.*Pomorstvo/Journal of Maritime Studies, 22*(1).
Gracan, D., Gregoric, M., & Martinic, T. (2016). Nautical tourism in Croatia: Current situation and outlook. In *Faculty of tourism and hospitality management in Opatija. Biennial International Congress. Tourism & Hospitality Industry* (p. 66). University of Rijeka, Faculty of Tourism & Hospitality Management.
Hall, B. H., & Khan, B. (2003). *Adoption of new technology.*
Haass, H. (2011). Marina management for super yachts and cruise vessels. *Journal of coastal research, 61*(10061), 475–475.
Heron, R., & Juju, W. (2012). The marina-sustainable solutions for a profitable business. Lulu. Com.
Ivanić, K., Perić Hadžić, A., & Mohović, Đ. (2018). Nautical tourism: Generator of Croatian economy development. *Pomorstvo, 32*(1), 59–66.
Jin, N., Lee, S., & Lee, H. (2015). The effect of experience quality on perceived value, satisfaction, image and behavioral intention of water park patrons: New versus repeat visitors. *International Journal of Tourism Research, 17*(1), 82–95.
Jovanovic, T., Dragin, A., Armenski, T., Pavic, D., & Davidovic, N. (2013). What demotivates the tourist? Constraining factors of nautical tourism. *Journal of Travel & Tourism Marketing, 30*(8), 858–872.
Kasum, J., Žanić Mikuličić, J., & Kolić, V. (2018). Safety issues, security and risk management in nautical tourism. *Transactions on Maritime Science, 7*(02), 184–188.
Kizielewicz, J., & Luković, T. (2013). The phenomenon of the marina development to support the European model of economic development. *TransNav, International Journal on Marine Navigation and Safety Od Sea Transportation, 7*(3).
Klarić, S., Afrić Rakitovac, K., & Trošt Lesić, K. (2015). Corporate social responsibility of Croatian Marinas. *Ekonomska Misao i Praksa, 1*, 81–101.
Klein, R., Božić, K., Dowling, R., Gibson, P., Gržetić, Z., Haahti, A., & Pekkala, S. (2013). Opportunities for market development of nautical tourism in Europe. In *Nautical tourism* (pp. 201–224). Wallingford UK: CABI.
Kopke, K., O'Mahony, C., Cummins, V., & Gault, J. (2008). *Assessment of coastal recreational activity and capacity for increased boating in Cork harbour*. Coastal and Marine Resources Centre Report. University College Cork.
Kovačić, M., Favro, S., & Mezak, V. (2016). Construction of nautical tourism ports as an incentive to local development.*Environmental Engineering & Management Journal (EEMJ), 15*(2).
Kuzma, J. R., Veltri, F. R., Kuzma, A. T., & Miller, J. J. (2003). Negative corporate sponsor information: The impact on consumer attitudes and purchase intentions. *International Sports Journal, 7*(2), 140.
Lam-González, Y. E., & De León-Ledesma, J. (2015). El perfil del turista nautico en el destino Cabo Verde. *Revista Perspectiva Empresarial, 2*(2), 23–35.

Lam-González, Y. (2019). Maritime Tourism: Modelling consumer behaviour and its managerial implications: Supervisors. *European Journal of Tourism Research, 23*, 208–211.
Lazarus, E. D., & Ziros, L. A. (2021). Yachts and marinas as hotspots of coastal risk. *Anthropocene Coasts, 4*(1), 61–76.
Lončar, G., Bekić, D., Carević, D., Bujak, D., Bartolić, I., & Beg Paklar, G. (2016). The impact of culverts on the seawater exchange and wave action in marina waters. *Hrvatske vode, 24*(98), 275–284.
Luković, T., & Bilić, M. (2007). Luke nautičkog turizma u Hrvatskoj i strategija lokalnoga razvoja. *NAŠE MORE: Znanstveni Časopis Za More i Pomorstvo, 54*(3–4), 114–122.
Luković, T. (2012). Nautical tourism and its function in the economic development of Europe. In *Visions for global tourism industry: Creating and sustaining competitive strategies*. InTech.
Luković, T., Piplica, D., & Hruška, D. (2021). Comparative performance analysis of the marina and charter service industries in Croatia. *Pomorstvo, 35*(2), 318–326.
Maglić, L., Grbčić, A., Maglić, L., & Gundić, A. (2021). Application of smart technologies in Croatian Marinas. *Transactions on Maritime Science, 10*(01), 178–188.
Margariti, A., Antonopoulos, K., Marini, K., & Skanavis, C. (2017). The case of the port of Linaria-Skyros, Greecee: The human factor as the main dimension of impression and other findings in the guestbook tourists' evaluation.
Martín, R., & Yepes, V. (2019). The concept of landscape within marinas: Basis for consideration in the management. *Ocean & Coastal Management, 179*, 104815.
Martín, R., & Yepes, V. (2022). Assessing the relationship between landscape and management within Marinas: The Managers' Perception. *Land, 11*(7), 961.
Mikulić, J., Krešić, D., & Kožić, I. (2015). Critical factors of the maritime yachting tourism experience: An impact-asymmetry analysis of principal components. *Journal of Travel & Tourism Marketing, 32*(sup1), 30–41. https://doi.org/10.1080/10548408.2014.981628
Mundula, L., Ladu, M., Balletto, G., & Milesi, A. (2020). Smart marinas. The case of metropolitan city of Cagliari. In *International Conference on Computational Science and Its Applications* (pp. 51–66).
Nebot, N., Rosa-Jiménez, C., Ninot, R. P., & Perea-Medina, B. (2017). Challenges for the future of ports. What can be learnt from the Spanish Mediterranean ports? *Ocean & Coastal Management, 137*, 165–174.
Nowacki, M. (2015). Turystyka żeglarska [Sailing tourism. In Stasiak.
Padovan, A. V., & Selan Voglar, M. (2019). Marina operator liability insurance in croatian and slovenian law and practice. *Transactions on Maritime Science, 8*(01), 109–122.
Paker, N., & Vural, C. A. (2016). Customer segmentation for marinas: Evaluating marinas as destinations. *Tourism Management, 56*, 156–171.
Pallis, A. A., & Lekakou, M. B. (2004). Cruising the Mediterranean Sea: Market structures and EU policy initiatives. *Journal of Transport and Shipping, 2*.
Palumbo, F. (2015). Developing a new service for the digital traveler satisfaction: The Smart Tourist App. *International Journal of Digital Accounting Research, 15*.
Parasuraman, A., Zeithaml, V. A., & Berry, L. L. (1985). A conceptual model of service quality and its implications for future research. *Journal of Marketing, 49*(4), 41–50.
Payeras, M., Jacob, M., Garcia, A., Alemany, M., Alcover, A., & Martínez-Ribes, L. (2011). The yachting charter tourism SWOT: A basic analysis to design marketing strategies. *Tourismos, 6*(3), 111–134.
Perdomo-Verdecia, V., Sacristán-Díaz, M., & Garrido-Vega, P. (2022). Lean management in hotels: Where we are and where we might go. *International Journal of Hospitality Management, 104*, 103250.
Pizam, A., Shapoval, V., & Ellis, T. (2016). Customer satisfaction and its measurement in hospitality enterprises: A revisit and update. *International Journal of Contemporary Hospitality Management*
Poljičak, A. M., Hinić, M. L., & Kalac, A. (2022). Nautical tourism–case study in the republic of Croatia. *LOGI–Scientific Journal on Transport and Logistics, 13*(1), 73–83.

Poon, A. (1993). *Tourism, technology and competitive strategies*. CAB international.
Rafał, K., Agnieszka, S., Grzegorz, G., Cezary, K., & Dariusz, T. (2022). Conceptual framework of a Global Yacht Positioning System in Poland. *Applied Geomatics, 14*(1), 79–91.
Ramey, J., & Rao, P. G. (2011). The systematic literature review as a research genre. *IEEE International Professional Communication Conference, 2011*, 1–7.
Rojas-Sánchez, M. A., Palos-Sánchez, P. R., & Folgado-Fernández, J. A. (2022). Systematic literature review and bibliometric analysis on virtual reality and education. *Education and Information Technologies.*
Rojo, I. (2009). Economic development versus environmental sustainability: The case of tourist marinas in Andalusia. *European Journal of Tourism Research, 2*(2), 162–177.
Santos, N., & Perna, F. (2018). Yachts passing by the west coast of Portugal–what to do to make the marina and the destination of Figueira da Foz a nautical tourism reference? *Pomorstvo, 32*(2), 182–190.
Schaller, A. A., Vatananan-Thesenvitz, R., Pulsiri, N., & Schaller, A. M. (2019). The rise of digital business models: An analysis of the knowledge base. In *2019 Portland International Conference on Management of Engineering and Technology (PICMET)* (pp. 1–13)
Shen, Y., Kokkranikal, J., Christensen, C. P., & Morrison, A. M. (2021). Perceived importance of and satisfaction with marina attributes in sailing tourism experiences: A kano model approach. *Journal of Outdoor Recreation and Tourism, 35.*
Silveira, L., & Santos, N. (2014). Marina impacts on the local population and on tourism development in Horta (Azores Islands Portugal). *Tourism in Marine Environments, 9*(3–4), 193–202.
Škorić, S., Bartoluci, M., & Čustonja, Z. (2012). Public financing in Croatian sport. *Financial Theory and Practice, 36*(2), 179–197.
Smythe, T. C. (2010). Can coastal management programs protect and promote water-dependent uses? *Coastal Management, 38*(6), 665–680.
Stipanović, C., & Gračan, D. (2005). Development strategies functioning as ACI marina Umag competitiveness. *NAŠE MORE: znanstveni časopis za more i pomorstvo, 52*(3–4), 163–172.
Stipanović, C., Gračan, D., & Bradetić, M. (2012). KONCEPCIJA RAZVOJA U FUNKCIJI KONKURENTNOSTI MARINE FRAPA ROGOZNICA. *NAŠE MORE: znanstveni časopis za more i pomorstvo, 59*(1–2), 61–69.
Toro-Sánchez, F. (2022). Comparison of memorable tourist experiences based on collections of reviews from trip advisor: Acropolis of Athens and Royal Alcazar of Seville. In *Transcending borders in tourism through innovation.*
Ugolini, G. M., & Ivaldi, E. (2017). Tourist ports and yachting: The case of Sardinia. *Revista De Estudios Andaluces, 34*(1), 429–452.
Vlasic, D., Poldrugovac, K., & Jankovic, S. (2019). The Competitive pricing in marina business: Exploring relative price position and price fluctuation. *Journal of Tourism, Heritage & Services Marketing, 5*(1), 3–8.
Vuković, A., & Mišura, M. (2019). Croatian maritime code reform–charter and contract of nautical berth. *NAŠE MORE: znanstveni časopis za more i pomorstvo, 66*(1), 43–49.
Yang, L., Ching-Min, & Jan-Jung. (2014). Assessing fishing port locations for adaption into YACHT Marinas in Taiwan. *Journal of Marine Science and Technology,* 22(Iss. 5). https://doi.org/10.6119/JMST-013-1029-1

Implication of Ukraine War on EU Migration Flows: Perspectives and Challenges

Mirela Ionela Aceleanu, Ioana Manuela Mîndrican, and Javier Ordóñez Monfort

Abstract Since the annexation of Crimea by Russia, Eastern Europe has experienced an unprecedented migration crisis. Most Ukrainians are internally displaced or have migrated to other countries, facing an uncertain and difficult future. Moreover, the outbreak of war in Ukraine has considerably affected various areas of the economic, social, and political life of European states, including on tourism. The main objective of this paper is to highlight the situation of the migrant crisis and the implications that this phenomenon has on the destination states. The implications of the current situation in Ukraine for mobility and migration are significant and have left a deep imprint on the socio-economic context of Europe. The first part of this paper highlights the contextual issues specific to the migration of Ukrainians, more precisely, the patterns of flows and the practices of their acceptance and reception. The second section outlines an overview of the stocks of immigrants that have been registered, and in the final section, it will be presented the main challenges facing European states from the perspective of destination states, but also of war refugees.

Keywords Migration · Socio-economic context · Tourism · Mobility · Crisis · War · Refugees

JEL Classification F22 · O15 · J 61 · G01

M. I. Aceleanu (✉)
Department of Economics and Economic Policies, Bucharest University of Economic Studies, Bucharest, Romania
e-mail: aceleanu_mirela@yahoo.com; mirela.aceleanu@economie.ase.ro

I. M. Mîndrican
Doctoral School Economics I, Bucharest University of Economic Studies, Bucharest, Romania

J. O. Monfort
Department of Economics, Jaume I University, Castelló, Spain

© The Author(s), under exclusive license to Springer Nature Switzerland AG 2023

V. Katsoni (ed.), *Tourism, Travel, and Hospitality in a Smart and Sustainable World*, Springer Proceedings in Business and Economics,
https://doi.org/10.1007/978-3-031-26829-8_38

1 Introduction

The limitation of relations between Russia and the West in recent years has gradually led to the possibility of a new major war on the territory of Europe (Legvold, 2016). The conflict triggered in 2014 over Ukraine determined the development and imposition of economic and financial sanctions on Russia, by the United States of America through the North Atlantic Treaty and the European Union, an action that materialized through the intensification of the deployment simultaneously with the planning of the military forces to Russia for a possible war.

At the moment, as a result of the outbreak of the war between Russia and Ukraine, the European Union is facing major threats from the perspective of security, but also of the global socio-economic context. The recent influx of Ukrainian citizens fleeing war zones to European states is undoubtedly an unprecedented case. The main destination countries considered by emigrants are the countries belonging to the former Soviet bloc, more precisely those that made the transition from a planned, centralized economy to a market economy, these being: Poland, Romania, the Republic of Moldova, Slovakia, Hungary and not lastly, Belarus. However, part of the Ukrainian citizens used the neighboring states only as transit locations, given the fact that they migrated from Poland, Romania, and Hungary to Germany, the Czech Republic, Turkey, France, and Great Britain. In other words, since March 3, 2022, about a week after the Russian invasion, more than 1 million refugees, representing 2.3% of Ukraine's population, have emigrated from Ukraine to neighboring states. Although the flow of refugees may cause the biggest migration crisis in Europe since the Second World War, the response of the European states is paradoxically different compared to the one generated by the Syrian refugee crisis in 2015 Martín (2022). Without a doubt, the influx of refugees is considerable, in the current context, bearing in mind that the United Nations Refugee Agency has estimated the number of refugees for the coming months at four to seven million people, this figure far exceeding the total number of emigrants who have crossed and taken refuge on European territory in the last seven years.

This time, the fate of the Ukrainian people seems to be much more troubling compared to the influx of refugees on European states. For this reason, the European Union aimed to manage the entire situation in optimal conditions and in real time, thus activating the provisions of the Temporary Protection Directive from 2001 for the first time on March 3, which refers to the unlimited admission of refugees during a crisis situation, with automatic acceptance of refugee status, avoiding the administrative process of recognition. Moreover, on June 23, 2022, in the framework of the European Council, the leaders of the European Union granted the Ukrainian state the right to be a candidate country for the European Union. Under these conditions, the leaders of the European Union reaffirmed their commitment to continue providing support to Ukraine from the point of view: economic, political, financial, humanitarian, and last but not least, military.

For this reason, the main objective of this paper is to highlight the crisis situation of migrants and the implications that this phenomenon has on the destination states. At the same time, this briefing analyzes the statistics regarding the impact of Ukrainian migration on European states, with the focus being on Ukraine's neighboring countries.

Tourism was particularly affected as a result of the state of uncertainty, but also of the increase in the number of refugees in certain areas.

2 The Relationship Between Migration and Tourism

Migration and tourism are two of the most important manifestations of globalization. The migration process brings significant economic and social contributions to the destination states, culturally enriching their society as a whole, at the same time increasing the tourism product and providing labor for the tourism, travel, hospitality, and catering sectors Poiarkova (2022). Moreover, tourism and migration are often interconnected, as there is currently a significant overlap between infrastructure used for tourism and migration. It should be noted that, in practice, the differences between migration and tourism are blurred. Tourism often generates various forms of mobility, and international labor is fundamental to maintaining functioning tourism economies. While tourism is often viewed in a positive light, migration is recurrently politicized and seen as affecting social systems and cultural values, despite tourism's reliance on migrant labor.

Migration and tourism have complex interactions, in some situations driving each other, while in others, they become competing practices and phenomena (Poiarkova 2022; Dwyer et al., 2014; Hall & Williams, 2002; Pappas & Papatheodorou, 2017; Williams & Hall, 2000). Both, the migration and tourism come from a combination of economic, social, and political factors, which are influenced by context, uncertainty, and mentality (Carling & Collins, 2018). Economists have increasingly recognized the disparate forms that international migratory processes can take, with transient, incomplete, liquid, circular, seasonal, and temporary manifestations (Collins, 2012; Hall & Williams, 2002). Long-term migrants, for example, may become tourists while traveling back to their home countries for short trips, visiting friends and relatives, but also consider "return" migration. At the same time, the "short-term" structure is also a relative expression in tourism, because some tourists become temporary or permanent migrants depending on their lifestyle (Benson & O'Reilly, 2009). Consequently, migration and tourism both involve large-scale movements of people who engage with destinations in different ways and move between distinct categories.

Bearing in mind that both migration and tourism are production or consumption oriented, mobility regimes are also interconnected with other markets such as retail, education, entrepreneurship, finance, and health. At the same time, tourism, like migration, is a particularly complex process, each destination having different histories, actors, and mechanisms and where movement possibilities are closely related to ethnicity, wealth, class, and profession (Eisenschitz, 2016). While migrants and tourists are alike, the global inequalities of border regimes and visa systems, respectively, ensure that their mobilities and experiences are, of course, different. Most destinations, at least until recently, have come to meet and support tourists from certain states, offering them visa-free or low-cost access and broad protections. Conversely, migration has been problematized by political debates and socio-political rhetoric. Therefore, migrant citizens often have fewer rights than tourists in many destinations due to deep inequalities (Abram et al., 2015).

In the current period, there are numerous studies that link the migration process and tourism in contemporary societies (Bianchi, 2000; Lugosi & Ndiuini, 2022). These examined topics are as follows: the role of globalized capitalism, border management, the welfare of migrant workers, social discrimination, inclusion, and settlement in host states (Bianchi, 2000; Choe et al., 2020; Ladkin, 2011; Eizenberg and Jabareen 2017). They derive the idea that migrant workers contribute significantly to economic development, innovation, entrepreneurship, and cultural diversity, along with knowledge transfer in tourist destinations. Tourism savings are mostly measured by tourist arrivals and receipts. Migration often generates tourist flows, through the "geographical extension of friendship and kinship networks" (Williams & Hall, 2002, p. 7). Moreover, migration, at the moment, is a determining factor of tourism, and the relationship between the two has increased progressively. An eloquent example can be represented by states such as Dubai and Macau, which cannot sustain their tourism sector without the involvement of migrant workers. As Macao's local labor force is not large enough to fill positions in the tourism sector, it has to attract a large number of foreign migrant workers (Choe et al., 2020). At the same time, migrant citizens conduct entrepreneurial activities using of course their cultural capital in the form of gastronomic knowledge, which often becomes part of the culture of the destination. As such, their contributions to destinations can gradually become heritage attractions and help destinations display "cosmopolitan cultures, which are then used in marketing and branding" (Lugosi & Ndiuini, 2022). Migration can thus transform a host society and impact its cultural heritage and lifestyle.

In conclusion, it can be stated that migration and tourism represent two fundamental concepts in the current economic context, closely interconnected and with similar forms, but which present different particularities.

3 The Context of the Increase in Emigration from Ukraine and the Measures Adopted at the European Level

Since the Russian invasion of Ukraine, the European Commission has had to have a prompt and rapid response to this situation, a fact that has materialized through the creation of the European Union Migration and Crisis Preparedness Plan. It provides an operational framework to ensure, on the one hand, awareness of the current context and better preparation, and on the other hand, effective governance and obtaining an eloquent response in a timely manner.

In response to the exodus from Ukraine, the European Union has drawn up and implemented a series of legislative proposals to provide immediate protection to refugees, bearing in mind that this is expected to be Europe's biggest refugee crisis this century. Other measures adopted by the European Union were the creation of a solidarity platform, the offering of humanitarian aid packages, and the increase in the level of flexibility for European states in terms of obtaining EU funding. The solidarity platform was created to ensure the implementation of the Temporary Protection Directive in close agreement with all parties involved in this process. Its main role is to monitor the identified needs and to coordinate the operational response at the level of the member states of the European Union, regarding the refugee crisis in Ukraine. The solidarity platform brings together: the member states of the European Union, EU Agencies, such as the European Union Agency for Asylum, Frontex, and Europol, the countries associated with the Schengen area and, last but not least, the Ukrainian authorities. The specific actions of the Solidarity Platform through which it differentiates itself in relation to the other instruments are the fact that it collects relevant information regarding the reception capacity, respectively, accommodation in the member states of the European Union, coordinates the matching offers received from these states regarding the acceptance of immigrants, and also aims at coordinating the transfer of persons from Moldova to the member states and those associated with the Schengen area. Moreover, the new proposals of the European Union regarding the management of borders are intended to create optimal conditions for entering the countries of destination and to accelerate the support offered to vulnerable people.

With the launch of the Russian invasion on February 24, 2022, approximately 2.5 million Ukrainian citizens had migrated from their country of origin by 11 March. Moreover, the European Commission supported the member states to use the financial funds belonging to the 2014–2020 cohesion in order to support refugees "in finding jobs, starting or continuing, depending on the context, education, as well as access to childcare". Also, the European aid fund for disadvantaged people will be used to provide basic material assistance and food. In addition to the aforementioned funds, the bloc is currently providing goods worth 100 million euros with the help of the European Union's Civil Protection Mechanism. Although the Republic of Moldova is not part of the European Union, it received approximately 20 million euros due to the high number of refugees it received. As a result of the intensification of the situation in Ukraine, the European Union proposed the consolidation of the budget for 2022 to the amount of 99.8 million euros from commitments. At the same time, the budget was increased by 176 million euros coming from payments. All these financial resources will be channeled through the Asylum, Migration and Integration Fund and the Instrument for Border and Visa Management. Taking into account the information presented previously, it can be highlighted the idea that the total amount of funding that will be made available to European states to support migration and manage the situation at the borders is 400 million euros. The funds will be useful, mainly, to the member countries most affected by the costs of receiving and registering citizens emigrating from Ukraine. In short, the European Union is trying to ensure that financial resources are sufficient to meet the needs of immigrants from

Ukraine. Through the resources that belong to the Asylum, Migration and Integration Fund of the European Commission, it is intended to create optimal conditions for the reception of immigrants, more precisely the provision of shelter, clothing, food, medical assistance, legal and translation assistance, psychological services social, and of course, other necessary services that allow them to integrate and register in another state.

The funds belonging to the instrument for border and visa management, BMVI, will allow the rapid and efficient application of the necessary procedures at or near external borders, such as identification, registration, fingerprinting, and security checks of third-country nationals Dijkstra et al. (2022). Another eloquent example to highlight the solidarity of people in relation to the situation generated by the war is the global employment event, entitled "Stand up for Ukraine" through which the sum of 9.1 billion euros was collected to support the vulnerable situation of refugees. At the same time, starting from March, the decision of the European Commission to grant temporary protection to people migrating from Ukraine entered into force. This measure will provide immediate protection to Ukrainian citizens and third-country nationals who have refugee status or permanent residence on the territory of Ukraine. This protection allows access to health care, schools, housing, and jobs, but it only lasts for one year with the possibility of extension.

4 Research Methodology. The Impact of Migration From Ukraine–The Situation of Refugees in Neighboring Countries

In this article, the following research methods were used: systematization, comparison, analysis, and synthesis. These methods were used to investigate the current situation from the perspective of the migration process and the effects resulting from the war on the member states of the European Union. The information base of the research consists of scientific articles and specialized studies belonging to European institutions.

According to statistical data published by the United Nations Refugee Agency, the number of Ukrainian citizens who emigrated is estimated at approximately 12 million people amid the Russian invasion launched on February 24, 2022. This figure constitutes approximately a quarter of Ukraine's total population of 44 million inhabitants. Part of the population fled to other regions of Ukraine, away from the fighting, and the other part, more than 5.7 million people, emigrated entirely to seek refuge abroad. It must be stated that this war led to the outbreak of the biggest humanitarian crisis in Europe since the end of the Second World War. The number of refugees crossing the border from Ukraine into a neighboring country each day peaked at around 200,000 on March 7, but has since fallen to around 50,000 in mid-July. Since the beginning of May, the number of people returning to Ukraine has been increasing and was around 30,000 per day in mid-July. The flow of refugees from Ukraine is very different compared to other migration flows in the past, this is due to the fact that the people involved in the process are women and children, while the adult men had to stay in Ukraine to continue fighting in war. According to the statistical data published by the United Nations, more than 5.2 million Ukrainian citizens have been registered throughout Europe, and more than 3.5 million have applied for temporary residence in another state United Nations (2022).

However, the number of refugees changes rapidly from one day to the next, taking into account the complexity and extent of the phenomenon; for this reason, the statistical data to be presented are a series of indicative estimates on the number of Ukrainian emigrants who are found in the states of destination from Europe. Moreover, movements from one state to another are not necessarily officially registered. In some countries, Ukrainians are allowed to travel without a visa but, in other countries, they must apply for asylum.

- *Poland*

  In mid-February, Poland was expecting a possible Russian attack on the Ukrainian state, so the Polish government asked communities to prepare to receive almost 1 million refugees. According to statistical information published by the Statistical Institute of Poland, by July 19, 2022, approximately 4.8 million Ukrainian refugees had entered Poland. For this reason, Poland has removed some of the usual formalities regarding the entry of foreign nationals into the state and has claimed that different identity documents will be accepted to facilitate the whole process. Moreover, collection points have been implemented in every district of Poland, through which free accommodation, food, and other necessary supplies are offered to immigrants. In the light of the previously discussed, it can be emphasized that the Polish government is preparing a series of legislative changes in order to simplify the procedure for obtaining a job for Ukrainian citizens, given the fact that a work visa is currently required, because many Ukrainians come from outside the European Union. Another cause that supports the increase in the number of Ukrainian immigrants on the territory of Poland is the labor market open to numerous opportunities, the cities very accessible to foreign citizens, and the pre-existing diaspora has made this state a much more attractive alternative for Ukrainians.

  So far, Poland has essentially faced two out of three possible migration waves. In the first wave, people arrived with family members or friends living in Poland or other EU member states. It should be noted that before the outbreak of the war, there were already approximately 800 thousand Ukrainians working or studying on the territory of Poland. As for the second wave, after the bombing of civilian facilities in major cities, people without family or friends living in Poland gradually began to arrive. For this reason, they need full assistance; thus, it is estimated that the next wave will be bigger compared to the previous ones. In the context in which the situation on the front is getting worse, and the repressions of the Russian troops are getting tougher, Poland could even face another two million people who would leave Ukraine, especially young people, women, children, and the elderly.

- *Romania*

  In May, the Romanian government reported that 989,357 Ukrainian citizens entered Romania, of which 80,000 remained in the country. Romania had a good organization and coordination of decision-making at all levels of government, in response to the refugee humanitarian crisis.

- *Russia*

  According to the information presented by the Ukrainian authorities, Russian troops in the territories of Ukraine occupied by the Russian state are forced to deport people from Ukraine to Russia, passing them off as refugees. According to Russian government statistics, 971,417 refugees left for Russia by May 26, 2022. At the end of April, the number of Ukrainian children from the east of the state who entered the Russian adoption system was reported to be around 150,000.

- *Republic of Moldova*

  Moldova was one of the first states to receive refugees from Odesa and Vinnytsia regions. It should be noted that the Moldovan authorities have activated a current crisis management center in order to facilitate humanitarian aid and accommodation for refugees. On July 19, 2022, approximately 541,323 Ukrainian refugees had entered the territory of Moldova. Moreover, Moldova has received the highest number of refugees per capita of any country, despite being one of the poorest countries in Europe. This considerable flow of emigrants led to the emergence of social tensions, and international aid was considered crucial to support Moldovan institutions in their management. In accordance with the statistical data published by the Statistical Institute of Moldova, in the current period 4% of the total population of Moldova are refugees, the state government requesting financial aid to be able to deal with this emergency situation. Consequently, the European Union agreed to grant 695.5 million EUR in aid to Moldova.

- *Hungary*

  From the beginning of the Russian invasion until July 19, 995,637 refugees from Ukraine arrived in Hungary. There are no border controls in the Schengen area. Hungary does not know how many people have moved to other Schengen countries, but 500 people from third countries, mostly students or migrant workers from Asia and Africa who had lived in Ukraine, asked for help in Budapest.

- *Slovakia*

  Since the outbreak of the war on February 24, 2022, approximately 327,000 Ukrainian citizens have crossed the Slovak-Ukrainian border. In April 2022, 301,772 were Ukrainians and 12,817 were third-country nationals residing in Ukraine.

At the border, the International Organization for Migration provided support to Ukrainian immigrants and conducted a face-to-face survey with them to find out

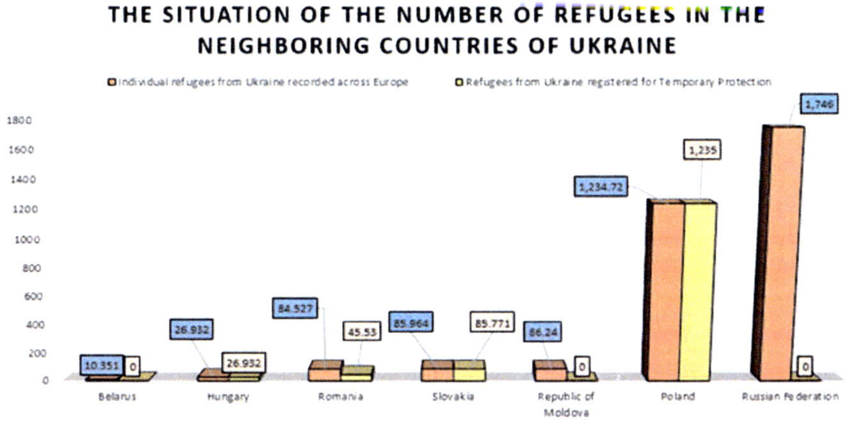

Fig. 1 The situation of the number of refugees in the neighboring countries of Ukraine. *Source* Processing based on data provided by the Operational Data Portal—Refugee situations—April 2022, European Commission 2022 a, b

the real context of the problem and the trend of the phenomenon as a whole. The survey was addressed to a number of 161 Ukrainian refugees on the territory of Slovakia. Among those displaced from Ukraine and arriving in Slovakia, women and girls are by far the vast majority, representing 88% of the sample surveyed. Most respondents traveled to Slovakia from Ukraine as part of a group (88%) and most (81%) with relatives. More than half (57%) of the refugees in the sample stated that Slovakia was their intended destination, especially Kosice (44%), Prešov (14%), Bratislava (5%). or Trenčín (4%). For 23% of those surveyed, the destination city was not known.

Other destinations considered by Ukrainian citizens were Germany (14%), Czech Republic (4%), Italy (4%), Poland (2%), and other European countries (12%). Some of those surveyed (7%) did not yet know their final destination (Fig. 1).

As can be seen from the graph presented above, a very large number of refugees were registered in Poland, this country being an easier platform for their movement to the developed countries of Western Europe.

5 The Consequences of the War in Ukraine—Economic, Social, Geopolitical

The war in Ukraine, in all its dimensions, is having alarming cascading consequences for the European economy already hit by the COVID-19 pandemic and climate change, with a remarkable impact on developing states. The conflict in Ukraine has taken its toll on people, infrastructure, and the economic and political context. According to statistics published by the United Nations, about 3,000 civilians have been killed or injured, 6.5 million people are internally displaced, and 4 million have been forced to migrate to neighboring states. The Ukrainian government has estimated the financial costs of the war at nearly $570 billion, while the Russian offensive shows no signs of ending the fighting any time soon. As previously mentioned, the conflict in Ukraine has led to the emergence of reverberating effects at the European level, from the perspective of increased commodity prices, energy shocks, and last but not least, the emerging geopolitical consequences of a war in the heart of Europe, which was instigated by a nuclear-armed country and a member of the UN Security Council.

Historically, periods of high geopolitical risk have been associated with notable negative consequences for the global economic context. This also materialized in the case of European states, given the fact that the war affected physical and human capital, there was a transfer of resources to inefficient and unproductive uses, there was a diversion of international trade and capital flows, and last but not least, they global supply chains have been disrupted. Forbes and Warnock (2012) Moreover, changing perspectives on the outcomes of this geopolitical event have also impacted the economic sector by delaying investment, eroding consumer confidence, and tightening financial conditions.

Economic consequences

To the extent that the war, for now, is limited to the territory of Ukraine, its effects of economic propagation are felt at the global level, implicitly by the European states. Anayi et al. (2022) Ukraine and Russia together account for only 2% of gross domestic product and global trade, while the stock of foreign investment in and from Russia reaches almost 1% of the global total Federle et al. (2022). Thus, the two states are large suppliers of goods: corn, sunflowers, fuel, metals, wheat, fertilizers, and the prices of which have increased considerably since the outbreak of the war. Using a conservative assumption of a 10 percentage point increase in global prices for food, wheat, fuel, and metal products during 2022, it is estimated that the negative impact of the war through the economic and trade channels of these products on low- and middle-income states it will have a value of 18 billion dollars. However, the main states that are strongly affected by the negative effects of the war are those in the geographical and historical proximity of Russia, such as Belarus, Moldova, Romania, and Armenia.

The war in Ukraine risks affecting Europe's economic recovery, as high energy prices and trade disruptions have gradually destabilized EU firms, which were already weakened by the effects of the pandemic International Monetary Fund (2022). According to a study published by the European Investment Bank (2022), the ever-increasing inflation rate will push many Europeans below the poverty line. Caldara et al. (2022) The economic situation in the European Union after the impact resulting from the pandemic was trying to recover when the war broke out, but the uncertainty increased simultaneously with the increase in food, energy, and commodity prices had a negative impact on the investment sector and on sustainable economic development and favorable to inclusion.

However, economic shocks are difficult to predict at the moment given the gradual expansion of the process, and the impact will be different for each member state of the European Union. In other words, after the situation generated by the COVID-19 pandemic, the Russian invasion of Ukraine still remains an external economic shock that brings with it asymmetric consequences throughout Europe. Undoubtedly, the unequal distribution of economic difficulties between the member countries of the European Union is not only an economic problem, but also a political one. Ensuring that the member states of the European Union remain united in the coming months or years in the face of Russian aggression will involve sharing the burden so that certain states are not affected more than others. The current situation focuses on the one hand on direct trade exposure, as well as the potential costs of energy dependence on Russia, and on the other hand on vulnerabilities to rising energy prices in general Council of the European Union (2022).

This first insight reveals a number of important and relevant patterns. First, states located in Central and Eastern Europe present a high level of vulnerability from a number of factors. This implies export dependence, dependence on Russian energy imports, as well as a pronounced exposure to the general increase in energy prices. Second, Italy and Germany are known to be heavily dependent on Russian gas, but could also be affected by supply shortages in the automotive industries. Third, the overall effect of higher energy prices differs from country to country. Although it would be very difficult to bear for the poorer states located in Eastern Europe, which have colder winters and implicitly, energy-consuming economies, states located in Northern and Western Europe would of course also suffer disproportionately. Therefore, the European Union will have to share the economic burden equally in order to preserve political unity.

Moreover, the war between Russia and Ukraine is disrupting the existing trade relations between the European Union and Russia. The European Union's direct trade sanctions are relatively limited to certain sectors, such as dual-use goods, weapons, and the space industry. However, financial sanctions have significantly increased trade costs outside the energy sector. A comparison to underline the aforementioned point is that a complete SWIFT ban on Iran reduced the state's trade by a third. Moreover, many companies have reduced or severed their relations with Russia, even in areas not subject to sanctions, possibly due to the skyrocketing reputational costs of doing business in Russia. In general, trade relations with Russia play a minor role for Europe. The EU is Russia's main trading partner and accounts for almost half of the country's total exports.

From the point of view of tourism, a conflict situation affects tourism in the countries involved. In the neighboring countries or in the destination countries for emigrants, their large wave has created some imbalances in terms of tourism as well. In the first stage, the need to support immigrants caused some accommodation spaces to be earmarked for immigrants. For a period, this affected tourism in the respective areas.

On the other hand, at the end of the conflict, immigrants can contribute to the development of tourism in different areas, by contributing as cheap labor and by promoting the integration areas to those left in the country of departure.

Social consequences

In the past decade, the world has faced several refugee crises, stemming from conflicts in the Americas, Africa, Asia, and the Middle East. However, in the last seven months of 2022, a new migrant crisis has emerged, with Europe once again in the spotlight. For this reason, it can be said that the war in Ukraine triggered the highest migration rate since the Second World War, with children and women having the largest share in the total number of refugees. According to statistics published by the United Nations, approximately seven million Ukrainians have entered Europe since the beginning of the Russian invasion. Without a doubt, this war is a tragedy for the people of Ukraine, who are being resolutely targeted by the Russian aggressors. It is worth noting that these events are quite shocking and have not been encountered in Europe since the Second World War. However, for the member countries of the European Union and their societies, this war represents a major social and humanitarian challenge, resulting from the unprecedented movement of the Ukrainian population.

First, European societies have unconditionally shown their support for Ukraine. Also, the political elites within the European Union are unanimous regarding the political and economic support of the Ukrainian state and the permanent provision of assistance in military equipment. However, migration was a contentious topic prior to the Russian invasion, given that the rise of a populist tide was associated with social and political discontent with the migration process, as most clearly evidenced by the rise of Brexit Grzymski (2022).

However, to date, Ukrainian migration has manifested itself in a reaction of empathy and support from the citizens of European states, since no one questions the refugee status of Ukrainians.

Second, the bleak outlook for a possible political outcome of the war in the near future indicates that refugees may not return to Ukraine quickly. This aspect is influenced by the existence of several factors. The chances of a complete withdrawal of Russian troops beyond the pre-war borders of Ukraine are quite small; consequently, the possibilities of a return to pre-war normality are equally low. Of course, this return to normality also brings with it a series of preconditions. On the one hand, some Ukrainian citizens would return to rebuild their home state, and on the other hand, others would need certainty regarding the country's stability, and this is unlikely in the near future. Moreover, Ukraine has proven to be effective in its defensive resistance. Therefore, it is possible that Ukraine to remain in an unstable situation with some kind of volatile ceasefire and partially occupied territory. This would be a disincentive for many Ukrainian refugees in their decision to return. The other factor that can prevent the return of refugees is the nature of this mobility, consisting mostly of mothers with children and elderly people. Men between the ages of 18 and 60 cannot leave Ukraine. This means that many refugees will stay longer in their new places with deeper integration because, for example, children have to go to school.

Geopolitical consequences

The invasion of Ukraine by the Russian state is essentially a direct challenge to the post-war order in Europe, being present through multilateral institutions such as NATO and the European Union and supported by the power of the United States of America. European states' policies on migration and refugees are being highlighted more than ever, raising both practical and moral questions about states' responsibilities under refugee law.

There is also a very real and very high risk that conflict-induced instability and economic disruption will provide an opportunity for organized crime and human trafficking groups to use an alternative smuggling route into Europe. It should be noted that organized crime and corruption have been a major challenge in Ukraine for years and were underlying issues that led to the Orange Revolution of 2004–2005.

Less clear are the implications of this war at the global level from the perspective of the conflict and the breaking of the West's relations with the Russian state. While geographically this war is being waged in Europe, how the geopolitics of the conflict will play out in the coming months is likely to reshape not just the European landscape, but the long-standing patterns of relations between Global North and Global South.

6 Conclusions

The human costs of war are clear and unsettling. In addition to its direct impact, the war came as an additional unwelcome shock to global and national economies still dealing with the impact of the pandemic. Unlike March 2020, when countries faced a common enemy, the crisis in Ukraine has global reverberations that affect different countries in different ways.

Moreover, the crisis in Ukraine may also create labor market disruptions in neighboring states, mainly Poland, Hungary, Romania, Moldova, and Slovakia. If hostilities continue, Ukrainian refugees would have to remain in other countries longer, putting more pressure on the labor market and social protection systems in these states, thus increasing unemployment in many of them. This situation in Ukraine also created a shock to the global economy, further complicating the recovery from the COVID-19 crisis.

In the short term, it is clear that tourism will be affected in conflict countries, or in the countries of destination of a large shoulder of refugees, as a result of the uncertainty and the pressure on the social protection and security systems.

The present research highlights the fact that the war in Ukraine brings with it a series of social challenges, such as increased mortality, migration, increased crime rate, increased unemployment, and exerting a negative impact on civil society. It is therefore particularly important to develop a constructive and pragmatic approach to help maintain peace and stability in Europe. In the long term, the integration of refugees in the reception areas can bring advantages, through the cultural integration of refugees and the creation of new forms of tourism. We believe that such a policy must be based on the existence of some economic and social pillars, which, on the one hand, will determine the realization of a permanent dialogue between policymakers and, on the other hand, will contribute to the creation of a sustainable basis for stability and cooperation at the European level. Moreover, this policy should mainly focus on reducing dependence on natural resources from Russia and on implementing non-political forms of cooperation beneficial to all parties involved. These pillars and alternative bridges should be intellectual elites, cooperation of middle-class members, strengthening of interstate relations, cultural and environmental cooperation, business cooperation, supporting the labor market integration of immigrants, and scientific and educational cooperation between states to facilitate access of foreign and Ukrainian students to studies.

Therefore, it can be stated that the war in Ukraine considerably affects the geopolitical security, the stability of Europe, the balance of the labor market, and the general economic context and tourism.

References

Abrams, K. M., Wang, Z., Song, Y. J., Galindo-Gonzalez, S. (2015). Data richness trade-offs between face-to-face, online audiovisual, and online text-only focus groups. *Social Science Computer Review,* 33.

Anayi, L., Bloom, N., Bunn, P., Mizen, P., Thwaites, G., Yotzov, I. (2022). The impact of the war in Ukraine on economic uncertainty. *VoxEU.org*, 16.

Benson, M., O'Reilly, K. (2009). Migration and the search for a better way of life: A critical exploration of lifestyle migration. *The Sociological Review, 57*(4).

Bianchi, R. V. (2000). Migrant tourist-workers: Exploring the 'contact zones' of post-industrial tourism. *Current Issues in Tourism, 3*(2).

Caldara and Iacoviello, 2022 Caldara, D., & Iacoviello, M. (2022). Measuring geopolitical risk. *American Economic Review, 112*, 1194–1225.

Caldara, D., Conlisk, S., Iacoviello, M., Penn, M. (2022). The effect of the war in Ukraine on global activity and inflation. FEDS Notes, 27 May 2022. Federal Reserve Board–Home.

Carling, J., & Collins, F. (2018). Aspiration, desire and drivers of migration. *Journal of Ethnic and Migration Studies, 44*(6).

Choe, J., O'Regan, M., & Kimbu, A. N. (2020). Filipino migrant workers' leisure and subjective quality of life in Macao. *World Leisure Journal, 62*(3).

Collins, F. (2012). Transnational mobilities and urban spatialities: Notes from the Asia-Pacific. *Progress in Human Geography*, 36(3).

Council of the European Union (2022). Retrieved July 16, 2022, from https://www.consilium.europa.eu/ro/policies/eu-response-ukraine-invasion/.

Dijkstra, H., Cavelty, M. D., Jenne, N., Reykers, Y. f. (2022). War in Ukraine. *Contemporary Security Policy, 43*(3). https://doi.org/10.1080/13523260.2022.2099085

Dwyer, L., Seetaram, N., Forsyth, P., & King, B. (2014). Is the migration-tourism relationship only about VFR? *Annals of Tourism Research, 46*(3), 130–143. https://doi.org/10.1016/j.annals.2014.03.009

Eisenschitz, A. (2016). Tourism, Class and Crisis. *Human Geography, Sage Journals, 9*(3), https://doi.org/10.1177/194277861600900308

Eizenberg, E., & Jabareen, Y. (2017). Social sustainability: A new conceptual framework. *Sustainability, 9*(1), 68.

European Central Bank (2022) Retrieved July 12, 2022, from https://www.ecb.europa.eu/pub/economicbulletin/focus/2022/html/ecb.ebbox202204_03~c9ddc08308.en.html

European Commission. (2022a). Retrieved Juyl 11, 2022, from https://ec.europa.eu/migrant-integration/news/slovakia-more-half-new-arrivals-ukraine-intend-stay_en

European Commission. (2022b). Retrieved July 12, 2022, from https://home-affairs.ec.europa.eu/policies/migration-and-asylum/migration-management/migration-management-welcoming-refugees-ukraine_en

European Investment Bank (2022). Retrieved July 14, 2022, from https://www.eib.org/en/press/all/2022-268-new-report-the-impact-of-the-war-on-europe-s-economic-recovery

Federle, J., Andre, M., Müller, G., Sehn, V. (2022). Proximity to war: The stock market response to the Russian invasion of Ukraine. CEPR Discussion Paper 17185.

Forbes, K. J., & Warnock, F. E. (2012). Capital flow waves: Surges, stops, flight, and retrenchment. *Journal of International Economics, 88*(2), 235–251.

Grzymski, J. (Centre of Migration Research, University of Warsaw). (2022). The Ukrainian Refugee Crisis: Consequences for Europe, Brexit Institute News. Retrieved July 24, 2022, from dcubrexitinstitute.eu

Hall, M. C., & Williams M. A. (2002). Tourism and Migration.New Relationships between Production and Consumption, *Geographies Journal Library* (GEJL, volume 65), Springer

International Monetary Fund. (2022). Retrieved July 21, 2022, from https://blogs.imf.org/2022/04/22/war-in-ukraine-is-serious-setback-to-europes-economic-recovery/

Ladkin, A. (2011). Exploring tourism labor. *Annals of Tourism Research, 38*(3), July 2011, https://doi.org/10.1016/j.annals.2011.03.010

Legvold, R. (2016). *Return to Cold War*, Oxford, Polity Press

Lugosi, P. & Ndiuini, A. (2022). Migrant mobility and value creation in hospitality labour, *Annals of Tourism Research 95*:103429, https://doi.org/10.1016/j.annals.2022.103429

Martín, I. (2022). The war in Ukraine: A migration crisis like no other. Policy Center For The New South. Retrieved July 24, 2022, from https://www.policycenter.ma/publications/war-ukraine-migration-crisis-no-other

Pappas, N.& Papatheodorou, A. (2017).Tourism and the refugee crisis in Greece: Perceptions and decision-making of accommodation providers, *Tourism Management, 63*(C), 31–41, http://www.sciencedirect.com/science/article/pii/S0261517717301279

Poiarkova, T. (2022). "Others" in the Ukrainian society as a result of the modern Russian-Ukrainian War. *European Journal of Sustainable Development, 11*(1). https://doi.org/10.14207/ejsd.2022.v11n1p20

United Nations. (2022). Global Impact of war in Ukraine on food, energy and financial systems, Brief No.1–13 April 2022. UNEP–UN Environment Programme.

Perception of Security by Tourism Service Providers: The Case of Acapulco, Mexico

Carmen Exchel Cliche Nájera, Karla Rosalba Anzaldúa Soulé, Mirella Saldaña Almazán, and Arely Adriana Almazán Adame

Abstract Tourism is one of the main economic sectors worldwide that, in times of crisis due to security incidents, configures a fight for businessmen's social and economic survival. The research determined the perception of tourist security in Acapulco based on the opinion of tourist service providers, to implement strategies that allow them to recover their competitiveness. The methodological process consisted of two phases: in the first, a documentary analysis of social and economic factors was carried out; in the second: the direct, flexible, and dynamic interview was structured. The investigation lasted two years, beginning in 2019 and ending in January 2021. About 60 interviews were conducted with men and women, in the tourist areas of Acapulco: diamond, traditional, and golden, aimed at tourism service providers, such as businessmen, presidents of associations, and managers of the small, medium, and large companies. The number of interviews was defined based on the theoretical saturation criterion, where the information became repetitive in the data collection and no new information emerged. The approximate duration of these interviews ranged between 25 and 50 min. Some findings are: the port was affected by security incidents such as organized crime, delinquency, extortion, kidnapping, payment of land, homicides, and drug trafficking; some causes of low security: corruption and non-enforcement of laws. Tourist service providers dedicate a good part of their income to maintaining your security with the hiring of private security personnel, alarms, security cameras, and resistant locks. Some proposals for improvement are: to fight corruption at all levels of government, promote local and citizen organizations, generate better-paid jobs, and renew the educational system. Finally, a proposed alternative is the creation of a comprehensive tourism security plan.

C. E. C. Nájera · K. R. A. Soulé (✉)
Faculty of Tourism, Autonomous University of Guerrero, Acapulco, México
e-mail: karlasoule88@gmail.com

M. S. Almazán
Center for Innovation Competitiveness and Sustainability, University of Guerrero, Acapulco, Mexico

A. A. A. Adame
Regional Development Science Center, University of Guerrero, Acapulco, México

© The Author(s), under exclusive license to Springer Nature Switzerland AG 2023

V. Katsoni (ed.), *Tourism, Travel, and Hospitality in a Smart and Sustainable World*, Springer Proceedings in Business and Economics,
https://doi.org/10.1007/978-3-031-26829-8_39

Keywords Tourist security · Tourist destination · Perception · Tourist service providers

JEL Classification Z32 Tourism and development. · Z39 Tourism · Others

1 Introduction

The importance of tourism activity is manifested in the world economy due to the social and economic relevance it represents for the development of regions and communities with tourism potential.

However, tourism is sensitive to negative events worldwide such as terrorist acts, local wars, natural disasters, epidemics, and pandemics, causing the perception of security to decrease.

Tourism security is an essential condition for the development of destinations since it represents the protection of life, health, physical, psychological and economic integrity of visitors, service providers, and members of receiving communities.

The insecurity that is observed mainly derived from violent acts in any destination has become a highly important factor in the construction of a negative image of the place.

Tourist destinations that aspire to be in the imagination of any issuing market must guarantee security since it is essential to satisfy the greatest number of expectations of potential tourists, future investors, businessmen, employees, and inhabitants about their ideal place to live.

However, some prestigious tourist sites have faced a crisis to guarantee tourist safety, as such is the case of the Acapulco destination, which has been susceptible to losses that constantly affect and deteriorate its prestige. These events usually manifest fear and panic in tourists, businessmen, and inhabitants and consequently cause a decrease in the potential of the destination, generating negative impacts on the sectors of the economy of the site due to the low influx of tourists. It should be noted that there are numerous investigations to know the perception of security focused on tourists, but incipiently on the opinion of tourism service providers. The case of security of tourism deserves a special treatment of analysis, taking into account the repercussions that its study may derive. In this sense, the objective of this research was to know the perception that tourism service providers have about security and to propose strategies that allow the repositioning of the Acapulco destination.

2 Literature Review

Tourism is recognized as a global phenomenon, one of the fastest growing sectors of the economy. The volume of tourism equals or even exceeds that of oil, food products, or automobile exports, in addition to the fact that it has become one of the main players in international trade, and represents one of the main sources of income (World Tourism Organization, 2018).

Every day different countries are committed to promoting their tourist destinations, promoting a high degree of competitiveness against others, and integrating different factors such as: guaranteeing efficient services, visitor satisfaction, accessibility to the destination, transport and accommodation infrastructure, willingness to implement development, maintain a good international image, and guarantee security (Salinas et al., 2022). Tourist safety in a destination is one of the most important aspects when it comes to offering tourists a quality experience (Otto & Ritchie, 1996).

Tourism security is understood as the protection of life, health, physical, psychological, and economic integrity of visitors, service providers, and members of receiving communities (Grünewald, 1989), or as Jorge Amonzabel puts it in his manual on tourist security as a series of protection conditions granted by a tourist destination or place of attraction, for the satisfaction and movement of the visitor; from another perspective it is the set of measures at the social, political, and economic levels that allow the tourist activity which is governed in an environment of tranquility and trust and involves a series of aspects that we can encompass in seven points (INISEG, 2019):

- Public security: a system that allows the free movement of tourists around the destination, reducing the percentage of conflict situations, mainly criminal acts (robbery and theft) and accidents.
- Social security: a system that allows the free movement of tourists around the destination in the face of social problems such as mobilizations, strikes, etc.
- Medical security: prevention, emergency, and protection system that allows medical assistance to the visitor before traveling or during their displacement in the tourist destination.
- Information security: a communication system that allows knowledge of the destination's offer.
- Economic security: traveler reinsurance system to prevent risks from an economic perspective.
- Security of tourist services: a system that allows the protection of the tourist while traveling through the different tourist and recreational service establishments (hotel, restaurant, travel agency, etc.) of the tourist destination.
- Security in events: a system that allows protection during the realization or participation of a sporting, musical, recreational event, etc. of the destination visitor.

Under this perspective, guarantee the physical, legal, and patrimonial security of all the actors involved in this activity, mainly tourists are aspirations of tourist destinations (González et al., 2019). The issue of tourist security is becoming more important every day in the tourism industry, even though some publications had already been published on the subject, such as Sonmez (1998) on the terrorist attacks in Egypt, in 2011 the Organization of the United Nations a watershed at the international level in addressing the issue due to the attack on the twin towers in New York.

However, tourism is exposed to constant events: cultural, a product of urban progress and the homogenization of the landscape; economic, due to fluctuations in demand and seasonality; and social, such as the increase in crime. Some authors argue that tourism is industry susceptible and fragile to any incident such as those related to nature: natural disasters, crime-related activities as well as plagues or epidemics, and it often has a quick and direct impact (Neumayer, 2011; Santana, 2003; Valencia & Crouch, 2008).

Consequences that can lead to a tourism crisis are understood as a situation that can endanger all businesses related to tourism and harm the image of a tourist destination in terms of safety and attractiveness of the place. Therefore, tourists are negatively influenced, and local tourism businesses are harmed as tourist arrivals and expenditures decrease (Fortini, 2013; Kurez & Prevolsek, 2015; Sommez et al., 1994).

In addition, destinations perceived as unsafe will tend to gradually disappear, since their image of the destination is tarnished which translates into a significant decrease in demand (Ceron and Silva, 2017; Korstanje, 2016; Mansfield and Pizam, 2006; McElroy, 2006).

The perception of security of the tourist place is frequently used as a mechanism to evaluate the performance and effectiveness of the government in different public affairs (Gélvez, 2018). It becomes a determining factor in the choice of a tourist destination (Peña & Contreras, 2019).

The complexity of this lies in being made up of two dimensions: objective and subjective. Objectivity is oriented toward statistical information: criminal situation, existing risks, and emerging entities available, while subjectivity is allied to the qualitative aspect where the perception of security of the destination in question is embedded. However, it is the subjective part that has a greater connotation in the first instance because it is the one that affects the decision to choose a tourist destination to visit and throughout the overnight stay (Grünewald, 2012; Yang & Nair, 2014).

On the other hand, in the construction of tourist security, various factors intervene in its application, such as the set of measures, objective conditions, and existing perceptions in the social, economic, and political spheres of a tourist destination, which allow the tourist experience to develop in an environment of freedom, trust, and tranquility and with the greatest physical, legal, or economic protection for tourists, their goods, and for those who offer services in said destination (Jiménez & Pérez, 2018).

3 Methodology

The research was qualitative; as Barrantes (2014) points out, it focuses on the study of the meanings of human actions and social life. In this sense, we sought to know the perception of tourism service providers in the Port of Acapulco about tourism security and its effects according to their experiences.

The study consisted of two phases: in the first phase, a documentary analysis of scientific journals was carried out, as well as journalistic notes and articles on social and economic factors.

In the second phase: the structured, direct, flexible, and dynamic interview was used as a technique. Its application lasted for two years and ended in January 2021. It should be noted that its application had a before and after the SARS-CoV2 coronavirus pandemic, which causes the COVID-19 disease. A total of 60 interviews were applied to men and women aimed at tourism service providers, such as entrepreneurs, managers of small, medium, and large companies, and presidents of associations. The number of interviews was defined based on the theoretical saturation criterion, where the information became repetitive in the data collection and no new information emerged. The approximate duration of these interviews ranged between 25 and 50 min. The geographic area of application was the tourist areas of Acapulco: Diamante, Tradicional, and Dorada.

The analysis process of the interviews consisted of 4 stages, which were integrated from the analysis of the methodologies of Erlingsson and Brysiewicz (2017) and Pieterse (2020):

1. Condensation: it was carried out to carry out an analysis and delimit the text while anticipating the central meaning.
2. Encoding: it is expressed as a label that accurately described the particular condenser unit.
3. Category: the codes are organized in categories when they describe different aspects, similar or differences, of the content of the text that belongs together.
4. Topic: an interpretive analysis of the content to obtain results.

4 Results, Capitalize the First Letter of Every Word in the Title

Social and economic factors that affect the construction of tourist security in the port of Acapulco.

The municipality of Acapulco, a tourist destination located in the State of Guerrero, Mexico, 2019 was declared the epicenter of organized crime in Mexico, with more groups fighting for control of the territory than in any other region (International Crisis Group, 2020). About 927 victims of intentional homicide were registered, occupying the national position of the number 6 victim rate, 10 victims of femicide, 144 victims of extortion (collection of flat), 1921 victims of robbery with violence, 31 victims of kidnapping, and 899 robberies of vehicles with violence, Executive Secretariat of the National Public Security System [SESNSP] (2020).

Acapulco has been positioned in the ranking of the 50 most dangerous cities in the world for several years (Citizen Council for Public Security and Criminal Justice A.C., 2022).

Alerts of recommendations issued by various tourist-issuing countries have increased, as is the case of the United States Department of State, which has suggested for more than 10 years to the present that citizens could not travel to the entity due to its high prices and crime levels (Embassy and Consulates of the United States in Mexico, 2022).

Some consequences of the criminal events that affect the tourist destination are: according to the National Chamber of Commerce [CANACO] (2019), about 400 micro-enterprises have closed their doors; the arrival of international tourists has declined; and in 2019 alone 12,602 international tourists arrived, the main issuing poles being the United States with 4982 visitors and Canada with 7186 (Secretaria de Turismo [SECTUR], 2019).

As the main source of jobs and economic benefits for a population of nearly 900,000 thousand inhabitants, National Institute of Statistics and Geography [INEGI] (2015), who live in the destination and who directly and indirectly depend on tourism for its multiplier effect, which makes the destination more vulnerable to the lack of tourist influx, since in the last fifteen years Acapulco has had an average occupancy of 50% of its 19,483 registered rooms in all its categories (Statistical Yearbook, 2019), which has become poverty, social marginalization, and delinquency due to the lack of employment opportunities stated intimidation of criminal groups, collection of flat.

In 2020, security incidents in the tourist destination Acapulco persisted, since 752 victims of intentional homicide were registered, occupying the national position of the number 8 victim rate, 10 victims of femicide, 129 victims of extortion (charge of the floor), 1325 victims of robbery with violence, 585 vehicle thefts with violence, and 9 victims of kidnapping (SESNSP, 2020). Different clandestine graves with human bodies were found in the port (State File of Guerrero, 2020).

Some factors that reflect the tourist situation in terms of the economic part that was experienced in the year 2020 in the port are:

During the second quarter of 2019, the tourist influx in Acapulco was 1,844,790 visitors, and in the first critical point which was in the second quarter of 2020, a stage where the economy was completely closed, the tourist influx was only 9605: a drop of 99.4%. For the second quarter of 2020, when infections were on the rise and reaching their first maximum peak, hotel occupancy decreased by 41%, going from 9,465 occupied rooms on average before, to 5601 rooms, and for the third quarter of 2020 to 4082 rooms.

The economic spill has unusual behavior; since before the start of the pandemic, it begins to fall rapidly and bottoms out in the second quarter of 2020, the date of the first wave of infections, passing from 9,851,055,652 million generated in the second quarter of 2019 to 423,009,300 a drop of 96%. In the second wave of infections that occurs in the first quarter of 2021, it reports a spill of 3,313,000,000 million, 68% less than that generated in the same period of 2019.

In the holiday season, it received just over 193 thousand tourists, with a hotel occupancy of 37.8%, which, compared to the previous year, when it was 83.4%, is practically equivalent to less than half. The estimated spill for Acapulco was 794 million pesos, against 2 thousand 765 million generated in December 2019, with the visit of 628 thousand tourists (La Jornada, 2021).

In 2020, the employed population decreased from 307,171 jobs in the second quarter of 2019 to 279,578 jobs, a loss of 27,593 jobs, equivalent to 9%, of which 20,858 were located in the tertiary sector and 11,723 in retail establishments, restaurants, and accommodation. In the fourth quarter of 2020, a slight increase in employment was observed, but in the second wave of infections that corresponded to the first quarter of 2021, employment again decreased, reaching 284,195 jobs, 34,031 jobs less than in the first quarter of 2020. In 2019, the shock was felt by the tertiary sector with 29,561 of them, 13,020 of which were located in restaurant and accommodation establishments.

The pandemic had very strong repercussions on the employment of tourism-related companies; in the second quarter of 2020, when the peak of infections reached its first maximum, 11,723 jobs had been lost, a drop of 22% compared to the same quarter of 2019, and by the third quarter, another 6202 jobs were lost again. Of this loss, women accounted for 55% and men 45% (Perez & Añorve, 2021).

However, what was observed was that during the SARS-CoV2 coronavirus pandemic, which causes the COVID-19 disease in 2020, security incidents such as homicides decreased, an action that is related to the quarantine that was kept by mandate of the World Health Organization.

Acapulco was a world-class tourist destination that positioned Mexico in the world; however, it has been affected by various social and economic elements such as insecurity, which is observed mainly as a result of violent acts, murders, armed robberies, rapes, drug trafficking, and the high rates of unemployment and business closures, facts that are reflected in the worn and abandoned infrastructure.

In tourism, security at the international level has permeated even in the choice of the international destination of the trip, through recommendations that are issued by government agencies to its citizens. One of the main factors that affect the perception of tourism safety in both citizens and tourists in Acapulco is insecurity. These events usually manifest fear and panic in tourists, thus creating a negative image of the place and consequently causing the potential of the destination to decrease significantly, generating negative impacts on all sectors of the economy reflected in a low influx of tourists to the destination.

The perception of service providers on tourism security in the year 2019–2020:

- Of the total number of interviewees, 57 people answered that they do not live in a safe destination and only 3 considered that they did.
- The tourist security of the Acapulco destination is weakened for the following reasons: delinquency, collusion between the government and organized crime, homicides, and drug trafficking; however, an interviewee answered that there is tourist security in the port, since feel safe coming and going to your home and the tourists keep coming.
- Companies in the tourist destination Acapulco have mainly faced incidents: robberies/thefts, disorderly conduct (excessive noise, vandalism), homicides and assaults, violence with weapons, and to a lesser extent sexual assault and prostitution.

- One hundred percent of those interviewed agree that the security protocols in the tourist destination coordinated at the municipal, state, and federal levels are not effective; likewise, the main reason stated is due to corruption between government agencies with organized crime. Likewise, some participants agreed that they currently do not dare to report any criminal act, for fear that the information will leak to criminal groups since the authority that provides security has lost credibility and trust.
- The causes that have generated insecurity in the port, according to your point of view, are the first position: corruption, non-application of laws, unemployment, poverty, poor quality education, and lack of values.
- They added that each month they spend approximately 30% of their income on maintaining the security of their businesses. They have had to choose.

Hire private security, security alarms, and resistant locks, reserve the right of admission to your businesses or jobs, cameras, cancelations, and closed circuits.

The main strategies to improve the tourist safety of the port are: to fight corruption, organization, and local unity, create paid jobs, renew social education systems, and ensure that governments comply with their mandates.

Some outstanding findings regarding the perception of tourism service providers on security in the year 2020–2021.

- It was identified that this year there was a minimally positive variation in the perception of security at the destination since forty people considered the site to be unsafe, and twenty interviewees as safe, highlighting some comments such as: despite all the security incidents, the tourism generally coming on weekends, considering that this is due to the pressure of confinement, the proximity to Mexico City, and the desires of visitors to travel.
- The actions that affect tourist security are corruption and organized crime. Some incidents that tourism service providers have faced are robberies, collection of flats, homicides, and kidnappings.
- The total number of interviewees stated that the tourist security protocols continue to be inefficient and are only projected as a promotion of the government that holds the position.
- The causes of insecurity in the port are corruption in society and collusion of the security authorities at the National, State, and Municipal levels with organized crime and unemployment.
- Added more than a third of participants have had to increase economic spending to maintain the safety of their businesses; on the other hand, the instability of tourism due to insecurity and confinement due to the pandemic declared by the World Health Organization forced many investors and companies in the tourism sector to close.

Finally, based on the analysis of the results and according to various success cases applied to guarantee tourist safety in destinations worldwide, one proposal is the creation of a Comprehensive Plan for Tourist Safety where citizens, providers of tourist services, the press, other officials, and the different security departments at their different levels of government, Municipal, State and Federal closely coordinate and promote a public policy that guarantees a safe tourist destination. However, the objective is to seek to create a sense of solidarity and shared benefits for the participants.

Peace and security are basic requirements for the success and development of tourism, not only image and price are the basis for selecting a vacation destination, but an essential factor is the perception of security (Carballo et al., 2021; Shin, 2005).

5 Conclusion

The perception of tourist safety was not positive on the part of the tourist service providers in Acapulco. It is considered an unsafe place, as a result of the various criminal incidents such as homicides, collection of flats, kidnappings, and robberies, which gradually increase due to crime. Some examples are drug trafficking, assaults, unemployment, and the weakness of public security institutions, highlighting the collusion between organized crime and authorities from different levels of government.

Brings with it some effects such as the deterioration of the image of the port due to the acts of violence broadcast in the media and in the alerts from the embassies, which have made the flows of international tourists from potential markets such as the United States and Canada which have gradually moved away the national tourist influx of high purchasing power.

This has had a strong impact on the economic sector, causing business closures, unemployment, and increased poverty in the port, influencing the lack of investment and abandonment of the infrastructure of companies in the tourism sector.

It is urgent to implement strategies that mitigate security incidents through risk prevention management, to mitigate current and future effects, create reflections, and identify proposals that allow generating a favorable environment for the development of tourism.

Finally, it is important to consider that the union between security and tourism is of vital importance to protect the life, physical and psychological integrity of tourism service providers, which must be assumed as the responsibility of all the main actors that manage tourism, tourism sector.

Acknowledgements We thank all those who joined in this investigation such as the different providers of tourist services in Acapulco, the Associations and Business Unions of the port, and the Autonomous University of Guerrero, Faculty of Tourism for supporting this investigation.

References

Barrantes, R. (2014). *Investigación: un camino al conocimiento, un enfoque cualitativo, cuantitativo y mixto* (No. 001.4 B268i2). San José, CR: EUNED.

Carballo, R. R., Carballo, M. M., León, C. J., & Gil, S. M. (2021). La percepción del riesgo y su implicación en la gestión y promoción de los destinos turísticos. El efecto moderador del destino. *Cuadernos de turismo,*(47), 23–36.

Ceron, H., & Silva, J. E. (2017). La relación entre un proxi de la dinámica de la inseguridad pública y el turismo internacional a México: un análisis econométrico. *El periplo sustentable,*(33), 105–131.

Donaldson, R., & Ferreira, S. (2009). (Re-) creating urban destination image: Opinions of foreign visitors to South Africa on safety and security? *Urban Forum, 20*(1), 1.

Erlingsson, C., & Rysiewicz, P. (2017). A hands-on guide to doing content analysis. *African Journal of Emergency Medicine, 7*(3), 93–99. https://doi.org/10.1016/j.afjem.2017.08.001

Fernández, J. A. F., Duarte, P. A. O., & Mogollón, J. M. H. (2011). Imagen del destino y marca turística: sinergias e impresiones. *Estudios De Gestión y Turismo, 1*, 904–914.

Ficha Estatal Guerrero. (2020). Causa en común. http://causaencomun.org.mx/beta/wp-content/uploads/2020/08/200814-Ficha-Guerrero-2020.pdf.

Fotini, M. (2013). *How safety and security influence the decision of tourists to visit another destination, A case study of Greece*. GRIN Verlag. https://www.grin.com/document/365399.

Gélvez, J. D. (2018). ¿Cuáles determinantes se relacionan con la percepción de inseguridad? Un análisis estadístico y espacial para la ciudad de Bogotá, D. C. Revista Criminalidad, *61*(1), 69–84. Recuperado de http://www.scielo.org.co/pdf/crim/v61n1/1794-3108-crim-61-01-00069.pdf. MINTIC. (10 de Enero de 2019). Plan Estratégico Institucional MINTIC 2019–2022.

González, G. H., Rodríguez, I. S., & Hernández, A. G. (2019). Turismo y Diversidad cultural en la huasteca potosina: Una propuesta de Análisis de seguridad turística.

Grünewald, L. (1989). La Seguridad de la actividad Turística. La percepción desde la óptica de la demanda. Argentina: Foro de Seguridad Turística, Departamento de Seguridad Turística Subsecretaría de Turismo y Deporte de Argentina.

Grünewald, L. (2012). La seguridad en el marco de la competitividad de los destinos turísticos. Revista de Ciencias Sociales. http://www.unq.edu.ar/advf/documentos/59399b905441b.pdf.

Hernández Sampieri, R., Fernández Collado, C., & Baptista Lucio, P. (2014). Metodología de la Investigación. México: Interamericana Editores S.A. DE C.V.

INISEG. (2019). Seguridad turística: su importancia y proyección. https://www.iniseg.es/blog/seguridad/seguridad-turistica-su-importancia-y-proyeccion/.

International crisis group. (2020). La guerra cotidiana: Guerrero y los retos a la paz en México. https://www.crisisgroup.org/es/latin-america-caribbean/mexico/80-mexicos-everyday-war-guerrero-and-trials-peace.

Jiménez, J., & Pérez, M. (2018). La seguridad como componente esencial del concepto de calidad turística. *Estudios y Perspectivas En Turismo, 27*, 921–943. https://doi.org/10.4102/hts.v76i1.5851

Korstanje, M. (2016). El espíritu del terrorismo: Turismo, sindicalización y terrorismo. *Pasos Revista de Turismo y Patrimonio Cultural, 14*(3), 239–250.

Kurež, B., & Prevolšek, B. (2015). Influencia de las amenazas a la seguridad en el desarrollo de destinos turísticos. Tims. Acta: naučni časopis za sport, turizam i velnes, *9*(2).

La Jornada. (2021). La Jornada. Obtenido de Estados: https://www.jornada.com.mx/notas/2021/01/04/estados/cayo-3-mil-millones-la-derrama-turisticadecembrina-en-guerrero/.

Lisowska, A. (2017). Crime in tourism destinations: A research review. *Central and Eastern European Online Library, 21*(1), 1–10. https://doi.org/10.1515/tour-2017-0004

Mansfeld, Y., & Pizam, A. (2006). Tourism and crime issues. In A. Pizam & Y. Mansfeld (Eds.), *Tourism, security, and safety: From theory to practice* (pp. 84–85). Elsevier.

McElroy, J. L. (2006). The growth of the Caribbean narcoeconomy: Implications for tourism. In *Tourism, security and safety,* (pp. 87–104). Routledge.

Neumayer, E. (2011). Sostenibilidad y desigualdad en el desarrollo humano. Documentos ocasionales de PNUD-HDRO, (2011/4).

Nohlen, D (2020). El método comparativo. Biblioteca Jurídica Virtual del Instituto de Investigaciones Jurídicas de la UNAM. Capítulo tercero. https://www.juridicas.unam.mx/.

Otto, J. & Ritchie, J. (1996). La experiencia de servicio en el turismo. *Gestión turística, 17*(3), 165–174.

Peña, A., & Contreras, J. (2019). Percepción de seguridad en el destino turístico. Universidad piloto de Colombia.

Peréz, U., & Añorve, N. (2021). Acapulco's tourism market crisis as a result of Covid-19, 2019–2021. *South Florida Journal of Development*. file:///C:/Users/Exchel/Downloads/ART.+035+SFJD.pdf.

Pieterse, H. J. C. (2020). La teoría fundamentada metodología para llevar a cabo análisis de contenido de sermones y entrevistas: crítica y respuesta. *HTS Estudios Teológicos/Estudios Teológicos, 76*(1), a5851. https://doi.org/10.4102/hts.v76i1.5851.

Salinas, J., Guaita, J. Y., & Martín, J. (2022). An analysis of the competitiveness of the tourism industry in a context of economic recovery following the COVID19 pandemic. *Technological Forecasting, and Social Change, 174*. https://doi.org/10.1016/j.techfore.2021.121301.

Santana, G. (2003). Gestión de crisis y turismo: más allá de la retórica. *Revista de marketing de viajes y turismo, 15*(4), 299–321.

Secretariado Ejecutivo del Sistema Nacional de Seguridad Pública. (2020). Ficha estatal Guerrero 2020, causa en común. http://causaencomun.org.mx/beta/wp-content/uploads/2020/08/200814-Ficha-Guerrero-2020.pdf.

Shin, S. (2005). Turismo de seguridad, protección y paz: El caso del área DMZ. *Asia Pacífico Revista de Investigación Turística, 10*(4), 411–426. https://doi.org/10.1080/10941660500363777.

Sonmez, S. F. (1998). Turismo, terrorismo, y e inestabilidad politica. *Anales De Turismo Investigación, 25*(2), 416–456.

Sönmez, S., Backman, K., & Allen, L. (1994). *Managing tourism crises: A guidebook*. Clemson University.

Valencia, J., & Crouch, G. I. (2008). Comportamiento de viaje en tiempos convulsos: el papel de la autoconfianza del consumidor. *Revista de marketing de viajes y turismo, 25*(1).

World Tourism Organization (2018). Tourism – an economic and social phenomenon, World Tourism Organization a UN Specialized Agency, https://www.unwto.org/why-tourism

Yang, E. C. L., & Nair, V. (2014). Turismo en riesgo: una revisión del riesgo y el riesgo percibido en el turismo. *Asia-Pacífico Journal of Innovation in Hospitality and Tourism (APJIHT), 3*(2), 1–21.

Sports Tourism an Opportunity for Urban Areas: Classifying Sports Tourists According to Their Level of Sports Engagement and Consumer Behavior

Maria Vrasida, Tatiana Chalkidou, and Despina Gavrili

Abstract Sports tourism is one of the fastest growing sectors in tourism by combining elements of culture, health, and wellness and recreation. More and more tourists are interested in sport activities during their trips whether sports are the main objective of travel or not. Following running or other popular sports events is acting as a strong motivator for the decision-making process in the choice of destination and greatly influences the decision-making phase of consumer behavior. Sport events of various kinds and sizes attract tourists as participants or spectators and destinations try to add local flavors to them to distinguish themselves and provide authentic local experiences. The aim of this paper is to identify and classify sports tourists according to their different level of engagement in the sporting activity. Furthermore, a hypothesis is made and tested that consumer behavior will present differences both in the motivational (deciding on a destination) and in the experiential phase (visiting and experiencing a destination). A multi input matrix will allow to determine proximity to the culturally oriented or health-oriented part of the scale and define and identify consumer behavioral patterns accordingly. In the post-COVID-19 era, it is very important for destinations to find ways to reignite their tourism product while placing safety as a topo priority. The link of sports tourism to health and wellness issues presents a valuable opportunity especially for urban areas to explore alternative visitation patterns.

Keywords Hotels · Athletes · Tourism · Urban · Reignite · Destinations

JEL Classification Z32 · Tourism and development

M. Vrasida (✉) · T. Chalkidou · D. Gavrili
Deree—The American College of Greece, Athens, Greece
e-mail: mvrasida@acg.edu

© The Author(s), under exclusive license to Springer Nature Switzerland AG 2023

V. Katsoni (ed.), *Tourism, Travel, and Hospitality in a Smart and Sustainable World*, Springer Proceedings in Business and Economics,
https://doi.org/10.1007/978-3-031-26829-8_40

1 Introduction

Sports tourism is one of the fastest growing sectors in tourism. More and more tourists are interested in sport activities during their trips whether sports are the main motivation for traveling or not. Following running, or other popular sports events, is acting as a strong motivator for the decision-making process in the choice of destination. Sport events of various kinds and sizes attract tourists as participants or spectators and destinations try to add local flavors to them to distinguish themselves and provide authentic local experiences. Teams are in constant search for location and destination that can host their preparatory training. Sport participation, especially in correlation to health and wellness, is a rising lifestyle trend that can't be ignored and athletes (professional, semi-professional, amateur, and more). Sports events can be a catalyst for tourism development if successfully leveraged in terms of destination branding, infrastructure development, and other economic and social benefits. The advantages of Sports Tourism are evident for destinations that are now in the early development phase but in the recent year even well-established destinations have being shifting their focus toward athletic events in an attempt to change their segmentation market, to extend their season or to become more sustainable, with Mykonos being the most characteristic example organizing its first running event in October 2022.

Being the first part of a wider research project, this paper aims to explore the potential for promoting sports tourism in the Greek tourism market, attempting to make a connection between sports-related tourism and the accommodation market by developing a categorization of hotels in line with a set of criteria according to their capacity to host sports tourists. In this paper, therefore, a literature review is conducted, hoping to shed light to the hypothesis that consumer behavior of tourists, and particularly sports tourists, will present differences both in the motivational and in the experiential phase of the journey. In other words, the literature review focuses on the existence of differences in the way sports tourists decide on which destination to visit, as to the way in which they experience a destination while visiting it. Ensuing, a discussion is made, stemming from the literature review themes.

2 Literature Review

2.1 Available Tools for Measurement

With about 1.5 billion international tourists in 2019, the direct tourism earnings topped $1.7 trillion (WTO, 2020). Based on the size of the tourism and hospitality industry and in conjunction with the dynamic growth and increasing competition in the sector at macro and micro levels, it is important for researchers to understand people's perceptions, attitudes, and behaviors in the sector and one of the ways to do so is by using scales. In particular, a scale is a mechanism or technique used in research to quantify constructs and to distinguish individuals in categories based on predetermined variables used in a study (Sekaran & Bougie, 2013). Various metrics, including nominal, ordinal, interval, or ratios, may be used to build scales based on scaling procedures and processes, including rating and ranking (Bruner, 2012; Sekaran & Bougie, 2013).

The design and application of scales are essential, particularly in the social sciences, where they may transform an abstract idea, an unobservable condition, or a phenomenon into a quantifiable, comparable concept (Kock et al., 2019). Scales may be seen as instruments that enable academics and practitioners to explore the causes and effects of a phenomenon in many settings, contexts, and sectors in order to provide new information. Additionally, two significant considerations make the creation of scales in the tourist and hospitality industries, particularly beneficial. Firstly, tourism is one of the largest industries in the world in terms of earnings and employment generated (WTTC, 2020). Secondly, the tourism and hospitality sector involve frequent and intensive social contact, interaction, and communication among service providers as well as between customers and staff. Therefore, it may be quite helpful to design and employ constructs that allow for differentiation and classification in the tourism and hospitality sector and allow researchers the ability to interpret how individuals interact with processes, with systems, and with each other.

As of 2020, there were 253 tourism and hospitality scales (Koc & Ayyildiz, 2021), divided into various categories, such as motivation scales (Category 1), residents' perceptions and attitudes scales (Category 2), and destination image scales (Category 3). One can argue that these categories tend to overlap because they measure more or less the same phenomena. Furthermore, the SERVQUAL model which was created by Parasuraman et al. (1988) consists of a multiple-item scale for assessing customer perceptions of service quality, which is one of the scales that focus on performance evaluation, quality assessment, loyalty, and satisfaction, which is another broad category of scales (Category 4). Furthermore, this particular scale (SERVQUAL) has been used for the development of the LODGSERV (Knutson et al., 1990), the LQI (Getty & Getty, 2003), and the HOLSERV (Mei et al., 1999) scales, which are all used to assess the quality of services provided by businesses in the lodging and hospitality sectors. Based on the above and the overlaps identified among different scales, it appears that opportunity exists when it comes to the development of scales in special interest tourism (SIT) fields and as such, there is a need for scales pertaining to less researched subfields like dark, religious or faith tourism, urban tourism, nature tourism, civic tourism, and sport-related tourism among others.

At this point, it is also important to note that there is an increasing demand for sports-led tourism, which is defined as journeys taken for leisurely purposes in order to take part in or watch sports activities that take place away from one's usual place of residence (Hall, 1992) and may or may not have a competitive purpose; a more detailed discussion of the term will be conducted later in this paper. At the national level, however, the rise of sports tourism in Greece could only be supported by the corresponding infrastructures. After organizing the 2004 Olympics and a number of metal positions to follow either in team sports (football and Euro 2004) or in individual sports (Wimbledon, 2021), Greece increased its sporting legacy. The country slowly shifted its focus and has managed to earn the trust of professional and national teams as a preferred location for team training as well as staging a number of athletic events in various destinations. The Swedish Olympic Committee has reached an agreement with two Greek resorts ("Sivota Retreat" in Sivota, Epirus, and "Porto Myrina—powered by Playitas" in Avlona, Lemnos) for hosting the training camps of its Olympic Team (Tryfona, 2022a). This deal was the outcome of a lot of effort put in the last five years to upgrade the tourist product and bring it to levels that meet the standards for Olympic team preparation. Indicative of this effort is the fact that Porto Myrina offers apart from excellent accommodation services access to a variety of sport-related facilities [Olympic-standard gym, WOD box (for CrossFit), 3 padel courts (combination of tennis and squash), multipurpose sports facilities, spa, water sports centers, access to biking and hiking trails, etc.]. It goes without saying that efforts are made by the Minister of Tourism, Mr. Kikilias, to support the development of sports-training tourism in Greece, and his meeting with the CEO of Apollo Northern Europe, which is the largest tour operator in Scandinavia, and member of the International Board of DER Touristik Leif Vase Larsen and Commercial Director, Nina Hornewall, which proves the interest to promote sport-training tourism (Tryfona, 2022b). Since most training occurs in the off-season, the growth and development of sports-training tourism may significantly extend the destinations' tourist seasons. Additionally, by providing packages that include sports activities, which are now in demand, hotels may profit from the sports infrastructure they have built or considering developing.

2.2 *Sports Tourism Definitions Discussion*

As early as 1998, there has been a connection established between sport and tourism as being equal forms of travel (Gibson, 1998); hence athletes, i.e., people traveling to an area to participate in a sport event, is also considered tourists in that area (ibid). Sports tourism is also a phenomenon with cultural ties and a strong socioeconomic background for both the host country and the country of origin. Through sports, a bond is formed between people and space, and participants are connected to a specific location, country of team. In recent marketing terms, the social bond created by athletic events (InFocus, 2020) was the unique selling point of this 2022 Euro Basket Advertisement of the Greek participation with the coach of the Greek team calling people at random and the central slogan "X you are on the team". Social integration, cohesion, respect, and acceptance of different cultures, cultural exchange, and finally civic pride are all very important elements of a successful sustainable community and can all be promoted and achieved through culture and sports.

Still the correlation between culture and sports, though obvious and well defined throughout history, is a more complicated matter. Many are the scholars that argue modern sports have given birth to vandalism, hooliganism, violence, and opposition. Sports in some of its extreme violent representations is viewed as the modern civilized evolution of war. It is clear that when an idea is abused or there is a small group of people following illegal, violent behaviors, that cannot and should not reflect back on the ideal but it should be treated as an isolated incident that requires better organization in order to be avoided.

Viewing sport competition as an event that provides an experience in its own merit (Getz, 2008), deriving a meaning out of the experience, especially when immersing in it through active participation (Pine & Gilmore, 1998), and sieving this experience and its derived meaning as affected by the sport event's setting, is also discussed (Funk & Bruun, 2007; Getz, 2008; Green & Chalip, 1998).

The discussion here below is not at all exhaustive, as the topic can be approached from a wide variety of angles, as the literature indicates (Gibson, 1998; and many more). The point of view of this particular paper is not to focus on destinations and the role of the public sector in order to develop sports tourism. The paper follows a marketing and operational approach in an attempt to include the private sector in the argument. Hotels and accommodation are facing intense competition due to the widening of the tourism and hospitality sector to new destinations, but also they are facing intense competition within the destination, in their external micro environment, by new entries and sharing economy platforms (VRBO, AirBnB, and more). It is now more relevant than ever, to examine the theoretical aspects at first, of sports tourism in an attempt to relate them to marketing opportunities for existing accommodation, hotels and the private sector.

2.3 *Categorization of Sports Tourism Analysis*

In 1998, Gibson (1998) identified three main expressions of sports tourism according to the main behavioral patterns or reasons for traveling. As expected, the first category is **active participation in sports** followed by **Sports Tourism as an Event** and finally **nostalgic sports tourism**. Active participation in sports tourism includes anyone traveling to participate as an athlete in sports. This definition does not make a distinction between professional athletes, organized tournaments, team sports of single athletes, and recreational sports activity. Professional NBA players traveling for a game are in the same category as a family of four visiting a ski resort (Gammon & Robinson, 1997).

Event related sports tourism includes all the people traveling to a destination to participate as athletes at the event or be spectators to the events or work in administrative or volunteering positions for the event (ibid). In marketing terms, this segmentation is very broad and does not entail specific behavioral patterns. Events related to tourism can prove to have the most benefits for the accommodation and business sector of a destination. The location of the event is predetermined competition and marketing can be limited to the micro external environment, contained within the destination. In reoccurring events, it creates a loyal clientele with very specific well -defined and well-known demands that are easily met by tourism professional. Events are usually off-season or in the outskirts of the peak season and can present an opportunity for initiating future visits. Sports events are one of the first tools employed by destinations in order to increase their tourism demand (POROSEA, 2019). Good example can be the triathlon in Poros, Zagori Mountain Running, Spartathlon, and even the Boston Marathon. It is worth mentioning that in terms of segmentation profiling people participating in sport events have an inclination toward a healthier lifestyle and demonstrate more sustainable behavioral patterns.

Passive participation in events tourism can include accompanying people for the athletes. The recreational requirements are enhanced since they do not have to follow a competitive athlete's lifestyle in full but they are semi-actively engaged in the sporting event (POROSEA, 2019). Again the destination is predetermined and hotels have to compete in the micro environment within the destination. Common behavioral patterns are difficult to identify and relate to the level of engagement with the event and the type of event.

In individual recreational sports tourism, the destinations is not a given and there is competition. In most cases, the competition is limited to places offering similar experiences (e.g., the Austrian and Italian Alps, Bansco, or rafting activities in rivers). Participants usually travel in small groups and the activity is highly seasonal (Dragovic & Pasic, 2020). Hotels need to compete among their immediate neighbors (micro external environment) and have a global reach as well in order to support the destination during the decision-making process of the potential tourists. Recreational needs are also a priority but again the healthy lifestyle supported by this type of tourism points toward spa treatments, relaxation, good food, nature appreciation, and sustainable environmentally friendly behavior (ibid). The interesting finding in the segmentation analysis is that individuals participating in such activities have above average education and income. A good example would be golfing, where all the supporting activities have to follow the same level of sophistication like for example wine tasting.

Nostalgic sports tourism is the only category that does not include athletes and involves traveling to a location to pay tribute to a sport-related field of hall of fame. However, Ramshaw and Gammon explain that the term "nostalgia" is too limited to include heritage, which should be addressed as a separate idea, so that the "complexity of the sporting past be revealed" (Ramshaw & Gammon, 2015, pp. 248–249). Although the discussion on Nostalgic Sport Tourism could easily deviate toward Heritage Tourism, or Sports Heritage Tourism, for the purpose of this paper, it will be contained to Nostalgic Sport Tourism, in the sense that, when experienced by tourists as such, it does not entail any special sport-related implication for hotels in the area.

It is clear that this very broad categorization cannot include all the different categories occurring ad hoc in relation to different sports, different motivations, and different level of engagement. In terms of accommodation and with the goal of providing personalized service according to individual needs, a more detailed segmentation is required. Still, even at this early stage, some common behavioral patterns among tourism can start occurring. The initial methodological approach was to run three separate very small focus groups on semi-informed and informed individuals in the field of tourism, marketing, and sports. The literature review and the very small focus groups will lead in the formulation of a research hypothesis. The initial findings will act as pilots in designing a more detailed and focused methodology approach which will focus on behavioral patterns and opportunities occurring for hotels in relation to sports tourism.

In 2004, Robinson and Gammon (2004) made an attempt to further divide sports tourism into secondary categories by developing the idea of a sports tourist—athlete including a hard and soft notion in the motivation and behavioral pattern. Hard tourist athletes are mainly motivated by the competitive side of sports and soft tourist athletes view sports as a recreational rather than a competitive activity. Both these categories are more relevant to active participation in sports and can be applied to a very limited extend to event sports tourism and not at all to nostalgic sports tourism It is clear from this categorization that soft tourist athletes would have their tourist side emphasized, thus requiring a mixture of other supplementary activities to complete their experience. They can be viewed as recreational tourists with a special interest in sports with the social elements being of importance in their overall experience (Weed & Bull, 2004). Hard tourist athletes are more interested in the competition side, therefore requiring support infrastructure like special meals or training ground and have less time for other supplementary tourism products. One can argue that they are business tourist with their main business being sports.

3 Discussion

The main hypothesis for this paper is that sports tourism can present an opportunity and a competitive advantage in marketing terms for hotels. The need for a classification system and a scale that would reflect the needs of sports tourist would act as a guide in the decision-making process when deciding on accommodation, which is evident. Sports tourism has been identified as a dynamic part of the tourism market and is growing rapidly as the lifestyle and mobility patterns change. Destinations have indeed identified this trend and are keener in bidding for mega sports events or even ghosting smaller sporting events. The private sector and hotels are only now beginning to realize the potential of this growing market with the exception of the hotels mentioned in the paper. The need for a classification model that will follow the current tourism trends for personalized service and will include the needs of sports tourists is evident. The marketing approach focusing more on identifying hotels that can offer services tailor-made to the requirements of different categories of sports—tourists or athlete—tourists will provide a competitive advantage for hotels and will allow visitors to make more informed choices, thus enhancing their experience and proving beneficial to the destination as well.

A more detailed analysis of the needs and requirements of sport tourist is therefore required based on their behavioral patterns. Simple motivation and main reason for traveling cannot cover the complexity of the issues, and more detailed indicators such as type of sports, level of engagement, seasonality, and life style are also important elements when attempting to create a badge or scale that will refer to amenities and services but will also reflect individual or group needs. Following this initial conceptual paper, the research will continue with detailed structured interviews in order to identify specific needs and requirements and attempt to co-relate them with hotel services and amenities with the overall goal of creating a multi-entry matrix leading to a sports hotel classification.

References

Bruner, G. (2012). *Marketing scales handbook* (6th ed.). GCBII Productions. ISBN 978-0-615-63068-7.

Dragovic, N. & Pasic, M., (2020). sustainable tourism in ski resorts of Europe and the world. In *SITCON 2020 Tourism and Sustainable Development Challenges, Opportunities and Contradiction* (pp. 108–116).

Funk, D. C., & Bruun, T. J. (2007). The role of socio–psychological and culture-education motives in marketing international sport tourism: A cross-cultural perspective. *Tourism Management, 28*(3), 806–819. https://doi.org/10.1016/j.tourman.2006.05.011

Gammon, S., & Robinson, T. (1997). Sport and tourism: A conceptual framework. *Journal of Sport & Tourism, 4*(3), 11–18.

Getz, D. (2008). Event tourism: Definition, evolution, and research. *Tourism Management, 29*(3), 403–428. https://doi.org/10.1016/j.tourman.2007.07.017

Getty, J. M., & Getty, R. L. (2003). Lodging quality index (LQI): Assessing customers' perceptions of quality delivery. *International Journal of Contemporary Hospitality Management, 15*(2), 94–104. https://doi.org/10.1108/09596110310462940

Gibson, H. J. (1998). Sport tourism: A critical analysis of research. *Sport Management Review, 1*(1), 45–76. https://doi.org/10.1016/S1441-3523(98)70099-3

Green, B. C., & Chalip, L. (1998). Sport tourism as the celebration of subculture. *Annals of Tourism Research, 25*(2), 275–291. https://doi.org/10.1016/S0160-7383(97)00073-X

Hall, C. M. (1992). Adventure, sport and health tourism. In C.M. Hall & B. Weiler (Eds.), *Special Interest Tourism.* Pluto Press.

InFocus (2020). *London Together Learning Guide 2: Approaches to sport and social integration.* www.london.gov.uk

Knutson, B., Stevens, P., Wullaert, C., Patton, M., & Yokoyama, F. (1990). Lodgserv: A service quality index for the lodging industry. *Hospitality Research Journal, 14*(2), 227–284.

Koc, E., & Yazici Ayyildiz, A. (2021). An overview of tourism and hospitality scales: discussion and recommendations. *Journal of Hospitality and Tourism Insights,* Vol. ahead-of-print No. ahead-of-print. https://doi.org/10.1108/JHTI-06-2021-0147

Kock, F., Josiassen, A., & Assaf, A. G. (2019). Scale development in tourism research: Advocating for a new paradigm. *Journal of Travel Research, 58*(7), 1227–1229.

Mei, A. W. O., Dean, A. M., & White, C. J. (1999). Analyzing service quality in the hospitality industry. *Managing Service Quality, 9*(2), 136–143.

Parasuraman, A., Zeithaml, V. A., & Berry, L. L. (1988). Servqual: A multiple-item scale for measuring consumer perception of service quality. *Journal of Retailing, 64*(1), 12–40.

Pine, J. B. & Gilmore, J. H. (1998). Welcome to the experience economy. *Harvard Business Review,* July–August, 97–105. https://hbr.org

Porosea (2019). Data Collected from the Event. https://why-n.gr/event/porosea-2019/

Ramshaw, G., & Gammon, S. (2005). More than just nostalgia? Exploring the heritage/sport tourism nexus. *Journal of Sport Tourism, 10*(4), 229–241. https://doi.org/10.1080/14775080600805416

Robinson, T., & Gammon, S. (2004). Revisiting and applying the sport tourism framework: A question of primary and secondary motives. *The Journal of Sport Tourism, 9*(3), 221–233.

Sekaran, U., & Bougie, R. (2013). *Research methods for business: A skill-building approach.* Wiley.

Tryfona, V. (2022a). *Sports Tourism Hotel Opening in Lemnos.*(found in Greek), Hotel & Restaurant Daily, (2/6/2022a). https://hotelandrestaurant.gr/hotel/xenodocheio-gia-athlitiko-tourismo-egkainiazetai-sti-limno/

Tryfona, V. (2022b). *Εγκαίνια για το αθλητικό ξενοδοχείο στη Λήμνο.* Hotel & Restaurant Daily, (2/6/2022b). https://hotelandrestaurant.gr/hotel/egkainia-gia-to-athlitiko-xenodocheio-sti-limno/

Weed, M., & Bull, C. (2004). *Sports tourism: Participants, Policy and providers* (2nd ed.). 2009. Elsevier Butterworth-Heinemann.

World Tourism Organization (2020). *International tourism growth continues to outplace the global economy*. https://www.unwto.org/international-tourism-growth-continues-to-outpacethe-economy

WTTC (2020). *World travel and tourism council. Tourism impact*. www.wttc.org/economic-impact/country-analysis.

Quality of Life and Health Tourism: A Conceptual Roadmap of Enhancing Cognition and Well-Being

Evgenia Gkinton, Giorgos Telonis, Constantinos Halkiopoulos, and Basilis Boutsinas

Abstract Health tourism is a unique form of tourism that mixes vacations with the prevention and treatment of physical and mental conditions. It refers to all relationships and phenomena that result from a change in location and residence of individuals and aims to promote, stabilize, and restore, as needed, their physical, mental, and emotional health and well-being using services in a location that is not their permanent place of residence or employment. Wellness tourism encompasses all interactions and phenomena resulting from the travel and stay of individuals whose primary objective is to restore, preserve, and improve their physical and mental health and social well-being. Wellness is a combination of attitudes and behaviors designed to prevent disease, improve health, and enhance the quality of life. This paper aims to investigate the contribution of wellness tourism to enhancing the quality of life. In addition, cognitive and emotional characteristics of the tourist traveler's personality that can be enhanced through wellness tourism and contribute to the general improvement of life quality are investigated. This article is an overview review of research conducted over the past ten years. The results underlined, among other things, the role of wellness tourism in enhancing the quality of life in its many facets, including physical and mental health, well-being, and social environment. In addition, enhancing cognitive function and emotional well-being, both of which can be attained through wellness tourism, are two additional determinants of a harmonious quality of life. Health tourism in general and wellness tourism can bring numerous benefits, both at the individual level to the physical and psychological health of the beneficiary of the tourist product and at the community level by contributing to their economic and social vibrancy. As a result, upgrading tourist facilities and infrastructure and attracting investment funds to strengthen health and wellness tourism in tourist-travel destinations is considered an unavoidable necessity, as it can bring substantial social, economic, and cultural benefits to the local communities of states.

E. Gkinton (✉) · G. Telonis · C. Halkiopoulos · B. Boutsinas
Department of Business Administration, University of Patras, Patras, Greece
e-mail: evigintoni@upatras.gr

C. Halkiopoulos
e-mail: halkion@upatras.gr

© The Author(s), under exclusive license to Springer Nature Switzerland AG 2023
V. Katsoni (ed.), *Tourism, Travel, and Hospitality in a Smart and Sustainable World*, Springer Proceedings in Business and Economics,
https://doi.org/10.1007/978-3-031-26829-8_41

Keywords Quality of life · Health tourism · Cognition · Well-being · Tourism industry · Data analytics

JEL Classifications D91 · Z32 · O32 · I12 · I31 · C80

1 Introduction

Tourism has become a significant socioeconomic factor in both developing and developed markets. Its efficacy as a source of economic development has made it an indispensable component of strategic planning in practically every nation. In addition to economic instability, recession, stagnation, social instability, war/terrorism, information technology, social media, environmental and sustainability issues, distribution of tourism benefits, and the non-economic value of tourism, the tourism industry faces numerous challenges today. There are evident repercussions of these issues for the tourism industry, and there has been abundant study examining certain facets of these challenges. The link between tourism activities, their repercussions, and the quality of life (QOL) of individuals involved in creating or consuming tourism goods and services is one of the research fields gathering traction and greater attention. This QOL research stream in the tourism industry is gaining momentum and will likely receive more significant focus. Close connections exist between health tourism and improving life quality and well-being. To address this issue, we attempt to develop an integrated theoretical framework that highlights the role of health tourism and, more specifically, wellness tourism in enhancing the quality of life in its many dimensions, including physical and mental health, well-being, public health (Katsoni & Koliopoulos, 2020), and social environment.

Moreover, the enhancement of cognitive function and emotional well-being, which can be obtained through wellness tourism, are two additional factors determining a harmonious quality of life. Therefore, we attempt to demonstrate these two factors' reciprocal influence in our study. Tourism is frequently considered a manifestation of human conduct. The tourism business encompasses transportation, loading, and entertainment. Furthermore, the experiences and activities of tourists tend to have favorable effects on various life areas, including familial, social, leisure, and cultural life (Uysal et al., 2016). All of those, as mentioned earlier, construct cognitive and emotional behavioral characteristics and improve quality of life and well-being (Harrill & Potts, 2003).

2 Literature Review

2.1 Definition of Health Tourism and Components

Health tourism is a unique form of tourism that combines vacations with the prevention and treatment of physical and mental-spiritual disorders. It refers to all relationships and phenomena that result from a change in location and residence of individuals and aim to promote, stabilize, and restore, as needed, their physical, mental, and emotional health and well-being by utilizing services in a location that is not their permanent place of residence or employment. The three categories of health tourism are medical tourism, spa-thermal tourism, and wellness tourism.

Medical tourism refers to the migration of individuals with various types of health problems within and across international borders and the provision of medical monitoring and care to prevent, diagnose, and maintain their health. Patients are typically joined by (healthy) friends or family members who remain close during treatment. As with other types of tourism, medical tourism includes medical care, transportation, information, sightseeing, catering, and hotel accommodations. In medical tourism, it is also possible for individuals requiring long-term medical treatment, such as those with kidney and heart failure and other chronic conditions, to travel quickly to their preferred tourist locations.

Spa-thermal tourism is a specific form of delivering tourist services in regions whose primary characteristic is utilizing recognized natural healing resources for therapeutic purposes in specialized facilities. It also comprises coordinated and mutually complementary acts, with thermal waters as the dominating element, aimed at preventing, maintaining, or restoring the individual's physical and mental health and well-being.

Wellness tourism is the total of all connections and phenomena resulting from the travel and stay of individuals whose primary objective is to restore, preserve, and enhance their physical and mental health and social well-being. Wellness is a combination of attitudes and behaviors designed to prevent disease, improve health, and enhance the quality of life. It encompasses both primary and secondary wellness visitors (Baralou et al., 2020). Generally, health tourism is offered in establishments with specialized tourist infrastructures, such as thermal treatment units, Thalassotherapy centers, Rejuvenation centers (spas), and thermal tourism centers.

2.2 Wellness Tourism and its Benefits

Wellness can be characterized in numerous ways. According to Saracci (1997), the frequently quoted World Health Organization definition of health is "a condition of total physical, mental, and social well-being and not only the absence of sickness or disability." Another definition of wellness is a way of life that emphasizes optimal health and well-being by integrating the body, mind, and spirit to live more completely within the natural human community (*Myers et al.*, 2001). This definition seems to incorporate many of the concepts discussed by the authors in this issue. Adams (2018) mentions four fundamental wellness principles: (1) Wellness is multidimensional; (2) wellness research and practice should focus on identifying the reasons of wellness as opposed to the causes of illness; (3) wellness is about balance; and (4) wellness is relative, subjective, or perceptual. This issue's editors and contributors agree that well-being is more of a psychological than a physical state, as implied by the final point, which is intriguing. In many ways, this is one of the oldest forms of tourism if one considers the Romans' and Greeks' meticulous attention to health, the Medieval pilgrims' journeys for spiritual enlightenment, and the medical seashore and spa tourism of the 18th-century European elite. In recent years, however, the desire for wellness has arguably intensified like never before in the history of travel. Never before have there been so many wellness facilities, holistic retreats, spas, spiritual pilgrimages, and supplementary and alternative therapies. As explanations for this exponential increase, numerous theories point to the cynicism of western capitalist cultures, the disintegration of traditional religions, and the fracturing of communities (Katsoni & Stratigea, 2016). Concurrent advances in science and medicine have resulted in improved bodily preservation and a decline in the prevalence of sickness, but mental, psychological, and emotional issues are frequently ignored. Depression is frequently listed as one of the most burdensome diseases of the 21st century, and suicide rates are on the rise, particularly among young men. However, assistance appears to be on the horizon in the shape of novel psychotherapies, complementary treatments, and an expanding wellness leisure and tourism industry. The scant data available implies that those who take advantage of the abundance of offered experiences are not only on a touristic voyage of physical activity but also on a path toward better self-awareness and happiness.

Wellness tourism dominates all other forms of tourism on a global scale. Wellness tourism draws affluent tourists from wealthy and developing countries; it boosts the economies of developed nations and tiny or disadvantaged communities rich in conventional health and wellness practices. In our Wellness Tourism Guide, we have highlighted wellness places for the convenience of our readers. Improving health is the essence of well-being, automatically leading to a higher quality of life. Several wellness programs at any wellness resort aim to eradicate a person's physical and mental "toxins" through relaxation and stress-reduction techniques. In addition, wellness centers and retreats promote healthy living through nutritious, wholesome cuisine, and fitness-related activities. They also emphasize spiritual and mental health, cosmetic treatments, and healthy sleeping practices to enhance and improve health.

Wellness tourism results in physical and spiritual development and enables tourists to discover the region's natural and historical sites, where they can spend a relaxing and enjoyable vacation. As a result, individuals are more attentive to their wellness needs and have established goals they are anxious to achieve through wellness tourism. With no hurried, exhausting schedule, only precious moments of rest and renewal through physical activities, nutritious cuisine, wellness spas, and treatments, people can meet new people and form new relationships. Away from the stresses of daily life, individuals could restart their lives. They can concentrate more on tiny personal improvements that will last even when they return home (Nopiyani & Wirawan, 2021).

Wellness holidays are focused on rejuvenating the body and mind. Only a vacation to a resort that offers body and mind-rejuvenating therapies, and practices can alleviate the constant tension associated with living in today's technology-driven society. Wellness holidays are the ideal way to enjoy a vacation and return home feeling revitalized and re-energized in a world where it can be difficult to maintain health and fitness due to the demands of daily life (Panas et al., 2017). Wellness tourism is the junction of two enormous and expanding industries: the $2.6 trillion tourism industry and the $4.5 trillion wellness industry (Anttiroiko, 2018). Increasingly, holistic health and prevention are at the forefront of consumer decision-making. As a result, people now anticipate maintaining their healthy lives and wellness habits when traveling (Luo et al., 2018).

In 2013, the Global Wellness Institute (GWI) published the first edition of the Global Wellness Tourism Economy report, a landmark study that defined the criteria and characteristics of the developing wellness tourism sector, calculated its global size, and outlined its far-reaching economic implications. Since then, this tourism subsegment has expanded globally. According to the Global Wellness Institute, wellness tourism was a $639 billion global sector in 2017, expanding more than twice as quickly as traditional tourism from 2015 to 2017. The Global Wellness Institute defines *wellness tourism* as travel to maintain or boost one's health. In this manner, wellness is a pursuit of life quality and well-being that enhances cognitive and emotional facets of individuals. Wellness tourists are often self-aware, active individuals who strive to improve their well-being, health, and happiness.

2.3 Quality of Life Promotion in Health Tourism

Quality of life is a complex concept that expresses more subjective experiences, perceptions, and demands than objective life circumstances. It shows the harmonious satisfaction of an individual's own aims, "beliefs," desires, and personal happiness and life satisfaction. It is directly tied to physical and mental health (Gkintoni et al, 2017), a sense of independence, positive social relationships, possibilities for enjoyment, and the body's capacity to not tire quickly from routine everyday tasks. According to the World Health Organization (1998), quality of life is a multifaceted notion that refers to human, social, economic, and environmental characteristics. The quality of life determines an individual's well-being in any condition of physical (disability issues and movement problems) and mental health (depression, anxiety, and other interpersonal issues) (Tzanos et al., 2019). In greater detail, the concept of quality of life is characterized by broad dimensions such as physical and mental health, the level of independence, social interactions, personal values, and relationships with the environment (WHO 1998). Specifically, family life, job, and life happiness, education, gender equality, economic and political status, and the environment are regarded as crucial factors (Calaminus & Barr, 2008).

The quality of a person's life is determined by their judgments of their place in life, within the context of their culture and value system, regarding their objectives, expectations, and standards (WHO 1998). Consequently, researchers frequently concentrate on "Health-Related Quality of Life," which examines the impact of sickness on a person's physical, psychological, and social health, emphasizing the significance of its evaluation for the field of health sciences. Its evaluation is conducted through questionnaires that contain scales and criteria relevant to a person's behavior or experiences, such as physical functioning and health.

Modern western culture is defined by the alienation of man from nature, the contamination of the environment, and the intense rhythms of daily life in situations of competition and uncertainty. The individual's low quality of life results from the factors mentioned above, with lack of free time, loneliness, stress, and the emergence of mental and physical disorders being the most significant (Matatolu et al., 2019). Health tourism, with the services provided within the framework of its three categories, aims to combat this morbid condition and assist man in regaining his equilibrium, given that the body's agility and vitality evoke positive emotions, which are directly related to the individual's happiness and sense of fulfillment in life. Health tourism, with the services provided within the framework of its three aforementioned categories, aims to combat this morbid condition and assist man in regaining his equilibrium, given that the body's agility and vitality evoke positive emotions, which are directly related to happiness and fulfillment that the individual derives from his or her life (Liburd et al., 2011). Good health and a cheerful disposition produce positive emotions. As a person's perception of his body is tied to his self-esteem and eventually has a substantial impact on his mental health and well-being, it is imperative to seek out ways and procedures that can contribute to this direction.

In 2018, the World Tourism Organization (UNWTO) and the European Travel Commission (ETC) defined health tourism as "the types of tourism that have as their primary motivation the contribution to physical, mental, and/or spiritual health through medical activities and wellness activities, enhancing individuals' capacity to meet their own needs and function better as individuals within the environment and society." The birth of this new sort of tourism from the intersection of three scientific fields—health, economics, and communication—complicates its conceptual definition. Moreover, it is a novel method for addressing health demands aimed at individuals willing to go to a different location to obtain health services, combining treatment, recuperation, or alternative medicine with recreation, vacations, and visits to destinations of their choosing (Markgraf et al., 2012). According to the World Tourism Organization, developing people's quality of life is one of tourism's most significant concerns. In conclusion, there is a closer relationship between health improvement, quality of life promotion, and tourism, and between health tourism and travel. Individuals who want to improve their quality of life through a destination of coordinated and mutually complementary actions with the goals of preventive, holistic approach, enhancing health, and enhancing the quality of life are targeted by health tourism (Dolnicar, et al., 2013).

2.4 Cognition Enhancement—A Privilege of Health Tourism

Cognitive psychology can serve in comprehending the mental processes linking stimulus perception to behavior (Gkintoni et al., 2021). Here, cognitive psychology is considered to include perceptual and attention processes, memory procedures (Gkintoni et al., 2022) such as visual taxonomy and classification, components of emotion elicitation, and states of consciousness such as mindfulness and flow (Antonopoulou et al., 2022; Hadinejad, et al., 2022).

Several prior studies have focused on the psychological implications of tourism. Considering holidays may elicit fantasies and emotions that have a significant effect in determining vacation location selections (Decrop & Snelders, 2004). Individuals tend to prefer enjoyable vacation experiences for enjoyment, imagination, and sensory stimulation. Important components of such encounters include positive and joyful emotions and sentiments (Lopes et al., 2021). Multisensory vacation experiences include tastes, sounds, smells, tactile sensations, and visual pictures. Emotions impact the memorability of holidays (Larsen & Jenssen, 2004). Clearly, psychology is a field that has important ramifications (Halkiopoulos et al, 2021b) for the study of all categories of tourism and particularly, health tourism.

Health tourism has discernible influence on the personal development of individuals, particularly cognitive development. Mainly, cognitive components such as consciousness, attention, and memory, together with emotion and sentiments, appear to be significantly enhanced and fostered (Skavronskaya et al., 2017).

2.5 Consciousness and Mindfulness

Consciousness is the experience of the current moment (Mannell & Iso-Ahola, 1987); hence, it is comparable to the term moment-by-moment experience used in tourist literature. In addition, mindfulness is a characteristic or state of conscious awareness in which a person is implicitly aware of the context and substance of information (Chen, 2015).

Since the mid-1980s, mindfulness has been regarded a strategy for controlling tourism experiences (Chen, 2015; Frauman & Norman, 2004; Pearce & Moscardo, 1986). Tung et al. (2016) underline the significance of awareness in memory encoding. Mindfulness is educational, offering travelers insight into their conscious state (Chen, 2015). Mindfulness is a trigger that can improve tourist memory retention, demonstrating the significance of attention and emotional states during memory encoding (Brown & Ryan, 2003; Chen, 2015; Lopes et al., 2021; Proceedings of the First Hospitality and Leisure, 2007).

Flow is a state of awareness utilized in health tourism study to comprehend the tourist experience. Flow is a state of emotional stimulation (wellness centers, spa treatments, yoga) that happens when one's abilities equal the degree of difficulty, resulting in a sense of oneness with the surrounding environment (Csikszentmihalyi, 2000; Filep, 2016). In a state of flow, a person is completely absorbed in the activity at hand and can ignore their immediate surroundings. Consequently, flow and awareness are separate states of consciousness associated with various experience settings in the context of health tourism activities.

2.6 *Attention*

Attention focuses on the mental processing of external stimuli or internal memories and ideas (Davenport & Beck, 2001) and influences perception, memory, and learning. Moreover, attention is associated with goal achievement, memory processing, and behavior. Additionally, focus is crucial for recalling events (Campos, 2022). In addition, a cognitive schema is a framework in a person's memory that conveys information about a concept or kind of stimulus, such as a person, event, or object, including its features and the relationships between those properties (Ghosh & Gilboa, 2014; Mercille, 2005). Cognitive schema allows the recovery of cognitively previously processed information and governs human behavior by allowing a person to function in a complicated context and comprehend new information. When assessing a particular person or thing, the same procedure is used. There are three degrees of homogeneity between a stimulus item and an activated schema, according to schema theory. Several tourist words, including destination image, are connected to form. A destination image is "a mental structure that integrates the components, impressions, and values that individuals project to a given place" and can be regarded as a particular instance of a more generic image schema (Halkiopoulos et al., 2021a). Understanding the cognitive processes involved in schema construction is crucial for destination marketing, particularly for medical tourism locations (Halkiopoulos et al., 2021b).

2.7 *Memory*

Memory is the mental process of encoding, storing, and retrieving information, but it may also refer to the information itself, the memory we recall and remember. Cognitive psychology identifies two fundamental forms of memory with distinct characteristics. Working memory (short-term memory) may hold information for a limited time but has a limited capacity. In contrast, long-term memory has a great capacity and stores acquired experiences and knowledge throughout life. There are two forms of long-term memory: semantic and episodic. Semantic memory is the recall of general information about the universe, whereas episodic memory relates to particular memories such as personal experience and facts. Autobiographical memory is a subset of episodic memory and is defined as the recalling of an individual's past-related experiences. Memories are pliable and susceptible to distortion by external sources such as commercials or stories or by subsequent events. Attention to novel stimuli influences the quantity of data encoded and stored (Kim et al., 2022).

In health tourism locations, tourist memories are significant because visitors' base future decisions on prior travel experiences. An important objective of a travel experience is to create lasting memories. Unforgettable tourist experiences at wellness and treatment centers account for a negligible proportion of thoughts processed in working memory and transmitted to long-term memory (Halkiopoulos et al., 2022).

Moreover, a health tourism location may be remembered primarily due to social engagement with friends, which is probable because social interaction is more likely to elicit emotional responses. Due to the structure of autobiographical memory, tales and narratives are crucial vehicles of communication and persuasion (Panteli et al., 2021). It has been observed that the story shapes travel experiences (LaMondia et al., 2010). Moreover, the multimodal encoding of memories is a rich area of health tourism psychology study based on the brain and connected to its mnemonic substrate.

2.8 Emotions

Emotions are mental events induced by the appraisal of relevant inputs and are accompanied by quantifiable bodily changes such as perspiration and heart rate rise. Emotions originate from an "emotional condition of the body" that not only the limbic system but also the somatosensory system communicates back to the brain (Gkintoni & Dimakos, 2022). Emotions are essentially motivating and illuminating due to their experience or emotional character. Experiences may have a greater personal and symbolic impact on customers than items. Therefore, emotions as summative evaluations of tourism experiences appear to warrant additional study. Most tourist studies of experiences have just sought to explain an experience's emotional impact without exploring why specific stimuli generate specific emotional reactions (Hosany & Gilbert, 2010). In conclusion, emotions greatly influence attitude by favorably or adversely altering the individual's ideas about a message. Emotions elicited by stimuli influence the extent to which people consider and their attitudes toward message content.

2.9 Well-Being Promotion

Well-being is a broad feeling, inner radiation of health, power, beauty, and physical health. It concerns the condition of man as a whole, as a living being, and not the absence of disease or pathology. Health is defined as the condition of complete physical, mental, and social well-being, not merely the absence of disease. From this, it follows that the concept of health is not solely linked to medicine, as it is commonly described today, but also to elements such as the environment, the economy, and work, and is based entirely on the subjective experience of well-being (WHO, 1998). According to the non-profit National Wellness Institute, whose sole mission is to promote wellness worldwide, wellness is defined as "an ongoing and active process in which each person realizes and learns that there are choices (healthy choices) that can lead him to a longer and better quality of life, to a more general "well-being." As an alternative to the traditional, medical model-based approach to wellness for treating mental and physical illnesses, a new wellness paradigm has recently arisen. Physical activity and wellness. Physical activity is inextricably linked to well-being. Physical activity promotes, sustains, and recovers not just physical but also mental and social health (Moore et al., 2012). It can assist everyone in approaching their well-being prescription through natural interventions and strategies, as well as a more general shift in his life outlook. It also teaches people how to be more functional and joyful in their daily lives, understanding what was already recognized in Greek medicine: that man is composed of spirit, body, and soul and that the balance between them determines his health level. Wellness is also a positive approach, not only to disease prevention but also to the quest for a more satisfying existence. It includes mental and physical abilities based on responsibility, dietary knowledge, physical fitness, stress management, and environmental consciousness. The first step to wellness is assessing one's current state. This can be accomplished through self-administered assessment questions, providing insight into areas requiring improvement. Understanding oneself and potential, knowing what serenity is and how to reach it, increasing self-confidence when necessary, enhancing the environment, and engaging in regular physical activity are emphasized (McCabe & Johnson, 2013).

Wellness tourism is regarded as the field of health tourism with the most rapid growth. It is geared toward travelers who have an excellent desire for physical and mental health through treatment and wellness treatments to attain personal achievement and happiness. Wellness tourism can be a journey of "self-knowledge," allowing the health tourist to set aside worries and distractions of daily life and delve into a thoughtful examination of oneself. In addition, wellness tourism aims at the psychosomatic rejuvenation of the individual in a pleasant, relaxing environment. Wellness tourism encompasses a wide range of tourism firms with significantly increased services in the current economic climate. They also comprise services with therapeutic properties for chronic conditions, aesthetic and body treatments, skincare, complementary and alternative therapies, and aromatherapy. In contrast, specialized spa hotels know to give proper care by integrating exercise services, beauty care, healthy nutrition, meditation, and other spiritual activities (Pyke et al., 2016).

In wellness tourism, the emphasis is centered on physical and spiritual refreshment and body care, with various techniques employed. Services such as special baths, massages, facial and aesthetic treatments, physical exercise, and sauna represent the broader term "wellness tourism." Wellness tourism is defined by the Ministry of Culture and Tourism's 2012 action plan for the development of health and wellness tourism as "the type of tourism chosen by the visitor for personal wellness and relaxation, but also relaxation with an alternative—paramedical methods of treatment, such as practical—traditional methods, exercise, etc."

To sum up, *wellness tourism* might be defined as the short-term travel of people to destinations beyond their place of home or employment to improve their mental state. According to Adams (2018), each stakeholder must adhere to the following four principles to approach the concept of well-being:

The concept of well-being has numerous interpretations and dimensions. In particular, the social, spiritual, mental, emotional, professional, and environmental dimensions. The concept of equilibrium is incorporated into the definition of wellness. In the third principle, it is acknowledged that the search for and implementation of wellness should concentrate on identifying the causes that lead to it rather than the causes that cause diseases perse.

In the fourth and final premise, it is acknowledged that the concept of well-being is not objective and has a direct association with psychological variables rather than physical characteristics. It is important to note that throughout the world's literature, well-being is described by multiple terms, such as wellness and well-being, which creates difficulty in comprehending scientific writings. The term well-being has a definite, qualitative connotation. It interprets the broad notion of well-being, whereas well-being is more concrete and quantifiable in that it specifies the variables that comprise the condition of mental upliftment—happiness (Tuohino & Konu, 2014).

Wellness tourism, health tourism, and spa tourism are alternative tourism whose motivating factors are the pursuit of self-development and inner calm, the reinterpretation of values, and the overall enhancement of daily life. Since the dawn of the civilized world, individuals have been offered vacations that have the potential to improve their health and well-being. Wellness tourism emerged as a dynamic idea in the tourism industry in the 1990s, as people began to seek a healthier living outside their daily lives and on vacations, with products and services that may meet their demands. The concepts of health and well-being may date back to antiquity. However, demographic changes, stress, trends, intense lifestyles, the modern reality, the cost of preserving health, etc., have led to their rapid development as forms of global tourism, emphasizing the need for holistic and alternative experiences and treatments (Cohen & Cohen, 2017).

In addition, the life expectancy of the world's working population may have increased, but man's everyday stress and worry are rising dramatically. Those mentioned above, coupled with the availability of a vast selection of wellness goods at varying rates, make it accessible to everybody. At the same time, modern tourists are more knowledgeable and seasoned than ever before, as they can be updated and broaden their horizons daily with the help of new technologies, which provide them with insight into their travel experience and how this experience—the trip—will improve their lives (Halkiopoulos et al., 2020). The proliferation of skepticism toward traditional medicine is attributable to the availability of online knowledge at any time of day and the rising cost of medical industry services; the globe is increasingly focusing on the prevention of various diseases through wellness and holistic philosophy.

It is necessary to note that the economic and health crisis contributed to those mentioned above since the collapse of the welfare state necessitated individuals taking personal health safety and prevention measures. It is impossible to ignore the new beauty standards offered on social media and their indirect or direct influence on individuals, whether consciously or subconsciously, continually generating an extraordinary virtual reality that many aspire to attain. Negative emotions like despair and anxiety arise as a result. By experimenting with holistic and alternative treatments and wellness, offered away from western medicine and its rigid health systems, individuals can approach their inner self, love him, and heal. All of the above is intended to justify the conditions that led to the proliferation of health and, by extension, wellness in tourism.

3 Discussion and Conclusion

This review has touched upon a number of key concepts for health tourism research. Specifically, the present study's main purpose was to investigate the contribution of wellness tourism to enhancing the quality of life. In addition, cognitive and emotional characteristics of the tourist traveler's personality that can be enhanced through wellness tourism and contribute to the general improvement of life quality are investigated.

The results underlined, among other things, the role of wellness tourism in enhancing the quality of life in its many facets, including physical and mental health, well-being, and social environment. In addition, enhancing cognitive function and emotional well-being, both of which can be attained through wellness tourism, are two additional determinants of a harmonious quality of life.

Cognitive psychology may also help to understand why holidays are beneficial to health and well-being. While traveling, people are away from normal environment, separate from family, friends, and colleagues, so that travelers are free to think and feel and behave in ways that they would not ordinarily consider (Kottler, 2002). Similarly, travel is accepted as a transformative experience is accepted, but the psychological processes supporting transformation and well-being are only beginning to be studied.

This study sheds light on the relationship between health tourism and quality of life. The research demonstrated that tourism has several effects on quality of life for instance two parameters have been highlighted: Cognition enhancement and well-being. The findings add to a deeper understanding of the relationship between health tourism and quality of life, with implications for tourism development (Aref et al., 2010). Tourism stakeholders and community planners may adopt tourism development plans based on people's views of tourism's influence on the quality of life in tourist locations with the help of the findings of this study. Furthermore, as individuals engage in self-development, health tourism, particularly holistic or spiritual tourism, can enhance their emotional and spiritual health (Choe, et al., 2021).

The social and cultural well-being of health tourists who typically travel alone can be improved via contact with other tourists and locals who face similar obstacles. This can help individuals share their experiences and recover more quickly. Typically, health tourists want comfort and security because they wish to cure their physical body, mind, or spirit without being distracted by worry. Health tourists, mainly medical tourists, may bring along family or friends as accompanying travelers for moral support. In addition, spas and wellness facilities (such as hotels) are attracting an increasing number of families with children and couples seeking a romantic retreat. Numerous holistic retreats provide a variety of activities that facilitate the attainment of specific desired moods. However, in order to attain total harmony and holism, these elements should be balanced. Many individuals attend holistic retreat centers to repair their lives or find new significance. Holistic denotes that they focus on the whole person and the integration of body, mind, and spirit, in addition to other areas (Smith & Kelly, 2006). Travel is genuinely transformative since it reveals one's actual or genuine self, as well as maybe a new and more adventurous or unconventional self that might not have existed if the individual had never traveled.

References

Adams, J. M. (2018). The value of wellness. *Public Health Reports, 133*(2), 127–129. https://doi.org/10.1177/0033354918754543.

Antonopoulou, H., Halkiopoulos, C., Gkintoni, E., & Katsibelis, A. (2022). Application of gamification tools for identification of neurocognitive and social function in distance learning education. *International Journal of Learning, Teaching and Educational Research, 21*(5), 367–400. https://doi.org/10.26803/ijlter.21.5.19.

Anttiroiko, A-V. (2018). Wellness city strategies in local-global dialectic. *Wellness City*. 99–137 https://doi.org/10.1007/978-3-319-75562-5_6.

Aref, F., Redzuan, M., Gill, S. S., & Aref, A. (2010). Community capacity building in tourism development in local communities. *Journal of Sustainable Development*, *3*(1). https://doi.org/10.5539/jsd.v3n1p81.

Baralou, E., Daskalaki, K., Georgiadis, E., Katsoni, V., Malliou, P., & Panagiotou, G. (2020). GoFit erasmus project: A transdisciplinary approach for exercise, health and tourism. *Springer Proceedings in Business and Economics*, 281–289. https://doi.org/10.1007/978-3-030-36342-0_23.

Brent Ritchie, J. R., Wing Sun Tung, V., & J.B. Ritchie, R. (2011). Tourism experience management research. *International Journal of Contemporary Hospitality Management*, *23*(4), 419-438. https://doi.org/10.1108/09596111111129968.

Brown, K. W., & Ryan, R. M. (2003). Mindful attention awareness scale. *PsycTESTS Dataset*. https://doi.org/10.1037/t04259-000.

Calaminus, G., & Barr, R. (2008). Economic evaluation and health-related quality of life. *Pediatric Blood & Cancer, 50*(S5), 1112–1115. https://doi.org/10.1002/pbc.21456.

Campos, A. C. (2022). Living labs in tourism. *Encyclopedia of Tourism Management and Marketing*. https://doi.org/10.4337/9781800377486.living.labs.

Chen, 2015 Chen, X. (2015). Surplus, tourism. *Encyclopedia of Tourism* (pp. 1–2). https://doi.org/10.1007/978-3-319-01669-6_628-1.

Choe et al., 2021 Choe, J., O' Regan, M., Kimbu, A., Lund, N. F., & Ladkin, A. (2021). Quality of life perspectives for different social groups in a world centre of tourism and leisure. *Tourist Studies, 21*(4), 615–637. https://doi.org/10.1177/14687976211038758.

Cohen, S. A., & Cohen, E. (2017). New directions in the sociology of tourism. *Current Issues in Tourism, 22*(2), 153–172. https://doi.org/10.1080/13683500.2017.1347151.

Csikszentmihalyi, M. (2000). *Flow. Encyclopedia of psychology* (Vol. 3, pp. 381–382). https://doi.org/10.1037/10518-188.

Davenport and Beck, 2001 Davenport, T. H., & Beck, J. C. (2001). The attention economy. *Ubiquity, 2001*(May), 1. https://doi.org/10.1145/376625.376626.

Decrop, A., & Snelders, D. (2004). Planning the summer vacation. *Annals of Tourism Research, 31*(4), 1008–1030. https://doi.org/10.1016/j.annals.2004.03.004.

Dolnicar, S., Lazarevski, K., & Yanamandram, V. (2013). Quality of life and tourism: A conceptual framework and novel segmentation base. *Journal of Business Research, 66*(6), 724–729. https://doi.org/10.1016/j.jbusres.2011.09.010.

Filep, S. (2016). *Positive Tourism*. https://doi.org/10.4324/9781315707129.

Frauman, E., & Norman, W. C. (2004). Mindfulness as a tool for managing visitors to tourism destinations. *Journal of Travel Research, 42*(4), 381–389. https://doi.org/10.1177/0047287504263033.

Ghosh, V. E., & Gilboa, A. (2014). What is a memory schema? a historical perspective on current neuroscience literature. *Neuropsychologia, 53*, 104–114. https://doi.org/10.1016/j.neuropsychologia.2013.11.010.

Gkintoni, E., & Dimakos, I. (2022). An overview of cognitive neuroscience in education. In *14th Annual International Conference on Education and New Learning Technologies, EDULEARN22 Proceedings*. https://doi.org/10.21125/edulearn.2022.1343.

Gkintoni, E., Halkiopoulos, C., & Antonopoulou, H. (2022). Neuroleadership an asset in educational settings: An overview. *Emerging Science Journal, 6*(4), 893–904. https://doi.org/10.28991/esj-2022-06-04-016.

Gkintoni, E., Meintani, P. M., & Dimakos, I. (2021). Neurocognitive and emotional parameters in learning and education process. In *14th Annual International Conference of Education, Research and Innovation, 8th-10th November*. Spain: Seville. https://doi.org/10.21125/iceri.2021.0659.

Gkintoni, E., Pallis, E., Bitsios, P., & Giakoumaki, S. (2017). Neurocognitive performance, psychopathology and social functioning in individuals at high-genetic risk for schizophrenia and psychotic bipolar disorder. *International Journal of Affective Disorders, 208*, 512–520. https://doi.org/10.1016/j.jad.2016.10.032.

Hadinejad, A., Gardiner, S., Kralj, A., Scott, N., & Moyle, B. D. (2022). Cognition, metacognition and attitude to tourism destinations: The impact of emotional arousal and source credibility. *Journal of Hospitality and Tourism Management, 51*, 502–511. https://doi.org/10.1016/j.jhtm.2022.05.005.

Halkiopoulos, C., Antonopoulou, H., Gkintoni E., & Giannoukou, I. (2021a). An expert system for recommendation tourist destinations: An innovative approach of digital marketing and decision-making process. *International Journal of Innovative Science and Research Technology, 6*(4), 398-404. ISSN: 2456-2165.

Halkiopoulos, C., Antonopoulou, H., & Gkintoni, E. (2021a). Consumer behavioral data analysis for decision making process: Cognitive and emotional factors. *International Journal of Innovative Science and Research Technology*. ISSN: 2456-2165.

Halkiopoulos, C., Antonopoulou, H., Gkintoni, E., & Aroutzidis, A. (2022). Neuromarketing as an indicator of cognitive consumer behavior in decision-making process of tourism destination—an overview. In V. Katsoni, & A. C. Şerban (Eds.), *Transcending Borders in Tourism Through Innovation and Cultural Heritage. Springer Proceedings in Business and Economics*. Cham: Springer. https://doi.org/10.1007/978-3-030-92491-1_41.

Halkiopoulos, C., Antonopoulou, H., Papadopoulos, D., Giannoukou, I., & Gkintoni, E. (2020). Online reservation systems in e-business: Analyzing decision making in e-tourism. *Journal of*

Tourism, Heritage & Services Marketing (JTHMS), *6*(1), 9–16. ISSN: 2529-1947. doi:https://doi.org/10.5281/zenodo.3603312.
Halkiopoulos, C., Dimou, E., Kompothrekas, A., Telonis, G., & Boutsinas, B. (2021b). The e-tour facilitator platform supporting an innovative health tourism marketing strategy. In V. Katsoni, C. van. Zyl (Eds), *Culture and Tourism in a Smart, Globalized, and Sustainable World. Springer Proceedings in Business and Economics*. Cham: Springer. DOI: https://doi.org/10.1007/978-3-030-72469-6_39.
Harrill, R., & Potts, T. D. (2003). Tourism planning in historic districts: Attitudes toward tourism development in charleston. *Journal of the American Planning Association, 69*(3), 233–244. https://doi.org/10.1080/01944360308978017.
Holbrook, M. B., & Hirschman, E. C. (1982). The experiential aspects of consumption: Consumer fantasies, feelings, and fun. *Journal of Consumer Research, 9*(2), 132. https://doi.org/10.1086/208906.
Hosany, S., & Gilbert, D. (2010). Measuring tourists "emotional experiences toward hedonic holiday destinations. *Journal of Travel Research, 49*(4), 513–526. https://doi.org/10.1177/0047287509349267.
Katsoni, V., & Koliopoulos, T. (2020). Innovative airport designs' infrastructures to support an integrated sustainable tourism policy for sustainable development and public health protection. *Turystyka i Rozwój Regionalny, 13*, 43–57. https://doi.org/10.22630/tirr.2020.13.4.
Katsoni, V., & Stratigea, A. (Eds.). (2016). Tourism and culture in the age of innovation. Springer Proceedings in Business and Economics. https://doi.org/10.1007/978-3-319-27528-4.
Kim, Y., Ribeiro, M. A., & Li, G. (2022). Tourism memory, mood repair and behavioral intention. *Annals of Tourism Research, 93*, 103369. https://doi.org/10.1016/j.annals.2022.103369.
Kottler, J. A. (2002). *International Journal for the Advancement of Counselling, 24*(4), 207–210. https://doi.org/10.1023/a:1023610514717.
LaMondia, J., Snell, T., & Bhat, C. (2010). Traveler behavior and values analysis in the context of vacation destination and travel mode choices. *Transportation Research Record, 2156*(1), 140-149.
Larsen, S., & Jenssen, D. (2004). The school trip: Travellingwith, not to or from. *Scandinavian Journal of Hospitality and Tourism, 4*(1), 43–57. https://doi.org/10.1080/15022250410006273.
Liburd, J. J., Benckendorff, P., & Carlsen, J. (2011). Tourism and quality-of-life: How does tourism measure up?. *Handbook of Tourism and Quality-of-Life Research*, 105–132. https://doi.org/10.1007/978-94-007-2288-0_7.
Lopes, T. M., Palrão, T., & Rodrigues, R. I. (2021). Creativity as an opportunity to stimulate a cognitive approach to tourist demand. *Frontiers in Psychology, 12*. https://doi.org/10.3389/fpsyg.2021.711930.
Mannell, R. C., & Iso-Ahola, S. E. (1987). Psychological nature of leisure and tourism experience. *Annals of Tourism Research, 14*(3), 314–331. https://doi.org/10.1016/0160-7383(87)90105-8.
Markgraf, I., Scheffer, D., & Pulkenat J., (2012). The needs of package tourists and travel agents–Neuromarketing in the tourism sector. In *Trends and Issues in Global Tourism* (pp. 79–91). https://doi.org/10.1007/978-3-642-27404-6_8#page-1.
Matatolu, I. (2019). Tourism and residents' quality of life: A critical examination. *The Journal of Pacific Studies, 39*(1). https://doi.org/10.33318/jpacs.2019.39(1)-06.
McCabe, S., & Johnson, S. (2013). The happiness factor in tourism: Subjective well-being and social tourism. *Annals of Tourism Research, 41*, 42–65. https://doi.org/10.1016/j.annals.2012.12.001.
Smith and (Senior Lecturer in Cultural Tourism) & Catherine Kelly (Senior Lecturer in Cultural Tourism), 2006 Smith, M., & Kelly, C. (2006). Wellness tourism. *Tourism Recreation Research, 31*(1), 1–4. https://doi.org/10.1080/02508281.2006.11081241.
Mercille, J. (2005). Media effects on image. *Annals of Tourism Research, 32*(4), 1039–1055. https://doi.org/10.1016/j.annals.2005.02.001.

Moore, K., Smallman, C., Wilson, J., & Simmons, D. (2012), Dynamic in-destination decision-making: An adjustment model. *Tourism management*, 635–645. doi:https://doi.org/10.1016/j.tourman.2011.07.005.

Myers, J. E., Sweeney, T. J., & Witmer, M. (2001). Wellness evaluation of lifestyle. *PsycTESTS Dataset*. https://doi.org/10.1037/t06502-000.

Nopiyani, N. M. S., & Wirawan, I. M. A. (2021). The impact of tourism on the quality of life of communities in tourist destination areas: A systematic review. *Open Access Macedonian Journal of Medical Sciences, 9*(F), 129–136. https://doi.org/10.3889/oamjms.2021.5966.

Panas, G., Heliades, G., Halkiopoulos, C., Tsavalia, D., & Bougioura, A. (2017). Evaluation of athens as a city break destination: Tourist perspective explored via Ddata mining techniques. In V. Katsoni, A. Upadhya, A. Stratigea (Eds), *Tourism, Culture and Heritage in a Smart Economy. Springer Proceedings in Business and Economics*. Cham: Springer. https://doi.org/10.1007/978-3-319-47732-9_6.

Panteli, A., Kompothrekas, A., Halkiopoulos, C., & Boutsinas, B. (2021). An innovative recommender system for health tourism. In V. Katsoni, van C. Zyl (Eds), *Culture and Tourism in a Smart, Globalized, and Sustainable World. Springer Proceedings in Business and Economics*. Cham: Springer. https://doi.org/10.1007/978-3-030-72469-6_42.

Pearce, P. L., & Moscardo, G. M. (1986). The concept of authenticity in tourist experiences. *The Australian and New Zealand Journal of Sociology, 22*(1), 121–132. https://doi.org/10.1177/144078338602200107.

Chen, J. S., Gherissi-Labben, T., & Mungall, A. (2007). *Proceedings of the first hospitality and leisure: Business advances and applied research conference*. Switzerland: Lausanne.

Pyke, S., Hartwell, H., Blake, A., & Hemingway, A. (2016). Exploring well-being as a tourism product resource. *Tourism Management, 55*, 94–105. https://doi.org/10.1016/j.tourman.2016.02.004.

Saracci, R. (1997). The world health organisation needs to reconsider its definition of health. *BMJ, 314*(7091), 1409–1409. https://doi.org/10.1136/bmj.314.7091.1409.

Skavronskaya, L., Scott, N., Moyle, B., Le, D., Hadinejad, A., Zhang, R., Gardiner, S., Coghlan, A., & Shakeela, A. (2017). Cognitive psychology and tourism research: State of the art. *Tourism Review, 72*(2), 221–237. https://doi.org/10.1108/tr-03-2017-0041.

Tung, V. W. S., Lin, P., Qiu Zhang, H., & Zhao, A. (2016). A framework of memory management and tourism experiences. *Journal of Travel & Tourism Marketing, 34*(7), 853–866. https://doi.org/10.1080/10548408.2016.1260521.

Tuohino, A., & Konu, H. (2014). Local stakeholders' views about destination management: who are leading tourism development? *Tourism Review, 69*(3), 202–215. https://doi.org/10.1108/tr-06-2013-0033.

Tzanos, I. A., Kyriakides, A., Gkintoni, E., & Panagiotopoulos, E. (2019). "Quality of life (QoL) of people with spinal cord injury (SCI) in western greece". *Rehabilitation Science, 4*(1), 7–12. ISSN: 2637-594X. doi: https://doi.org/10.11648/j.rs.20190401.12.

Uysal, M., Sirgy, M. J, Woo, E., & Kim (Lina), H. (2016). Quality of life (QOL) and well-being research in tourism. *Tourism Management, 53*, 244–261. https://doi.org/10.1016/j.tourman.2015.07.013.

WHO. (1998). World health organization quality of life-100. *PsycTESTS Dataset*. https://doi.org/10.1037/t59581-000.

Luo, Y., Lanlung (Luke), C., Kim, E., Tang, L. R., & Song, S. M. (2018). Towards quality of life: the effects of the wellness tourism experience. *Journal of Travel & Tourism Marketing, 35*(4), 410–424. https://doi.org/10.1080/10548408.2017.1358236.

Pilot Survey to Investigate Participation in Exercise for All Programs and Sport Events and Their Impact on the Evaluation of the Attica Region

Christina Avloniti, Georgia Yfantidou, and Charilaos Kouthouris

Abstract Sport events are one of the most important and constantly developing parts of events industry. Municipalities organize exercise for all programs, and nowadays, they are involved in the organization of local sport events (Balaska, 2007, 2013). The ability of sport events to attract visitors—tourists, their positive economic impact, but also their wide coverage by the media contribute to the development of destination tourism promotion strategies (Balaska, 2013). Also, sport events improve personal well-being and the quality of life of the participants, as well as they promote an active lifestyle (Alexandris et al, 2012; Balaska & Kouthouris, 2014). Furthermore, sport tourism events are used to enhance the image of the destination (Keller, 1993). suggested that brand image consists of elements with features, benefits, and attitudes toward the "product." Understanding these characteristics would help to identify the elements that determine the image of sport tourism event, which seemed to consist of images related to organization, environment, physical activity, socialization, self-fulfillment, and emotions (Kaplanidou & Vogt 2007). This study attempted to estimate and understand the relationship between the image of a small-scale sport tourism event, its impact on destination image, and the profile of participants. The study involved 203 individuals, who participated in municipal exercise for all programs and sport events in various municipalities of Attica. Participation was voluntary, and participants were informed by public invitation. The questionnaire, which was used, was translated from (Kaplanidou and Kaplanidou, 2006) scale about the effect of sport event image on destination image. The results of this research could help to provide destination marketers and sport event marketers with insight into how various entities should be best combined according always to the participants' profile.

Keywords Sport tourism · Sport city · Sport events · Destination image · Municipalities

C. Avloniti (✉) · G. Yfantidou
Democritus University of Thrace, Komotini, Greece
e-mail: cavlonit@phyed.duth.gr

C. Kouthouris
University of Thessaly, Trikala, Greece

© The Author(s), under exclusive license to Springer Nature Switzerland AG 2023

V. Katsoni (ed.), *Tourism, Travel, and Hospitality in a Smart and Sustainable World*, Springer Proceedings in Business and Economics,
https://doi.org/10.1007/978-3-031-26829-8_42

JEL Classifications L83 · Sports · Gambling · Restaurants · Recreation · Tourism

1 Introduction

The implementation of mass sports programs (Sport for All) is carried out by Local Authorities (Code of Municipalities and Communities, N.3463/2006, Government Gazette A114, No.75). The programs include various sports and recreational activities that take place in outdoor or indoor areas and are aimed at general and special population of all ages. The philosophy of mass sport defines exercise with systematic guidance as an essential social investment because it improves the quality of human life, helps prevent degenerative tendencies, develops interpersonal relationships, and contributes to a better relationship between man and the natural environment. The aim of the programs is to improve the biological level of the participants, to cultivate a sporting character and sporting awareness, to optimize their leisure time, and to provide recreation for the participants (www.gss.gov.gr).

In order to fulfill the need of citizens for sport participation, the General Secretariat for Sports designed the Sport for All Programmes. These programs are implemented under the central responsibility of the Ministry of Culture and Sports through the General Secretariat for Sports (G.G.A.), in cooperation with the Local Authorities and other bodies such as sports federations, sports clubs, educational associations, cultural associations, sports centers, company sports programs, and higher education institutions, and are addressed to all citizens. The participation of adults in exercise programs is limited, as a large proportion of them do not show interest in exercise (Balaska et al., 2008) or are reluctant to start participating in physical exercise activities (Alexandris & Carroll, 1997a, b). For the participants, the two most important reasons for encouraging participation in exercise and recreation programs were to maintain health and fitness and to reduce body weight (Yfantidou et al., 2004; Fakriadou et al., 2009) and furthermore, for the emotions gained by participating in exercise (Balaska et al., 2018).

One of the fastest-growing forms of sport tourism internationally, and in Greece, are city sports events (Horne, 2007; Kaplanidou et al., 2013a, 2013b). It is a form of sports tourism that has not been recorded and studied in depth in Greece, although its contribution to the economic development of local communities can be very important (0). In general, sport activities are useful both for promoting physical activity as a means of strengthening the local economy (Matheson, 2009; Tien et al., 2011) and enhancing the lives of local residents (Bretherton et al., 2016; Frawley & Cush, 2011; Girginov & Hills, 2008; Ramchandani et al., 2015; Veal et al., 2012; Wicker et al., 2013). They can promote the area by developing the brand of the destination, contributing to tourism development and economically strengthening the local community, enhancing tourism in all seasons, and enhancing social and environmental benefits (Kaplanidou & Gibson, 2012; Kaplanidou et al., 2013a, 2013b). The main target of the municipalities that organize sporting events and local events is the citizens and residents of the local communities. Their benefits are multiple for local communities because they reinforce positive attitudes toward sport, promote mass participation in sport and unity of residents, and give them prestige. There is a significant increase in sports participants/tourists these days, who travel to participate in major and smaller sporting events and combine their travel with their sports participation (Funk, 2008; Gibson et al., 2012; Kaplanidou & Vogt, 2007; Ridinger et al., 2012; Ziakas & Costa, 2011).

The purpose of this research is to investigate the relationship between the image of a sporting event on the destination image and the influence of social and demographic factors (such as age and marital status, occupation, and economic level) on the evaluation of destination and sporting event images.

2 Literature Review

The involvement of local authorities in the provision of sport and leisure services has become even more pronounced and massive in recent decades. For this reason, their effectiveness and quality have attracted the interest of the scientific community. A study by Balaska et al. (2017) examined the relationship between the services provided by recreational sports and leisure programs and the degree of involvement of participants in these programs. The modified SERVQUAL scale of Alexandris et al. (2004a) was used to measure service quality, while the Kyle & Chick (2002) scale was used to measure involvement. The strong involvement resulting from their satisfaction with the services provided determines and influences the decision of individuals to participate in the programs. In general, there was low satisfaction with the quality of services provided, particularly in the quality of facilities, exercise conditions, variety, timing and type of programs, and issues related to responsiveness and problem-solving. However, despite the low satisfaction with services, the level of engagement with these sport recreation programs was satisfactory. The researchers concluded that the services provided can be improved through the adoption of new strategies and suggestions to solve specific problems and upgrade services with the ultimate goal of increasing participant satisfaction.

Regarding the factors of participation in the sport and recreation programs of municipal organizations, there were gender differences (Afthinos et al., 2005; Fakriadou et al., 2009), with women considered as important participation factors the functionality of the facilities, variety of programs and easy access, while men considered the interaction with other people as important factors. In contrast, in research by Yfantidou et al. (2011a, 2011b); Giles-Corti & Donovan (2002), there were no differences in men's and women's perceptions of participation factors such as location, appearance of the facilities, and safety. Older people perceived the responsibility of the staff working in the programs as more important (Afthinos et al., 2005; Fakriadou et al., 2009) and educational level did not affect the overall perception of participation factors (Afthinos et al., 2005; Fariadou et al., 2009; Yfantidou et al., 2011a, 2011b).

In addition to the regular sports and leisure programs, sports events are held in sports facilities, which are one of the most important and growing segments of the events industry. In recent years, local authorities, alongside mass sport activities, have started to get involved in the organization of local sports events (Balaska, 2007, 2013). A variety of sports are included in today's sporting events, and their impact is large and multifaceted. They are important for the strategic development of tourism destinations due to their ability to attract visitors, their positive economic impact, and their wide media coverage (Balaska, 2013). They also promote more active living and increase the personal wealth and standard of living of participants (Alexandris et al., 2012; Balaska & Kouthouris, 2014).

Sports tourism events are used to improve the image of the destination, but the brand image elements of an event have not been thoroughly studied. To determine the brand image dimensions of sporting events, research has been conducted in which it was found that consumer interest was increased by conveying specific images, by providing experiences that enhance or alter the image of the event, and by creative use of the brand image of the event's products (event history, uniforms, trophies, teams, stadiums, finals, sports ambassadors) (Chadwick & Holt, 2008; Keller, 1993; Papadimitriou et al., 2016). Existing literature suggests that images of sporting events and destinations could interact with the behavior of sports tourists. In a study by Beerli and Martin (2004), it was argued that motivation influences the affective component of the image, the travel-holiday experience has a significant relationship with cognitive and affective images, and the social and demographic characteristics influence cognitive and affective appraisal of the image. Several researchers aimed to investigate the variables (behavior, motivation, destination characteristics, socio-demographic data) of affective and cognitive image for branding of sports tourism destination. From the results of these studies, it was found that the impact of event image on destination image was significant, but not vice versa. The factors that increased the intention to repeat participation in the event were (a) the tourists' feeling of satisfaction, which was related to the degree of cooperation between destination marketing managers and sport event organizers, and (b) destination image (Kaplanidou & Gipson, 2010; Kaplanidou & Vogt, 2007; Keller, 1993; Yfantidou et al., 2011a, 2011b).

"Sport cities are important for the economic transformation as well as the industrialization of sport. They also provide opportunities for the development of sport and leisure tourism (Gibson, 1998). Modern cities are considered areas that contain "consumption-related activities and tourism, sport, culture and entertainment" (Hannigan, 2003). Some cities, such as Melbourne (Francis & Murphy, 2005) and Birmingham and Sheffield (Smith, 2005), use sport to promote their images. In addition, some cities, such as Manchester, Cardiff, and Dubai, are considered as 'sports cities' due to the special themed sports zones in which their governments create sports facilities (Smith, 2010)" (Avloniti et al., 2022).

3 Methodology

3.1 Sample

The study involved people who participated in municipal sports and recreation programs in various municipalities of Attica. Participation was voluntary and the participants were informed by public invitation at the places where they were exercising. The total sample size was 203 citizens (83.7% women, 15.3% men, and 1% other). The questionnaires were distributed at the sports facilities and were completed in hard copy (n = 110) and via google forms (n = 93).

3.2 Measurement Scales

The questionnaires used were as follows: (a) the negotiation strategies scale of Balaska et al. (2019), derived from the questionnaires of Alexandris et al. (2007) and Marcus et al. (1992c), which includes eleven dimensions: (1) Knowledge Increase, (2) Understanding Negative Effects of Non-Exercise, (3) Self-Motivation, (4) Activation, (5) Socialization, (6) Empowerment, (7) Engagement, (8) Stimulus Creation, (9) Relationship Development for Encouragement, (10) Time, and (11) Financial (Responses recorded on a seven-point Likert-type scale, 1 = never, 7 = forever), (b) Schiffman and Kanuk's (2007) three-dimensional model scale, as translated into Balaska's (2013) doctoral thesis, for capturing attitudes toward sport recreation, consisting of the dimensions: Cognitive, Affective and Intention to Participate, and (c) Kaplanidou's (2006) scale concerning the impact of the image of sporting events on the image of the destination.

3.3 Process

This study was approved by the Research Ethics Committee of the Democritus University of Thrace. The contact with the participants of the sport programs was conducted after approval by the municipal authorities, and outreach was conducted during their participation in the sport and recreation programs from February to April 2022. Questionnaires were distributed and collected at the facilities where the sports and recreation programs were implemented. A total of 180 paper questionnaires were distributed of which 110 were returned completed, 6 were returned blank and 64 were not returned at all. Also, some participants preferred to complete the questionnaire through google forms and provided their email address, where the link of the questionnaire was sent to them and 93 responded.

4 Results

Total Reliability of Scale

The overall reliability was tested through Cronbach α analysis for all variables of the questionnaire and scored a very satisfactory value of 0.91.

Reliability of sub-scales

All sub-scales that were used in the research revealed satisfactory values. In particular:

Three-dimensional model scale for capturing attitudes toward sport recreation Cronbach $\alpha = 0.83$
The scale of intention to participate or continue to participate in the exercise in future Cronbach $\alpha = 0.89$
The negotiation strategies scale Cronbach $\alpha = 0.92$
The image of sporting events on the image of the destination Cronbach $\alpha = 0.895$
The scale of evaluation of the municipality as a destination for sports activities Cronbach $\alpha = 0.87$.

Frequencies

Regarding gender, 83.7% of the survey participants were female (n = 170), 15.3% were male (n = 31), and 1% people did not answer which gender they belong to (n = 2).

Regarding educational level, 42.9% were university graduates, 21.7% were holding master, 17.7% were lyceum graduates, 9.4% were college graduates, 5.4% had Ph.D and 2% were high school graduates, and 0.5% were technical public school and 0.5% other.

As far as occupation, 24.6% of the total sample were civil servants, 21.7% were pensioners, 20.2% were private employees, 11.8% were freelancers, 11.3% were engaged in domestic work, 5.9% were unemployed, 2.5% stated that they did not belong to any of the above-mentioned professional categories and 2% were students.

Concerning income, 53.2% noted that they had a total family annual income of 20.000–60.000€, 38.4% below 20.000€, 1% 60.000–100.000€, and 7.4% did not answer about their income (Fig. 1).

Furthermore, 55.7% were in the 40–59 age group, 32.5% were 60 and over, 10.8% were 17–39 years old and 1% did not respond (Fig. 2).

Also, 61.1% of the total sample were married, 14.8% divorced, 12.3% single/never married, 8.4% widowed, and 3% cohabiting (Fig. 3).

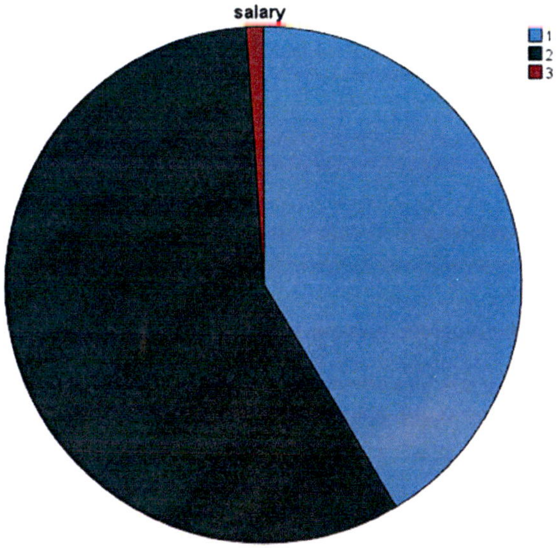

Fig. 1 Total family income: 1. below 20.000 €, 2. between 20.000–60.000 €, 3. over 60.000–100.000 €

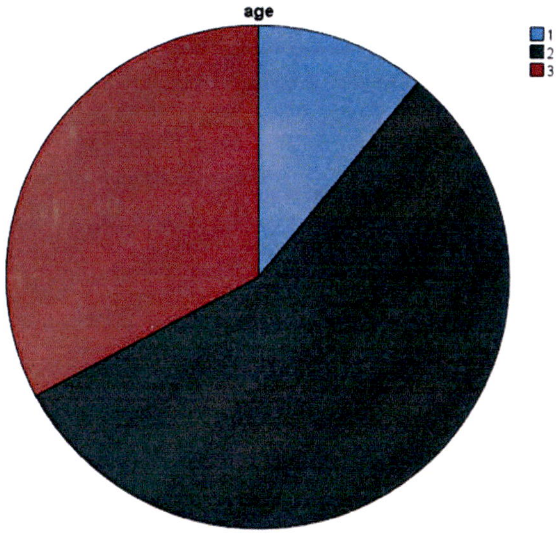

Fig. 2 Age: 1. 17–39 years, 2. 40–59 years, 3. 60 and over

Descriptive Analysis of Municipality's Image Evaluation

Using the descriptive analysis of the questions related to the image evaluation of the municipality developing sports events and sports programs where the participants exercised, the following table shows the mean values and standard deviations (Table 1).

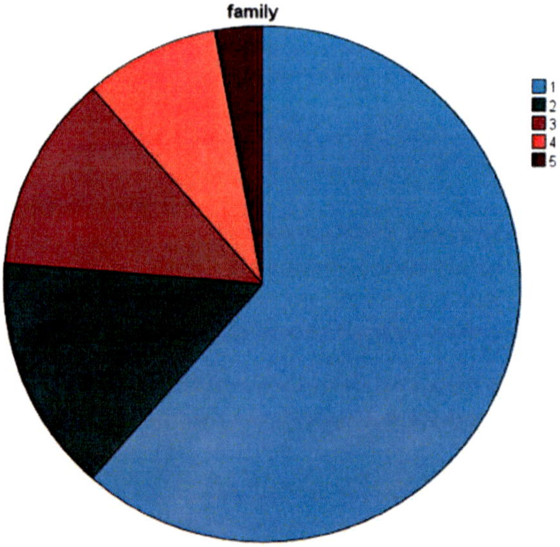

Fig. 3 Family status: 1. Married, 2. Divorced, 3. Singe/never married, 4. Widowed, 5. Cohabiting

Table 1 Mean values and std of municipality's image evaluation

	N	Min	Max	Mean	Std
Opportunities for sport activities	197	1	7	5,56	1,218
Good climate	194	1	7	5,30	1,352
Friendly people	196	1	7	5,15	1,286
Standard hygiene and cleanliness	199	1	7	5,13	1,495
Quality of infrastructure	196	2	7	5,08	1,244
Personal safety	197	1	7	5,05	1,393
Great trails	192	1	7	5,03	1,709
Suitable accommodations	192	1	7	5,02	1,631
Beautiful scenery/natural attractions	192	1	7	4,96	1,776
Good value for money	191	1	7	4,92	1,378
Shopping facilities	190	1	7	4,56	1,565
Good quality restaurants	190	1	7	4,55	1,657
Unpolluted/unspoiled environment	192	1	7	4,52	1,659
Good nightlife and entertainment	194	1	7	3,85	1,664
Interesting historical attractions	190	1	7	2,33	1,875
Great beaches	188	1	7	1,79	1,796
Great museums	187	1	7	1,78	1,433
Valid N (listwise)	171				

Table 2 Mean values and std of municipality's evaluation as a sport destination

	N	Minimum	Maximum	Mean	Std. Deviation
Friendly	190	1	7	4,86	1,640
Pleasant	192	1	7	4,72	1,680
Arousing	196	1	7	4,68	1,483
Exciting	190	1	7	4,33	1,636
Relaxing	188	1	7	4,27	1,837
Valid N (listwise)	184				

Descriptive Analysis of Municipality's Evaluation as a Sport Destination

Using the descriptive analysis of the questions related to the municipality's evaluation as a sport destination, the following table shows the mean values and standard deviations (Table 2).

T Test Between Municipality's Image Evaluation and Gender

According to the results of the questions related to the evaluation of the image of the municipality, the t test of independent variables showed only one effect of gender on the question about the quality of infrastructure with $t_{192} = 3.117$ and $p < 0.05$. The mean values of the answers given by men and women were $M = 5.72$ and $M = 4.96$, respectively.

The t test of the independent variables did not show the effect of gender on the following questions ($p > 0.05$):

- Good nightlife and entertainment with $t_{190} = -0.409$
- Personal safety with $t_{193} = 1{,}392$
- Standard hygiene and cleanliness with $t_{195} = 1{,}892$
- Suitable accommodations with $t_{188} = 1{,}178$
- Good quality restaurants with $t_{186} = 0{,}364$
- Great beaches with $t_{184} = 0{,}243$
- Friendly people with $t_{192} = 0{,}739$
- Great museums with $t_{183} = 1{,}958$
- Interesting historical attractions with $t_{186} = 1{,}329$
- Beautiful scenery/natural attractions with $t_{188} = 0{,}020$
- Good value for money with $t_{187} = 0{,}772$
- Unpolluted/unspoiled environment with $t_{188} = 0{,}788$
- Good climate with $t_{190} = 0{,}184$
- Opportunities for sport activities with $t_{193} = -0{,}108$
- Shopping facilities with $t_{186} = -0{,}064$
- Great trails with $t_{188} = 0{,}014$.

T Test Between Municipality's Evaluation as a Sport Destination and Gender

According to the results of the questions concerning municipality's evaluation as a sport destination, the t test of the independent variables did not show the effect of gender in any of the questions of this scale ($p > 0.05$):

- Exciting, $t_{186} = 0.924$
- Pleasant, $t_{188} = -0{,}252$
- Arousing, $t_{192} = 0.833$
- Relaxing, $t_{184} = -0.416$
- Friendly, $t_{186} = -0.852$.

Interaction and Main Effect of the Factors Age and Family Status on the Evaluation of the Image of the Municipality

Testing the interaction of age and marital status did not reveal any statistically significant difference in any of the variables examined (Table 3).

Testing the main effect of the age factor revealed no statistically significant difference in any of the variables examined (Table 4).

The test of the main effect of the marital status factor revealed no statistically significant difference in any of the variables examined (Table 5).

Table 3 Tests of between-subjects effects—age*family

Questions	$F_{(6,158)}$	p
1. Good nightlife and entertainment	0,407	0,874
2. Quality of infrastructure	1,103	0,363
3. Personal safety	0,972	0,446
4. Standard hygiene and cleanliness	1,540	0,168
5. Suitable accommodations	1,960	0,074
6. Good quality restaurants	0,915	0,486
7. Great beaches	0,373	0,895
8. Friendly people	1,474	0,190
9. Great museums	0,570	0,753
10. Interesting historical attractions	0,492	0,813
11. Beautiful scenery/natural attractions	0,810	0,564
12. Good value for money	1,293	0,263
13. Unpolluted/unspoiled environment	1,196	0,311
14. Good climate	0,828	0,550
15. Opportunities for sport activities	0,876	0,514
16. Shopping facilities	1,148	0,337
17. Great trails	1,244	0,287

Table 4 Tests of between-subjects effects—age

Questions	$F_{(2,158)}$	p
1. Good nightlife and entertainment	1,175	0,312
2. Quality of infrastructure	0,178	0,837
3. Personal safety	0,222	0,801
4. Standard hygiene and cleanliness	0,013	0,987
5. Suitable accommodations	0,551	0,577
6. Good quality restaurants	0,184	0,832
7. Great beaches	0,727	0,485
8. Friendly people	1,025	0,361
9. Great museums	1,892	0,154
10. Interesting historical attractions	0,280	0,756
11. Beautiful scenery/natural attractions	0,330	0,720
12. Good value for money	0,543	0,582
13. Unpolluted/unspoiled environment	1,453	0,237
14. Good climate	2,225	0,111
15. Opportunities for sport activities	0,465	0,629
16. Shopping facilities	0,186	0,830
17. Great trails	1,459	0,236

Table 5 Tests of between-subjects effects—family

Questions	$F_{(4,158)}$	p
1. Good nightlife and entertainment	0,487	0,745
2. Quality of infrastructure	1,137	0,341
3. Personal safety	0,970	0,426
4. Standard hygiene and cleanliness	0,779	0,540
5. Suitable accommodations	0,281	0,890
6. Good quality restaurants	0,597	0,665
7. Great beaches	0,230	0,921
8. Friendly people	0,355	0,841
9. Great museums	0,241	0,915
10. Interesting historical attractions	2,495	0,050
11. Beautiful scenery/natural attractions	0,790	0,533
12. Good value for money	1,223	0,303
13. Unpolluted/unspoiled environment	0,647	0,630
14. Good climate	1,017	0,400
15. Opportunities for sport activities	0,728	0,574
16. Shopping facilities	0,619	0,649
17. Great trails	0,190	0,943

Table 6 Tests of between-subjects effects—job*salary

Questions	$F_{(7,146)}$	p
1. Good nightlife and entertainment	1,695	0,114
2. Quality of infrastructure	0,794	0,593
3. Personal safety	1,225	0,293
4. Standard hygiene and cleanliness	2,348	0,027
5. Suitable accommodations	2,013	0,057
6. Good quality restaurants	1,913	0,071
7. Great beaches	0,629	0,731
8. Friendly people	2,159	0,041
9. Great museums	1,232	0,289
10. Interesting historical attractions	2,142	0,043
11. Beautiful scenery/natural attractions	1,076	0,381
12. Good value for money	0,625	0,734
13. Unpolluted/unspoiled environment	0,488	0,842
14. Good climate	1,218	0,297
15. Opportunities for sport activities	0,864	0,537
16. Shopping facilities	1,367	0,223
17. Great trails	1,302	0,253

Interaction and Main Effect of the Factors Occupation and Income on the Evaluation of the Image of the Municipality

The test of the interaction of the factors occupation and income revealed statistically significant differences with $p < 0.05$ (Table 6):

- In the question on standard hygiene and cleanliness, a statistically significant interaction was found between those who answered that they belonged to the occupational category of private employees with Mean = 5.640 and unemployed with Mean = 3.250, belonging to the category with income of 20,000–60,000 euro.
- In the question on friendly people, a statistically significant interaction was found between civil servants with Mean = 4.667 and freelancers with Mean = 5.750, belonging to the income category below 20,000 euro.
- In the question on friendly people, a statistically significant interaction was found between private employees with Mean = 5,520 and pensioners with Mean = 4,471 who belong to the income category below 20.000–60.000 euro.
- In the question on friendly people, a statistically significant interaction was found between civil servants with Mean = 5.300 and pensioners with Mean = 4.471 who belong to the income category below 20.000–60.000 euro.

- In the question on interesting historical attractions, a statistically significant interaction was found between those who answered that they belonged to the professional category of civil servants with Mean = 1,933 and freelancers with Mean = 4,091 who belonged to the income category of 20.000–60.000 euro.

From the test of the main effect of the occupation factor, there was a statistically significant difference in the question on good nightlife and entertainment and the question on unpolluted/unspoiled environment (Table 7). The main effect analysis identified differences in the question on good nightlife and entertainment between civil servants and those who were engaged in domestic work, with means of 4.81 and 2.39, respectively. In the question about the unpolluted/unspoiled environment, differences were found between civil servants and those who were engaged in domestic work, with averages of 3.82 and 6.27, respectively.

The test of the main effect of the income factor found no statistically significant difference in any question except one. The difference was in the question on great museums between the income category of less than 20.000 euro with Mean = 2.237 and 20.000–60.000 euro with Mean = 1.575 (Table 8).

Table 7 Tests of between-subjects effects—occupation

Questions	$F_{(7,146)}$	p
1. Good nightlife and entertainment	3,399	0,002
2. Quality of infrastructure	1,518	0,166
3. Personal safety	1,369	0,222
4. Standard hygiene and cleanliness	1,683	0,117
5. Suitable accommodations	0,917	0,495
6. Good quality restaurants	1,879	0,077
7. Great beaches	0,904	0,505
8. Friendly people	0,598	0,757
9. Great museums	1,180	0,318
10. Interesting historical attractions	0,558	0,789
11. Beautiful scenery/natural attractions	0,604	0,752
12. Good value for money	1,515	0,166
13. Unpolluted/unspoiled environment	2,115	0,045
14. Good climate	1,083	0,377
15. Opportunities for sport activities	0,862	0,539
16. Shopping facilities	1,048	0,400
17. Great trails	1,376	0,220

Table 8 Tests of between-subjects effects—income

Questions	$F_{(2,146)}$	p
1. Good nightlife and entertainment	0,313	0,732
2. Quality of infrastructure	0,714	0,491
3. Personal safety	0,894	0,411
4. Standard hygiene and cleanliness	0,650	0,524
5. Suitable accommodations	2,152	0,120
6. Good quality restaurants	0,961	0,385
7. Great beaches	1,340	0,265
8. Friendly people	1,064	0,348
9. Great museums	3,645	0,029
10. Interesting historical attractions	2,067	0,130
11. Beautiful scenery/natural attractions	1,349	0,263
12. Good value for money	1,397	0,251
13. Unpolluted/unspoiled environment	0,315	0,730
14. Good climate	0,441	0,644
15. Opportunities for sport activities	0,102	0,903
16. Shopping facilities	1,336	0,266
17. Great trails	1,733	0,180

5 Conclusion

From the averages of the responses relating to the image evaluation of the municipality developing sports events and sports programs where participants practiced, it appeared that there were several opportunities for sporting activities, a fairly good climate, friendly people, standard hygiene and cleanliness, quality of infrastructure, personal safety, great trails, suitable accommodation, beautiful scenery/natural attractions, and good value for money. For shopping facilities, quality restaurants, and unpolluted/unspoiled environment, they responded that it was offered neither very much nor a little to somewhat, and for good nightlife and entertainment, they responded that it was offered a little to neither very much nor a little. Finally, for interesting historical attractions they answered that they were offered very little and the great beaches and great museums were offered very little to none at all. The answers concerning the questions on beaches and large museums were due to the limitation of the survey to municipalities close to the mountainous area of Attica and not to the historical center of Athens where the majority of historical attractions are located.

From the averages of the responses regarding the evaluation of the municipality as a sports destination, it appeared to be quite friendly, pleasant, and arousing and somewhat exciting and relaxing. According to the results of the questions related to the evaluation of the image of the municipality, the effect of gender on the question about the quality of infrastructure was evident. The mean values of the answers given by men were higher than the mean values given by women in terms of the assessment of the infrastructure offered.

The test of the interaction between the factors of occupation and income revealed statistically significant differences in the question on standard hygiene and cleanliness between those belonging to the occupational category of private employees and the unemployed, in the income category of 20,000–60,000 euro. The mean values of the answers given by private employees were higher than the mean values of the unemployed in terms of the assessment of the standard hygiene and cleanliness services offered, maybe because they invested more and they seek for value for money. Also, in the question on friendly people, a statistically significant interaction was found between civil servants and freelancers, who belong to the income category below 20,000 euro, with the latter having higher average values in their answers and this may arise from the social skills that freelancers acquire due to the demands of their work. In the same question, a statistically significant interaction was found between private employees and pensioners in the income category of 20.000–60.000 euro, with the former having higher average values in their responses. In addition, in the same question, a statistically significant interaction was found between civil servants and pensioners in the income of 20.000–60.000 euro, with the former having higher average values in their responses. Maybe pensioners feel difficult to socialize with others due to health reasons or lack of social activities. Finally, statistically significant differences were found in the question on interesting historical attractions between those belonging to the professional category of civil servants and freelancers with an income of 20.000–60.000 euro. The mean values of the answers given by civil servants were lower than the mean values of freelancers with regard to the assessment of interesting historical attractions. Maybe civil servants demand more from the municipalities due to their position at the public sector.

From the test of the main effect of the occupation factor, there was a statistically significant difference in the question on good nightlife and entertainment and the question on unpolluted/unspoiled environment. From the main effect analysis, there were differences in the question on good nightlife and entertainment between civil servants and those who were engaged in domestic work, with higher mean scores in the responses of civil servants. Those who do domestic work is difficult to invest money in nightlife as they fight to survive the years of economic and health crisis. In the question on the unpolluted/unspoiled environment, differences were found between civil servants and domestic workers, with higher averages in the responses of domestic workers.

Finally, the test of the main effect of the income factor revealed a statistically significant difference in one question. The difference was in the question on great museums between the income category of less than 20.000 euro and 20.000–60.000 euro, with higher mean scores in the responses of those in the income category of less than 20.000 euro. The prices in Greek museums are not expensive, so people with less income appreciate more the opportunity to visit museums.

We expect the results of this survey will contribute to improve the quality of municipal sport and leisure programs and events, meeting the needs of participants more effectively and timely. The social benefits are multiple, as it will help to set standards for the extroversion of municipalities to other visitors and to develop more attractive destinations through sport and recreation. The findings can be used by sports and tourism marketers and managers to assess intentions and organize relevant programs and events to attract more tourists and visitors to sport destinations/cities.

References

Avloniti, Ch., Yfantidou, G., & Kouthouris, Ch., & Costa, G. (2022). The development of a sports city as a tourist attraction. In E. Christou & A. Fotiadis, *Reviving tourism in the post-pandemic era* (Vol II, pp. 578-591). International Hellenic University, ISBN: 978-618-5630-06-5.

Afthinos, Y., Theodorakis, N. D., & Nassis, P. (2005). Customers' expectations of service in Greek fitness centers: Gender, age, type of sport center, and motivation differences. *Managing Service Quality: An International Journal, 15*(3), 245–258.

Alexandris, K., & Carroll, B. (1997). An analysis of leisure constraints based on different recreational sport participation levels: Results from a study in Greece. *Leisure Sciences, 19*(1), 1–15.

Alexandris, K., & Carroll, B. (1997). Demographic differences in the perception of constraints on recreational sport participation: results from a study in Greece. *Leisure Studies, 16*, 107–125.

Alexandris, K., Douka, S., & Balaska, P. (2012). Involvement with active leisure participation: Does service quality have a role? *Managing Leisure, 17*(1), 54–67.

Alexandris, K., Zahariadis, P., Tsorbatzoudis, C., & Grouios, G. (2004). An empirical investigation into the role of the outcome dimension in measuring perceived service quality in a health club context. *International Journal of Sport Management, 5*, 281–294.

Alexandris, K., Kouthouris, C., & Girgolas, G. (2007). Investigating the relationships among motivation, negotiation, and alpine skiing participation. *Journal of Leisure Research, 39*(4), 648-667.

Balaska, P., & Kouthouris, C. (2014). Promoting sport participation in Greece: Issues and challenges. *International Journal of Sport Management, Recreation & Tourism, 13*.

Balaska, P., Yfantidou, G., Kenanidis, Th., Spyridopoulou, E., & Alexandris, K. (2019). *Exploring how recreational sport participants with different motivation levels use leisure negotiation strategies*. Lisbon: European Academy of Management "EURAM 2019".1039 (ISSN 2466-7498).

Beerli, A., & Martın, J. (2004). Tourists' characteristics and the perceived image of tourist destinations: a quantitative analysis–a case study of Lanzarote, Spain. *Tourism Management, 25*, 623–636. https://doi.org/10.1016/j.tourman.2003.06.004.

Bretherton, P., Piggin, J., & Bodet, G. (2016). Olympic sport and physical activity promotion: The rise and fall of the London 2012 pre-event mass participation 'legacy.' *International Journal of Sport Policy and Politics, 8*(4), 609–624.

Chadwick, S., & Holt, M. (2008). Releasing latent brand equity: The case of UEFA's champions league. *The Marketing Review, 16*, 147–162.

Francis, S., & Murphy, P. (2005). Sport tourism destinations: The active sport tourist perspective. In J. Higham (Ed.), *Sport tourism destinations* (pp. 73–92). Elsevier.

Frawley, S., & Cush, A. (2011). Major sport events and participation legacy: The case of the 2003 rugby world cup. *Managing Leisure, 16*(1), 65–76.

Funk, D. C. (2008). *Consumer behaviour for sport & events: Marketing action*. Elsevier.

Gibson, H. J.,(1998). Active sport tourism: who participates? *Leisure Studies, 17*(2).

Gibson, H. J., Kaplanidou, K., & Kang, S. J. (2012). Small-scale event sport tourism: A case study in sustainable tourism. *Sport Management Review, 15*(2), 160–170.

Giles-Corti, B., & Donovan, R. J. (2002). The relative influence of individual, social and physical environment determinants of physical activity. *Social Science & Medicine, 54*(12), 1793–1812.

Girginov, V., & Hills, L. (2008). A sustainable sports legacy: Creating a link between the London olympics and sports participation. *The International Journal of the History of Sport, 25*(14), 2091–2116.
Hannigan, J. (2003). Symposium on branding, the entertainment economy and urban place Building. *International Journal of Urban and Regional Research, 27*(2), 352–360.
Horne, J. (2007). The four 'Knowns' of sports mega events. *Leisure Studies, 26*(1), 81–96.
Kaplanidou, K. (2006). *The impact of sport tourism event image on destination image and intentions to travel: A structural equation modeling analysis.* Dissertation thesis, Michigan State University.
Kaplanidou, K., & Gibson, H. (2010). Predicting behavioral intentions of active event sport tourists: The case of a small-scale recurring sports event. *Journal of Sport & Tourism, 15*(2), 163–179. https://doi.org/10.1080/14775085.2010.498261
Kaplanidou, K., & Gibson, H. (2012). Event image and traveling parents' intentions to attend youth sport events: A test of the reasoned action model. *European Sport Management Quarterly, 12*, 3–18.
Kaplanidou, K., & Vogt, C. (2007). *Do sport tourism events have a brand image?.* In: R Burns, K. Robinson, comps. *Proceedings of the 2006 Northeastern Recreation Research Symposium; 2006 April 9-11.* General Technical Report: Bolton Landing, NY. NRS-P-14. Newtown Square, PA: US
Kaplanidou, K., Karadakis, K., Gibson, H., Thapa, B., Walker, M., Geldenhuys, S., & Coetzee, W. (2013). Quality of life, event impacts, and mega-event support among South African residents before and after the 2010 FIFA world cup. *Journal of Travel Research, 52*, 631–645.
Kaplanidou, K., Kerwin, S., & Karadakis, K. (2013b). Understanding sport event success: exploring perceptions of sport event consumers and event providers. *Journal of sport & Tourism, 18*(3), 137−159. *Department of Agriculture, Forest Service, Northern Research Station, 14*, 2−7.
Keller, K. L. (1993). Conceptualizing, measuring, and managing customer-based brand equity. *Journal of Marketing, 57*(January), 1–22.
Kyle, G., & Chick, G. E. (2002). Enduring leisure involvement: The importance of relationships. *Leisure Studies, 23*, 243–266.
Marcus, B. H., Rossi, J. S., Selby, V. C., Niaura, R. S., & Abrams, D. B. (1992). The stage and processes of exercise adoption and maintenance in a worksite sample. *Health Psychology, 11*, 386–395.
Matheson, V. A. (2009). Introduction to the symposium on mega events. *Eastern Economic Journal, 35*(4), 479–480.
Papadimitriou, D., Apostolopoulou, A., & Kaplanidou, K. (2016). Participant-based brand image perceptions of international sport events: The case of the Universiade. *Journal of Convention & Event Tourism, 17*(1), 1–20. https://doi.org/10.1080/15470148.2015.1046100
Ramchandani, G., Davies, L. E., Coleman, R., Shibli, S., & Bingham, J. (2015). Limited or lasting legacy? The effect of non-mega sport event attendance on participation. *European Sport Management Quarterly, 15*(1), 93–110.
Ridinger, L. L., Funk, D. C., Jordan, J. S., & Kaplanidou, K. (2012). Marathons for the masses: Exploring the role of negotiation-efficacy and involvement on running commitment. *Journal of Leisure Research, 44*(2), 155–178.
Schiffman, L. G., & Kanuk, L. L. (2007). *Consumer behavior*. Prentice Hall.
Smith, A. (2005). Reimaging the city: The value of sport initiatives. *Annals of Tourism Research, 32*(1), 217–236.
Smith, A. (2010). The development of "sports-city" zones and their potential value as tourism resources for urban areas. *European Planning Studies, 18*(3), 385–410.
Tien, C., Lo, H. C., & Lin, H. W. (2011). The economic benefits of mega events: A myth or a reality? A longitudinal study on the Olympic Games. *Journal of Sport Management, 25*(1), 11–23.
Veal, A. J., Toohey, K., & Frawley, S. (2012). The sport participation legacy of the sydney 2000 olympic games and other international sporting events hosted in Australia. *Journal of Policy Research in Tourism, Leisure and Events, 4*(2), 155–184.

Wicker, P., Hallmann, K., & Breuer, C. (2013). Analyzing the impact of sport infrastructure on sport participation using geo-coded data: Evidence from multi-level models. *Sport Management Review, 16*(1), 54–67.

Yfantidou, G., Costa, G., & Michalopoulou, M. (2011). Sports tourist destination brands: the case of Greece. *Tourism Today, 11*, 96–109.

Yfantidou, G., Tsernou, A., Michalopoulos, M., & Costa, G. (2011). Criteria of participation at the sport and recreation services of municipal organizations. *International Journal of Sport Management, Recreation & Tourism, 7*, 1–10.

Ziakas, V., & Costa, C. (2011). The use of an event portfolio in regional community and tourism development: Creating synergy between sport and cultural events. *Journal of Sport and Tourism, 16*(2), 149–175.

Ελληνική Βιβλιογραφία

Alexandris, K. (2011). *Principles of Management and Marketing of Organizations and Enterprises of Sports and Leisure.* Thessaloniki: Christodoulides Publications (in Greek).

Balaska, P. (2007). *Application of the Cross Frame Model and the Constraint Factors Model in sports and leisure activities.* Master Thesis, Department of Physical Education and Sport Science. Aristotle University of Thessaloniki (in Greek).

Balaska, P. (2013). *Recreational sports in local government organizations: development factors and strategies for promoting programs.* Doctoral dissertation, Aristotle University of Thessaloniki (Aristotle University of Thessaloniki), School of Physical Education and Sport Science. Department of Physical Education and Sport Science) (in Greek).

Balaska, P., Alexandris, K., Kouthoeris, Ch. (2008). Application of the Diatheoretical model against the participation of individuals in recreational sport activities. *Sport and Recreation Management*, 5(1), 2–16 (in Greek).

Balaska, P., Yfantidou, G. & Kaimakamis, D. (2017). Investigating the relationship between perception of Quality of Service and Participation in Recreational Sports and Leisure Programs. *Hellenic Journal of Sport & Recreation Management*, 14(2), 28–42 (in Greek).

Balaska, P., Yfantidou, G., Papaioannou, A., & Spyridopoulou, E. (2018). Investigating the relationship between attitudes and negotiation strategies in sports recreation activities. *International Review of Services Management (IRSM),* Special, (1), 28–53 (in Greek).

Fakriadou, S., Yfantidou, G., Kosta, G., Mavromatis, G. & Laios, A. (2009). Recording, investigation and evaluation of the services of the sports organization of the municipality of Alexandroupolis. *Sport and Society*, 48, 46–59.

General Secretariat of Sports *http://gga.gov.gr/*

Yfantidou G., Kosta G., Michalopoulou M., Taxildaris K. (2004). Inventory, investigation and evaluation of the services of the Municipal Organization of Youth and Sport of Komotini. *Proceedings of the 12th International Conference on Physical Education and Sport.* Komotini, (May 2004), 231 (in Greek).

Wellness Tourism in the Post COVID-19 Era in Blue Zones and Insular Areas. The Case of the Island of Ikaria

Vakoula Eleonora and Constantoglou Mary

Abstract This article presents a type of tourism, that of wellness tourism, which is developing rapidly in Greece and more specifically aims to examine how wellness tourism evolved during the COVID-19 pandemic, how it is located on the island of Ikaria, and also what is the perspective of potential and current consumers of tourist services on this island. The research indicated that although the pandemic damaged severely the tourism sector, tourists are willing to travel, showing also a preference in the country. The study also showed that the island of Ikaria is a well-known destination with a high percentage of the sample having visited it but without receiving any wellness services and a quite significant percentage not knowing about them. Moreover, the majority of the sample expressed their intention to visit the island and to promote the destination to others and also their strong belief that the staff in the tourism industry will meet the individuals' needs regarding the measures taken due to the pandemic.

Keywords Wellness tourism · Health tourism · CoVid-19 · Blue Zones · Ikaria · Longevity

JEL Classification Z30 · Z32

1 Introduction

Wellness tourism, the phenomenon of traveling to destinations offering a good natural environment in order to improve and/or preserve health, undergoes a rapid development during the last decades, and many researchers have focused on the demand for wellness tourism as well as on strategies for its development. The recent pandemic

V. Eleonora (✉)
MSc Tourism Business Administration, Hellenic Open University, Patras, Greece
e-mail: eleonora-vak@hotmail.com

C. Mary
Department of Economics and Tourism Management, University of the Aegean, Chios, Greece

© The Author(s), under exclusive license to Springer Nature Switzerland AG 2023

V. Katsoni (ed.), *Tourism, Travel, and Hospitality in a Smart and Sustainable World*, Springer Proceedings in Business and Economics,
https://doi.org/10.1007/978-3-031-26829-8_43

of COVID-19, however, has changed the way of traveling and traveling in general as it was known.

According to World Travel & Tourism Council (WTTC), tourism globally accounted for a total of 10,3% of GDP and in Greece specifically, tourism accounted for 20,8% of the country's economy (World Travel & Tourism Council (WTTC), 2020). The implementation of lockdowns, in order to slow down the spread of the coronavirus, eroded the tourism sector, with many countries refusing to accept visitors and many companies going bankrupt. However, the tourism industry seems to recover, since international tourist arrivals during the first five months of 2022 tripled (+221%), although remained 54% below 2019 levels (UNWTO, 2022). Europe and the United States of America lead the recovery of tourism, even though the arrivals are still below those of 2019 (Table 1).

In Greece, in 2021, 96% of the hotels operated as opposed to 2020 when the 60% of them operated and during August the occupancy reached 67,9% as opposed to 27,7% in 2020 (RIT, 2021).

The tourism sector has been hit hard, and it is estimated that the tourists' decisions on the way they travel will change, since from now on they will seek services in trustworthy environments. Instead of returning to previous operating model, the pandemic is forcing experts to rethink the unsustainability of the pre-crisis travel and tourism industry (Constantoglou & Klothaki, 2021). Health and safety will be of top priority in the new era and wellness services will grow rapidly in the post COVID-19 era. Health Tourism Worldwide (HTW) and Wellness Tourism Association (WTA) state that one of the most critical consequences of the pandemic is that personal well-being will be an important issue and travelers will look for suitable destinations and wellness services. Wellness remains an important incentive for travel, while tourism services that will adopt wellness service features will become more important than wellness travel (WTA & Health Tourism Worldwide, 2021).

2 Wellness Tourism

According to World Health Organization (WHO), health does not refer only to the absence of illness, but is a state of physical, mental, and social well-being and integrates the need for beauty, consciousness raising, longevity, and spiritual sensibility (Wang et al., 2020). Wellness is an antithesis of illness, and wellness tourism targets healthy people with an interest in maintaining and enhancing their health (Page et al., 2017). Wellness tourists are attracted by natural environments offering alternative wellness services and activities, such as SPAs, thermal spring resorts, fitness centers, and sacred sites providing spiritual restoration (Lowry, 2017).

The rapid development of wellness tourism has caught the attention of researchers in recent years, who have focused on the factors influencing the decision-making regarding a travel seeking wellness. These factors include socio-demographic and behavioral characteristics, previous experiences of wellness tourists along with other factors enhancing their well-being (Thal & Hudson, 2019). The term well-being is

Table 1 International Tourist Arrivals by (sub) region

Monthly/quarterly data series	(Million)			Share	Change (%)			Change (%)											
				(%)				2022 over 2021*						2022 Over 2019					
	2019	2020*	2021*	2021*	20/19	21/20*	21/19*	YTD	Jan	Feb	Mar	Apr	May	YTD	Jan	Feb	Mar	Apr	May
World	**1465**	**406**	**429**	**100**	**−72.3**	**5.8**	**−70.7**	**221**	**143**	**232**	**218**	**274**	**229**	**−53.9**	**−65.4**	**−58.7**	**−53.8**	**−49.0**	**−45.8**
Advanced economies[1]	*777*	*218*	*227*	*52.8*	*−71.9*	*4.0*	*−70.8*	*401*	*265*	*418*	*414*	*541*	*370*	*−49.9*	*−62.5*	*−52.8*	*−48.7*	*−45.3*	*−44.4*
Emerging economies[1]	*689*	*188*	*203*	*47.2*	*−72.8*	*8.0*	*−70.6*	*122*	*81*	*130*	*116*	*144*	*135*	*−58.1*	*−67.9*	*−64.1*	*−58.8*	*−53.1*	*−47.6*
By UNWTO regions:																			
Europe	***745.2***	***238.1***	***288.4***	***67.2***	***−68.0***	***21.1***	***−61.3***	***350***	***244***	***339***	***356***	***458***	***336***	***−36.4***	***−48.1***	***−36.4***	***−34.2***	***−33.3***	***−33.7***
Northern Europe	83.7	23.3	22.4	5.2	−72.1	−4.1	−73.3	801	392	643	923	↑	847	−35.6	−65.1	−46.1	−32.7	−21.2	−23.1
Western Europe	204.2	79.8	72.4	16.9	−60.9	−9.3	−64.6	545	392	539	545	785	469	−32.1	−50.6	−39.8	−37.1	−21.2	−21.2
Central/Eastern Eur	153.3	46.7	54.7	12.7	−69.5	17.2	−64.3	124	130	146	143	117	99	−51.3	−52.5	−47.0	−47.6	−53.1	−54.4
Southern/Medit. Eur	304.0	88.3	138.9	32.4	−70.9	57.3	−54.3	366	242	362	347	523	362	−31.3	−36.1	−22.3	−24.8	−35.0	−34.4
UNWTO, July 2022 of which EU-27	540.5	181.4	209.5	48.8	−66.4	15.5	−61.2	430	289	427	458	611	384	−36.8	−47.2	−34.9	−33.2	−35.1	−36.1
Asia and the Pacific	***359.6***	***59.2***	***20.6***	***4.8***	***−83.5***	***−65.1***	***−94.3***	***94***	***39***	***96***	***56***	***97***	***179***	***−90.4***	***−93.3***	***−93.9***	***−91.7***	***−88.3***	***−84.2***
North-East Asia	170.3	20.3	11.3	2.6	−88.1	−44.1	−93.3	−9	−14	23	−33	−12	8	−94.4	−94.5	−96.6	−95.6	−93.5	−91.8

(continued)

Table 1 (continued)

Monthly/quarterly data series

	(Million)			**Share** (%)	**Change (%)**			**Change (%)**											
								2022 over 2021*						2022 Over 2019					
	2019	2020*	2021*	2021*	20/19	21/20*	21/19*	YTD	Jan	Feb	Mar	Apr	May	YTD	Jan	Feb	Mar	Apr	May
South-East Asia	138.0	25.5	2.9	0.7	−81.5	−88.8	−97.9	280	109	95	134	381	651	−93.5	−96.5	−97.2	−95.8	−91.9	−84.8
Oceania	17.5	3.6	0.7	0.2	−79.2	−80.2	−95.9	462	301	573	↑	538	296	−82.1	−92.4	−91.6	−83.0	−71.3	−67.6
South Asia	33.7	9.7	5.7	1.3	−71.1	−41.3	−83.1	215	119	143	171	215	677	−59.4	−74.3	−70.8	−59.3	−47.5	−35.4
Americas	***219.3***	***69.8***	***82.4***	***19.2***	***−68.2***	***18.1***	***−62.4***	***112***	***99***	***150***	***117***	***126***	***86***	***−40.1***	***−51.6***	***−44.9***	***−39.6***	***−32.9***	***−32.1***
North America	146.6	46.5	57.0	13.3	−68.3	22.8	−61.1	98	77	123	107	113	80	−39.2	−47.6	−44.8	−39.4	−33.4	−33.3
Caribbean	26.3	10.3	15.0	3.5	−60.8	44.9	−43.2	89	122	156	86	86	38	−18.4	−28.0	−15.6	−21.7	−11.9	−13.9
Central America	10.9	3.1	4.9	1.1	−71.6	58.1	−55.2	157	188	238	144	163	99	−27.4	−46.1	−30.7	−24.8	−17.4	−12.7
South America	35.4	9.9	5.5	1.3	−72.0	−44.5	−84.5	349	198	302	337	524	464	−63.8	−77.3	−69.4	−61.1	−54.2	−48.1
Africa	***68.1***	***18.8***	***19.4***	***4.5***	***−72.5***	***3.3***	***−71.5***	***156***	***56***	***156***	***170***	***198***	***215***	***−49.9***	***−65.8***	***−54.7***	***−47.8***	***−49.1***	***−30.5***
North Africa	25.6	5.6	7.0	1.6	−78.2	25.8	−72.6	238	−20	173	224	459	547	−48.2	−82.3	−55.8	−44.3	−49.9	−10.7
Subsaharan Africa	42.5	13.2	12.4	2.9	−69.0	−6.2	−70.9	127	88	149	147	135	125	−50.7	−59.0	−54.2	−49.6	−48.6	−40.8
Middle East	***73.0***	***19.8***	***18.6***	***4.3***	***−72.9***	***−6.2***	***−74.5***	***157***	***55***	***126***	***172***	***236***	***200***	***−54.4***	***−69.4***	***−63.5***	***−51.9***	***−45.1***	***−43.5***

Source UNWTO (2022)

closely linked to life satisfaction and depends upon life choices, varying between individuals, which influence people's perception of happiness and researchers argue that the concept of well-being is an aspect of a much wider concept, that of Quality of Life (Page et al., 2017).

In tourism terms, Quality of Life is linked to the sustainability, authenticity, attractiveness, and competitiveness of a destination and is evolved around three parts: the tourists themselves, the host community, and the tourism workers. The literature suggests that apart from the life-improving experiences offered to tourists, the carrying capacity of the destination and the enhancement of lives of those working in the tourism industry should be considered (Higgins-Desbioles, 2018). Wellness tourism seems to make a positive remark on the sustainability of the destinations, since it supports the values of the host community, promoting new forms of economic development and managing activities that lead to mental and physical harmony and prosperity (Quintela et al., 2017).

3 Wellness Tourism Economic Figures

Wellness tourism constitutes a rapid developing tourism sector, regarding the consumers' intention to strengthen their mental and physical health, with the global wellness economy valued at $4,9 trillion in 2019 as opposed to $4,4 trillion in 2020 due to the pandemic of COVID-19, since well-being as an endeavor is integrated into all the activities of a person's lifestyle (Global Wellness Institute, 2021).

Globally, wellness tourism shows a growth rate of 6,6% from 2017 to 2019, which is twice as much as the growth rate of general tourism, experiencing a fall by 11% in 2020. Asia–Pacific was the region with the largest wellness spending in 2020, followed by North America and Europe (Table 2).

Due to the pandemic, mental wellness showed evident growth (+7,2%), since consumers requested for products and services that would help them cope with the stress-facing, and also healthy eating, nutrition, and weight loss (3,6%) grew since they sought ways to strengthen their immune system. On the contrary, due to the overall decline in consumption, personal care and beauty services showed a decline (−13%) and also spas (−38,6%) and thermal springs (−38,9%) were adversely affected by COVID-19 due to the restrictions imposed. However, despite the harsh effects of the pandemic, it is estimated that during the period of 2020–2025, the wellness economy will grow at a rate of 9,9% annually, reaching $7 trillion in 2025 (Global Wellness Institute, 2021).

Greece is known as a sea and sun destination. However, the need for the diversification of the product offered, as well as the market demand for the sustainability of the destinations, led to the development of special interest tourism products, where the primary goal was to provide experiences to travelers (Vasileiou et al., 2016). Wellness tourism in Greece is a steadily growing sector of the country's tourism industry, since the resources, the climate, and the coastal and lakeside locations favor a growth focused on the body and mind well-being. It is worth mentioning

Table 2 Wellness economy by region, 2017, 2019, and 2020

	Wellness economy (US$ billions)			Average annual growth rate		
	2017*	2019	2020	Per capita 2020	2017–2019 (%)	2019–2020 (%)
Asia-Pacific	$1,370.5	$1,602.8	$1,500.2	$359.74	8.1	−6.4
North America	$1,288.1	$1,514.0	$1,310.8	$3,566.93	8.4	−13.4
Europe	$1,168.3	$1,288.6	$1,141.5	$1,235.83	5.0	−11.4
Latin America-Caribbean	$311.7	$302.1	$235.4	$360.31	−1.5	−22.1
Middle East-North Africa	$104.7	$120.9	$107.7	$214.53	7.4	−10.9
Sub-Saharan Africa	$74.0	$81.7	$73.7	$64.74	5.1	−9.8
World	$4,317.3	$4,909.9	$4,369.3	$563.44	6.6	−11.0

Source Global Wellness Institute (2021)

that 1 out of 5 people working in the tourism sector is employed in wellness tourism and the country is seventh among the European countries regarding wellness travels (Global Wellness Institute, 2015).

4 Case Study Area: The Island of Ikaria

The island of Ikaria belongs to the North Aegean region, with an area of 254.41 sq.km. and a population of 8.555 inhabitants, regarding the 2021 census. Its capital is Agios Kirikos, named after the Greek myth of Daedalus and Icarus. It is one of the biggest islands in the North Aegean region with a remarkable historical and cultural civilization, including archeological findings, traditional architecture, and a local dialect. The island is mainly mountainous with many rare species of flora and fauna (Fig. 1).

The inhabitants' main occupations are tourism, agriculture, and fishing, and some of the products for which the island is known are red wine with the name "Pramneios Oinos", olive oil, and honey. Ikaria is also known worldwide for the longevity of its inhabitants and for the "slow" rhythm of everyday life—a pole of attraction for the tourists who seek peace of mind.

According to Buhalis et al. (2022), destinations are the raison d'être for travel and tourism and by amalgamating cultures, nature, traditions, and humanity with private and public organizations, they offer transformative tourism experiences based on environmental, sociocultural, and economic resources. Ikaria is a wellness tourism destination due to the existence of thermal springs, with one of the natural resources recognized officially for its healing properties. Ikaria has a total rent capacity of 24 hotels, 1* to 3* hotels, and specifically 6 1-star hotels, 10 2-star hotels, and 8 3-star hotels and a total amount of 570 rooms and 1.096 beds. In Ikaria, there are also 89 units of rental rooms varying from 1st to 3rd category, and specifically 14 units of

Fig. 1 The island of Ikaria, Greece. *Source* Google Earth (2022)

1st category, 68 of 2nd and 7 of 3rd, and a total amount of 589 rooms and 1.235 beds and the average tourist expenditure per visit is 559€.

During 2021, the arrivals of inbound tourists were 2.896 people and those of domestic tourists were 5.539 people, with 11.309 overnight stays for the first group and 15.102 overnight stays for the second group, respectively, and an occupancy percentage of 18.4% (INSETE, 2022) (Fig. 2).

5 Blue Zones and the Ikarian Longevity

There are few regions in the world, regarded as places with a high percentage of centenarians, compared with other surrounding regions. These regions are labeled Blue Zones, and they are isolated or nearly isolated places where the inhabitants share homogeneous genetic background, traditional lifestyle, and dietary habits (Poulain, Buettner & Pes, 2020). Up until now, five regions are considered as Blue Zones and are accepted by scientists as an effective model of healthy aging transferable to more societies: the island of Okinawa in Japan, the island of Ikaria in Greece, a mountainous area on the island of Sardinia in Italy, the peninsula of Nicoya in Costa Rica, and the city of Loma Linda in California (Marston et al., 2021; Poulain, Buettner & Pes, 2020). The high percentage of people that live healthy and lead active lives beyond 100 years of age without serious disease and with a close relationship to

Fig. 2 Ikaria island capacity. *Source* INSETE (2022)

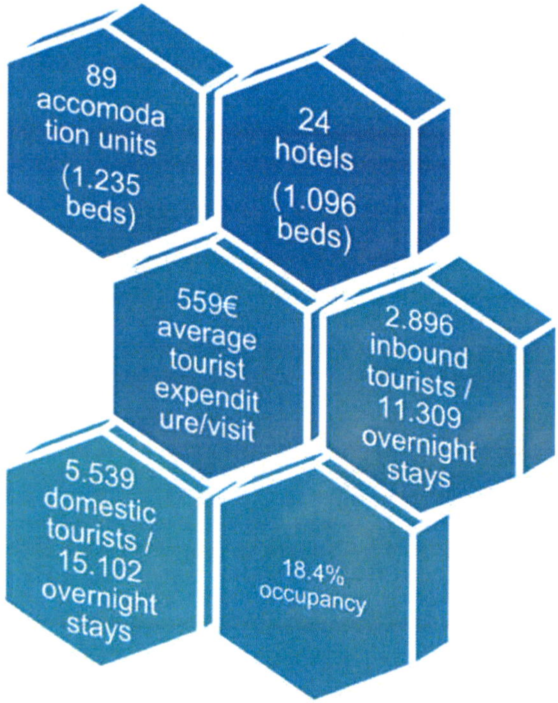

friends and families led researchers trying to identify the factors associated with the longevity of the population in these specific areas (Poulain, Buettner & Pes, 2020), and they came up with 9 principles contributing to these areas unique characteristics, forming a triangle with natural movement at the top and dietary habits and the sense of belonging at the bottom (Marston et al., 2021).

Natural movement is considered as extremely important since immobility fosters decline whereas 20 min of exercise benefits a person a lot. Good physical health is equally important as emotional well-being and eliminating stress in life are found to be vital for centenarians, who keep a balance between having a purpose in life and not letting life become so serious. The eating habits of the population in these regions were investigated thoroughly, and it was found that the dietary habits contribute positively to their overall well-being. People in these areas seem to follow the Mediterranean diet: they use olive oil which is found that benefits the heart, they have a moderate intake of alcohol, and they seem to consume food only until they are 80% full. The sense of belonging is crucial and the centenarians have an active social life, perform their religious duties, and enjoy spending time with friends and family, whereas not a small percentage of them live under the same roof with their grandchildren (Beer, 2022).

In the island of Ikaria in Greece, people live more than the age of 90 years and more than 30% of the deaths are above that age and the death from natural causes is ten years greater than other parts of Greece and the world (Stefanadis, 2011). Researches have shown that the traditional way of living of its inhabitants, as well as the healthy Mediterranean diet, the physical activity status, the active social life, and an optimistic attitude of life, contributes to the longevity of the people there.

According to studies, the longevity of people in Ikaria relies also on the commonest occupation there, farming, which made it possible for oldest aged people to continue to work. Apart from that, researchers also found a relationship between the marital status and family solidarity with the low rates of depression among people there. The high level of social interaction is considered also important, since it appears to have beneficial effects on health, reducing mortality risk and improving mental health. The island of Ikaria is also known for the "Panigiria", festive gatherings related to religious events, organized throughout the year in the different villages of the island, where the participants celebrate and dance all together, the participation which could have a positive influence on health (Legrand et al., 2021).

Ikaria is characterized by the presence of radioactive hot springs, an environmental factor influencing human health (Chrysohoou et al., 2013). The natural radioactivity in the surface of the island was measured, and it was found that the levels of radioactivity together with the consumption of potable water, the spring called "immortal water", contribute to the radiation overdose for the population, an overdose though, which is not linked to mortality but to the longevity of the population living in the area.

As mentioned above, locals follow the Mediterranean diet, which is associated with protection from coronary heart disease and is characterized by a high intake of vegetables and olive oil, together with fish and dairy products intake and a moderate consumption of res wine. What is more, the inhabitants also show a moderate or high level of physical activity with the benefits of the regular physical activity in older adults being extensive (Legrand et al., 2021).

6 Methodology

The rationale of this research was to process data regarding the sector of wellness tourism on the island of Ikaria and attempt a prediction for the future, after the pandemic of COVID-19. The purpose of the sample examination was to clarify whether the field of wellness tourism is known to those tourists who have visited the island and to its residents, and the use of its properties before and after the appearance of the pandemic.

Regarding the local wellness tourism, the thermal springs, the existing facilities, and the possible use by visitors, as well as any facility and/or accommodation that falls under the category of wellness, were taken into account.

Not only that, but also the sample's view was asked, as to whether and how the pandemic would affect the tourism product and to what extent its choices will be affected in future in relation to the new situation.

In order to undertake this research, the quantitative approach was chosen. This type of survey is based on the collection of information resulting from data measurements, which can be either quantitative or qualitative (Constantoglou & Klothaki, 2021). The main survey instrument in a quantitative survey is the structured questionnaire (Creswell, 2011), which was used in this survey and will be presented in the results section. The majority of the questions were closed-ended, and especially belonging to the Likert scale. The reasons that led to the choice of this type of questions were the possibility they provide to the researchers for easy comparison of the answers and their ease of coding. The fact that they do not provide the freedom of answers to the user is mentioned as an important disadvantage of this type of questions (Creswell, 2011) and this is one of the limitations of this study. Prior to the drafting of the final questionnaire, a pilot questionnaire was distributed to 5 visitors of the island, who did not participate in the final survey, in order to identify possible ambiguities and duplications in the questions. Finally, the statistical analysis of the data was implemented with the statistical program SPSS v22., using appropriate Descriptive Statistics instruments.

The questionnaire was created in google forms and sent to the study's participants by email. Respondents were asked to respond exactly to the same questions, as part of a research strategy, in order to collect descriptive and explanatory data regarding views, attitudes, characteristics, etc. The sample consisted of all age groups, since wellness tourism concerns an extremely large age range, in order to be achieved a broader picture in the examination of results. The limitations encountered relate to the approach of the sample due to the limitations in interpersonal relationships. In the approach of the sample, the contribution of the local tourism market players was considered significant, since they acted as a communication channel on the one hand, and as facilitators of the research on the other, during a time of restrictions, such as the quarantine imposed for health reasons. Considering the aforementioned factor, a sample of 70 people was formed through random sampling using questionnaires, out of a total of 150 questionnaires sent via email.

7 Results

The survey examined a mixed sample of men (37,1%) and women (62,9%), of all age groups (Fig. 3) with a higher percentage (37,1%) of those between 31 and 40 years old. The 45,7% of the sample are married with children (Fig. 4) and the 37,1% of the respondents are single, whereas the majority of the sample (44,3%) have received higher education, while the minority (1,4%) received compulsory education (Fig. 5).

Forty people among the respondents believe that tourism has been severely affected by the pandemic (Fig. 6), while none of them thinks that the industry has not been affected at all.

In the question regarding whether the respondents are aware of wellness tourism, a quite significant percentage of them (34,3%) are not aware of what wellness tourism includes—a fact demonstrating the need for communication and advertising to the

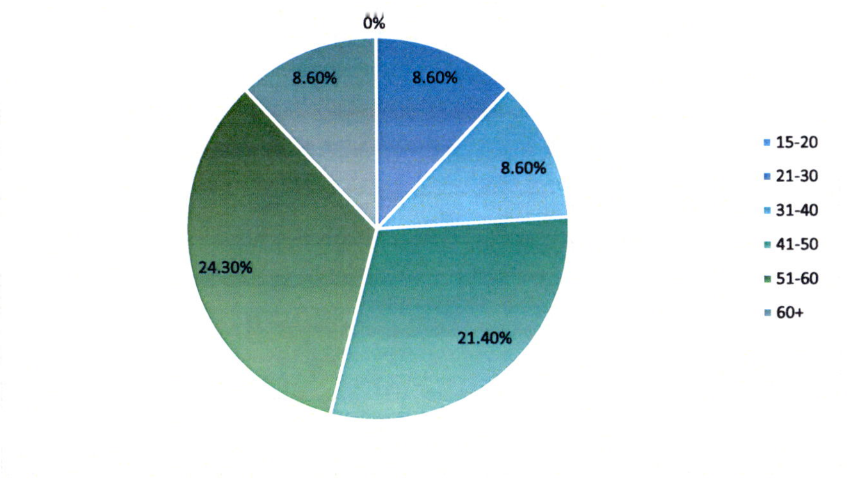

Fig. 3 Age group

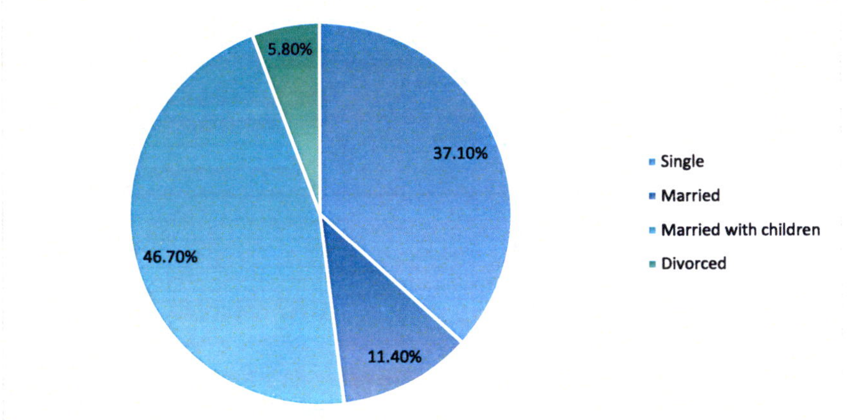

Fig. 4 Marital status

general public. Of the 65,7%, who are aware of wellness tourism, the 26,8% of them have opted for such services in the past and those who have no experience of wellness tourism services so far intend to choose a destination providing services such as thermal springs, spas, fitness centers, and resorts promoting both mental and physical health in future (89,2%).

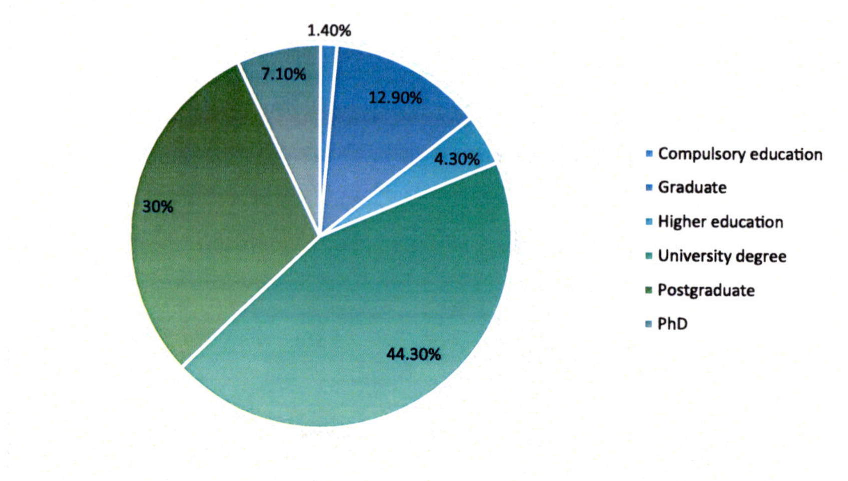

Fig. 5 Education

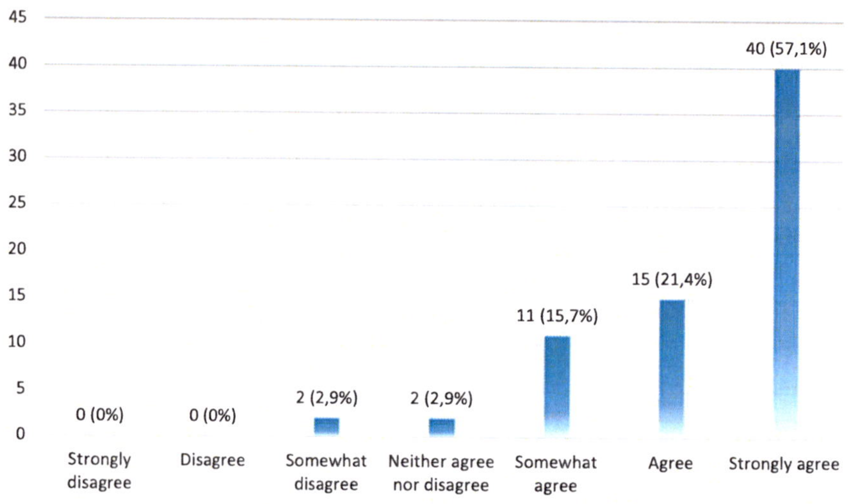

Fig. 6 The tourism industry has been affected negatively by the pandemic of COVID-19

An interesting fact is that tourists intend to travel to Greece by 84,3%, compared to 15,7% that would prefer a destination abroad (Fig. 7)—a fact that highlights the country's dynamic in the tourism sector, as well as in future tourism development.

The public is ready and willing to receive wellness services, without specific requirements in mind from the destination (41,4%), while 24 people stated that they prefer a remote destination for this purpose (34,3%) (Fig. 8).

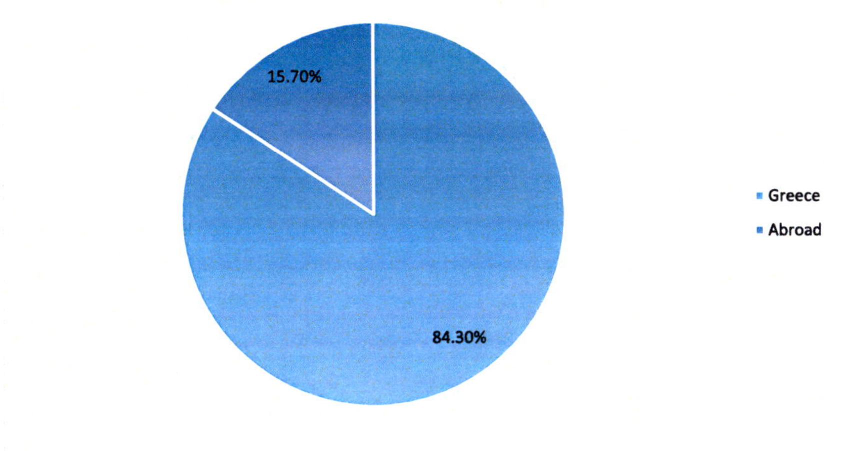

Fig. 7 Greece as a wellness destination

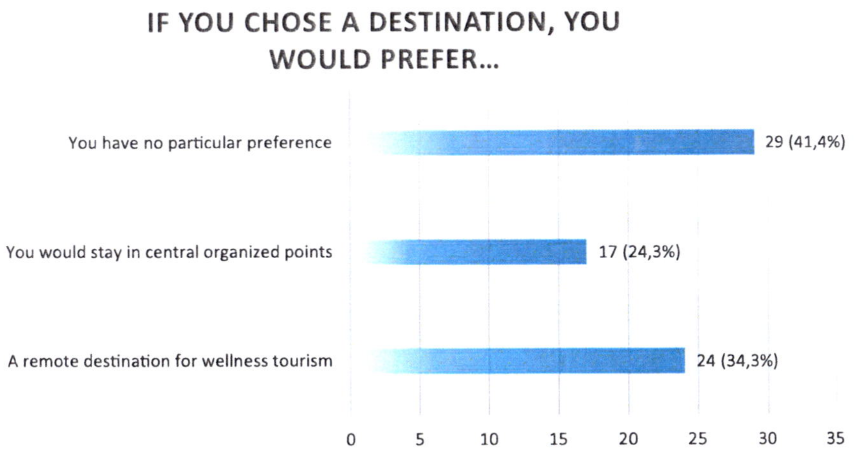

Fig. 8 Choosing a destination

Regarding the intention of travel during the season of 2021 (2nd year of the pandemic), 77,1% of the sample intend to travel (Fig. 9).

As far as the island of Ikaria is concerned, 80% of the sample (Fig. 10) have already visited sites providing wellness tourism services; however, 42,9% are not aware of the facilities existing on the island and only 7,1% knowing them very well (Fig. 11). The need of promoting the island and its services is evident here, in order to strengthen its brand awareness.

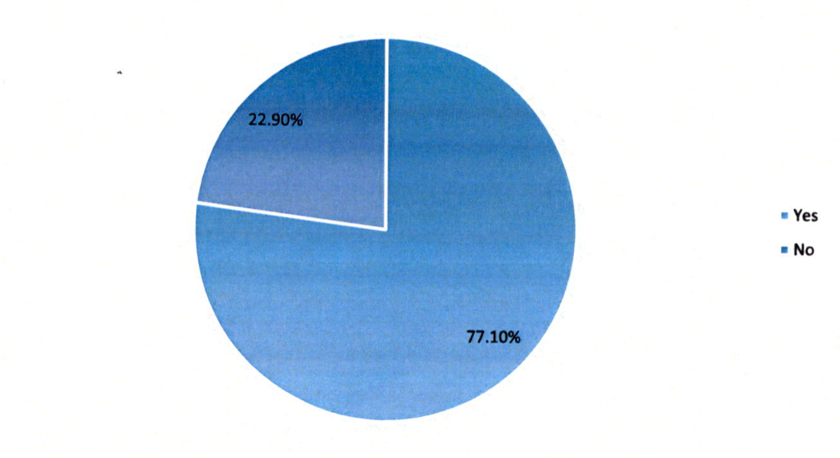

Fig. 9 Intention of traveling in 2021

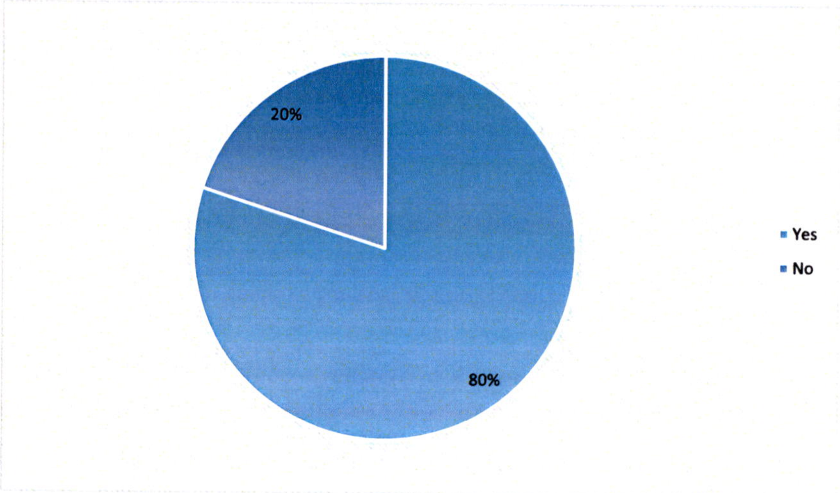

Fig. 10 Previous visit in wellness-related sites on the island of Ikaria?

About 76,8% agree that Ikaria is a well-known destination, and 44,9% confirm that the island has a good reputation, while 37,7% admit knowing the special features and activities someone can deal with there. However, only 8,7% agree that the level of services provided there is high (Fig. 12).

When it comes to whether the respondents would suggest others to visit the island (Fig. 13), 58,3% intend to do so, and the 38,3% would visit the island again, while

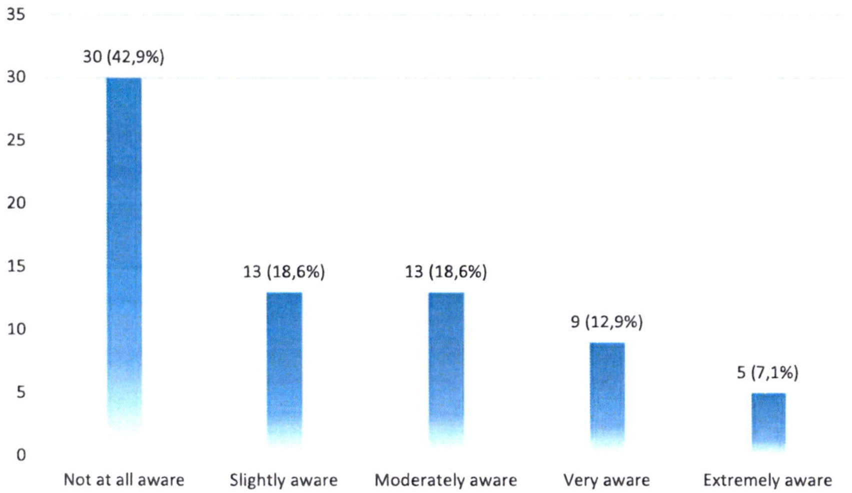

Fig. 11 Awareness of the tourist facilities on the island of Ikaria

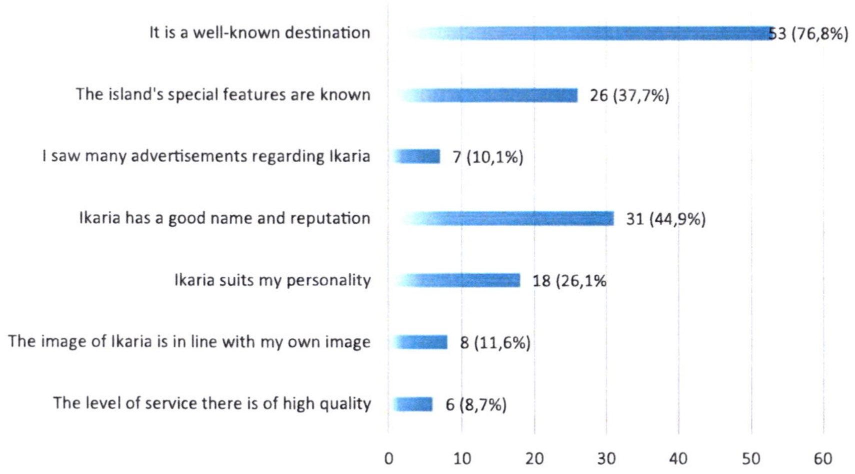

Fig. 12 Beliefs about the island of Ikaria

43,3% express their intention to visit Ikaria even if prices increase. The latter case shows a competitive advantage of this place, as it gathers unique features of interest to a part of the public willing to pay more if they have to in order to visit the island again.

Regarding the sample's thoughts about how the next tourism season will end up on the island, 77,1% of them are optimistic that during the next season, tourism on the island will recover.

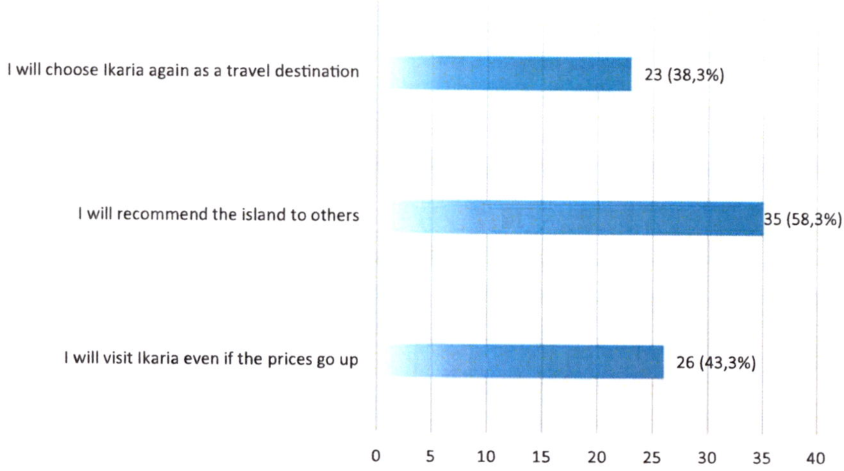

Fig. 13 Intention to repeat the visit in Ikaria in future

About 40% of the respondents express a moderate view on whether the health protocols set by the State affect negatively their accommodation, whereas 5,7% believe that it will not be affected and 14,3% that it will be affected a lot (Fig. 14).

The majority of the respondents believe that hotel owners will comply to the measures set by the State, in order to prevent the spread of COVID-19, with an only small percentage of them (5,7%) expressing the opposite view (Fig. 15).

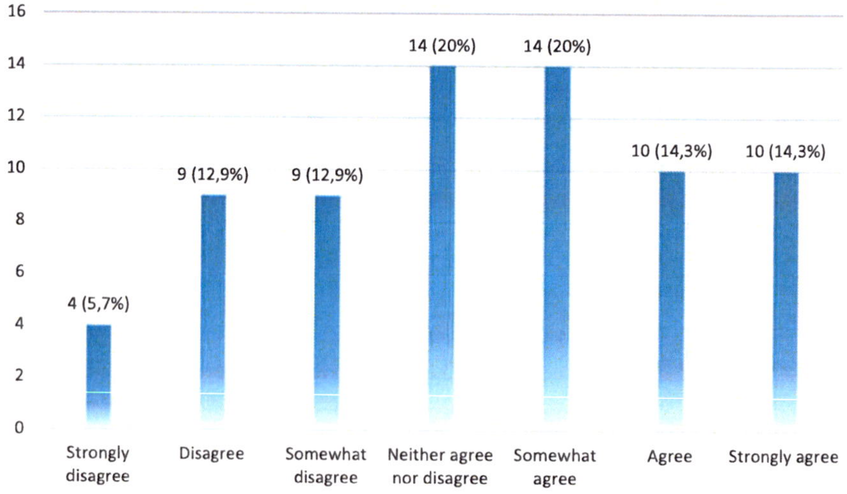

Fig. 14 Beliefs whether the health protocols set by the State affect the accommodation

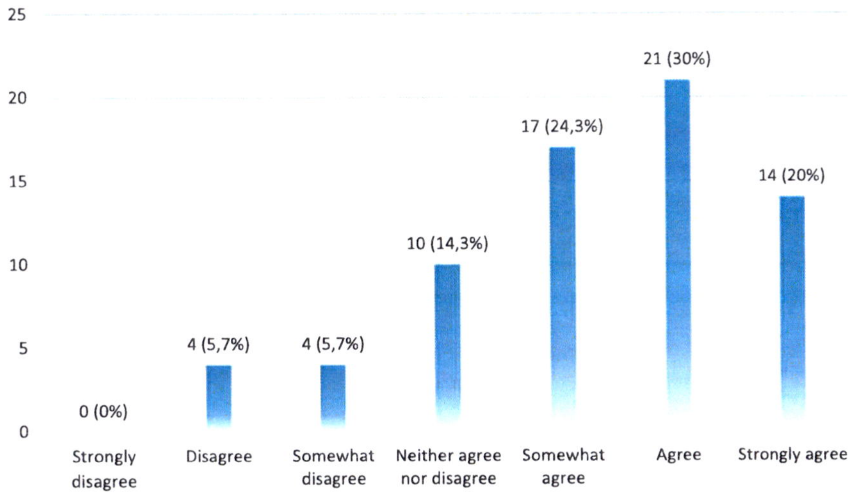

Fig. 15 Compliance of the hoteliers to the health protocols set in order to prevent the spread of COVID-19

Considering the staff's response to the special health and safety conditions, the respondents believe that the staff will respond to the fullest extent (Fig. 16).

Not only that, but also the vast majority of the sample believes that the staff will meet quickly the customers' needs, so that they do not waste extra time due to the procedures imposed by the pandemic (Fig. 17).

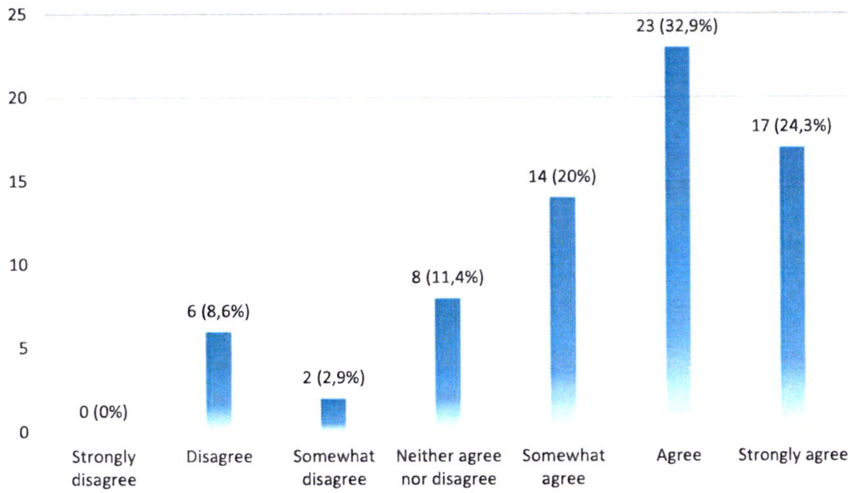

Fig. 16 Compliance of the staff to the health protocols set in order to prevent the spread of COVID-19

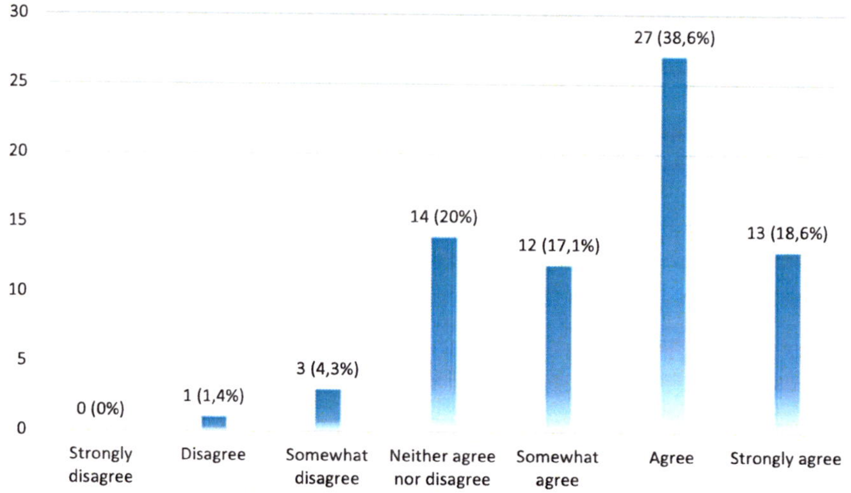

Fig. 17 The staff will meet the customers' needs so as not to waste extra time

The respondents think highly of the staff since they also think in a great percentage that the staff will be able to respond to individuals' needs while at the same time will meet the requirements of the pandemic (Fig. 18).

Finally, a mediocre view is expressed regarding whether the facilities of Ikaria have an up-to-date character and whether they meet the expectations of wellness

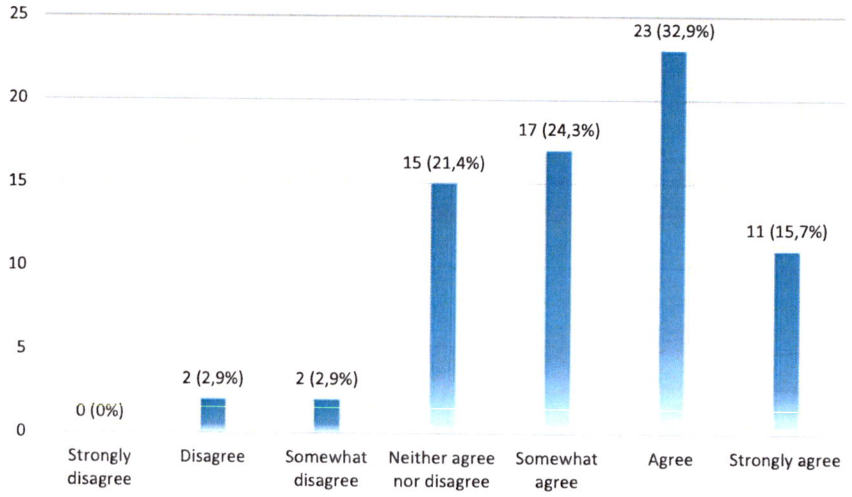

Fig. 18 Response of the staff to individuals' needs

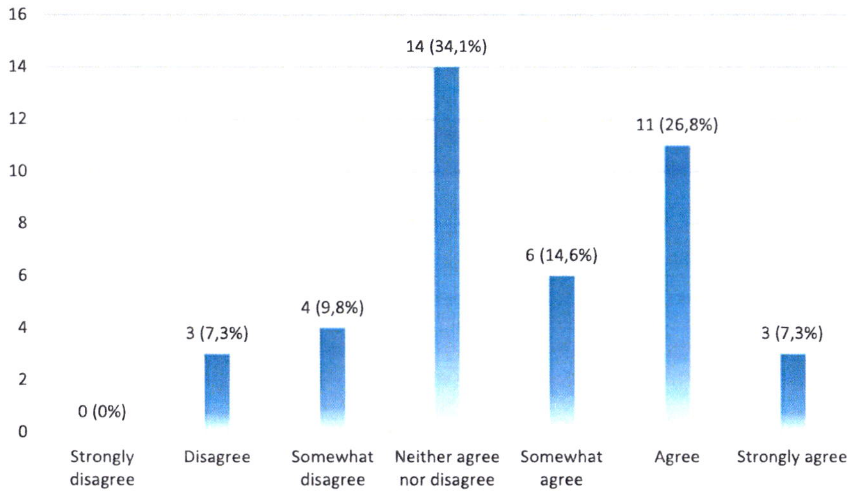

Fig. 19 The facilities on the island of Ikaria are up to date

tourists, with only the 7,3% agreeing that facilities on the island of Ikaria are up to date (Fig. 19).

The island of Ikaria has several thermal springs. One of them, "Apollo", in the area of Therma, was recognized by the Greek Government as medicinal in 2019, and although the Government voted for a law that promotes the modernization of the hot springs with a simultaneous attraction of private investments, the respondents who are familiar with the tourist facilities of Ikaria do not completely agree or disagree that the infrastructure of the thermal springs has been entirely renovated (Fig. 20).

8 Conclusion

The research carried out on the island of Ikaria demonstrated the need for modernization, communication, and promotion of the island to the general, potential tourist public, and also the need to upgrade the services, so that the tourists enjoy a complete travel experience, which will motivate them to travel again and increase the tourist expenditure, contributing to the overall prosperity and development.

The sample thinks that the pandemic of COVID-19 affects tourism, but believes firmly that the people working in the tourism industry will try to meet the unprecedented needs and requirements.

The island of Ikaria has several natural thermal springs, and one of them, "Apollo", is recognized by law by the Green government as medicinal, a recognition that adds value to the island. However, the infrastructure of the thermal springs, apart from being functional, could have been more attractive for the general public and compete

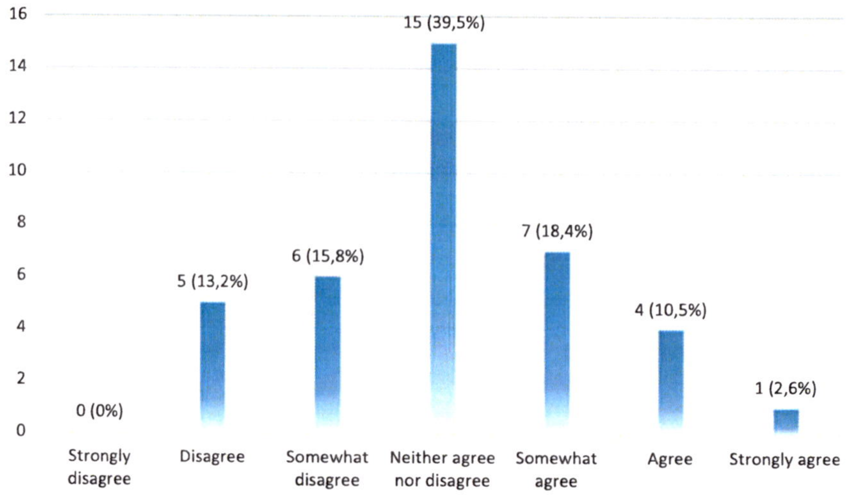

Fig. 20 The infrastructure of the thermal springs is completely renovated

other destinations that invested in their tourism product. Their landscaping would add value, as the tourist perceives wellness as an overall experience—physical, mental, and visual—concepts completely intertwined.

Recently, the Greek government, taking into consideration that tourism yields a lot of income to the country and that the natural springs are a valuable asset to it, has voted for the establishment of a limited liability company under the name "Thermal Springs of Greece" with the main purpose being the exploitation of the natural resources and springs of the country, as part of the property of the Green State. This company will be responsible for the administration, management, and exploitation of the natural thermal resources and springs, as well as their facilities and the environment around, in a radius of five hundred (500) meters, and the government with this law seeks to promote wellness tourism, a rapidly growing sector with multiple benefits for the tourism economy.

An organized tourism campaign supported by the power of social media, as well as with the help of the state's promotion of local destinations, will strengthen the tourism economy, which strives to offer quality services again.

The island of Ikaria is able to compete other well-known and well-publicized tourist destinations, and with its unique characteristics, it constitutes an excellent place with unique elements, fully exploitable for tourist purposes.

Finally, it would be of interest for further investigation regarding Ikarian hoteliers and their intention to promote tourism on their island and how familiar they are with the use of the new marketing tools, as well as what the intention of local actors is, in order to modernize and upgrade the tourist infrastructure.

References

Beer, G. O. (2022). *Blue zones: Unlocking key themes in the centenarian's life*. Honors College Theses, p. 115.

Buhalis, D., Constantoglou, M. (2022). Destination typologies. In D. Buhalis (ed.), *Encyclopedia of tourism management and marketing*. Edward Elgar Publishing, Incorporated.

Chrysohoou, C., Panagiotakos, D. B., Pitsavos, C., Bassiakos, Y., Lazaros, G., Tousoulis, D., & Stefanadis, C. (2013). Exposure to low environmental radiation and longevity. Insights from the Ikaria Study. *International journal of cardiology, 169*(6), e97–e98.

Constantoglou, M., & Klothaki, T. (2021). How much tourism is too much? Stakeholder's perceptions on overtourism, sustainable destination management during the pandemic of COVID-19 Era in Santorini Island Greece. *Journal of Tourism and Hospitality Management, 9*(5), 288–313. https://doi.org/10.17265/2328-2169/2021.05.004

Global Wellness Institute. (2015). *Wellness tourism will be crucial to Greece no matter how crisis plays out*. Retrieved January 10, 2021, from https://globalwellnessinstitute.org/global-wellness-institute-blog/2015/07/08/2015-7-7-wellness-tourism-will-be-critical-to-greece

Global Wellness Institute. (2021). *The global wellness economy: Looking beyond COVID*. Retrieved August 10, 2021, from https://globalwellnessinstitute.org/industry-research/the-global-wellness-economy-looking-beyond-covid/

Health Tourism Worldwide. (2021). *Wellness Travel 2030, Post-COVID19: A pioneering study*. Retrieved April 22, 2021, from https://htww.life/download/wellness-travel-2030-a-pioneering-study-full-report/

Higgins-Desbiolles, F. (2018). Sustainable tourism: Sustaining tourism or something more? *Tourism Management Perspectives, 25*, 157–160.

INSETE. (2022). *Tourist statistics of the North Aegean region*. Retrieved August 10, 2022, from https://insete.gr/perifereies/

Legrand, R., Nuemi, G., Poulain, M., & Manckoundia, P. (2021). Description of lifestyle, including social life, diet and physical activity, of people≥ 90 years living in Ikaria, a Longevity Blue Zone. *International Journal of Environmental Research and Public Health, 18*(12), 6602.

Lowry, L. (2017). The SAGE international encyclopedia of travel and tourism. SAGE Publications, Inc.

Marston, H. R., Niles-Yokum, K., & Silva, P. A. (2021). A commentary on blue Zones®: A critical review of age-friendly environments in the 21st century and beyond. *International Journal of Environmental Research and Public Health, 18*(2), 837.

Page, S. J., Hartwell, H., Johns, N., Fyall, A., Ladkin, A., & Hemingway, A. (2017). Case study: Wellness, tourism and small business development in a UK coastal resort: Public engagement in practice. *Tourism Management, 60*, 466–477.

Poulain, M., Buettner, D., & Pes, G. (2020). Blue zones. In *Encyclopedia of biomedical gerontology* (vol. 1, pp. 296–305). Elsevier, Academic Press.

Quintela, J. A., Costa, C., & Correia, A. (2017). The role of Health and wellness tourism in sustainable territorial development. *Tourism and Hospitality International Journal, 9*(2), 113–121.

RIT. (2021). Performance of Greek Hotels in 2021. Retrieved August 10, 2022, from https://www.itep.gr/wp-content/uploads/2022/01/itep-annual-survey-2021-public-en.pdf

Stefanadis, C. I. (2011). Unveiling the secrets of longevity: The Ikaria study. *Hellenic Journal of Cardiology, 52*(5), 479–480.

Thal, K. I., & Hudson, S. (2019). A conceptual model of wellness destination characteristics that contribute to psychological well-being. *Journal of Hospitality & Tourism Research, 43*(1), 41–57.

UNWTO. (2020). Tourism back to 1990 levels as arrivals fall by more than 70%. Retrieved December 27, 2020, from https://www.unwto.org/news/tourism-back-to-1990-levels-as-arrivals-fall-by-more-than-70

UNWTO. (2022). World tourism barometer and statistical annex. World Tourism Organization, 20 (4).

Vasileiou, M., Tsartas, P., & Stogiannidou, M. (2016). Wellness tourism: Integrating special interest tourism within the Greek tourism market. *Tourismos, 11*(3), 210–2206.

Wang, K., Xu, H., & Huang, L. (2020). Wellness tourism and spatial stigma: A case study of Bama, China. *Tourism Management, 78*, 104039.

World Travel & Tourism Council (WTTC). (2020). To recovery & beyond, the future of travel & tourism in the wake of CoVid-19. Retrieved January 9, 2021, from https://wttc.org/Portals/0/Documents/Reports/2020/To%20Recovery%20and%20Beyond-The%20Future%20of%20Travel%20Tourism%20in%20the%20Wake%20of%20COVID-19.pdf?ver=2021-02-25-183120-543

Brand and Competitiveness in Health and Wellness Tourism

Eleonora Santos, Cátia Crespo, Jacinta Moreira, and Rui Alexandre Castanho

Abstract Health and wellness tourism is growing worldwide, triggering a resurgence of competition in this segment and a clear need for the supply to develop new products and services, increasingly creative and exclusive, capable of attracting tourists. Thermal waters in Portugal are considered very rich in terms of chemical composition, which makes their use highly important in various wellness and rejuvenating therapies like balneotherapy, thermalism, etc. This enables the country to pursue a differentiated wellness-related experience product, which is a key determinant of customer loyalty and can play an important role in regional development. Natural resources contribute to the strengths of territories and can be leveraged by marketing tools such as brands. This article intends to draw inferences about the effect of the brand on the competitiveness of the wellness tourism supply, across NUTS II regions of the continent, during the period of ten years, with a special focus on the recent period of the COVID-19 pandemic. With a sample of 10 spas in Portugal, retrieved from SABI database, we analyze the evolution of four competitiveness indicators: market share, productivity, return on investment and talent retention, for 2010–2020, by clusters of firms with and without brand. The correlation analysis between brand and competitiveness indicators suggests a strong positive and significant relationship between brand and market share, in the Centro and Lisbon regions, starting in 2017 and increasing over time. Finally, some recommendations on tourism management are made to improve the competitiveness of the supply.

Keywords Health and wellness tourism · Regional policy · Tourism management

JEL Classification M21–Business Economics · Z32–Tourism and Development

E. Santos (✉) · C. Crespo · J. Moreira
CARME—Centre of Applied Research in Management and Economics, Polytechnic Institute of Leiria, Leiria, Portugal
e-mail: eleonora.santos@ipleiria.pt

R. A. Castanho
Faculty of Applied Sciences, WSB University, Dabrowa Górnicza, Poland

College of Business and Economics, University of Johannesburg, Johannesburg, South Africa

© The Author(s), under exclusive license to Springer Nature Switzerland AG 2023

V. Katsoni (ed.), *Tourism, Travel, and Hospitality in a Smart and Sustainable World*, Springer Proceedings in Business and Economics,
https://doi.org/10.1007/978-3-031-26829-8_44

1 Introduction

Health and wellness tourism (HWT) is growing all over the world, which has caused changes in its supply, mainly due to the new needs, interests and demands of tourists (Buhalis and Amaranggana, 2015; Vujko & Gajic, 2014). Thus, competition in this segment is also increasing, showing a clear need to develop new products and services, increasingly creative and exclusive, capable of attracting tourists. The current trend in demand is focused on the holistic (body–mind–spirit) and preventive health, to the detriment of the curative aspect.

In this context, the water of naturally occurring hot springs usually possesses a very high quantity of mineral content, which makes their use highly important in various wellness and rejuvenating therapies such as balneotherapy, thermalism.

Among the natural hot spring's components, some of the most famous destinations are the Boiling River in Yellowstone Park and the Challis Hot Springs in Idaho, in the USA; the blue lagoon in Iceland; Terme di Saturnia in Italy; Pamukkale in Turkey; and Takaragawa Onsen in Japan. Natural resources play an important role on the competitiveness of HWT destinations since they represent a differentiation factor and harness the potential for new business opportunities (Pessot et al., 2021). Marketing strategies to address HWT competitiveness usually focus on the therapies based on differentiated natural resources as thermal waters. The richness entailed in the natural resources of specific regions may differentiate a wellness-related experience product (Romão et al., 2018) which is a key determinant of customer loyalty to a HWT destination (Lee & King, 2010; Shablii, 2018). Schalber and Peters (Schalber & Peters, 2012) highlight the role of sustainable management of natural resources when determining the strengths of territories to be leveraged from a marketing perspective. Successful marketing of HWT destinations is essential to value and differentiate local tourism products. Marketing includes the promotion of a HWT destination by using effective communication media as websites (Untari & Satria, 2019), and activities aimed at bestowing strategies with wellness-related meanings (Bočkus et al., 2021). Thermal waters in Portugal are considered very rich in terms of chemical composition and suitable for medicinal treatment. Most spas are located in the North and Centro regions of Portugal. This enables the country to pursue a differentiated wellness-related experience product, which is a key determinant of customer loyalty and can play an important role in regional development. Natural resources contribute to the strengths of territories and can be leveraged by marketing tools such as brands. Thus, this article intends to draw inferences about the effect of the brand on the competitiveness of the wellness tourism supply, across NUTS II regions of the continent, during the period of ten years, with a special focus on the recent period of the COVID-19 pandemic. With a sample of 10 spas in Portugal, retrieved from the SABI database, we analyze the evolution of four competitiveness indicators: market share, productivity, return on investment and talent retention, for 2010–2020, by clusters of firms with and without brand.

Besides the introduction, Sect. 2 describes the factors of competitiveness; Sect. 3 presents the trends in wellness tourism in Portugal; Sect. 4 describes data and methodology; Sect. 5 presents and discusses the results; and Sect. 6 draws conclusions and policy implications.

2 SWOT Analysis

Increased competition is a threat to tourism supply and prompts firms to implement collaborative strategies that can contribute to increasing productivity and competitiveness (see Fig. 2). Competitiveness is the ability of firms to use strategies that understand the external (environment) and the internal contexts (organization) to maintain or surpass market share and generate added value (Ali & Anwar, 2021; McDougall et al., 2021). Other threats include seasonality and the segmentation of demand (Ghasemi et al., 2021). In addition, a major weakness is the fact that tourism activities comprise many small-sized enterprises (SMEs) with limited resources that prevent them to achieve sustainable development alone (Anaya-Aguilar et al., 2021; Luederitz et al., 2021; Santos & Khan, 2018). However, tourism and the territory have a symbiotic relationship in the sense that resources can be changed or become tourist attractions (Buhalis and Amaranggana, 2015; Ridderstaat et al., 2019). Thus, cultural and spatial specificities are major strengths since they create opportunities to product and process differentiation through marketing techniques, namely branding (Dieguez & Conceição, 2020; Buhalis and Park, 2021; Liu et al., 2022; Santos et al., 2022), and thus are sources of innovations and may facilitate the surge of new market segments and niches (e.g., medical tourism) (Alves et al., 2021; Lubowiecki-Vikuk & Sousa, 2021) (Fig. 1).

Other opportunities include the improvement of service quality and communication with the various segments, cross-selling (Al Idrus et al., 2021; Dam & Dam, 2021; Ali et al., 2021) and the establishment of alliances and collaborative practices. These collaborative forms can be informal (networks) and formal (alliances) and assume the form of simple trading agreements to full integration (Moore et al., 2021), and can be horizontal—with rivals, or vertical—with suppliers or customers.

Spas can help promoting local identity and increase the demand for cultural products contributing to the differentiation of the tourism destination (Pinos-Navarrete et al., 2022; Zanten & Tulder, 2021). Collaboration can generate crucial changes, namely on perceptions about competition, and therefore, it is key to promote new markets and facilitate synergies between companies aiming at the sustainability of tourism destinations. Hence, the analysis of the competitiveness factors of health and wellness tourism in Portugal aims to understand the design of development strategies adjusted to the local sustainable promotion.

These collaborative forms allow SMEs to access resources, markets and technologies; share ideas and risks; promote innovation strategies; acquire skills, knowledge and reputation; and reduce costs and eliminate losses (Fernandes et al., 2021; Pilving et al., 2021; Scheyvens & Cheer, 2021).

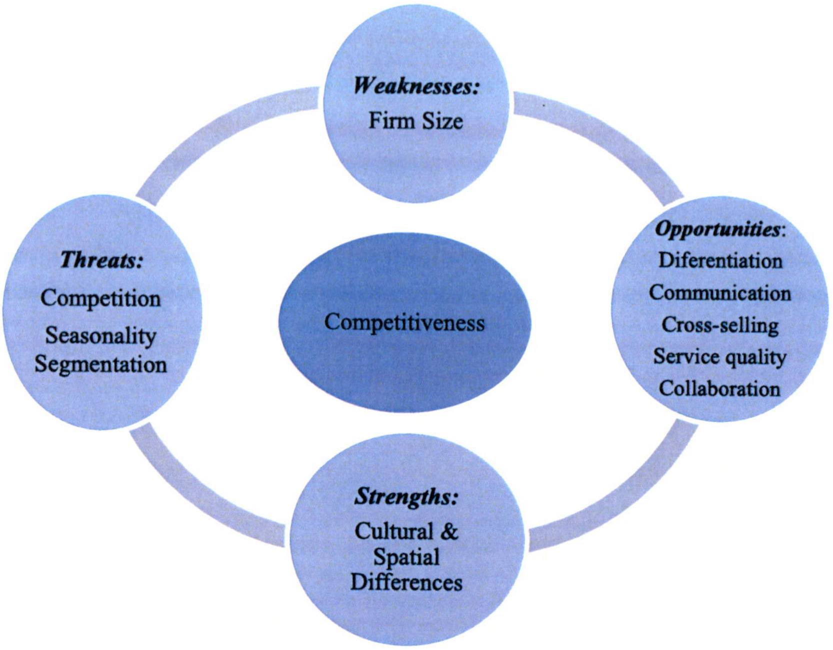

Fig. 1 SWOT analysis

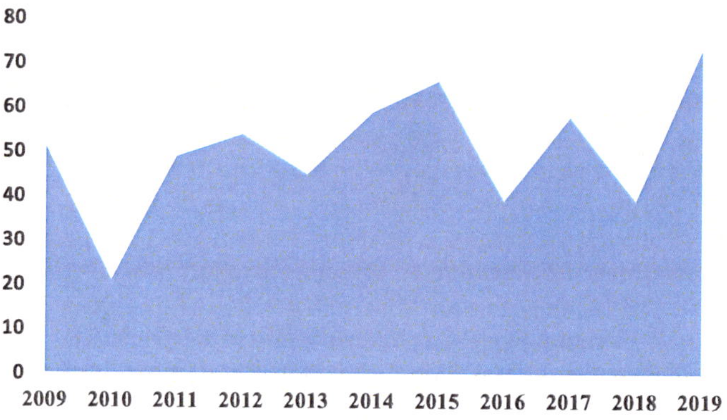

Fig. 2 Number of trips by residents in Portugal for health reasons. *Source* PORDATA

Table 1 Objectives and measures to strengthen competitiveness

Goals	Measures
Restructuring and differentiating the supply and promoting the attraction and growth of demand	• Creation of a network of thermal spas for the socioeconomic dynamism of the territory • Requalification of the surroundings • Integration of the various attractive elements (thermal spas, accommodation, gastronomy, entertainment activities, heritage, culture and nature) • Quality certification • Innovation, applied research and specialized training
Strategic marketing plan	• Improving the effectiveness and profitability of distribution, communication and marketing activities through: → Strategies based on value proposition → Strategic vision by market segment → Focus on the Internet channel → Proactive management

For Portugal, the objectives and measures to strengthen competitiveness (Gonçalves & Guerra, 2019) are described in Table 1.

3 Trends in Demand for Spas

In the past, traveling for medical reasons implied a flow from less developed to more advanced economies (Hoz-Correa & Muñoz-Leiva, 2019). Currently, the trend is the opposite as several less or more developed countries resort to competitive strategies in medical tourism based on service quality and price (Singh, 2019).

According to Fig. 2, the number of trips by residents in Portugal for health reasons has changed significantly over the period (2009–2019), with the years 2010, 2016 and 2018 being the ones that recorded lower values. The year of 2010 recorded the worst scenario for the tourism sector, with only around 20 trips per resident, which may have reflected the economic crisis experienced in that period. In contrast, in 2015 and 2019, the number of trips per resident was always above 60.

The access of the most disadvantaged populations to spas, because of social support from welfare state, has increased the average age of tourists, instituting the senior tourist profile. However, in recent decades, spas are adapting to a new reality by offering products and services oriented toward aesthetics, the body and healthy lifestyles. However, spas continue to show some lack of competitiveness, possibly explained by the neglect of the importance of medical hydrology and the lack of modernization since they are in government-protected areas.

According to data from Associação das Termas de Portugal, after a period of decline between 2010 and 2014, since 2015, there has been an increase in customers seeking health and wellness programs at the spas in operation.

The demand for health and well-being services has become a promising tourism sub-sector, as there is today a greater desire to travel to improve the health and well-being of the individual (Kim et al., 2021; Quintela, 2021).

Classical spa treatments are oriented toward therapeutic goals associated with specific pathologies, whereas in the case of wellness spas, it orients its supply toward a type of demand with recreational and tourist motivations that may have a therapeutic component. Examples of classic thermal establishments include Caldas da Felgueira, which today supplies complementary services related to prevention, nutrition and relaxation therapies; the Caldas da Rainha thermal hospital is the oldest in the world; Pedras Salgadas spa has a historic spa that combines the quality of water from four world-renowned springs with a set of complementary activities such as fitness treatments and cosmetic treatments. Classic spas have a high demand, tourist profile being associated with females over 45 years old, with the most representative age groups being over 60 years old. Their socioeconomic condition is medium or low average, and they are generally retired people who have acquired the habit of going to spas, the main motivation of this segment is to alleviate their chronic medical condition and their consumption is limited to the spa and accommodation. Among the most treated pathologies, in 2017, half were related to rheumatic and musculoskeletal diseases, followed by airway pathologies with just under a third of demand. In the latter case, the number of children and young people who sought, in thermal treatments, the therapeutic solution for different types of respiratory allergies, namely rhinitis, sinusitis, and asthma, among others, grew.

As for the wellness tourist, he is usually between 25 and 45 years old, although there are older tourists in a higher socioeconomic stratum, who are in good physical condition and usually seek prevention and relaxation treatments, as well as aesthetic ones. In this type of spa, it is more common to find a more demanding foreign population with greater economic power, and the stays are usually short, just a few days or even just at the weekend (Table 2).

In 2010, classic spa demand was seasonal in all regions, with the 3rd quarter recording 46% of total demand in this segment. Portugal had 36 thermal establishments in operation in 2010, of which 19 were in the Centro region (53% of total establishments), 13 in the North (36%) and 4 in the regions of Lisbon, Alentejo and Algarve (11%).

The North region, with 8 thousand users of well-being and leisure spa services (26% of the country's total), was the one that showed the only increase (+38%, corresponding to + 2 thousand customers). The Centro region concentrated 65% of classic spa practitioners, of which 51% preferred wellness and leisure programs. The spa of S. Pedro do Sul represented 22% of the country's total demand and 35% of the Centro region. The remaining five main spas showed year-on-year increases, except for Termas da Felgueira which registered a slight decrease (−253 registrations).

Despite the predominance of the classical spa in the North (71% of the total) and Centro (73%) regions, it decreased by 5% in the country. Lisbon, Alentejo and

Table 2 Main features of spas in Portugal, 2010 and 2020

	2010	2020
No Spas:	36	46
North	13	20
Centro	19	23
Lisbon, Alentejo and Algarve	4	3
Staff:	962	387
Blue collar	576	275
Technical/Administrative	208	34
Physicians	96	35
Top managers	82	43
Demand:		
Classic spa	62,000	12,498
Wellness and leisure	29,176	24,514

Notes Only full-time employees
Source Associação de Termas de Portugal

Algarve have the highest share of wellness and leisure spas (71%). Many users of classic spa and wellness and leisure programs mainly chose the holiday months for their stays. The Easter holidays and June are peak seasons. In 2010, each thermal establishment earned, on average, €243.12 per registration and treatments, which corresponded to a slight year-on-year decrease of €2.72 (−1%).

Registrations made by foreigners always have a very small share of the total. Spain, with 1,181 enrollments and a share of 73%, in 2010, returned to the first place in the ranking of foreign markets that opted for thermal treatments in Portugal. France, with 219 entries, ranked in second place (14% of the total). Switzerland, with 37 registrations (−7 than in 2009), ranked in third position, with a representation of 2% in total. Luxembourg and the USA, with 29 and 26 entries, respectively, ranked in fourth and fifth place.

In 2010, the North region received 26,000 spa practitioners. The 3rd quarter was the period that clearly concentrated the largest number of users, both for thermal treatments (46% of demand, which translated into 9,000 spa people) and for enjoying well-being and leisure (39%, that is, 3 thousand users). The 2nd quarter was the second choice for 5,000 classic thermal spas (29% of the total) and 2,000 users of the "well-being and leisure" aspect (22% of the total).

In 2010, the spas in the Centro region attracted 56 thousand users. Classic spa, with 41 thousand clients, decreased by 4% (−2 thousand spa users). The 3rd quarter was the period in which the largest number of spa practitioners was involved, for both modalities (45% of classic spa practitioners and 38% of well-being and leisure practitioners). The well-being and leisure programs showed a more regular activities throughout the year, with the 3rd quarter concentrating 38% of total demand. Classical spas in the Centro showed consecutive decreases in demand, from 53,189 inscriptions in 2005 to 41,773 in 2009. In 2010, each spa in this region earned, on average, €242.02 per registration and treatments, which corresponded to a decrease of €18.41 (−7%) compared to 2009.

In 2010, the regions of Lisbon, Alentejo and Algarve received 10 thousand spa users which corresponded to a decrease of 8% compared to 2009. Customers who opted for the spa for well-being and leisure preferentially chose the 3rd quarter to make their stay (53% of total demand, that is, 4,000 users). This period showed a year-on-year growth of customers of 31%. The 3rd quarter was also the option to stay for 49% of users of classic spa treatments (1,400 users). In 2010, each spa establishment in these regions earned, on average, €302.11 per registration and treatments, which corresponded to a year-on-year growth of 33% (+€74.65).

In 2020, the spas received 12,498 clients for thermal treatments (34% of the total) and 24,514 for well-being and leisure (66%). As a result of the pandemic caused by COVID-19, sharp drops in demand were also felt in the spa sector. The classic spa showed from 2019 to 2020 a decrease of 68% in the number of clients, while the wellness and leisure one decreased by 67%. For both classic spa and wellness and leisure, the 3rd quarter was the period with the highest demand, with 67 and 51% of total customers, respectively. The 2nd quarter presented the lowest values because it was in this period that the most severe lockdowns took place.

The Centro region attracted 57% of clients for treatments and 51% for a wellness and leisure stay. Around 64% of the clients who used the spas for treatments are 65 years old or older and 56% of customers for well-being and leisure are between 35 and 64 years old.

In 2020, the turnover of classic spas reached €3,200.5 thousand and that related to well-being and leisure reached €957.3 thousand. Receipts in classic spas showed a decrease of 72% from 2019 to 2020, while those for well-being and leisure fell less sharply (−57%). In 2020, each classic spa client paid €256.08 for treatments, on average (−14% compared to 2019, that is, −€42.83). When referring to the stay in well-being and leisure, it increased €39.05 (+30% compared to 2019). The foreign market (3,071 registrations) represented 8.3% of total demand.

In 2020, the North region, with 20 spas registered 4,573 clients in classic spas (−60% compared to 2019) and 11,269 in wellness and leisure stays (−48%). Classical spas represented 29% of the total number of clients; and wellness and leisure, 71%.

Around 59% of spa practitioners chose the 3rd trimester to spend a few days at the spa. Classical spas earned €1,174.7 thousand (76% of the total), 64% less than in 2019 and well-being and leisure earned €602.1 thousand (24%), which translated into a 39% decrease. Each classic spa client paid, on average, €256.88 in 2020 and €286.45 in 2019 (−10%). In relation to wellness and leisure spa services, the average amount paid by each customer was €32.83, compared to €27.76 in 2019, which translates into a year-on-year increase of 18%.

In 2020, the Centro region, with 23 spas registered 7,084 clients in classic spas (−71% compared to 2019) and 12,386 in wellness and leisure stays (−72%). Classic spas represented 77% of the total number of clients; and wellness and leisure, 23%. The 3rd quarter concentrated 52% of the practitioners. Classical spa treatments had a turnover of €1,784.6 thousand (77% of total revenues), 76% less than in 2019, and well-being and leisure had a turnover of €541.7 thousand (23%), which represented a decrease of 60%. Each classic spa client paid, on average, €251.92 in 2020 and €305.57 in 2019 (−18%). In relation to wellness and leisure spas, the average amount paid by each customer was €43.73, compared to €30.86 in 2019, which translates into a year-on-year growth of 42%.

In 2020, the three spas open in the regions of Lisbon, Alentejo and Algarve received 841 clients in classic spas (−71% compared to 2019) and 859 in wellness and leisure stays (−90%). Classical spas represented 49.5% of the total number of clients at the spas and wellness and leisure, 50.5%. The 3rd quarter concentrated 91% of the thermal practitioners. Classic spas had a turnover of €241,200, 70% less than in 2019, and wellness and leisure €45,700, which represented a decrease of 83%. Each classic spa client paid, on average, €286.85 in 2020 and €275.73 in 2019 (+4%). Regarding wellness and leisure spas, the average amount paid by each customer was €53.26, compared to €31.12 in 2019, which translates into a year-on-year growth of 71%.

4 Data and Methodology

Using panel data from SABI database, the sample of ten hot springs comprises four competitiveness indicators in 2010–2020. We calculated four performance indicators: market share, return on investment, productivity and talent retention.

Market share is the percentage that a given hospital has gained over competitors over a period. The larger this share, the more competitive the hospital is. This indicator helps managers to grow in the market. Losing market share requires strategic adjustments. However, having a very high market share can also imply a greater risk of having to face more aggressive competition. Consequently, hospitals must seek a lower-than-desired market share to avoid being attacked by competition and strict regulations. The calculation formula is

$$\text{Market share} = \text{sales}/\text{Total market sales} \tag{1}$$

If it is low, it is a sign that patients prefer the competition. Based on the results presented by this indicator, it is possible to investigate the causes of low invoicing such as pricing policy and the final quality of services.

Return on investment (ROI) shows how much the hospital can recover the money invested in projects or in the financing of the business model:

$$\text{ROI} = (\text{Profit} - \text{Investment})/\text{Investment} \tag{2}$$

Productivity. Hospitals that manage to produce a lot with as little resources as possible and without compromising the final quality tend to stand out in the market. We use labor productivity as measured by:

$$\text{Labor productivity} = \text{Turnover/Number of workers} \quad (3)$$

Talent Retention. Since we do not have qualitative data, we use

$$\text{Employment growth rate} = \text{No. Employees}_{\text{n}} - \text{No.Employees}_{\text{n}-1}/\text{No.Employees}_{\text{n}-1} \quad (4)$$

5 Results and Discussion

We perform the analysis by year, region and cluster of firms with brands. The analysis of the market share indicator suggests the presence of a strong heterogeneity in the spas sample. Considering both the mean and standard deviation obtained, the coefficient of variation for the market share indicator is considerably above 30%, indicating strong data heterogeneity (Favero & Belfiore, 2019). The mean of return on investment (ROI) is negative in the period from 2010 to 2020, indicating that the investment costs surpass the revenue generated by the companies in the health and wellness tourism sector. Both productivity and talent retention mean and standard deviation analysis support the existence of a strong level of heterogeneity in the data sample, since both coefficients of variation for productivity and talent retention surpass the 30% threshold (Favero & Belfiore, 2019) (Table 3).

The correlation analysis between brand and the four considered performance indicators, respectively, market share, return on investment, productivity and talent retention, indicates the presence of a strong positive and significant relationship between brand and market share, as well as between brand and productivity (Table 4).

A detailed analysis by year and region suggests a strong positive and significant relationship between brand and market share, only in the Centro and Lisbon regions, starting in 2017 and increasing over time.

Figure 3 shows that productivity changed very little over the period under analysis, although 2019 stands out slightly (56.80) compared to other years. This evolution is not reflected in other indicators such as talent retention, for example. Indeed, regarding the latter indicator, large variations can be observed between 2010 and 2020, with the years 2010, 2013 and 2020 being those in which the indicator presented

Table 3 Basic statistics for the competitiveness indicators Obs = 104

	Mean	Std. dev	Min	Max
Market share	0.11	0.13	0	0.51
Return on investment	−0.92	0.48	−2	3
Productivity	37.09	25.62	2	146
Talent retention	4.74	65.85	−94.12	500

Table 1 Correlation between brand and competitive factors, whole sample

	Brand	Market share	Return on investment	Productivity	Talent retention
Brand	1.00				
Market share	0.64**	1.00			
Return on investment	−0.11	−0.12	1.00		
Productivity	0.35*	−0.05	−0.04	1.00	
Talent retention	−0.02	−0.05	−0.10	−0.14	1.00

Notes * $p < 0.05$, ** $p < 0.01$, *** $p < 0.001$
Source own calculations in Stata 17.0

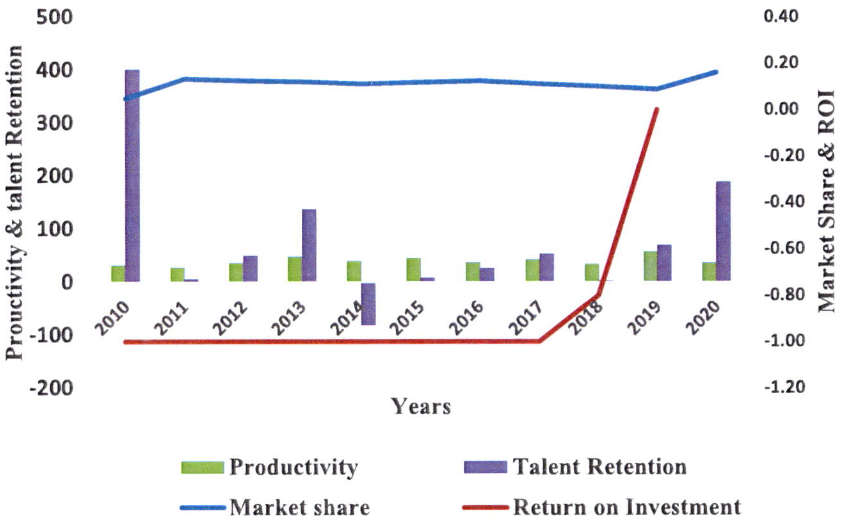

Fig. 3 Evolution of average competitiveness indicators, 2010–2020

higher values, respectively of 400.45, 136.90 and 187.84, which reveals the importance that companies have given to human capital as a factor that enhances growth and competitiveness. On the other hand, 2014 recorded a negative value (−82.04), which reveals that firms experienced an employment reduction and suggests that talents were not retained. Regarding market share as an indicator of competitiveness, there are essentially two periods of growth: 2010–2011, in which it increased from 0.05 to 0.13, and 2019–2020, where the value grew by 0.07. Finally, when analyzing the return on investment (ROI), it is constant and negative in the period from 2010 to 2017, increasing slightly between 2017 and 2018, although still negative, which reveals that over this period, the investment costs were higher than the revenue

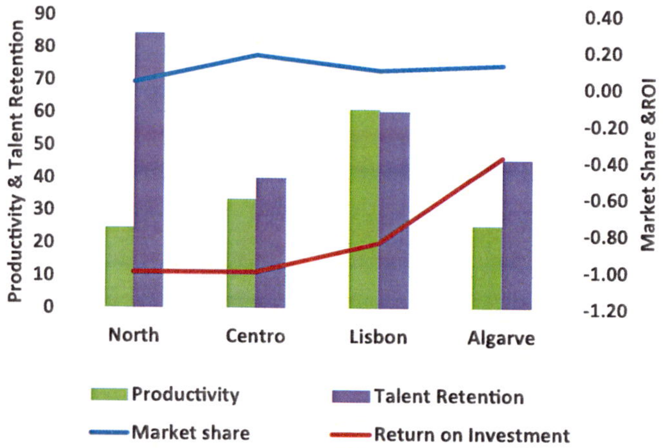

Fig. 4 Average competitiveness indicators across NUTs II regions, 2010–2020

generated by companies in the sector. From 2018 to 2019, there is a recovery, with the indicator assuming the value "0," which means that the amount invested (costs) equaled the revenue generated.

Figure 4 shows the competitiveness indicators across NUTs II regions. Regarding productivity, in 2010–2020, it was higher in the Lisbon region (61.12), followed by the Centro region (33.46), Algarve (25.38) and the North region (24 0.74). Regarding talent retention, the North region is the one with a higher value for this indicator (84.44), which suggests that the most productive companies are not always those that retain talents or that invest more in this productive factor. The region of Lisbon ranks second, followed by Algarve and Centro, with values of 60.52, 45.75 and 40.10, respectively.

Concerning the market share by region, during the period under analysis, the Centro region was the one that registered a higher value for the indicator (0.18), followed by Algarve (0.13), Lisbon (0.10) and North (0.04); while in terms of ROI, although it is always negative, it is in the Algarve that there is a value closer to "0," which means that in this region, the revenues equal the costs of investment.

These regional differences propose that strategic management guarantees better results when performed with the understanding of the local/ regional context (Santos, 2020b).

Also, one should bear in mind that tourists adopt a proactive behavior to fulfill their needs and, thus, are an active element when it comes to creating experiences while traveling (Santos, 2020a). Moreover, brands can perform a crucial role in the wellness tourism competitiveness, by defining positioning and transmitting the uniqueness of each wellness tourism destination. Brands also help to minimize consumers' perceived risk (Boga & Weiermair, 2011), contributing for consumers' creation of a distinctive image of a service or product, which ultimately can influence decision. Therefore, brands can enhance customers' attitudes and emotional connections toward wellness tourism destinations, allowing differentiation from competitors. These aspects should be regarded when searching for explanations of tourism supply performance across regions.

6 Conclusion

This study intended to investigate the effect of the brand on the competitiveness of the wellness tourism supply, across NUTS II regions of Portugal. Our results demonstrate the presence of a strong positive and significant relationship between brand and market share, as well as between brand and productivity.

The health and wellness tourism destinations face several threats that arise from increasing competition, seasonality and the segmentation of demand. Additionally, the fact that tourism activities include many small-sized enterprises with limited resources complicates the ability to achieve competitiveness advantages. However, cultural and spatial specificities can be a source of important opportunities to product and process differentiation through marketing techniques and can stimulate the creation of new market segments. Moreover, the improvement of service quality and communication with the various segments, the development of cross-selling practices and the establishment of alliances and collaborative agreements can contribute to the differentiation of the tourism destination.

Many of the health and wellness destinations are in non-urban areas, most affected by unemployment and threatened by depopulation. Health and wellness tourism can also mitigate regional asymmetries and generate more investment, whether direct in core activity or indirect in sport activities, creating a multiplier effect that benefits local economic agents.

For Portuguese spas, the restructuration and differentiation of the supply can be achieved through the creation of a network of thermal spas, the requalification of the surroundings complemented by the integration of different attractive elements and the implementation of quality certification and innovation. Moreover, to strengthen competitiveness, it is crucial to develop and implement proactive and integrated communication and marketing activities, customized to each market segment.

Consequently, Portuguese spas should enhance integrated marketing activities able to communicate the brand positioning to all target groups and to express the uniqueness of each wellness tourism destination. To increase its competitiveness, the brand should incorporate the service promise and capture the core value of the wellness tourism experience to differentiate each wellness tourism destination from competitors.

Acknowledgements This paper is financed by National Funds of the FCT—Portuguese Foundation for Science and Technology within the project «UIDB/04928/2020»

References

Al, S., Abdussakir, A., Djakfar, M., & Al, S. (2021). The effect of product knowledge and service quality on customer satisfaction. *The Journal of Asian Finance, Economics and Business, 8*(1), 927–938.

Ali, B. J., & Anwar, G. (2021). Business strategy: The influence of strategic competitiveness on competitive advantage. *International Journal of Electrical, Electronics and Computers, 6*(2).

Ali, B. J., Gardi, B., Othman, B. J., Ahmed, S. A., Ismael, N. B., Hamza, P. A., Aziz, H. M., Sabir, B. Y., Anwar, G. (2021). Hotel service quality: The impact of service quality on customer satisfaction in hospitality. *International Journal of Engineering, Business and Management, 5*(3), 14–28.

Alves, G. M., Sousa, B. B., & Belino, M. (2021). Understanding the brand management and rebranding processes in specific contexts of medical tourism. In *New Techniques for Brand Management in the Healthcare Sector* (pp. 124–141). IGI Global.

Anaya-Aguilar, R., Gemar, G., & Anaya-Aguilar, C. (2021). Challenges of spa tourism in Andalusia: Experts' proposed solutions. *International Journal of Environmental Research and Public Health, 18*(4), 1829.

Bočkus, D., Sulkinoja, E., Kolesnikova, N., & Komppula, R. (2021). Differences in the concept of wellness and its materialisation in service offering: a multiple case study from Eastern Finland, Russian Karelia, and Lithuania. *Tourism Recreation Research,* 1–16.

Boga, T. C., & Weiermair, K. (2011). Branding new services in health tourism. *Tourism Review.*

Buhalis, D., & Amaranggana, A. (2015). Smart tourism destinations enhancing tourism experience through personalisation of services. In *Information and communication technologies in tourism 2015* (pp. 377–389). Cham: Springer.

Buhalis, D., & Park, S. (2021). Brand management and cocreation lessons from tourism and hospitality: introduction to the special issue. *Journal of Product & Brand Management.*

Dam, S. M., & Dam, T. C. (2021). Relationships between service quality, brand image, customer satisfaction, and customer loyalty. *The Journal of Asian Finance, Economics and Business, 8*(3), 585–593.

de la Hoz, A., & Muñoz-, F. (2019). The role of information sources and image on the intention to visit a medical tourism destination: A cross-cultural analysis. *Journal of Travel & Tourism Marketing, 36*(2), 204–219.

Dieguez, T., & Conceição, O. (2020). Innovative destination branding:"Porto.". In *Advances in Tourism, Technology and Smart Systems* (pp. 131–140). Singapore: Springer.

Favero, L., & Belfiore, P. (2019). *Data science for business and decision making.* Academic Press.

Fernandes, G., Daniel, A., & Almeida, H. (2021). Collaborative strategies and tourist competitiveness in medium mountain destinations: Study on the positioning of tourist actors in the Natural Park of Serra da Estrela. *European Journal of Tourism, Hospitality and Recreation, 11*(1), 121–134.

Ghasemi, P., Mehdiabadi, A., Spulbar, C., & Birau, R. (2021). Ranking of sustainable medical tourism destinations in Iran: An integrated approach using fuzzy Swara-Promethee. *Sustainability, 13*(2), 683.

Gonçalves, E. C. C., & Guerra, R. C. (2019). O Turismo De Saúde E Bem-Estar Como Fator De Desenvolvimento Local. *Pasos Revista De Turismo y Patrimonio Cultural, 17*(2), 453–472.

Kim, J. J., Han, H., & Ariza, A. (2021). The impact of hotel attributes, well-being perception, and attitudes on brand loyalty: Examining the moderating role of COVID-19 pandemic. *Journal of Retailing and Consumer Services, 62*, 102634.

Lee, C. F., & King, B. (2010). International competitiveness in hot springs tourism: An application of the analytical hierarchy process approach. *Tourism Analysis, 15*, 531–544.

Liu, H., Hasan, M., Cui, D., Yan, J., & Sun, G. (2022). Evaluation of tourism competitiveness and mechanisms of spatial differentiation in Xinjiang. *China. Plos One, 17*(2), e0263229.

Lubowiecki-Vikuk, A., & Sousa, B. (2021). Tourism business in a vuca world: Marketing and management implications. *Journal of Environmental Management & Tourism, 12*(4), 867–876.

Luederitz, C., Caniglia, G., Colbert, B., & Burch, S. (2021). How do small businesses pursue sustainability? The role of collective agency for integrating planned and emergent strategy making. *Business Strategy and the Environment, 30*(7), 3376–3393.

McDougall, N., Wagner, B., & MacBryde, J. (2021). Leveraging competitiveness from sustainable operations: frameworks to understand the dynamic capabilities needed to realise NRBV supply chain strategies. *Supply Chain Management: An International Journal.*

Moore, E. M., Dau, L. A., & Mingo, S. (2021). The effects of trade integration on formal and informal entrepreneurship: The moderating role of economic development. *Journal of International Business Studies, 52*(4), 746–772.

Pessot, E., Spoladore, D., Zangiacomi, A., & Sacco, M. (2021). Natural resources in health tourism: A systematic literature review. *Sustainability, 13*(5), 2661.

Pilving, T., Kull, T., Suškevics, M., & Viira, A. H. (2021). Creating shared collaborative tourism identity in a post-communist environment. *Scandinavian Journal of Hospitality and Tourism, 21*(3), 313–340.

Pinos, A., Abarca-Álvarez, F. J., & Maroto-Martos, J. C. (2022). Perceptions and profiles of young people regarding spa tourism: A comparative study of students from Granada and Aachen Universities. *International Journal of Environmental Research and Public Health, 19*(5), 2580.

Quintela, J. A. F. D. S. (2021). *Contribution of health and wellness tourism to happiness and quality of life: a comparative study between Portugal and Hungary.* Doctoral dissertation, Universidade de Aveiro.

Ridderstaat, J., Singh, D., & DeMicco, F. (2019). The impact of major tourist markets on health tourism spending in the United States. *Journal of Destination Marketing & Management, 11*, 270–280.

Romão, J., Machino, K., & Nijkamp, P. (2018). Integrative diversification of wellness tourism services in rural areas–an operational framework model applied to east Hokkaido (Japan). *Asia Pacific Journal of Tourism Research, 23*(7), 734–746.

Santos, E. (2020). Promoting firms' sustainability in tourism planning through employment. In *Advances in Tourism, Technology and Smart Systems* (pp. 345–355). Singapore: Springer.

Santos, E. (2020). Do cultural tourism firms perform better than their rivals?. In *Advances in Tourism, Technology and Smart Systems* (pp. 383–393). Singapore: Springer.

Santos, E., Lisboa, I., Crespo, C., Moreira, J., & Eugenio, T. (2022). Evaluating economic sustainability of nautical tourism through brand equity and corporate performance. In *Transcending Borders in Tourism Through Innovation and Cultural Heritage* (pp. 105–118). Cham: Springer.

Santos, E., & Khan, S. (2018). Determinant factors of pecuniary externalities. *The International Journal of Business & Management, 6*(8), 180–198.

Schalber, C., & Peters, M. (2012). Determinants of health tourism competitiveness: An Alpine case study. *Tourism: An International Interdisciplinary Journal, 60*(3), 307–323.

Scheyvens, R., & Cheer, J. M. (2021). Tourism, the SDGs and partnerships. *Journal of Sustainable Tourism*, 1–11.

Shablii, O. I. (2018). The main problems of healthcare and wellness tourism in Ukraine. *J. Geol. Geol. Geoecol., 27*, 337–345.

Singh, L. (2019). Medical tourism motivations: The driving force. *Journal of Multidisciplinary Academic Tourism, 4*(2), 77–86.

Untari, D. T., & Satria, B. (2019). Measuring website effectiveness in communicating tourism destinations in Jakarta, Indonesia. *African Journal of Hospitality, Tourism and Leisure, 8*(4), 1–16.

van Zanten, J. A., & van Tulder, R. (2021). Improving companies' impacts on sustainable development: A nexus approach to the SDGS. *Business Strategy and the Environment, 30*(8), 3703–3720.

Vujko, A., & Gajic, T. (2014). Opportunities for tourism development and cooperation in the region by improving the quality of tourism services–the 'Danube Cycle Route'case study. *Economic Research-Ekonomska Istraživanja, 27*(1), 847–860.

Investigation into the Contribution of Social Tourism Programmes to the Satisfaction of Entrepreneurs in the Hospitality Sector

Despotaki N. C. Georgia and Doumi P. Maria

Abstract Social tourism, as a form of special interest tourism, gives the opportunity not only to members of society who are unable to meet their tourist needs by their own means but also to entrepreneurs in the hospitality sector to participate in social tourism programmes aiming at the financial growth of their businesses. The purpose of this paper is to investigate the value of the contribution of Greek Social Tourism programmes given to the hospitality sector by evaluating the level of satisfaction of entrepreneurs according to the effectiveness of their implementation and with respect to both their socio-demographic and business' profile. The empirical research of this study was conducted among all entrepreneurs in the hospitality sector of Greece, who have implemented social tourism programmes during the last ten years. The inventory method was used for the collection of the data, and a sampling in proportion to the population was collected, i.e., 287 questionnaires. Descriptive (numerical data summaries) and inductive statistics (cluster analysis, chi-square test) were used for data analysis. Through this research, the significance of the implementation of social tourism programmes in hospitality businesses was highlighted. This is because according to the respondents, social tourism contributes both to financial development and to the mitigation of seasonality. The majority of the respondents expressed their satisfaction with both the implementation of the programmes and the benefits and conditions of their participation. This research is the springboard for future planning of social tourism programmes not only in Greece but also worldwide. In several cases, in which Greece faced crises, such as the refugee crisis or the more recent health crisis (COVID-19), social tourism has been considered to be a "panacea" for tourism development.

Keywords Social tourism · Hospitality sector · Entrepreneurs' satisfaction · Greece

---

D. N. C. Georgia (✉)
Department of Business Administration, University of the Aegean, Chios, Greece
e-mail: g.despotaki@chios.aegean.gr

D. P. Maria
Department of Tourism Economics and Management, University of the Aegean, Chios, Greece
e-mail: mdoumi@aegean.gr

© The Author(s), under exclusive license to Springer Nature Switzerland AG 2023
V. Katsoni (ed.), *Tourism, Travel, and Hospitality in a Smart and Sustainable World*, Springer Proceedings in Business and Economics,
https://doi.org/10.1007/978-3-031-26829-8_45

JEL Classification Z000 Other Special Topics: General · Z390 Tourism · Other

1 Introduction

In1936, the International Labour Convention (ILO) agreed on the Holiday with Pay Convention (Convention n°52) and the Universal Declaration of Human Rights (1948) also refers to the right to rest and leisure and periodic holidays with pay. Despite the fact that tourism has been recognized as a human right, there are members of society who are unable to meet tourist needs by their own means; therefore, social tourism could be considered as an extension of the right to rest and holidays with pay.

The European Economic and Social Commission, in its declaration in Barcelona (2006), refers to the set of advantages resulting from the implementation of social tourism, which include the improvement of well-being, the social development of the beneficiaries, the socio-economic development of the host society's (social tourists), improving the mental health of participants as well as increasing employment. In the same declaration, the positive contribution that the form of social tourism can have to a more sustainable tourism was emphasized.

Social tourism programmes contribute to the country's wider economy and sustainable tourism development. The benefits of tourism can be wide ranging, extending to benefits to the economy, social life for people living in tourist destinations as well as personal benefits to tourists (Gilbert & Abdullah, 2004; World Tourism Organization (WTO), 1999). Furthermore, social tourism has a humanitarian character and contributes to social stability as it gives the opportunity to vulnerable social groups who are unable for various reasons such as health, financial or social, to be able to enjoy the benefits of a vacation (Diekmann, McCabe & Ferreira, 2018). Consequentially, there is now much greater evidence on the importance of social tourism which is providing significant benefits for the people supported through social tourism schemes. These benefits have also been shown to extend to wider sections of society as well as to destinations, businesses, governments and communities, who welcome social tourists and the resulting increase in income they bring.

In this article, the contribution of social tourism programmes to the satisfaction of entrepreneurs in the hospitality sector will be studied. The article aims to highlight the significance of the implementation of social tourism programmes in hospitality businesses according to their socio-demographic and business profile. The paper presents findings from application of social tourism to businesses in the hospitality sector in North Greece who are contracted with social tourism programmes and specifically the degree of entrepreneurs' satisfaction resulting from their implementing of these programmes.

2 Literature Review

The milestone year of the global development of social tourism was 1920 when the International Labour Organization (ILO) organized the first conference concerning the advantage of worker's leisure time, recognizing the right of vacation (Kokkosis, Tsartas & Griba 2011; Sfakianakis, 2000; Lytras, 1998).

According to the main institution of social tourism, the International Bureau of Social Tourism (BITS), which was founded in 1963, and renamed as International Organization of Social Tourism (ISTO), in 2010 (Belanger & Jolin, 2011, p.475) and the pioneer of Social Tourism who was the first president of ISTO, Professor W. Hunziker, the social tourism referred to "the set of relationships and phenomena with touristic elements facilitating the participation of weaker economic classes in tourism" (Hunziker, 1951:1).

Social tourism in Greece was implemented for the first time in 1982 with the cooperation of the Greek National Tourism Organization (GNTO), the Worker's Social Benefits Organization and the General Secretariat for Youth (Board Resolution 7631/3–11-1982). Over the years, many other institutions have implemented social tourism programmes:

- 1995: The Legal Insurance (Presidential Degree: 225/1995).
- 2002: The Organization of Agricultural Insurance (Law Number 3050/2002).
- 2013: Hellenic Manpower Employment Organization "OAED" (Law Number: 4144/2013).

According to the above legislations of social tourism institutions adapted the following table in which, the main criteria for the participation of the beneficiaries and the social groups who benefit are described (Table 1).

The biggest upset in the history of Greek Social Tourism came in 2012 with the abolition of social tourism programmes from the Workers' Social Fund (OEE), which was one of the institutions with the greatest involvement in social tourism programmes, as a result of the first memorandum for reducing public debt (Law Number:4046/2012, 2022). In these programmes, 1,154 hotels and 1,477 room rentals were involved, and its abolition resulted to the loss of more than 6,320

Table 1 The criteria for the participation in social tourism and the social groups who benefit

Participation criteria	Beneficiaries
Financial (taxable income)	Employees and their dependents with low taxable income
Age	Senior citizens
Duration of unemployment	Unemployed
Main insurance	Farmers
Rate of disability	People with a disability of more than 67%
Marital status	Families with four or more children, single-parent families

jobs in the accommodation area (Despotaki, Tsartas & Doumi, 2015). In Greece during a financial depression, social tourism is a breath of fresh air for the citizens of the country, not only for the vulnerable social groups, by providing them with the opportunity for economical holidays, but also for the hospitality businesses that implemented the said social tourism programmes. The solution to the problem came in 2013 with the welcomed implementation of social tourism by the Hellenic Manpower Employment Organization "OAED".

Apart from the economic crisis, Greece was faced with other forms of crisis which had a significant impact on the development and operation of tourist accommodations. The most important ones are the refugee crisis and the COVID-19 crisis.

In 2015, the regions of the Northern and Southern Aegean received an unprecedented wave of immigration and an influx of refugees (Tsartas et al, 2020, UNCHR, 2015). The significance of this phenomenon, which still exists today, greatly affected the economic activities of these areas, and its impact on the tourism industry became evident at a very early stage, most importantly concerning cancellations of overnight stays in tourist accommodations. The social tourism institutions of OAED and the Organization of Agricultural Insurance, showing a sense of social responsibility towards the residents and businesses that are on the front line of managing refugee flows, specifically the islands of Lesbos, Chios, Samos Leros, Kos as well as in the Prefecture of Evros, proceeded with a series of changes regarding the social tourism programmes offered in these areas. The aim was to strengthen the tourism businesses and to increase the tourism demand in the affected areas (Hellenic Public Employment Service "OAED", 2022, OPEKA, 2021). The changes made by the respective social tourism institutions concern:

- Increase in the number of overnight stays from 5 to 10.
- Zero private financial participation of the beneficiaries
- Increase in the accommodation subsidy.

Finally, in the context of strengthening tourism, due to the reduced tourist demand caused by the COVID-19 pandemic, the institutions of Greek Social Tourism (Hellenic Public Employment Service OAED, 2020, 2021, OGA, 2021) implemented the following:

- Subsidy of ferry tickets.
- Tripling of the budget from 10 million euros until 2019 (Hellenic Public Employment Service "OAED", 2019) to 30 million euros for the years 2020 and 2021.
- Double points and free ferry tickets for people with disabilities.
- Increase in subsidy in 2-, 3-, 4- and 5-star hotels with breakfast.
- Reduction of private participation in all accommodations.
- More than doubling of the number of beneficiaries, from 140,000 to 300,000.
- Increasing the length of overnight stays from 5 to 6 nights, except the islands of Chios, Samos, Lesbos, Leros and Kos, where the overnights remain at 10.
- Reduction of mandatory bed availability to 25 from 40%.

Social tourism, in whether or not in times of crisis, has been used as a panacea to boost tourism in general and specifically the tourist accommodation businesses that have contracts with social tourism institutions.

3 Methodology

3.1 Sampling Method

The objectives of the survey were carried out by using a quantitative research method, i.e. by developing a structured questionnaire to identify the satisfaction levels of owners of tourist accommodation facilities (hotels and rental apartments) derived from social tourism programmes and policies adopted in Greece. Hence, each questionnaire included closed-ended and scale questions, because they are credible in measuring people's perceptions (Oppenhein, 1992), while the participants ranked each of sixteen satisfaction items on a 5-point scale, with "1" indicating "strongly dissatisfied" and "5" indicating "strongly satisfied". The questionnaire was divided into three sections. The first section gathered information on the socio-demographic factors which were collected, including gender, age and educational level (Sánchez-Pérez et al., 2021). The second section aimed at obtaining information on the profile of the tourism businesses, hotel and rental apartments that implement social tourism programmes, including the type of the tourist accommodation and its category in stars for the hotels and keys for the rental apartments, the number of participations in social tourism programmes and the institutions with whom they are cooperating. The third section collected information on the degree of satisfaction of the respondents, with regard to the policies of social tourism adopted in Greece. A pilot survey was conducted among 20 participants, as piloting can help both with the wording of question sequences and the reduction of non-response rates (Oppenhein, 1992). Using stratified sampling by geographical region of Greece, 287 questionnaires were successfully completed by entrepreneurs adopting and promoting social tourism programmes and strategies. The study was conducted from February to May 2019, and the data collected were based on descriptive and inductive statistics analysed by SPSS v.26. The reliability of the Likert variables was controlled using Cronbach's alpha, which is a suitable method that can be used for Likert scale items (Ercan et al., 2007). The coefficient of reliability of the alpha Cronbach questionnaire was calculated and equalled 0.804, which means the results of the questionnaire were extremely reliable (Bonett & Wright, 2015).

3.2 Data Analysis Methods

For the purposes of the exploratory analysis, a K-means cluster analysis was employed to identify structures within the data. Cluster analysis is a statistical method that aims to classify existing observations into groups, using the information contained in some variables. In particular, it examines how similar some observations are in terms of a certain number of variables, with the aim of creating groups of observations that are similar to each other (Karlis, 2005). As the authors predefine the number of extracted clusters in advance, K-means cluster was deemed suitable due to the size of the sample and was carried out in order to display similar groups of observations among social tourism entrepreneurs' satisfaction levels.

A chi-square test was used to examine the significance of the differences between two separate samples regarding some particular characteristics. It should be noted that the main prerequisites for the use of this test are fulfilled (Angelis, 2009). The carrying out of specialized conclusions was completed with the creation of correlation tables, between the resulting clusters and the main demographic characteristics of the respondents (age and level of education), as well as between the characteristics of their business profile (type of tourist accommodation and its category in stars and keys, the number of participations in social tourism programmes and the institutions with whom they are cooperating).

4 Findings and Analysis

4.1 Social Tourism Entrepreneurs' Profile

Based on their gender, social tourism entrepreneurs are equally represented in the sample, while 46.8% represent females and 53.20% refer to males. Regarding age, participants aged from 40 to 61 consist of the largest group accounting for 71.2% of the sample, followed by the group aged from 18 to 39 (28.8%). With regard to the level of education, it was observed that the majority (50.2%) have a university degree, while 28.7% were educated in vocational and technical institutions and one quarter has completed up to the level of secondary education (21.1%).

4.2 Social Tourism Accommodation Businesses' Profile

Based on the type of tourist accommodation that is contracted with social tourism programmes, the majority refers to hotels (55.2%), while 44.8% concerned the rental apartments. With respect to the classification of the accommodation facilities, hotels with 2*or 3* consist of the largest group accounting for 47.1% of the sample. Similarly, in the case of keys classification of rental apartments, the majority, i.e., 51.2%

of the sample, refer to 2 or 3 keys. With respect to the number of participated in social tourism programmes, the majority (87.2%) have participated more than 7 times and have collaborated with all social tourism institutions, but mainly with OAED (43.2%) and OGA (50.1%).

4.3 The Satisfaction of Entrepreneurs in Hospitality Sector

In order to investigate the degree of satisfaction of tourist accommodation entrepreneurs with the implementation and policies of social tourism programmes, we proceeded to apply cluster analysis. K-means analysis was based on sixteen satisfaction items, and its results are summarized in Table 2.

From the cluster analysis, three groups of entrepreneurs emerged:

- **Cluster A: "Pessimists".** Despite their higher education and strong contact with social tourism programmes, the means of pessimists are below the mean scores of all satisfaction levels. Pessimists expressed a negative satisfaction level, mainly related to inability to reduce the own participation rate of tourists in the price of accommodation and to the limited options provided to tourists, so that they are able to choose accommodation from several tourist destinations.
- **Cluster B: "Realists".** Realists tend to be more satisfied than pessimists, as eleven in sixteen satisfaction items displayed higher mean scores than the mean scores of the sample size. Realists expressed their satisfaction towards the reduction of bureaucracy regarding the participation of accommodations in social tourism programmes and highlighted the increase of subsidy into tourist accommodation.
- **Cluster B: "Supporters".** Supporters presented higher mean scores than the means of sample size in all satisfaction categories. As the findings of the study revealed, supporters presented higher mean scores than those presented by realists, only in seven satisfaction statements, while "realists" and "supporters", tend to express a more positive attitude towards entrepreneurs' satisfaction.

In order to investigate the satisfaction of entrepreneurs resulting from the effectiveness of their implementation, cross-tabulations and chi-square tests were made between the cluster analysis data and the data collected from the respondents' demographic profile and their business profile.

Table 3 contains the cluster data as obtained in the previous section and two of the main demographic characteristics of the respondents regarding age and education level.

The socio-demographic profile of three exacted clusters is of particular interest, as females and males are almost equally represented in each group, and respondents from 40 to 61 consist of the largest group for each cluster (62.0, 58.8 and 65.7%). The majorities of entrepreneurs in each cluster have university degree (54.6, 48.7 and 38.7%).

The chi-square tests show a correlation between the demographic data and the resulting data from the cluster analysis.

Table 2 K-means cluster analysis

Variables[a]	Total	Cluster			
	n:287	A	B	C	F[b]
Facilities of social tourism programmes	3.13	2.94	2.46	3.87	149.832
Cooperation with social tourism stakeholders	3.54	3.35	2.81	4.34	101.093
Educational and training levels of social tourism stakeholders	3.47	3.40	2.58	4.31	103.968
Terms and conditions for the participation of property accommodation in social tourism programmes	3.23	3.02	2.66	3.91	68.354
Subsidy into property accommodation	3.08	2.88	2.49	3.74	71.728
Increase of subsidy into tourist accommodation	4.35	3.69	4.73	4.53	74.884
Reduce of bureaucracy for the participation of property accommodation in social tourism programmes	4.23	3.51	4.86	4.25	12.569
Organization of programmes, seminars for the education and training of social tourism entrepreneurs	3.71	2.78	4.33	3.92	68.233
Promotion of social tourism programmes through media	4.10	3.14	4.51	4.49	101.864
Discounts on transfer tickets	3.68	2.72	4.06	4.10	46.865
Discounts on tickets/granting of vouchers of cultural interest	3.58	2.45	4.04	4.07	80.536
Inclusion and participation of persons accompanying beneficiaries with disabilities who participate in social tourism programmes	4.12	3.12	4.58	4.50	100.234
Reduce the own participation rate of tourists in the price of accommodation	2.08	1.65	2.41	2.14	12.569
Options provided to tourists, so that they can choose accommodation from a wider range of tourist destinations	3.10	2.14	3.53	3.48	64.469
Increase overnight stays	3.25	2.51	3.61	3.53	28.045
Provision of BB/HB	2.73	1.76	3.23	3.05	36.190

[a]df = 361, level of probability 0.05, sig < 0.05
[b]Variables were measured on a 5-point scale from 1: Strongly Dissatisfied 5: Strongly Satisfied

However, the table reveals that the respondents aged between 40 and 61 tend to be more supportive, while the younger aged between 18 and 39 are more realistic and less supportive. Respectively, the 40 and 61 age group is almost evenly distributed between the groups of pessimists, realists and supporters. Also, based on the analysis of the above table, it is found that the people who have completed higher education in the largest percentage declare themselves to be pessimists and realists. Therefore, respondents who hold university degrees and are aged between 18 and 61 do not express a high degree of satisfaction regarding the implementation of social tourism programmes and tend to be more pessimistic.

Table 3 Cross-tabulations and chi-square test among the demographic data and the cluster data

Cross-tabulations and Chi-square test						
			Clusters			
			%			
			A:"Pessimists"	B:"Realists"	C:"Supporters"	Chi Square Test*
Socio-demographic characteristics	Age	18–39	30.6	36.1	21.9	$X^2 = 15.654$, $p = 0{,}114$
		40–61	62.0	58.8	65.7	
		> 62	7.4	5.0	12.4	
	Education level	Secondary level	19.4	18.5	32.8	$X^2 = 10.317$, $p = 0{,}135$
		Technical institution	25.9	32.8	28.5	
		University degree	54.7	48.7	38.7	

*a = 0.05

Table 4 contains the cluster data and the profile characteristics of the respondents with respect to the type of the tourism business they own and its classification, the number of the participations in social tourism programmes and the institutions with which they cooperated.

The chi-square tests show a correlation between the profile business data and the resulting data from the cluster analysis.

According to the table above, an even distribution is observed between hotels and rooms for rent regarding their satisfaction with social tourism. Similarly, it shows an even distribution of pessimists, realists and supporters between hotels regardless of their star rating. A sharp differentiation appears among the participating rental apartments where those who own 4 keys in their largest percentage, 42.9%, are supporters, while those who have 2 keys accommodation facilities are more realistic

Table 4 Cross-tabulations and chi-square test among the profile business data and the cluster data

Cross-tabulations and chi-square test

			Clusters			
			%			
			A:"Pessimists"	B:"Realists"	C:"Supporters"	Chi-square test*
Profile Business	Type of tourist accommodation	Hotel	47.9	51.6	45.3	$X^2 =$ 0.936, p = 0,626
		Rental apartments	52.1	48.4	54.7	
	Star classification of hotels	5*	0.0	1.2	0.0	$X^2 =$ 4.911, p = 0,767
		4*	14.0	13.4	8.1	
		3*	33.3	34.1	29.7	
		2*	43.9	42.7	43.2	
		1*	8.8	8.5	18.9	
	Keys classification of rental apartments	4	28.6	28.5	42.9	$X^2 =$ 10.942, p = 0,090
		3	39.3	38.2	22.5	
		2	27.7	47.0	25.3	
	Number of participations in social tourism programmes	1–3	10.1	13.8	9.3	$X^2 =$ 7.155, p = 0,520
		4–6	19.3	22.0	30.2	
		7–9	25.2	17.7	17.4	
		10–12	18.5	17.6	14.0	
		> 13	26.9	28.9	29.1	
	Cooperating social tourism institutions	OAED	37.4	40.4	44.0	$X^2 =$ 3.848, p = 0,664
		Lawyer Insurance Fund	2.1	1.6	0.7	
		OGA	54.5	51.4	52.5	
		All the above	5.9	6.7	2.8	

*a = 0.05

(47%). On the other hand, three-key room rental owners tend to be more pessimistic (39.3%) and realistic (38.2%).

The number of participations of tourist accommodations in social tourism programmes affects the degree of satisfaction of their owners. Thus, those who have participated up to 6 times in the programmes are identified as supporters and realists, while those who have participated in the programmes between 7 and 12 times tend to be realistic and pessimistic. Finally, entrepreneurs who have participated in social tourism programmes more than 13 times tend to be more supportive of the programmes.

The policy of the social tourism programme of the Hellenic Manpower Employment Organization (OAED) enjoys a high level of satisfaction from the respondents, who declare themselves supporters (44%) unlike the programme of the Agricultural Insurance Organization (OGA), where the distribution among the respondents is even with the pessimists being slightly more (54.5%).

5 Conclusion and Implications

Through this research, the degree of satisfaction of tourist accommodation entrepreneurs who are contracted with social tourism programmes was highlighted. According to the research, factors such as demographic characteristics (age and level of education) as well as the characteristics of the profile of tourist accommodations contracting with social tourism (category of accommodation and its classifications, the number of participations in social tourism programmes and the contracted social tourism institutions) tend to influence the degree of satisfaction of the entrepreneurs of tourist accommodations regarding the implementation policy of social tourism programmes.

Through a cluster analysis, three groups of entrepreneurs emerged: the pessimists, the realists and the supporters of the social tourism programmes. Pessimists who expressed a negative satisfaction level mainly referred to inability to reduce the own participation rate of tourists in the price of accommodation and to the limited options provided to tourists, in order for them to choose their accommodation from several tourist destinations. "Realists" and "supporters" tend to express a more positive attitude towards entrepreneurs' satisfaction. Thus, the perceptions of the aforementioned groups set a common ground that needs to be taken into account both for the segmentation of social tourism stakeholders and the planning of sustainable tourism development strategies.

This research is the springboard for future planning of social tourism programmes not only in Greece but also worldwide. In several cases, in which Greece faced crises, such as the financial crisis or the more recent health crisis (COVID-19), social tourism was used as a "panacea" for tourism development. For a more effective implementation of social tourism programmes, it is proposed to have a central coordination between institutions providing social tourism programmes and the Ministry of Tourism to cooperate directly with hospitality sector entrepreneurs.

References

Angelis, V. (2009). *Non parametrc Statistics*. University notes.

Belanger, C. H., & Jolin, L. (2011). The International Organisation of Social Tourism (ISTO) working towards a right to holidays and tourism for all. *Current Issues in Tourism, 14*(5), 475–482. https://doi.org/10.1080/13683500.2011.568056

Bonett, G. D., & Wright, A. T. (2015). Cronbach's alpha reliability: Interval estimation. hypothesis testing. and sample size planning. *Journal of Organizational Behavior, 36*, 3–15. https://doi.org/10.1002/job.1960

Despotaki, G., Tsartas, P., Doumi, M. (2015). Social tourism as a breath of the crisis in greece: research on social tourism institutions. In *1st International Conference on Experiential Tourism*, (pp. 9–11). Santorini: Greece, 9–11 October 2015. Retrieved July 15, 2022, from http://imic2015.conferences.gr/themata/parousiaseis/

Diekmann, A., McCabe, S., & Ferreira, C. C. (2018). Social tourism: research advances, but stasis in policy. Bridging the divide. *Journal of Policy Research in Tourism, Leisure and Events, 10*(3), 181–188. https://doi.org/10.1080/19407963.2018.1490859

Ercan, I., Yazici, B., Sigirli, D., Ediz, B., & Kan, I. S. (2007). Examining cronbach alpha, theta, omega reliability coefficients according to sample size. *Journal of Modern Applied Statistical Methods, 6*(1), 291–303. https://doi.org/10.22237/jmasm/1177993560

European Committee. (2006). Opinion of the European Economic and Social Committee on Social Tourism in Europe. *Official Journal of the European Union C318/12*, 67–77. Retrieved July 15, 2022, from https://eur-lex.europa.eu/legal-content/EN/TXT/PDF/?uri=CELEX:52006IE1155&from=EL

Gilbert, D., & Abdullah, J. (2004). Holiday taking and the Sense of Wellbeing. *Annals of Tourism Research, 31*(1), 103–121. https://doi.org/10.1016/j.annals.2003.06.001

Hellenic Public Employment Service "DYPA". (2022). Notice to tender 6/2022. Programme for the subsidized vacation of employees, unemployed persons and their families with the use of social tourism vouchers for the period 2022–2023. Digital Publication Number, ADA": 9ΩΓ6491Ω2-ΥΤΧ. (Protocole No.: 477780/15.06.2022). Retrieved July 16, 2022, from https://www.dypa.gov.gr/storage/koinoniki-politiki/koinonikos-toyrismos-2022-2023/9og64691o2-ytx.pdf

Hellenic Public Employment Service "OAED". (2019). Notice to tender 5/2019. Programme for the subsidized vacation of employees, unemployed persons and their families with the use of social tourism vouchers for the period 2019–2020. Digital Publication Number. (Protocole No.: 45207/31.05.2019). Retrieved July 16, 2022, from https://www.dypa.gov.gr/documents/10195/1794012/6%CE%A01%CE%9E4691%CE%A92-%CE%9D9%CE%A7.pdf

Hellenic Public Employment Service "OAED". (2020). Notice to tender 6/2020. Programme for the subsidized vacation of employees, unemployed persons and their families with the use of social tourism vouchers for the period 2020–2021. Digital Publication Number (Protocole No.: 36335/24.06.2020). Retrieved July 16, 2022, from https://www.dypa.gov.gr/documents/10195/2107146/%CE%A9%CE%A3%CE%9F04691%CE%A92-%CE%996%CE%9F-%CF%80%CF%81%CE%BF%CF%83%CE%BA%CE%BB%CE%B7%CF%83%CE%B7.pdf

Hellenic Public Employment Service "OAED". (2021). Notice to tender 8/2022. Programme for the subsidized vacation of employees, unemployed persons and their families with the use of social tourism vouchers for the period 2021–2022. Digital Publication Number. (Protocole No.: 30268/09.06.2021). Retrieved July 16, 2022, from https://www.dypa.gov.gr/storage/koinoniki-politiki/kinonikos-turismos-2021-2022/ps7mps4691o2-ag0.pdf

Hunziker, W. (1951). *Social tourism: Its nature and problems International Tourists Alliance*. Scientific Commission. Impr. fédérative

International Labour Organization, ILO. (1936). *Convention 052–holidays with pay convention*. Retrieved July 18, 2022, from https://www.ilo.org/dyn/normlex/en/f?p=NORMLEXPUB:12100:0::NO::P12100_ILO_CODE:C052

Karlis, D. (2005). *Multivariate statistical analysis*. Stamoulis. [in Greek].

Kokkosis, C., Tsartas, P., & Griba, E. (2011). *Special and alternative forms of tourism.* Demand and supply of new tourism products. Athens: Kritiki (in Greek).

Law No 3050/2002. (2022). Organization of Agricultural Insurance started to implement the Social Tourism for farmers and pensioners' farmers. Retrieved August 11, 2022, from https://www.e-nomothesia.gr/kat-agrotike-anaptukse/n-3050-2002.html

Law No 4046/2012 (2022) *Abolition of social tourism programs from the workers' social fund to reduce public debt.* Retrieved July 11, 2022, from https://www.e-nomothesia.gr/kat-oikonomia/n-4046-2012.html

Law No 4144/2013. (2022) *Holiday subsidy for workers, unemployed and their families from Manpower Employment Organization.* Retrieved August 11, 2022, from https://www.taxheaven.gr/law/4144/2013

Lytras, P. (1998). *Sociology of Tourism* (2nd ed.). Athens: Interbooks (in Greek).

Oppenheim, A. N. (1992). *Questionnaires Design, Interviewing and Attitude Measurement.* Continuum.

Organisation of Welfare Benefits and Social Solidarity "OPEKA". (2021). Rural housing programme for the year 2021. Protocole Number: 29736/13–05–2021. Retrieved September 5, 2022, from http://www.koinonikostourismos.gr/pdfs/opeka2021_plirofories_dikaiouxon.pdf

Presidential Degree: 225/1995. (2022). Holiday subsidy for pensioners of Legal Insurance who not receive pension from any other social fund. Retrieved September 5, 2022, from http://www.et.gr/api/DownloadFeksApi/?fek_pdf=19950100130

Sanchez-Perez, M., Teran-Yepez, E., Marin-Carillo, M. B., Marin-Carillo, G. M., & Illescas-Manzano, M. D. (2021). The impact of the COVID-19 health crisis on tourist evaluation and behavioural intentions in Spain: Implications for market segmentation analysis. *Current Issues in Tourism, 24*(7), 919–933. https://doi.org/10.1080/13683500.2021.1889481

Sfakianakis, M. (2000). *Alternative forms of tourism.* Ellin. (in Greek).

Tsartas, P., Kyriakaki, A., Stavrinoudis, T., Despotaki, G., Doumi, M., Sarantakou, E., & Tsilimpokos, K. (2020). Refugees and tourism: A case study from the islands of Chios and Lesvos, Greece. *Current Issues in Tourism, 24*(11), 1311–1327. https://doi.org/10.1080/13683500.2019.1632275

UNHCR. (2015). The sea route to Europe: The Mediterranean passage in the age of refugees. UNHCR, The UN Refugee Agency, p. 20. Retrieved September 1, 2022, from https://www.unhcr.org/protection/operations/5592bd059/sea-route-europe-mediterranean-passage-age-refugees.html

United Nations. (1948). Universal Declaration of Human Rights. Retrieved September 1, 2022, from https://www.un.org/en/about-us/universal-declaration-of-human-rights

World Tourism Organization (WTO). (1999). Global Code of Ethics for Tourism. Retrieved September 1, 2022, from http://www.worldtourism.org.pressrel/CODEOFE.htm

Hybrid Events and Congresses and Their Impact on the Greek Meetings Industry

Athina Papageorgiou and Alexandra Chalkia

Abstract Both the economic crisis and the pandemic outbreak significantly changed the meetings industry and the organization process of conferences and other events. Mobility restrictions and social distancing, necessary to reduce and subsequently minimize the pandemic and turn it into an endemic, initially led to the cancellation of the vast majority of conferences and events worldwide. However, shortly after the restrictions' induction, the industry used all tools provided by online platforms to organize virtual conferences that dominated the pandemic era as they met at a certain extent the industry needs. After the gradual de-escalation of the health protocols, hybrid conferences emerged as the new norm of the COVID-19 era. Online conferences, although a necessity, became disastrous for the meetings industry. Thus, hybrid conferences are an intermediate solution, as there is no sign of immediate return to traditional conferences in the near future. The aim of this paper was to investigate how hybrid conferences affected the Greek Meetings Industry during the COVID-19 pandemic. For this, we conducted both a primary and secondary research; primary data were collected by interviewing the administrative personnel of hotels, PCOs, DMCs and audiovisual companies, and secondary data were obtained by reviewing journals, relevant studies conducted by tourism associations and also UNWTO statistics. Our results confirmed that hybrid events are currently the preferred type of the Hellenic Conferences industry and will play a major role in the years to come. We believe that this study provides evidence that hybrid conferences will set a new standard for the conduct of conferences in the post-COVID-19 era.

Keywords Hybrid conferences · Event organization · COVID-19 · Professional event organizers · Audiovisual equipment · Rental companies · Hotels

JEL Classification Z31 · Z32

A. Papageorgiou · A. Chalkia (✉)
University of West Attica, Egaleo, Greece
e-mail: alexchalkia@uniwa.gr

© The Author(s), under exclusive license to Springer Nature Switzerland AG 2023
V. Katsoni (ed.), *Tourism, Travel, and Hospitality in a Smart and Sustainable World*, Springer Proceedings in Business and Economics,
https://doi.org/10.1007/978-3-031-26829-8_46

1 Introduction

According to Mistilis and Dwyer (2000), MICE tourism generates a variety of economic and social benefits and promotes the sustainable use of all resources available (Mistilis & Dwyer, 2000). In addition, it enables the tourism industry to grow in a balanced manner (Aburumman, 2020). Furthermore, the meetings industry has emerged as a hub of technology innovation in recent years, as many businesses in the sector use digital technologies for their marketing and operation strategies. This use of technology enhanced after the COVID-19 global pandemic outbreak, as many event organizers tried to cope with the challenges of the new condition. Technology was greatly used to conduct meetings in a virtual form at first, and meetings were held via Internet platforms providing an alternative solution to the market.

Even though virtual meetings became a major component of the meetings industry market, they are considered to be an "emergency" solution, as they affect many businesses in the meetings industry and, in a way, pose as a threat to their existence.

As soon as the circumstances allowed it, hybrid meetings started to emerge. Conducting meetings in a hybrid way combined the best features of both virtual and traditional meetings, as they are more flexible for attending, reduce costs for the organizers and still achieve to provide the full meeting experience to on-site delegates (social program, in-person interaction, etc.).

2 Meetings Industry Shift During COVID-19 Era

The coronavirus disease outbreak (COVID-19), shortly after its onset at the end of 2019, had developed to be a global pandemic, as declared on 11 March 2020 by the World Health Organization (WHO), causing chaos around the world and vastly impacting every aspect of life, including health, social life and the economy (Haleem et al., 2020; Mohanty et al., 2021). One of the sectors majorly impacted was travel and tourism, which suffered losses in both supply and demand for tourism-related goods. According to UNWTO (2022a), the year 2020 was disastrous for tourism as international travel suffered a 72% decrease, which resulted to 1.1 billion less international tourists travelling worldwide (overnight visitors), bringing the number of travellers back to levels seen thirty years ago. In addition, compared to 2020, there were 22 million more international tourist arrivals (overnight tourists) as a result of a 5% rise in international tourism in 2021 (427 million versus 405 million). However, foreign arrivals were still 71% lower in 2019 than in the year prior to the epidemic.

According to UNWTO (2022a), there are substantial signs of recovery as international tourism experienced a significant upswing, with visitor numbers achieving almost half (46%) of the levels of the same period in 2019. Europe and the Americas are leading the recovery amongst other regions, but there is still a long way to cover for the losses of the last two years. However, in the first five months of 2022, several destinations have shown signs of full recovery as preliminary international arrival and receipts data exceed pre-pandemic levels.

These encouraging results are anticipated to be further strengthened by strong summer season tourism demand in the northern regions of the globe; thus, the rising economic and geopolitical risks continue to endanger the recovery process (UNWTO, 2022a). The conflict between Russia and Ukraine, apart from being a human tragedy, poses a danger to global travel and economy in general. The conflict highly affected the oil prices and the transportation costs to rise excessively in the past few months (UNWTO, 2022b).

The meetings industry, as anticipated, was greatly affected by the pandemic. An ICCA (2021a) study states that in 2020, the majority of association meetings were postponed (44%), and 14% of the meetings were cancelled. In addition, virtual meetings, the second-largest segment, accounted for 30% of the total meetings held during the same period, even though it started as the smallest percentage. Hybrid meetings accounted for 2% of the total meetings held attracting the most attention from August onwards.

In 2021, it was becoming evident that virtual and hybrid meetings held the reins, as they accounted for 43 and 11%, respectively, of the total association meetings. Virtual meetings experienced a growth of 13%, whereas hybrid meetings share percentage grew by 9%. In addition, there was a significant upward trend of 25% in the meetings market compared to 2020 (ICCA, 2021b).

3 Hybrid Events and Congresses

After the COVID-19 outbreak, the meetings industry was compelled to review the manner in which events are conducted in the light of the changing circumstances. Mobility restrictions and social distancing, which were necessary conditions for the subsequent containment of the pandemic, initially led to the cancellation of almost all conferences and events. However, there was still the need of conducting events and conferences, as they are the main components for the dissemination and exchange of scientific knowledge, as well as ideas, opinions and strategies, which are essential in today's societies. The meeting industry's main focus was on the implementation of new technologies to conduct virtual and hybrid events due to the restrictions placed on the physical conduct. The Internet provided the vast majority of the necessary sources and tools needed. Online platforms such as Zoom and Microsoft Teams were used to conduct virtual and hybrid meetings and conferences along with augmented reality (AR), virtual reality (VR) and Green Box technology tools.

Virtual meetings are easily accessible, environmental-friendly (reduced carbon footprint) and provide lower participation costs (Puccinelli et al., 2022). On the other hand, they do not allow social interaction between participants, and technical difficulties are prone to occur and also depend on the Internet quality. (Roos et al., 2020; Standaert et al., 2022). In addition, virtual meetings have a major impact on travel-related businesses, such as airlines, hotels, PCOs as they pose as a threat to their high profitability on business travel (Standaert et al., 2022).

In Greece, the MICE sector suffered great losses due to COVID-19 pandemic. More specifically, until July 2020, 43 out of 122 PCO companies have reported losses of over 67 million Euros, while by April 2021, exhibition organizers have reported losses of over 200 million Euros (HAPCO, 2020; HELEXPO, 2021).

Additionally, if one considers that for every 1.00 Euro spent on an exhibition or conference, 3.13 Euros are generated in indirect and induced additional economic activity (fiscal multiplier effect); we understand the magnitude of the damage the pandemic has induced to the meetings industry and the Greek economy in general (Ikkos & Koutsos, 2021).

Gradually recovering from the pandemic made possible to conduct hybrid conferences (combination of traditional and virtual meetings with both in-person and online attendance). This type of meetings gave delegates the ability to choose their preferred way to attend meetings and conferences and provides many benefits for both attendees and organizers (attendance flexibility, wider audience reach, reduced conference costs, etc.) (Chalkia & Papageorgiou, 2022).

Advantages and Disadvantages of Hybrid Congresses and Events

The main advantages of hybrid events are the destination, the in-person attendance, live-streaming and the quality of the content. More specifically:

- Destination. Destination in hybrid conferences continues to provide a comparative advantage, as delegates who decided to attend will naturally choose a conference amongst many similar ones based on the attractiveness of the destination.
- Attendance. Attendance options are extended in a hybrid conference. Delegates can select if they wish to attend physically or online. Furthermore, hybrid events provide delegates who cannot attend physically due to restrictions imposed in their country, lack of mobility, financial barriers, etc. with a flexible solution. At the same time, keynote speakers can address the delegates via online platforms, which adds value to the scientific part while reducing costs.
- Live-streaming. Although many meetings in the past already offered the possibility of "on demand" viewing, the feeling of a "canned" video presentation is an inhibiting factor for watching, especially when costs are imposed for such viewing. Live-streaming is an advantage, particularly when combined with the use of online media that attendees can use to ask questions. Online attendance at a hybrid event is also likely to be greater than physical, since not all sessions attract all registered delegate on-site.
- Content. The speaker and the content of the speech are known to be key elements in attracting delegates, both in physical and online conferences. Lower fees and the absence of travel and accommodation costs for qualified speakers are a significant advantage of a hybrid conference, which can secure speakers that would not have been attracted in the past due to finances, distance or personal time to attend.

Accordingly, the main disadvantages of hybrid events are:

- Communication. Many online participants find it challenging and time-consuming to communicate with the chair and other delegates, and many are unable to do so during the sessions' limited time. The use of response systems, chat rooms or social media for interaction between online participants and also the possibility to ask questions has limitations, because it is very challenging to gather, group and answer delegate's questions by the speakers in real-time during meetings, given the time constraints and especially when the process of asking questions from the physical participants in the room is ongoing.
- Contact. As opposed to the past, online delegates cannot communicate with speakers, sponsors or advertisers. They are also deprived of social interaction during ceremonies, meals and other events. In addition, they are unable to participate in the customary extra-conference activities like sightseeing, visiting museums and shopping. Social interaction, making acquaintances and catching up with old friends and colleagues are very important parts of the social side of a conference event.
- Overall experience. Attending a conference is a multifaceted experience that involves more than just fulfilling professional obligations, learning objectives and skill requirements. Emotional satisfaction is a very important issue which cannot be satisfied remotely, since there is no participation in sessions, events, briefings and visits. Additionally, online attendance offers no chance of rest or holiday.

4 Research Aims and Methodology

This paper investigates the impact of hybrid conferences on the Greek Meetings Industry during the COVID-19 outbreak. The main goal was to assess the development of hybrid congresses and events during the COVID-19 era and the way the business sectors were involved, how they adapted their operations and practices to fit this new form of meeting and what were the main factors considered to provide similar experiences for both physical and virtual participants.

We use the qualitative approach to achieve the goals of this investigation. Key informants, i.e., professionals from meeting and event planning firms, audiovisual rental providers and hotels with meeting spaces, were chosen by researchers for personal interviews. This method documents the participants' empirical knowledge as well as their opinions and perspectives (Zafiropoulos, 2015). The purpose of the study was to ascertain how the meetings industry in Greece reacted to these novel circumstances and the way it altered.

The selection of the key informants was based on their relevance to the study's goals and was determined by factors such as their expertise in the meetings sector, job experience and involvement in MICE coordinating activities. Twelve interviews with project managers, event coordinators, sales managers and other officials were obtained and analysed. It included three audiovisual companies, three hotel managers and six PCOs and DMCs.

To encourage comprehensive interviews for the key study, semi-structured interviews were used. Interviews, commenced in April 2022 and completed in May 2022, had a duration of approximately 30–40 min each and were made via online platforms. Some interviewees chose to provide a written response. Using an open-ended question approach, interviewees first provided basic information about their job title and experience and then discussed how technology was utilized in their business during the COVID-19 pandemic and what was the effect on the way they organized and conducted the Hybrid events in particular.

5 Results

Most of the participants in the study were prompted to re-evaluate their work and its future prospects under the light of the opportunities provided by new technology advancements and their implementation. Digital transformation in the meetings industry is considered to be essential, and many respondents were already upgrading their systems and encouraging remote access for their staff, as they conducted virtual events, even before the pandemic outbreak. The lack of digitalization policies was mentioned as a factor of slow-paced technology integration.

Within the first months of the pandemic, it became evident that digital transformation was the only way to attract business, even though scarce, through virtual events. Major investments were made in terms of software and hardware in order to conduct online congresses and events. However, many barriers had to be surpassed, such as emerging technical issues due to low Internet speed and its reliability, educating the users (audience and speakers) on the operation of the various platforms, educating staff on the use of event organizing and meeting platforms and cultivating a remote collaboration environment.

As soon as hybrid events were possible, it was evident that both staff and delegates became more familiar with the use of new technologies and their utilization. Furthermore, hybrid events' benefits were obvious to all interviewees: higher yield of income (compared to virtual events), in-person interaction and full "congress experience", provision of advanced online experiences to the remote participants and a wider audience (compared to in-person events). This model is considered to be more appealing to the interviewees in contrast to the virtual one, although it is considered to be rather challenging to combine the elements of both traditional and virtual events effectively to create similar congress experiences for both on-site and online attendees. Some of the hybrid event platform challenges mentioned are delays in live-streaming and videos, while time zone of remote participants and Internet security issues are considered to be important. In this context, the majority of the participants in this study mentioned that excessive efforts have been made, in order to close the gap between on-site and online participants in terms of experience. However, the notion that hybrid events are in an early stage and there is plenty of room for advancement for this form of meeting was expressed by many participants, who were optimistic regarding their future implementation and further integration in the meetings market.

On a more specific note, PCO and DMC professionals mentioned that they incorporated the distribution of meeting platform manuals for the delegates prior to events, as well as technical assistance on this concept in their practices. In addition, they emphasized the importance of the IT professionals in the organization and conduct of hybrid events. Conference organizing companies also agreed that there must be equal efforts in terms of fulfilling the needs of both on-site and online delegates, by creating as much as similar experiences as possible. In that context, they stated that they have invested in upgrading their systems and Internet connection capabilities, in order to provide a higher service quality. The need of enhancing the experience of online delegates through short and focused sessions in order to maintain their attention was also mentioned, as well as implementing live chat tools (including social media platforms), online break-out room sessions, private virtual 1 to 1 meetings or even customized event applications, which include sponsor information, delegate virtual business cards, etc. in order to address the needs of both on-site and online audiences. Furthermore, congress organizers consider that venues have to comply with the health protocols and provide detailed crisis management plans.

Audiovisual rental companies professionals similarly promoted the distribution of meeting platform manuals and technical assistance to delegates. They also mentioned that IT professionals are of paramount importance in terms of organizing and operating Hybrid events. In terms of service quality, all companies stated that have improved their Internet connections and equipment to fulfil the delegate needs and advance the quality of services provided. The online experience enhancement was also important for AV companies, a finding that corresponds well with the notions expressed by the Congress Organizers.

Hotel professionals usually rely on audiovisual rental companies for the IT support, and the importance of IT professionals was highlighted in this study. Similarly, to Congress Organizers and AV companies, hotels have also invested in the quality and speed of the Internet provided and usually use a dedicated network for the conference facilities. Similarly, to Congress Organizers, hotel managers agree in venue compliance to strict health protocols regardless of the COVID-19 pandemic, as well as having a crisis management plan available henceforth.

6 Conclusion

This study showed that the Greek Meetings Industry has evolved significantly in order to comply with the novel circumstances of hybrid events and congresses. Even though there was some interest in integrating technology in the organization and conduct of events prior to the pandemic, the first months after the COVID-19 outbreak made it evident for businesses to invest in technology in order to continue their operation.

Virtual conferences and events have a catastrophic impact to the meeting industry, as they exclude key businesses of the MICE sector. Hybrid conferences and events are an intermediate solution which enrols all MICE sector-related businesses, since the traditional form of conferencing is not yet considered as a viable solution. Hybrid events have many benefits for both organizers and attendees, as well as for the destination itself.

The first couple of months of the pandemic provided valuable time for businesses to adapt to the new norm and introduce their staff to new technologies. In turn, participants were given instructions and training on how to utilize the conferencing platforms. This facilitated the introduction of hybrid meetings, when circumstances permitted the physical presence of delegates in meetings.

In order to enhance the engagement of delegates in hybrid events, the duration of sessions was reduced, and their content became more condensed. Several online tools were implemented in order to provide a sense of inclusion and a perception of providing a full "congress experience" mainly for virtual attendees. In addition, further spending on advancing Internet speed was vital in order to provide high-quality attendance and enhance delegate satisfaction.

This study aimed to assess the status of hybrid congresses and events during COVID-19 pandemic period. It became evident that major investments in technology were made by key players in the meetings industry, in order to meet the requirements of the emerging meeting models, as they are expected to continue to exist for several years after the end of the pandemic. The meetings industry businesses adapted quickly to the new norm by implementing new practices and online tools in the organization and conduct of conferences. Furthermore, emphasis was given in the quality of experience for both on-site and online participants and the way new technology was utilized to provide similar results for all audiences. It seems that hybrid events tend to represent a significant element of the MICE market; however, their resilience is yet to be tested with time and especially when circumstances will allow congress participation without restrictions and health protocols.

References

Aburumman, A. A. (2020). COVID-19 impact and survival strategy in business tourism market: The example of the UAE MICE industry. *Humanities and Social Sciences Communications, 7*(1), 141. https://doi.org/10.1057/s41599-020-00630-8

Chalkia, A., & Papageorgiou, A. (2022). The management of conferences and business events in periods of crisis. The new digital paradigm. In V. Katsoni & A. C. Şerban (Eds.), *Transcending borders in tourism through innovation and cultural heritage* (pp. 463–473). Springer International Publishing. https://doi.org/10.1007/978-3-030-92491-1_28

Haleem, A., Javaid, M., Vaishya, R., & Deshmukh, S. G. (2020). Areas of academic research with the impact of COVID. *The American Journal of Emergency Medicine, 38*, 1524–1526. https://doi.org/10.1016/2Fj.ajem.2020.04.022

HAPCO. (2020). *The data of the first nationwide survey on the effects of the global health crisis on the Greek market of conferences and events*. HAPCO. Retrieved from https://hapco.gr/index.php/nea/deltia-typou-hapco/830-ta-stoixeia-tis-protis-panelladikis-erevnas-gia-tis-epiptoseis-tis-pagkosmias-ygeionomikis-krisis-stin-elliniki-agora-synedrion-kai-ekdiloseon

HELEXPO. (2021). *Scream of anguish from the exhibition industry*. HELEXPO. Retrieved from https://www.helexpo.gr/pressoffice/el/press20210412

ICCA. (2021a). *ICCA annual statistics study 2020*. ICCA. Retrieved from https://www.iccaworld.org/cnt/Research/ICCA%20Statistics%20Study%202020_270521_Final.pdf

ICCA. (2021b). *ICCA association meetings outlook abstract*. ICCA. Retrieved from https://www.iccaworld.org/cnt/Research/2021_ICCA_Association_Meetings_Outlook_.pdf

Ikkos, A., & Koutsos, S. (2021). *Contribution of tourism in the Greek economy in 2020*. INSETE. Retrieved from https://insete.gr/wp-content/uploads/2021/11/21_11_Tourism_and_Greek_Economy_2019-2020.pdf

Mistilis, N., & Dwyer, L. (2000). Information technology and service standards in MICE tourism. *Journal of Convention & Exhibition Management, 2*(1), 55–65. https://doi.org/10.1300/J143v02n01_04

Mohanty, P., Dhoundiyal, H., & Choudhury, R. (2021). [Preprint] Events tourism in the eye of the COVID-19 storm: Impacts and implications. In S. Arora & A. Sharma (Eds.), *Event tourism in Asian countries: Challenges and prospects*.

Puccinelli, E., Zeppilli, D., Stefanoudis, P., Wittische-Helou, A., Kermorgant, M., Fuchs, S., Sarrazin, J., Easton, E. E., & Anh-Thu, A. (2022). *Hybrid conferences: Opportunities, challenges and ways forward*. 17. https://doi.org/10.1101/2022.03.18.484941

Roos, G., Oláh, J., Ingle, R., Kobayashi, R., & Feldt, M. (2020). Online conferences—Towards a new (virtual) reality. *Computational and Theoretical Chemistry, 1189*, 112975. https://doi.org/10.1016/j.comptc.2020.112975

Standaert, W., Muylle, S., & Basu, A. (2022). Business meetings in a postpandemic world: When and how to meet virtually. *Business Horizons, 65*(3), 267–275. https://doi.org/10.1016/j.bushor.2021.02.047

UNWTO. (2022a). *Impact assessment of the Covid-19 outbreak on international tourism*. Retrieved from https://www.unwto.org/impact-assessment-of-the-covid-19-outbreak-on-international-tourism

UNWTO. (2022b). *Impact of the Russian offensive in Ukraine on international tourism*. Retrieved from https://www.unwto.org/impact-russian-offensive-in-ukraine-on-tourism

Zafiropoulos, K. (2015). *How is a scientific paperwork done? Scientific research and writing of papers*. (2nd ed.). Kritiki Publishing.